6th International Conference on 3D Radiation Dosimetry 2010

(IC3DDose)

Journal of Physics: Conference Series, Volume 250

Hilton Head, South Carolina, USA
22-26 August 2010

Editors:

Mark Oldham

ISBN: 978-1-61782-126-4
ISSN: 1742-6588

Printed from e-media with permission by:

Curran Associates, Inc.
57 Morehouse Lane
Red Hook, NY 12571

Some format issues inherent in the e-media version may also appear in this print version.

Printed by Curran Associates, Inc. (2010)

For permission requests, please contact the Institute of Physics
at the address below.

Institute of Physics
Dirac House, Temple Back
Bristol BS1 6BE UK

Phone: 44 1 17 929 7481
Fax: 44 1 17 920 0979

techtracking@iop.org

Additional copies of this publication are available from:

Curran Associates, Inc.
57 Morehouse Lane
Red Hook, NY 12571 USA
Phone: 845-758-0400
Fax: 845-758-2634
Email: curran@proceedings.com
Web: www.proceedings.com

TABLE OF CONTENTS

QA in Radiation Therapy: The RPC Perspective 1
G S Ibbott

Dosimetry Challenges for Implementing Emerging Technologies 8
Fang-Fang Yin, Mark Oldham, Jing Cai, Qiuwen Wu

A Practical Use for FXG Gel Dosimetry 12
T Olding, G Salomons, J Darko, L J Schreiner

Evaluation of the Breast Absorbed Dose Distribution Using the Fricke Xylenol Gel 17
C Czelusniak, L S Del Lama, M V Moreira, A De Almeida

Development of PDRESS (patient Specific Dose Real Evaluation Systems) Using a Tenomag Gel and Optical CT (VISTA TM) in Clinical IMRT Prostate Case 20
Suk Lee, Jang Bo Shim, Kyung Hwan Chang, Yuanjie, Cao, Jaeyoun Yi, Jinwoo Park, Sam Ju Cho, Sang Hoon Lee, HyunDo Huh, Kwang Hwan Cho, Chul Kee Min, Dae Sik Yang, Young Je Park, Won Seob Yoon, Chul Yong Kim

Toward 3D Dosimetry of Intensity Modulated Radiation Therapy Treatments with Plastic Scintillation Detectors 24
M Guillot, L Gingras, L Archambault, S Beddar, L Beaulieu

Fast, Large Field-of-view, Telecentric Optical-CT Scanning System for 3D Radiochromic Dosimetry 29
A Thomas, M Oldham

The Characterisation of a Genipin-gelatin Gel Dosimeter 34
JB Davies, S Bosi, C Baldock

Dose Rate Dependency of Micelle Leucodye 3d Gel Dosimeters 39
J Vandecasteele, S Ghysel, Y De Deene

The "push-pull" Dosimeter: When Two Pigments Are Better Than One 44
Stephen G Bosi, Clive Baldock, Robert Smee

Preliminary Investigation of the Dosimetric Properties of 'RadGel' 49
J R Newton, A Thomas, A Appleby, C Marsden, E A Christman, J G Wolodzko, M Oldham

Dosimetric Properties of Magic-f Polymer Gel Assessed to Radiotherapy Clinical Beams 54
T Marques, M Schwarcke, CE Garrido, O Baffa, P Nicolucci

Influence of Gelling Agents on the Dosimetric Performance of the Turnbull Blue Gel Dosimeter 59
Jaroslav Šolc, Vladimír Sochor, Václav Spevacek

Polymer Gel Dosimetry 64
L J Schreiner, T Olding, K B McAuley

How to Scan Polymer Gels with MRI? 69
Yves De Deene

Improved Dose Sensitivity of Normoxic Polyacrylamide Gelatin Gel Dosimeter with Sucrose 79
M Yoshioka, S Hayashi, S Usui, K Haneda, T Kondo, H Numasaki, T Teshima, T Tominaga

Impact of Oxygen on the Accuracy and Precision of Normoxic Polymer Gel Dosimeters 83
Mahbod Sedaghat, Rachel Bujold, Martin Lepage

Effect of the Exothermal Polymerization Reaction on Polymer Gel Dosimetric Measurements 88
Mahbod Sedaghat, Rachel Bujold, Martin Lepage

An Investigation of the Pre-irradiation Temporal Stability of PAGAT Gel Dosimeter 93
S Khoei, J B Moorrees, C M Langton, J V Trapp

2d and 3d Dose Verification at the Netherlands Cancer Institute–antoni Van Leeuwenhoek Hospital Using EPIDs 97
Ben Mijnheer, Anton Mans, Igor Olaciregui-Ruiz, Jan-Jakob Sonke, Rene Tielenburg, Marcel Van Herk, Ron Vijlbrief, Joep StroomZ

Use of Electronic Portal Image Detectors for Quality Assurance of Advanced Treatments 103
Jean M Moran

An Evaluation of Cine-mode 3D Portal Image Dosimetry for Volumetric Modulated Arc Therapy 108
W Ansbacher, C-L Swift, P B Greer

Phase Space Modulation Method for EPID-based Monte Carlo Dosimetry of Imrt and Rapidarc Plans 112
Avery Berman, Reid Townson, Karl Bush, Sergei Zavgorodni

Preliminary Investigations with a Photodiode-based Fan-beam Optical CT Scanner 117
W G Campbell, A Jirasek, D Wells

Development of a New Laser-line and Ccd Based Optical-CT Scanner for the Readout of 3D Radiation Dosimeters 122
A E Papadakis, T G Maris, G Zacharakis, J Ripoll, C Varveris, J Damilakis

Dosimetry and Quality Assurance for Static and Rotational Fields 125
Thomas Rockwell Mackie

Novel Approaches to Improve the Combined Geometric and Dosimetric Performance of Existing IG-IMRT Systems 126
D A Jaffray, M Sharpe, A Rink, D Letourneau

Effective Management of FXG Gel Dosimetry 127
T Olding, J Darko, L J Schreiner

Ionization Chamber Array for Patient Specific Vmat, Tomotherapy and Imrt QA 132
Sotiri Stathakis

Development of a Radiochromic Ferric Oligomer Hydrogel 136
Kevin Jordan, Masaya Sekimoto

Non-diffusing Radiochromic Micelle Gel 140
Kevin Jordan, Masaya Sekimoto

Feasibility of Quantitative Pet/ct Dosimetry for Proton Therapy Using Polymer Gels 143
O A Zeidan, W C Hsi, O Lopatiuk-Tirpak, S I Sriprisan, S L Meeks, P A Kupelian, Z Li, J R Palta

Initial Study of 3d Dose Verification of Multi-field Proton Therapy Treatment Along Match Line with Polymer Gel Detectors 148
S Avery, C Freeman, K Shahnazi

An Investigation of the Response of the Radiochromic Dosimeter Presagetm to Irradiation by 62 MeV Protons 153
Shamsa Al Nowais, Andrzej Kacperek, John N H Brunt, John Adamovics, Andrew Nisbet, Simon J Doran

Determination of the Depth Dose Distribution of Proton Beam Using Presage TM Dosimeter 158
L Zhao, I J Das, Q Zhao, A Thomas, J Adamovics, M Oldman

An Evaluation of Genipin Gel As a Water Equivalent Dosimeter for Megavoltage Electron Beams and Kilovoltage X-ray Beams 162
Tina Gorjiara, Robin Hill, Zdenka Kuncic, Stephen Bosi, Clive Baldock

MAGIC-f Gel Dosimetry for Clinical Electron Beam 167
T A Pianoschi, M Alva, M Santanna, O Baffa, P Nicolucci

A Liquid Scintillator System for Dosimetry of Photon and Proton Beams 171
S Beddar

Dosimetry of a New P-32 Ophthalmic Applicator 174
Chang Heon Choi, Hyon Soo Han, Kwang-Jae Son, Ul Jae Park, Sung-Joon Ye

Investigation on Tissue Equivalent Normoxic Polymer Gel Dosimeter Using In-house Laser CT Scanning System 177
D Senthil Kumar, E James Jebaseelan Samuel

Cone Beam Optical CT Investigation on Tissue Equivalent Normoxic Polymer Gel Dosimeter 182
D Senthil Kumar, E James Jebaseelan Samuel

Preliminary Dose Calibration Results Using a TENOMAG Polymer Gel Dosimeter and Optical CT (vistaTM) 187
Suk Lee, Jang Bo Shim, Kyoung Hwan Chang, Yuanjie Cao, Jaeyoun Yi, Jinwoo Park, Sam Ju Cho, Sang Hoon Lee, HyunDo Huh, Kwang Hwan Cho, Chul Kee Min, Dae Sik Yang, Young Je Park, Won Seob Yoon, Chul Yong Kim

Review of Recent Advances in Radiochromic Materials for 3D Dosimetry 191
Kevin Jordan

How to Perform Dosimetry with Optical CT 198
CT Cheng-Shie Wuu, Y Xu

Achieving Accurate Radiochromic Optical-CT Imaging When Using a Polychromatic Light Source 208
A Thomas, M Pierquet, M Oldham

Creation of Sophisticated Test Objects for Quality Assurance of Optical Computed Tomography Scanners 213
A T Abdul Rahman, Elke Bräuer-Krisch, Thierry Brochard, John Adamovics, Steve Clowes, David Bradley, Simon Doran

An Investigation Into a New Re-useable 3d Radiochromic Dosimetry Material, PresageREU 218
Michael Pierquet, Andrew Thomas, John Adamovics, Mark Oldham

Effects of Adding Glycerol and Sucrose to Ferrous Xylenol Orange Hydrogel 222
Kevin Jordan, Masaya Sekimoto

Diode Arrays and QA of Advanced Techniques 225
Alonso N Gutiérrez, Oscar Calvo

Patient Specific Quality Assurance: Transition from IMRT to IMAT 229
Jennifer O'Daniel, Shiva Das, Jackie Wu, Fang-Fang Yin

Tumor-tracking Radiotherapy of Moving Targets; Verification Using 3D Polymer Gel, 2D Ion-chamber Array and Biplanar Diode Array 233
Sofie Ceberg, Marianne Falk, Per Munck Af Rosenschöld, Herbert Cattell, Helen Gustafsson, Paul Keall, Stine S Korreman, Joakim Medin, Fredrik Nordström, Gitte Persson, Amit Sawant, Michelle Svatos, Jens Zimmerman, Sven ÅJ Bäck

3D Analysis of Intensity-modulated Radiation Therapy Quality Assurance Measurement Using a 2D Diode Array 238
M Lawrence, T Cullip, L Potter, J Lian, S Wang, Z Xu, K Burkhardt, S Chang

Study of Dosimetric Water Equivalency of Presage® for Megavoltage and Kilovoltage X-ray Beams 242
Tina Gorjiara, Robin Hill, Jung-Ha Kim, Zdenka Kuncic, John Adamovics, Clive Baldock

Commissioning of Brachytherapy TPS Using a 2D-Array of Ion Chambers 247
Mammo Yewondwossen, Jim Meng

3D Dosimetry Fundamentals: Gels and Plastics 251
M Lepage, K Jordan

Small Field: Dosimetry in Electron Disequilibrium Region 259
Timothy C Zhu

SRT and SBRT: Current Practices for QA Dosimetry and 3D 269
S H Benedict, J Cai, B Libby, M Lovelock, D Schlesinger, K Sheng, W Yang

Dose Verification of Stereotactic Radiosurgery Treatment for Trigeminal Neuralgia with Presage 3D Dosimetry System 280
Z Wang, A Thomas, J Newton, G Ibbott, J Deasy, M Oldham

3D Geometric Gel Dosimetry Verification of Intraprostatic Fiducial Guided Hypofractionated Radiotherapy of Prostate Cancer 284
Fredrik Nordström, Sofie Ceberg, Sacha af Wetterstedt, Per Nilsson, Crister Ceberg, Sven ÅJ Bäck

On the Use of Polymer Gels for Assessing the Total Geometrical Accuracy in Clinical Gamma Knife Radiosurgery Applications 289
A Moutsatsos, P Karaiskos, L Petrokokkinos, K Zourari, E Pantelis, L Sakelliou, I Seimenis, C Constantinou, A Peraticou, E Georgiou

MAGAT Gel Dosimetry for Its Application in Small Field Treatment Techniques 294
N Gopishankar, S Vivekanandhan, S S Kale, G K Rath, S Senthil Kumaran, Sanjay Thulkar, V Subramani, M A Laviraj, R K Bisht, A K Mahapatra

Multiple Slot Array Collimator to Minimize Stray Light in Optical Cone Beam CT 299
Kevin Jordan, Jonatan Snir, Jerry Battista

Light Scattering in Optical CT Scanning of Presage Dosimeters 303
Y Xu, J Adamovics, J C Cheeseborough, K S Chao, C S Wuu

ScanSim: A Tool for Simulating Optical-CT Imaging 308
Mark Oldham

Mathematical Modelling of Depth-dose Response of Polymer Gel Dosimeters 313
J N M Chain, L J Schreiner, K B McAuley

Mathematical Modelling of Response of Polymer Gel Dosimeters to Brachytherapy Radiation 318
A T Nasr, J N M Chain, L J Schreiner, K B McAuley

Monte Carlo Simulation of MAGIC-f Gel for Radiotherapy Using PENELOPE 321
M Alva, T Pianoschi, T Marques, M Santanna M, O Baffa, P Nicolucci

High Performance Dosimetry Calculations Using Adapted Ray-tracing 325
Lancelot Perrotte, Guillaume Saupin

3d Polymer Gel Dosimetry and Geant4 Monte Carlo Characterization of Novel Needle Based X-ray Source 330
Y Liu, E Sozontov, V Safronov, G Gutman, E Strumban, Q Jiang, S Li

Alternative Imaging Modalities for Polymer Gel Dosimetry 335
Andrew Jirasek

Gamma Dose Distribution Evaluation Tool 346
Daniel A Low

Isopropanol-based Polymer Gel Dosimeters for Use with X-ray CT Imaging 357
A Jirasek, M Hilts, K McAuley

Polymer Gel Dosimetry Using X-ray Computed Tomography: Investigation of the Effect of Reconstruction Technique 362
Jonathan Hindmarsh, Roger Fulton, Lyn Oliver, Clive Baldock

Dose-rate Dependence of PAGAT Polymer Gel Dosimeter Evaluated Using X-ray CT Scanner 367
P Sellakumar, E James Jebaseelan Samuel, D Senthil Kumar

Investigation of Ultrasonic Properties of MAGIC Gels for Pulse-echo Gel Dosimetry 372
T J Atkins, V F Humphrey, F A Duck, M A Tooley

Improvement in the Accuracy of Polymer Gel Dosimeters Using Scintillating Fibers 377
Nicolas M Tremblay, Vincent Hubert-Tremblay, Rachel Bujold, Luc Beaulieu, Martin Lepage

Femtosecond Laser Pulse Filamentation Characterized by Polymer Gel Dosimetry and Fricke Dosimetry 382
Ridthee Meesat, Jean-François Allard, Daniel Houde, Luc Tremblay, Abdelouahed Khalil, Jean-Paul Jay-Gerin, Martin Lepage

Preliminary Commissioning Investigations with the DMOS-RPC Optical-CT Scanner 387
J Newton, A Thomas, G Ibbott, M Oldham

Initial Experience with Optical-CT Scanning of RadBall Dosimeters 391
M Oldham, C Clift, A Thomas, E Farfan, T Foley, T Jannik, J Adamovics, C Holmes, S Stanley

RadBallTM Technology Testing in the Savannah River Site's Health Physics Instrument Calibration Laboratory 395
Eduardo B Farfán, Trevor Q Foley, G Timothy Jannik, Larry J Harpring, John R Gordon, Ronald Blessing, J Rusty Coleman, Christopher J Holmes, Mark Oldham, John Adamovics, Steven J Stanley

RadBallTM Technology Testing and MCNP Modeling of the Tungsten Collimator 400
Eduardo B Farfán, Trevor Q Foley, J Rusty Coleman, G Timothy Jannik, Christopher J Holmes, Mark Oldham, John Adamovics, Steven J Stanley

MAGIC-f Gel in Nuclear Medicine Dosimetry: Study in an External Beam of Iodine-131 405
M Schwarcke, T Marques, C Garrido, P Nicolucci, O Baffa

Verification of Synchrotron Microbeam Radiation Therapy Using a Purpose-built Optical CT Microscope 410
A T Abdul Rahman, Elke Bräuer-Krisch, Thierry Brochard, John Adamovics, David Bradley, Simon Doran

Gel Dosimetry Analysis of Gold Nanoparticle Application in Kilovoltage Radiation Therapy 415
T Marques, M Schwarcke, C Garrido, V Zucolot, O Baffa, P Nicolucci

Data Analysis Tools for 3D Dosimetry: The Use of CERR as a Platform to Integrate and Compare Measurements and Treatment Planning Information 420
Joe Deasy, Aditya Apte

Imaging and 3-D Dosimetry: Top Tips for MRI and Optical CT 421
Simon J Doran

Radiotherapy Delivery During Motion 429
Sofie Ceberg, Sven Å J Bäck

3D Dosimetry for Bachytherapy and Heterogeneities 433
G S Ibbott, M P Heard

Dosimetry in HDR Brachytherapy with Fricke-gel Layers and Fricke-gel Catheters 439
G Gambarini, M Carrara, A Negri, M Invernizzi, C Tenconi, A Scotti, L Pirola, M Borroni, S Tomatis, C Fallai

Electron Dosimetry in the Presence of Small Cavities 444
Simon Doran, Russell Thomas, Rachel Hollingdale, John Adamovics, Andrew Nisbet

On the Feasibility of Verification of 3D Dosimetry Near Brachytherapy Sources Using PRESAGE/Optical-CT 449
M Pierquet, O Craciunescu, B Steffey, H Song, M Oldham

Dose Enhancement Caused by Gold Foils on Polymer Gels 454
L C Afonso, M Greiter, F Schoefer, C Hoeschen

Water Equivalency Evaluation of PRESAGE® Dosimeters for Dosimetry of Cs-137 and Ir-192 Brachytherapy Sources 459
Tina Gorjiara, Robin Hill, Zdenka Kuncic, Clive Baldock

High Resolution Polymer Gel Dosimetry for Small Beam Irradiation Using a 7T Micro-MRI Scanner 464
Xuanfeng Ding, John Olsen, Ryan Best, Marcus Bennett, Inna McGowin, Jennifer Dorand, Kerry Link, J Daniel Bourland

Three-Dimensional Dosimetry of a Beta-Emitting Radionuclide Using PRESAGE® Dosimeters 467
R L Grant, M L Crowder, G S Ibbott, J Simon, R K Frank, J Rogers, H M Loy, J Adamovics, J Newton, M Oldham, S Stearns, R E Wendt

Fricke Dosimetry of Irradiated Sugar by 1H NMR Relaxation 470
J Y C Lau, T Olding, G Wu, L J Schreiner

A 'Quad-phantom' Film Dosimeter for use as a Multi-planar Verification Tool for PRESAGE/Optical-CT 473
L Stunja, A Thomas, J Adamovics, J Deasy, M Oldham

Absorbed Dose Distribution Visualization for Superficial Treatments Through the Fricke Xylenol Gel Dosimeter (FXG) 478
M Alva, F G A Sampaio, M V Moreira, P C D Petchevist, A de Almeida

Dose Overshoot Reduction by Tetrakis Hydroxymethyl Phosphonium Chloride in Polyacrylamide Gel Dosimeter 481
K Vávru, J Šemnická, V Spevacek
Effect of Irradiation and Storage Temperature on PRESAGETM Dose Response 485
P S Skyt, P Balling, J B Petersen, E S Yates, L P Muren
Experience Using DosimetryCheck Software for IMRT and RapidArc Patient Pre-treatment QA and a New Feature for QA During Treatment 487
Arthur Pinkerton, Michael Hannon, Jae Kwag, Wendel Dean Renner
Author Index

Journal of Physics: Conference Series **250** (2010) 011001 doi:10.1088/1742-6596/250/1/011001

Chairman's Introduction and Welcome Message:

Every clinical physicist I have asked readily acknowledges the great desirability of a 3D dosimetry system for the verification of advanced radiation therapy treatments. An accurate and practical 3D dosimetry system would greatly strengthen the foundation of quality assurance in radiation therapy by enabling a rigorous and comprehensive whole system test. Such systems are now emerging, and the innovations and progress that led to them are remarkably captured in the proceedings of five prior DOSGEL conferences, the last three of which are freely available in the Journal of Physics: Conference Series. These meetings included a focus on the technical challenges of various approaches to 3D dosimetry. When considering plans for the present 6th meeting, the scientific committee recognized that the field has matured, and a broader focus was desirable, including a strengthening of the clinical and applications component, while preserving a strong technical component. There was also the desire to embrace a variety of other semi-3D techniques which have also recently emerged to implementation in the clinic. In accordance with these sentiments, the committee approved changing the name of the conference from the International Conference on Radiotherapy Gel Dosimetry (DOSGEL) to the International Conference on 3D Radiation Dosimetry (IC3DDose) and to the following objectives -

Conference Objectives:

1. ***To provide a forum*** *to discuss the latest research and developments in 3D and advanced radiation dosimetry.*
2. ***To elevate the quality*** *of radiation therapy treatments (quality assurance QA) through improved clinical dosimetry.*
3. ***To explore*** *the dosimetric challenges posed by modern radiation treatment techniques*
4. ***To energize and diversify*** *dosimetry research and clinical practice by encouraging interaction and synergy between advanced, 3D, and semi-3D dosimetry techniques*

The scientific program of the IC3DDose2010 meeting has been crafted to meet the objectives listed above, and we were fortunate that many leading speakers shared their experience and perspectives to help achieve these aims.

On a more personal note it was a true pleasure to extend a warm welcome to all conference attendees. We have a very diverse group of clinical and research physicists from many parts of the world, and it is truly a pleasure to welcome you all.

Finally I would like to acknowledge the many people who have made the meeting possible. Special thanks go to the Scientific Organizing Committee (listed opposite) who reviewed all the conference abstracts and participated in many planning conference calls, and meeting activities. The Local Organizing Committee shouldered a significant load finalizing the conference abstract book and ensuring smooth meeting logistics. We are also very grateful to our sponsors, both academic and industrial, whose support was vital to this meeting.

Mark Oldham, PhD, FAAPM
Associate Professor, Radiation Oncology and Biomedical Engineering,
Duke University Medical Center, Durham, NC, USA
mark.oldham@duke.edu

Journal of Physics: Conference Series **250** (2010) 011002 doi:10.1088/1742-6596/250/1/011002

Proceedings: IC3DDose 2010

6th International Conference on 3D Radiation Dosimetry
www.IC3DDose.org

Edited by

Mark Oldham
Assisted by Joseph R. Newton

"All papers published in this volume of *Journal of Physics: Conference Series* have been peer reviewed through processes administered by the proceedings Editors. Reviews were conducted by expert referees to the professional and scientific standards expected of a proceedings journal published by IOP Publishing".

Academic Sponsors:

American Association of Physicists in Medicine (AAPM)
South Eastern Chapter of the AAPM (SEAAPM)
The Duke Medical Physics Graduate Program
The Department of Radiation Oncology, Duke University Medical Center

Scientific Organizing Committee:

Sven Back (Sweden), Clive Baldock (Australia), Cheng-Shie Wuu (USA), Yves De Deene (Belgium), Simon Doran (UK), Geoffrey Ibbott (USA), Andrew Jirasek (Canada - President), Kevin Jordan (Canada), Martin Lepage (Canada), Thomas Maris (Greece), Mark Oldham (USA - Chair), Evangelos Pappas (Greece), John Schreiner (Canada)

Local Organizing Committee:

Mark Oldham, Carolyn Crank, Joseph Newton, Andrew Thomas, Matthew DeLorenzo
Fang-Fang Yin, Duke University Medical Center, Durham, NC, USA

MGS RESEARCH, INC.

ScandiDos

SUN NUCLEAR corporation

PTW

iba Dosimetry

VARIAN medical systems

A partner for life

Journal of Physics: Conference Series **250** (2010) 011002 doi:10.1088/1742-6596/250/1/011002

TomoTherapy

CIVCO
MEDICAL SOLUTIONS

modusQA

Best®
Theratronics

ACCURAY®
Our Business Begins with Patients™

2010 IC3D DOSE
HILTON HEAD ISLAND

Journal of Physics: Conference Series **250** (2010) 011005 doi:10.1088/1742-6596/250/1/011005

Peer review statement

All papers published in this volume of *Journal of Physics: Conference Series* have been peer reviewed through processes administered by the proceedings Editors. Reviews were conducted by expert referees to the professional and scientific standards expected of a proceedings journal published by IOP Publishing.

Journal of Physics: Conference Series **250** (2010) 012001 doi:10.1088/1742-6596/250/1/012001

QA in Radiation Therapy: The RPC Perspective

G. S. Ibbott

UT M. D. Anderson Cancer Center

gibbott@mdanderson.org

Abstract. The Radiological Physics Center (RPC) is charged with assuring the consistent delivery of radiation doses to patients on NCI-sponsored clinical trials. To accomplish this, the RPC conducts annual mailed audits of machine calibration, dosimetry audit visits to institutions, reviews of treatment records, and credentialing procedures requiring the irradiation of anthropomorphic phantoms. Through these measurements, the RPC has gained an understanding of the level of quality assurance practiced in this cohort of institutions, and a database of measurements of beam characteristics of a large number of treatment machines. The results of irradiations of phantoms have yielded insight into the delivery of advanced technology treatment procedures.

1. Introduction

The requirements of quality assurance (QA) were brought to the attention of the US public in 2000 when a report from the Institute of Medicine highlighted the frequency of patient deaths from medical errors. [1] While the magnitude of the problem was shocking – as many as 98,000 patients dying each year from such errors – practitioners of radiation therapy were probably better aware of both the problem and potential solutions than those of many other medical fields. QA has a lengthy and solid foundation in radiation therapy, resulting at least in part from its quantitative nature.

During the past few years, a number of significant radiation therapy errors have been reported. [2-4] Several reasons are postulated, including the introduction of, and increasing dependence upon, advanced technologies. [5-8] Several publications reported that the introduction of advanced technology equipment permitted errors to occur that might otherwise have been detected. [9] The demands of advanced technologies on department resources might have drawn resources from simpler or basic functions. [10]

2. The Components of a QA Program

Quality assurance (QA) is defined by the Joint Commission on Accreditation of Healthcare Organizations (JCAHO) as "All planned or systematic actions necessary to provide adequate confidence that a service or product will satisfy given requirements for quality." [11] In radiation oncology, this means that a QA program must be (a) planned, meaning that it must be prospective; (b) systematic, meaning that it must be well organized and intended to address all appropriate aspects of performance; (c) and necessary, meaning that it should include only those activities that contribute meaningfully to maintaining quality. Ultimately, the QA program must provide "adequate confidence" that the delivery of radiation therapy will meet published and reasonable expectations for quality, accuracy, and safety.

A critical aspect of a QA program is independence; that is, the QA procedures conducted to assure the quality and accuracy of the product or process (in this case the delivery of radiation therapy) must be independent of the product or process itself. The failure to establish independence can lead to the risk that the QA device merely mimics the performance of the parameter being measured, masking an error or change.

3. The Role of the Radiological Physics Center

The RPC was established in 1968 to contribute to the development, conduct, and QA of multi-institutional cooperative group clinical trials. A description of clinical trials and the role of several of the QA offices can be found in a report published by the AAPM. [12] The mission of the Radiological Physics Center is to assure NCI and the Cooperative Groups that institutions participating in clinical trials deliver prescribed radiation doses that are clinically comparable and consistent.

In 1999, the RPC joined with several other QA offices to form the Advanced Technologies Consortium. [13] The ATC currently consists of the RPC, the Image-Guided Therapy QA Center (ITC) in St. Louis, the Quality Assurance Resource Center (QARC) in Providence, RI and the headquarters dosimetry group of the Radiation Therapy Oncology Group (RTOG), located in Philadelphia. The role of the ATC is to support the development and conduct of advanced-technology clinical trials, and to facilitate communications among the four QA offices.

3.1. Remote Audits

The RPC initiated a program of mailed thermoluminescent dosimeters (TLDs) in 1977. [14, 15] The uncertainty of the TLD system to measure output of accelerators remotely has been evaluated and found to be 1.5%. [15] Consequently, the RPC's measurement of an institution's output can be stated at an uncertainty of less than 5% using a 99% confidence interval. When the TLD measurement disagrees with an institution's stated dose by more than 5%, the RPC initiates a series of activities to resolve the discrepancy. If the discrepancy cannot be resolved through telephone calls and the review of procedures and documentation, an on-site dosimetry visit is scheduled. The RPC recently replaced the TLD system with one based on optically stimulated luminescence dosimetry (OSLD). [16]

3.2. On-site Dosimetry Review Visits

An on-site audit has been recommended by several organizations, including the AAPM and the IAEA. [17, 18] An independent audit is especially important for solo practitioners but is a valuable exercise for all practicing clinical medical physicists. It need not be extensive, but should address key activities such as basic calibrations, the overall QA program and documentation.

The RPC visit procedure consists of a review of the institution's QA procedures and documentation; a review of treatment records to ascertain the consistency of the procedures used for treatment planning and monitor unit calculations; and measurements of the radiation beams and radioactive sources. The measurements include mechanical alignment and accuracy of position readout devices, light versus radiation field congruency, calibration of treatment machine output and brachytherapy source strength, relative field-size dependence, percent depth dose, off-axis ratio, some asymmetric jaw and irregular field parameters, and accessory transmission factors. Several measurements are designed to evaluate the basic data required for delivery of IMRT. Procedures for evaluating image-guided radiation therapy (IGRT) are under development, and procedures for visits to proton-beam facilities are currently being implemented.

3.3. Reviews of Patient Treatment Records

In some cases, the RPC (or another QA office) reviews the treatment plans prepared by participating institutions for patients registered on the clinical trial. When the number of patients to be registered is relatively small, and the protocol is complex, the study chair may require that institutions submit their

Journal of Physics: Conference Series **250** (2010) 012001 doi:10.1088/1742-6596/250/1/012001

treatment plans for review before the patients are treated. The RPC often participates in such "rapid reviews" to ensure that the treatment plans meet the dosimetric requirements of the protocol. In other cases, the RPC performs retrospective reviews. The RPC relies on measurement made at the institution through the TLD program and on-site dosimetry reviews, or if a visit has not yet been made, its database of measured "standard" data. Using these data and the treatment parameters (field size, depth, MU setting, etc.) the RPC can independently calculate the dose received by the patient.

3.4. Credentialing for Advanced Technology Clinical Trials

Clinical trials that require the use of advanced technologies such as IMRT and prostate brachytherapy are considered sufficiently challenging that institutions are required to demonstrate their ability to use these technologies before being permitted to register patients. Credentialing for such clinical trials generally involves the completion of questionnaires, review of patients treated previously, computation and submission of a benchmark treatment plan, and in some cases, irradiation of an anthropomorphic phantom.

4. Results from RPC's QA Audits

4.1. Annual Calibration Checks

Over the years, 5% to 6% of the US megavoltage beams audited with TLD have fallen outside of the RPC's ±5% dose or 5 mm electron depth dose criteria on the first measurement. (Figure 1) With 3200 machines being monitored today, this represents approximately 700 beams. The RPC has determined that the approximately 750 institutions that have been visited have contributed about 85% of all clinical trial patients. Among these institutions, approximately 20% or 150 have one or more beams outside of the RPC's criteria on an annual basis that require an investigation by the RPC. (Figure 2)

An annual independent audit coincides with the quality assurance standard requirement in most states that each beam be calibrated on an annual basis to ascertain that the output has not changed. The precedent for performing the TLD audit annually was established by the RPC many years ago and all current trial data and results are based on having this level of QA. A review of EORTC trial results indicated that decreases in tumor control probability were associated with discrepancies in the beam calibration, as measured by a TLD audit program. [19] At the same time, increases in normal tissue morbidity were associated with discrepant high TLD measurements. The article also indicated that sequential TLD audits improved the uniformity of the clinical outcome and that small deviations in beam output might lead to clinically important variations in outcome. Mailed TLD audits were deemed to be an integral part of quality assurance for trials.

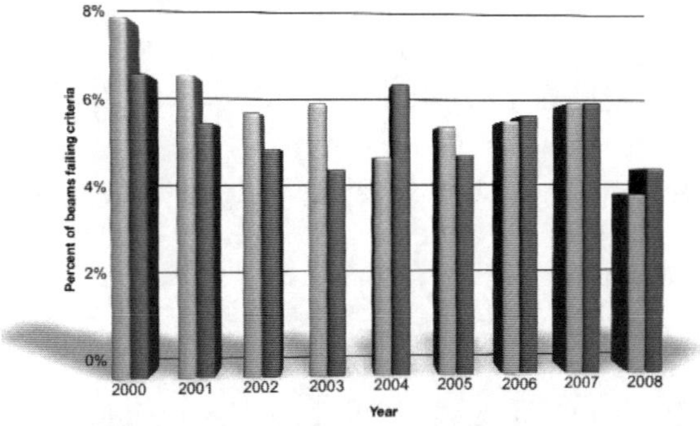

Figure 1: The percent of megavoltage radiation beams at US institutions that fail to meet the RPC's 5 %/5 mm criteria for acceptability. The blue bars indicate the proportions at institutions visited by the RPC.

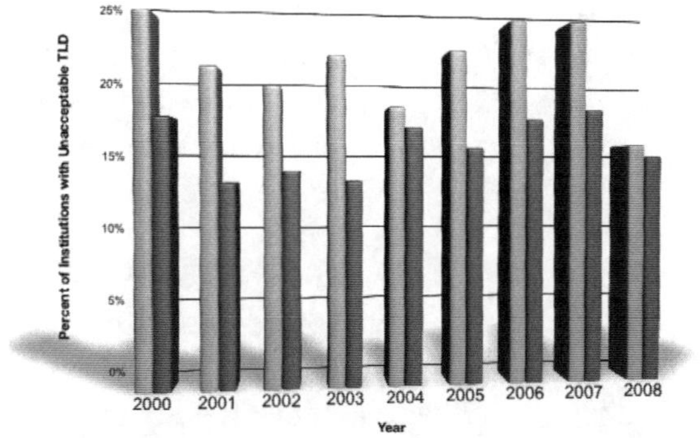

Figure 2: The percent of institutions irradiating TLD in any year that have at least one beam failing the RPC's 5 %/5 mm criteria for acceptability.

4.2. Measurements of Beam Parameters

Following a dosimetry review visit, the RPC generates a detailed report describing the observations and measurements that were made and the level of agreement with the institution's planning data. The report clearly indicates any measurements that disagree by a significant amount with the institution's data, and when the disagreements exceed appropriate thresholds, the report includes recommendations to the institution for improvement. These recommendations then demonstrate areas that require attention by the institution. When considered in aggregate, these recommendations form an indication of the areas of general concern at the visited institutions. The common recommendations and the frequency with which institutions receive them are shown in Table 1. Of note are the recommendations indicated by asterisks; these are considered important dosimetry parameters. Overall, 70% of visited institutions received one or more of these recommendations.

Table 1: Selected discrepancies detected during 2004-2008 during RPC dosimetry review visits to 165 institutions. The parameters indicated by asterisks are considered significant dosimetry parameters. During this 5-year period, 70% of the visited institutions received one or more recommendations to address discrepancies in these dosimetry parameters.

Errors Regarding:	Number of Institutions (%)
Review QA Program	127 (84%)
*Wedge Transmission	53 (32%)
*Photon FSD (small fields)	46 (28%)
Off-axis Factors, Beam Symmetry	42 (25%)
*Photon depth dose	34 (21%)
*Electron Calibration	25 (15%)
*Photon Calibration	22 (13%)
*Electron Depth Dose	19 (12%)

4.3. Reviews of QA Programs

The RPC judges the quality of an institution's QA program against the AAPM's TG-40 and TG-142 recommendations. [20, 21] A list of common failures or lapses in QA programs found by the RPC appears in Table 2.

Table 2. Common QA lapses and deficiencies found at institutions during RPC visits.

- QA records not available or maintained
- Annual calibrations or monthly checks not performed timely
- No record of comparison to clinical values on annual report
- No record of comparison of daily and monthly checks against annual "baseline" values
- Physicist review of daily checks not documented
- No record of corrective actions and repeat measurements
- Daily check of electron beam energy not performed
- Output or field flatness constancy with gantry angle not checked during annual calibration

4.4. Observations from Reviews of Patient Records

Over a 5-year period from 2005-2009 the RPC found systematic errors in 1% of charts, individual errors in 8% of charts, and transcription errors in 27% of the charts. In each case, the error was corrected by the RPC and reported to the study group so that correct information could be used for evaluation of the clinical trial. The results of these reviews were also reported to the institutions promptly, to enable the institutions to take corrective action.

4.5. Results of Anthropomorphic Phantom Reviews

During the time period 2001 to 2009 the RPC mailed IMRT head-and-neck phantoms to 472 distinct institutions. See Table 3 and Figure 3. A total of 752 irradiations were analyzed. Of these, 585 irradiations or 78% successfully met the irradiation criteria. More than 350 institutions failed to meet the irradiation criteria on the first attempt and had to repeat the phantom irradiation. Of those failing to meet the accreditation criteria, the majority failed only the dose criterion. The remaining unsuccessful irradiations failed the distance-to-agreement (DTA) criterion or both the dose and DTA criteria.

Table 3. Passing rates for the RPC's anthropomorphic phantoms. The criteria for agreement are 7 % and 4 mm for all phantoms except the lung phantom, for which the criteria are 5 % and 5 mm.

Phantom	H&N	Prostate	Spine	Lung	Liver
Irradiations	752	174	19	174	23
Pass	585	143	13	124	12
Pass %	78%	82%	68%	71%	52%
Criteria	7%/4mm	7%/4mm	5%/3mm	5%/5mm	7%/4mm
Year introduced	2001	2004	2009	2004	2005

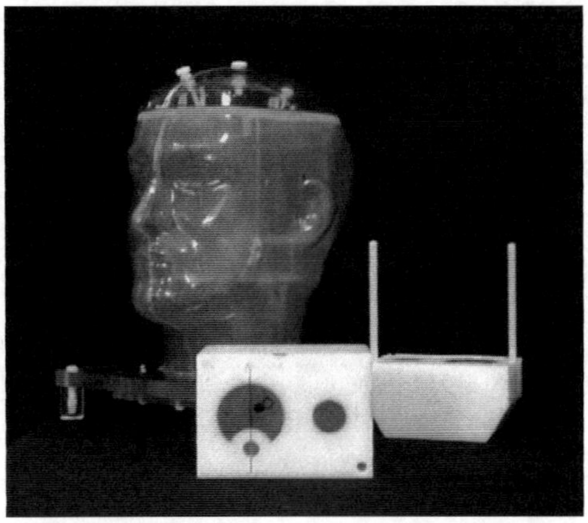

Figure 3. The RPC's head-and-neck phantom. The phantom is representative of all RPC phantoms and consists of a water-filled shell containing inserts with imageable structures representing target volumes and organs at risk. The inserts also contain radiochromic film and TLD dosimeters to evaluate the delivered dose.

The average TLD/Institution ratio for the PTV structures was 0.99 with a standard deviation of 5%. The range of the measurements was large, with ratios extending from 0.44 to 1.26. The distance to agreement (DTA) for the high dose gradient region between the primary PTV and the OAR averaged 0.2 mm with a standard deviation of 3.1 mm. The majority of the phantoms met the 4 mm criterion. The range of the DTAs was from -15 mm to +17 mm. A negative DTA meant that an institution delivered dose posteriorly beyond the planned distribution and delivered a higher than intended dose to the OAR.

5. Summary and Conclusions

5.1. State of QA in the US

As described above, the majority of institutions audited by the RPC meet well-accepted criteria established by organizations such as the AAPM and RTOG. However, a significant number of institutions fail to meet these standards. The RPC's mailed TLD program found reference calibration discrepancies at 15% to 20% of the audited institutions each year of the last eight. During dosimetry audit visits, measurements with an ionization chamber in a water phantom were made of several significant dosimetry parameters. Seventy percent of the institutions visited had discrepancies in one or more of these parameters. In particular, the RPC disagreed with the institution's calibration of one or more photon beams during audit visits to 13% of the institutions visited during the past 5 years. During reviews of treatment records, the RPC has disagreed with calculations performed by the audited institutions in 12% of the reviewed charts. Measurements of tumor dose delivered to anthropomorphic phantoms demonstrated that institutions delivered the intended tumor dose within the 7% / 4 mm DTA criteria 75% to 80% of the time.

References

[1] Kohn L. T., J. M. Corrigan, M. S. Donaldson. (2000). "To err is human: Building a safer health system". Washington, DC: National Academy Press.

[2] Donaldson, Sir Liam, *Towards Safer Radiotherapy*. The Royal College of Radiologists, Society and College of Radiographers, Institute of Physics and Engineering in Medicine, National Patient Safety Agency, British Institute of Radiology. London, The Royal College of Radiologists, 2008.

[3] International Atomic Energy Agency. Safety Reports Series No. 17. Lessons Learned from Accidental Exposures in Radiotherapy. Vienna: IAEA, 2000.

[4] International Commission on Radiation Protection. "Prevention of Accidental Exposures to Patients Undergoing Radiation Therapy". Annals of the ICRP, Volume 30, Issue 3, September 2000

[5] Barthelemy-Brichant N, Sabatier J, Dewe W, et al. (1999) Evaluation of frequency and type of errors detected by a computerized record and verify system during radiation treatment. *Radiother Oncol*;**53**:149 –154.

[6] Perrow, Charles "Normal Accidents: Living with High Risk Technologies," Princeton University Press, Princeton, New Jersey, 1999.

[7] Macklis RM, Meier T, Weinhous MS. Error rates in clinical radiotherapy. *J Clin Oncol* 1998; 16: 551–556.

[8] Huang G, Medlam G, Lee J et al. (2005) "Error in the delivery of radiation therapy: results of a quality assurance review." *Int J Radiat Oncol Bio Phys* ; 61: 1590–1595.

[9] Patton G. A., D. K. Gaffney, J. H. Moeller. (2003). "Facilitation of radiotherapeutic error by computerized record and verify systems. *Int J Radiat Oncol Biol Phys*. 56: 50-57

[10] Marks L. B., K. L. Light, J. L. Hubbs, D. L. Georgas, E. L. Jones, M. C. Wright, C. G. Willett, F. F. Yin. (2007). "The impact of advanced technologies on treatment deviations in radiation treatment delivery." *Int J Radiat Oncol Biol Phys*. **69**: 1579-1586

[11] O'Leary, M. R. *Lexikon: Dictionary of health care terms, organizations and acronyms for the era of reform.* Oak Brook Terrace, IL. Joint Commission on Accreditation of Healthcare Organizations. 1994

[12] Olch A., R. Kline, G. S. Ibbott, J. R. Anderson, J. Deye, T. J. Fitzgerald, D. S. Followill, M. T. Gillin, M. S. Huq, J. R. Palta, J. A. Purdy, M. M. Urie. (2004) "Quality Assurance for Clinical Trials: A Primer for Physicists, Prepared by AAPM Subcommittee on QA for Clinical Trials". AAPM Report No. 86. American Institute of Physics, College Park, MD.

[13] Purdy J, Palta J, Ibbott G. (2004) The Advanced Technology QA Consortium (ATC). *Med. Phys*. **31**:1833.

[14] Kirby, T. H., W.F. Hanson, R.J. Gastorf, C.H. Chu, R.J. Shalek. (1986) Mailable TLD system for photon and electron therapy beams. Int. J. Radiat. Oncol. Biol. Phys. 12, 261-265.

[15] Kirby, T.H., W.F. Hanson, D.A. Johnston. (1992). Uncertainty analysis of absorbed dose calculations from thermoluminescensce dosimeters. Med. Phys. 19, 1427-1433.

[16] Homnick J, Ibbott G, Springer A, Aguirre J. Optically Stimulated Luminescene (OSL) Dosimeters Can Be Used for Remote Dosimetry Services. Med Phys 35(6):2994, 6/2008.

[17] Halvorsen PH, Das IJ, Fraser M, Freedman DJ, Rice RE, 3rd, Ibbott GS, Parsai EI, Robin TT, Jr, Thomadsen BR. (2005). "AAPM Task Group 103 report on peer review in clinical radiation oncology physics." *J Appl Clin Med Phys* 6:50-64.

[18] International Atomic Energy Agency "Comprehensive Audits of Radiotherapy Practices: A Tool for Quality Improvement". IAEA Publications, Vienna, 2007

[19] Bentzen et al, Clinical impact of dosimetry quality assurance programmes assessed by radiobiological modeling of data from the thermoluminescent dosimetry study of the European Organization for Research and Treatment of Cancer, European J. Cancer (36), 615-620, 2000.

[20] Kutcher, Gerald J., L. Coia, M. Gillin, W. F. Hanson, S. Leibel, R. J. Morton, J. R. Palta, J. A. Purdy, L. E. Reinstein, G. K. Svensson, M. Weller, L. Wingfield. "Comprehensive QA for Radiation Oncology". *Med. Phys.*, 21, 581-618, 1994

[21] Klein EE, Hanley J, Bayouth J, Yin FF, Simon W, Dresser S, Serago C, Aguirre F, Ma L, Arjomandy B, Liu C, Sandin C, Holmes T. Task Group 142 Report: Quality Assurance of Medical Accelerators. Med Phys 36:4197-4212, 2009.

Journal of Physics: Conference Series **250** (2010) 012002 doi:10.1088/1742-6596/250/1/012002

Dosimetry challenges for implementing emerging technologies

Fang-Fang Yin, Mark Oldham, Jing Cai, and Qiuwen Wu

Department of Radiation Oncology
Duke University Medical Center, Durham, NC 27516, USA

Corresponding author email: Fangfang.yin@duke.edu

Abstract. During the last 10 years, radiation therapy technologies have gone through major changes, mainly related introduction of sophisticated delivery and imaging techniques to improve the target localization accuracy and dose conformity. While implementation of these emerging technologies such as image-guided SRS/SBRT, IMRT/IMAT, IGRT, 4D motion management, and special delivery technologies showed substantial clinical gains for patient care, many other factors, such as training/quality, efficiency/efficacy, and cost/effectiveness etc. remain to be challenging. This talk will address technical challenges for dosimetry verification of implementing these emerging technologies in radiation therapy.

1. Introduction

ASTRO regards any technology, yet to be optimally implemented, as having the potential to improve the effectiveness and efficiency of current radiation therapy practice by providing better definition of the treatment target, more conformal dose distribution, more accurate treatment delivery, and modification of dose scheme. The rapid implementation of emerging technology in radiation therapy has provided an excellent opportunity to provide high quality radiation therapy to patients. However, it also posts substantial challenge to clinic professionals who wish to rapidly implement these technologies in an accurate and safe manner. In this presentation we highlight some of the key clinical techniques where more comprehensive, and 3D dosimetry techniques, may be able to facilitate accurate clinical implementation.

2. SRS/SBRT: small field dosimetry (Refs 1-4)

Treatment of SRS and SBRT requires dosimetry of small fields down to a few mm beam apertures. Measurement of beam characteristics with such small fields are very challenging, giving the effect of partial volume of detector responses, radiation lateral scatter, etc. Errors are often devastating due to the nature of high prescription dose. Such serious treatment errors have been reported in the media resulting in compromised patient care. One effective measure to avoid these problems is to measure the dosimetry of the treatment planning system with 3D dosimeter. That equally requires high resolution, high sensitivity and stable 3D dosimetry measurement device, and robust dosimetric verification protocol. Currently this is not available although SRS/SBRT is widely available clinically.

3. IMRT/VMAT: complicated dosimetry (Refs 5-10)

IMRT/VMAT involves not only small fields, but also sophisticated mechanical movement of MLC and gantry, as well as dynamic change of dose rate. In addition to complicated treatment modification from the conventional treatment unit, some new developments use very special techniques for

Journal of Physics: Conference Series **250** (2010) 012002 doi:10.1088/1742-6596/250/1/012002

radiation delivery, such as robotic moving beam with Cyberknife and spiral delivery using Tomotherapy. Any deviation of these parameters could have substantial dosimetry impact on the planned treatment. Due to its complicated dose distribution and high modulated nature, typical 2D dosimetry may have difficulty to detect these variations at all regions. It is desirable to have high resolution and 3D dosimetric information, which is not yet available for routine clinical practice.

4. IGRT: imaging dosimetry (Refs. 11-23)

A variety of imaging systems are being introduced into the radiation oncology for real-time target localization and verification, such as stereoscopic imaging, cone-beam CT with both kV and MV sources, etc. Some clinical trials also require repeated imaging during treatment course. Over exposure to patient during imaging has been reported in the media. Those technologies have substantially improved the localization accuracy. However, imaging dose is also becoming an increasing concern especially because the imaging volume is much larger than the treatment volume. For example, dose to the eye and other components when imaging for treating head and neck patients, dose to the lung and to the contra-lateral breast when imaging chest are concerning factors for treatment complication and increased probability of radiation induced cancer. Accurate measurement of that dose would be very critical for developing imaging protocols and for imaging dose calculations.

5. Organ motion: 4D dosimetry (Refs. 24,25)

Dosimetry of moving target is not well known for many reasons. Not only because the target is moving, but also the target involve substantial non-uniform tissue organs. Motion is often not uniform in all directions, the pattern of motions of different organs is typically different, and the organ may be deformable. Dose measurement of moving target using 2D dosimetry is extremely challenging if not possible. Precise measurement of such dosimetry will need to be at least at three dimensional and high resolution. Tracking the dose along different treatment fractions is also challenging.

6. Dosimetry tools

The sections above outline the many areas of emerging technologies where comprehensive 3D dose measurements would be highly desirable. At present many clinics are using 2D dosimeters to verify complex distributions, including film, diode/chamber arrays, and EPIDS. This applies even to the commissioning phase, where extensive measurements and more time and resources are allocated for verification. Many of these devices have the major advantage of convenience, and many provide near real-time feedback with results. Much can be achieved with these devices. Two basic limitations that accompany many such devices however, include poor spatial resolution, and limited coverage of the whole distribution to be verified. The 3D techniques described in this meeting (including optical-CT, MRI, x-ray-CT and other approaches) can certainly improve on the limitations. The factors of convenience and real-time feedback of results continue to improve, and we shall see during the meeting the current state of the art.

An important factor in the transition of these technologies into the clinic is determining a clear context of use. At present gamma criteria are commonly used to evaluate agreement on 2D planar dosimetry measurements. Dose distributions are not 2D entities however, and extending the analysis to three dimensions is an important goal. This subject is addressed in a dedicated session in this meeting. Determining accepted threshold for agreement is not consistently applied at present, and the clinical implementation needs strengthening.

Another important aspect for any dimensional dosimetry is related to the quality management. Dosimetry tools are primary used for quality assurance of treatment delivery accuracy. However, the accuracy of dosimetry devices is equally important to ensure the measurement accuracy. Therefore, quality assurance and proper calibration of dosimetry tools should be part of routine practices in all clinics. The accurate dosimetry information is also dependent on the measurement methods and proper use of dosimetry tools.

7. Summary

Dosimetry could be regarded as a classical topic because it has been applied and used for decades. However, it is also an emerging technology because we yet to reveal techniques and devices to measure challenging dosimetry listed above and to reveal many unknown factors and mechanisms of different measurement devices. In addition to the dosimetric topics above, some other emerging dosimetry problems are also need to be explored, such as particle therapy, brachytherapy both using by-product and/or electronic brachytherapy, as well as sophisticated combined modality treatment. The dosimetry for fast growing proton therapy is specially challenging due to no exit dosimetry information. Therefore, it is fair to say; 3D and 4D dosimetry coupled with effective and accurate measurements will be one of most challenging topics for implementing emerging technologies in modern radiation therapy. We not only need to perform accurate dose measurement but also need to secure accurate measurement devices.

References

[1] R.K. Rice, J.L. Hansen, G.K. Svensson, *et al.* Measurements of dose distributions in small beams of 6 MV x-rays *Phys. Med. Biol.* 1987 (32);1087-1099.

[2] I.J. Das, G.X. Ding GX, A. Ahnesjo. Small fields: nonequilibrium radiation dosimetry, *Med. Phys.* 2008 (35); 206-215.

[3] J. Fan, K. Paskalev, L. Wang, *et al.* "Determination of output factors for stereotactic radiosurgery beams". *Med Phys* **36**, 5292-5300 (2009).

[4] C. Clift, A. Thomas, J. Adamovics, *et al.* Toward acquiring comprehensive radiosurgery field commissioning data using the PRESAGE/optical-CT 3D dosimetry system. Phys Med Biol. 2010 Mar 7;55(5):1279-93. Epub 2010 Feb 4.

[5] J.L. Bedford, Y.K. Lee, P. Wai, *et al*, Evaluation of the Delta4 phantom for IMRT and VMAT verification, *Phys Med Biol* (2009) 54; N167–N176.

[6] A. van Esch, C. Clermont, M. Devillers, *et al*, On-line quality assurance of rotational radiotherapy treatment delivery by means of a 2D ion chamber array and the Octavius phantom, *Med. Phys.* (2007) 34; 3825–3837.

[7] S. Korreman, J. Medin and F. Kjaer-Kristoffersen, Dosimetric verification of RapidArc treatment delivery, *Acta Oncol* (2009) 48; 185–191.

[8] C.F. Chuang, L.J. Verhey, and P. Xia, Investigation of the use of MOSFET for clinical IMRT dosimetric verification, *Med. Phys.* (2002) 29; 1109–1115.

[9] A. Mans, P. Remeijer, I. Olaciregui-Ruiz, *et al*, 3D Dosimetric verification of volumetric-modulated arc therapy by portal dosimetry; *Radio. Oncol.* (2010) 94; 181–187.

[10] H. Sakhalkar, D. Sterling, J Adamovics, *et al,* Investigation of the feasibility of relative 3D dosimetry in the Radiologic Physics Center Head and Neck IMRT phantom using presage/optical-CT. Med Phys. 2009 Jul;36(7):3371-7

[11] M.K. Islam, T.G. Purdie, B.D. Norrlinger *et al*, Patient dose from kilovoltage cone beam computed tomography imaging in radiation therapy, *Med. Phys.* (2006) 33; 1573–1582.

[12] W.K. Kan, H.T. Leung, W. Wong, *et al*, Radiation dose from cone beam computed tomography for image-guided radiation therapy. *Int. J. Radiation Oncology Biol. Phys.* (2008) 70; 272–279.

[13] P. Alaei, G.X. Ding, H. Guan. "Inclusion of the dose from kilovoltage cone beam CT in the radiation therapy treatment plans". *Med Phys* **37**, 244-248 (2010).

[14] J.C. Chow, M.K. Leung, M.K. Islam, *et al.* "Evaluation of the effect of patient dose from cone beam computed tomography on prostate IMRT using Monte Carlo simulation". *Med Phys* **35**, 52-60 (2008).

[15] G.X. Ding, C.W. Coffey, "Radiation dose from kilovoltage cone beam computed

tomography in an image-guided radiotherapy procedure". *Int J Radiat Oncol Biol Phys* **73**, 610-617 (2009).

[16] G.X. Ding, D.M. Duggan, C.W. Coffey, "Accurate patient dosimetry of kilovoltage cone-beam CT in radiation therapy". *Med Phys* **35**, 1135-1144 (2008).

[17] E.K. Osei, B. Schaly, A. Fleck, *et al,* "Dose assessment from an online kilovoltage imaging system in radiation therapy". *J Radiol Prot* **29**, 37-50 (2009).

[18] C. Beltran, R. Lukose, B. Gangadharan, *et al,* "Image quality & dosimetric property of an investigational imaging beam line MV-CBCT". *J Appl Clin Med Phys* **10**, 3023 (2009).

[19] O. Gayou, D.S. Parda, M. Johnson, *et al,* "Patient dose and image quality from mega-voltage cone beam computed tomography imaging". *Med Phys* **34**, 499-506 (2007).

[20] E.K. Hansen, D.A. Larson, M. Aubin, *et al,* "Image-guided radiotherapy using megavoltage cone-beam computed tomography for treatment of paraspinous tumors in the presence of orthopedic hardware". *Int J Radiat Oncol Biol Phys* **66**, 323-326 (2006).

[21] L.C. Peng, C.C. Yang, S. Sim, *et al,* "Dose comparison of megavoltage cone-beam and orthogonal-pair portal images". *J Appl Clin Med Phys* **8**, 10-20 (2007).

[22] J. Pouliot, A. Bani-Hashemi, J. Chen, *et al,* "Low-dose megavoltage cone-beam CT for radiation therapy". *Int J Radiat Oncol Biol Phys* **61**, 552-560 (2005).

[23] J. Sillanpaa, J. Chang, G. Mageras, *et al,* "Developments in megavoltage cone beam CT with an amorphous silicon EPID: reduction of exposure and synchronization with respiratory gating". *Med Phys* **32**, 819-829 (2005).

[24] J. Sillanpaa, J. Chang, G. Mageras, *et al,* "Low-dose megavoltage cone-beam computed tomography for lung tumors using a high-efficiency image receptor". *Med Phys* **33**, 3489-3497 (2006).

[25] A. Cherpak, W. Ding, A. Hallil, *et al,* Evaluation of a novel 4D *in vivo* dosimetry system *Med. Phys.* (2009) 36; 1672-1679.

[26] S.L. Brady, W.E. Brown, C.C Clift, *et al,* Investigation into the feasibility of using PRESAGE/optical-CT dosimetry for the verification of gating treatments. Phys Med Biol. 2010 Apr 21;55(8):2187-201. Epub 2010 Mar 26.

Journal of Physics: Conference Series **250** (2010) 012003 doi:10.1088/1742-6596/250/1/012003

A Practical Use for FXG Gel Dosimetry

T Olding[1], G Salomons[2], J Darko[2] and L J Schreiner[1,2]

[1]Department of Physics, Queen's University, Kingston, Ontario, Canada, K7L3N6
[2]Cancer Centre of South Eastern Ontario at Kingston General Hospital, 25 King Street West, Kingston, ON, Canada, K7L5P9

E-Mail: Tim.Olding@krcc.on.ca

Abstract. In-phantom Fricke-xylenol orange-gelatin (FXG) gel dosimetry yields three dimensional (3D) dose data for intensity modulated radiation therapy (IMRT) treatment plan verification within 18-24 hours from the point of request. The information obtained from a 3% dose difference, 3 mm distance-to-agreement gamma function comparison between treatment plan dose and gel-measured dose then provides a useful secondary 3D quality assurance check of the treatment plan prior to delivery.

1. Introduction

Quality assurance (QA) in IMRT is important. While two dimensional (2D) and electronic portal image device (EPID) 3D techniques are quick and powerful, they can fail. Then what? Do we use gels? This paper presents an example of the application of FXG gel dosimetry [1] in combination with cone beam optical computed tomography (CT) for IMRT treatment plan evaluation [2] when a standard IMRT QA has failed. The investigation makes use of previous work related to scanner characterization [3]. A calibration basis for this FXG-gel based IMRT evaluation (megavoltage range energy independence, dose calibration linearity and intra-batch reproducibility of the FXG dosimeter) is outlined in a companion paper in these proceedings. With this background, a head-and-neck IMRT plan evaluation using a standard FXG gel-filled 1 L polyethylene terephthlate (PETE) jar-in-phantom is described.

2. Experiment

2.1. IMRT Treatment Plan Quality Assurance

Figure 1a shows a CT slice of a selected head-and-neck case at the Cancer Centre of Southeastern Ontario (CCSEO), including the overlaid planning target volume (PTV) isodose contours from the 7 field, 6 MV IMRT treatment plan calculated in Eclipse™ (Varian Medical, Palo Alto, CA, USA). Two checks were employed for treatment plan QA. First, the individual fields of the plan were cumulatively delivered at a gantry angle of 0° to an ImRT Matrixx running software version 1.6 (Scanditronix Wellhöfer, Bartlett, TN, USA). The Matrixx consists of 1020 vented ion chambers in a 32 by 32 square grid, with 7.6 mm spacing between chambers, and the treatment unit was a Varian Clinac 21iX linear accelerator with cone beam CT (CBCT) On-Board Imaging (Advanced OBI). The aggregate Matrixx measured dose was then compared to the aggregate Eclipse plan dose using a two dimensional (2D) Low's gamma function evaluation [4] with 3% dose difference and 3 mm distance-to-agreement criteria. The 2D gamma map (figure 1b) indicated full agreement (i.e. gamma value < 1)

Journal of Physics: Conference Series **250** (2010) 012003 doi:10.1088/1742-6596/250/1/012003

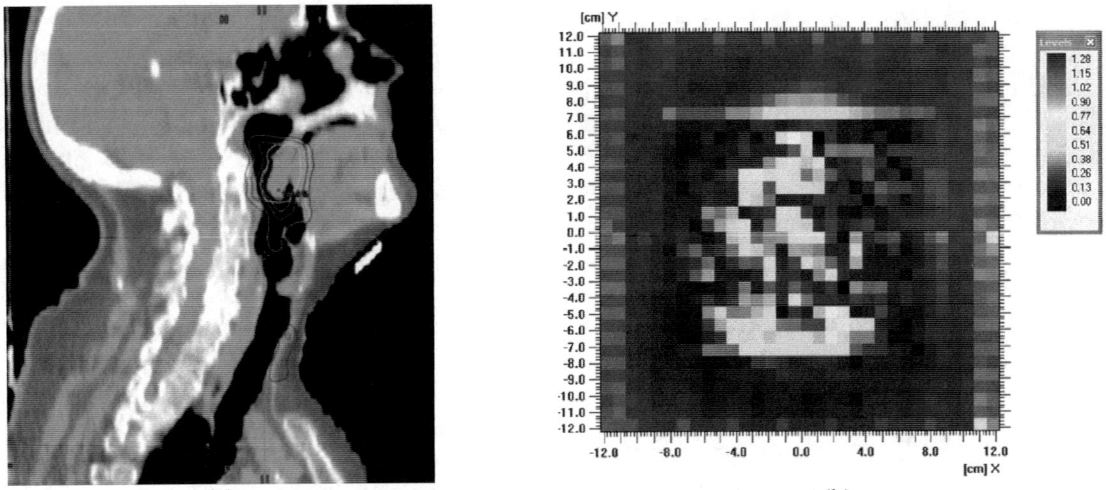

(a) (b)

Figure 1: (a) A sagittal slice of a selected head-and-neck cancer case, showing the PTV isodose contours from the calculated IMRT treatment plan in Eclipse. (b) A gamma map comparison using 3%, 3mm evaluation criteria of the aggregate Eclipse plan and Matrixx measurement dose for the treatment plan fields delivered at a gantry angle of 0°.

between the plan and the measurement within the specified criteria.

In the second QA check, portal image data was acquired from the delivery of the IMRT plan at the treatment unit without the patient in place. A back-projection technique-based software package (termed 'Epidose') [5] was then used to reconstruct 3D dose to a virtual phantom on the treatment couch (see figure 2a). The plan dose and Epidose-measured dose were compared within the software using the weighted dose difference chi (χ) test described by Bakai *et al* [6], with 4% dose difference and 4 mm distance-to-agreement criteria. Referring to figure 2b, the agreement between plan and measured dose in the high dose region ($\geq$ 80% of the normalized prescription dose) was 78.3%, falling short of the QA target value of 80%. The agreements within the phantom PTV and body contours were 31.8% and 98.7% respectively. When the criteria were tightened to the preferred metrics of 3% and 3 mm, the agreement within the high dose region was reduced to 46.0%.

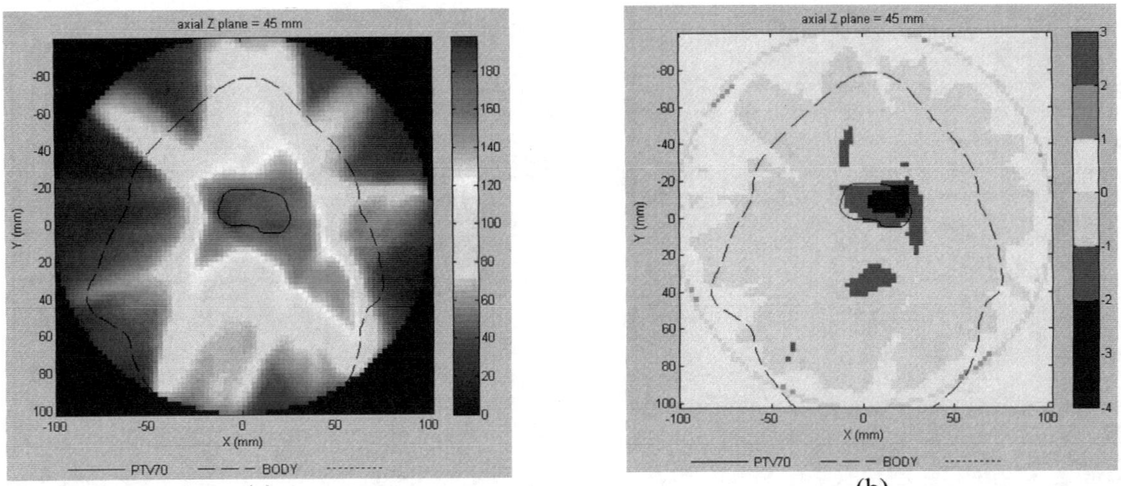

(a) (b)

Figure 2: (a) An axial slice of the Epidose reconstructed dose distribution. The PTV and body contours are indicated on the slice and the colour bar indicates the measured dose in cGy. (b) The 4%, 4 mm χ test results for the axial slice in (a).

Journal of Physics: Conference Series **250** (2010) 012003 doi:10.1088/1742-6596/250/1/012003

2.2. *FXG Gel Dosimetry*

FXG gel dosimeters were prepared according to the recipe outlined by Babic *et al* (2008), with final concentrations of 0.3 mM ferrous ammonium sulphate (Cat.No.203505, Sigma-Aldrich Ltd., Oakville, ON, Canada), 0.05 mM xylenol orange (Cat.No.398187, Sigma-Aldrich Ltd), 65 mM sulfuric acid (Cat.No.258105, Sigma-Aldrich Ltd), and 6 wt% gelatin (Cat.No.G2500, Sigma-Aldrich Ltd) in distilled, de-ionized water. After preparation, the heated formulations were poured into 1 L PETE jars and allowed to set overnight in a refrigerator prior to irradiation.

Three dimensional imaging of the dosimeters was performed at a wavelength of 590 nm using the Vista™ cone beam optical CT scanner (Modus Medical Devices, Inc., London, Canada) [3]. Depth fiducials were marked on the surface of the PETE jars in permanent red marker, serving as reference points for registration of depth dose to attenuation in the reconstructed optical CT image.

A wax-filled reproduction of the head-and-neck from a Kyoto SBU-4 Rando anthropomorphic phantom (Capintec, Ramsey, NJ) was prepared at the CCSEO for the IMRT study, with a custom cavity designed to fit the standard 1 L PETE jar-sized dosimeter (figure 3a). Two 1 L PETE containers were filled from a single batch of gel. One container was irradiated with a well characterized 12 MeV electron beam using the Varian Clinac 21iX linear accelerator (figures 3b and 3c) and used for dose-to-attenuation calibration. The other container was inserted into the wax-filled Rando reproduction as the IMRT measurement dosimeter. A CT data set was acquired with a FXG gel dosimeter 'blank' inserted in the phantom cavity on a Picker PQ 5000 Philips CT scanner (Philips Medical Systems, Andover, MA, USA). Fiducial marks were added to the exterior surface of the wax-filled reproduction as reference points for positioning at the treatment unit.

The wax Rando CT was first registered to the patient CT in Eclipse. The Matrixx and Epidose-evaluated IMRT treatment plan was then applied to the phantom CT data set and a single fraction (prescription dose of 200 cGy) scheduled to the treatment unit. CBCT imaging of the phantom was completed at 1 mm cubic voxel resolution according to the standard head protocol (100kVp/145mAs, half scan with full bowtie filter), and 3D-3D planning CT-onboard CBCT matching was performed. The planned treatment fraction was then delivered to the phantom.

Optical CT gel-measured dose data were registered to the Eclipse plan data in MatLab (Mathworks, Newark, NJ, USA), for viewing in the Computational Environment for Radiotherapy Research (CERR) developed at Washington University (St Louis, MO, USA) [7]. Prior to irradiation, a red marker was used to indicate fiducials on the surface of the PETE jar, to fix the spatial position of the optical CT volume data. Two sets of three fiducials marked on separate parallel planes were adequate

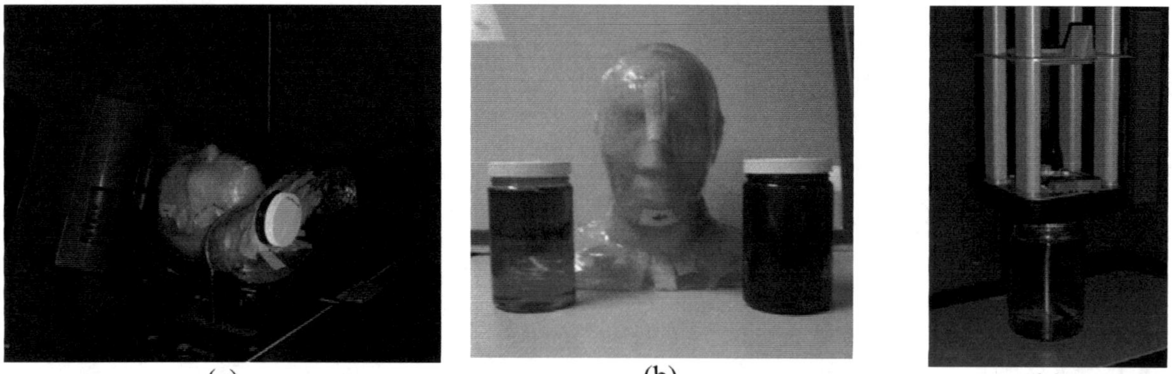

(a) (b) (c)

Figure 3: (a) A wax-filled reproduction of the head and neck from a Kyoto SBU-4 Rando anthropomorphic phantom (Capintec, Ramsey, NJ) prepared at the CCSEO, with a cavity for insertion of a gel-filled 1 L PETE jar. (b) Another view of the wax Rando reproduction, with a measurement jar for a test IMRT plan on the right, and an electron beam-irradiated FXG gel calibration jar on the left. (c) The calibration gel dosimeter electron beam setup at the treatment unit.

Journal of Physics: Conference Series **250** (2010) 012003 doi:10.1088/1742-6596/250/1/012003

for this purpose. Steel 1 mm diameter beads were affixed with tape on top of the permanent marker fiducials to provide positioning information from the 3D-3D planning CT matched onboard CBCT scan. The beads were removed from the jar after the irradiation, before optical CT scanning. An in-house registration routine written in MATLAB [8] was used to align the optical CT and CBCT fiducial data points through an affine transformation solid body rotation, and register the optical CT dose data to the planning CT. The mean CBCT scan dose to the gel dosimeter was determined by ion chamber measurements as 0.8 ± 0.1 cGy, and was appropriately accounted for. Dose evaluations between the reference Eclipse planning dose and optical CT gel-measured dose were completed using in-house developed routines in MatLab and contouring tools in CERR for full 3D voxel-by-voxel Low's gamma function analysis [8].

Figure 4a shows 3D visualizations (in CERR) of the treatment plan (top) and gel-measured dose distributions (bottom) overlaid on the contoured planning CT images for the 7 field, 6 MV, 200 cGy (single fraction) IMRT treatment plan delivery. Figure 4b presents the results of a 3D Low's gamma function voxel-by-voxel comparison of gel-measured dose against the reference Eclipse dose for the IMRT delivery, employing 3% dose difference and 3 mm distance-to-agreement criteria. The percentage of treatment plan dose and measured gel dose voxels in agreement (i.e. gamma value less than 1) within the dose evaluation volume covering the central 8 cm in diameter and 10 cm in height of the jar was 99.7%.

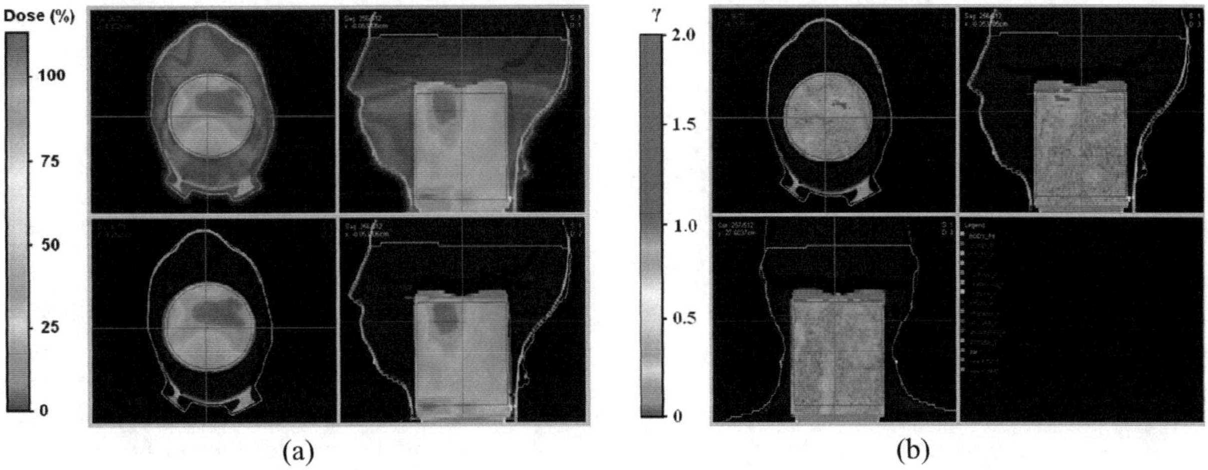

Figure 4: (a) Eclipse treatment planning dose (top) and gel-measured dose (bottom) for the 7 field, 6MV 200 cGy IMRT plan delivered to the wax-filled Rando reproduction. (b) A Low's gamma function comparison (3%, 3mm) between reference treatment planning dose calculated in Eclipse™ and FXG gel-measured dose.

3. Discussion & Conclusions

More than 90% of the IMRT treatment plan evaluations completed at the CSSEO to date have satisfied the agreement criteria between Eclipse plan dose and Epidose-measured dose, and hence have not required additional QA and/or re-planning. In this particular head-and-neck case, however, the χ test indicated significant disagreement between the plan and the measurement. The alternative 3D dose check using FXG gel dosimetry indicated, on the other hand, that the plan delivery yields an acceptable dose distribution in the phantom within 3%, 3 mm gamma comparison criteria. Given a total gel preparation time of 12 hours or less, CT acquisition and plan registration less than 1 hour, scan and reconstruction time under 30 minutes and analysis time of 3-4 hours, the 3D dose verification was easily completed within 18-24 hours from the point of request.

This illustrative example underscores the relevance of 3D dose verification for complex IMRT treatment plans. The 3D Epidose data indicated a problem with the delivery of this particular plan that did not show up in the 2D Matrixx check. Following the initial evaluation, the Epidose QA process was repeated on a separate occasion, with a comparable failed outcome. While the cause of disagreement in this particular case has not as yet been determined, similar occasional unexplained failures of Epidose tests in clinical practice have been reported [5].

The in-phantom FXG gel dose measurement subsequently indicated (in a timely manner) that the IMRT treatment plan was acceptable. These results lead to some questions. First of all, do we believe the gel measurement? And, if we do believe the gel measurement, how do we explain the Epidose data? Finally, and perhaps most importantly, does the 3D dose data from gel dosimetry lend the necessary confidence to proceed with this course of treatment? The experience reported here has prompted further validation of the Epidose analysis beyond the ion chamber and film commissioning already performed. Phantoms have been built to enable direct gel-based measurement of complex 3D distributions in the Epidose cylindrical geometry.

The authors suggest that FXG gel dosimetry can gain the necessary degree of confidence to make a significant contribution to IMRT QA, as it builds up a track record of careful usage in clinically relevant cases. The data in this report indicates that FXG gel dosimetry is capable of providing a useful *secondary* 3D measure of IMRT treatment plan dose. But a single case study is not enough to warrant further conclusions. We need to regularly engage in the sideline activity of using gel dosimetry for selected clinical plan evaluations with the clear understanding that a deeper case history will help to establish gel measurement as one of the available tools in the quality assurance decision-making process.

Acknowledgements
Research funding for this work has been provided by the Canadian Institutes of Health Research (CIHR) and the Cancer Centre of Southeastern Ontario.

References

[1] Kelly B G, Jordan K J and Battista J 1998 Optical CT reconstruction of 3D dose distributions using the ferrous-benzoic-xylenol (FBX) gel dosimeter *Med. Phys.* **25** 1741-50
[2] Babic S, Battista J and Jordan K 2008 Three-dimensional dose verification for intensity-modulated radiation therapy in the radiological physics centre head-and-neck phantom using optical computed tomography scans of ferrous xylenol-orange gel dosimeters *Int. J. Radiat. Oncol. Biol. Phys.* **70** 1281-91
[3] Olding T, Holmes O and Schreiner L J 2010 Cone beam optical computed tomography for gel dosimetry I: scanner characterization *Phys. Med. Biol.* **55** 2819-40
[4] Low D A and Dempsey Evaluation of the gamma dose distribution comparison method J F 2003 *Med. Phys.* **30** 2455-64
[5] Ansbacher W 2006 Three-dimensional portal image-based dose reconstruction in a virtual phantom for rapid evaluation of IMRT plans *Med. Phys.* **33** 3369-82
[6] Bakai A, Alber M and Nusslin F 2003 *Phys. Med. Biol.* **48** 3543-53
[7] Deasy J O, Blanco A I and Clark CERR: a computational environment for radiotherapy research V H 2003 *Med.Phys.* **30** 979-85
[8] Holmes O 2008 *Validation of Conformal Radiation Therapy Treatments in 3D Using Polymer Gel Dosimeters and Optical Computed Tomography* M.Sc. thesis (Kingston: Queen's University)

Journal of Physics: Conference Series **250** (2010) 012004 doi:10.1088/1742-6596/250/1/012004

Evaluation of the breast absorbed dose distribution using the Fricke Xylenol Gel

C Czelusniak[1], L S Del Lama[1], M V Moreira[2], A de Almeida[1]

[1]Departamento de Física e Matemática, Universidade de São Paulo, SP, Bandeirantes, 3900, DFM/FFCLRP, CEP 14040-901, Ribeirão Preto, São Paulo, Brazil
[2]Instituto de Radioterapia e Megavoltagem – IRMEV, Ribeirão Preto, SP, Brazil

E-mail:dalmeida@ffclrp.usp.br

Abstract. During a breast cancer radiotherapy treatment, several issues have to be taken into account, among them, hot spots, gradient of doses delivered over the breast, as well as in the lungs and the heart. The present work aims to apply the Fricke Xylenol Gel (FXG) dosimeter in the study of these issues, using a CCD camera to analyse the dose deposited distribution. Thus, the CCD was used to capture the images of different cuvettes that were filled with FXG and irradiated considering analogous setups employed in breast cancer radiotherapy treatments. Thereafter, these pictures where processed in a MatLab routine and the spatial dose distributions could be evaluated. These distributions were compared with the ones that were obtained from dedicated treatment planning's softwares. According to the results obtained, the FXG, allied with the CCD system, has shown to be a complementary tool in dosimetry, helping to prevent possible complications during breast cancer treatments.

1. Introduction

In 1921, Fricke introduced a solution which consisted of ferrous sulfate, sulphuric acid and water, that once irradiated, conducted the oxidation of Fe^{2+} to Fe^{3+} ions, whose concentration was directly proportional to the energy deposited. The original dosimeter recipe was improved with the addition of xylenol orange indicator and swine skin gelatin, which provided better sensitivity for lower absorbed doses, as well as better signal stability [1, 2]. These modifications leaded to the currently known Fricke Xylenol Gel (FXG), a dosimeter that has effective atomic number of 7.75 and density of 1.05 g/cm^3 (close to those of water), high spatial resolution, linear dependence with the absorbed dose (up to 30 Gy) and maximum absorption peak at 585 nm [3,4]. In this work, the Fricke Xylenol Gel was used to evaluate the dose distributions in breast cancer radiotherapy and to verify doses deriving from scattering, which could contribute to increase the risks of heart and lung diseases [5, 6]. In this sense, the dose distributions obtained with a developed CCD system were compared with those obtained with the software MEVIS CAT3D, used for a Siemens MEVATRON VI linear accelerator (LINAC), and with the isodose charts of an Atomic Energy of Canada Limited (AECL) THERATRON 780C ^{60}Co.

2. Materials and Methods

A dedicated thorax PMMA phantom was manufactured, as shown in Figure (1a), which was composed of five 200 x 300 x 50 mm^3 pieces, simulating a human thorax. The breast, Figure (1b), was composed of three pieces: a FXG cuvette (black part) surrounded by two breast-like materials (not shown in Figure 1b). The doses, besides being computed in the breast, were also evaluated up to 50 mm below

Journal of Physics: Conference Series **250** (2010) 012004 doi:10.1088/1742-6596/250/1/012004

the thorax surface. The phantoms were irradiated with both sources cited earlier, considering a routine breast treatment (two parallel and opposite fields) and different wedge filters.

Figure 1. Thorax (a) and breast (b) phantoms.

In order to confirm the linear behavior between the light absorption and the absorbed dose, standard cuvettes (10 x 10 x 35 mm3) were filled with FXG and irradiated with 2, 4, 6, 8 and 10 Gy, in 10 x 10 cm^2 fields. The isodose patterns were evaluated using rectangular, 150 x 150 x 10 mm^3, cuvettes for different filters (15°, 30°, 45° and 60°) and the breast shaped cuvette was irradiated considering the 30° filter. The images once obtained, were processed with a MatLab especially developed routine and, at last, were compared to the isodose charts available.

3. Results and Discussion

The FXG showed linear behaviors in the calibration curves. This is an important feature of the FXG-CCD dosimeter system developed, allowing it to discriminate the absorbed dose values. The CCD processed images are presented below.

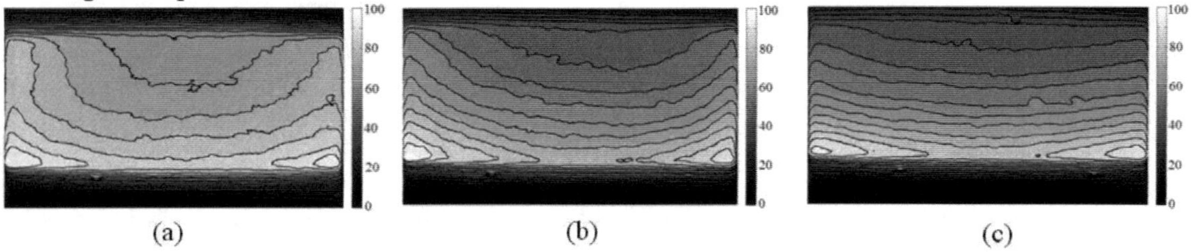

Figure 2. Isodoses curves obtained from the rectangular cuvettes, irradiated with 8 Gy and different wedge filters for the ^{60}Co photons. (a) 15°, (b) 30° and (c) 45° wedge filters. The absorbed doses were normalized to 100%, considering the highest pixel intensity in the irradiated field.

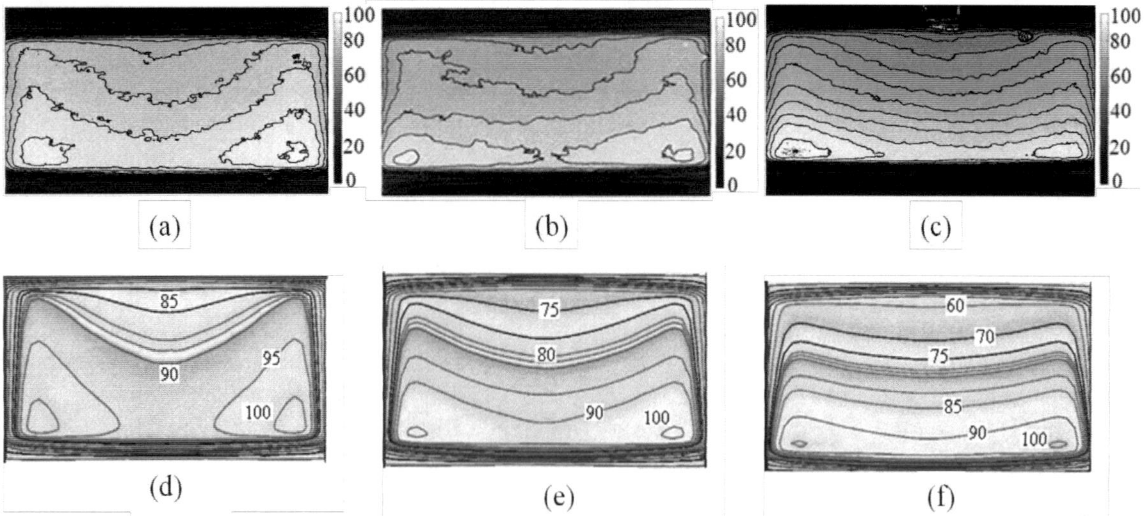

Figure 3. Isodoses curves obtained from rectangular cuvettes, irradiated with 8 Gy and different wedge filters for the LINAC, (a) 15°, (b) 30° and (c) 45°wedge filters. The absorbed doses were normalized to 100%, considering the highest pixel intensity in the irradiated field. In figures (d), (e) and (f) the respective isodoses obtained with the MEVIS software are presented.

As noticed in Figure (2) and (3), the FXG-CCD system was able to detect subtle changes in absorbed doses for each cuvette-filter configuration. One can see through the images how the doses gradients change according to the filter applied. As shown in ^{60}Co and LINAC isodoses, as the wedge filters were increased, the dose gradients became wider, although, for each filter, the gradient of the ^{60}Co isodoses were always higher when compared with the LINAC. From these results, it was possible to notice that the isodoses obtained with the FXG-CCD system corroborated with the isodoses obtained with the software. Considering the breast shaped cuvette, it was seen that the doses were also influenced by the wedge filters applied. This was reflected by more homogeneous isodoses presented in Figure (4b). According to this image, the dose delivered to the breast was more suitable for a treatment planning, considering the size and shape of the breast.

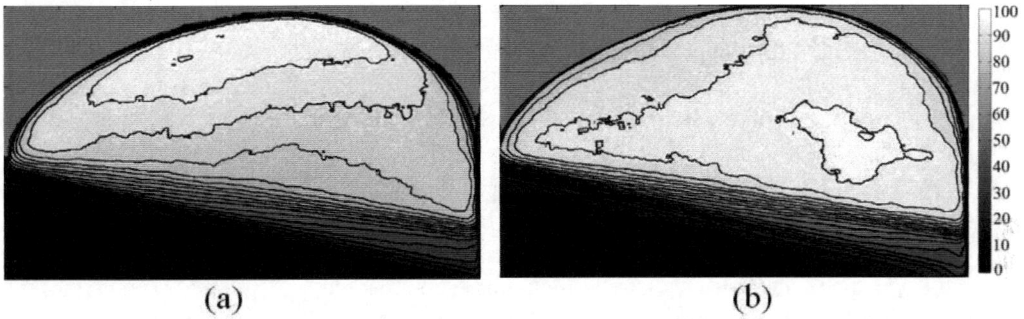

Figure 4. Breast shaped LINAC irradiated cuvettes. (a) Without and (b) with a 30° wedge filter.

4. Conclusion

In this work, it was possible, using thorax and breast phantoms, to evaluate the absorbed doses distributions in the breast in a routine irradiation treatment. The FXG-CCD system developed demonstrated to be sensitive, effective and practical, being suitable for this type of study. Mapping the absorbed doses with different filters, it was able to investigate how isodoses and hot spots behave. Such measurements are interesting, once they simulated the dose distributions delivered in a real treatment, contributing to understand and reduce undesirable doses in the pacient. Therein, FXG showed to be a powerful tool for quality control in breast treatment planning's.

References

[1] Bero M A, Gilboy W B, Glover P M, and El-Masri H M 2000 *Instr. and Meth.* **166-167** 820-825
[2] Gambarini G, Birattari C, Mariani M, Marchesini R, Pirola L, Prestini P, Sella M and Tomatis S 2004 *Instr. and Meth.* **213** 321-324
[3] Schreiner L J 2004 *Inst. Phys. Publ. J. Phys.: Conf. Ser.* **3** 9-21
[4] Calcina C S G, De Oliveira L N, De Almeida C E and De Almeida A 2007 *Phys. Med. Biol.* **52** 1431-1439
[5] Darby S C, McGale P, Taylor C. W and Peto R 2005 *Lancet Oncol.* **6** 557-565
[6] Giordano S H, Kuo Y-F, Freeman J L, Buchholz T A, Hortobagyi G N and Goodwin J S, 2005 *Journal of the National Cancer Institute* **97** 419-424

IC3DDose: The 6th International Conference on 3D Radiation Dosimetry IOP Publishing
Journal of Physics: Conference Series **250** (2010) 012005 doi:10.1088/1742-6596/250/1/012005

Development of ^{P}DRESS (Patient Specific Dose Real Evaluation Systems) using a TENOMAG Gel and Optical CT (VISTA^{TM}) in Clinical IMRT Prostate Case

Suk Lee[1], Jang Bo Shim[1], Kyung Hwan Chang[1], Yuanjie, Cao[1], Jaeyoun Yi[2], Jinwoo Park[2], Sam Ju Cho[3], Sang Hoon Lee[4], HyunDo Huh[5], Kwang Hwan Cho[6], Chul Kee Min[7], Dae Sik Yang[1], Young Je Park[1], Won Seob Yoon[1], Chul Yong Kim[1]

[1] Department of Radiation Oncology, College of Medicine, Korea University, Seoul, Korea
[2] INFINITT Healthcare, Seoul, Korea
[3]Department of Radiation Oncology, Eulji University, Seongnam, Korea
[4]Department of Radiation Oncology, Cheil General Hospital & Women's Healthcare Center, Kwandong University College of Medicine, Seoul, Korea,
[5]Department of Radiation Oncology, Inha University Hospital, Incheon, Korea
[6]Department of Radiation Oncology, Soonchunhyang University Boocheon Hospital, Boocheon, Korea
[7]Department of Radiation Oncology, Konyang University Hospital, Daejeon, Korea

sukmp@korea.ac.kr

Abstract. The aims of this study, we present the preliminary results of 3 dimensional dose evaluation software (^{P}DRESS, patient specific dose real evaluation systems). In this work, we compared planned 3D dose distribution with measured 3D dose distribution using a novel normoxic polymer gel dosimeter (TENOMAG) and a commercial cone-beam optical CT scanner (VISTA^{TM}, Modus Medical Devices, Inc., London, ON, Canada) to verify the 3D dose distribution in intensity-modulated radiation therapy (IMRT) prostate case. And we developed ^{P}DRESS using the Xelis Flatform which is developed by INFINITT Corporation is used to display the 3D dose distribution by loading the DICOM RT Data which is exported from RTP and optical-CT reconstructed VFF file. Data analysis is achieved by comparing the RTP data with the VFF data using profile, gamma map, and DTA. The profiles showed good agreement between RTP data, gel dosimeter, and gamma distribution and the precision of the dose distribution is within ± 5%. The results from this study show that there are no significantly discrepancies between the calculated dose distribution from treatment plan and the measured dose distribution from a TENOMAG gel scanned with an optical CT scanner. The 3D dose evaluation software (^{P}DRESS) which is developed in this study evaluates the accuracy of the three dimensional dose distributions.

1. Introduction

A study on 3D dose verification is still lacking in clinical cases. In this work, we compared planned 3D dose distribution with measured 3D dose distribution using a novel normoxic polymer gel dosimeter (Tetrakis hydroxymethly phosphonium chloride-Normoxic-Methacrylic acid-Gelatin, TENOMAG) and a commercial cone-beam optical CT scanner (VISTA^{TM}, Modus Medical Devices, Inc., London, ON, Canada) to verify the 3D dose distribution in intensity-modulated radiation therapy (IMRT) prostate case [1-3]. So, we present the preliminary results of 3 dimensional dose evaluation software (^{P}DRESS, patient

Journal of Physics: Conference Series **250** (2010) 012005 doi:10.1088/1742-6596/250/1/012005

specific dose real evaluation systems).

2. Materials and Methods

TENOMAG dosimeter was consisted of water, gelatin, THPC and HQ. The process and the condition for preparing the polymer gel used in this study can be summarized as follows: Gelatin (300 blooms, Sigma-Aldrich, USA) was mixed with high-purity distilled water (HPLC) in a reaction flask. The reaction flask was heated by using a hot-plate magnetic stirrer to stir the mixture after the gelatin had been. The heating was continued until the gelatin melted completely, and the stirring speed was maintained such that there was no bubble formation in the flask. The heating was continued when the temperature of the gelatin solution approached 50℃, after which a clear gelatin solution was obtained due to the complete melting of the gelatin in the water. Then, the temperature of the gelatin solution was cooled down to 41℃ in a water bath. In addition, a specific amount of methacrylic acid (Sigma-Aldrich, USA) was added to the reaction flask by using a micropipette. Also, a small amount of hydroquinone (Sigma-Aldrich, USA) was added to the flask. Finally, a type of antioxidant, THP (Tetrakis hydroxymethyl phosphonium, Sigma-Aldrich, USA), was added to the reaction flask. Then, the solution was stirred using the hot-plate magnetic stirrer until the mixture had fully melted and had become a clear solution [4, 5]. Scanned CT (Philips) images of dosimeter were transferred to the radiation treatment planning system (Varian ECLIPSE, Varian Medical System, Palo Alto, CA) where the IMRT prostate plan was designed. We used a 10 MV photon beam (iX, Varian Medical System, Palo Alto, CA) to deliver the prostate treatment plan. After irradiation, the TENOMAG dosimeter was scanned in the VISTA^{TM} scanner. The optical CT scanner was composed with 590 nm LED diffuse light, CCD camera, stepper motor, matching tank, gel container and holder. The 180 projections data was acquired during 360° rotation and 1 mm resolution. Scanning time was about 4 min. The scanned data were reconstructed using VistaRecon software (Feldkamp reconstruction algorithm with Hamming filter) to obtain 3D dose distribution of OD. The gel dosimeter is taken through an entire IMRT treatment procedure and irradiated by an external beam which is delivered by iX treatment system. Optical-CT scanner was used to readout the dose distribution in gel dosimeter (Figure 1.).

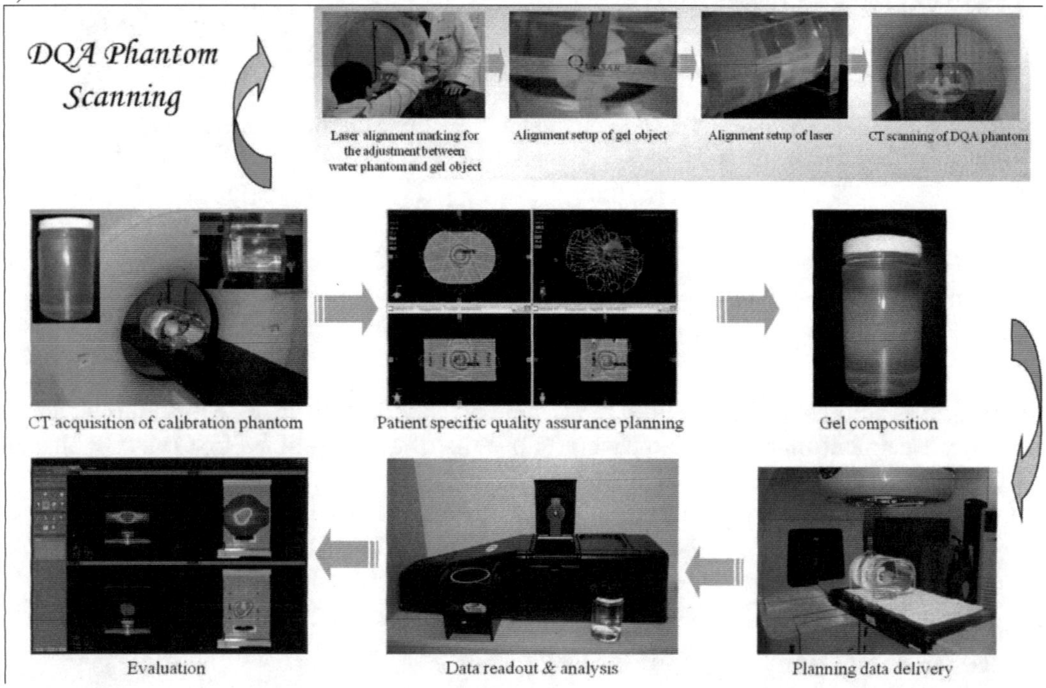

Figure 1. It shows that schematic diagram of experimental process of 3D dosimetry using gel dosimeter and optical CT.

Moreover we developed ^{P}DRESS using the Xelis Flatform which is developed by INFINITT Corporation is used to display the 3D dose distribution by loading the DICOM RT data which is exported from RTP and optical-CT reconstructed VFF file. Independent ^{P}DRESS using ionizing chamber and EBT film is performed to compare the dose distribution of RTP with gel dosimeter. After calibration, the data analysis is achieved by comparing the RTP data with the VFF data using profile, gamma map, and DTA.

3. Results and Discussion

The agreement between the normalized EBT, gel dosimeter and RTP data was evaluated using both qualitative and quantitative methods like profile, gamma map and DTA. The profiles showed good agreement between RTP data, gel dosimeter, and gamma distribution and the precision of the dose distribution is within ± 5% (Figure 2.).

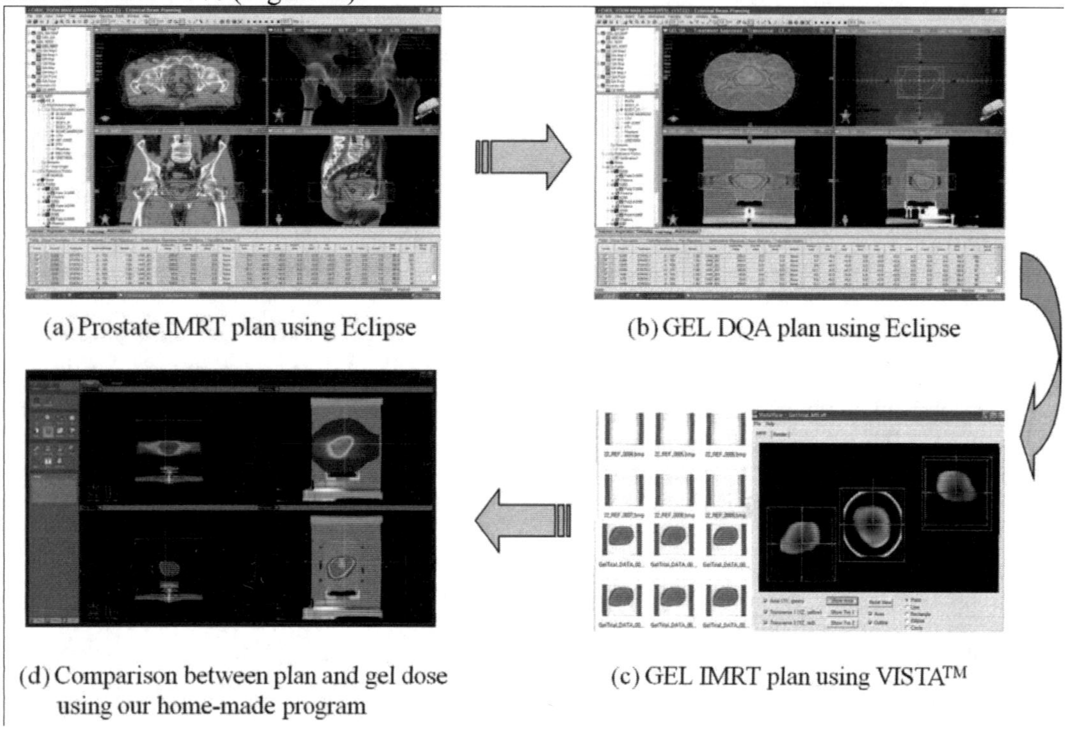

Figure 2. It shows that 3D dosimetry of prostate IMRT case using optical CT: (a) Prostate IMRT plan using Eclipse; (b) GEL DQA plan using Eclipse; (c) GEL IMRT plan using VISTA^{TM}; (d) Comparison between plan and gel dose using our home-made program.

4. Conclusion

The results from this study show that there are no significantly discrepancies between the calculated dose distribution from treatment plan and the measured dose distribution from a TENOMAG gel scanned with an optical CT scanner. The 3D dose evaluation software (^{P}DRESS) which is developed in this study evaluates the accuracy of the three dimensional dose distributions. Further more comfortable clinical implications could be expected in further study.

5. Acknowledgements:

The work was supported by a Korea University Grant. The authors thank Dr. Kevin Jordan, Ph.D. from the Department of Physics and Engineering, London Regional Cancer Program at London Health

Sciences Centre, London, ON, Canada. And the authors thank Dr. John Miller from the Modus Medical Devices, Inc., London, ON, Canada.

References

[1] Oldham M, Guo P, Gluckman G and Adamovics J. 2006 IMRT verification using a radiochromic/optical-CT dosimetry system *J. Physics.: Conf. Ser.:* **56** 221-224
[2] Sakhalkar H S, Adamovics J, Ibbot G and Oldham M 2009 A comprehensive evaluation of the PRESAGE/optical-CT 3D dosimetry system *Med. Phys.* **36** 71-82
[3] S Babic, J Battista and Jordan K 2008 Three-dimensional dose verification for intensity-modulated radiation therapy in the radiological physics centre head-and-neck phantom using optical computed tomography scans of ferrous xylenol-orange gel dosimeters *Int J. Radiat. Oncol. Biol. Phys.* **70** 1281-1291
[4] SJ Cho, DO Shin and Suk Lee et al. 2007 Development of a novel normoxic polymer gel dosimeter (TENOMAG) Journal of Korean Physical Society. **52** 1798-1805
[5] SJ Cho, Su Jung Shim and Suk Lee et al. 2010 Dose Response Characteristics in Polymer Gel for the Composition Ratio Journal of Korean Physical Society. **56** 147-152

Journal of Physics: Conference Series **250** (2010) 012006 doi:10.1088/1742-6596/250/1/012006

Toward 3D dosimetry of intensity modulated radiation therapy treatments with plastic scintillation detectors

M Guillot[1, 2], L Gingras[1, 2], L Archambault[1, 2], S Beddar[3] and L Beaulieu[1, 2]

[1] Département de Physique, de Génie Physique et d'Optique, Université Laval, Québec, Québec, G1V 0A6, Canada

[2] Département de Radio-Oncologie, Hôtel-Dieu de Québec, Centre Hospitalier Universitaire de Québec, Québec, Québec, G1R 2J6, Canada

[3] Department of Radiation Physics, The University of Texas M. D. Anderson Cancer Center Houston, Texas, 77030, USA

E-mail: mathieu.guillot.1@ulaval.ca

Abstract. In this work, we present a novel two Dimensional Plastic Scintillation Detector (2D-PSD) array designed to measure dose distributions generated by high energy photon beams from medical linear accelerators. This study aim to demonstrate that the dose distribution in the irradiated volume is not modified by the presence of several hundred plastic scintillation detectors (PSDs). The 2D-PSD consists of 781 PSDs inserted in a plastic water slab. The dose distributions measured with the 2D-PSD were compared to calculations from a treatment planning system (Pinnacle3, Philips Medical Systems) and with measurements taken with an ionization chambers array (MatriXX Evolution, IBA Dosimetry). Furthermore, a clinical head and neck IMRT plan was delivered on the 2D-PSD. A good agreement is obtained between the measured and planned dose distributions. The results show that the 2D arrangement presented in this work is water equivalent and transparent to x-ray radiation. As a consequence, our design could be extended to multiple detection planes, opening the possibility for 3D dosimetry with PSDs.

1. Introduction

Advanced radiation treatment modalities such as step and shoot intensity modulated radiation therapy (IMRT) and arc techniques (RapidArc, VMAT, Tomotherapy) can generate complex 3D dose distributions with high conformity to the target volume in the patient. These dose distributions have to be accurately measured for machine commissioning and during the frequent patient plan verifications to ensure a complete control of the dose delivered. Plastic Scintillation Detectors (PSDs) are a promising new technology for the measurement of complex dose distributions. They are characterized by an unique set of properties including: high spatial resolution, angular independence, energy independence and real-time readout [1][2][3]. PSDs use plastic scintillators to convert the absorbed dose to photon emitted in the visible spectral region. Optical fibers are commonly used to guide the scintillation photon to a photodetector. One Dimensional Plastic Scintillation Detector (1D-PSD) arrays has already been presented in the literature [3] [4].

Two main aspects will characterize the development of 3D-PSD arrays. Firstly, because PSDs have small sensitive volumes, 3D-PSD arrays are expected to be composed of many hundreds of PSDs to ensure a good sampling of the dose distribution. Secondly, the radiation beam will necessarily pass through many PSDs in such device. It is therefore important that the dose distribution is not modified by the presence of PSDs. This will requires using detectors made from nearly water equivalent materials.

In this work, we present a novel two Dimensional Plastic Scintillation Detector array (2D-PSD) designed to measure dose distributions generated by high energy photon beams from medical linear accelerators. The aim of this study is to demonstrate that the dose distribution in the irradiated volume is not modified by the presence of several hundred plastic scintillation detectors (PSDs). We also evaluate the capacity of the 2D-PSD prototype to measure complex dose distributions generated from a clinical head and neck IMRT plan.

2. Methods and Materials

2.1. The 2D-PSD prototype

A picture of the 2D-PSD is presented in figure 1(a). The prototype is built entirely from nearly water equivalent plastic materials. It consists of 781 PSDs inserted vertically in a 3 x 30 x 30 cm^3 plastic water slab (Plastic Water DT, CIRS, Norfolk, VA, USA) with a uniform spacing of 1 cm, covering a 26 x 26 cm^2 region. The 2D-PSD also has two 5 mm spacing detection lines positioned in a cross shape centered on the detection plane. Each detector consists of a 1 mm diameter and 3 mm long polystyrene plastic scintillating fiber (BFC-12, Saint-Gobain Inc., Paris, France) coupled to a clear plastic optical fiber of the same diameter. The plastic water slab containing the 781 PSDs is fixed in a specially designed acrylic water tank. The user has to fill the acrylic tank with water and add build-up materials on the top of the array prior to measurements. An image from a Computed Tomography (CT) scan of the 2D-PSD is presented in figure 1(b). The image was taken in a slice containing 53 PSDs spaced by 5 mm. Individual PSDs can not be distinguished from others water-equivalent materials on the image. For comparison, an image from a CT scan of an ionization chambers array (MatriXX Evolution, IBA Dosimetry, Louvain-La-Neuve, Belgium) is presented in figure 1(c). The presence of the 32 ionization chambers (air cavities and electrodes) can be clearly seen on the image.

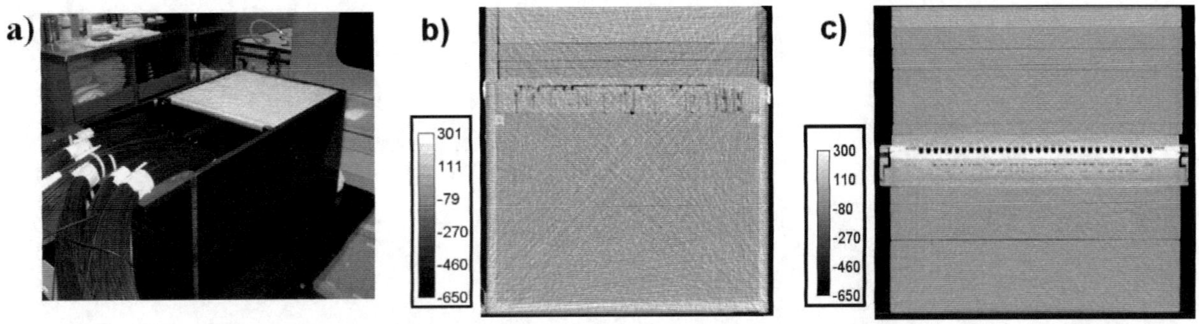

Figure 1. a) The two Dimensional Plastic Scintillation Detector array (2D-PSD). b) An image from a CT scan of the 2D-PSD. The slice contains 53 PSDs spaced by 5 mm. c) An image from a CT scan of an ionization chambers array (MatriXX Evolution, IBA Dosimetry, Louvain-La-Neuve, Belgium). The scale bar on the CT images is expressed in Hounsfield units.

Journal of Physics: Conference Series **250** (2010) 012006 doi:10.1088/1742-6596/250/1/012006

2.2. Characterization of the water-equivalence

In current dosimetry protocols for external beams radiotherapy, the measured dose always refers to dose in water to emulate the conditions encountered in the patient. To characterize the degree of water-equivalence of the 2D-PSD prototype, the detector array was irradiated with a square field size 10x10 cm^2 using the beam incidences 90° and 120°. Those beam incidences are interesting because the radiation beam is passing through many lines of detectors. The perturbation of the beam fluence by the detector materials can therefore be characterized. The same procedure was performed using the MatriXX ion chambers array. Both detector arrays where imbedded in an identical phantom made from Plastic Water slabs to reproduce the same scattering conditions. The dose distributions in the detection plane were also calculated with a Treatment Planning System (TPS) (Pinnacle^3 Version 8.0m, Philips Medical Systems, Andover, MA, USA) with no heterogeneities corrections (dose to water). No corrections to account for the volume averaging of the ionization chambers in the MatriXX were applied to the calculations of the TPS.

2.3. Validation of an IMRT plan

A clinical head and neck IMRT plan (step and shoot) was delivered to the 2D-PSD to verify its capacity to measure complex dose distributions. The IMRT plan consists of six beam incidences (0°, 103°, 257°, 309°, 165° and 195°), each beam having five or seven different MultiLeaf Collimator (MLC) configurations. The dose distribution of the IMRT plan was also calculated on a CT scan of the 2D-PSD using the TPS.

3. Results

3.1. Characterization of the water-equivalence

Dose profiles on the central beam axis measured with the 2D-PSD, the MatriXX ion chambers array and calculated with Pinnacle^3 are plotted on figure 2 for beam incidences 90° and 120°. The close agreement between the dose distributions measured with the 2D-PSD and calculated with Pinnacle^3 (dose to water) is a clear indication that the dose is minimally perturbed by the presence of the 781 PSDs. The discrepancy observed between the dose measured with the MatriXX and calculated with Pinnacle^3 is caused by the beam fluence perturbation from the presence of air cavities and electrodes in the irradiated phantom.

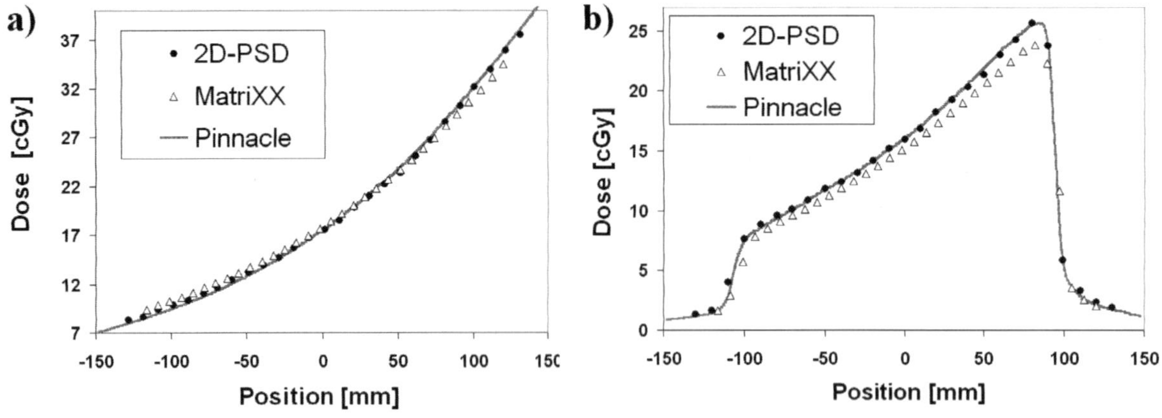

Figure 2. Dose profiles on the central beam axis measured with the 2D-PSD, the MatriXX

ionization chambers array and calculated with Pinnacle[3] for various beam incidences : a) 90° b) 120°.

3.2. Validation of an IMRT plan

The total dose distribution measured with the 2D-PSD for the IMRT plan is presented in figure 3 (a). The results of the gamma evaluation performed on the measured and calculated dose distributions with Pinnacle[3] are presented on figure 3 (b) [5]. For the total dose distribution of the treatment, the gamma evaluation was successful for 96 % and 99 % of the 781 detectors for dose tolerances and distances to agreement of (3%, 3mm) and (4%, 4mm) respectively. The gamma evaluation was successful for 92 % and 98 % of the detectors in the high dose region (dose higher than 50 % of the maximum dose) for the previously mentioned tolerances.

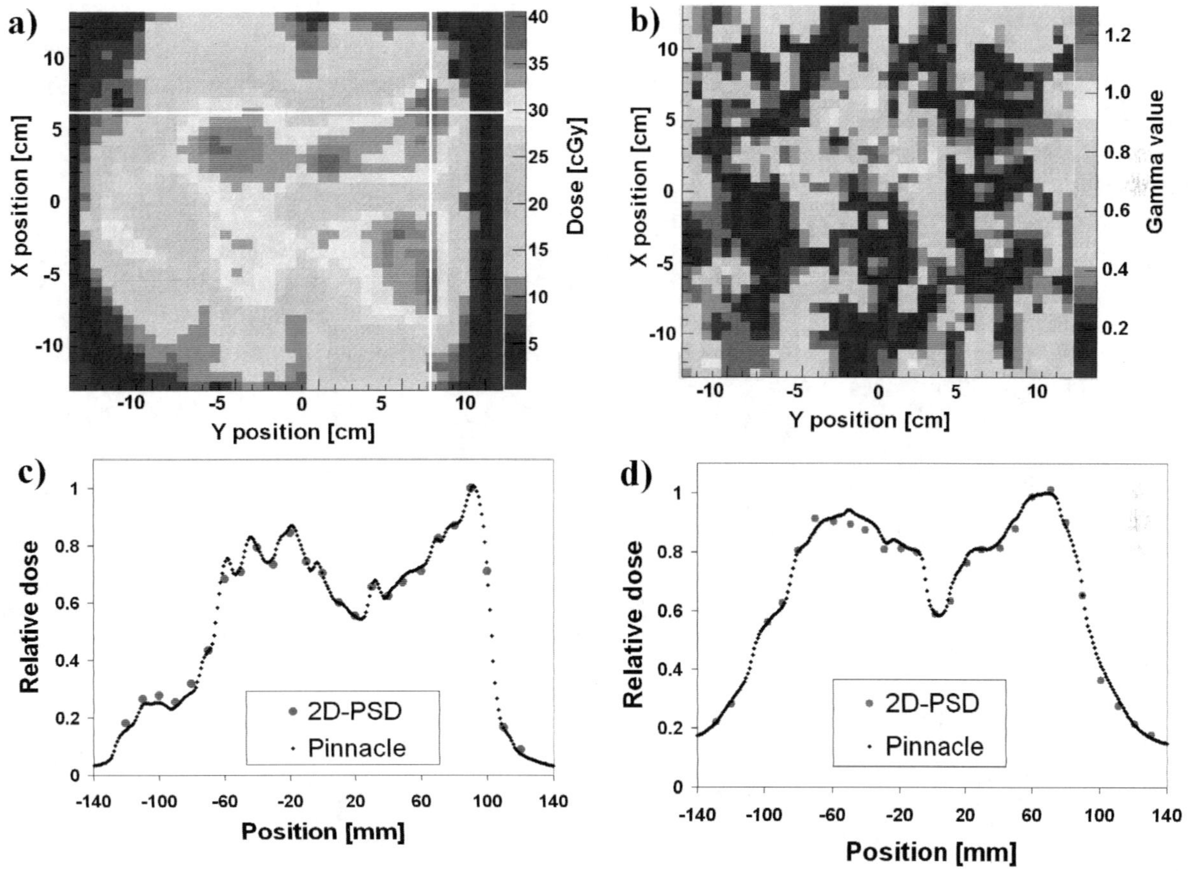

Figure 3. Results for a clinical head and neck IMRT treatment. Six beam incidences are used: 0°, 103°, 257°, 309°, 165° and 195°. a) The total dose distribution measured with the 2D-PSD. b) A color map of the gamma value (3 %, 3 mm) for the comparison of the measured and planned dose distributions. c) Comparison of the measured and planned dose profiles along the Y axis crossing the maximum value of the dose distribution (white line on figure 3(a)). d) Comparison of the measured and planned dose profiles along the X axis crossing the maximum value of the dose distribution (white line on figure 3(a)).

4. Discussion and Conclusions

In this work, we have presented a two dimensional plastic scintillation detector array technology composed of 781 sensors for quality assurance of IMRT treatments. The dose distributions measured with the 2D-PSD was compared to measurement taken with an ionization chamber array and calculations from a TPS. The presented results have demonstrated that the dose distribution in the irradiated volume is not modified by the presence of several hundred PSDs. This property is essential for the development of a future three Dimensional Plastic Scintillation Detector (3D-PSD) array because the radiation beam will pass through many PSDs in such device. As a consequence, our design could be extended to multiple detection planes, opening the possibility for 3D dosimetry with PSDs.

Acknowledgments

This work was supported by a National Sciences and Engineering Research Council of Canada (NSERC) reseach grant. M. Guillot would like to acknowledge NSERC for their financial support through a doctoral scholarship.

References

[1] Beddar A S, Mackie T R and Attix F H 1992 *Phys. Med. Biol.* **37** 1883-90
[2] Beddar A S, Mackie T R and Attix F H 1992 *Phys. Med. Biol.* **37** 1901-13
[3] Archambault L, Beddar A S, Gingras L, Lacroix F, Roy R and Beaulieu L 2007 *Med. Phys.* **34** 1583-92
[4] Lacoix F, Archambault L, Gingras L, Guillot M, Beddar A S and Beaulieu L 2008 *Med. Phys.* **35** 3682-90
[5] Low D A and Dempsey J F 2003 *Med. Phys.* **30** 2455-64

Journal of Physics: Conference Series **250** (2010) 012007 doi:10.1088/1742-6596/250/1/012007

Fast, large field-of-view, telecentric optical-CT scanning system for 3D radiochromic dosimetry

A Thomas*, M Oldham

Dept. of Radiation Oncology, Duke University Medical Center, Durham, NC, USA

*Corresponding author: ast5@duke.edu

Abstract. We describe initial experiences with an in-house, fast, large field-of-view optical-CT telecentric scanner (the **D**uke **L**arge field of view **O**ptical-CT **S**canner (DLOS)). The DLOS system is designed to enable telecentric optical-CT imaging of dosimeters up to 24 *cm* in diameter with a spatial resolution of 1 mm^3, in approximately 10 minutes. These capabilities render the DLOS system a unique device at present. The system is a scaled up version of early prototypes in our lab. This scaling introduces several challenges, including the accurate measurement of a greatly increased range of light attenuation within the dosimeter, and the need to reduce even minor reflections and scattered light within the imaging chain. We present several corrections and techniques that enable accurate, low noise, 3D dosimetery with the DLOS system.

1. Introduction

Telecentric scanners consist of a light source, rotation stage, telecentric lens, and an imaging array, typically CMOS or CCD. Telecentric based area scanners have been suggested by several authors and have potential to speed up the scanning process an order of magnitude over the 1st generation scanning laser systems [1-4]. Currently such systems have limited field of view (FOV) (e.g. < 10 cm diameter [3, 5]). A variety of challenges arise when attempting to scale up the relatively small telecentric scanner to a larger scanner with FOV up to 24 *cm*. First, telecentric lenses have an inherent internal reflection from lenses comprising the telecentric lens which produce cross talk on the entire array amounting to approximately .1 % of the maximum intensity incident on the array. If unaccounted for, this can lead to severe errors in the image reconstructions. Equally important is the increase in dynamic range required to image larger dosimeters. The detector is responsible for capturing accurate signal levels from the unattenuated light source simultaneously with the most attenuating pathlengths of the dosimeter, where attenuation follows Beer's Law. Typical CCD detectors have inherent dynamic range of ~60 *dB*, which is sufficient for dosimeters up to 10 *cm*. This work presents our first experiences with the scanner, and considers methods to overcome the two primary challenges described above for imaging larger dosimeters.

2. Methods

2.1. Scanner

A schematic of the DLOS system is shown in figure 1. An LED light source is collimated and projected onto a dosimeter placed on a rotating stage within an anti-reflection coated glass aquarium. The aquarium is filled with a mixture of mineral oil, octyl salicylate, and octyl cinnamate such that the

Journal of Physics: Conference Series **250** (2010) 012007 doi:10.1088/1742-6596/250/1/012007

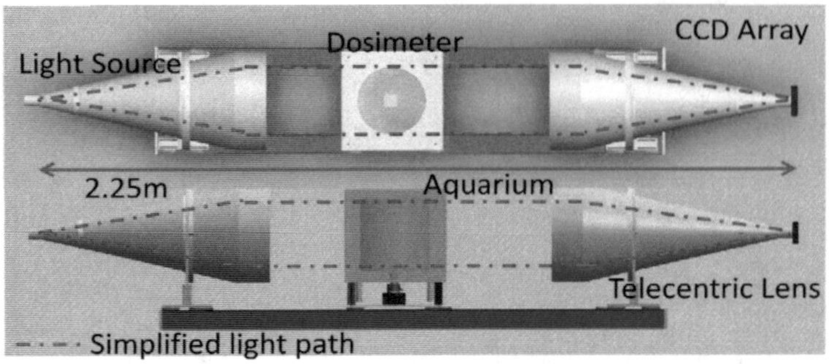

Figure 1: A schematic of the DLOS system in a top and side view showing the collimated LED light source, acquarium, rotation stage, telecentric lens and detector array.

indices of refraction of the fluid and the dosimeter are matched to minimize reflection and refraction distortions of incident light. As the sample rotates, projection images are mapped to a CCD array via a telecentric lens, which only utilizes light rays nearly parallel (~0.1°) to the optical axis. Light rays thus pass from the source and through the dosimeter where they are differentially attenuated and continue through to the telecentric lens where they are imaged onto the 12 bit monochrome CCD array. Generally, 360 projection images with 25 averages each are acquired in 1° increments. Each projection is then flood and dark noise corrected.

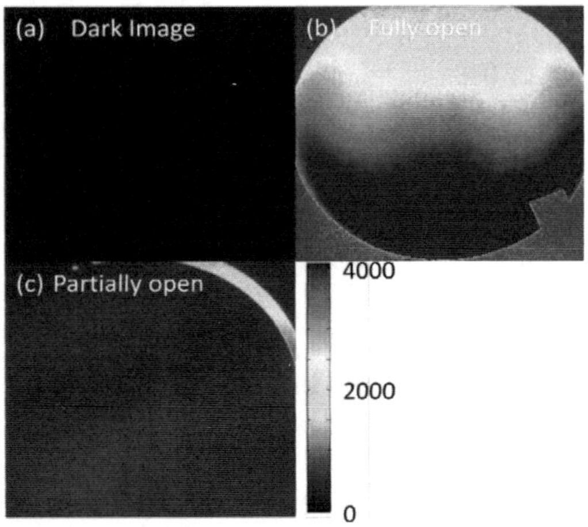

Figure 2: series of images showing the internal scatter present within the lens. (a) Image acquired with the lens cover on, (b) an image acquired without the lens cover and (c) an image captured with the lens cover in position to let a "sliver" of light through. Notice the contrast between (a) & (b).

2.2. Corrections

When imaging highly attenuating media, a stray light correction is required. This correction is for internal scatter caused by internal reflections within the imaging lens. Any light collected by the telecentric lens within the acceptance angle is mapped to the appropriate location. Due to internal reflections from AR coated lenses that comprise the lens, however, an approximate -50 *dB* signal can also be detected in the arrays other pixel locations. Figure 2 illustrates this effect. A "sliver" of light on one side of the aperture is observed to 'spill' over to the entire CCD array. This effect can be reduced by subtracting a scatter map of the low intensity regions created by placing an opaque structure in beam representing the higher attenuating regions of the dosimeter. This procedure achieves full dynamic range (~60 *dB*) from the detector and yields more accurate images.

When the dosimeter has a signal drop higher in magnitude than the dynamic range which the camera allows, ie > 60 *dB*, a separate correction needs to be introduced in combination with the stray light correction to get high dynamic range images from a low dynamic range device. The correction aims to increase the dynamic range of the system above that of the detector alone through a series of images taken with increasing exposure times, all other aspects remaining unchanged [6]. The images are converted such that pixel values are proportional to exposure rate by taking the ratio of grey scale values to exposure time. Each new image in the series replaces the previous images lower 25% exposure rates. Figure 3 shows a series of images with the intention of illustrating how the technique (combined with the previously discussed subtraction correction to account for internal scattering) can extend the inherent dynamic range of the detector. The image in figure 3a is the calculated OD of known neutral density filters placed within a filter wheel after a single exposure with no corrections applied. The image in figure 3b is an image of the same filter wheel after 3 exposures were combined to give accurate OD's to within 3% of the expected values.

Journal of Physics: Conference Series **250** (2010) 012007 doi:10.1088/1742-6596/250/1/012007

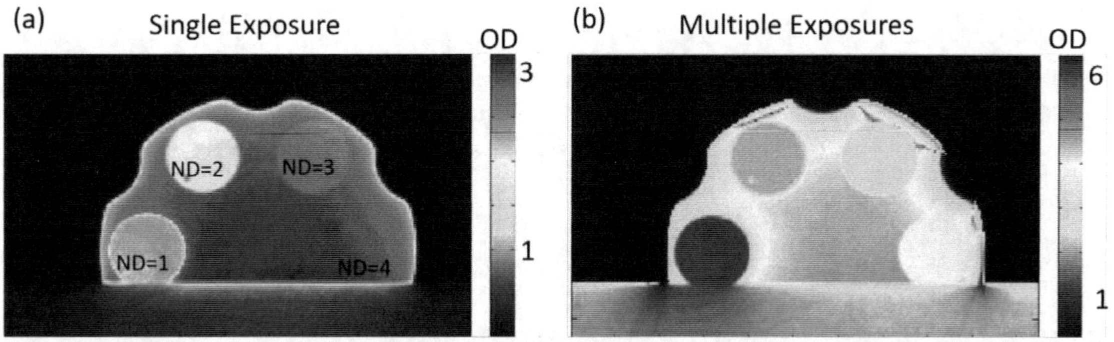

Figure 3: (a) Image of a filter wheel with known optical densities ranging from 1 – 4 converted to OD. The OD values of ND = 1 & 2 match the known filter values, however, the higher values, ND = 3 & 4 are not near their known values. (b) An image of the same filter wheel taken with a series of 3 images with increasing exposure times to create an accurate (within 4%), high dynamic range image. This represents an image accurately covering 80 *dB* of dynamic range.

2.3. Irradiations, imaging & analysis

To show the DLOS system is capable of returning dosimetricly accurate images the output was compared to the images generated from MGS's OCTOPUS scanner via 4%, 3 *mm* gamma analysis with a 5% threshold.

A 15 *cm* diameter PRESAGE dosimeter, 12 *cm* tall was irradiated with a simple two field plan of orthogonal 6 *cm* beams as shown in figure 4. The dosimeter has a sensitivity that yields a ΔOD of .06 */Gy/cm*, so to match the dynamic range of the detector the dosimeter was irradiated to 2 *Gy*. Before irradiation, a prescan was acquired in both systems using the same refractive index matching fluid, a mixture of octyl salicylate and octyl cinnamate. Images were acquired in the DLOS system by taking an average of 16 images at each projection angle for 360 projections in one degree increments. To mimic these conditions in the MGS scanner, 360 projections were acquired in one degree increments as well. Slices were acquired 5mm apart over 15 slices covering the central 7.5 *cm* of the dosimeter.

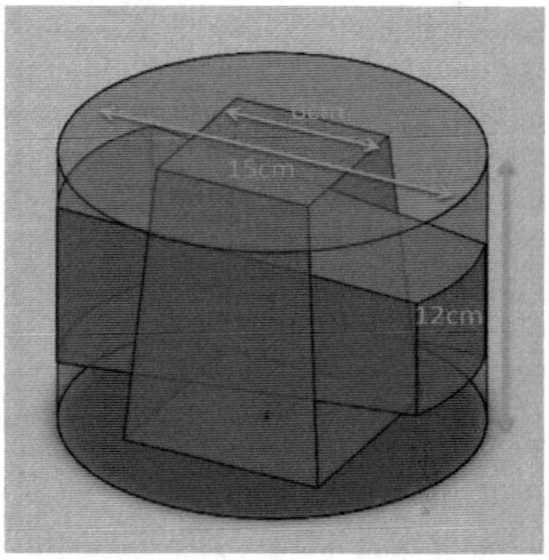

Figure 4: Sketch of the two field 6 *cm* x 6 *cm* plan delivered to the 15 *cm* x 12 *cm* dosimeter.

After irradiation the dosimeter was scanned in the DLOS system (~10 minutes) followed by a scan in the MGS scanner all with the same acquisition parameters as their respective pre scans. All images were reconstructed using the Matlab iradon function to 1 *mm* x 1 *mm* pixels. Upon image reconstructions, post and pre scans of the images were manually registered and subtracted to create a ΔOD image which isproportional to the dose delivered to the dosimeter. The two 3D data sets were then compared on a slice by slice basis with 2D gamma criteria of 4%, 3 *mm*. A 3D gamma analysis with identical criteria was also carried out by interpolating the MGS 5 *mm* gaps to 1 *mm* increments to match the DLOS system data set.

3. Results & Discussion

Ideally the two scanners would produce identical images. In practice this is challenging due to the differences involved in the acquiring optics among the two; however, general agreement between the two scanners is evident from figures 5a & 5b showing a representative slice from each scanner.

Figure 4c shows a 2D gamma comparison of the two slices showing more differences than a first glance of figures 5a & 5b may detect. The image from the laser scanner gives results that are expected by the physical setup of the system. Figures 6a & 6b show the sinograms of the slice from figure 5 for the prescans from both scanners. The sinograms are in the form of $\log(I_0/I)$ and thus the grey scale values represent the amount of attenuation for each path length for a given projection angle.

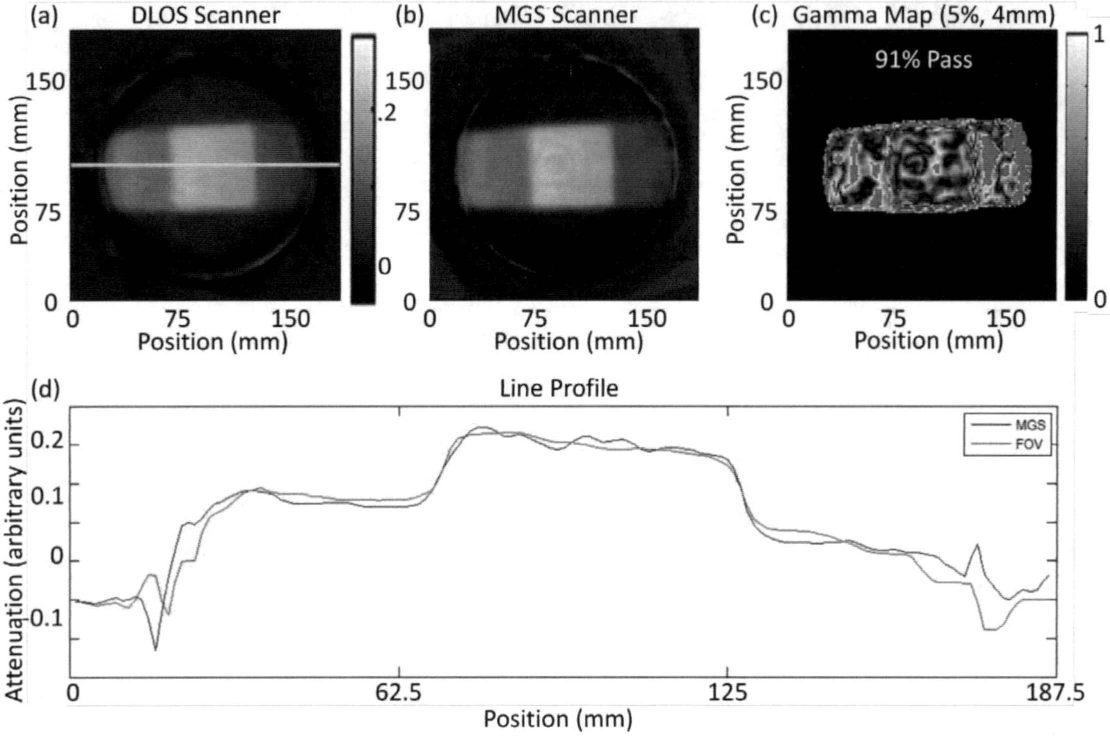

Figure 5: (a) Reconstructed slice from the DLOS system. (b) The same reconstructed slice from the MGS scanner. Both scans were acquired with 360 projections over 360 degrees. (c) A 5%, 4 *mm*, 5% threshold for the slices shown in (a) and (b) with a 91% pass rate calculated only from pixels within the dosimeter. (d) Line profiles taken from the green line in slices (a) and (b).

Two obvious differences are present in the sinograms: (1) the edge artefact is significantly more pronounced in the DLOS system and (2) the magnitude of attenuation observed is slightly higher (~10%) in the MGS scanner. Observation (1) shows that the DLOS system is much more sensitive to refractive index matching than the MGS scanner is. This is due to the relatively large collection mirror bending the laser beam after traversing the dosimeter to a collection lens which focuses the beam onto a photodiode for detection. Large changes in direction (up to ~2° can still be fully collected by the photodiode at each laser position. Conversely, when the DLOS system has the same matching fluid there is no mechanism to correct for shifts in the direction of the photons. In fact, the telecentric lens *rejects* any light bent more than 0.1° off the optical axis creating artificially high attenuation paths along the edges of the sinogram. The absolute attenuation difference seen in the sinograms is most likely attributable to the differences in light sources. The MGS scanner is using a very narrow bandwidth (.1 *nm*) 633.1 *nm* HeNe laser while the DLOS system is utilizing light from a 1 *nm* FWHM 633 *nm* filter from a red LED. The actual spectrum from the 1 *nm* is more on the order of 5 *nm* FWHM which can account for differences in the measured attenuation as the absorption peak for PRESAGE is 633 *nm*, so the laser will be preferentially attenuated.

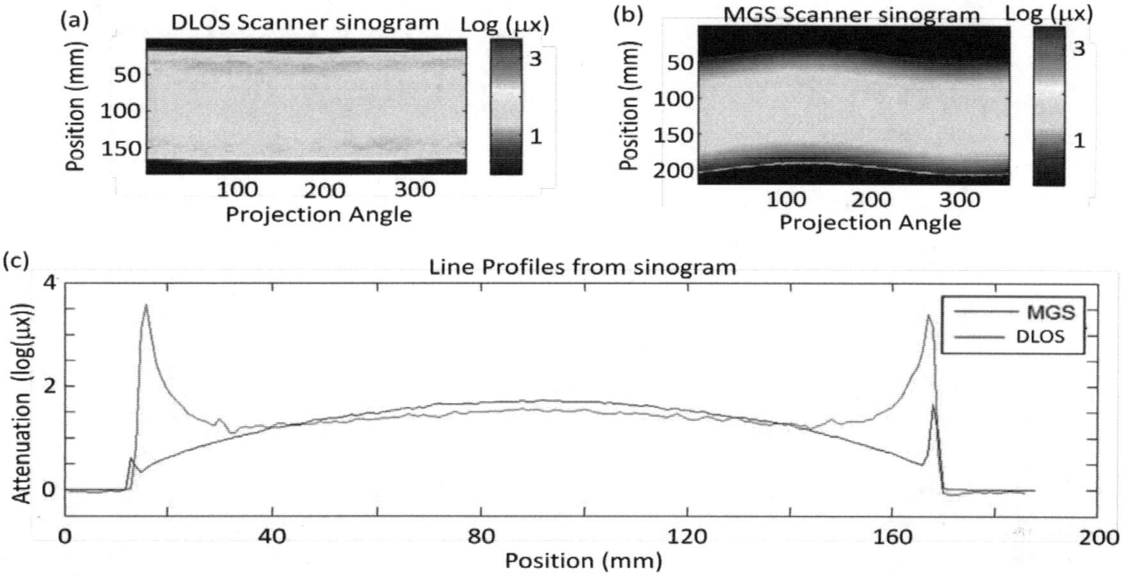

Figure 6: (a) Flood corrected sinogram from prescan captured with the large DLOS system. (b) Flood corrected sinogram from the prescan acquired with the MGS scanner & (c) a set of line profiles from each sinogram in (a) and (b) showing the differences in the scanners. Most evident are the edge artefacts present in each scanner, however, those in the DLOS system are much more pronounced and alters the shape of the curve for the 1st and last 25 *mm* of each projection angle. The same fluid was used in each scan and shows how the MGS scanner is much less sensitive to the refractive index match.

4. Conclusion

The DLOS system was designed and built with the purpose of measuring dosimeters up to 24 *cm* in diameter and 20 *cm* in height in a short time. Initial results demonstrate a scanner with excellent spatial resolution (down to 688 *μm* voxels) and low noise (~10 *cGy* for a 2 *Gy* irradiation) for large dosimeters. Good agreement is observed when compared to the MGS OCTOPUS laser scanning system, when appropriate corrections are applied. Ongoing scanner commissioning work will compare DLOS 3D dosimetry for simple plans, with calculated dose distributions from a commissioned treatment planning. This work demonstrates the DLOS /Presage system has strong potential as the first large-field-of view telecentric 3D dosimetry system capable of providing high resolution 3D dosimetry in minutes.

Acknowledgement: Supported by NIH R01 CA100835

References

[1] Doran, S.J., et al., A CCD-based optical CT scanner for high-resolution 3D imaging of radiation dose distributions: equipment specifications, optical simulations and preliminary results. Phys Med Biol, 2001. **46**(12): p. 3191-213.
[2] Oldham, M., et al., High resolution gel-dosimetry by optical-CT and MR scanning. Med Phys, 2001. **28**(7): p. 1436-45.
[3] Sakhalkar, H.S. and M. Oldham, Fast, high-resolution 3D dosimetry utilizing a novel optical-CT scanner incorporating tertiary telecentric collimation. Med Phys, 2008. **35**(1): p. 101-11.
[4] Wolodzko, J.G., C. Marsden, and A. Appleby, CCD imaging for optical tomography of gel radiation dosimeters. Med Phys, 1999. **26**(11): p. 2508-13.
[5] Clift, C., et al., Toward acquiring comprehensive radiosurgery field commissioning data using the PRESAGE/optical-CT 3D dosimetry system. Phys Med Biol. **55**(5): p. 1279-93.
[6] M Robertson, S.B., L Stevenson, DYNAMIC RANGE IMPROVEMENT THROUGHMULTIPLE EXPOSURES. IEEE Image Processing, 1999.

Journal of Physics: Conference Series **250** (2010) 012008 doi:10.1088/1742-6596/250/1/012008

The characterisation of a genipin-gelatin gel dosimeter

Davies JB[1,2], Bosi S[1,3], Baldock C[1]

[1] Institute of Medical Physics, School of Physics, University of Sydney, NSW 2006

[2] Australian Nuclear Science and Technology Organisation, Lucas Heights, NSW 2234

[3] Radiation Oncology Prince of Wales Hospital, High Street Randwick NSW 2031

jbd@ansto.gov.au

Abstract. Genipin cross links gelatin to slowly form a blue colour that bleaches upon irradiation. Spectrophotometric measurements of the absorbance change following irradiation to doses up to 100 Gy gives a linear dose response for certain concentrations of the gel ingredients; genipin, gelatin and sulphuric acid. Dose sensitivity increases with increasing concentrations of sulphuric acid and genipin and is also strongly dependent on the time allowed for the genipin-gelatin cross linking reaction (referred to here as blending) to take place. The optimum formulation of this gel was found for genipin concentration between 0.3 – 0.5 mM and blending time of at least 4 h.

1. Introduction

Gel dosimeters, having the ability to retain dosimetry information in three-dimensions (3D) have significant advantages in various radiotherapy applications [1]. In the Fricke-type gel dosimeter, however, ferric ions will diffuse post-irradiation where there are dose gradients within the gel [2, 3], leading to a loss in dosimetric information. Genipin, a fruit extract from *Gardenia jasminoides Ellis*, is a natural cross-linker of gelatin [4] and slowly reacts during preparation to give a blue colour that bleaches quantitatively upon irradiation. It has been demonstrated that a genipin-gelatin hydrogel does not diffuse post-irradiation [5] and could also be useful in measuring surface doses [6]. The aim is to develop a non-diffusing gel dosimeter for optical CT 3D dosimetry [6, 7]. Ideally a gel dosimeter for 3D dosimetry will respond linearly with dose from 0 Gy to a chosen maximum useful for radiotherapy, allowing the dose distribution to be calibrated. This paper investigates the dose response characteristics of a genipin-gelatin gel dosimeter.

2. Methods

Gelatin (Sigma-Aldrich, G2500, ~300 bloom) was dissolved in water and then genipin (Wako Chemicals 078-03021) added from a stock solution. The mixture was continuously stirred in a water bath at 45.0 ± 0.5 °C. The solution slowly changed from colourless to blue and steadily darkened over time. The time allowed for genipin and gelatin to blend prior to adding sulphuric acid is referred to here as the blending time. After sulphuric acid was mixed the solution was pipetted into 1 cm plastic cuvettes and allowed to set overnight at about 4 °C. Samples were irradiated in a calibrated cobalt-60 Gammacell with doses up to 100 Gy. The radiation induced colour change in the gel was measured at its peak absorbance of 600 nm in a spectrophotometer (Cary 100, Varian).

Journal of Physics: Conference Series **250** (2010) 012008 doi:10.1088/1742-6596/250/1/012008

As the genipin-gelatin mixture was darkening, samples were taken at regular intervals to initially observe the colour change over time. Further experiments were carried out by pouring 50 ml from a batch into a separate beaker, immediately adding sulphuric acid and preparing cuvettes. This was repeated at regular intervals to evaluate and compare the dose response as a function of blending time and a range of concentrations of genipin, gelatin and sulphuric acid.

3. Results and discussion

Figure 1 shows how the genipin-gelatin reaction darkens over time. Irrespective of the genipin concentration, the colour change reaches a plateau after approximately 48 hours.

It was found that for low genipin concentration (0.2 mM or less), a second order polynomial produced a better fit to the absorbance change readings against dose (Figure 2). At 0.1 mM, the correlation coefficients were 1.000 (polynomial) and 0.980 (linear), and at 0.2 mM, the correlation coefficients were 0.999 (polynomial) and 0.995 (linear). As the genipin concentration was increased up to 0.5 mM a linear relationship was found between 0 – 100 Gy with correlation coefficients greater than 0.999.

Dose sensitivity increases with increasing sulphuric acid concentration (Figure 3) and genipin concentration (Figure 4). As the concentration of either sulphuric acid or genipin is increased, the dose response also becomes more sensitive to the blending time. Previous work showed gel sensitivity for a linear fit of doses up to 50 Gy of approximately -2.3×10^{-3} cm^{-1} Gy^{-1} [8]. In these experiments the sensitivities found ranged from -0.5 to -3.5 ($\times 10^{-3}$ cm^{-1} Gy^{-1}). While the dose sensitivity is strongly dependent on the concentrations of genipin and sulphuric acid, only a small effect of the gelatin content was found, although results were only obtained for gelatin in the range 3 – 5 % (Figure 5). The blending time is also a dependent variable so careful control of these parameters is required to obtain a reproducible dose response of this gel.

Accurate 3D dosimetry requires optimization of the gel formulation including influence quantities such as the blending time. The optimum formulation of this gel for high sensitivity and linear dose response up to 100 Gy was found for genipin concentration between 0.3 – 0.5 mM and blending time of at least 4 h.

Journal of Physics: Conference Series **250** (2010) 012008 doi:10.1088/1742-6596/250/1/012008

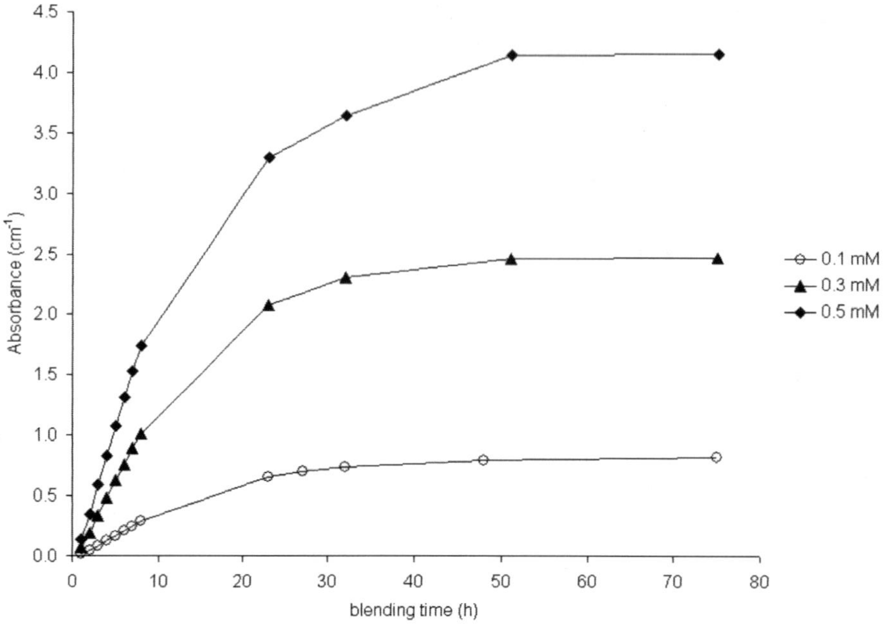

Figure 1. Darkening of the genipin-gelatin reaction for genipin concentrations 0.1, 0.3 and 0.5 mM. Gelatin (4.0 %) was constant and there was no acid added.

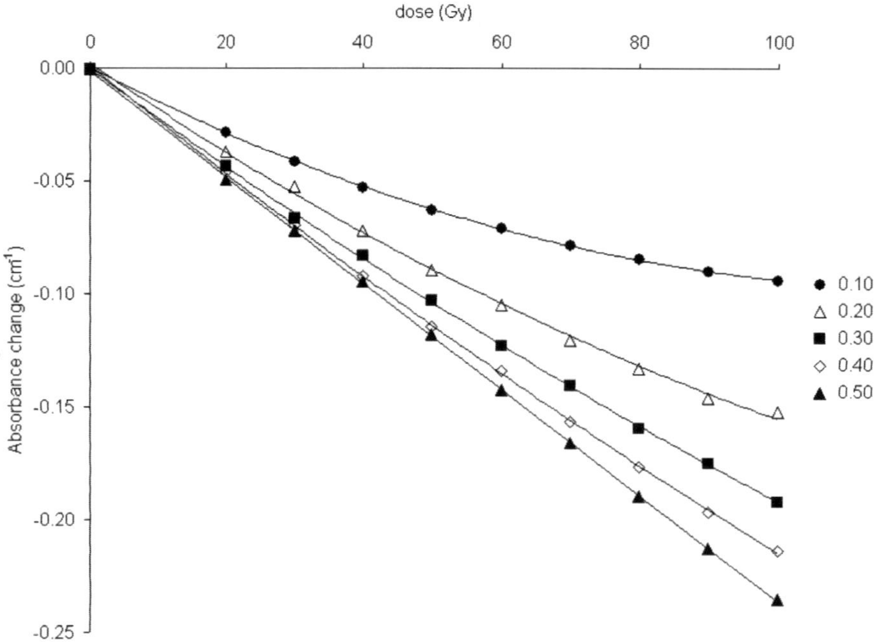

Figure 2. Dose response for various [genipin] (mM). The gel bleaches upon irradiation resulting in a decrease in optical absorbance. Gelatin (4.0 %), [H_2SO_4] (100 mM) and blending time (6.5 h) were constant.

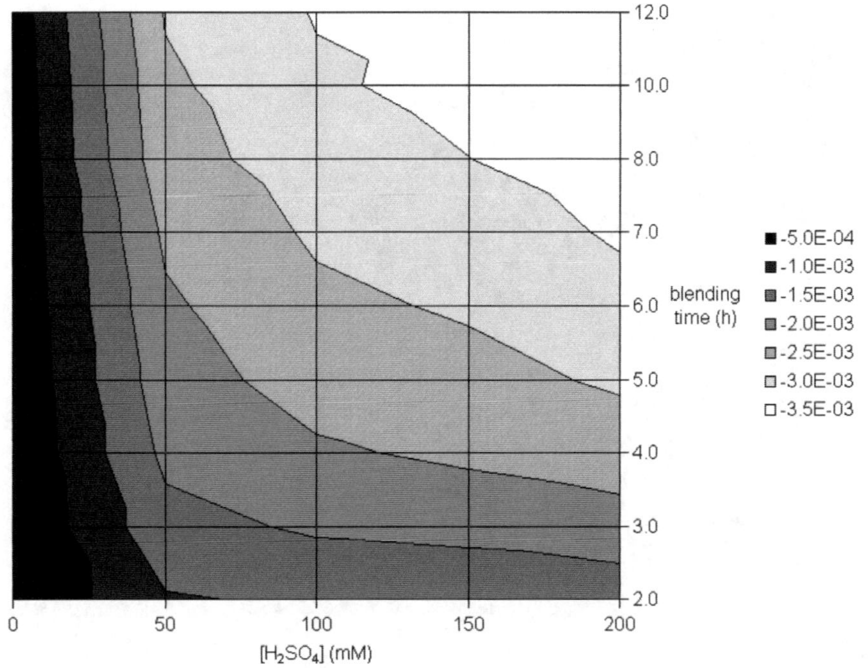

Figure 3. Variation of dose sensitivity (cm^{-1} Gy^{-1}) with sulphuric acid concentration and blending time. [genipin] (0.3 mM) and gelatin (4.0 %) were constant.

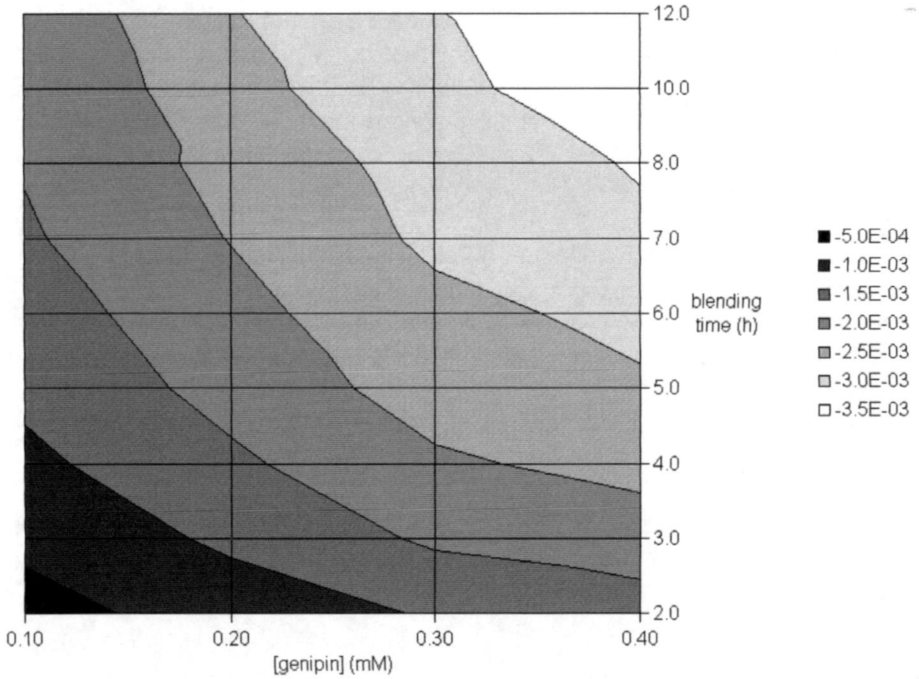

Figure 4. Variation of dose sensitivity (cm^{-1} Gy^{-1}) with genipin concentration and blending time. Gelatin (4.0 %) and [H_2SO_4] (100 mM) were constant.

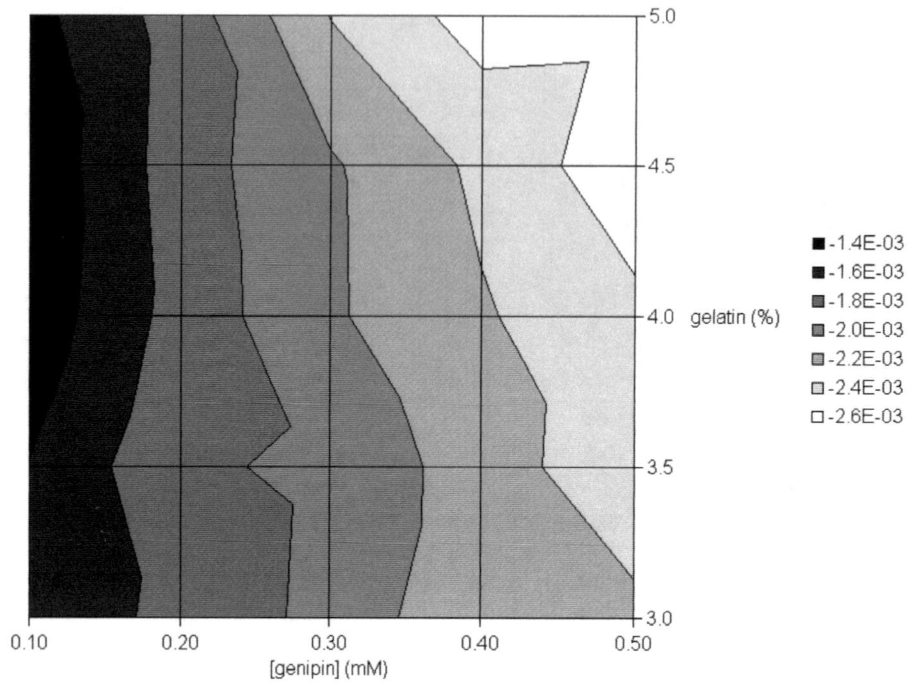

Figure 5. Variation of dose sensitivity (cm^{-1} Gy^{-1}) with genipin concentration and gelatin content. [H_2SO_4] (100 mM) and blending time (6.5 h) were constant.

References

[1] Baldock C, De Deene Y, Doran S, Ibbott G, Jirasek A, Lepage M, McAuley K B, Oldham M and Schreiner L J. 2010. Polymer gel dosimetry. Phys. Med. Biol., **55**(5): p. R1-R63.

[2] Baldock C, Harris P J, Piercy A R and Healy B 2001. Experimental determination of the diffusion coefficient in two dimensions in ferrous sulphate gels using the finite element method. Australas. Phys. Eng. Sci. Med. **24**, 19–30.

[3] Davies J B and Baldock C, 2008. Sensitivity and stability of the Fricke-gelatin-xylenol orange gel dosimeter. Rad. Phys. Chem. **77**, 690-696.

[4] Yao C, Liu B, Chang C, Hsu S, Chen Y. 2004. Preparation of networks of gelatin and genipin as degradable biomaterials. Mat. Chem. Phys. **83**, 204-208.

[5] Jordan K. 2008. Zero diffusion radiochromic genipin-gelatin dosimeter. Med. Phys. **35**, 3413.

[6] Jordan K. 2009. Optical CT scanning of cross-linked radiochromic gel without cylinder wall. J. Phys.: Conf. Ser. **164**, 012029.

[7] Bosi S, Naseri P, Puran A, Davies J and Baldock C. 2007. Initial investigation of a novel light-scattering gel phantom for evaluation of optical CT scanners for radiotherapy gel dosimetry. Phys. Med. Biol. **52**, 2893–2903.

[8] Jordan K. June 2008. Zero diffusion radiochromic genipin-gelatin dosimeter for three-dimensional radiation dosimetry. Conference Proceedings COMP annual meeting.

Journal of Physics: Conference Series **250** (2010) 012009 doi:10.1088/1742-6596/250/1/012009

Dose rate dependency of micelle leucodye 3D gel dosimeters

J Vandecasteele, S Ghysel and Y De Deene

Ghent University, Department of radiation oncology and experimental cancer research, De Pintelaan 185, Ghent, Belgium

Jan.Vandecasteele@UGent.be

Abstract. Recently a novel 3D radiochromic gel dosimeter was introduced which uses micelles to dissolve a leucodye in a gelatin matrix. Experimental results show that this 3D micelle gel dosimeter was found to be dose rate dependent. A maximum difference in optical dose sensitivity of 70% was found for dose rates between 50 cGy min^{-1} and 400 cGy min^{-1}. A novel composition of 3D radiochromic dosimeter is proposed composed of gelatin, sodium dodecyl sulphate, chloroform, trichloroacetic acid and leucomalachite green. The novel gel dosimeter formulation exhibits comparable radio-physical properties in respect to the composition previously proposed. Nevertheless, the novel formulation was found to be still dose rate dependent. A maximum difference of 33% was found for dose rates between 50 cGy min^{-1} and 400 cGy min^{-1}. On the basis of these experimental results it is concluded that the leucodye micelle gel dosimeter is still unsatisfactory for clinical radiation therapy dose verifications. Some insights in the physico-chemical mechanisms were obtained and are discussed.

1. Introduction

3D radiochromic dosimeters change color upon radiation. The radiochromic dosimeter investigated in this study changes color upon irradiation due to the oxidation of leucodyes by halogen radicals. In commercially available products such as Presage™, the leucodyes and halogens are dissolved in a polyurethane matrix [1]. In recently published papers by Jordan *et al.* and Babic *et al.* another approach was taken: In their proposed dosimeter the color dye and halogen are dissolved in a gelatin gel [2,3]. As the color dye and halogen do not readily dissolve in the gelatin hydrogel, the dye and halogen are embedded in micelles. The authors concluded that the micelle gel dosimeters are potential 3D dosimeters that had no dose rate dependence. The authors conclusion on dose rate dependence was based on a comparison of the shape of a gel measured dose depth curve for electron beams and a dose depth curve acquired with an ionization chamber. In our study in which the dose rate dependency of the leucodye micelle gels were investigated independently with photon beams, however a significant dose rate dependency was found. The dose rate dependency of different kinds of leucodye gel dosimeters were investigated, including the gel dosimeter proposed by Jordan *et al.* [2]. A novel leucodye gel dosimeter is proposed that has a similar dose sensitivity and improved dose rate dependency in respect to the composition proposed by Jordan *et al.* [2].

2. Materials and methods

2.1. Gel fabrication

The leucodye micelle gel dosimeter proposed by Jordan *et al.* (further referred to as LMD1) was fabricated according to the fabrication procedure described in their publication [2] and poured into small cuvettes (1x1x4.5 cm^3, PMMA, Sigma Aldrich) and placed in a refrigerator overnight at 4°C. These cuvettes are further referred to as spectroscopic samples.

A novel composition of leucodye micelle gel is proposed that consists of 6% gelatin, 80 mM chloroform ($CHCl_3$), 50 mM sodium dodecyl sulphate (SDS), 5 mM trichloroacetic acid (CCl_3COOH) an 0.37 mM leucomalachite green (LMG) all dissolved in deionised water (further referred to as LMD2). Gelatin is dissolved in 60% (w/w) of total water volume at room temperature and is left to swell for 10 minutes. Thereafter the gelatin-water solution is heated to 50°C. The remaining 32% of total water volume is used to dissolve SDS and CCl_3COOH. $CHCl_3$ is used to dissolve LMG after which this solution is carefully added to the water-SDS-CCl_3COOH-solution. The solution is wrapped in aluminum foil to minimize any photochemical induced reaction. After cooling of the gelatin-water solution to approximately 40°C, the two solutions are added together and stirred for 15 minutes in a dark room. Finally the gel is poured into small cuvettes, small PMMA phantoms (dimensions 1x1x8 cm^3) and a large PMMA phantom (dimensions 1x1x40 cm^3).

2.2. Dose sensitivity

Spectroscopic samples were irradiated with doses ranging between 0 Gy and 60 Gy in steps of 5 Gy and between 60 Gy and 100 Gy in steps of 10 Gy. Afterwards all spectroscopic samples were placed in a refrigerator overnight at 4°C. The spectroscopic samples were read out approximately 12 hours post irradiation by use of an USB 4000 spectrometer (Ocean Optics, Dunedin, USA). The ΔOD dose plot was fitted in Matlab and the slope was used in further comparisons.

2.3. Dose rate dependence

The dose rate dependency for the LMD1 type and LMD2 type dosimeter is investigated for photon beams of 6 MV by changing the photon beam pulsation rate. Spectroscopic samples were irradiated at dose rates of 50 cGy min^{-1}, 100 cGy min^{-1}, 200 cGy min^{-1} and 400 cGy min^{-1} at isocentre.

2.4. Depth dose profiles

Depth dose profiles were measured for electron beams of 6 MeV at an SSD of 95 cm and field size 10x10 cm^2 (using an electron applicator and appropriate cut-out) for dose rates set at 50 cGy min^{-1}, 200 cGy min^{-1}, and 400 cGy min^{-1} at isocentre. The position of the central axis of the small PMMA phantoms was along the beam direction. Measured depth dose profiles were rescaled to dose maximum for the PMMA phantom irradiated at 400 cGy min^{-1}. Ionization chamber measurements for a 10x10 cm^2 electron field of 6 MeV were obtained from the treatment planning system. Depth dose profiles were compared with depth dose profiles acquired with the gel.

A depth dose profile was measured for a photon beam of 6 MV at an SSD of 95 cm and field size 10x10 cm^2 for a dose rate set at 400 cGy min^{-1} at isocentre. The position of the central axis of the large PMMA phantom was along the beam direction. The measured depth dose profile was rescaled to dose maximum. Ionization chamber measurement was performed with a 0.016 cm^3 pinpoint ionization chamber (PTW 31016) and automated water phantom (PTW). The depth dose profile was compared with the depth dose profile acquired with the gel.

All PMMA phantoms were read out by the USB 4000 fiber optic spectrometer along the depth of the recorded dose distribution using a linear stage (PTW, Freiburg, Germany) which travelled in spatial increments of 1 mm. From all measured spectral data, optical density differences (ΔOD) were calculated with Matlab scripts. A relative calibration method of ΔOD to dose was used to calibrate the ΔOD data to dose data.

2.5. Spectroscopic measurements

All spectroscopic samples were taken out of the refrigerator about 45 minutes prior to readout and left to equilibrate at room temperature. Subsequently, the samples were read out by an USB 4000 fiber optic spectrometer using a in-house constructed white led light source. The readout parameters were: 100 ms integration time, 7 nm boxcar width of running average, 50 averages at 632 nm.

Journal of Physics: Conference Series **250** (2010) 012009 doi:10.1088/1742-6596/250/1/012009

3. Results

3.1. Dose sensitivity

The dose sensitivity of the LMD2 dosimeter amounted to 4.375 x 10^{-3} $cm^{-1}Gy^{-1}$ (dose rate 400 cGy min^{-1}). The dose sensitivity curve shows a linear dependency. The dose sensitivity of the LMD1 dosimeter amounted to 0.8925 x 10^{-3} $cm^{-1}Gy^{-1}$ (dose rate 400 cGy min^{-1}).

3.2. Dose rate dependence

A clear dose rate dependency is seen for all radiochromic dosimeters investigated in this study. Figure 1 shows that the deviation of the optical dose sensitivity between 50 cGy min^{-1} and 400 cGy min^{-1} of LMD1 type dosimeter reached 74.6% relative to 400 cGy min^{-1} as compared to 33.7% for the LMD2 type dosimeter.

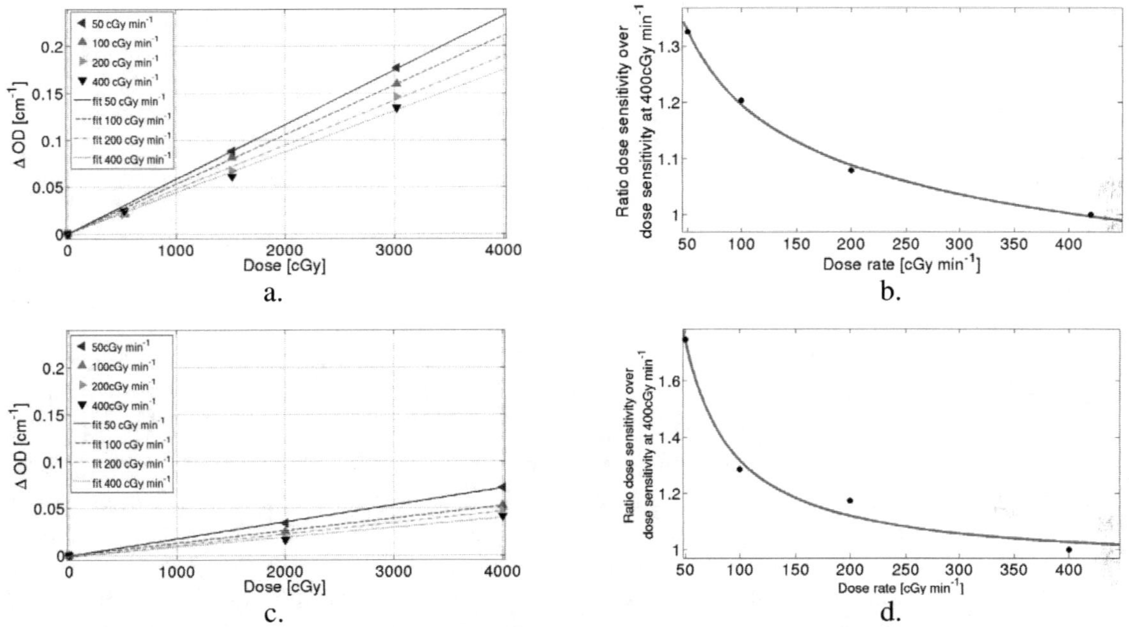

Figure 1. Comparison of dose response for LMD2 dosimeter a and LMD1 dosimeter c irradiated at different dose rates (50 cGy min^{-1} (full line), 100 cGy min^{-1} (dashed line), 200 cGy min^{-1} (dot-dashed line), 400 cGy min^{-1} (dotted line)). Plot b and d show the dose sensitivity as function of dose rate relative to 400 cGy min^{-1}. An exponential fit is shown as a full line.

3.3. Depth dose profiles

In figure 2a, depth dose profiles of 6 MeV electron beams are plotted all scaled to the dose maximum measured at 400 cGy min^{-1}. In figure 2c all depth dose profiles are individually scaled to the maximum and minimum dose demonstrating the high apparent correspondence between the three curves.

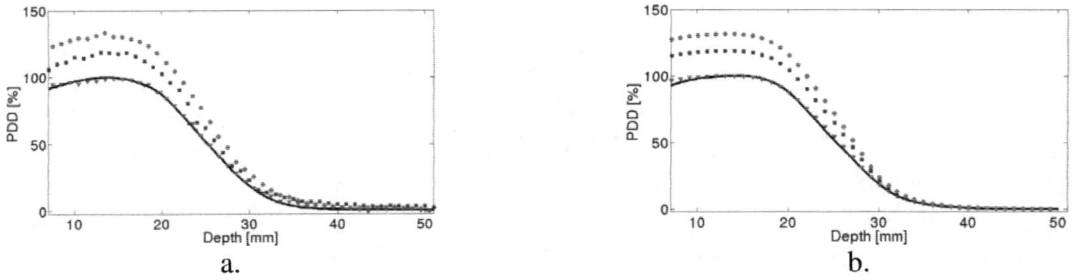

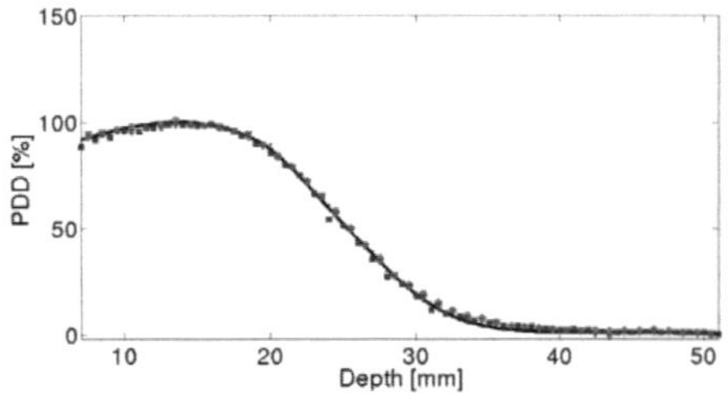

c.

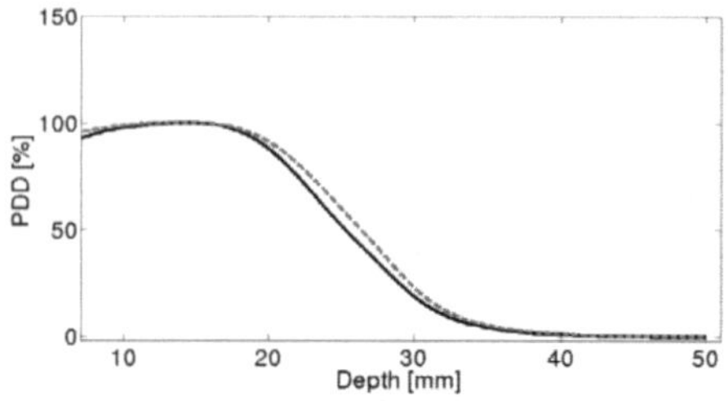

d.

Figure 2. Plot a and c show electron depth dose profiles (6 MeV) measured with the LMD2 type dosimeter at three dose rates: 50 cGy min^{-1} (red dotted line), 100 cGy min^{-1} (blue dotted line) and 400 cGy min^{-1} (green dotted line) compared with an ionisation chamber measured PDD (black full line). The PDD's are scaled with the same normalisation factor for the PDD recorded at 400 cGy min^{-1} in a. The PDD's are scaled with their individual normalization factor in c. Plot b and d show a simulation of the influence of measured dose rate dependence upon the depth dose profile (dashed line) compared to a PDD recorded with an ionisation chamber (full line).

Additionally, in figure 2b and 2d simulated depth dose profiles are shown as calculated from an ionization chamber measured depth dose distribution and based on the dose rate dependency for electrons.

4. Discussion

4.1. Dose sensitivity

The maximum optical dose sensitivity for LMD2 was 4.375 x 10^{-3} $cm^{-1}Gy^{-1}$ (dose rate 400 cGy min^{-1}) which is 5% lower than the value of the leucodye dosimeter proposed by Jordan *et al.* (4.6 x 10^{-3} $cm^{-1}Gy^{-1}$) [2]. In order to compare the characteristics of the LMD2 dosimeter, the leucodye dosimeter proposed by Jordan *et al.* (LMD1) was fabricated according to an identical fabrication procedure as described in their publication [2]. The maximum optical dose sensitivity for LMD1 amounted to 1.031 x 10^{-3} $cm^{-1}Gy^{-1}$ (dose rate 400 cGy min^{-1}) which is four times lower than the value reported by Jordan *et al.* [2]. An explanation for the difference in dose sensitivity was not found. An significant increase of the background color is seen during the first hour post fabrication although the dosimeter was stored in a refrigerator at 4°C. Attempts to reduces this background color by fabricating the gel in a dark room have not been successful.

4.2. Dose rate dependence

Both LMD1 type and LMD2 type dosimeters are shown to be strongly dependent on the dose rate. To illustrate the effect of the dose rate dependency on a dose distribution, a PDD of a 6 MeV electron beam was recorded with a LMD2 type dosimeter irradiated at different dose rates. It is clear that lower dose rate regimes result in higher dose sensitivity (figure 2a). However when all PDD's are rescaled to their dose maximum and minimum (figure 2b) no significant deviation from the ionization chamber measurement is found.

The dose rate dependency effects were also simulated on a depth dose distribution measured with an ionization chamber. The maximum deviation between the dose rate dependent and independent PDD amounts to 8% of dose maximum between 20 mm and 30 mm depth (figure 2d.). However a positional shift of 0.9 mm results in an almost complete correspondence of both curves between dose maximum and 5 cm depth indicating that this method of identifying dose rate dependency is very sensitive to positional errors.

To extend the previous results, the effects of the dose rate dependent optical dose response was also investigated for photon beams. A 6 MV photon beam depth dose distribution was recorded with the LMD2 type dosimeter. Deviations up to 4.9% relative to the dose maximum are found for the LMD2 type dosimeter (figure 3). A dose rate dependent PDD was also simulated based on the measured dose rate dependency of the LMD2 type dosimeter. The results are consistent with a measured PDD.

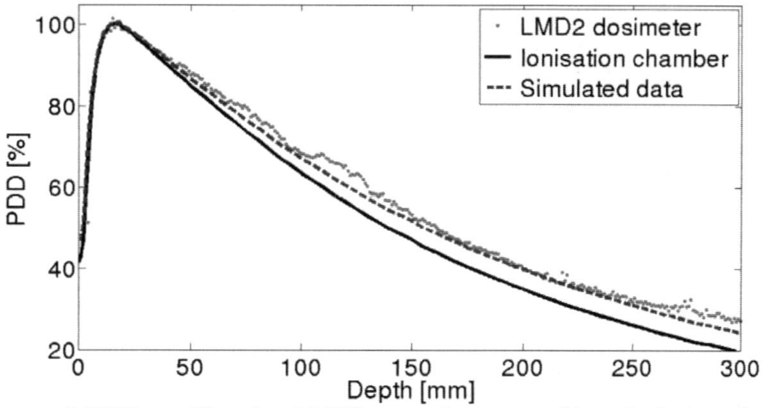

Figure 3. Comparison of PDD profiles for LMD2 type dosimeter (dotted plot) versus ionisation chamber measurements (full line) of 6 MV photon beams. The dashed line represents the simulated dose rate dependent PDD curve as calculated from the measured dose rate dependency of the LMD2 type dosimeter.

The predominant chemical that influences the dose rate dependency is believed to be the surfactant. Changing the concentration and type of surfactant (SDS versus Triton X100) results in substantial differences in dose rate dependency. A change in gelatin concentrations however does not avoid a dose rate dependent response.

The large number of chemical components and the chemical heterogeneity of the micelles in the LMD type dosimeters results in a high complexity of radiation induced reaction kinetics which cannot be completely disentangled at this moment.

5. Conclusions

The increasing need for dedicated real 3D quality assurance tools in modern radiation therapy is the main motivation behind the development of accurate and user-friendly 3D dosimeters. Recently a novel 3D radiochromic gel dosimeter was introduced which uses micelles to dissolve a leucodye in a gelatin matrix. Experimental results show that the previously introduced micelle dosimeter was found to be significantly dose rate dependent. A novel composition of 3D radiochromic dosimeter is proposed composed of gelatin, sodium dodecyl sulphate, chloroform, trichloroacetic acid and leucomalachite green. The novel gel dosimeter formulation exhibits comparable radio-physical properties and a reduced dose rate dependency, however it is still unsatisfactory for clinical radiation therapy dose verifications. Future research should focus on minimizing the dose rate dependency through physic-chemical analysis of the dosimetry systems. A robust calibration method for 'absolute' dosimetry is under development.

Acknowledgements

This research is supported by the Institute for the Promotion of Innovation though Science and Technology in Flanders (IWT -Vlaanderen).

References

[1] Adamovics J and Maryanski M J. "Characterisation of Presage: a new 3D radiochromic solid polymer dosemeter for ionizing radiation", *Rad. Pro. Dosim.* 2006; 120: 107-112.

[2] Jordan K and Avaakumov N. "Radiochromic leuco dye micelle hydrogels: I. Initial investigation", *Phys. Med. Biol.* 2009; 56: 141-153.

[3] Babic S, Battista J, Jordan K. "Radiochromic leuco dye micelle hydrogels: II. Low diffusion rate leuco crystal violet gel", *Phys. Med. Biol.* 2009; 54: 6791-6808.

Journal of Physics: Conference Series **250** (2010) 012010 doi:10.1088/1742-6596/250/1/012010

The "push-pull" dosimeter: When two pigments are better than one.

Stephen G. Bosi*[1,2], Clive Baldock[2] and Robert Smee[1]

[1]Dept of Radiation Oncology, Prince of Wales Cancer Centre, Randwick NSW 2031 Australia

[2]Institute of Medical Physics, School of Physics, University of Sydney, NSW 2006 Australia

*Corresponding author s.bosi@physics.usyd.edu.au

Abstract. A new kind of gel dosimeter (the "push-pull" dosimeter) is proposed which would contain two spectrally complementary pigments, one which darkens with increasing dose and another which bleaches. The bleaching pigment would be optimised for high sensitivity and the darkening pigment for low sensitivity. By employing dual pigments optimised independently, the usual requisite compromises between sensitivity at low dose and accuracy at high dose would be relaxed. Such a gel, after exposure would be read using two successive optical CT scans employing two different scan wavelengths. The use of dual pigments could also reduce the occurrence of regions of high optical attenuation which generate artefacts in optical CT. This paper also presents results of simulations of the behaviour of such a gel when scanned using optical cone beam CT.

1. Introduction

The choice of dynamic range for any dosimeter involves a compromise between sensitivity and detection limit at the low end, and saturation or loss of accuracy at the high dose end. This compromise is typically resolved by having a range of dosimeters and choosing one with a dynamic range matched to the task. However, in an application such as 3D dosimetry for IMRS/IMRT QA where there is a wide range of dose confined to a small volume, it is challenging to address that compromise.

Moreover, in the case of a 3D gel dosimeter being read optically using either optical cone beam or scanning laser CT, even if saturation of the dosimeter can be avoided, then the highest doses produce regions of high optical density in the gel which can produce shadow and streaking artefacts (Figure 1) and other artefacts such as cupping [1] in reconstructions of the dose map, especially in the presence of noise and stray scattered light.

In this paper, a new design of gel dosimeter is proposed, the "push-pull gel" which relaxes the compromise between accuracy and sensitivity at either end of the dosimeter's dynamic range. This proposed gel would include two separate dosimetric pigments. A mathematical model of such a system has been developed to analyse the optical and dosimetric parameters required for such a pair of pigments. This paper presents some of the initial results yielded by this model.

Journal of Physics: Conference Series **250** (2010) 012010 doi:10.1088/1742-6596/250/1/012010

2. Rationale

The two pigments would be chosen so the maximum dose-induced change in extinction coefficient of the pigments occurs at well-spaced wavelengths, and the wavelength widths of their respective absorption peaks should be narrow enough to minimise spectral overlap. These conditions would allow one to interrogate each pigment separately using a separate optical scan at these two wavelengths, with minimal cross-talk (Figure 2).

One pigment would be optimised to measure the high-dose part of the gel's dynamic range, while the other would be used for low-dose range. Assuming the sensitivity of the two pigments could be independently optimised (for example by varying their concentrations), then the possibility arises that a gel with enhanced dose dynamic range can be produced. The increase in dynamic range by separation of the dynamic range into two separate components, serviced by distinct mechanisms, is reminiscent of the "push-pull" amplifier in electronics, hence the adoption of the name "push-pull" to denote a gel employing two pigments to deal with opposite ends of the dynamic range.

There is another important aspect of this design. One pigment would (as is most typical for dosimeters) darken or *stain* with dose. If however, the second pigment is chosen to *bleach* with dose, then it opens another possibility.

The staining pigment would be optimised for low sensitivity. In that way, it would not saturate in high dose regions. Also, the extinction coefficient in those regions would never be excessive, thus avoiding the optical CT artefacts that result from high attenuation. However, this pigment would necessarily provide poor information in the low-dose regions.

If the bleaching pigment is optimised for maximum sensitivity, then it would serve as the low-dose dosimeter, complementing the function of the staining pigment. The reason for this choice is that in regions of high dose, the bleaching pigment would simply bleach away. In high dose regions, the bleaching pigment would provide no information but also its minimal attenuation would avoid the possibility of high attenuation artefacts.

The possibility that reading a *single pigment* gel dosimeter at multiple wavelengths might improve dynamic range [2] or improve control of image contrast [3] has been suggested before. These suggestions were based on properties of the spectrum of Fricke-xylenol orange gel dosimeter (FXB) which exhibits both bleaching and staining optical wavelength regions. At wavelengths less than ~470 nm, the gel bleaches with dose, while for wavelengths above this, it stains [3]. However, being a single dosimeter pigment, one could not separately optimise both the sensitivity and saturation doses in these two wavelength ranges. The optical extinction coefficients at wavelengths above and below 470 nm in FXB are not independent parameters. Only by using two separately optimisable pigments are the sensitivity and saturation doses at the two ends of the dynamic range truly independently optimisable.

The recent development of a dosimeter based on the radio-bleaching pigment genipin [4] turns this hypothetical design into a distinct possibility.

3. Method

A mathematical model was developed in order to test the underlying optical principles required for the realisation of this new dosimeter and its behaviour in a dual-wavelength optical CT scanner. Both pigments were assumed to be nearly linear in dose in the low-dose range, but to saturate above a certain high dose range.

For each hypothetical pigment, the wavelength of maximum absorption, the width of the absorption peak, the slope and height of the spectrum background, the sensitivity (slope of the attenuation versus dose curve in the linear region), the saturation range and the sharpness of the saturation knee are adjustable. In the current version of the mode, each pigment is assumed to possess a single gaussian absorption peak, although it is easily generalised to include other forms.

Hypothetical dose maps can be input into the model to obtain an extinction coefficient map at the two scan wavelengths which can then be projected numerically to produce simulated optical projections one would obtain in an optical CT scanner. Noise, stray light and other non-ideal behaviours can also be added. The open-source cone-beam reconstruction software OSCaR2 [5] is then used to reconstruct these projections and the resulting reconstructed extinction coefficient map can then be converted back to dose using the calibration curves and this compared with the original dose map.

For this paper, a dose map (maximum dose 55 Gy) was taken from the dose plan of an actual intensity modulated radiosurgery treatment of the head. The treatment was planned using Radionics XKnife 4. A 2D slice of the treatment plan was then converted into several optical extinction coefficient maps using the model, assuming different sensitivities for the staining pigment. The effect on the reconstructed map, of adding a uniform 10 PPM of stray light to the resulting projections, was examined. The optical scan wavelengths were 455 nm and 617 nm.

4. Results

Figure 3 reveals the reconstructed extinction coefficient map (scanned at 455 nm) resulting from a simulated exposed push-pull gel with a staining gel of moderate sensitivity. Moderate sensitivity can be accommodated because the bleaching pigment (scanned at 617 nm) would be responsible for measuring the lower doses. The two images and accompanying plots in Figure 3 reveal that the addition of 10 PPM of stray light has a negligible effect on the extinction coefficient profile.

Figure 4 shows two profiles taken through extinction coefficient map similar to Figure 3, except that now the sensitivity of the staining pigment is increased by 44%. Without the bleaching pigment, the sensitivity of the staining pigment might need to be increased to improve detection at low dose. Figure 4 reveals the limit to this approach - the increased optical attenuation renders the reconstruction sensitive to stray light.

5. Conclusions

The reduction in sensitivity of the staining dye, allowed for by the presence of the bleaching dye, renders the reconstructed optical attenuation coefficient map (and hence dose map) less sensitive to attenuation-induced artefacts.

Although the push-pull concept was devised for 3D gel dosimetry, a similar principle could also be applied to 2D film dosimeters to increase dynamic range.

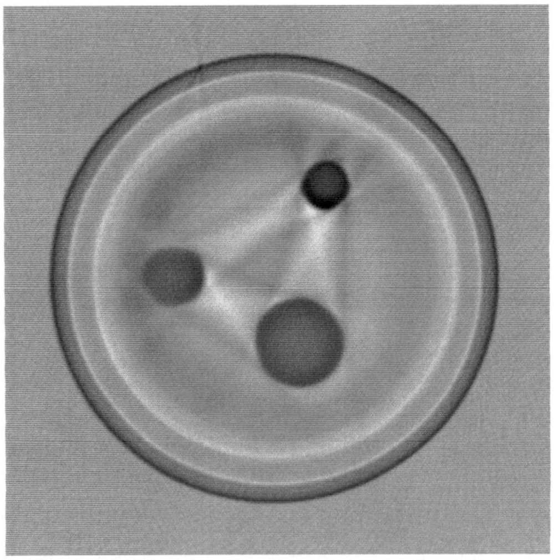

Figure 1. Example of shadow/streaking artefacts in an axial slice of a scanned laser optical CT image of a Dettol gel "finger phantom" with three strongly attenuating fingers. The region surrounding and between the fingers should be uniform grey.

Journal of Physics: Conference Series **250** (2010) 012010 doi:10.1088/1742-6596/250/1/012010

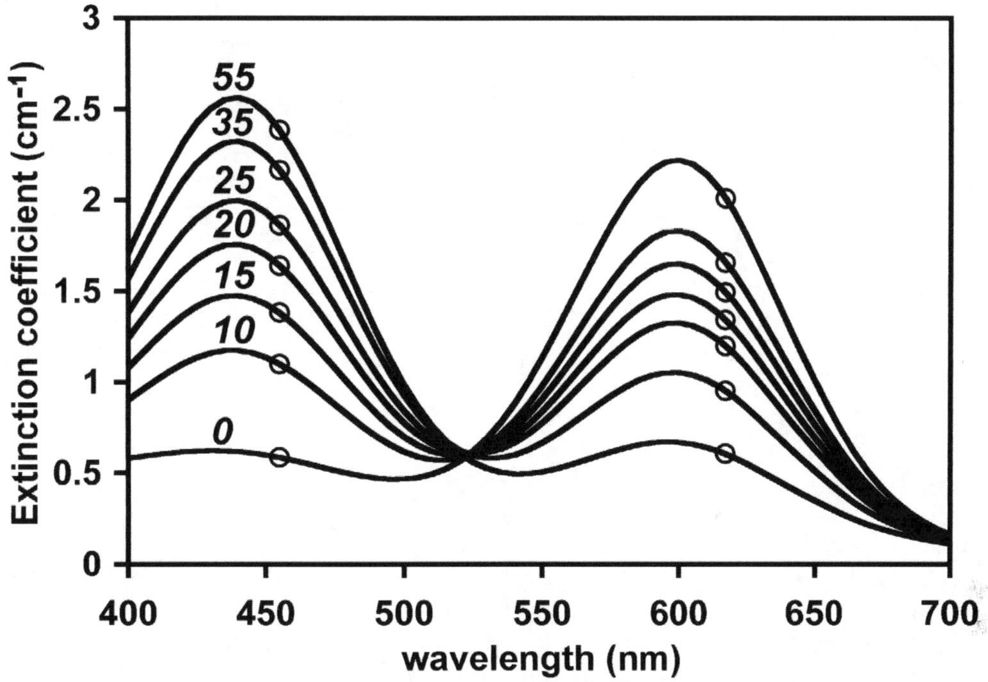

Figure 2. Modelled spectra for a hypothetical gel containing a "push-pull" pair of dosimetric pigments. Each spectrum is labelled by the hypothetical dose in gray. The open circles indicate extinction coefficients measured at two widely-spaced LED wavelengths available for use in the diffuse light source of a Vista optical cone-beam CT scanner (Modus Medical Devices™).

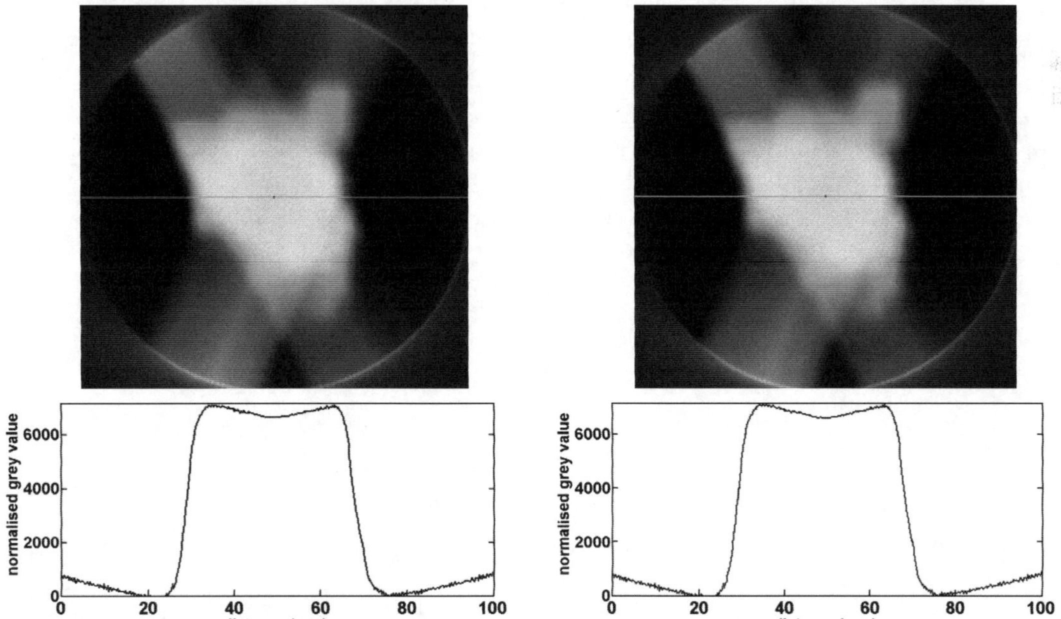

Figure 3. Left: Reconstructed extinction coefficient map (455 nm) for the staining pigment at lower sensitivity, without stray light. Right: Similar reconstruction but with 10 PPM stray light added to optical CT projections. Plots underneath are profiles through the maps, grey values normalised against dosimeter sensitivity. The similarity of the two plots reveals that the effect of stray light is negligible for this dosimeter sensitivity.

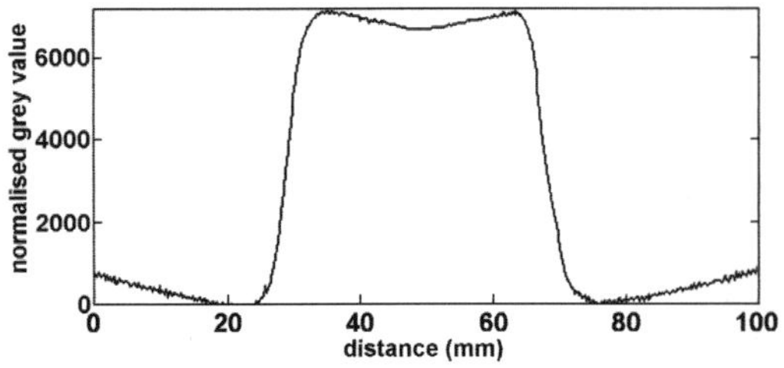

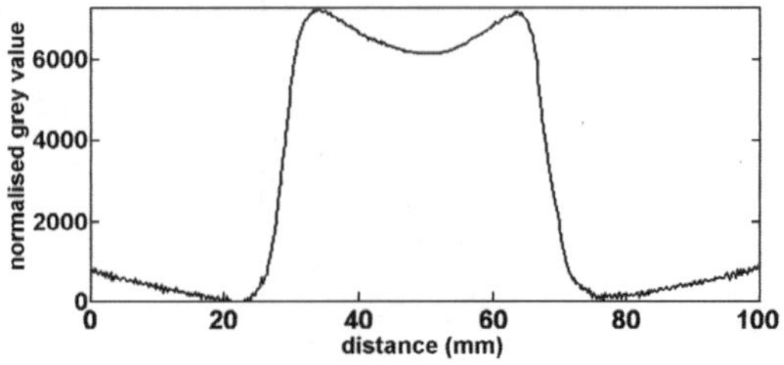

Figure 4. Profiles through reconstructed extinction coefficient maps (455 nm) for a staining pigment 44% more sensitive than in Figure 3. Left: Without stray light. Right: 10 PPM stray light added to optical CT projections. Evident increased cupping in the middle of the bottom profile reveals that even a moderately more sensitive staining pigment renders the gel more seriously affected by stray light.

References

[1] Bosi S.G., Naseri P. and Baldock C. Light scattering-induced artifacts in a complex polymer gel dosimetry phantom. Appl. Opt. **48** 2427-2434 (2009).
Bosi S.G., Naseri P., Puran A., Davies J. and Baldock C. Initial investigation of a novel light-scattering gel phantom for evaluation of optical CT scanners for radiotherapy gel dosimetry. Phys. Med. Biol. **52**, 2893-2903 (2007)

[2] Bero M A, Jenneson P M, Gilboy W B, Glover P M and Doran S J 1999b Faster optical tomography with parallel-beam white light for three-dimensional dosimetry Proc. 1st Int. Workshop on Radiation Therapy Gel Dosimetry July 21-23, 1999, Lexington, Kentucky USA.

[3] Doran S.J., Koerkamp K.K., Bero M.A., Jenneson P., Morton E.J. and Gilboy W.B. A CCD-based optical CT scanner for high-resolution 3D imaging of radiation dose distributions: equipment specifications, optical simulations and preliminary results. Phys. Med. Biol. **46** 3191–3213 (2001).

[4] Jordan K. Optical CT scanning of cross-linked radiochromic gel without cylinder wall. J. Phys: Conf. Series **164** 012029 (2009)

[5] Rezvani N., Aruliah D., Jackson K., Moseley D. and Siewerdsen J., SU-FF-I-16: OSCaR: An Open-Source Cone-Beam CT Reconstruction Tool for Imaging Research. 2007 AAPM Annual Meeting Minneapolis, July 22-26, 2007. Med. Phys. **34**(6), 2341 (2007).
Permalink: http://dx.doi.org/10.1118/1.2760393

Journal of Physics: Conference Series **250** (2010) 012011 doi:10.1088/1742-6596/250/1/012011

Preliminary Investigation of the Dosimetric Properties of 'RadGel'

J. R. Newton[1], A. Thomas[1], A. Appleby[2], C. Marsden[2], E.A. Christman[2], J.G. Wolodzko[2], M. Oldham[1]

[1]Duke University, Durham, North Carolina
[2]Dosimetrics Inc., Middlesex, NJ

*Corresponding autor: joseph.newton@duke.edu

Abstract. A preliminary investigation into the efficacy of a new 3D dosimetry material, RadGel™, for verification of radiation therapy dose distributions is presented. Small volumes of RadGel™ were found to exhibit a linear, reproducible response to dose. A gradual increase in optical-density (OD) with time was observed, suggesting scanning should be completed within 18 hours to keep a linear correlation of $R^2 > 0.99$. A larger 10 cm diameter volume of RadGel™ was irradiated with a rotationally symmetric "spoke" plan designed to rigorously evaluate scanner/dosimeter combined performance. The dosimeter was imaged with the **D**uke **M**id-sized **O**ptical-CT **S**canner (DMOS). Promising OD and corresponding dose maps were obtained. Edge artefacts were observed and are suspected to be exacerbated by the particular container used in this early study. Further studies will evaluate new containers and methods for refractive matching at the gel-container-fluid interface.

1. Introduction

The use of gel dosimetry has been the subject of significant interest recently as a result of the necessity for a plausible 3 dimensional dosimetry technique for verification and quality assurance of complex radiation therapy techniques, such as IMRT and radiosurgery. The purpose of the current study is to characterize the basic dosimetric properties of a new formulation of gel, called RadGel™, which is a near tissue equivalent material with a CT number of 30 HU. A characterization of a previous formulation can be found in [1]. The properties investigated in this work include absorption spectrum, optical dose response (sensitivity), temporal stability of optical response, dose-rate effects, reproducibility, water equivalence, and volume effects (i.e. whether large volumes respond in a similar manner to small volumes).

2. Methods

Methods are organized into two sections. Section 2.1 details extensive studies on small volumes of the RadGel™ material in order to efficiently investigate a wide range of basic dosimetric properties. Section 2.2 details experiments on a larger volume of RadGel™, to investigate any volume effect as well as the consistency of behavior between large and small volumes.

Journal of Physics: Conference Series **250** (2010) 012011 doi:10.1088/1742-6596/250/1/012011

2.1. Small Volume Cuvette Studies

A series of standard (1 cm x 1 cm) optical cuvettes filled with RadGel™ were irradiated with 6 MV photons in the same beam geometry, consisting of a 10 cm x 10 cm field with an SSD of 100 cm. The cuvettes were individually placed below a 5 cm slab of solid water with 10 cm of solid water below, for backscatter. The center of the cuvette was at a depth of 5.5 cm in solid water and was surrounded by a tissue equivalent bolus. The optical density of individual cuvettes was measured with a Thermo Scientific Genesys 20 spectrophotometer both before and after irradiation.

The first dosimetric property of interest is the absorption spectrum, in order to determine the wavelength at which the maximum ΔOD is observed after irradiation. The radiation induced ΔOD of a cuvette irradiated to 4Gy was measured with the spectrophotometer over the wavelengths 500 - 700 nm at intervals of 20 nm, with intervals of 2 nm over the range 600 - 612 nm. This higher resolution data coincided with the peak OD change reported in a previous study of ~600 nm (Guo et al. [2]). The next property, optical-dose-response, was measured using 6 cuvettes irradiated to different doses ranging from 1 - 20 Gy. The optical density of each cuvette was measured prior to irradiation and then immediately after irradiation.

The temporal response of optical density was studied by two separate methods. In the first method, three cuvettes were irradiated to a dose of 8 Gy and the immediate change in optical density measured. One cuvette was then scanned by the spectrophotometer at 15 minute intervals for 12 hours, beginning one hour after irradiation. The second cuvette was stored in the refrigerator and the third cuvette was left at room temperature. The ΔOD of all three was then measured 19 hours after irradiation. In the second method, the optical density of the 6 cuvettes used for the dose response study was re-measured at intervals of 17.5, 41.7 and 113.5 hours after irradiation. These 6 cuvettes were stored in the refrigerator between measurements to minimize any temperature effects on the optical density. The cuvettes were allowed to acclimate to room temperature before each measurement in the spectrophotometer. Dose rate effects were investigated by irradiating separate cuvettes to 8 Gy at dose rates of 100, 300 and 600 Monitor Units per minute.

2.2. Volume effect Study

A larger volume 10 cm dosimeter consisting of a plastic bottle filled with RadGel™ was irradiated with a spoke plan modeled after that used by Sakhalkar et al. [2]. The plan was created in the planning system with the AAA algorithm. The spoke plan geometry, consisting of a total of 9 alternating high dose, medium dose and low dose spokes, provides an excellent test of the radial uniformity of dose response of the dosimeter. The spoke plan was created to extend outside the outer diameter of the dosimeter, in order to separate any penumbra effects from edge effects of the dosimeter and bottle. The high-dose spoke was prescribed 6.5 Gy and the medium and low-dose spokes were prescribed 4.7 Gy (72%) and 2.1 Gy (32%) respectively. The gantry was positioned at an angle of 180° with an SSD of 94.2 cm. The Radgel™ dosimeter was placed in a water bath and the plan was delivered with 6 MV radiation from a Varian 600c. The ΔOD of the RadGel™ dosimeter after irradiation was measured using the **D**uke **M**idsized **O**ptical **S**canner (DMOS), which is a telecentric optical-CT system. More details of the principle of the DMOS scanner are described in an accompanying abstract by Thomas et. al.[3] and elsewhere [4-6]. A solution of ~7% (by volume) ethylene glycol in water was used for refractive index matching of the dosimeter. The measured ΔOD was then compared to the calculated ECLIPSE dose distribution using profiles and gamma maps.

3. Results and Discussion

3.1. What is the optimal scanning wavelength?

The change in optical density as a function of wavelength for the first cuvette irradiated to 4 Gy is shown in Figure 1(a). The peak of this response curve occurs at a wavelength of 608 nm and has a width of ~100 nm. A wavelength of 608 nm was therefore selected for all further measurements of the optical characterization of RadGel™.

Journal of Physics: Conference Series **250** (2010) 012011 doi:10.1088/1742-6596/250/1/012011

3.2. The Response of Optical Density as a Function of Dose

The immediate OD response of the six cuvettes irradiated with doses ranging from 1 - 20 Gy is displayed in Fig. 1(b). A linear regression of the data is also included resulting in an optical density response of 0.010 (Gy cm)$^{-1}$ with a correlation of $R^2 = 0.997$, showing a remarkably linear response.

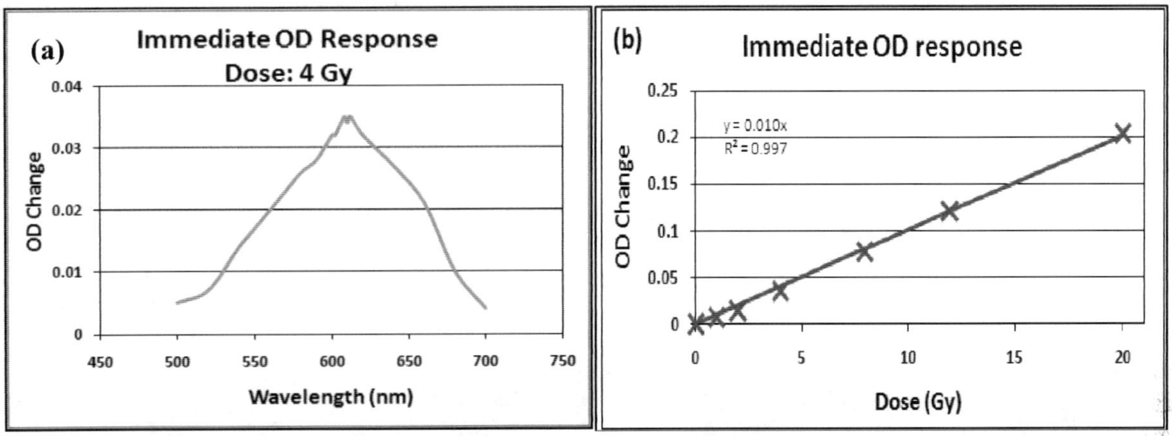

Fig. 1 (a) the spectral response of RadGel™ cuvette after irradiation dose of 4 Gy. Maximum response was found at 608 nm. (b) The immediate measured change in optical density of the RadGel™ cuvettes after irradiation with 6 MV photons of doses ranging from 1 to 20 Gy.

3.3. Temporal Response of Optical Density

The results of the 12 hour, 15 minute interval study of the temporal response of OD in a single cuvette are shown in Fig. 2(a). A linear OD darkening of the RadGel™ of 0.009 (hour cm)$^{-1}$ was observed. Also included in this plot are three data points showing the OD response of cuvettes stored at different temperatures 19 hours after irradiation. A clear increase in OD change is observed with increasing temperature, suggesting a significant temperature dependence of the temporal response. The results of the temporal response of OD with dose for the 6 cuvettes irradiated to differing doses are shown in Fig. 2(b). There are two interesting effects to notice. The first is that the rate of darkening is significantly higher for higher dose points and the second is that the linearity of response is breaking down over time.

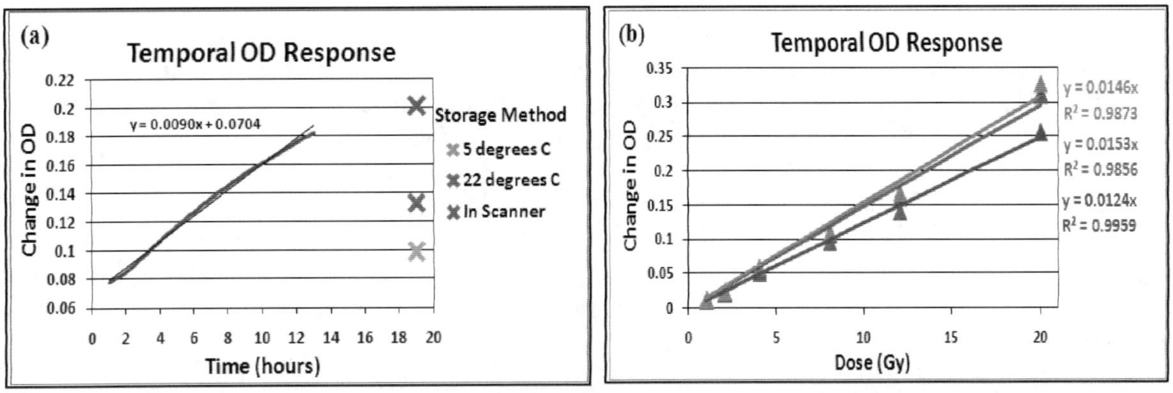

Fig. 2 (a) The change in OD of a representative RadGel™ cuvette irradiated to 8Gy. Also included are the values 19 hours after irradiation for 3 cuvettes (cross marks) stored at different temperatures. (b) The measured change in OD of the 6 RadGel™ cuvettes after irradiation to different doses determined at 17.5 hours (blue), 41.7 hours (red) and 113.5 hours (green) after irradiation.

Journal of Physics: Conference Series **250** (2010) 012011 doi:10.1088/1742-6596/250/1/012011

3.4. Dose Rate Response and Reproducibility

The ΔOD after irradiation of RadGel™ cuvettes was found to be extremely reproducible with no effects attributable to the delivered dose rate. The immediate ΔOD's were unaffected by dose rates of 85 - 500 cGy/minute. Dose rate effects below this range are expected to be indistinguishable from the temporal effects discussed previously.

4. Large Volume Studies

A qualitative comparison of a central slice of the calculated and measured spoke plan dose distribution is shown in figure 3. The alternating intensities of the spokes are clearly visible in both distributions.

The edge of the dosimeter is very bright suggesting a difference in index of refraction of the plastic bottle and the RadGel™ inside the bottle. This leads to significant edge effects which should be remedied by selection of an appropriate container for the RadGel™.

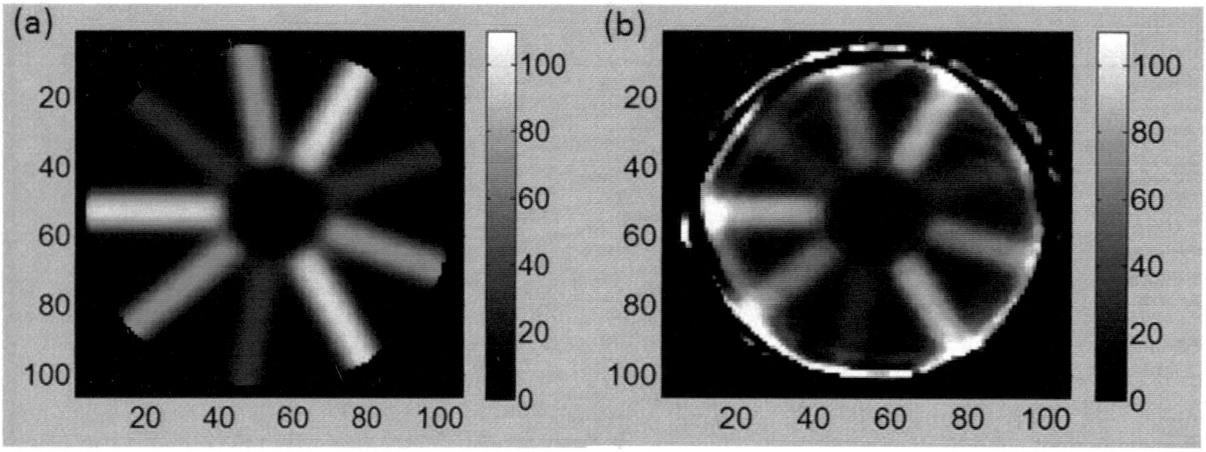

Fig. 3 (a) Calculated dose distribution in the central axial slice of the Spoke irradiation delivered to the 10 cm diameter RadGel™ dosimeter. (b) Reconstructed Optical-CT image of the change in OD in the same slice.

Figure 4 (a) shows the gamma map (8% difference, 3mm DTA) calculated for the difference of the images from Figure 3. As expected, the outer edge of the dosimeter is in poor agreement with the Eclipse calculations due to pronounced edge artifacts arising from the difficulty in matching refractive-indices at the fluid/plastic/RadGel™ interface. However, a reasonable pass rate of 83% is found for the inner 3.5 cm radius of the dosimeter. If the gamma criteria are relaxed to a 10% difference, the pass rate increases to 87% for the inner 3.5 cm radius. Figure 4 (b) shows a CT scan of the dosimeter placed in a water bath. The line profile shows the CT number of RadGel™ to be ~ 30 HU, consistent with the surrounding water. The spikes in CT number correspond to the walls of the plastic container.

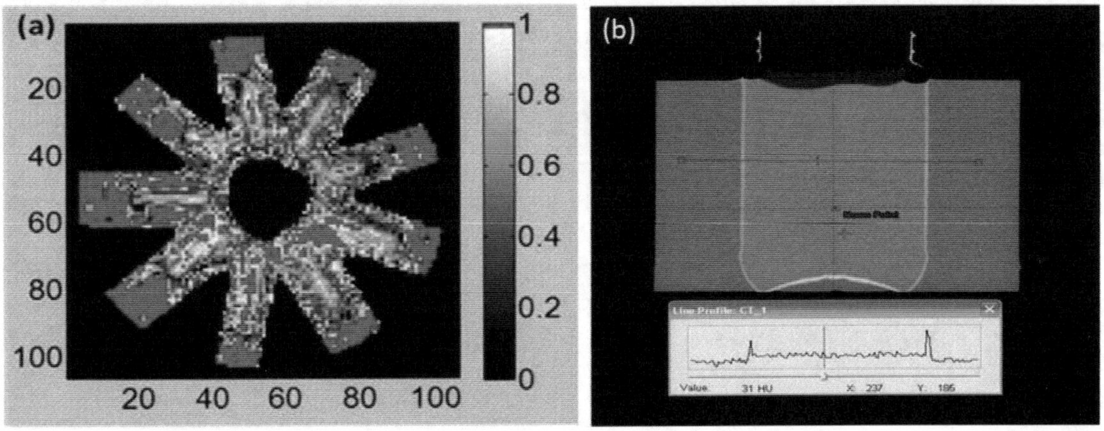

Fig. 4 (a) Gamma map calculated for the 8% difference, 3mm distance to agreement between the Optical-CT measurement and the Eclipse calculation shown in Fig. 2. A pass rate of 83% is found for the inner 3.5 cm radius of this dosimeter. (b) CT scan of plastic container filled with Radgel™ immersed in water.

5. Conclusions

This preliminary investigation of the dosimetric properties of RadGel™ has produced promising results. The linearity of response found in the cuvette studies and the reproducibility of response are attractive features. The temperature dependence of the temporal response suggests that storing the RadGel™ in a refrigerator after irradiation will significantly slow the OD darkening and allow for an 18 hour period in which the dose response remains linear. The results of the large volume study produced good images of the OD changes in the dosimeter. Strong edge artifacts were observed, however, which are attributed partly to a degree of light scattering in the container walls, and partly to the challenge of achieving a good refractive match at the gel-plastic-fluid interfaces. Further work is anticipated to provide improvements in these areas.

Acknowledgements: Supported by NIH R01CA100835.

References

[1] H. S. Sakhalkar, J. Adamovics, G. Ibbott and M. Oldham, *A comprehensive evaluation of the PRESAGE/optical-CT 3D dosimetry system*, Med. Phys., 2009. **36** (1): p. 71-82.

[2] P. Guo, A. Appleby and M. Oldham, *SU-FF-T-257:The study of the Dosimetric Properties of 'RadGel', a New Dosimeter for Three-Dimensional Gel Dosimetry*, Med. Phys., 2005. **32**(6).

[3] A. Thomas, *An investigation into the feasibility of a fast, large field-of-view, telecentric optical-CT scanning system for 3D radiochromic dosimetry*, IC3D Dose 2010.

[4] N. Krstajic and S. J. Doran, *Focusing optics of a parallel beam CCD optical tomography apparatus for 3D radiation gel dosimetry*, Phys. Med. Biol., **51** (2006): p. 2055-2075.

[5] R. G. Kelly, K. J. Jordan and J. J. Batista, *Optical CT reconstruction of 3D dose distributions using the ferrous-benzoic-xylenol (FBX) gel dosimeter*, Med. Phys., 1998. **25** (9): p. 1741-1750.

[6] J. G. Wolodzko, C. Marsden and A. Appleby, *CCD imaging for optical tomography of gel radiation dosimeters*, Med. Phys., 1999. **26** (11): p. 2508-2513.

[7] J. C. Gore, M. Ranade, M. J. Maryanski and R. J. Schulz, *Radiation dose distributions in three dimensions from tomographic optical density scanning of polymer gels: I. Development of an optical scanner*, Phys. Med. Biol., **41** (1996): p. 2695-2704.

Journal of Physics: Conference Series **250** (2010) 012012 doi:10.1088/1742-6596/250/1/012012

Dosimetric properties of MAGIC-*f* polymer gel assessed to Radiotherapy clinical beams

Marques T., Schwarcke M., Garrido C.E., Baffa O., Nicolucci P.

Departamento de Física e Matemática, Universidade de São Paulo.
Av. Bandeirantes, 3900 - 14040-901. Ribeirão Preto – SP, Brazil.

tatiana.marques@usp.br

Abstract. Dosimetric properties of MAGIC-*f* gel were studied for a wide range of beams used in Radiotherapy. The MAGIC-*f* tissue-equivalence was theoretically verified using PENELOPE Monte Carlo code and experimentally by percentage depth dose curves for water and gel (maximum differences of 2.0% and 0.5% for 6MV and 10MV, respectively). Energy and dose-rate dependency were evaluated in a range from 60 kV to 10 MV and 0.44 Gy/min to 10 Gy/min respectively. MAGIC-*f* presents linearity to all energies studied and its sensitivity presents maximum and average variation of 8.6% and 7.7% respectively in the range of energy considered. The study of dose rate dependency evidence that MAGIC-*f* presents no significant variation of response in clinical intervals of dose rate: lower than 0.7% when x-ray are considered and lower than 1.9% when all beams are considered. Its dosimetric characteristics indicate that MAGIC-*f* is a very suitable dosimeter for Radiotherapy.

1. Introduction

Gel dosimetry has been extensively investigated aiming to provide accurate tri-dimensional (3D) information of dose to complex Radiotherapy treatment plans. The idea of using gelatin matrix simulating soft tissue to record dose deposition historically underlies the scenario of dose verifications and, nowadays, is the flagship of 3D dosimetry [1]. Matrix integrity is directly proportional to the ability of precisely record dose depositions and spatial resolution and the capability of absorbing the same dose as biological tissue directly reflects gel's applicability in Radiotherapy dosimetry [2].

Mass attenuation coefficient is possibly one of the more complete dosimetric quantities to indicate radiological tissue-equivalence of a dosimeter. The stability of polymer chains inside gelatin matrix implies in its dosimetric reproducibility [1-2]. From the radiation physics point of view, free radicals are created inside matter in a spatial range proportional to the electron energies in the initial beam or of those created after photon interactions. This may indicate that polymer chains can be structured and created in different sizes, directions and rates depending on the energy and dose rate of the initial beam.

Formaldehyde was added to MAGIC gel (*Methacrylic Ascorbic acid in Gelatin Initiated by Copper added formaldehyde)* aiming to increase matrix stability due to the large number of hydrogen bonds formed between gelatin and formaldehyde chemical sites. Melting point of MAGIC-*f* gel has its value doubled when 1% of formaldehyde is used [3-5]. In this work, MAGIC-*f* gel (*Methacrylic Ascorbic acid in Gelatin Initiated by Copper with added formaldehyde*) is investigated to determine its dosimetric characteristics of matrix integrity, tissue-equivalence, dose rate and energy dependency.

2. Material and Methods

2.1. MAGIC gel with formaldehyde

The polymer gel investigated in this work is MAGIC-*f*. This gel has an excellent stability of response due to gelatin integrity preservation during long time after irradiation [3-4]. Temperature stability is also reported in terms of its melting point of 69° C [5]. Gel composition is presented in table 1. Effective atomic number of 7.53 and density of 1.0 g/cm³ indicate gel equivalence with tissue.

Table 1 – MAGIC-*f* gel composition

Component	Concentration (w/w)	Atomic component	Mass fraction (w/w)
Mili-q water	82.7 %	**H**	10.33 %
Gelatin	8.4 %	**O**	62.68 %
Copper	0.02 %	**C**	23.21 %
Ascorbic acid	0.03 %	**N**	2.52 %
Methacrylic acid	5.9 %	***Others*** (Concentrations lower than 1% were omitted in table)	0.81 %
Formaldehyde	1.0 %		

Abbreviations: Concentrations of chemical components and atomic compositions are expressed in terms of mass fraction (w/w).

2.1.1. Gel manufacture.

Gelatine is added to water at 45°C and mixed without interruption. Temperature is kept until gelatine is completely dissolved and then heating is stopped. Other components are added when temperature reach 30°C in the same order presented in table 1. Gel's PH was mapped during manufacture process to identify free radicals presence. It was possible to observe that gelatine itself provides radicals to solution (PH changes from 6.04 to 5.43). Hydroquinone was not used to consume these radicals given the decrease in sensitivity previously reported for its use [6]. After gel solution homogenization, it was stored in 5mL glass tubes, in a vacuum condition. Glass surrounding the samples has a thickness of 1mm that was considered in the irradiation procedures. Samples were kept in 10°C during 24 hours before irradiation and 12 hours inside MRI tomography room before NMR readings aiming thermal equilibrium.

2.1.2. Irradiation procedures in high-energy beams

Radiotherapy clinical photon beams were used in this study. Varian® 2100C linear accelerators from Hospital do Câncer de Barretos (HCB) were used to produce photon beams of 6MV and 10MV. A Gammatron ^{60}Co unit from Hospital das Clínicas de Ribeirão Preto (HCFMRP) was used to study high energy γ-rays. Irradiations were made in a 10x10 cm² beam size with source-skin distance of 100 cm and 80 cm to x-rays and ^{60}Co, respectively. Samples were positioned at corresponding build-up depth of each beam in a water phantom of 50x50x50 cm³. Doses were delivered homogeneously to all samples. The range of dose rate investigated was from 0.8 Gy/min to 10 Gy/min.

2.1.3. Irradiation procedures in low-energy beams

Siemens Stabillipan ortovoltage clinical beams from HCFMRP were studied. Beams of nominal potential equal to 60kV, 100kV, 120kV and 250kV were investigated. Samples were positioned outside a water box 20cm high and 40cm length to guarantee backscattering conditions. Dose rate ranged from 0.5 Gy/min to 1.68 Gy/min. Irradiations with Iridium-192 (Varian GammaMed plus HDR source) were performed at HCFMRP, similar as previously described [7]. Dose-rate range considered in Iridium irradiations varies from 0.44Gy/min to 10Gy/min. Ionization chamber measures were made in each distance studied from the source to assurance dose values.

2.1.4. NMR reading.
Gel samples were read by relaxometry in a Philips 3.0 T MRI tomograph at HCFMRP. A head coil and a multi-spin echo sequence of 5 echoes, with echo-time of 20ms and repetition-time of 4000ms, were used. Total scan-time was 4.6 minutes for each set of two slices. Pixel size and slice thickness were 0.5mm and 2.0mm respectively. Three acquisitions were averaged to each scan aiming to increase reading accuracy. Transversal relaxation rates (R2) correlated to doses were calculated by Philips MRI software, normally used to medical diagnostic analysis. An acrylic box containing manganese chloride (0.04%) and sodium chloride (0.6%) solution was used to avoid abrupt differences in magnetic susceptibility. All NMR readings were made 12 hours after irradiation assuring thermal equilibrium. Data were analyzed using relaxometry process [8].

2.2. Monte Carlo simulations
Radiation transport in matter using Monte Carlo method as described in PENELOPE – 2008 code [9] was used to determine photoelectric, Compton, pair-creation and Rayleigh cross-sections of MAGIC-*f* polymer gel. Electronic transition probabilities of nine levels, from K to M5 atomic edges are considered in PENELOPE – 2008 code. Mass attenuation coefficients were also calculated and compared to water. Relative depth dose curves (PDD) were obtained experimentally and by simulation using the cross-sections determined in PENELOPE code to both 6MV and 10MV clinical beams.

3. Results and Discussion

MAGIC-*f* presented R2 equal to 0.47 s^{-1} with a standard deviation of 0.06 s^{-1} when not irradiated. This result indicates low levels of polymerization in non-irradiated gel 12 hours after being in MRI room temperature. Tissue equivalence of MAGIC-*f* gel was investigated theoretically comparing mass attenuation coefficient (μ/ρ) of gel and water and experimentally comparing PDD curves obtained to 6MV and 10MV beams, as shown in figure 1.

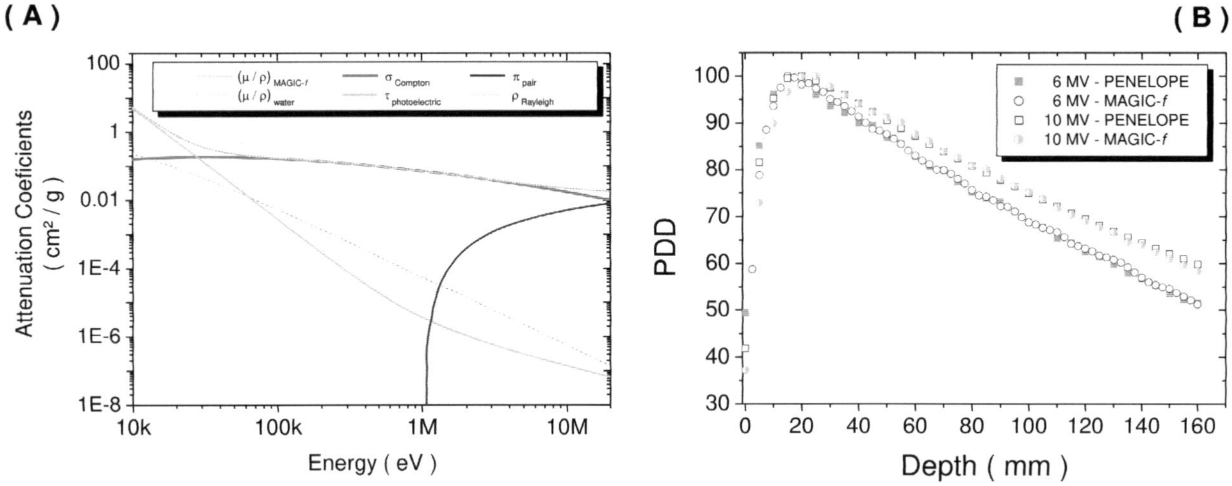

Figure 1 – (a) Mass attenuation coefficients of MAGIC-*f* gel and water determined by PENELOPE code. Cross-sections to Compton, photoelectric, pair-production and Rayleigh scattering are referred by $\sigma_{Compton}$, $\tau_{photoelectric}$, π_{pair} and $\rho_{Rayleigh}$ symbols respectively; (b) Relative depth dose curves experimentally obtained for 6MV and 10MV beams compared to PENELOPE Monte Carlo simulations.

Journal of Physics: Conference Series **250** (2010) 012012 doi:10.1088/1742-6596/250/1/012012

Sensitivity of gel ($\Delta R2/\Delta D$) varies on average 7.7% with energy, being 1.31 s^{-1}/Gy to the lower energy beam (60kV) and 1.42 s^{-1}/Gy to the higher energy beam studied (10MV). However, linearity with dose is present to all energies as can be seen in figure 2.a. Figure 2.b presents MAGIC-*f* sensitivity dependency with energy to all clinical beam studies. It is possible to observe that the standard deviation is significant grater to low energies than to high energies. This effect appears because the homogeneous area analyzed inside the tubes is different to each energy: grows with the energy. This methodology was adopted to guarantee that all average values of R2 were obtained in regions of gel that received doses between 99% and 100%. The corresponding regions were determined either by observing PDD curves simulated by Monte Carlo to each beam. Simulations consider monoenergetic beams with effective energy of each spectrum and attenuation caused by 1mm of glass positioned before gel.

Dose rate dependency (DRD) was investigated in all clinical beams studied using 2 Gy as reference dose. DRD was evaluated quantifying the mean signal variation corresponding to a same dose applied in gel with different dose rates. Figures 2.c and 2.d shows the DRD obtained to all beams studied. In a range from 0.44 Gy/min to 1.68 Gy/min MAGIC-*f* presents no significant variation between x-ray beams measurements (< 0.7% around 1.38 s^{-1}). In the range from 0.8 Gy/min to 10 Gy/min studied with energies between Iridium source and the 10 MV linear accelerator, gel presents mean variation of 1.8% around 1.40 s^{-1}.

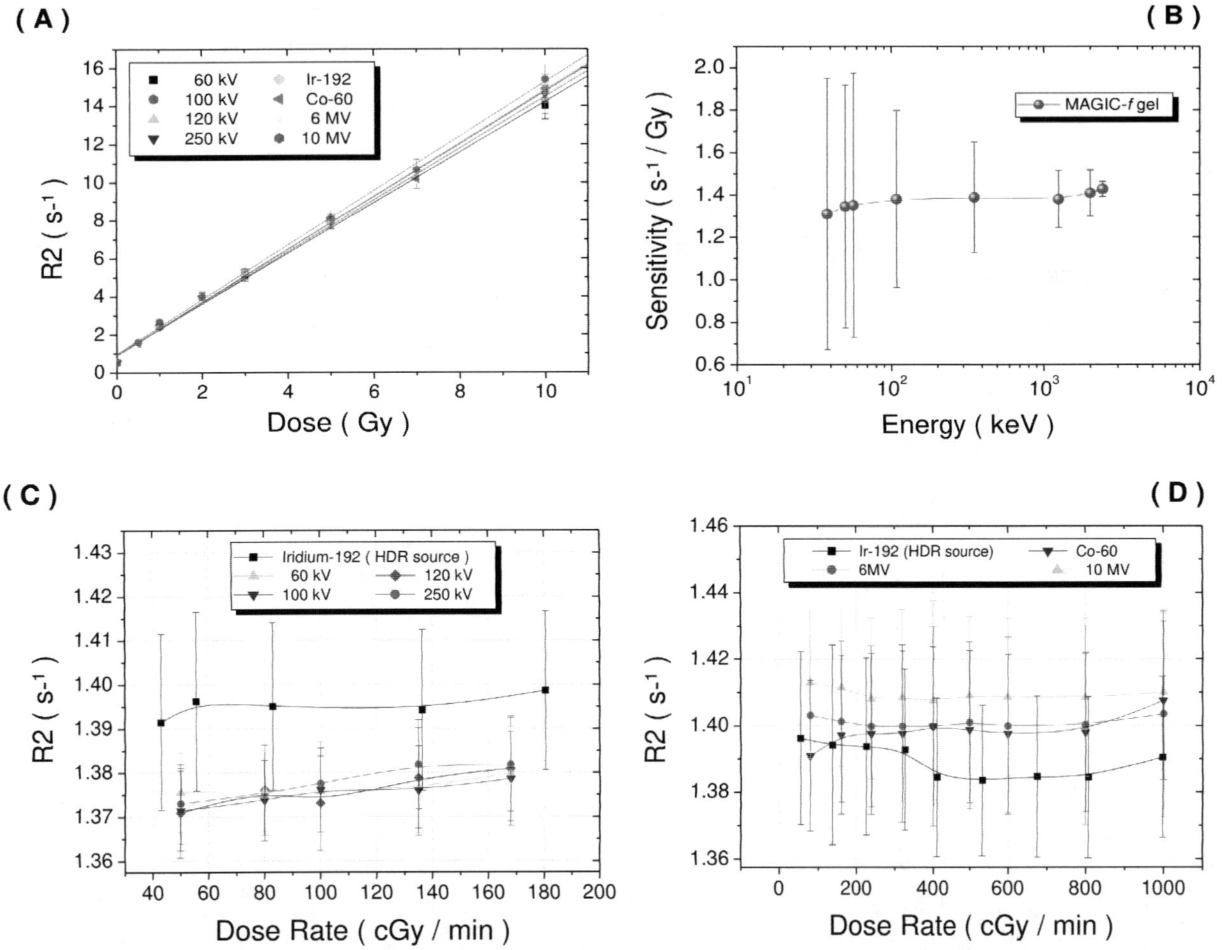

Figure 2 – (A) Linearity of response evaluated to photon clinical beams of nominal energies of 60kV, 100kV, 120kV, 250kV, 6MV, 10MV and clinical sources of Ir-192 (HDR) and Co-60; (B) MAGIC-*f* sensitivity variation with energy; (C) and (D) Dose rate dependency assessed in a range from 0.44 Gy/min to 10 Gy/min.

Analyzing the linear behavior of gel and sensitivity variation with energy, it is possible to observe that proper calibration curves must be previously defined to each clinical beam. Gel response varies a maximum of 8.6% in the energy range studied and a maximum of 1.9% with the dose rate between 0.44 Gy/min and 10 Gy/min. These dosimetric dependencies evidence MAGIC-*f* as a suitable gel dosimeter to megavoltage Radiotherapy clinical beams. Applications in dosimetry of clinical kilo-voltage beams and brachytherapy sources requires more carful, being viable when calibration curves were previously defined under the same irradiation conditions of the desired 3D dosimetric study.

Acknowledgments

Authors are grateful to JL Aziani, to HCB and to HCFMRP for the essential technical contribution. This work was financially supported by grants from CNPq and CAPES/NANOBIOMED.

References

[1] Baldock C., De Deene Y., Doran S., Ibbott G., Jirasek A., Lepage M., McAuley K. B., Oldham M. and Schreiner L. J.: "Polymer gel dosimetry", Phys. Med. Biol. 55, R1-R63 (2010).

[2] De Deene Y., Hurley C., Venning A., Vergote K., Mather M., Healy B. J. and Baldock C.: "A basic study of some normoxic polymer gel dosimeters", Phys. Med. Biol. 42, 3441-3463 (2002).

[3] Fong P., Keil D.C., Does M.D., Gore J.C.: "Polymer gels for magnetic resonance imaging of radiation dose distributions at normal room atmosphere", Phys. Med. Biol. 46, 3105–3113 (2001).

[4] Pavoni J. and Baffa O.: "Estabilidade temporal do dosímetro", In: Dosimetria tridimensional por imagens de ressonância magnética com gel MAGIC modificado. PhD Thesis (2009), Universidade de São Paulo.

[5] Fernandes J., Pastorello B., Araújo D. and Baffa O.: "Formaldehyde increases MAGIC gel dosimeter melting point and sensitivity", Phys. Med. Biol. 53, N53-N58 (2008).

[6] Luci J. J., Whitney H. M. and Gore J. C.: "Optimization of MAGIC gel formulation forthree-dimensional radiation therapy dosimetry", Phys. Med. Biol. 52, N241–N248 (2007).

[7] Marques T., Fernandes J., Barbi G., Nicolucci P. and Baffa O.: "MAGIC with formaldehyde applied to dosimetry of HDR Brachytherapy source", J. Phys. 164, 012067 (2009).

[8] De Deene Y., Van de Walle R., Achten E. and De Wagter C., "Mathematical analysis and experimental investigation of noise in quantitative magnetic resonance imaging applied in polymer gel dosimetry", Sig. Proces. 70, 85-101 (1998).

[9] Salvat F., Fernández-Varea and Sempau J.: "PENELOPE – 2008: a code system for Monte Carlo simulation of electron and photon transport", OECD - NEA nº 6416 (2009).

Journal of Physics: Conference Series **250** (2010) 012013 doi:10.1088/1742-6596/250/1/012013

Influence of gelling agents on the dosimetric performance of the Turnbull Blue gel dosimeter

Jaroslav Šolc[1], Vladimír Sochor[1] and Václav Spěváček[2]

[1]Czech Metrology Institute - Inspectorate for Ionizing Radiation, Radiová 1, CZ-102 00 Prague 10, Czech Republic

[2] Czech Technical University in Prague, Faculty of Nuclear Sciences and Physical Engineering, Břehová 7, CZ-11519 Prague 1, Czech Republic

jsolc@cmi.cz

Abstract. Gelling agents such as agarose, phytagel, and several types of gelatin were used for preparation of Turnbull Blue radiochromic gel dosimeter. Their influence on gel dose response and background value was assessed. It was found that all gelatins cause significant increase of background in a short period of time after gel preparation therefore gelatin is not a suitable gelling agent for this dosimeter. Phytagel and agarose gels exhibit low and stable background and higher dose sensitivity than gelatin gels; however, the disadvantage is increased scattered light intensity in the gel in comparison to gelatin gels. A simple measurement was done demonstrating that the scattered light intensity significantly increases in phytagel and agarose gel in comparison to gelatin gels.

1. Introduction

The main advantages of radiochromic gel dosimeters in comparison to polymer gels are easy preparation procedure and the possibility to evaluate gel response using optical methods, particularly optical cone beam computed tomography (optical CBCT; [1,2]). Although new types of radiochromic gels were recently presented [3,4], the only one widely used for 2D and 3D dose distribution measurement is a Fricke-infused gel with xylenol orange Fe^{3+} ion indicator (the FXG dosimeter; [5]). Some researchers also use the radiochromic solid polymer dosimeter PRESAGE™ [6]. Recently, our group presented a new radiochromic gel based on radiation-induced creation of blue dye Turnbull Blue (the TBG dosimeter or the TB gel; [7]). Its main advantage - besides easy preparation - is the suppressed diffusion of dose patterns recorded in the irradiated gel. However, the gel exhibits lower dose sensitivity and higher background level in comparison to the FXG dosimeter. In this study we prepared this gel with different gelling agents in order to find a better gelling agent, resulting in higher gel sensitivity and lower background level. The final goal is a transparent solid gel with low and stable background, whose response could be easily evaluated using the optical CBCT.

2. Materials and Methods

2.1. Original gelling agents for the TBG dosimeter

The most suitable gelling agent was assumed to be gelatin because it is easy to prepare and forms a clear, colorless solid gel. However, former experiments with gelatin (gelatin form porcine skin, type A; Sigma, G2500) revealed [8] that the TB gel with gelatin exhibits a very high background. The reason was unclear, but it was suspected it could be caused by some additives in gelatin. That is why several types of gelatin were bought to use for preparation of the TB gel.

The other already tested gelling agent was phytagel. It creates nearly as clear and colorless gel as gelatin, however it was observed that phytagel was forming lumps during addition of chemicals required for the TB gel [8].

2.2. Examined gelling agents

Eight gelling agents were chosen to study their influence on sensitivity and background level of the TBG dosimeter: gelatin from porcine skin, type A, for electrophoresis, 300 bloom (Sigma, G8150) - abbreviated as GP-1; gelatin from porcine skin, high gel strength (Fluka, 48724) - GP-2; gelatin from porcine skin (Fluka, 04055) - GP-3; gelatin from bovine skin, type B (Sigma, G9391) - GB; gelatin from cold water fish skin (Sigma, G7041) - GF; edible food gelatin (natura, Czech Republic) - GE; phytagel, plant cell culture (Sigma, P8169) - PHY; and agarose, low EEO, for routine use (Sigma, A5093) - AGA.

2.3. Gel preparation

Gel composition was as follows: 4 % w/w gelatin (or 0.25 % w/w phytagel or agarose), 1 mM potassium ferricyanide if gelatin gels or 0.5 mM if phytagel or agarose gels, 1 mM ferric chloride, and 1 mM sulphuric acid. Gelling agents were dissolved in deionized water (10 % w/w gelatin solution, 0.6 % w/w agarose or phytagel solution) placed into a water bath (60°C if gelatin, 90°C otherwise). Ferric chloride was dissolved in diluted sulphuric acid (approximately 2/5 of the total gel volume). Before gel mixing, this solution was warmed to ~35°C, while the gelling agent was cooled down to ~35°C as well. Diluted potassium ferricyanide was then added to the gelling agent, followed by ferric chloride. Well mixed gel was filled in PMMA $1.0\times1.0\times4.5$ cm^3 spectrophotometric cuvettes (1.0 cm optical path) and refrigerated.

2.4. Irradiation and evaluation of gel response

Gel samples were irradiated by Co-60 photons (mean energy 1.25 MeV) to several dose values reaching up to 80 Gy. Two cuvettes per every batch were taken aside as background samples.

Evaluation of gel response was performed using an optical method. Gel samples were illuminated with red diodes light (emission maximum at the wavelength of 640 nm) and the intensity of light passed through the gel in terms of gray scale image intensity was captured by a pixelated CCD camera. Linearity of the CCD camera response and other details were described elsewhere [9]. So-called CCD absorbance is then calculated for each gel sample as ln(A/B), where A and B are median gray scale pixel values of the background cuvette and the gel sample, respectively. This quantity is linearly dependent on the gel optical density.

A calibration curve was determined between the CCD absorbance measured at the given optical conditions and spectrophotometric absorbance measured at 690 nm (maximum of Turnbull Blue absorption peak) in order to obtain and state the results in familiar quantity - the spectrophotometric absorbance. The calibration curve has the equation $A_{sp} = (0.487 \pm 0.004) \times A_{CCD}$ where A_{sp} is the spectrophotometric absorbance per 1 cm optical path and A_{CCD} is the CCD absorbance.

To obtain quantitative information about the intensity of scattered light in different gelling agents, dummy samples in PMMA cuvettes were prepared containing only the gelling agent in concentration used in the gel. Measurement of intensity of scattered light was inspired by [10]. Samples were inserted into a black cuvette holder, and a black color dot (4 mm diameter) glued onto a transparent plastic foil was placed just behind the sample. From the photographs, median gray scale pixel values were calculated at the central part of the dot, I_{dot}, and at open area, I_0, of the sample. Values I_{dot}/I_0, subtracted from that for cuvette with water, were calculated to assess the contribution of scattered light by different gelling agents to the total intensity of light passed through the gel.

3. Results and Discussion

3.1. Gelling agents

Basic dosimetric characteristics of the TB gel prepared with different gelling agents are presented in Table 1.

Table 1: Sensitivity, background level and intensity of scattered light of the TBG dosimeter prepared with different gelling agents

Gelling agent [1)]	Sensitivity [2)] ($Gy^{-1}cm^{-1}$)	Background level [3)] (cm^{-1})	Scattered light intensity [4)]
Porcine gelatin, GP-1	$(5.5 \pm 0.5)\times10^{-3}$ (22 h)	0.58 (54 h)	1.6 %
Porcine gelatin, GP-2	$(5.0 \pm 0.5)\times10^{-3}$ (22 h)	0.48 (54 h)	Not measured
Porcine gelatin, GP-3	$(4.1 \pm 0.6)\times10^{-3}$ (2 h)	0.28 (8 h)	Not measured
Bovine gelatin, GB	-	1.25 (54 h)	5.7 %
Fish gelatin, GF	-	Did not solidify	Not measured
Edible gelatin, GE	$(3.6 \pm 0.6)\times10^{-3}$ (22 h)	0.23 (24 h)	0.8 %
Phytagel, PHY	$(7.7 \pm 0.2)\times10^{-3}$ (22 h)	0.12 (66 h); 0.15 (7 days); 0.19 (18 days); 0.22 (45 days)	3.5 %
Agarose, AGA	$(9.7 \pm 0.5)\times10^{-3}$ (22 h)	0.12 (8 h); 0.14 (24 h)	4.3 %

1) For detailed specification of gelling agents see paragraph 2.2
2) Slope of the linear part of the dose response curve at the stated time after gel irradiation (1σ confidence level)
3) Absorbance of a background sample against gelatin GP-1 at the stated time after gel preparation; stored at 5°C
4) Intensity of scattered light after 1 cm optical path in the gelling agent

A TBG dosimeter suitable for optical CBCT evaluation should fulfill three requirements: 1) it has to solidify, 2) it should be fully transparent, and 3) it has to exhibit reasonably low background at least for a sufficient period of time, enabling irradiation and evaluation of the gel.

The first requirement was not met by the gel made of gelatin from cold water fish skin. Gels with 4 % and even 30 % of this gelatin did not solidify in a refrigerator. Porcine skin gelatin should be represented by at least 3.5 % of gel weight to sufficiently solidify.

Gel transparency is indicated in the last column of Table 1. Scattered light causes fogging of the gel and adds inaccuracy to the optical CBCT evaluation. Mild fog was observed for agarose and bovine skin gelatin gels, milder for phytagel dosimeter. Bovine skin gelatin exhibited very high background as well, therefore it was not studied further. On the other hand agarose gel possessed very low background, which was stable during a long period of time if stored in a refrigerator. That is why this gelling agent could be recommended preferably for non-optical CBCT evaluation methods. It was found out that, besides the gelling agent itself, the gel fogging is generally caused when potassium ferricyanide is present in an insufficiently acidic gel. The fogging was sometimes also observed in gelatin gels with high potassium ferricyanide and gelatin concentration. Lowering the concentration of potassium ferricyanide from the initial 1.5 mM [7] up to 0.5 mM or adding more sulphuric acid decreases fogging of the phytagel dosimeter and gelatin gels become transparent.

The third requirement of "reasonably low background" was not met for all gelatin gels. Generally, all gels prepared with gelatin exhibit significant increase of background and over 1.5× lower dose sensitivity than phytagel dosimeter. However, gelatin gels are colored yellow just after mixing, indicating that no reduced Fe^{2+}, which would create Turnbull Blue, is present at that time in the gel. After preparation, the gel darkens slowly and continuously (gel changes color from yellow to green). The reason of this effect is unknown but it is assumed that some reducing radicals or impurities are present in gelatin. To suppress them, one batch of gel made with gelatin GP-3 was prepared with 0.15 % w/w of hydrogen peroxide added into the diluting gelatin. However, the background value of this gel was even higher than without the additive.

3.2. TBG dosimeter with phytagel gelling agent

Gel with phytagel gelling agent proved to be the most perspective variant of the TBG dosimeter, therefore this gel will be used for next studies. In addition to the already published characteristics [7], Figure 1 shows the dose dependence of the phytagel TB gel presented in this paper. The dose response is linear from 0 Gy up to 90 Gy with the sensitivity of (7.7 ± 0.3) $Gy^{-1}cm^{-1}$ (measured 22 h after irradiation). Post-irradiation behavior is pictured in Figure 2. A gel sample was irradiated to 7.5 Gy in 4.5 min and immediately photographed in 5 s intervals. It can be seen that the stable response is reached approximately 40 min after irradiation.

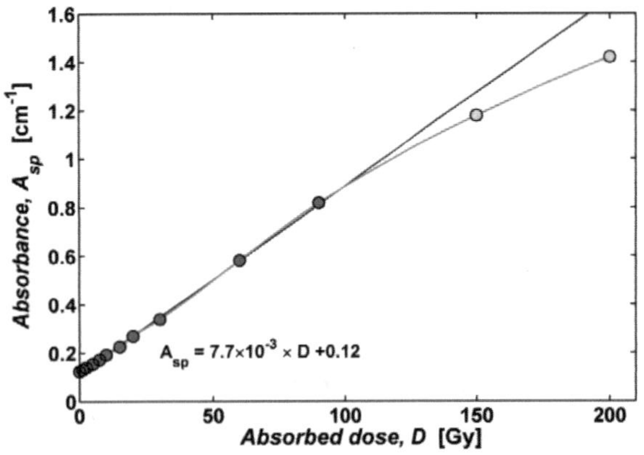

Figure 1: Dose response of the TB gel with phytagel gelling agent. Green dots show values beyond the gel linear response.

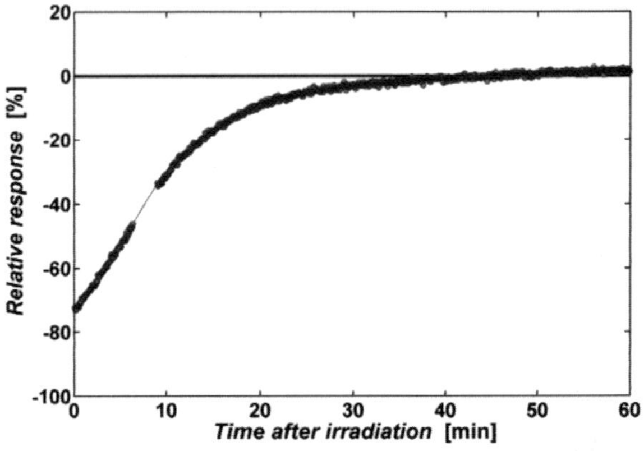

Figure 2: Post-irradiation behavior of the TB gel with phytagel gelling agent.

4. Conclusions

Presented paper describes dosimetric performance of the Turnbull Blue radiochromic gel dosimeter prepared with different gelling agents. It was found out that every type of gelatin tested causes continuous increase of gel background values of optical density, making the evaluation of 3D dose distributions recorded in gelatin gels of large dimensions very difficult. Agarose gel possesses low background, however it is not fully transparent because it scatters light, which also limits the use of optical cone beam CT for gel response evaluation. The best performance is exhibited by a gel with phytagel gelling agent, although the concentrations of chemicals can only vary by a narrow margin, otherwise phytagel forms lumps. Until the cause of spontaneous reduction of Fe^{3+} in gelatin is found and eliminated, phytagel remains the most suitable gelling agent for Turnbull Blue dosimeter.

Acknowledgements

This work was carried out within the scope of the project EMRP TP2:J06 Brachytherapy, supported by the European Commission, in accordance with Grant Agreement No. 217257 between the EC and EURAMET e.V. (European Association of National Metrology Institutes) under the Seventh Framework Programme.

References

[1] Doran, S.J. *The history and principles of optical computed tomography for scanning 3-D radiation dosimeters: 2008 update*. Journal of Physics: Conference Series **164** (2009), doi: 10.1088/1742-6596/164/1/012020

[2] Wolodzko J.G.; Marsden C. and Appleby A. *CCD imaging for optical tomography of gel radiation dosimeters*. Med. Phys., Vol.**26**, No.11 (1999), p.2508-13

[3] Babic S.; Battista J. and Jordan K. *Micelle hydrogels for three-dimensional dose verification*. J. Phys.: Conf. Ser. 164 012044 (2009); doi: 10.1088/1742-6596/164/1/012044

[4] Jordan K. *Sci-Fri PM: Planning-08: Zero diffusion radiochromic genipin-gelatin dosimeter*. Med. Phys. Volume **35**, Issue 7, pp. 3413-3413 (July 2008), http://dx.doi.org/10.1118/1.2965980

[5] Bero M.A.; Gilboy W.B.; Glover P.M. and El-masri H.M. *Tissue-equivalent gel for non-invasive spatial radiation dose measurements*. NIM-B, Vol.166, No. (2000), p.820-825

[6] Adamovics J. and Maryanski M.J. *Characterisation of PRESAGE™: A new 3-D radiochromic solid polymer dosemeter for ionising radiation*. Radiat Prot Dosimetry (September 2006) 120 (1-4): 107-112. doi: 10.1093/rpd/nci555

[7] Šolc J. and Spěváček V. *New Radiochromic Gel for 3D Dosimetry Based on Turnbull Blue: Basic Properties*. Phys. Med. Biol. **54** (2009) 5095-5107. doi: 10.1088/0031-9155/54/17/002

[8] Šolc J. and Spěváček V. *New 3D radiochromic gel dosimeters with inhibited diffusion*. J. Phys.: Conf. Ser. 164, pp.012047-51, 2009. doi: 10.1088/1742-6596/164/1/012047

[9] Šolc J., Sochor V., Kačur M. and Šmoldasová J. *3D dose distribution measurements in brachytherapy using radiochromic gel dosimeters*. Nucl. Instr. and Meth. A, in print, available onlie (2009), doi:10.1016/j.nima.2009.10.145

[10] Olding T., Holmes O. and Schreiner L.J. *Scatter Corrections for Cone Beam Optical CT* J. Phys.: Conf. Ser. 164 012031 (2009), doi: 10.1088/1742-6596/164/1/0120

Journal of Physics: Conference Series **250** (2010) 012014 doi:10.1088/1742-6596/250/1/012014

Polymer gel dosimetry

L J Schreiner[1], T Olding[1] and K B McAuley[2]

[1] Cancer Centre of South Eastern Ontario at Kingston General Hospital and Depts. of Oncology and Physics, Queen's University, Kingston, ON, Canada,
[2] Department of Chemical Engineering, Queen's University, Kingston, ON, Canada

e-mail: John.Schreiner@krcc.on.ca;

Abstract: Polymer gels are chemical dosimeters based on dose dependent radiation-induced polymerization and cross-linking of monomers in an irradiated volume. The changes are spatially localized in the volume by incorporating the initial monomers in an aqueous gel matrix in the dosimeter and can be probed by various imaging techniques such as magnetic resonance imaging (MRI), x-ray computed tomography (CT), and optical CT. As they are chemical dosimeters, polymer gels are sensitive to preparation conditions. The three dimensional dose readout is sensitive to the imaging modality and also to the technical conditions in use during specific scans. This brief article is intended to present an introduction to these points which need to be taken into account as one attempts to establish this dosimetry in the clinic.

1. Introduction

Gel dosimetry[1] is a class of chemical dosimetry in which aqueous gels spatially store radiation-induced changes in chemical species. The dose-dependent changes can subsequently be probed in three dimensions (3D) by various techniques. Diffusion based limitations with early Fricke-infused gel dosimeters led to the development of polymer gel systems based on radiation-induced polymerization and cross-linking of monomers infused in gel[2-4]. The localized radiation-induced copolymerization enables dose readout by MRI[3,5,6], optical scanning[4,7], x-ray computed tomography[8,9], and ultrasound[10]. It is this marriage of the imaging with the dosimetry [2] that can be considered to fix the birth of gel dosimetry; polymer-based dosimetry had been established decades earlier[11-13] but had not come into standard practice, mainly because of the limits in dose readout. All of the current imaging modalities yield 3D data with good spatial resolution. Polymer gels systems integrate the radiation effects over time and can be used when the dose is delivered dynamically, for example, when the radiation source, the treated volume, or both move throughout the delivery. Therefore, gel dosimeters enable verification of the dose delivery achieved even in extended intensity modulated and adaptive radiation therapy deliveries brachytherapy[14-17].

Polymer gels share some features common to all gel dosimeters. On the plus side, polymer gels are typically soft-tissue equivalent through appropriate preparation and since it is the material changes in the gel that provide the dosimetry, the dosimeter and the surrounding medium are identical[18,19]. The dosimeters are liquid at some point in the preparation and can be poured into complex vessels and phantoms including anthropomorphic phantoms of clinical interest[6,20,21]. Some common constituents of polymer gels in use are presented in Table 1. Other characteristics can be problematic. In particular, as with all chemical dosimeters, the dose response is sensitive to the preparation and can change if the manufacturing conditions are even slightly altered, for example by a change in chemical suppliers or a change in timing of a given stage at a given temperature[22-25].

Other characteristics of the dosimetry are governed primarily by the nature of the readout techniques and imaging. Therefore it is important, for example, to optimize the pulse sequences used for MRI readout[26]or to limit optical CT measurements of polymer gels to regimes in which opacity and scatter do not drive the response to non-linear regimes[27,28]. These topics will be addressed by others during this meeting.

While 3D polymer gel dosimetry has been discussed for 15 years it has not come into broad practice to date. This is still the major barrier to adoption in the local setting. In the last few years, a number of advances have reduced impediments to adoption. Firstly, polymer gel systems have been developed that can be prepared in oxygenated environments (called normoxic polymer gels)[25,29-31] and with less toxic constituents[32,33]. Therefore, the preparation of the dosimeters is possible in the clinical setting. Secondly, physical properties of irradiated polymer gel, such as opacity and density, enable

Component	PAGAT	MAGIC	MAGAT	NIPAM	VIPAR	Formula
	(weight percent)					
Water	88.9	82.8	85.9	90.9	80.5	H_2O
Gelatin	5	8	8	5	7.5	*
Bis	3	-	-	2	4	$C_7H_{10}N_2O_2$
Acrylamide	3	-	-	-	-	C_3H_5NO
Methacrylic acid	-	9	6	-	-	$C_4H_6O_2$
N-isopropyl acrylamide	-	-	-	2	-	$C_6H_{11}NO$
N-vinyl pyrrolidone	-	-	-	-	8	C_6H_9NO
Ascorbic acid	-	0.035	-	-	0.0007	$C_6H_8O_6$
Hydroquinone	-	0.02	-	-	-	$C_6H_4(OH)_2$
Copper sulphate	-	0.002	-	-	0.0008	$CuSO_4.5H_2O$
	(mM)					
THP[1]	-	-	2	-	-	$[(CH_2OH)_4P]_2SO_4$
THPC[2]	5	-	-	6.67	-	$C_4H_{12}ClO_4P$

*The constituents for the 300 bloom porcine gelatin has been averaged from the amino acid breakdown of some representative gelatins.

[1]THP stand for bis-tetrakis(hydroxymethyl)phosphonium sulphate

[2]THPC stands for tetrakis(hydroxymethyl)phosphonium chloride

Table 1: Constituents and formulations of some common polymer gel dosimeters[21].

dose imaging using optical or x-ray computed tomography (CT) imaging. These alternative imaging schemes are attractive since access to MRI, the gold standard imaging, is often limited. The research and development of optical computed tomography systems for polymer gel dosimetry is that economical optical CT systems are now available for accessible dose imaging in the clinic. X-ray CT is ubiquitous to clinics with CT simulation as the standard of practice for treatment planning.

The objective of this paper is to provide a summary and outline of the various characteristics of polymer dosimeters, a guide to the issues related to their use, and a bibliography to enable the reader to easily become familiar with the points that will be presented in the course of this conference. It is not intended to provide a complete description of the basic physics and chemistry of the dosimetry nor a detailed review of the growing literature dedicated to the various polymer systems that have been developed. These points have been very well presented in the last few years in a number of detailed descriptions[22,34-39] from previous gel dosimetry conferences available in the *Journal of Physics Conference Series*, in a review of the basic science and practical issues of polymer gel dosimeters given in the general context of gel dosimetry in an AAPM summer school Monograph[21], and in a comprehensive topical review published in *Physics in Medicine and Biology* earlier this year[23].

References

[1] Gore, J. C., Kang, Y. S., and Schulz, R. J. 1984 Measurement of radiation dose distributions by nuclear magnetic resonance (NMR) imaging. Phys Med Biol **29** 1189-1197.
[2] Maryanski, M. J., Gore, J. C., Kennan, R. P., and Schulz, R. J. 1993 NMR relaxation enhancement in gels polymerized and cross-linked by ionizing radiation: a new approach to 3D dosimetry by MRI. Magn Reson Imaging **11** 253-258.
[3] Maryanski, M. J., Schulz, R. J., Ibbott, G. S., Gatenby, J. C., Horton, D., and Gore, J. C. 1994 Magnetic resonance imaging of radiation dose distributions using a polymer-gel dosimeter. Phys Med Biol **39** 1437-1455.
[4] Maryanski, M. J., Zastavker, Y. Z., and Gore, J. C. 1996 Radiation dose distributions in three dimensions from tomographic optical density scanning of polymer gels: II. Optical properties of the BANG polymer gel. Phys Med Biol **41** 2705-2717.
[5] Baldock, C., Lepage, M., Back, S. A., Murry, P. J., Jayasekera, P. M., Porter, D., and Kron, T. 2001 Dose resolution in radiotherapy polymer gel dosimetry: effect of echo spacing in MRI pulse sequence. Phys Med Biol **46** 449-460.
[6] De Deene, Y., De Wagter, C., Van Duyse, B., Derycke, S., De Neve, W., and Achten, E. 1998 Three-dimensional dosimetry using polymer gel and magnetic resonance imaging applied to the verification of conformal radiation therapy in head-and-neck cancer. Radiother Oncol **48** 283-291.
[7] Oldham, M., Siewerdsen, J. H., Kumar, S., Wong, J., and Jaffray, D. A. 2003 Optical-CT gel-dosimetry I: basic investigations. Med Phys **30** 623-634.
[8] Hilts, M., Audet, C., Duzenli, C., and Jirasek, A. 2000 Polymer gel dosimetry using x-ray computed tomography: a feasibility study. Phys Med Biol **45** 2559-2571.
[9] Hilts, M., Jirasek, A., and Duzenli, C. 2005 Technical considerations for implementation of x-ray CT polymer gel dosimetry. Physics in Medicine & Biology **50** 1727-1745.
[10] Mather, M. L., De Deene, Y., Whittaker, A. K., Simon, G. P., Rutgers, R., and Baldock, C. 2002 Investigation of ultrasonic properties of PAG and MAGIC polymer gel dosimeters. Phys Med Biol **47** 4397-4409.
[11] Andrews, Howard L., Murphy, Raymond E., and LeBrun, Edward J. 1957 Gel Dosimeter for Depth-Dose Measurements. Review of Scientific Instruments **28** 329-332.
[12] Hoecker, F. E. and Watkins, I. W. 1958 Radiation polymerization dosimetry. Int J Appl Radiat Isot **3** 31-35.
[13] Boni, A. L. 1961 A Polyacrylamide gamma dosimeter. Radiat Res **14** 374-380.
[14] Ceberg, S., Gustavsson, H., Korreman, S., Medin, J., Kjaer-Kristoffersen, F., and Back, S. A. 2008 RapidArc™ treatment verification using polymer gel dosimetry. 323-327.

[15] Ceberg, S., Karlsson, A., Gustavsson, H., Wittgren, L., and Back, S. A. 2008 Verification of dynamic radiotherapy: the potential for 3D dosimetry under respiratory-like motion using polymer gel. Phys Med Biol **53** N387-N396.

[16] Pantelis, E., Antypas, C., Petrokokkinos, L., Karaiskos, P., Papagiannis, P., Kozicki, M., Georgiou, E., Sakelliou, L., and Seimenis, I. 2008 Dosimetric characterization of CyberKnife radiosurgical photon beams using polymer gels. Med Phys **35** 2312-2320.

[17] Papagiannis, P., Pappas, E., Kipouros, P., Angelopoulos, A., Sakelliou, L., Baras, P., Karaiskos, P., Seimenis, I., Sandilos, P., and Baltas, D. 2001 Dosimetry close to an Ir-192 HDR source using N-vinylpyrrolidone based polymer gels and magnetic resonance imaging. Med Phys **28** 1416-1426.

[18] Brown, S., Venning, A., De Deene, Y., Vial, P., Oliver, L., Adamovics, J., and Baldock, C. 2008 Radiological properties of the PRESAGE and PAGAT polymer dosimeters. Appl Radiat Isot **66** 1970-1974.

[19] Rogers D W 2009 Stopping-Power Ratios, Ratios of Mass-Energy Absorption Coefficients and CSDA Ranges of Electrons. In: Rogers, D. W. and Cygler, J. E. AAPM Medical Physics Monograph No. 34: Clinical Dosimetry Measurements in Radiotherapy. Madison, WI, USA, Medical Physics Publishing. pp1097.

[20] Scheib, S. G. and Gianolini, S. 2002 Three-dimensional dose verification using BANG gel: a clinical example. J Neurosurg **97** 582-587.

[21] Schreiner L J and Olding T 2009 Gel Dosimetry. In: Rogers, D. W. and Cygler, J. E. AAPM Medical Physics Monograph No. 34: Clinical Dosimetry Measurements in Radiotherapy. Madison, WI, USA, Medical Physics Publishing.979-1025.

[22] Jirasek, A., McAuley, K. B., and Lepage, M. 2009 How does the chemistry of polymer gel dosimeters affect their performance? J Phys : Conf Ser **164** 22-35.

[23] Baldock, C., De Deene, Y., Doran, S., Ibbott, G., Jirasek, A., Lepage, M., McAuley, K. B., Oldham, M., and Schreiner, L. J. 2010 Polymer gel dosimetry. Phys Med Biol **55** R1-63.

[24] De Deene, Y., Hanselaer, P., De Wagter, C., Achten, E., and De Neve, W. 2000 An investigation of the chemical stability of a monomer/polymer gel dosimeter. Phys Med Biol **45** 859-878.

[25] De Deene, Y., Hurley, C., Venning, A., Vergote, K., Mather, M., Healy, B. J., and Baldock, C. 2002 A basic study of some normoxic polymer gel dosimeters. Phys Med Biol **47** 3441-3463.

[26] De Deene, Y. and Baldock, C. 2002 Optimization of multiple spin-echo sequences for 3D polymer gel dosimetry. Phys Med Biol **47** 3117-3141.

[27] Oldham, M. and Kim, L. 2004 Optical-CT gel-dosimetry. II: Optical artifacts and geometrical distortion. Med Phys **31** 1093-1104.

[28] Olding, T., Holmes, O., and Schreiner, L. J. 2010 Cone beam optical computed tomography for gel dosimetry I: scanner characterization. Phys Med Biol **55** 2819-2840.

[29] Fong, P. M., Keil, D. C., Does, M. D., and Gore, J. C. 2001 Polymer gels for magnetic resonance imaging of radiation dose distributions at normal room atmosphere. Phys Med Biol **46** 3105-3113.

[30] Bayreder, C., Georg, D., Moser, E., and Berg, A. 2006 Basic investigations on the performance of a normoxic polymer gel with tetrakis-hydroxy-methyl-phosphonium chloride as an oxygen scavenger: reproducibility, accuracy, stability, and dose rate dependence. Med Phys **33** 2506-2518.

[31] De Deene, Y., Vergote, K., Claeys, C., and De Wagter, C. 2006 The fundamental radiation properties of normoxic polymer gel dosimeters: a comparison between a methacrylic acid based gel and acrylamide based gels. Phys Med Biol **51** 653-673.

[32] Pappas, E., Maris, T., Angelopoulos, A., Paparigopoulou, M., Sakelliou, L., Sandilos, P., Voyiatzi, S., and Vlachos, L. 1999 A new polymer gel for magnetic resonance imaging (MRI) radiation dosimetry. Phys Med Biol **44** 2677-2684.

[33] Senden, R. J., De Jean, P., McAuley, K. B., and Schreiner, L. J. 2006 Polymer gel dosimeters with reduced toxicity: a preliminary investigation of the NMR and optical dose-response using different monomers. Phys Med Biol **51** 3301-3314.
[34] De Deene, Y. 2004 Essential characteristics of polymer gel dosimeters. J Phys : Conf Ser **3** 34-57.
[35] Doran, S. and Krstajic, N. 2006 The history and principles of optical computed tomography for scanning 3-D radiation dosimeters. J Phys : Conf Ser **56** 45-57.
[36] Jirasek, A. and Hilts, M. 2009 An overview of polymer gel dosimetry using x-ray CT. J Phys : Conf Ser **164** 246-256.
[37] McAuley, K. B. 2006 Fundamentals of polymer gel dosimetry. J Phys : Conf Ser **56** 35-44.
[38] McAuley, K. B. 2004 The chemistry and physics of polyacrylamide gel dosimeters: why they do and don't work. J Phys : Conf Ser **3** 29-33.
[39] Oldham, M. 2004 Optical scanning of polymer gels. J Phys : Conf Ser **3** 122-135.

Journal of Physics: Conference Series **250** (2010) 012015 doi:10.1088/1742-6596/250/1/012015

How to scan polymer gels with MRI?

Yves De Deene

Department for Radiotherapy and Experimental Cancer Research, Ghent University, Belgium.

Yves.DeDeene@UGent.be

Abstract. The absorbed radiation dose fixated in a polymer gel dosimeter can be read out by several methods such as magnetic resonance imaging (MRI), optical CT, X-ray CT and ultrasound with MRI being the first method that was explored. Although MRI was considered as an elegant scanning technique, readily available in most hospitals, it was later found that using a non-optimized imaging protocol may result in unacceptable deviations in the obtained dose distribution. Although most medical physicists have an understanding of the basic principles of magnetic resonance imaging (MRI), the optimization of quantitative imaging sequences and protocols is often perceived as the work of MRI experts. In this paper, we aim at providing the reader with some easy guidelines in how to obtain reliable quantitative MRI maps.

1. Introduction

In the early days of the development of gel dosimetry, magnetic resonance imaging (MRI) was suggested as thé method to read-out gel dosimeters. The use of MRI as a non-destructive imaging method of a dosimeter gel was first proposed in 1984 by Gore *et al* [1] who showed that ferrous sulfate chemical dosimeters initially developed in 1927 [2] could be probed by nuclear magnetic relaxometry and hence by magnetic resonance imaging (MRI) [1]. Gore *et al* investigated the nuclear magnetic resonance (NMR) relaxation properties of irradiated Fricke or ferrous sulfate dosimetry solutions showing that radiation-induced changes, in which ferrous (Fe^{2+}) ions are converted to ferric (Fe^{3+}) ions, could be quantified using MRI and subsequently showed that Fricke dosimetry solutions dispersed throughout a gel matrix could be used to obtain three-dimensional (3D) spatial information using MRI. It was subsequently shown that irradiated Fricke-type gel dosimeters did not retain a spatially stable dose distribution due to ion diffusion within the irradiated dosimeters [3]. Fricke solutions with various gelling agents such as gelatin, agar, sephadex™ and polyvinyl alcohol (PVA) were investigated. Chelating agents to reduce diffusion in Fricke gels, such as xylenol orange (XO), had only limited success [4] and diffusion remained a significant problem in the advancement of gel dosimetry. Different models have been described that explain the mechanism of how the relaxation rates are affected by the paramagnetic substances [1, 5-7]. The spin-lattice relaxation rate ($R_1 = 1/T_1$) and the spin-spin relaxation rate ($R_2 = 1/T_2$) in Fricke gels is altered significantly upon irradiation. As R_1 of a non-irradiated Fricke gel dosimeter is small as compared to the R_2 of a non-irradiated Fricke gel dosimeter, the dynamic range of the Fricke gel dosimeter in relative terms is higher for R_1 than R_2. For this reason, R_1 mapping is preferred to R_2 mapping for Fricke gel dosimeters. Also R_1 maps (at

Journal of Physics: Conference Series **250** (2010) 012015 doi:10.1088/1742-6596/250/1/012015

least in the early days) can be obtained with a shorter acquisition time than R_2 maps which is of crucial importance in avoiding diffusion related blurring of the dose distribution.

Polymer systems for the use of radiation dosimetry were first proposed as early as 1954, where Alexander *et al* discussed the effects of ionizing radiation on polymethylmethacrylate [8]. In 1992, Kennan *et al* reported on NMR longitudinal relaxation studies performed on an irradiated aqueous solution of N,N'-methylene-bis-acrylamide (Bis) and agar, which showed that the relaxation rates increased with absorbed dose [9]. Polymer gel dosimeters are based on the conversion of co-monomers to polymer aggregates upon irradiation. This reaction alters the mobility of surrounding water molecules which also results in a change in R_1 and R_2 [10]. The dose-response of R_2 in gelatin based polymer gel dosimeters however is more pronounced than of R_1. To explain the effect that the radiation-induced polymerization has on the R_2 relaxation rate, a model of fast exchange [11] is adopted [12–14]. It is shown in later studies that not only the relaxation rate can be used as an imaging parameter but also other MR contrasts such as magnetization transfer [15–17] and chemical shift [18]. For a further explanation of these mechanisms, the reader is also referred to other papers in these proceedings. Different imaging sequences can be used to acquire quantitative images. These imaging sequences may differ in performance in terms of accuracy, precision and speed. It will be shown that for a specific imaging sequence, these three properties are interconnected.

The target figure of accuracy that is aimed in gel dosimetry for high-precision radiotherapy is about 3-5% of the maximum dose in regions of homogeneous dose and a spatial error of less than about 2-3 mm in regions of high-dose gradients. This figure of accuracy encompasses the overall dosimetry experiment. The problem in evaluating the overall accuracy of the dose maps obtained with gel dosimetry is that there is no "golden dosimetric standard" to compare with. The most reasonable strategy is to compare doses obtained with gel dosimetry with doses obtained by the most reliable dosimetry techniques that apply to a certain spatial dimension. For example, dose profiles of a single field (photons and electrons) can be compared with dose profiles obtained with an ionization chamber or diamond detector [19–20]. In two dimensions, gel dosimetry can be compared with film dosimetry [19, 21–22]. Dose distributions obtained with gel dosimetry have been compared with those calculated with treatment planning software [21, 23–29]. Errors that compromise the accuracy may occur at different stages of the dosimetry procedure [30].

At the stage of imaging the dosimeter, several imaging artifacts may cause errors in the final dose map. These errors may be classified in dose inaccuracies or in deformations of the dose maps. Studies of these different artifacts have resulted in different compensation strategies. The verification of the treatment plan can be seen as the major purpose of gel dosimetry in radiotherapy quality assurance. Besides the possibility of systematic errors, the dose maps will also contain stochastic noise. To minimize the stochastic noise in the images, the imaging sequence parameters should be optimized.

2. A step-by-step guide for MRI scanning of polymer gel dosimeters.

Although there is a vast amount of scientific literature available on different quantitative scanning methods and compensation of imaging artifacts, for the medical physicist who is planning to start using MRI as a readout technique for polymer gel dosimeters, the implementation of a reliable scanning protocol may seem very complex. Here is a step-by-step procedure that may guide you in implementing a quantitative MRI protocol to scan polymer gels.

2.1. Get yourself familiar with the basics behind magnetic resonance imaging (MRI)

Without studying the quantum mechanical description of nuclear magnetic resonance, it is essential that you familiarize yourself with the basic principles behind magnetic resonance imaging (MRI). Knowledge is the best medicine against beginners fear. There are some nice primers on the basics of MRI [31-35]. It may also be beneficial to allow yourself to play around with the parameters of a basic spin-echo sequence. It should be noted that in order to make the theory of NMR more accessible, some of these works have made compromises on the preciseness of the quantum mechanical model. More advanced textbooks on MRI physics and sequence development will give a more precise description of the quantum mechanical model [36-38].

2.2. Choose a quantitative imaging sequence

Theoretically, any kind of imaging sequence that generates an image in which the pixel intensity is related to absorbed dose in a monotonic fashion is a possible candidate. A dose map can be obtained by calibrating every pixel intensity to the absorbed dose by use of a set of calibration vials that have been irradiated with known doses. In general, due to the inhomogeneity of the radiofrequency (B_1-) field, contrast weighted images (T_1w, T_2w, MTw) suffer from severe image non-uniformity which results in poor accuracy. Eventually, this can be compensated by acquiring an additional B_1-field map and using the B_1-field map to compensate for the non-uniformity before calibration. However, other artifacts such as eddy currents and B_0-field inhomogeneity may also lead to image non-uniformity. A better approach to compensate for image non-uniformity is obtained by using quantitative parametric maps such as T_1-, T_2- and MTR-maps. The contrast mechanism (T_1, T_2, MTR) should be chosen on the basis of the kind of gel dosimeter that is scanned. In Fricke based gel dosimeters, both T_1 and T_2 are significantly affected upon irradiation and thus both T_1 and T_2 maps can be used. In polymer gel dosimeters T_2 is more affected than T_1 and thus T_2-maps and MTR-maps will yield the highest dose resolution. A comprehensive list of possible imaging sequences to acquire parametric maps is provided in [39] and listed in table 1. The parameter of interest is calculated from a set of images that are acquired where one imaging parameter is varied (see table 1).

Table 1. Overview of important quantitative MR imaging sequences for R_1 (= $1/T_1$), R_2 (= $1/T_2$) and magnetization transfer ratio (MTR) imaging. The variable that is changed to acquire different contrast weighted images is shown in the third column. If the variable is varied automatically within the sequence, the parameter is shown between squared brackets. The meaning of the availability is as follows: (*) The imaging sequence is not provided by the manufacturer. The sequence should be developed in house. (**) The sequence is available but significant changes to the imaging parameters are required. (***) The imaging sequence is readily available on all clinical MRI scanners.

Sequence type	Conditions	Variable	Post-processing	Availability	Spatial accuracy
1. Quantitative R_1 imaging sequences ($R_1=1/T_1$)					
Single spin-echo (SE)	TE short	TR (×2/×N)	Fit	***	Very good
Saturation recovery (SRGE/SRSE)	TR long, TE short	TM (×2/×N)	Fit	***	Good / Very good
Inversion recovery (IRGE/IRSE)	TR long, TE short	TI (×2/×N)	Fit	***	Good / Very good
Driven Equilibrium Single Pulse Observation of T_1 (DESPOT)	-	FA (×2/×N)	Fit	*	Good
Look-Locker (LL, TOMROP)	FA small	TI (×2/×N)	Fit	*	Good
Steady-State Free Precession (SSFP)	TR >> T_2	FA (×2/×N)	Anal. / Fit	**	Good

IR - Very fast acquisition (EPI, GRASE, HASTE)	TR long	TI (×2/×N)	Fit	**	Poor
2. Quantitative R_2 imaging sequences ($R_2 = 1/T_2$)					
Single spin-echo (SE)	TR long	TE (×2/×N)	Anal. / Fit	***	Very good
Fast spin-echo (FSE, TSE, RARE)	TR long	TE (×2/×N)	Anal. / Fit	***	Good
Multiple spin-echo (MSE, MC-SE)	TR long	[ΔTE (N)]	Fit	***	Very good
Steady-State Free Precession (SSFP)	TR << T_1 FA = 90°	[2 echoes]	Anal.	**	Good
3. Quantitative magnetization transfer (MT) imaging sequences					
MT pulse prepared spin echo imaging sequence	TR long	MT pulse amplitude	Anal.	**	Very good
Pulsed MT steady state	TR short	MT pulse amplitude	Anal.	**	Good
Stimulated echo preparation	-	TM	Anal.	*	Very good

2.3. Develop image processing software for calculating parametric maps from a set of base images

The parametric maps are calculated from a set of base images with different image contrast. This is achieved by an adequate software program which is most often developed ‘in house’. The importance of the choice of an appropriate cost-function to be minimized has been emphasized in [40]. It has been shown that the use of a chi-square minimization results in a more precise R_2 map as compared to a least-square minimization. In order to test the image processing software, a synthetic base image data set may be generated on which stochastic (or structural) noise may be added in order to test its performance.

2.4. Optimize the imaging sequence parameters for optimal precision

If the range of R_1, R_2 or MTR values is known in advance, the variable imaging parameters (column 3 in table 1) can be optimized to yield the highest achievable dose resolution. Tables of optimal scanning parameters for different R_2 ranges for a single-echo sequence and a multi-spin echo sequence are provided in [40] and [41] respectively. Theoretical derivations can be made for other imaging sequences using similar mathematical formalisms as provided in [40]. In most applications, a dose resolution $D_{\Delta\%}^{p=95\%}$ of at least 2% is required which results in a signal-to-noise ratio (SNR) in the dose map of approximately 138 ($D_{\Delta\%}^{p=95\%} = 1.96\sqrt{2}\,\sigma_D/(D_{max} - D_{min}) = 0.02$).

The dynamic range of the parametric map can be easily assessed using a series of standard contrast agent solution. The R_1 and R_2 are linearly proportional with the molar concentration of the contrast agent. To change both R_1 and R_2 of the test phantoms independently, a mixture of two contrast agents (e.g. Gd-DTPA and $FeSO_4$) or a mixture of a contrast agent and a gelling agent can be used.

2.5. Test the spatial and dosimetric accuracy of the imaging sequence

It has been well described that imaging artifacts may compromise the accuracy of the acquired dose maps [30, 42-45] both spatially as in dose. It is therefore vital that before a gel dosimeter is scanned, the performance of the imaging sequence in terms of accuracy is investigated. A simple test consists of scanning a 'blank' phantom (phantom with a non-irradiated gel) with similar spatial dimensions as the gel dosimeter phantom. It is important to leave the 'blank' phantom in the scanner room for at least 24 hours before scanning in order to equilibrate at the scanner room temperature. Quantitative parametric maps should be calculated from the 'blank' images and the signal homogeneity should be assessed. To isolate stochastic and structural deviations in the image, a voxel based analysis method can be used as described elsewhere [39]. The structural deviations should be below the tolerated error in the quantitative parametric MR maps. The tolerated error is obtained from the tolerable dose error as

$$\varepsilon_P = \frac{\partial P}{\partial D}\varepsilon_D \tag{1}$$

where P = R_1, R_2 or MTR and ε_P and ε_D are the tolerated error in the parameter P and in dose respectively.

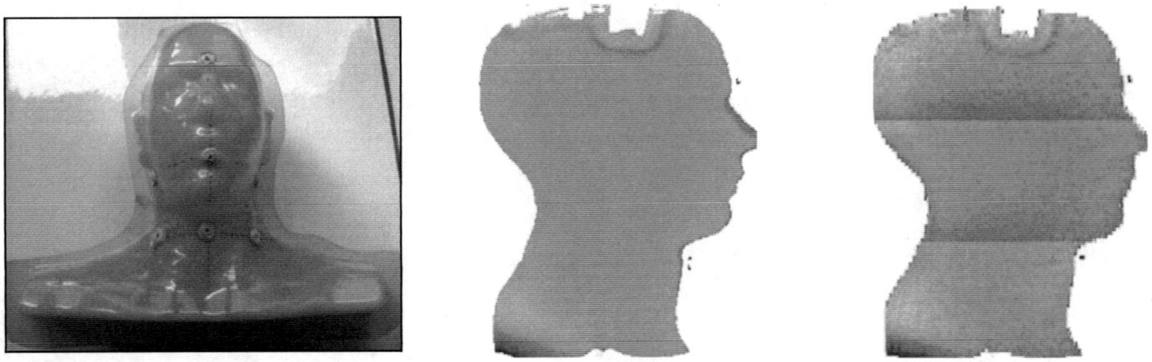

(a) **(b)** **(c)**

Figuur 1. Antropomorphic 'blank' gel dosimetry phantom with fiducial markers (a) and corresponding sagital slice (b) A sagital reconstructed from a stack of 105 transverse slices (c) reveals the effect of temperature drift during scanning. Transverse scanning was performed in three interleaved blocks. The ring shaped cranial artifact is attributed to oxygen effects.

At this point, it is advisable to test the image uniformity for several RF coils and check if the addition of a water load improves the homogeneity. Also, it is recommended to perform this uniformity test with the same imaging parameters as in the actual dosimetry experiment, as some artifacts may be phantom related. Also temperature drift as a result of the RF energy from the imaging sequence is dependent on the number of imaging slices and number of acquisitions [44]. If temperature drift results in non-uniformity in the imaged volume, either the repetition time should be increased to decrease the overall specific absorption rate (SAR) or the sequence should be modified to obtain central k-space ordering [44]. Only if the error in the parametric map is below the tolerable error, it is recommended to proceed to the next step.

A geometric quality control QC phantom can be used to check the geometric accuracy [30,39]. It should be emphasized that geometric distortions originating from magnetic field distortions caused by susceptibility differences are phantom dependent. A magnetic field map can be acquired and used to correct the geometric image distortions [46-47].

If dose maps are to be acquired in different image orientations, it is also advisable to scan the 'blank' phantom with different slice orientations. Although at most modern MR scanners, eddy currents are minimized by actively shielded gradient coils, with some magnetic field gradient intensive imaging sequences, eddy currents may still render the dose-R2 response curve dependent on the slice orientation and other sequence parameters [42].

2.6. Preparation of the gel dosimeter experiment

The gel dosimeter phantom and calibration vials are constructed. To minimize any systematic error introduced by the calibration vials, it is recommended to irradiate at least 15 to 20 calibration vials. The actual gel dosimeter phantom is provided with fiducial markers and scanned with CT or MRI to obtain the image data set on which the treatment is planned. The gel dosimeter phantom may be anthropomorphically shaped or may consist of a gel recipient inserted in an anthropomorphic cast. The isocenter on the treatment planning can be located with respect to the fiducial markers which facilitates the localization of the isocenter laser markers onto the antropomorphic phantom. Once the isocenter is determined, the isocenter laser lines are marked on the gel dosimeter phantom and the gel dosimeter phantom is irradiated according to the treatment plan. The calibration vials are irradiated with known doses of which the maximum dose is higher than the maximum dose in the treatment plan.

2.7. Transfer of the gel dosimeter to the MR scanner and scanning

Immediately after irradiation, the gel dosimeter phantom is transferred to the MR scanner room and left there for at least 24 hours in order to equilibrate at the MR scanner room temperature. The calibration vials are fixed onto the gel dosimeter phantom and scanned together with the gel dosimeter phantom. It may be helpful to make a special holder for the calibration vials.

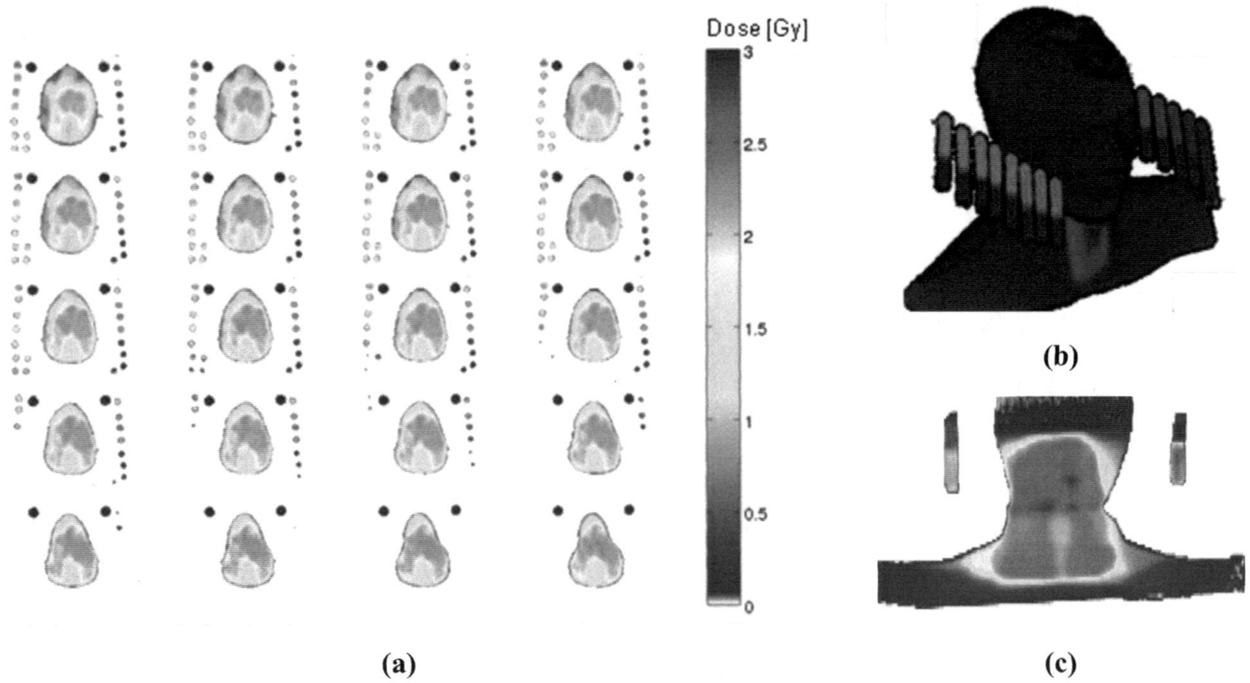

Figure 2. A collection of 20 gel measured dose maps out of a stack of 105 transverse dose maps and a 3D reconstructed rendered volume showing the position of the test tubes (b). A coronal dose map demonstrates the sparing of the brain stem (c).

Journal of Physics: Conference Series **250** (2010) 012015 doi:10.1088/1742-6596/250/1/012015

2.8. Post-processing and data interpretation

If all base images are acquired, they are converted in parametric maps. Subsequently, the parametric maps are converted into dose maps using the average values collected in regions-of-interest (ROIs) within the calibration vials. The acquired dose maps are then co-registered with the original (CT) reference image data set (treatment planning) using a rigid transformation on the basis of the fiducial markers that are visible in both data sets. When both image data sets are co-registered, a comparison of the treatment planning dose grid and the gel measured dose distribution can be performed.

Based on clinically relevant dose tolerance and distance to agreement criteria, gamma-maps [48] can be created and dose volume histograms (DVHs) can be generated for both the treatment planning dose distribution and the gel measured dose distribution.

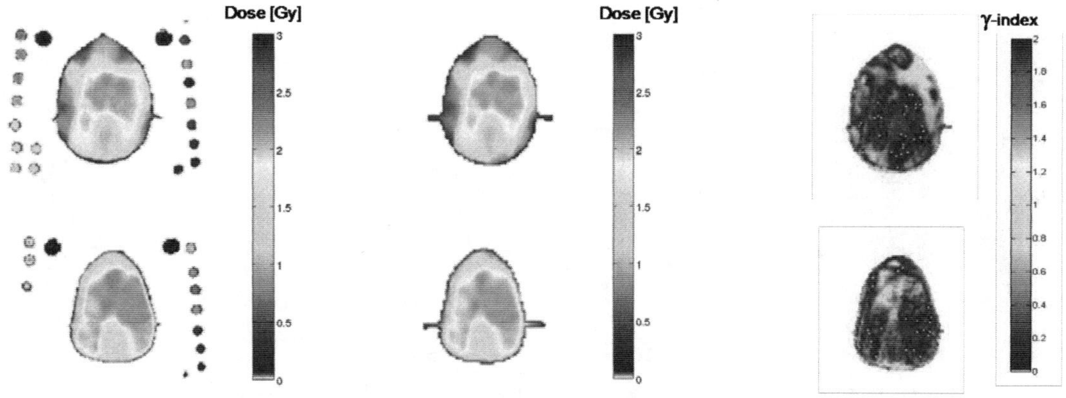

(a) **(b)** **(c)**

Figuur 3. Gel measured (a), calculated (treatment planning) dose maps (b) and corresponding gamma plots (c).

3. Conclusion

In contrast to other imaging modalities, MR scanning of polymer gel dosimeters provides many degrees of freedom. Several quantitative MR properties (such as T_1, T_2 and MTR) can be imaged and several different MR sequences can be used to acquire these properties. Whatever property or sequence is used to generate a quantitative parametric MR image data set, the accuracy and precision should be assessed and optimized using 'blank' phantoms (i.e. phantoms similar to the actual gel dosimeter phantom but filled with a non-irradiated gel).

The actual gel dosimetry experiment should be performed in a similar way as the patient treatment. The treatment planning should be performed on a reference image data set obtained from the gel dosimeter phantom and calibration vials should be irradiated with known doses. The gel measured dose distribution can be co-registered with the simulated (treatment planning) dose distribution on the basis of fiducial markers.

To obtain a picture of the overall accuracy of polymer gel dosimetry, also the fabrication and irradiation have to be included in the analysis (figure 3). This can be achieved through a reproducibility study of the complete gel dosimetry experiment from gel fabrication to dose distribution analysis with a well-characterized dose distribution (such as from a single square profile).

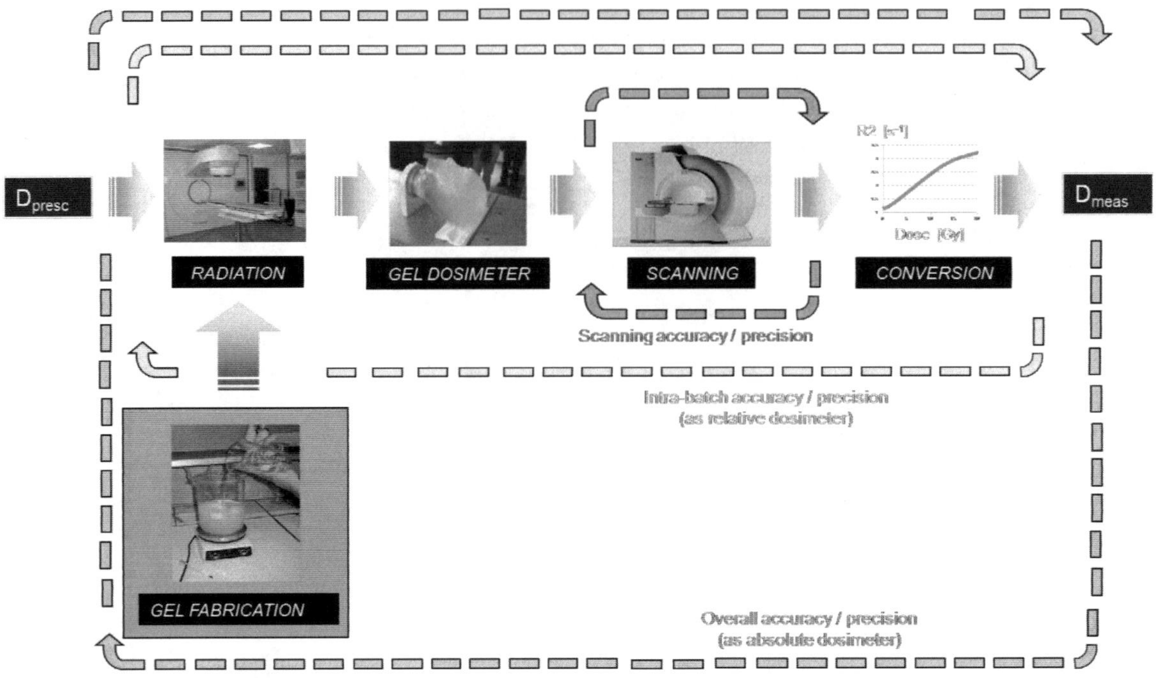

Figure 3. Gel dosimetry is performed in different steps. Errors can occur in each of the steps, leading to a decrease in the overall precision and accuracy. The accuracy and precision of each step can be determined independently.

References

[1] Gore J C, Kang Y S and Schulz R J 1984 Measurement of radiation dose distributions by nuclear magnetic resonance (NMR) imaging *Phys. Med. Biol.* **29** 1189–97

[2] Fricke H and Morse S 1927 The chemical action of Roentgen rays on dilute ferrosulphate solutions measure of dose *Am. J. Roentgenol. Radium Therapy Nucl. Med.* **18** 430–2

[3] Olsson L E, Petersson S, Ahlgren L and Mattsson S 1989 Ferrous sulphate gels for determination of absorbed dose distributions using MRI technique: basic studies *Phys. Med. Biol.* **34** 43–52

[4] Baldock C, Harris P J, Piercy A R and Healy B 2001b Experimental determination of the diffusion coefficient in two-dimensions in ferrous sulphate gels using the finite element method. *Australas Phys Eng Sci Med* **24** 19-30

[5] Podgorsak M B and Schreiner L J 1992 Nuclear magnetic relaxation characterization of irradiated Fricke solution *Med. Phys.* **19** 87–95

[6] Duzenli C, Sloboda R and Robinson D 1994 A spin-spin relaxation rate investigation of the gelatin ferrous sulphate NMR dosimeter *Phys. Med. Biol.* **39** 1577–92

[7] Audet C and Schreiner L J 1997 Multiple-site fast exchange model for spin-latttice relaxation in the Fricke-gelatin dosimeter *Med. Phys.* **24** 201–9

[8] Alexander P 1954 The degradation of solid polymethylmethacrylate by ionizing radiation *Proceedings of the royal society of london series A - Mathematical and physical sciences* **223** 392

[9] Kennan R P, Richardson K A, Zhong J, Maryanski M J and Gore J C 1996 The effects of cross-link density and chemical exchange on magnetization transfer in polyacrylamide gels *J*

Magn Reson B **110** 267-77

[10] Maryanski M J, Gore J C, Kennan R P and Schulz R J 1993 NMR relaxation enhancement in gels polymerized and cross-linked by ionizing radiation: A new approach to 3D dosimetry by MRI *Magn. Reson. Imaging* **11** 253–8

[11] Zimmerman J R and Brittin W E 1957 Nuclear magnetic resonance studies in multiple phase systems: lifetime of a water molecule in an adsorbing phase on silica gel *J. Phys. C: Solid State Phys.* **61**

[12] Audet C 1995 NMR dose response of the gels used for 3-D radiation dosimetry by magnetic resonance imaging (*Ph.D. thesis, McGill University, Montreal, Quebec*)

[13] De Deene Y, Hanselaer P, De Wagter C, Achten E and De Neve W 2000 An investigation of the chemical stability of a monomer/polymer gel dosimeter *Phys. Med. Biol.* **45** 859–78

[14] Lepage M, Whittaker A K, Rintoul L, Back S A J and Baldock C 2001 The relationship between radiation-induced chemical processes and transverse relaxation times in polymer gel dosimeters *Phys. Med. Biol.* **46** 1061–74

[15] Kennan R P, Richardson K A, Zhong J, Maryanski M J and Gore J C 1996 The effects of crosslink density and chemical exchange on magnetization transfer in polyacrylamide gels *J. Magn. Reson. – series B* **110** 267–77

[16] Gochberg D F, Fong P M and Gore J C 2001 Studies of magnetization transfer and relaxation in irradiated polymer gels – interpretation of MRI-based dosimetry *Phys. Med. Biol.* **46** 799–811

[17] Lepage M, McMahon K, Galloway G J, De Deene Y, Back S A J and Baldock C 2002 Magnetization transfer imaging for polymer gel dosimetry *Phys. Med. Biol.* **47** 1881–90

[18] Murphy P S, Cosgrove V P, Schwarz A, Webb S and Leach M O 2000 Proton spectroscopic imaging of polyacrylamide gel dosimeters for absolute radiation dosimetry *Phys. Med. Biol.* **45** 835–45

[19] De Deene Y, De Wagter C, Van Duyse B, Derycke S, De Neve W and Achten E 1998 Three-dimensional dosimetry using polymer gel and magnetic resonance imaging applied to the verification of conformal radiation therapy in head-and-neck cancer *Radiother. Oncol.* **48** 283–91

[20] Haraldsson P, Bäck S A J, Magnusson P and Olsson L E 2000 Dose response characteristics and basic dose distribution data for a polymerization-based dosemeter gel evaluated using MR *Br. J. Radiol.* **73** 58–65

[21] De Deene Y, De Wagter C, Van Duyse B, Derycke S, Mersseman B, De Gersem W, Voet T, Achten E and De Neve W 2000 Validation of MR-based polymer gel dosimetry as a preclinical three-dimensional verification tool in conformal radiotherapy. *Magn. Reson. Med.* **43** 116–25

[22] Pappas E, Seimenis I, Angelopoulos A, Georgolopoulou P, Kamariotaki-Paparigopoulou M, Maris T, Sakelliou L, Sandilos P and Vlachos L 2001 Narrow stereotactic beam profile measurements using N-vinylpyrrolidone based polymer gels and magnetic resonance imaging *Phys. Med. Biol.* **46** 783–97

[23] Ibbott G S, Maryanski M J, Eastman P, Holcomb S D, Zhang Y, Avison R G, Sanders M and Gore J C 1997 Three-dimensional visualization and measurement of conformal dose distributions using magnetic resonance imaging of BANG polymer gel dosimeters *Int. J. Radiat. Oncol. Biol. Phys.* **38** 1097–103

[24] Oldham M, Baustert I, Lord C, Smith T A D, McJury M, Warrington A P, Leach M O and Webb S 1998 An investigation into the dosimetry of a nine-field tomotherapy irradiation using BANG-gel dosimetry *Phys. Med. Biol.* **43** 1113–32

[25] Meeks S L, Bova F J, Maryanski M J, Kendrick L A, Ranade M K, Buatti J M and Friedman W A 1999 Image registration of BANG gel dose maps for quantitative dosimetry verification. *Int. J. Radiat. Oncol. Biol. Phys.* **43** 1141–53

[26] Cosgrove V P, Murphy P S, McJury M, Adams E J, Warrington A P, Leach M O and Webb S

2000 The reproducibility of polyacrylamide gel dosimetry applied to stereotactic conformal radiotherapy *Phys. Med. Biol.* **45** 1195–210

[27] Ramm U, Weber U, Bock M, Kramer M, Bankamp A, Damrau M, Thilmann C, Bottcher H D, Schad L R and Kraft G 2000 Three-dimensional BANG gel dosimetry in conformal carbon ion radiotherapy *Phys. Med. Biol.* **45** N95–102

[28] Farajollahi A R, Bonnett D E, Tattam D and Green S 2000 The potential use of polymer gel dosimetry in boron neutron capture therapy *Phys. Med. Biol.* **45** N9-N14

[29] Grebe G, Pfaender M, Roll M and Luedemann L 2001 Dynamic arc radiosurgery and radiotherapy: Commissioning and verification of dose distributions *Int. J. Radiat. Oncol. Biol. Phys.* **49** 1451–60

[30] De Deene Y 2006 On the accuracy and precision of gel dosimetry *J. Phys: Conf Series* **56** 72-85

[31] Hornak J P 2007 The basics of MRI [web document] http://www.cis.rit.edu/htbooks/mri/

[32] Magnetic Resonance Technology – Information Portal 2008 MR-Tip [web document] http://www.mr-tip.com/

[33] Campus Medica 2007 e-MRI, Magnetic Resonance Imaging physics and technique course on the web [web document] http://www.e-mri.org/

[34] Kuperman V 2000, Magnetic resonance imaging – Physical principles and applications, academic press, San Diego, USA

[35] McRobbie D W, Moore E A, Graves M J, Prince M R 2006, MRI from picture to proton, Cambridge University Press, Cambridge, UK

[36] Slichter C P 1978 Principles of magnetic resonance, Springer-Verlag Berlin Heidelberg

[37] Levitt M H 2002 Spin Dynamics: Basics of nuclear magnetic resonance, John Wiley and Sons, West Sussex, England

[38] Bernstein M A, King K F and Zhou X J 2004 Handbook of MRI pulse sequences, Elsevier Academic Press, Burlington, Massachutes, USA

[39] De Deene Y 2009 Review of quantitative MRI principles for gel dosimetry *J. Phys: Conf Series* **164** 012033, 1-42.

[40] De Deene Y, Van deWalle R, Achten E and DeWagter C 1998 Mathematical analysis and experimental investigation of noise in quantitative magnetic resonance imaging applied in polymer gel dosimetry *Signal Processing* **70** 85-101

[41] De Deene Y and Baldock C 2002 Optimization of multiple spin-echo sequences for 3D polymer gel dosimetry *Phys Med Biol* **47** 3117-41

[42] De Deene Y, De Wagter C, De Neve W, Achten E 2000, Artefacts in multi-echo T2 imaging for high-precision gel dosimetry: I. Analysis and compensation of eddy currents *Phys Med Biol* **45** 1807-23

[43] De Deene Y, De Wagter C, De Neve W 2000, Artefacts in multi-echo T2 imaging for high-precision gel dosimetry: II. Analysis of B1 field inhomogeneity *Phys Med Biol* **45** 1825-39.

[44] De Deene Y and DeWagter C 2001 Artefacts in multi-echo T2 imaging for high-precision gel dosimetry: III. Effects of temperature drift during scanning *Phys Med Biol* **46**, 2697-711.

[45] De Deene Y, Reynaerdt N and De Wagter C 2001 On the accuracy of monomer/polymer gel dosimetry in the proximity of a high-dose-rate ^{192}Ir source *Phys Med Biol* **46** 2801-25.

[46] Park H W, Ro Y M and Cho Z H 1988 Measurement of the magnetic susceptibility effect in high-field NMR imaging *Phys. Med. Biol.* **33** 339-49

[47] Cho Z H, Kim D J and Kim Y K 1988 Total inhomogeneity correction including chemical shifts and susceptibility effect in high-field NMR imaging *Med. Phys.* **15** 7-11

[48] Low D A, Harms W B, Mutic S and Purdy J A 1998 A technique for the quantitative evaluation of dose distributions *Med Phys* **25** 656-61.

Journal of Physics: Conference Series **250** (2010) 012016 doi:10.1088/1742-6596/250/1/012016

Improved dose sensitivity of normoxic polyacrylamide gelatin gel dosimeter with sucrose

M Yoshioka[1,2], S Hayashi[1], S Usui[1], K Haneda[1], T Kondo[1,3], H Numasaki[2], T Teshima[2], T Tominaga[1]

[1]Department of Clinical Radiology, Faculty of Health Sciences, Hiroshima International University, Higashi-Hiroshima, Hiroshima, 739-2695, Japan
[2]Department of Medical Physics & Engineering, Osaka University Graduate School of Medicine, Suita, Osaka, 565-0871, Japan
[3]Futaba Hospital, Kure, Hiroshima, 737-0143, Japan

m-yoshi@hs.hirokoku-u.ac.jp

Abstract. To improve the dose sensitivity of normoxic polyacrylamide gelatin gel (nPAG) dosimeter, the effect of sucrose as an additive is investigated. The dose-transverse relaxation rate (R_2) characteristics of the samples of nPAG with different sucrose concentrations were examined, and temperature increases due to exothermic polymerization reaction in the irradiated gel were also measured. As the result, the dose-R_2 sensitivity increased (~3 times) with increasing sucrose concentrations (0-25%), while other characteristics of dose response, such as dose integral property, dose rate dependence and temporal stability, were consistent with those of typical nPAG dosimeter. And as the sucrose concentrations increased, a larger temperature increase was observed. These results clearly indicate that the polymerization rate is increased with increasing sucrose concentrations.

1. Introduction

Polymer gel dosimeter has been used for measuring the three-dimensional dose distribution delivered by recent radiation therapy techniques. It is a device utilizing the radiation-induced polymerization in the gel, and the results of the polymerization reaction increase the gel's solidity and opacity with the increase of irradiation dose. The three-dimensional dose distributions are imaged by MRI or Optical CT measurements [1, 2].

Many gel dosimeters having different vinyl monomers, gelling agents and deoxygenated techniques have been developed so far [3, 4]. Among them, the normoxic polyacrylamide gelatin (nPAG) gel is one of the promising dosimeters [5-7], which is composed of two monomers, acrylamide (Aam) and *N,N′*-methylenebisacrylamide (Bis), and gelatin. The nPAG dosimeter has advantages of the dose integral property and the less dose rate dependence, however the dose sensitivity for the transverse relaxation rate (R_2) is unsatisfactory (ca. 0.1 $Gy^{-1}s^{-1}$).

To improve this property, Koeva *et al.* have proposed the gels with co-solvents (glycerol or isopropanol) which dissolves more Bis (7%) in water [8, 9]. The dose sensitivity was improved to 2.7 times higher than that of typical nPAG gel. In their studies, co-solvents (carbohydrates) play an important role to enhance the dose sensitivity even in a case of 3% Bis concentration [9, 10], though

the mechanisms are not clear. It has been also reported that saccharide additive enhanced with the dose sensitivity of methacrylic acid-based gel [11] and Fricke type gel dosimeters [12].
According to these facts, we focused on sucrose as carbohydrate additive for nPAG gel dosimeter. In this paper, we report on the dose-R_2 characteristics of nPAG with different sucrose concentrations and temperature increases due to exothermic polymerization reaction in the irradiated gel [13].

2. Materials and methods

2.1. Gel fabrication

In this study, the samples of nPAG with different sucrose concentrations were used (Table). The details of procedure of gel fabrication are as follows. Gelatin (300 bloom, Sigma Chemical co.) was added in water at room temperature and then it was heated at 60°C on a hot magnetic stirrer. After the solution got clear, it was cooled to 50°C and Bis (Wako Pure Chemical co.) was added. After Bis dissolved completely, sucrose was added. The solution was cooled to 40°C and Aam (Wako Pure Chemical co.) was then added. Finally, tetrakis(hydroxymethyl)phosphonium chloride (THPC 80%, Sigma Chemical co.) was added at 35°C. The solution was filled in 10 ml test tubes for MRI measurements or in 50 ml glass vials for temperature measurements. The samples were stored at 3-5°C in a refrigerator.

Table. Compositions of gel.

Chemical	Concentration
Water	
Acrylamide	3 %
N,N'-methylenebisacrylamide	3 %
Gelatin	5 %
Sucrose	0 - 25 %
THPC	5 mM

2.2. Irradiation

The samples were irradiated between 10 and 16 hours after gel preparation with a medical linear accelerator (Clinac iX, Varian) using 6 MV photon beams at 5 cm depth in a water phantom. A source sample distance was 100 cm, a field size was 10 × 10 cm^2 and constant dose rate was 300 cGy/min at dose maximum depth. Firstly, the samples were irradiated to known dose up to 40 Gy to compare the dose-R_2 response of each samples. Secondly, for examining the dependence on the dose integration in the dose response, the fractional irradiations up to 10 Gy of total dose were performed. The irradiation dose per fraction was 1 Gy and the time interval of each irradiation was one minute. And finally, various doses rate from 100 and 600 cGy/min at dose maximum depth were delivered. After irradiation, all samples were stored at 23°C in an incubator to obtain MR room-temperature equilibrium.

2.3. Temperature measurement

The samples were left at the irradiation site for at least 1h before the irradiation to stabilize their temperature. The samples were wrapped in polyethylene sheet to avoid heat loss to air and were set in the beam axis center on Solid Water. The top surface of 50 ml vials was set at the same level at 100 cm source surface distance. A fields size was 20 × 20 cm^2. The samples were irradiated to dose up to 20 Gy. Temperature probe was inserted vertically into the center part of 50 ml glass vial. Temperature histories were recorded from the start to after irradiation with digital thermo-recorders (TR-52, T&D).

2.4. MR measurement

All images were acquired with 0.3 Tesla MRI system (AIRIS-II comfort, Hitachi) using multiple spin echo sequence with the following parameters, TE = 20 and 250 ms, TR = 4000 ms, thickness = 5 mm

Journal of Physics: Conference Series **250** (2010) 012016 doi:10.1088/1742-6596/250/1/012016

(at a central beam axis of the irradiation), matrix size = 128 × 128, FOV = 140 × 140 mm^2, number of excitations = 3. All samples except that for the study on temporal stability of dose response were measured at after three days post-irradiation. The R_2 values were calculated from two images (TE = 20 and 250 ms). The R_2 value of each sample was determined by averaging above 50 pixels of circular region of interest in images. The samples for the temporal stability study were measured successively at the time interval between one and five days after irradiation.

3. Results and discussion

The dose-R_2 response relations of the samples with different sucrose concentrations are shown in Fig. 1(a). The gradient of the dose-R_2 response (0-5 Gy) increases with increasing sucrose concentrations [Fig. 1(b)]. The sensitivity (0.37 Gy^{-1}s^{-1}) of nPAG with 25% sucrose was ca. 3 times higher than that (0.12 Gy^{-1}s^{-1}) of nPAG without sucrose.

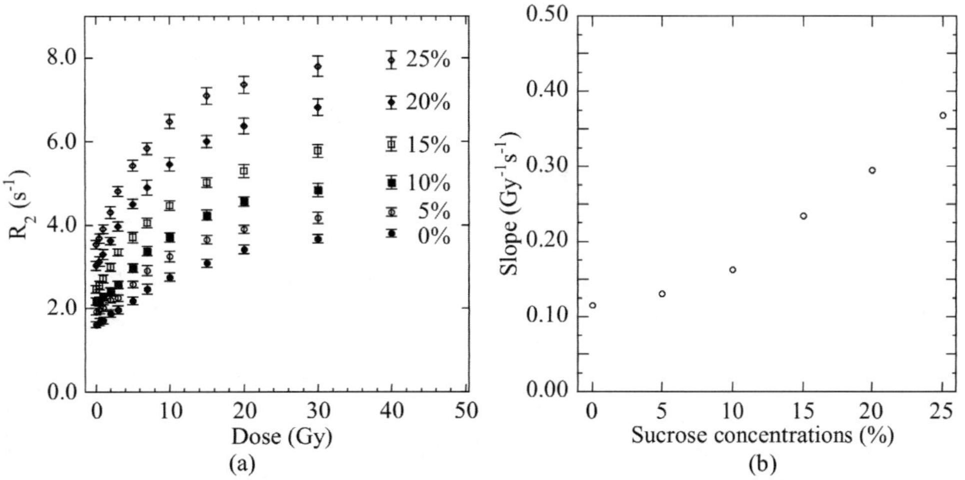

Figure 1. The dose-R_2 relation of nPAG with different sucrose concentrations (a). The plot of the dose-R_2 sensitivity at the linear part (0-5 Gy) versus sucrose concentrations (b).

The temperature increase due to radiation-induced polymerization is shown in Fig. 2. It was observed just after the irradiation start. With increasing sucrose concentrations, the temperature increases proportionally due to the exothermic reaction.

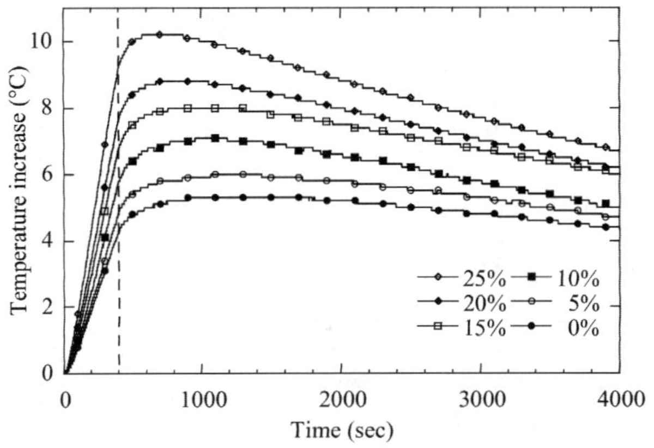

Figure 2. The temperature histories of nPAG with different sucrose concentrations due to the exothermic radiation-induced polymerization reaction. The vertical broken line shows the time at the end of irradiation.

The result indicates that the polymerization rate increases with sucrose concentrations so that the dose-R_2 sensitivities increase. The reasons of the higher sensitivity due to the presence of sucrose are not clear at this stage, but some possibilities are considered. One is due to the decreasing of water content that causes the higher concentration of monomers in the gel. Another is due to chemical interactions between carbohydrates and monomers/ polymer molecules which are postulated in the previous studies [9-11]. It is interesting that the enhancement in sensitivity is observed in the gels with a variety of water-soluble carbohydrates (e.g. sucrose, alcohols [9, 10], glucose [11]) even though each gradient of an enhancement is not consistent. The change of the water environment (structure) due to the hydrates formed may promote the polymerization of monomers in the gel.

The maximum difference of R_2 between the fractional irradiation and the single shot irradiation was less than 2% below 10 Gy. The sample with above 20% sucrose showed slightly lower R_2 at higher dose rate (600 cGy/min). The temporal R_2 changes after irradiation were significant for less than one day and slightly increased after two days post irradiation. These characteristics are consistent with those reported for typical nPAG gels [6,7].

4. Conclusion

We have examined the characteristics of sucrose-containing nPAG system and found that the dose-R_2 sensitivity are significantly increased with increasing sucrose concentrations even in the case of 3% Bis concentration.

References

[1] Maryanski M J, Gore J C, Kennan R P and Schulz R J 1993 NMR relaxation enhancement in gels polymerized and cross-linked by ionizing radiation: a new approach to 3D dosimetry by MRI *Magn. Reson. Imaging* **11** 253-8
[2] Gore J C, Ranade M, Maryanski M J and Schulz R J 1996 Radiation dose distributions in three dimensions from tomographic optical density scanning of polymer gels: I. Development of an optical scanner *Phys. Med. Biol.* **41** 2695-704
[3] Jirasek A 2006 Experimental investigations of polymer gel dosimeters *J. Phys.: Conf. Ser.* **56** 23-34
[4] Baldock C, De Deene Y, Doran S, Ibbott G, Jirasek A, Lepage M, McAuley K B, Oldham M and Schreiner L J 2010 Polymer gel dosimetry *Phys. Med. Biol.* **55** R1-R63
[5] Maryanski M J, Schulz R J, Ibbott G S, Gatenby J C, Xie J, Horton D and Gore J C 1994 Magnetic resonance imaging of radiation dose distributions using a polymer-gel dosimeter *Phys. Med. Biol.* **39** 1437-55
[6] De Deene Y, Vergote K, Claeys C and De Wagter C 2006 The fundamental radiation properties of normoxic polymer gel dosimeters: a comparison between a methacrylic acid based gel and acrylamide based gels *Phys. Med. Biol.* **51** 653-73
[7] Karlsson A, Gustavsson H, Månsson S, McAuley K B and Bäck S A 2007 Dose integration characteristics in normoxic polymer gel dosimetry investigated using sequential beam irradiation *Phys. Med. Biol.* **52** 4697-706
[8] Koeva V I, Csaszar E S, Senden R J, McAuley K B and Schreiner L J 2008 Polymer Gel Dosimeters with Increased Solubility: A Preliminary Investigation of the NMR and Optical Dose-Response Using Different Crosslinkers and Co-Solvents *Macromol. Symp.* **261** 157-66
[9] Koeva V I, Olding T, Jirasek A, Schreiner L J and McAuley K B 2009 Preliminary investigation of the NMR, optical and x-ray CT dose-response of polymer gel dosimeters incorporating cosolvents to improve dose sensitivity *Phys. Med. Biol.* **54** 2779-90
[10] Jirasek A, Hilts M, Berman A and McAuley K B 2009 Effects of glycerol co-solvent on the rate and form of polymer gel dose response *Phys. Med. Biol.* **54** 907-18
[11] Zhu X, Reese T G, Crowly E M and Fakhri G E 2010 Improved MAGIC gel for higher sensitivity and elemental tissue equivalent *Med. Phys.* **37** 183-88
[12] Healy B J, Zahmatkesh M H, Nitschke K N and Baldock C 2003 Effect of saccharide additives on response of ferrous-agarose-xylenol orange radiotherapy gel dosimeters *Med. Phys.* **30** 2282-91
[13] Salomons G J, Park Y S, McAuley K B and Schreiner L J 2002 Temperature increases associated with polymerization of irradiated PAG dosimeters *Phys. Med. Biol.* **47** 1435-48

Journal of Physics: Conference Series **250** (2010) 012017 doi:10.1088/1742-6596/250/1/012017

Impact of oxygen on the accuracy and precision of normoxic polymer gel dosimeters

Mahbod Sedaghat[1], Rachel Bujold[2] and Martin Lepage[1]

[1]Centre d'imagerie moléculaire de Sherbrooke, Département de médecine nucléaire et de radiobiologie, Université de Sherbrooke, Sherbrooke, QC, Canada and [2]Service de radio-oncologie, Centre hospitalier universitaire de Sherbrooke, Sherbrooke, QC, Canada

e-mail: Mahbod.Sedaghat@Usherbrooke.ca

Abstract. Diffusion of oxygen in normoxic polymer gel dosimeters was studied. Gel samples in glass tubes were exposed to oxygen for different periods and then irradiated. Oxygen inhibited polymerization at the surface and polymerization occurred at larger depths. However the interface between these two regions revealed a local increase of polymerization. R_2 values in these distinct regions are analyzed in terms of the concentration of oxygen and of antioxidant. Perhaps counter-intuitively, a small concentration of oxygen increases the sensitivity of the gel by consuming antioxidant molecules. The presence of antioxidant molecules is shown to cause a decrease of sensitivity, presumably through interactions with water free radicals and/or polymer radicals. Mixing and filling procedures as well as oxygen penetration from plastic caps or walls of gel containers are potential sources of oxygen-related discrepancies in normoxic polymer gel dosimeters. This effect may cause dosimetric inaccuracies of up to 60%.

1. Introduction

A comprehensive 3D dose distribution evaluation is crucial requirement for modern 3D radiation therapy treatments. The first three-dimensional dosimetry technique was proposed over a quarter of a century ago when quantitative 3D medical imaging was combined with a novel chemical dosimetry system [1]. It consisted of the well known Fricke solution dispersed in a hydrogel. The utility of this gel dosimeter was limited as the ferric ions generated by ionizing radiation could diffuse rapidly in the gel matrix, distorting the spatial integrity of the measurements. The next generation of gel dosimeters using water soluble monomers was proposed a few years later [2]. The monomers polymerize in chain reactions initiated by the radiation-induced production of water free radicals in the gel. The extent of this chain polymerization is dose-dependent. Unfortunately, oxygen was (and still is) a major concern in the preparation and in the accuracy of polymer gel dosimeters [3]. The reactions between radiation-induced free radicals and oxygen are so effective that polymerization is completely inhibited in polymer gel dosimeters prepared under normal atmospheric conditions. Therefore, these dosimeters must be prepared, cast and sealed under anoxic conditions. It was later proposed that adding a chemical oxygen scavenger (*e.g.*, an antioxidant) to the gel mixture during preparation would yield a radiosensitive gel [4]. These dosimeters were called normoxic, since they could be prepared in normal atmospheric environment.

The salient characteristics of normoxic polymer gel dosimeters have been investigated. Although no dose threshold due to oxygen inhibition is observed in normoxic gel dosimeters, addition of an oxygen scavenger reduces their R_2-dose sensitivity compared to their anoxic counterparts.

Discrepancies between the results of different research groups [5] and inaccuracies between the dose delivered to a phantom and the dose determined by normoxic gel dosimeters using calibration vials [6] have been reported. Here we report on the interplay between oxygen and antioxidant molecules within normoxic gel dosimeters. Specifically we demonstrate that oxygen scavengers can also scavenge water free radicals and/or polymer radicals, decreasing the dose sensitivity of normoxic gel dosimeters. Oxygen penetrating the dosimeter by diffusion actually increases the dose sensitivity to some extent, but higher concentration of oxygen eventually decreases the dose sensitivity.

2. Materials and Methods

Diffusion of oxygen was studied in three acrylamide-based and three methacrylic acid-based gel compositions. Tetrakis hydroxymethyl phosphonium chloride (THPC) and ascorbic acid (AscA) were used as antioxidants. Because the initial dissolved oxygen (DO) concentration may depend upon atmospheric conditions, gel fabrication was performed in a nitrogen-filled glove box. The level of oxygen in the glove box was monitored using a Fisher Scientific Accumet® XL60 DO meter. Oxygen level was 0.01 mg/l in the glove box and 0.00 mg/l in nitrogen-bubbled water. Acrylamide-based gels were mixed using concentrations of 5%, 3% and 3% (w/w) of gelatin, acrylamide (AA) and N,N'-methylene-bis-acrylamide (BIS), respectively. Eight glass tubes (KIMAX culture tubes with rubber-lined screw caps, length of 12 cm and volume of 25 ml) were filled. An appropriate volume of THPC was added to half of the remaining solution for a final concentration of 5 mM and a second series of tubes was filled. To the other half, 0.1 mM AscA and 0.02 mM copper(II) sulphate were added. This gives three sets of 8 tubes filled with acrylamide-based gel dosimeters. Methacrylic acid-based dosimeters were prepared from 8% gelatin and 7% methacrylic acid. As above, samples were prepared with 5 mM THPC and 2 mM AscA with 0.08 mM copper(II) sulphate. Three sets of glass tubes were filled and the caps were tightly closed. All the tubes were removed from the glove box and left for 3 h in a fridge for the gels to set. After this delay, caps of the tubes in each set were removed one by one in 12 hours intervals to expose the gels to air. The tubes were irradiated to 5 Gy, 84 h after gel fabrication, with a Gammacell 220 irradiator (Atomic Energy of Canada Ltd.). The machine is equipped with ^{60}Co sources and had a dose rate of 1.68 Gy/min. Tubes for each type of gel dosimeter were imaged simultaneously 24 h after irradiation with a Siemens Sonata 1.5 T clinical scanner (Siemens Medical Solutions Erlangen, Germany) with a multi-echo spin echo protocol. R_2 maps were calculated using in-house software running in MATLAB® environment.

3. Results

Figure 1(A) shows the position of a gel tube and the dose distribution in the Gammacell 220 sample chamber [7]. It is noted that the dose at the bottom and top of the tube is reduced by roughly 5%. Figure 1(B) illustrates spin-spin relaxation rate (R_2) values derived from gel tubes containing the acrylamide-based gel dosimeters exposed to air for 24 h. Oxygen diffusion has inhibited polymerization of the PAG dosimeter for the first 33 mm from the surface of the gel. This is seen as a uniform baseline with an R_2 value around 1 s^{-1}. From this point on, the concentration of oxygen declines to values below the amount required for complete inhibition and R_2 rapidly rises to a maximum of 2.4 s^{-1}, which is the expected R_2-dose response of PAG at 5 Gy. At the bottom of the tube (right of the figure), R_2 values decrease by about 4.3% which is consistent with the dose distribution of the Gammacell as shown in figure 1(A). As the time of exposure to air increases the oxygen inhibition front moves deeper into the gel (data not shown).

Less polymer is formed in gels containing antioxidant, which results in lower maximum R_2 values along their profiles. Progress of the oxygen inhibition front into the acrylamide-based gels containing THPC (PAGAT) and AscA (PAGIC) is slowed compared to the PAG dosimeter. This is expected due to the presence of antioxidant that reacts with oxygen, preventing it from further diffusion. Interestingly, the behavior of these gels at the oxygen inhibition front is clearly different from that of PAG. Their R_2 values rise from the inhibition baseline, reach a maximum and decrease to a relatively steady value roughly 20 mm deeper, presumably after the maximum range of penetration of oxygen.

Percentage differences between maximum R_2 values at the top of these maxima and those after the maximum penetration of oxygen is 12.6% for PAGAT and 18.8% for PAGIC.

Figure 1(C) compares oxygen diffusion in an anoxic MAG dosimeter with its two derivatives containing antioxidant. Oxygen has fully inhibited polymerization in the MAG up to 40 mm from the surface of the gel after 24 h. However, for the gels containing antioxidant complete oxygen inhibition was restricted to the first 8 mm from the surface. After the oxygen inhibition front, R_2 values of the MAGAT dosimeter sharply increase to a maximum of 19.6 s^{-1} and decrease down to a relatively steady value of 8.7 s^{-1} after a distance of 27 mm from the surface of the gel. The value of the maximum R_2 is more than twice of the steady R_2 reached after the oxygen diffusion range. In the MAGIC dosimeter, a very sharp peak is observed around 10 mm on the profile followed by a smooth

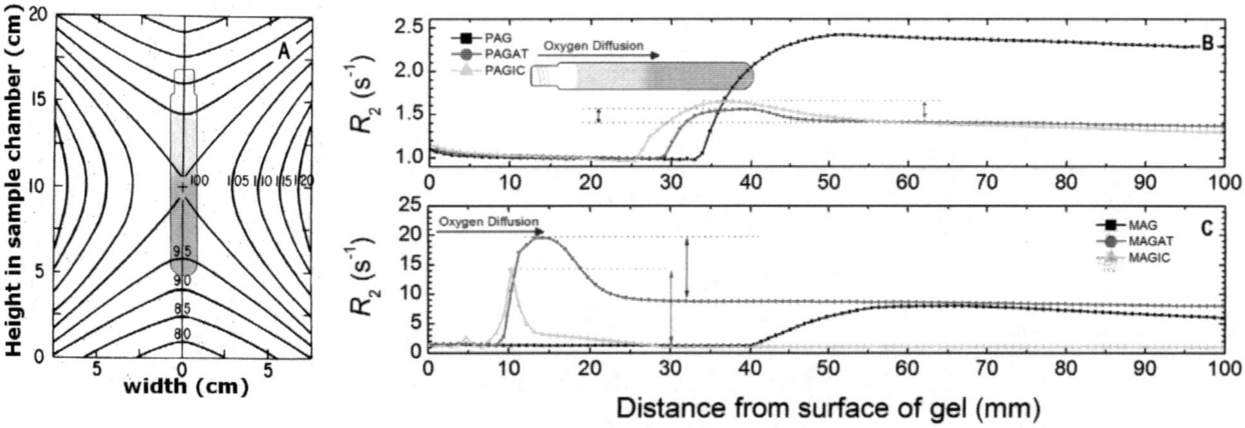

Figure 1. (A) Dose distribution in the Gammacell sample chamber and position of the gel tube within the chamber. A dose gradient exists along the tube axis. R_2 profiles of the acrylamide-based (B) and methacrylic acid-based (C) gel dosimeters in tubes exposed to air for 24 h. The profiles are derived by averaging 8 pixel columns along the tube axis. Each tube received a dose of 5 Gy and was scanned 24 h after irradiation. Note that y-axes on panels B and C are adjusted for clarity.

decrease to baseline values. Interestingly, no polymerization is observed after the maximum range of oxygen penetration (30 mm) in this tube. For acrylamide- and methacrylic acid-based gels the features described above were progressively observed deeper into the gel as the duration of exposure to air was increased.

4. Discussion

Reactions of antioxidants have not been specifically studied in polymer gel systems. However, antioxidant activity generally correlates well with radical scavenging activity. It has been shown that oxidation of ascorbic acid proceeds via production of ascorbyl radicals. Ascorbyl radicals react with other free radicals and effectively terminate the propagation of free radical reactions [8]. This is the reason for which AscA is not efficient in acrylamide-based gel dosimeters. We observed no radiation dose response for PAGIC dosimeters containing 0.5 mM of AscA. Interestingly, no polymerization is observed in MAGIC dosimeter after the maximum range of oxygen diffusion into the gel tube. This indicates that the dose sensitivity of MAGIC strongly relies on the presence of oxygen in the solution.

It was suggested that THPC reacts with gelatin in an acrylamide-based dosimeter [5] increasing its coagulation and stiffness and decreasing the R_2-dose sensitivity. The coagulation of gelatin is presumed to be uniform in the tubes and we assume that the gelatin network is not affected by the presence of oxygen after the network is formed. Yet, we clearly observe an increase in R_2 after the oxygen inhibition front in the gel dosimeters. To explain this observation, we suggest that THPC scavenges radicals during irradiation of the dosimeter, but whether these are water free radicals or polymer radicals, is not known.

Figure 2 shows the R_2 profile of the 5 mM THPC PAGAT with the maximum in R_2 values after the inhibition baseline highlighted in grey. On the left of this region the concentration of oxygen is high enough to fully inhibit propagation of polymerization and a baseline R_2 is measured. The oxygen concentration decreases along the hatched region to presumably zero. It follows that the concentration of THPC molecules (or their normal fragmentation in solution) that have not reacted with oxygen ("unreacted") is minimal in the first 28 mm of the tube and then increases to a maximum after the maximum range of oxygen penetration. Oxygen and antioxidant strike a balance in the grey region and this results in a larger extent of polymerization. In other words, both unreacted antioxidant and oxygen can lower the extent of polymerization, the absence of these species results in a maximum in the extent of polymerization.

Under ambient atmospheric conditions, the balance between oxygen and antioxidant will depend on the amount of oxygen that penetrates into the solution during stirring, filling and storage of the dosimeter. Any diffusion of oxygen from plastic caps or walls of the containers will further modify the amount of unreacted antioxidant. It should be noted that a variation of 10% on R_2 values may translate to a dose discrepancy of up to about 60% when using absolute calibration. As the gel is not unduly exposed to air in conventional gel dosimetry studies, such large discrepancies have seldom been reported. However oxygen-related discrepancies for normoxic gels can lead to poor reproducibility and precision in and between different studies.

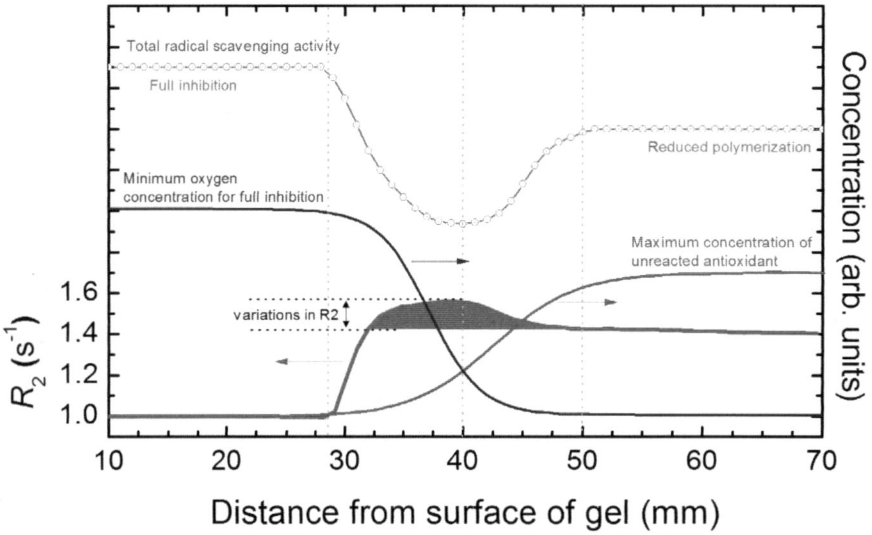

Figure 2. Schematic representation of the concentration of oxygen and antioxidant across the oxygen diffusion front in normoxic polymer gels. The interplay between these chemicals influences the R_2 value in polymer gel dosimeters.

5. Conclusion

Normoxic polymer gel dosimeters are *not* immune to oxygen-related discrepancies. The effect of oxygen may be more visible when using external calibration methods (*i.e.*, when calibration is done with a separate phantom or vials other than the original dosimeter) or when comparing the results of separate gel measurements. Oxygen diffusion through container walls alters the dose response of the gel within a single phantom. This effect may be restricted to the surface or may affect the whole sample, depending on the phantom size and time between preparation and irradiation.

References

[1] Gore J C, Kang Y S and Schulz R J 1984 Measurement of radiation dose distributions by nuclear magnetic resonance (NMR) imaging *Phys. Med. Biol.* **29** 1189–97

[2] Maryanski M J, Gore J C, Kennan R P and Schulz R J 1993 NMR relaxation enhancement in gels polymerized and cross-linked by ionizing radiation: a new approach to 3D dosimetry by MRI *Magn. Reson. Imaging* **11** 253-58

[3] Hepworth S J, Leach M O and Doran S J 1999 Dynamics of polymerization in polyacrylamide gel (PAG) dosimeters: (II) modeling oxygen diffusion *Phys. Med. Biol.* **44** 1875–84

[4] Fong P M, Keil D C, Does M D and Gore J C 2001 Polymer gels for magnetic resonance imaging of radiation dose distributions at normal room atmosphere *Phys.Med. Biol.* **46** 3105–13

[5] Jirasek A, Hilts M, Shaw C and Baxter P 2006 Investigation of tetrakis hydroxymethyl phosphonium chloride as an antioxidant for use in x-ray computed tomography polyacrylamide gel dosimetry *Phys. Med. Biol.* **51** 1891-1906

[6] De Deene Y, 2006 On the accuracy and precision of gel dosimetry *J. Phys.: Conf. Ser* **56**, 72–85

[7] Christensen R C and Song C-M 1981 Magnitude and position dependence of Gammacell 220 drawer transit doses *Health Phys.* **41** (3) 527-33

[8] Bendich A, Machlin L J, Scandurra O, Burton G W and Wayner D D M 1986 The antioxidant role of vitamin C *Adv. Free Radical Biol. Med.* **2**, 419-444.

Journal of Physics: Conference Series **250** (2010) 012018 doi:10.1088/1742-6596/250/1/012018

Effect of the exothermal polymerization reaction on polymer gel dosimetric measurements

Mahbod Sedaghat[1], Rachel Bujold [2] and Martin Lepage[1]

[1]Centre d'imagerie moléculaire de Sherbrooke, Département de médecine nucléaire et de radiobiologie, Université de Sherbrooke, Sherbrooke, QC, Canada and [2]Service de radio-oncologie, Centre hospitalier universitaire de Sherbrooke, Sherbrooke, QC, Canada

E-mail: mahbod.sedaghat@Usherbrooke.ca

Abstract. Discrepancies in polymer gel dosimetric measurements have been observed between containers of different sizes receiving the same radiation dose. We hypothesized that these deviations are caused by a change in the rate of polymerization due to internal heat increase in the gel containers resulting from the exothermic polymerization of monomers. Here, we test this hypothesis in a polyacrylamide gel dosimeter by recording the temperature in glass phantoms of different sizes during and after irradiation. The dose response of the samples was determined with magnetic resonance imaging. The difference of R_2 values along the depth of the containers was below ±1%. We discuss that this small difference can be attributed to variations in the rate of gelatin cooling during manufacture rather than to the measured heat increase during irradiation.

1. Introduction

Other authors and ourselves have already shown that a significant internal temperature increase can be measured during irradiation of polymer gel dosimeters due to release of energy from carbon-carbon double bond breaks (C=C) in exothermal polymerization reactions [1,2,3]. The temperature increase in the gel containers depends on three major factors: *i*) the ratio of volume to surface area (V/A) of the containers, *ii*) the volume of gel that receives radiation and *iii)* the type of monomers used. Other factors that influence the temperature increase are concentration of monomers, dose rate as well as material and thickness of the container walls.

Our hypothesis is that heat inside a gel container of interest can favour either propagation or inhibition reactions, leading to a different amount of polymer, which in turn modulates the dosimeter dose response as a function of gel volume. This becomes important not only for absolute dose determination but also relative dose evaluation since the relationship between transverse relaxation rates (R_2) and dose will consequently be modified by this effect.

Our previous attempts to pinpoint the effect of heat in normoxic gel dosimeters were not conclusive as the gel containers we had constructed were made of either Plexiglas® or Delrin® [1,4]. Oxygen can accumulate within these materials and may be returned to the gel before irradiation, hence interfering with the potential temperature effect on the dose response. We have reported elsewhere in these proceedings that traces of oxygen can increase the sensitivity of normoxic polymer gel dosimeters, thus creating ambiguity in the interpretation of the results. We re-investigated the possible effect of temperature increase in an anoxic PAG dosimeter.

2. Materials and methods

Four cylindrical gel containers were constructed from borosilicate glass. The containers were 20 cm long with internal diameters of 5.65, 3.65, 2.35 and 1 cm as shown in figure 1(a). The thickness of the glass was 4 mm. In order to insert temperature probes into the gel, four entries on the side of the cylinders were designed by sealing tiny glass tubes at different heights from the bottom of the cylinders. Eight MRI-compatible optical temperature sensors (NeOptix Inc, Quebec, Canada, accuracy ±1°C, resolution 0.1°C, length 5 m) were used to monitor temperature variations inside the containers

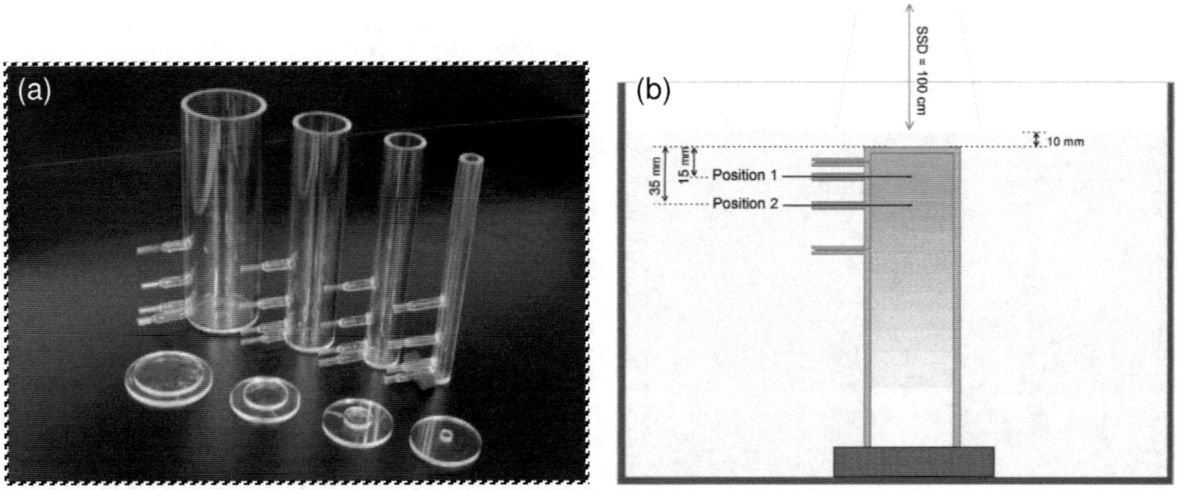

Figure 1. Four glass gel containers were specifically constructed with side entries for the temperature probes as shown in (a). To fill the containers plastic tubes were attached to all side entries and temperature probes were inserted to the gel through the tubes. In (b), irradiation set up in the water tank for one of the containers is sketched.

during irradiation. Before gel fabrication, two temperature probes were inserted into the entries at 15 mm and 35 mm from the bottom of each container. The probes were passed through plastic holder pipes which were fixed to the side entries. The containers were transferred into a nitrogen-flushed glove box and filled with a PAG dosimeter prepared with 5% gelatin (type A, 300 bloom, Sigma-Aldrich, Ontario, Canada), 3% acrylamide (AA) (electrophoresis grade, Sigma-Aldrich, Ontario, Canada) and 3% N,N'-methylene-bis-acrylamide (BIS) (99+%, Acros Organics BVBA, Geel, Belgium). A set of glass calibration vials (2.5 ml) was also filled with the gel. The containers were left in the glove box overnight for the gel to set in an anoxic environment. All gel samples were transferred to the radiotherapy department the day after gel fabrication and placed into a water tank. A second water tank was also filled with water and both were left to equilibrate to the ambient temperature for 12 h. The gel containers and calibration vials were irradiated with a Siemens Oncor™ linear accelerator (Siemens Medical Solutions AG, Germany). Irradiation set up for the gel containers is depicted in figure 1(b). The uniformity of the depth dose profiles in all four glass phantoms had been verified to be within 1% by Gafchromic® EBT2 (International Specialty Products, NJ, USA) film dosimetry. Doses of 10 and 15 Gy to D_{max} in water were delivered using 6MV photons with a dose rate of 3 Gy/min and a 10×10 cm^2 field. The SSD was set to 100 cm on the surface of water. Two probes at two different depths from the surface of water recorded the temperature changes in each container. The probe in position 1 was 15 mm lower than the point of maximum dose (D_{max}) and the probe in position 2 was 35 mm lower. This corresponds to a dose of 96% and 87.4% of D_{max} at the position of each probe, respectively.

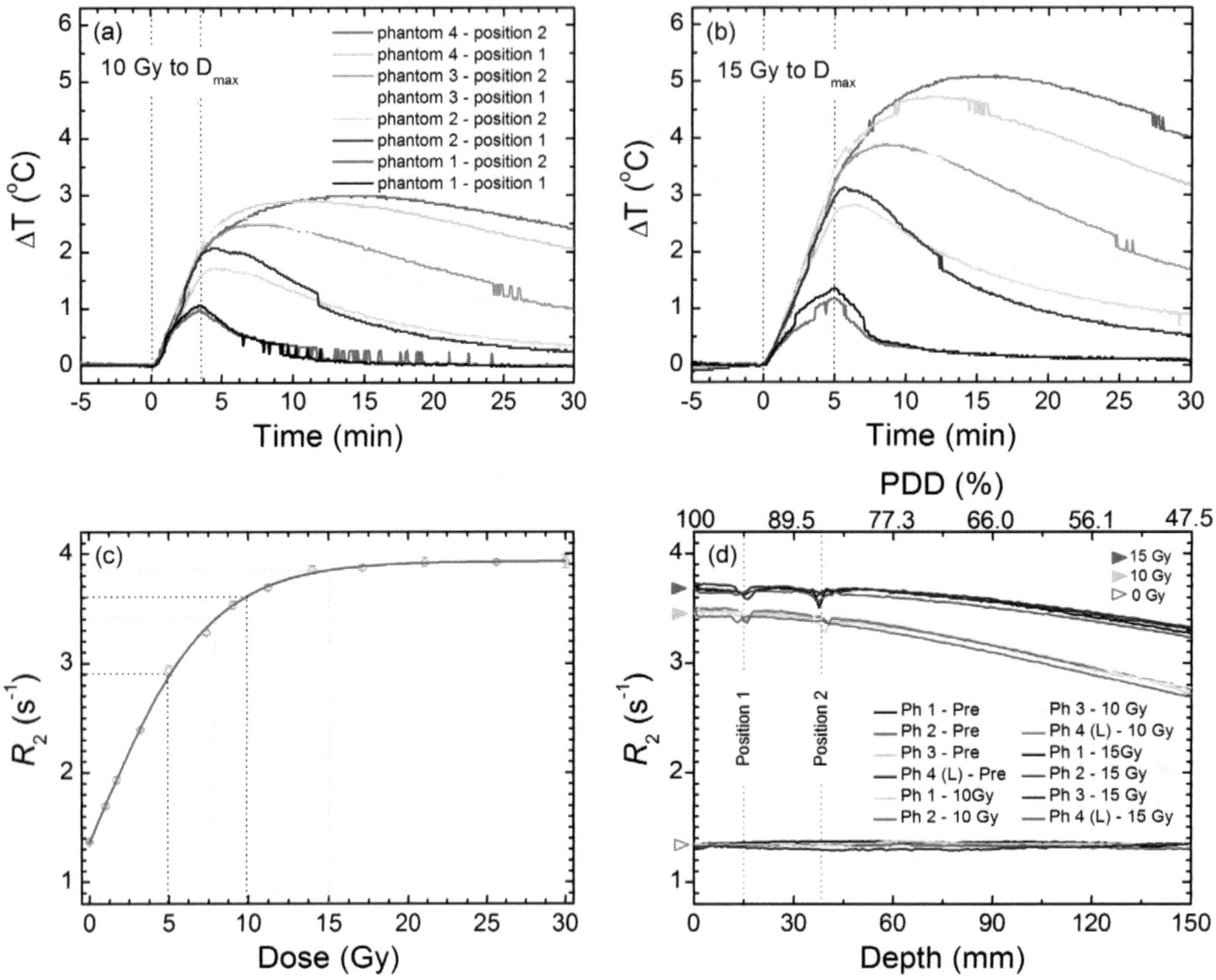

Figure 2. Temperature increase in the four glass containers and corresponding R_2 values. Irradiation to 10 Gy (a) or 15 Gy (b) started at time zero and ended at the second vertical dotted line. The R_2-dose response curve from calibration vials is shown in (c). Panel (d) illustrates profiles of R_2 values along the depth of the containers before irradiation and after 10 and 15 Gy. Positions of the probes are shown by vertical dotted lines. The upper axis provides percentage depth dose (PDD) data of the Siemens linac in water.

Temperature measurements started 30 min before irradiation and continued for 1 h after irradiation. Each gel container was transferred from the first water tank to the second water tank under the radiation field and was transferred back to the first water tank 20 min after the end of irradiation. The gel containers were scanned 24 h after irradiation using a Siemens Sonata (Siemens Medical Solutions Erlangen, Germany) 1.5 T clinical imager with a multi-echo spin echo protocol. The temperature probes were not removed to avoid distortions to the surrounding gel. Phantoms were scanned before irradiation to determine that the $R_2(0)$ values did not depend on phantom size. Data were analyzed with software developed for radiotherapy gel dosimetry in the MATLAB® environment.

3. Results

A summary of the results is presented in figure 2. Figure 2(a) shows internal temperature changes in the four gel containers for a dose of 10 Gy prescribed to D_{max} in water. Start and end of irradiation is illustrated by two dotted vertical lines. The temperature increased very rapidly after the start of irradiation. The time to reach a maximum depended on the container volume and the position of the probe inside the container. The temperature increased by ~ 1.0°C, 2.0°C, 2.6°C and 3.0°C for increasing container volume. The difference between the temperature in the smallest and largest container is 2°C for this radiation dose.

Figure 2(b) shows the internal temperature increase in the four gel containers when a dose of 15 Gy was delivered. In the smallest gel container the temperature increased by 1.3°C during the course of irradiation. As the gel container volume increased, the maximum temperature rise was of 3.1, 4.0 and 5.0°C. The internal temperature difference between the smallest and largest gel container in this case was 3.7°C. While the temperature inside the largest container is still increasing after 10 min, the smallest container has almost reached equilibrium with the ambient temperature after this time.

Figure 2(c) shows the R_2-dose response of the PAG dosimeter as determined from small calibration vials. Figure 2(d) shows the profiles of R_2 along the depth of the gel containers scanned before irradiation and the gel dosimeters irradiated to 10 and 15 Gy. The upper axis in figure 2(d) shows the percentage depth dose (PDD) of the linac in water. The PDD of the Siemens linac in water decreases by 52.5% at a depth of 15 cm from the surface. As our glass phantoms were 20 cm long, the gel at this depth was well preserved from the diffused oxygen from the caps of the containers. We discarded the data from the last 5 cm gel near the caps due to oxygen inhibition effects. The position of the probes in the gel containers is also shown by vertical doted lines. R_2 profiles of the gel dosimeters before irradiation are uniform at about 1.30 s^{-1}. R_2 profiles of the gel dosimeters irradiated to10 Gy start at 3.46 s^{-1} and end at 2.73 s^{-1}. For the dosimeters irradiated to 15 Gy, the profiles start from 3.69 s^{-1} and end at 3.28 s^{-1}. Profiles of each group have a similar shape and slope but the profile of the largest gel container is lower than the other 3 gel containers by roughly 1.5%. Comparing R_2 values in figure 2(c) and 2(d) reveals that figure 2(c) systematically shows higher R_2 of about 3% for all doses.

4. Discussion

The highest temperature was recorded in the largest container in position 2, which was 35 mm below the point of maximum dose. This suggests that the maximum temperature is not at the same position as D_{max}, suggesting that heat is rapidly lost at the extremity of the phantoms and/or that heat is flowing from D_{max} to larger depths. If the values of R_2 were directly correlated with the temperature rise, then we would expect a distortion of the profiles as a function of the container size, since the location of the maximum temperature rise depends on the container size. Such a trend is not observed in our results. It is important to note that the R_2 values in the largest container are actually *lower* by roughly 1.5%. This is consistent with the R_2 values of the small calibration vials that are higher by about 3% compared to the four larger containers. We recall that both the calibration vials and the gel dosimeters were irradiated with the same photon beams and identical setup in one session. Our results suggest that the R_2 values are most influenced by differences in the cooling rate of the gel in the phantoms and the calibration vials during and after manufacture.

To our knowledge, the effect of cooling history has not been studied for anoxic gel dosimeters. For a normoxic PAG dosimeter the R_2 values were about 7% higher in samples cooled at room temperature compared to those cooled in a refrigerator [5]. Although the study in [5] shows deviations between the dosimeters that have been subjected to different temperature history during storage, a consistent trend was not established between cooling rate and the R_2-dose response. In our experiments with an anoxic PAG dosimeter, faster cooling (*i.e.*, for smaller samples) always produces higher R_2 values of up to about 3%. This explains the deviation observed in figure 2(d) between the four gel containers and between small calibration vials in figure 2(c) and phantoms in figure 2(d).

For the anoxic PAG dosimeter, a temperature increase of up to 12°C has been reported for a volume of 160 ml of gel receiving a high dose of 52 Gy [3]. However, the changes of R_2 values at such a high dose would be below the uncertainty of the R_2 estimation, such that any temperature-induced variations could not be observed. The situation is different for normoxic PAG dosimeters and methacrylic acid-based dosimeters. The R_2 values of normoxic PAG dosimeters prepared with 10 mM of tetrakis(hydroxymethl) phosphonium chloride (THPC) saturate at about 30 Gy such that

Journal of Physics: Conference Series **250** (2010) 012018 doi:10.1088/1742-6596/250/1/012018

temperature effects, if any, should be observed in the dose range below saturation. In separate experiments, we have observed that PAGAT dosimeters prepared in the same conditions as PAG dosimeters cooled slower, suggesting that heat conductivity of PAGAT dosimeter is lower. This may contribute to a detectable temperature induced effect. On the other hand, methacylic acid-based dosimeters produce much higher heat during polymerization [1]. We have already measured temperature increases of up to 15°C in PAGAT and MAGAT dosimeters. Whether these internal temperature variations affect the final dose determination in PAGAT and MAGAT dosimeters is still under study.

5. Conclusion

The effect of internal temperature increase was studied in anoxic PAG dosimeter. For the range of doses between 10 – 15 Gy, no variations in the R_2 values could be attributed to the temperature increase resulting from the exothermal polymerization reaction in the gel dosimeter. An average discrepancy of 1.5% between the R_2 values of the four gel containers and discrepancies of up to 3% between the calibration vials and the dosimeter gel were instead attributed to differences in cooling history of the gel after filling the containers. Faster cooling always results in slightly higher R_2 values in anoxic PAG dosimeters. This may lead to dose inaccuracies of about 5%.

References

[1] Sedaghat M, Hubert-Tremblay V, Tremblay L, Bujold R and Lepage M 2009 Volume-dependent internal temperature increase within polymer gel dosimeters during irradiation *J. Phys. Conf. Ser.* **164** 012009

[2] Hayashi S, Yoshioka M, Usui S, Haneda K and Tominaga T 2009 The role of gelatin in a methacrylic acid based gel dosimeter *J. Phys. Conf. Ser.* **164** 012019

[3] Salomons G J, Park Y S, McAuley K B and Schreiner L J 2002 Temperature increases associated with polymerization of irradiated PAG dosimeters. *Phys. Med. Biol.* **47**, 1435–1448

[4] Dumas E-M., Leclerc G and Lepage M 2006 Effect of container size on the accuracy of polymer gel dosimetry. *J. Phys. Conf. Ser.* **56** 239

[5] De Deene Y, Pittomvils G and Visalatchi S 2007 The influence of cooling rate on the accuracy of normoxic polymer gel dosimeters *Phys. Med. Biol.* **52** 2719–2728

Journal of Physics: Conference Series **250** (2010) 012019 doi:10.1088/1742-6596/250/1/012019

An investigation of the pre-irradiation temporal stability of PAGAT gel dosimeter

S. Khoei, J. B. Moorrees, C. M. Langton, J. V. Trapp

Queensland University of Technology (QUT), Brisbane, Australia

s.khoei@qut.edu.au

Abstract. In this work we examine two aspects of the PAGAT gel dosimeter. The first aspect studied is determination of a stable range of concentrations of the anti-oxidant Tetrakis Hydroxy Phosphonium Chloride (THPC). Once the desired THPC concentration is determined, we proceed to an investigation into the effect of pre-irradiation storage time and how this affects the dose response of the gel.

1. Introduction

In earlier work, it was shown that the optimal concentration of Tetrakis Hydroxy Phosphonium Chloride (THPC) in PAGAT gels is approximately 5 mM [1]. Due to this relatively small concentration of THPC, inadvertent imprecision can occur in measurements during the manufacturing process and may cause a significant alteration in the sensitivity of the gel. For example, Venning *et al* showed that below 5 mM the dose response reduces rapidly with THPC concentration [1]; if the measurement of THPC was slightly below the intended amount it could lead to failure of the gel to respond sufficiently to radiation. We investigate concentrations above 5 mM to determine if there is a stable region where slight imprecision in measurements will not lead to a failure of the gel (albeit with a slight sacrifice of sensitivity). Once a stable concentration of THPC is determined, we investigate whether the PAGAT gel can be stored up to several weeks prior to irradiation and retain its usefulness as a dosimeter.

2. Method

2.1. THPC concentration

To investigate the effect of THPC concentration on the dose response of PAGAT gel dosimeters, gels were manufactured based on Venning *et al* [1] formulation with varying amounts of THPC. The 'many vials' method of irradiation has been shown to be of sufficient accuracy for the purposes of this measurement [2, 3], and hence the gels were poured into glass vials and refrigerated at 4° C for 24 hours priors to irradiation. Each gel was irradiated to various doses utilizing a Gammacell 220 Co-60 source with dose rate of 1759 Gy/hour. The irradiated gels were then refrigerated for 24 hours before imaging. Readout of the gels was performed using a MGS Research IQScan Optical CT Scanner. All acquisitions performed 360 projections over 180 degrees, A/D rate of 65000, and a field of view of 110 mm.

Journal of Physics: Conference Series **250** (2010) 012019 doi:10.1088/1742-6596/250/1/012019

2.2. Temporal stability measurements

A bulk batch (1 lit) of PAGAT gel dosimeter was manufactured according to method described by Venning *et al* [1], however THPC was not included. The bulk gel was then refrigerated in a large Pyrex container. At various times 250 ml of the bulk gel was extracted by placing the Pyrex container in a warm water bath until the gel melted. The extracted gel was then mixed with 8mM of THPC and after 24 hours the samples were irradiated using a Gammacell200 (Co-60) source. Imaging was performed for all samples in an MGS Research IQScan Optical CT Scanner at various post-irradiation times as shown in Table 1.

Batch	Time between gel manufacture and THPC addition	Time between irradiation and scanning
1	24 Hours	24 Hours 49 Days
2	28 Days	24 Hours

Table 1: Time between gel manufacture and THPC addition

3. Results and Discussion

3.1. THPC concentration

Figure 1 illustrates the dose response for PAGAT gels manufactured at different THPC concentration. Results show that at 4.46 mM the dose response is much less than for greater concentrations, which is consistent with the results of Venning *et al*. Analysis shows that the greatest response for the concentrations examined occurs at 5.10 mM, and with greater concentrations the expected decrease in sensitivity occurs. Importantly however, although there is a slight decrease in sensitivity when exceeding the optimal 5 mM concentration, this decrease is relatively small in comparison with the decrease shown when concentrations are below 5 mM.

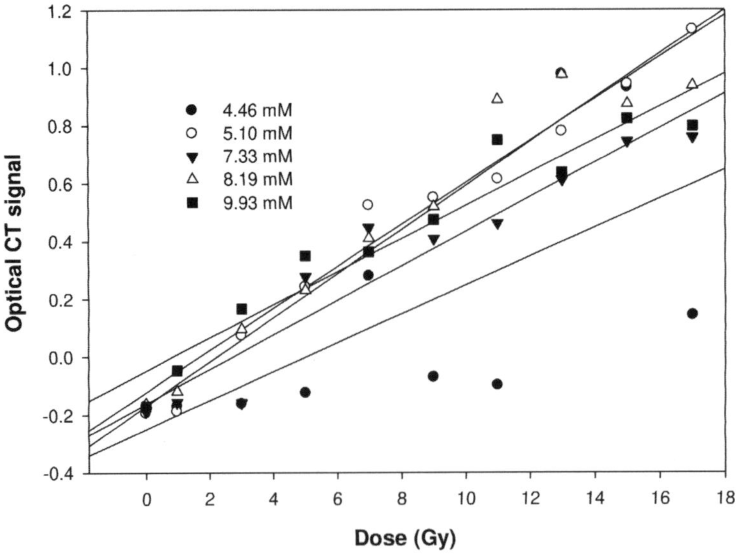

Figure 1: Optical CT scanner signal measured for all THPC concentrations across a variety of doses

3.2. Temporal stability

Figure 2 shows the dose response of Batches 1 and 2, when imaged 24 hours post-irradiation. Lines of best fit are shown to guide the eye. It can be seen that the sensitivity of the gel is not significantly affected by the time between manufacture and irradiation (for these gels where THPC is added 24 hours prior to irradiation), although there is an overall decrease in the optical CT response across all vials.

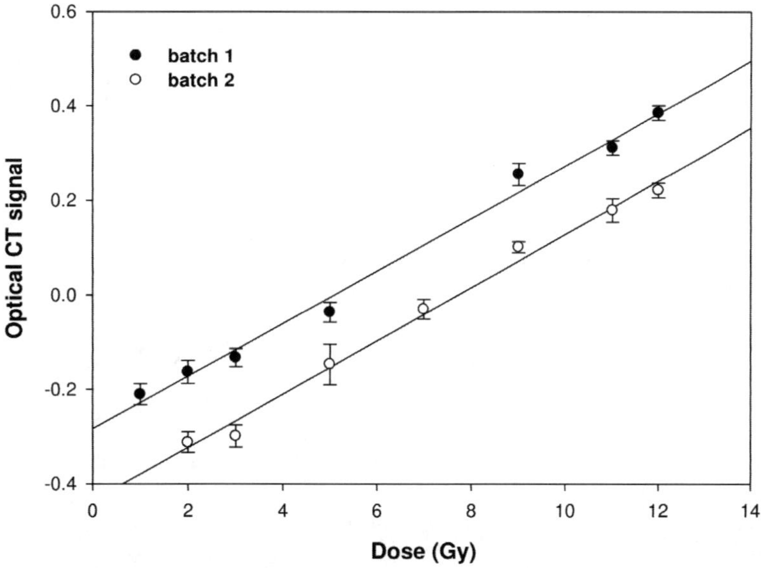

Figure 2: Dose response for PAGAT gel irradiated 1 day and 28 day post manufacture

Figure 3 shows the results for Batch 1 when scanned 1 day and 7 weeks post irradiation. The dose response indicates that there is little change in the sensitivity of the gel over time post-irradiation, however there is an overall decrease in optical CT dose response.

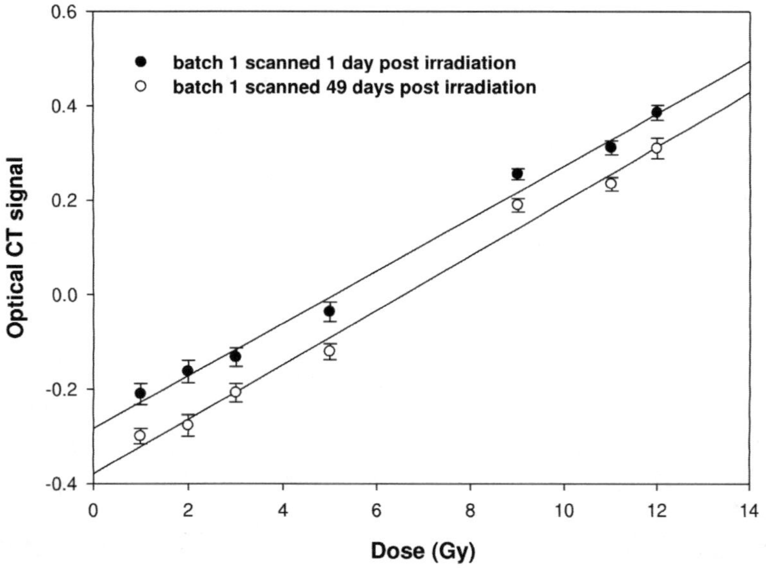

Figure 3: Dose response for Batch 1 scanned 1 day and also 7 weeks after irradiation

4. Conclusion

The concentration of THPC in PAGAT gels indicate a slight decrease in the sensitivity of the PAGAT gel when increased above the recommended 5 mM level, however a much greater decrease occurs when decreasing the concentration below 5 mM, even slightly. Therefore, although 5 mM THPC concentration produces the greatest sensitivity, users can choose to sacrifice some of this sensitivity to avoid possible failure of the gel due to measurement inaccuracies during the manufacturing process. As a result, this group now manufactures PAGAT gels with 7-8 mM of THPC.

Dose responses at different time of addition THPC to pre manufactured gel exhibits that gels can be stored without THPC in the mixture and THPC can be added 24 hours prior to irradiation. The advantage of this is that a large batch of PAGAT can be manufactured and stored for greater time efficiency, and users can simply extract the required amount of gel and add THPC 24 hours prior to use.

References

[1] A. J. Venning, B. Hill, S. Brindha, B. J. Healy, C. Baldock, "Investigation of the PAGAT polymer gel dosimeter using magnetic resonance imaging," Physics in Medicine and Biology, vol. 50, pp. 3875-3888, 2005.

[2] M. L. Taylor, R. D. Franich, P. N. Johnston, R. M. Millar, J. V. Trapp, "Systematic variations in polymer gel dosimeter calibration due to container influence and deviations from water equivalence," Physics in Medicine and Biology, vol. 52, pp. 3991-4005, 2007.

[3] M. L. Taylor, R. D. Franich, J. V. Trapp, Johnston, P. N. Johnston, "A Comparative Study on the Effect of Calibration Conditions on the Water Equivalence of a Range of Gel Dosimeters," Nuclear Science, vol. 56, 2009.

Journal of Physics: Conference Series **250** (2010) 012020 doi:10.1088/1742-6596/250/1/012020

2D AND 3D dose verification at The Netherlands Cancer Institute-Antoni van Leeuwenhoek Hospital using EPIDs

Ben Mijnheer, Anton Mans, Igor Olaciregui-Ruiz, Jan-Jakob Sonke, ReneTielenburg, Marcel van Herk, Ron Vijlbrief, Joep Stroom

Department of Radiation Oncology, The Netherlands Cancer Institute - Antoni van Leeuwenhoek Hospital, Plesmanlaan 121, 1066 CX Amsterdam, The Netherlands

b.mijnheer@nki.nl

Abstract. A review is given of the clinical use of EPID dosimetry in the Department of Radiation Oncology of The Netherlands Cancer Institute-Antoni van Leeuwenhoek Hospital. All curative plans (almost all IMRT or VMAT) are verified with EPID dosimetry, mostly *in vivo*. The 2D approach for IMRT verification and the 3D method for VMAT verification are elucidated and their clinical implementation described. It has been shown that EPID dosimetry plays an important role in the total chain of verification procedures that are implemented in our department. It provides a safety net for advanced treatments such as IMRT and VMAT, as well as a full account of the dose delivered.

1. Introduction

In January 2005, the demand for an efficient and accurate dose verification increased in our department with the use of a new treatment planning system (Pinnacle V8.0h, Philips Medical Systems, Eindhoven, The Netherlands), the large-scale introduction of IMRT on our 9 Elekta SL20i linear accelerators (Elekta, Crawley, UK), higher dose prescriptions (*e.g.,*78 Gy for IMRT prostate treatments) and the gradual transfer of data management systems from paper to digital format. Since January 2008, we verify all curative treatment plans (almost all IMRT) with EPID dosimetry, mostly *in vivo; i.e.,* the dose distribution is reconstructed within the patient from images acquired for each field during treatment. In August 2009 volumetric-modulated arc therapy (VMAT) has been introduced in our department for prostate treatments and later that year for lung stereotactic body radiotherapy. Also all VMAT treatments are verified by means of EPID dosimetry.

Measurements made with aSi EPIDs (PerkinElmer RID 1680 AL5/Elekta iView GT) are converted to dose by a back-projection algorithm. For IMRT verification, dose distributions are verified in 2D per field [1], while for VMAT verification, or if further analysis of an IMRT check is necessary, 3D verification is performed by summing back-projected images to a volume [2, 3].

Our philosophy of patient-specific QA is to use fast or automatic pre-treatment checks to prevent large errors, and in addition to have an independent end-to-end check of planned versus delivered dose distribution in the patient. Such an end-to-end test should not be limited to testing individual steps in the treatment process, but rather evaluate the complete process from image-based treatment design through the dose calculation to measured dose delivery, and should preferably be performed by the same persons who treat patients. This philosophy is translated into three-fold patient-specific QA. First, an independent monitor unit check is performed. Second, a consistency check of the basic plan properties is done by the therapists at the treatment machine of the data in the treatment planning system and the record-and-verify system. Third, a 2D verification of individual IMRT beams, or a 3D verification of the total dose distribution of a VMAT plan, is executed by EPID dosimetry, if possible *in vivo*, and otherwise pre-treatment. In our opinion, such a three-stage verification process is adequate and efficient for safe IMRT and VMAT delivery.

2. 2D IMRT verification

All our static beam IMRT treatments are verified with EPID dosimetry using a back-projection model [1]. The algorithm accounts for the inverse-square law, attenuation of the beam between the exit plane and the reconstruction plane (estimated from CT data), and three types of scattered radiation (within the EPID, from the patient to the EPID, and within the patient). The algorithm also takes couch attenuation and changes in EPID sensitivity with time into account. The planned and reconstructed dose distributions are compared for each beam in a plane through the isocentre perpendicular to the beam direction. The dose at the isocentre is checked for each beam, as well as the total dose per fraction. In addition a 2D gamma analysis is performed for each beam. The results of this analysis are given as a "dosimetry report" for each patient (see figures 1-3). Phantom measurements showed that for prostate, rectum, and head-and-neck cases, the percentage of points within γ-agreement are 99.9%, 98.1%, and 96.6%, respectively. At the dose prescription point, the dose values based on EPID measurements and the TPS agree generally within 1% [1, 3].

2.1. Clinical implementation

Within a few weeks of clinical use, EPID dosimetry replaced film measurements in a phantom and within 8 months *in vivo* EPID dosimetry replaced pre-treatment verification for all prostate IMRT checks. Currently EPID dosimetry is the only routine IMRT plan verification performed in our hospital. *In vivo* EPID dosimetry was performed for the following sites and number of patients in 2009: breast (779), lung (279), head-and-neck (241), prostate (233), rectum (206), and others, *e.g.,* skull and liver, (285). Pre-treatment EPID dosimetry is performed for single fraction brain treatments (20) and plans with fields larger than the detector area (160). Several patients had multiple plans (*e.g.,* a boost plan) that were checked. The total number of plans verified in 2009 was 2835.

One fte therapist (actually 2 part-time people) analyses these IMRT plans. Physicists are responsible for the initial calibration of each EPID for the various photon beams on all our accelerators, and are available to resolve problems that arise. For each field, a treatment image and an "open image" (no patient in the beam, acquired outside treatment time) is required to determine the dose inside the patient. These "open images" are measured by the therapists, which takes about 5 to 10 minutes per plan. When images have been acquired, the total analysis time per plan is about 10 min required for 3 fractions *in vivo* compared to about one hour for pre-treatment verification. In the near future we will replace the measured transmission data by calculated values, thus saving about 400 hours/year spent on measurement of "open images".

An alert is raised to potential clinical errors when criteria based on the γ-index (3.0% of the maximum dose per field and 3.0 mm distance) are surpassed. The mean γ-value (γ_{mean}), the maximum 1% γ value ($\gamma_{max1\%}$) and the percentage of points with γ>1 ($P_{\gamma>1}$) are calculated within the 20% isodose line of each field. The alert criteria depend on the tumour site and are for prostate cancer verification:

- $\gamma_{mean} > 0.5$, i.e. when average differences exceed ± 1.5% or 1.5 mm
- $\gamma_{max1\%} > 2.0$, i.e. when maximum differences exceed ± 6.0% or 6.0 mm
- $P_{\gamma>1} > 15\%$, i.e. when more than 15% of points exceed 3.0% or 3.0 mm
- $\Delta D_{isoc} > 3.0\%$, i.e. when the total dose at the isocentre deviates by more than 3.0% from the prescribed dose

2.2. Errors prevented

Since the clinical introduction of EPID dosimetry in January 2005 to July 2009 treatment plans of 4337 patients were verified. In 17 cases, major deviations detected gave rise to an intervention. In seven cases patient anatomy changes resulted in substantial dosimetric deviations due to weight loss (2), (recovery from) atelectasis (2), patient contour change (2), and emptying of a post-operative cavity that was filled during the planning CT (see figure 1). In four cases, the plan was flawed during transfer from the TPS to the treatment machine (one human error and three system communication errors). In two instances the fine-tuning of parameters in the TPS were found to be sub-optimal. Two plans were accidentally modified in the R&V system just prior to delivery of the first fraction. In one case, planning protocols were not obeyed, resulting in an undeliverable plan. Finally, the linear accelerator failed to deliver one of the segments of a step-and-shoot IMRT beam. A detailed analysis of the errors detected by means of our EPID dosimetry programme has been published recently [4]. In addition to these major deviations, a number of minor deviations resulted in re-planning of the patient.

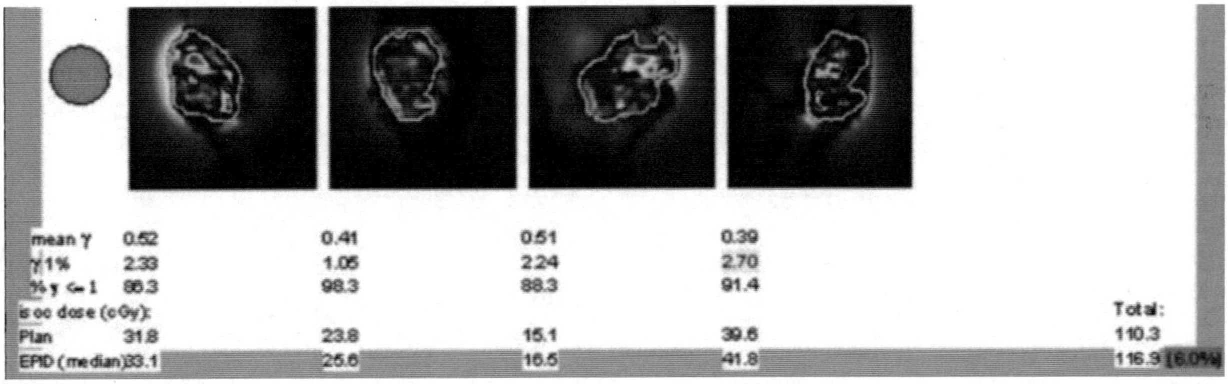

Figure-1. Results of the *in vivo* EPID dosimetry verification of an IMRT treatment of a head-and-neck cancer patient. The deviation between the planned and delivered dose at the isocentre is outside the action criterion. The red spot in the gamma images indicates that the post-operative cavity, which was filled up with fluid when making the planning CT scan, was removed during treatment. As a result of these measurements a new treatment plan was created for this patient.

3. 3D VMAT verification

For VMAT verification it was necessary to modify our software to incorporate gantry-angle resolved image acquisition. Furthermore, the 3D back-projection model applied for IMRT verification [3] was adapted to include a calculated transmission from CT data instead of measured transmission data, *e.g.*, no "open arc images" are needed for VMAT verification. The total 3D dose distribution is reconstructed and compared with the planned dose distribution using a 3D gamma evaluation, with 3% of the maximum planned dose and 3 mm as dose difference and distance-to-agreement criteria, respectively. Details of the adaptation of our back-projection portal dosimetry method for accurate 3D dose verification of VMAT can be found elsewhere [5].

3.1. Clinical implementation

The results of the 3D verification of each arc of a clinical VMAT plan are also presented as an EPID dosimetry report, in the same way as for 2D IMRT verification. Figures 2 and 3 show, as an example, dosimetry reports for a pre-treatment and an *in vivo* verification of the delivery of a single arc prostate cancer VMAT treatment. The pictures illustrate the results of the gamma evaluation in a sagittal, axial and coronal plane through the isocentre, while the data for the various gamma evaluation parameters within the 50% isodose surface, as well as the total dose at the isocentre, are also given. If the total dose is deviating more than 3.0% a warning is given (tolerance level exceeded), while for a deviation larger than 5% an error report is sent to a physicist (action level exceeded). Gamma evaluation statistics are reported with tolerance/action levels: 0.50/1.00 for the mean gamma value, 2.0/4.0 for the maximum gamma value, and 85%/70% for the percentage of gamma values below one, respectively.

The average difference of the dose at the isocentre was for the first 10 prostate cancer VMAT plan verifications (-0.2±1.6) % and (+0.4±1.6) %, for the pre-treatment and *in vivo* results, respectively. A preliminary analysis of the parameters applied for the 3D gamma evaluation showed almost the same results for pre-treatment and *in vivo* verification of the 3D dose distribution for prostate cancer VMAT (see figures 2 and 3). For instance, the average of the mean gamma values of these 10 patients is 0.42±0.12 and 0.48±0.06 for the pre-treatment and *in vivo* results, respectively.

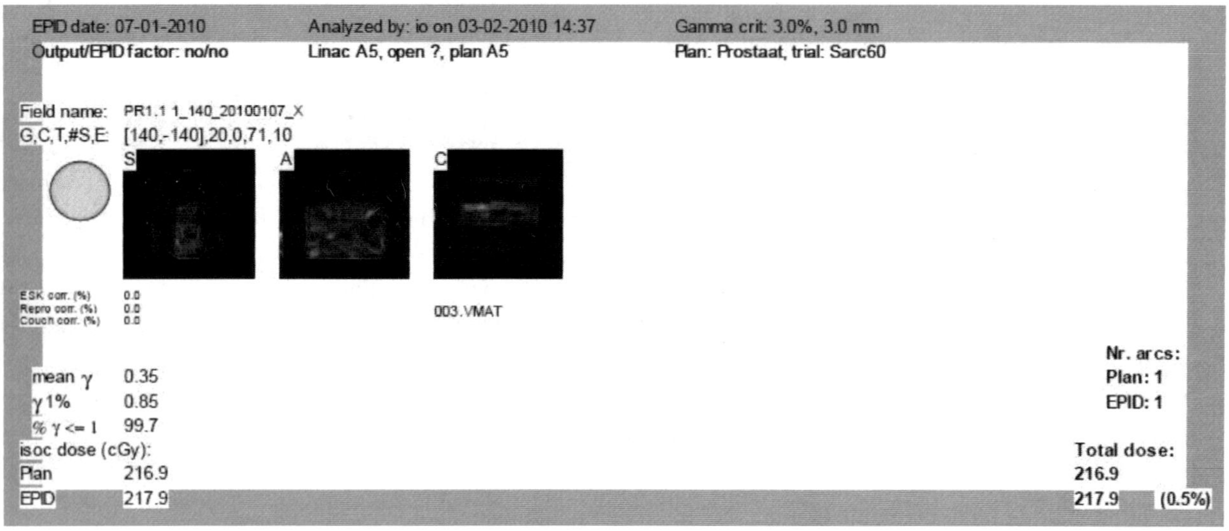

Figure-2. EPID dosimetry report of a 3D pre-treatment verification of a VMAT treatment of a prostate cancer patient.

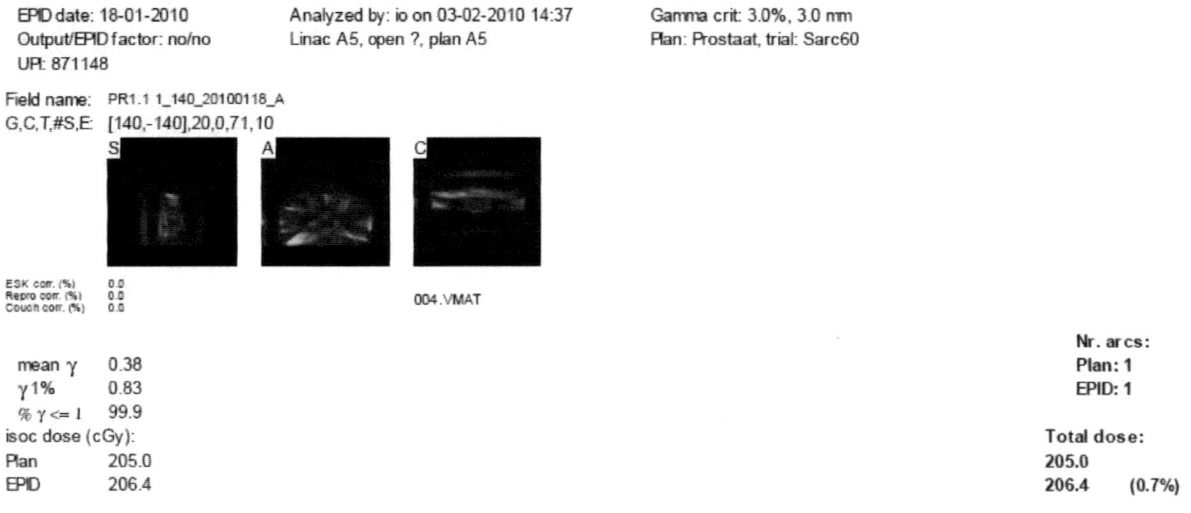

Figure-3. EPID dosimetry report of a 3D *in vivo* verification of a VMAT treatment of the same prostate cancer patient.

3.2. 2D versus 3D comparison

Our current γ-evaluation for IMRT applies a 2D comparison of measured and planned dose distributions in a plane through the isocentre perpendicular to each IMRT beam. The criteria for IMRT verification are based on the γ-index as well as the dose at the isocentre and differ for the various treatment sites. At present we are using the same criteria for the 3D dose comparison of VMAT checks. Because the 3D dose distribution in the target volume is generally intended to be uniform, the dose difference criterion will be dominant. However, dose deviations in a single beam or segment will be diluted in the total dose distribution and serious errors in the dose calculation or in the position of the leaves may no longer be detected. The distance-to-agreement criterion will mainly contribute to the γ-value in the dose gradient region at the border of target volumes. With the implementation of IMRT and VMAT combined with position verification techniques, increased dose conformality is achieved and the distance-to-agreement criterion will gain importance. Therefore, we are currently considering stricter γ-criteria for 3D dose verification in the clinic, such as 2%, 2 mm or 2%, 1.5 mm, also since identical criteria always yield smaller γ-values in 3D dose verification than in 2D [6].

More clinical data are needed to assess the values applied for our gamma evaluation criteria for 3D VMAT verification. Currently, our clinic does not have strict criteria for plan adaptation when EPID dosimetry data (or IGRT procedures) show deviations exceeding action levels. In such cases, the physicist informs the responsible physician, who decide together if plan adaptation is necessary.

4. Discussion and conclusions

The use of EPIDs for dosimetry purposes has matured and is now a reliable and accurate dose verification method that can be used in a large number of situations [7]. From our clinical results it can be concluded that EPID dosimetry is a valuable tool for 2D verification of IMRT and 3D verification of VMAT delivery, both pre-treatment and *in vivo*.

The drawback of applying *in vivo* verification only is that a small part of the treatment has already been given before an error is detected. The advantages, on the other hand, are verification of the actual patient treatment and (for our method) workload reduction. Certain types of errors can also easily be detected by pre-treatment methods. However, these methods are not capable of detecting a number of other errors due to, for instance, considerable variation in patient anatomy or patient position, and obstructions from table arms or immobilization devices. Furthermore, treatment parameters can be (accidentally) modified between pre-treatment verification and the first treatment, as we have detected several times. For these reasons, the choice was made in our department to use *in vivo* EPID dosimetry for all treatments with curative intent. When making this decision, verification of the entire radiotherapy chain, including the actual patient treatment was a strong argument.

The additional workload involved with EPID dosimetry is limited and little or no extra time is needed at the accelerator. Analysis of the results of EPID dosimetry during IMRT made it possible to identify small systematic errors related to the dose calculation in the TPS. Methods to integrate 3D *in vivo* dosimetry and IGRT procedures, such as the use of kV cone-beam CT, are under development [2, 7]. The acquisition and analysis of EPID dosimetry data can be performed by therapists if simple measurement and pass/fail tools are available. Compared to pre-treatment verification, *in vivo* EPID dosimetry is not only capable to detect errors but can also assess quantitatively the dosimetric impact of these errors and the need for plan adaptation (see figure 1). In combination with delineated structures from the plan, dose-area or dose-volume information can be obtained for the target volume as well as for organs at risk, based on 3D EPID-based reconstructed dose distributions (see figure 4).

Journal of Physics: Conference Series **250** (2010) 012020 doi:10.1088/1742-6596/250/1/012020

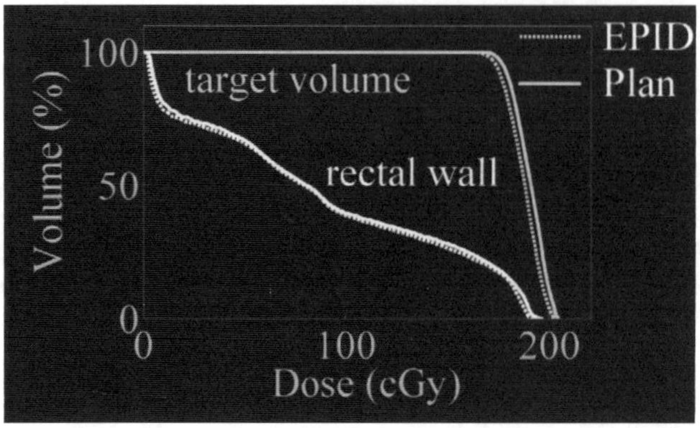

$D_{50\%}$ target volume:
EPID = 188 cGy
Plan = 191 cGy

$D_{50\%}$ rectal wall:
EPID = 78.9 cGy
Plan = 80.6 cGy

Figure-4. Dose-volume histogram of a 3D *in vivo* verification of an IMRT treatment of a prostate cancer patient.

In this way clinical judgment of deviations between planned and actual dose distributions is possible. Furthermore, in a growing number of countries *in vivo* dosimetry is required by law or strongly recommended by (inter)national organizations to avoid serious errors.

It can be concluded that EPID dosimetry plays an important role in the total chain of verification procedures that are implemented in our radiotherapy department. It provides a safety net for advanced treatments such as IMRT and VMAT, as well as a full account of the dose delivered.

References

[1] M. Wendling, R.J.W. Louwe, L.N. McDermott, J.-J. Sonke, M. van Herk, and B.J. Mijnheer, "Accurate two-dimensional IMRT verification using a back-projection EPID dosimetry method". Med Phys 33: 259-273, 2006.

[2] L.N. McDermott, M. Wendling, J. Nijkamp, A. Mans, J.-J. Sonke, B.J. Mijnheer and M. van Herk, "3D *in vivo* dose verification of entire hypo-fractionated IMRT treatments using an EPID and cone-beam CT". Radiother Oncol 86: 35-42, 2008.

[3] M. Wendling, L.N. McDermott, A. Mans, J.-J. Sonke, M. van Herk and B. Mijnheer, "A simple back-projection algorithm for 3D EPID dosimetry of IMRT treatments". Med Phys 36: 3310-3321, 2009.

[4] A. Mans, M. Wendling, L.N. McDermott, J.-J. Sonke, B. Mijnheer, M. van Herk and J.C. Stroom, "Catching errors with *in vivo* dosimetry". Med Phys 37: 2638-2644, 2010.

[5] A. Mans, P Remeijer, I. Olaciregui-Ruiz, M. Wendling, J-J Sonke, B. Mijnheer, M. van Herk and J. Stroom, "3D dosimetric verification of volumetric-modulated arc therapy by portal dosimetry". Radiother Oncol 94: 181-187, 2010.

[6] M. Wendling, L. Zijp, L.N. McDermott, E.J. Smit, J.-J. Sonke, B.J. Mijnheer, and M. van Herk, "A fast algorithm for gamma evaluation in 3D". Med Phys 34: 1647-1654, 2007.

[7] W. van Elmpt, L. McDermott, S. Nijsten, M. Wendling, P. Lambin, B. Mijnheer, "A literature review of electronic portal imaging for radiotherapy dosimetry". Radiother Oncol 88: 289-309, 2008.

Journal of Physics: Conference Series **250** (2010) 012021 doi:10.1088/1742-6596/250/1/012021

Use of Electronic Portal Image Detectors for Quality Assurance of Advanced Treatments

Jean M. Moran, Ph.D.

Department of Radiation Therapy, University of Michigan, 1500 E. Medical Center Drive, Ann Arbor MI 48109-0010

jmmoran@med.umich.edu

Abstract. As the complexity of radiation therapy has increased, the need for quantitative dosimetric evaluation of treatment delivery has also increased. A growing number of investigations have expanded the use of EPIDs from anatomic applications to dosimetric verification. This work focuses on the applications of EPIDs for pre-treatment dosimetric verification of IMRT and intensity modulated arc therapy techniques. The advantages and disadvantages of these techniques are discussed along with methods to extrapolate to 3D dose verification applications.

1. Introduction

The evolution of radiation therapy treatment planning and delivery methods has required a similar evolution for the dosimetric methods for evaluation. Delivery techniques such as intensity modulated radiation therapy (IMRT) and intensity modulated arc therapy (IMAT) require more detailed spatial information. As the need for pre-treatment patient-specific quality assurance has increased, so have investigations into the use of electronic portal imager detectors (EPIDs). EPIDs offer advantages compared to other systems in that they are attached to the gantry and are sold with the majority of new accelerator purchases. They are also efficient to use and do not require the set up of any additional equipment once the system has been commissioned for such a purpose. However, the response of EPIDs to radiation is complex due to their composition. These systems were designed and optimized to require as little delivered dose as possible to obtain a high quality image that can be used for patient localization.

EPIDs have been investigated for both pre-treatment and transit dosimetry applications. These different types of portal dosimetry methods have been comprehensively reviewed by van Elmpt et al.[1] Transit or in vivo dosimetry methods have been developed and reported on by investigators at the Netherlands Cancer Institute (see conference paper by Mijnheer et al).[1-5] These methods typically utilize the commercial configuration of the device. Additional software tools and acquisition methods have been developed with EPIDs to verify the accuracy of a delivered treatment pre-treatment without a patient and with or without a phantom present. The majority of these pre-treatment investigations have been performed utilizing commercial EPID systems in an indirect detection configuration where the system includes a scintillation-layer as part of the manufacturer supplied configuration. Other investigators have used direct detection configurations where no scintillation layer is present. This work focuses on methods developed for amorphous silicon-based systems for pre-treatment quality assurance methods.

2. EPID Dosimetry with Standard Configurations

Amorphous silicon EPIDs have been investigated for dosimetric applications for IMRT and arc therapy techniques. Reliable use of the system requires commissioning of the planning and delivery system prior to development of the EPID-based program. Then, it may be appropriate to establish a pre-treatment QA program based on calculations in the EPID compared with the EPID response. These methods typically rely on determining the appropriate parameters of operation of the EPID and accurate determination of correction factors and the accuracy of the calculation of a predicted portal dose image (PDI).[6-12] The features of the measurement are affected by the frequency of the data acquisition and communication speeds. The response itself can be affected by ghosting and lag. The response of these systems depends on the field size.

The presence of the scintillator results in an increased sensitivity to low energies. For example, the effect of the different layers of material and backscatter on the detector response as a function of energy has been quantified using Monte Carlo calculations.[13] Therefore the response of the system is dependent on the energy spectrum of the delivered beam. In some investigations, additional buildup is added to provide additional attenuation of the beam. Initial investigations demonstrated a clear dose rate effect due to saturation of the detector.[8]

2.1. Calculating the Predicted Portal Dose Image (PDI)

For pre-treatment QA, the EPID response is measured and then compared to the predicted portal dose image (PDI). This method is used to verify the accuracy of the intended fluence. The measured response is compared to a calculation in the imager rather than converting the imager response to dose in water. Initial methods used pencil beam methods to predict the portal dose image.[8] Other methods involved calculating separate fluences for the primary and scatter components.[6, 7] More recently, investigators have used Monte Carlo to create EPID-specific dose kernels. For example, one method uses Monte Carlo to incorporate details of the multileaf collimator design such as transmission, tongue-and-groove, and curved leaf ends.[9] In another study, Monte Carlo was used to generate imager-specific dose kernels to investigate different thicknesses of backscattering.[14] Other considerations included determination that the effect of backscatter due to the metallic components of the support arm can be as much as 6% for large fields.[15]

With the corrections applied, the majority of these studies found agreement of >90% of the points to be within criteria 2%/2 mm for dose and distance, respectively for IMRT and IMAT deliveries. These studies demonstrate the complexity of modeling required for robust use of EPIDs for IMRT pre-treatment QA.

2.2. Measuring the Portal Dose Image (PDI)

When operated in an integrative mode during delivery, amorphous silicon EPIDs have been found to have a reproducible response which is essential for dosimetric applications.[7, 8] Investigators have characterized the response for a range of field sizes, doses, dose rates, and source-to-detector distances. Continuous acquisition mode has also been investigated for EPID verification for IMRT and IMAT delivery methods to allow for discrimination of time-based information. In this mode, some information is lost during the readout and the system was determined to have a non-linear response to dose for low doses.[12] The system was found to be reliable for the IMRT and arc therapy fields that were evaluated. For arc and IMRT applications, there are discrepancies for deliveries of low monitor units.[8, 10, 12] These need to be further evaluated to ensure that EPID techniques are only used in validated situations.

3. Dosimetry with Direct Detection Configurations

Due to the difficulties in modeling scatter and the fact that the measured response differs substantially from dose in water, work has also been done using amorphous silicon-based systems in a direct detection configuration.[16-20] Because these systems do not have a scintillator, the overall signal is much lower. The measured signal is converted to dose more simply than with indirect detection

Journal of Physics: Conference Series **250** (2010) 012021 doi:10.1088/1742-6596/250/1/012021

systems although there the dark field must be adequately measured. The devices can be operated in continuous or integrated modes with considerations similar to commercial systems about loss of data in the continuous acquisition mode.[16, 18] The configuration results in measured results that are much more similar to those in water, even with minimal backscatter.[20]

4. 3D Applications

In vivo dosimetry methods rely on the backprojection of the measured response to a plane in the patient or phantom.[1] With cone beam CT data more readily acquired as part of a patient's treament, 3D verification of the actual delivered dose is becoming feasible. Methods for 3D dose verification have been developed for hypofractionated lung treatments[4, 21] and prostate, rectum, and head-and-neck treatments[5]. Methods are being developed to account for artifacts in the cone beam CT images.[22, 23] Monte Carlo methods have been used to reconstruct the dose in a phantom as a function of the measured exit dose for an IMRT treatment.[24] These methods promise must more information about the accuracy of the delivered dose to patients.

5. Summary

As radiation therapy continues to increase in complexity, it is crucial to have validated and easy to use methods to verify that the correct treatment plan was delivered to the patient. Information that has previously been obtained solely for adaptive therapy purposes can be used to develop more sophisticated models of the patient over the course of treatment. Methods to use commercial EPID systems for verification of IMRT and IMAT treatments have been developed and are in use in some centers. When used in the standard configuration (indirect mode), a number of correction factors must be applied to the measurement and to properly calculate the predicted portal dose image. Transit dosimetry methods have also been developed and the derived information is being coupled successfully with cone beam CT information to develop models of the 3D delivered doses. The accuracy of these methods must be validated with dosimetry systems of known performance. From a safety perspective, the use of EPID information to reconstruct 3D doses provides much more information about the quality of a patient delivery. This area is expected to mature over the next few years.

Table 1. Summary of investigations into the use of EPIDs for pre-treatment verification in a standard configuration for advanced technology treatments.

First Author/Year	Method	Conditions	Limitations
van Esch et al. 2004	Measurement: integrated acquisition mode; averaged gray scale image converted to portal dose image (PDI) Calculation: predicted PDI based on a pencil beam method	3x3 cm^2 to 25 x25 cm^2 for SDD=105 cm; SMLC/DMLC	Adversely affected by saturation of the detector at 600 mu/min dose rate
Chytyk et al. 2009	Improved calculation: Two-source fluence model with focal and extra-focal sources; more detailed modeling of jaws and MLC; Monte Carlo used to derive EPID-specific	Derived parameters with 1x1 cm^2up to 20x20 cm^2 fields Tested: Fields for 10 prostate and 10 oropharyngeal patient plans	Patient scatter not modelled and therefore not implemented for transit dosimetry

	dose kernels		
McCurdy et al. 2009	Measurement: continuous acquisition mode	IMRT and IMAT techniques	Difficulties with verification of low MU; significant discrepancies from ion chamber measurements below 50 MU
Iori et al. 2010	Measurement: intensity-modulated arc therapy; accessory mount clamped between EPID and collimator housing to keep EPID position stable during rotation Calculation: pencil beam kernels derived from measurements	23 Treatment fields delivered with both IMRT and IMAT techniques	Compared to an ion chamber 10% disagreement for ~15 MU; (3.3% discrepancy ≥100 MU)

References

[1] W. van Elmpt, L. McDermott, S. Nijsten, M. Wendling, et al., "A literature review of electronic portal imaging for radiotherapy dosimetry". Radiother Oncol. 88: 289-309, 2008.
[2] L.N. McDermott, M. Wendling, J. Nijkamp, A. Mans, et al., "3D in vivo dose verification of entire hypo-fractionated IMRT treatments using an EPID and cone-beam CT". Radiother Oncol. 86: 35-42, 2008.
[3] L.N. McDermott, M. Wendling, J.J. Sonke, M. van Herk, and B.J. Mijnheer, "Replacing pretreatment verification with in vivo EPID dosimetry for prostate IMRT". Int J Radiat Oncol Biol Phys. 67: 1568-77, 2007.
[4] W. van Elmpt, S. Nijsten, B. Mijnheer, A. Dekker, and P. Lambin, "The next step in patient-specific QA: 3D dose verification of conformal and intensity-modulated RT based on EPID dosimetry and Monte Carlo dose calculations". Radiother Oncol. 86: 86-92, 2008.
[5] M. Wendling, L.N. McDermott, A. Mans, J.J. Sonke, et al., "A simple backprojection algorithm for 3D in vivo EPID dosimetry of IMRT treatments". Med Phys. 36: 3310-21, 2009.
[6] B.M. McCurdy and S. Pistorius, "A two-step algorithm for predicting portal dose images in arbitrary detectors". Med Phys. 27: 2109-16, 2000.
[7] B.M. McCurdy, K. Luchka, and S. Pistorius, "Dosimetric investigation and portal dose image prediction using an amorphous silicon electronic portal imaging device". Med Phys. 28: 911-24, 2001.
[8] A. Van Esch, T. Depuydt, and D.P. Huyskens, "The use of an aSi-based EPID for routine absolute dosimetric pre-treatment verification of dynamic IMRT fields". Radiother Oncol. 71: 223-34, 2004.
[9] K. Chytyk and B.M. McCurdy, "Comprehensive fluence model for absolute portal dose image prediction". Med Phys. 36: 1389-98, 2009.
[10] M. Iori, E. Cagni, M. Paiusco, P. Munro, and A.E. Nahum, "Dosimetric verification of IMAT delivery with a conventional EPID system and a commercial portal dose image prediction tool". Med Phys. 37: 377-90, 2010.

[11] A. Kavuma, M. Glegg, M. Metwaly, G. Currie, and A. Elliott, "A novel method for patient exit and entrance dose prediction based on water equivalent path length measured with an amorphous silicon electronic portal imaging device". Phys Med Biol. 55: 435-52,

[12] B.M. McCurdy and P.B. Greer, "Dosimetric properties of an amorphous-silicon EPID used in continuous acquisition mode for application to dynamic and arc IMRT". Med Phys. 36: 3028-39, 2009.

[13] H. Gustafsson, P. Vial, Z. Kuncic, C. Baldock, and P.B. Greer, "EPID dosimetry: effect of different layers of materials on absorbed dose response". Med Phys. 36: 5665-74, 2009.

[14] S. Wang, J.K. Gardner, J.J. Gordon, W. Li, et al., "Monte Carlo-based adaptive EPID dose kernel accounting for different field size responses of imagers". Med Phys. 36: 3582-95, 2009.

[15] P. Rowshanfarzad and B.M. McCurdy, "Measurement and modeling of the effect of support arm backscatter on dosimetry with a Varian EPID". Med Phys. 37: 2269-2278, 2010.

[16] Y. Chen, J.M. Moran, D.A. Roberts, Y. El-Mohri, et al., "Performance of a direct-detection active matrix flat panel dosimeter (AMFPD) for IMRT measurements". Med Phys. 34: 4911-22, 2007.

[17] Y. El-Mohri, L.E. Antonuk, J. Yorkston, K.W. Jee, et al., "Relative dosimetry using active matrix flat-panel imager (AMFPI) technology". Med Phys. 26: 1530-41, 1999.

[18] J.M. Moran, D.A. Roberts, T.S. Nurushev, L.E. Antonuk, et al., "An Active Matrix Flat Panel Dosimeter (AMFPD) for in-phantom dosimetric measurements". Med Phys. 32: 466-72, 2005.

[19] P. Vial, P.B. Greer, L. Oliver, and C. Baldock, "Initial evaluation of a commercial EPID modified to a novel direct-detection configuration for radiotherapy dosimetry". Med Phys. 35: 4362-74, 2008.

[20] P. Vial, H. Gustafsson, L. Oliver, C. Baldock, and P.B. Greer, "Direct-detection EPID dosimetry: investigation of a potential clinical configuration for IMRT verification". Phys Med Biol. 54: 7151-69, 2009.

[21] W. van Elmpt, S. Petit, D. De Ruysscher, P. Lambin, and A. Dekker, "3D dose delivery verification using repeated cone-beam imaging and EPID dosimetry for stereotactic body radiotherapy of non-small cell lung cancer". Radiother Oncol. 94: 188-94, 2010.

[22] S.F. Petit, W.J. van Elmpt, P. Lambin, and A.L. Dekker, "Dose recalculation in megavoltage cone-beam CT for treatment evaluation: removal of cupping and truncation artefacts in scans of the thorax and abdomen". Radiother Oncol. 94: 359-66, 2010.

[23] S.F. Petit, W.J. van Elmpt, S.M. Nijsten, P. Lambin, and A.L. Dekker, "Calibration of megavoltage cone-beam CT for radiotherapy dose calculations: correction of cupping artifacts and conversion of CT numbers to electron density". Med Phys. 35: 849-65, 2008.

[24] I.J. Yeo, J.W. Jung, M. Chew, J.O. Kim, et al., "Dose reconstruction for intensity-modulated radiation therapy using a non-iterative method and portal dose image". Phys Med Biol. 54: 5223-36, 2009.

An evaluation of cine-mode 3D portal image dosimetry for Volumetric Modulated Arc Therapy

W Ansbacher[1], C-L Swift[2], P B Greer[3]

[1]BC Cancer Agency, Vancouver Island Centre, Victoria BC, Canada
[2]BC Cancer Agency, Vancouver Centre, Vancouver BC, Canada
[3]Calvary Mater Newcastle Hospital, Newcastle NSW, Australia

Email: wansbach@bccancer.bc.ca

Abstract: We investigated cine-mode portal imaging on a Varian Trilogy accelerator and found that the linearity and other dosimetric properties are sufficient for 3D dose reconstruction as used in patient-specific quality assurance for VMAT (RapidArc) treatments. We also evaluated the gantry angle label in the portal image file header as a surrogate for the true imaged angle. The precision is only just adequate for the 3D evaluation method chosen, as discrepancies of 2° were observed.

1. Introduction

The Electronic Portal Image dosimeter (EPID) has become a standard tool for evaluation of complex radiation treatments [1]. The most common configuration has been as an integrating dosimeter, but more recently, cine-mode acquisition has become viable. This mode provides a potentially useful tool for Volumetric Modulated Arc Therapy (VMAT), a novel treatment modality that involves simultaneous modulation of dose rate, field apertures and gantry speed.

Although the integrating mode has been adapted for VMAT dosimetry [2] the so-called "collapsed" 2D distribution is considered less useful than in conventional IMRT because the angular information is entirely absent, even if the gantry is rotated during acquisition. Cine acquisition opens up the possibility of true 3D dose reconstruction, as in a method recently developed by Mans *et.al.*.[3] for transmission dosimetry using the Elekta VMAT delivery and imaging system.

Cine mode presents some significant dosimetric challenges; Varian's IAS2 system, for example, has an unacceptably long dead-time [4]. McCurdy and Greer [5] performed the first definitive study of the current IAS3 system. Their study was limited to single frame images and dose rates that were varied by changing the nominal monitor unit (MU) rate. They discovered a deficiency of 1-2 frames per acquisition and speculated that this occurred at the end of irradiation. Beyond dosimetry, the accurate recording of the gantry angle during each image acquisition is also a major concern.

In this work we first evaluated Varian's IAS3 cine mode acquisition, paying particular attention to dose rate linearity and accuracy of the gantry angle recorded in the Dicom header of each image (the "header angle"). We then used this mode to deliver VMAT (RapidArc) plans for head & neck treatments, and evaluated them using an adaptation of a 3D virtual-phantom reconstruction technique that has proven very successful in 3D IMRT QA [6]. In this new technique, "Camera" (Composite Acquisition Method for Evaluating Radiotherapy Arcs), the header angle is a critical element because it enables the acquisition process to be self-contained. We show that Camera provides a rapid and robust method for determining whether the dose is out of tolerance compared to a calculated 3D distribution in a cylindrical phantom,

Journal of Physics: Conference Series **250** (2010) 012022 doi:10.1088/1742-6596/250/1/012022

and examples of successful and incorrect treatment deliveries are discussed. We also present data that suggests that 2D techniques may be inadequate to detect these incorrect deliveries.

2. Methods and Materials

All radiation deliveries were performed on a Varian Trilogy linear accelerator at the Vancouver Centre using a beam energy of 6MV.

2.1. Dosimetric linearity.

Linearity checks were performed at the nominal rate of 600MU/min, which is the mode used exclusively in RapidArc delivery. The actual MU rate was varied by changing the MU per degree, as in clinical operation, where the gantry rotation rate is limited to a maximum of 4.8°/s. Effective MU rates of 80 to 600MU/min were investigated, using 3 to 12 frames per image at an acquisition rate of about 12 frames/s.

2.2. Angular precision.

A "radiographic inclinometer" (Figure 1) was constructed that allowed comparison of the header angle to the actual start and stop angles of each image acquisition. It consisted of a pair of lead-solder wires wound in an open spiral around a 20cm long, 13cm diameter acrylic tube, coaxial with the gantry rotation axis. As the gantry rotated, the apparent intersection point of the wires moved proportionally along the axis. Cine sequences of up to 12 frames per image were recorded. The number of frames was limited by the blurring of the wire, to the point where the start and end points of the acquisition could not be determined. The inclinometer was designed to resolve the dead frame effect [5]. To achieve the required high spatial resolution, the inclinometer covered a range of only 40°; and it was calibrated with static fields 5° apart over that range.

We also delivered conformal arcs up to 360° (where the dose and gantry rotation rates are assumed to be constant), without the inclinometer in place. We made use of the constant frame acquisition rate, at 6 to 12 frames per image, to estimate the accuracy of the header angles.

2.3. RapidArc plan evaluation with Camera

Representative clinical dose distributions were optimized in RapidArc (Eclipse 8.6.17, Varian Medical Systems). These were converted to verification plans on a 20.4 cm diameter phantom and delivered with 6 to 12-frame cine imaging enabled, corresponding to 2-5° per image. Camera uses the same method of registration and calibration as its predecessor [6]; but for consistency with the arc acquisitions, the 10x10cm^2 calibration and flood fields were also acquired in cine mode.

The ability to validate patient plans and discriminate against invalid delivery or incorrect plans is based on χ–scores [7], with a dose/distance-to-agreement tolerance of 3%/3mm. This metric is evaluated within a high-dose volume (dose above 80% of the prescription) approximating the PTVs, and an intermediate dose volume (40-80%) representing the critical organ locations.

Some RapidArc plans were transformed into conformal arc deliveries with the same MLC apertures and the same total MU, but without dose rate or gantry speed variations, and these served as the incorrectly-delivered plans.

The cine images were also summed to represent integrated images. These were compared in 2D with "collapsed" verification plans at gantry angle 0°, in a plane normal to the beam axis.

3. Results

3.1. Dosimetry

Relative dose measurements reproduced the frame deficiency [5], and showed that it was independent of both the actual dose rate and MU delivered. The EPID signal was independent of frames per image to better than 1.5%, when the missing frames were accounted for. There was a small signal loss compared to the normal integrating mode, but all measurements for Camera were performed in cine mode so this effect was not investigated further.

Journal of Physics: Conference Series **250** (2010) 012022 doi:10.1088/1742-6596/250/1/012022

3.2. Angular precision

The inclinometer calibration was approximately 10 pixels/degree in the vertical image direction, with the exact scale being dependent on source to imager distance. The linearity and resolution was about ± 0.1° and showed that the 1.5-2 frame deficiency occurred during the final image, where acquisition terminated prematurely about 1° before the actual end of an arc delivery (In Figure 2, the gantry rotated from 2° to 34°). This effect was independent of the number of frames and dose rate. By comparing successive stop and start angles, it appeared that there was no dead time between images, for up to 12 frames per image.

The phase relationship between the header angle and the start or end of image acquisition was found to be rather complex. It is on average closer to the mid-point of the acquisition than either the start or end (mean difference ±0.9°); however the variation is significant ($\sigma = 0.6°$) and depends on the acquisition rate. The phase lags during 3-frame acquisition but tends to advance for 12 frames; and, from the number of repeated header angle instances, it seems that the angle may be updated in the Varian "4D-treat" application no more than 2-3 times per second.

Over the 360° range where the gantry rotation rate was assumed to be constant, we found that the header angle increment can vary by several degrees from its nominal value, ($\sigma = 1°$) and again, in several instances the increment was zero – i.e., the same angle appears in successive images.

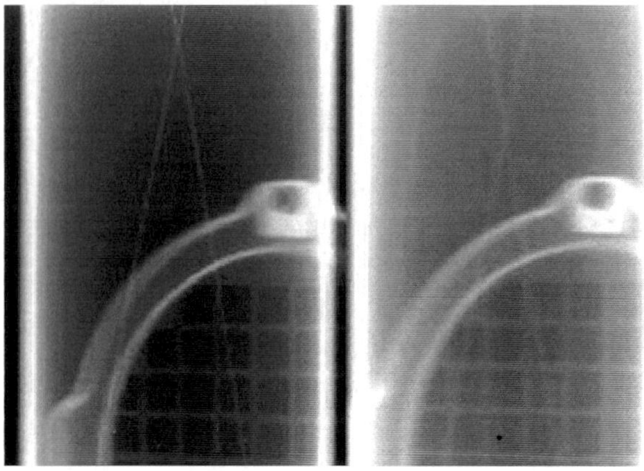

Figure 1

Portal images acquired with the cylindrical inclinometer: (Left) calibration image, (Right) 12-frame arc image. **Scale**: image height corresponds to 28°

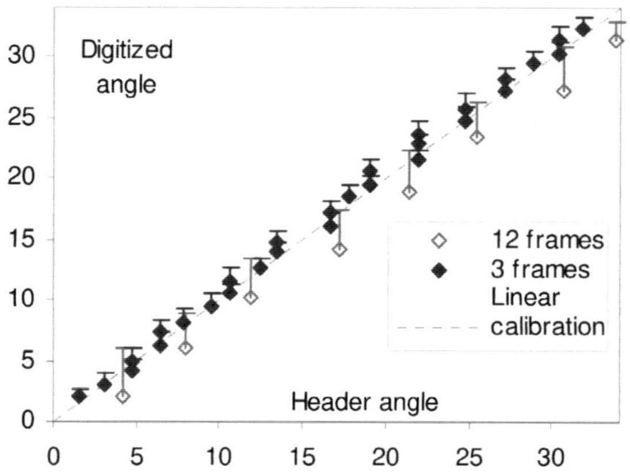

Figure 2

Inclinometer-digitized angles, from the start (◊,♦) to the end (–) of each acquisition, for 3 & 12 frame cine imaging, versus the EPID header angle. The arc delivery was from 2° to 34°.

3.3. Summary of Camera validations

Figure 3 shows one plan (a) delivered according to expectations, and one (b) delivered as a conformal arc so that the total MU and isocentric dose was correct, but the angular distribution was not. The χ-scores for case a) were 84% and 96% of points within tolerance for the high- and low-dose volumes respectively, while for case b) the corresponding scores were only 55% and 57%. Visual inspection confirmed the expected "hourglass"-shaped regions of discrepancy. Our previous experience with 3D IMRT validation (nearly 500 plans) has shown that an 80% score is a valid minimum pass criterion.

The integrating-mode comparisons with "collapsed" verification plans, which also used the χ metric in 2D, were unable to differentiate between the correct and incorrect treatment deliveries.

4. Discussion and Conclusion

We confirmed that Varian's cine mode is sufficiently linear during a typical VMAT delivery for EPID dosimetry to be viable. We identified issues with the EPID header angle that may confine its use to specific applications where these limitations have been properly assessed. Within Camera 3D verification, the header angle has been defined as mid-way through the image acquisition, although it is known to vary by up to 2° from this point. The cause is currently unknown but may be due to communication limitations between the linac controller and the treatment workstation, or even within the treatment/acquisition application itself. We are exploring this issue in more detail with real-time measurement of gantry angle during arc delivery, using a gantry-mounted inclinometer.

Given the limited precision of the header angle, we have shown that the maximum useful acquisition rate is about one image per 3-4° of gantry rotation, or about 10 frames (just under 1 s) per image. For the virtual phantom used in the initial version of Camera, this translates to a positional resolution at the cylinder surface of about 6 mm. As this is more than two pixels at the calculated dose resolution of 2.5 mm, higher angular precision would definitely be an advantage. Nevertheless, the current technology appears sufficiently accurate to detect *clinically-significant* 3D dose discrepancies in RapidArc deliveries where the high-dose volume is less than about 20cm in diameter.

Finally, by synthesizing integrated images, we were able to demonstrate that 2D QA methods would probably not have been able to detect the delivery errors that were introduced.

Acknowledgement

Partial funding for this work was provided by Varian Medical Systems.

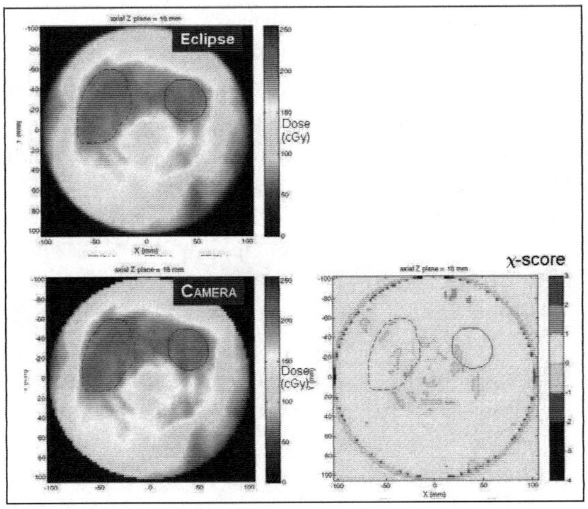

Figure 3a

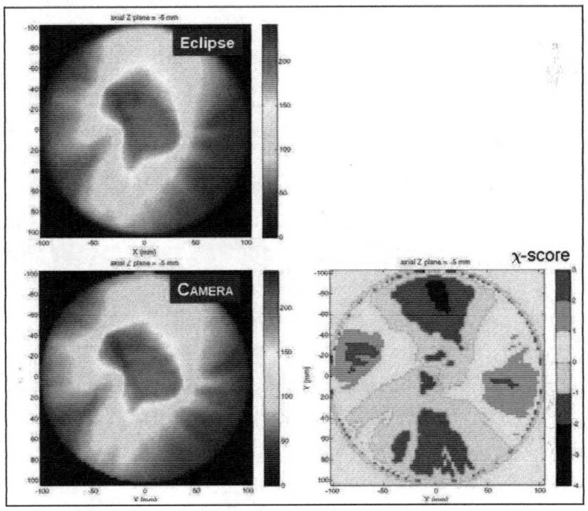

Figure 3b

Two RapidArc plans calculated in Eclipse, the corresponding "Camera" 3D reconstructions and χ-score differences for one plane in **a)** a complex Head and Neck plan that was delivered according to the plan, and **b)** an incorrect delivery where the dose rate and gantry speed were not modulated. Legend for χ-scores: Green areas are points within tolerance, |χ|<1; indigo, blue, orange and red: χ<-2, <-1, >1 >2 respectively.

References

[1] W. van Elmpt, L. McDermott, S. Nijsten, M. Wendling, P. Lambin and B. Mijnheer, *Radiother. Oncol.* **88,** 289-309 (2008).

[2] G. Nicolini, E. Vanetti, A. Clivio, A. Fogliata, S. Korreman, J. Bocanek and L. Cozzi, *Radiat. Oncol.* **3,** 24 (2008).

[3] A. Mans, P. Remeijer, I. Olaciregui-Ruiz, M. Wendling, J. Sonke, B. Mijnheer, M. van Herk and J. C. Stroom, *Radiother. Oncol.* **94,** 181-187 (2010).

[4] A. Piermattei, A. Fidanzio, L. Azario, F. Greco, A. Mameli, S. Cilla, L. Grimaldi, G. D'Onofrio, B. G. Augelli, G. Stimato, D. Gaudino, S. Ramella, R. D'Angelillo, F. Cellini and L. Trodella, *Med.Biol.Eng.Comput.* **47**, 425-33 (2009)

[5] B. M. McCurdy and P. B. Greer, *Med.Phys.* **36,** 3028-3039 (2009).

[6] W. Ansbacher, *Med. Phys.* **33,** 3369-3382 (2006).

[7] A. Bakai, M. Alber and F. Nusslin, *Phys.Med.Biol.* **48,** 3543-3553 (2003).

Phase space modulation method for EPID-based Monte Carlo dosimetry of IMRT and RapidArc plans

Avery Berman[1,2], Reid Townson[1,2], Karl Bush[1] and Sergei Zavgorodni[1,2]

[1]BC Cancer Agency–Vancouver Island Centre, Victoria, British Columbia, Canada

[2]Physics and Astronomy Department, University of Victoria, British Columbia, Canada

szavgorodni@bccancer.bc.ca

Abstract. Quality assurance for IMRT and VMAT require 3D evaluation of the dose distributions from the treatment planning system as compared to the distributions reconstructed from signals acquired during the plan delivery. This study presents the results of the dose reconstruction based on a novel method of Monte Carlo (MC) phase space modulation. Typically, in MC dose calculations the linear accelerator (linac) is modelled for each field in the plan and a phase space file (PSF) containing all relevant particle information is written for each field. Particles from the PSFs are then used in the dose calculation. This study investigates a method of omitting the modelling of the linac in cases where the treatment has been measured by an electronic portal imaging device. In this method each portal image is deconvolved using an empirically fit scatter kernel to obtain the primary photon fluence. The Phase Space Modulation (PSM) method consists of simulating the linac just once to create a large PSF for an open field and then modulating it using the delivered primary particle fluence. Reconstructed dose distributions in phantoms were produced using MC and the modulated PSFs. The kernel derived for this method accurately reproduced the dose distributions for 3x3, 10x10, and 15x15 cm^2 field sizes (mean relative dose-difference along the beam central axis is under 1%). The method has been applied to IMRT pre-treatment verification of 10 patients (including one RapidArc™ case), mean dose in the structures of interest agreed with that calculated by MC directly within 1%, and 95% of the voxels passed 2%/2mm criteria.

1. Introduction

For complex modern treatments, such as intensity modulated radiation therapy (IMRT) and volumetric modulated radiation therapy (VMAT), quality assurance (QA) and treatment verification are critical to ensure accurate dose delivery. However, presently there are few reliable three-dimensional (3D) dosimeters for performing QA. Two-dimensional (2D) dosimeters can instead be used to measure properties of the delivered treatment. An electronic portal imaging device (EPID) is one of such 2D dosimeters. The amorphous silicon (aSi) detector has been recognized as suitable for EPI dosimetry because it has a linear response to dose, is stable with time, and acquires images in real time at a high frame rate [1, 2].

When using portal images for 3D dosimetry, the images can be convolved with a kernel that represents a combined effect of the EPID scatter deconvolution and the convolution of the primary beam with phantom scatter to generate the dose in a phantom through a single convolution operation

Journal of Physics: Conference Series **250** (2010) 012023 doi:10.1088/1742-6596/250/1/012023

[3]. However, typically the images are deconvolved with a scatter kernel in order to obtain the primary photon fluence incident on the EPID[4-7].

After obtaining the primary photon fluence, the fluence can be convolved with a dose deposition kernel [7] to obtain the dose. Alternatively, the fluence can be substituted back into the treatment planning system [5, 8]. This allows the dose to be calculated in any phantom but the dose calculations themselves tend to be inaccurate near inhomogeneities [9, 10]. Further modifications of this method involve using the fluence to create Monte Carlo (MC) phase spaces which then get incorporated into dose calculations. As an example of this last method, van Elmpt et al.[6] deconvolved portal images with an empirically derived scatter kernel then created their own particle phase space (located just below the target) using ray-tracing from the EPID to the target to determine the particle positions and directions of travel. Particle energies were sampled from a measured and simulated on-axis spectrum and an off-axis correction factor was applied. The XVMC code [11] was used for dose calculations. When compared with film measurements, dose reconstruction in a homogeneous water phantom was within 1.5 % for square fields ranging from 5 to 20 cm width. For an IMRT field, 99 % of the pixels in the irradiated region had a gamma value less than one (for a 3 mm distance-to-agreement and 3 % maximum dose difference criterion) [6].

The method proposed here, Phase Space Modulation (PSM), is similar to van Elmpt et al.'s in that deconvolved portal images are used to reconstruct phase spaces, however, as it will be shown, it is much simpler, involves fewer assumptions and requires far less data manipulation. Instead of creating a phase space, a pre-existing phase space is modified by projecting the particles down onto the EPID and modulating their weights based on the deconvolved fluence values. No approximations are required in calculating the particles' energies and no extra processing is required to restore the flattening filter horns that get removed during the portal image flood field correction [5, 7].

2. Materials and methods

2.1. Image acquisition

The EPID used in this study was a Varian aS500 amorphous silicon device. The active imaging area is a 384x512 pixels^2 array of aSi light sensitive photodiodes with a 0.784 mm pitch. The acquired images are automatically stored as DICOM files containing the number of frames taken, gantry position, source-to-image distance (SID), collimator positions, and the average pixel values, among other parameters.

The process of image acquisition follows closely that described by Ansbacher [3].The EPID was positioned at an SID of approximately 110 cm and then repositioned longitudinally in order to optimally cover the entire beam. 6 MV photon beams were used at a delivery rate of 400 MU/min. Four types of images were acquired: a dark field (DF) image, a flood field (FF) image, a calibration (10x10 cm^2) image (CI), and the image(s) of interest. The DF, FF, and CI were all acquired with the gantry at 0°, and the patient-specific images were acquired at the gantry angles specified in their plans. Once all the images were acquired, the measured signals and required beam parameters were extracted from the DICOM files using an in-house developed DICOM library [12].

2.2. Image Deconvolution

Two forms of the scattering kernel were tested; both were the sum of multiple exponentials:

$$K(r) = \sum_{i=1}^{N} a_i \exp(-b_i r) \text{, with } \sum_{i=1}^{N} a_i = 1$$

Comparisons were made for variations on the Renner model [5] and the Kirkby model [4], which consisted of the sum of five and three exponentials, respectively. Kernel parameters were determined in an iterative process of deconvolving the portal images and then comparing the fluence profiles to those obtained from MC calculations. The goal was to match open field MC profiles for 3x3, 10x10, and 15x15 cm^2 fields to the profiles obtained from deconvolved images. Image deconvolution calculations were done using fast Fourier transforms.

2.3. Calculating modulation factors

After deconvolving the image of interest and the calibration image, the ratio between the image of interest's primary fluence $D^P(x,y)$ and the average fluence at CAX of the calibration image (CAL_{CAX}) was calculated for each pixel,

$$g_{\text{plan}}(x,y) = D^P(x,y) / CAL_{\text{CAX}}$$

Particle fluence modulation occurred in two stages: in a one-time calculation, particles were transported through the components of the linac head with an open field and scored as a phase space B (PS-B) [13]. BEAMnrc software [14] was used to create this phase space file. In the second stage, the trajectory of each particle was projected down to the level of the EPID using the particles' coordinates and direction cosines in the PS-B file and the position of the EPID specified in each DICOM file. If a particle struck the EPID at a point with coordinates (x,y), then the modulation factor of the particle f_{part} was given by $f_{part} = g_{\text{plan}}(x,y)$, provided (x,y) was within the active EPID area; otherwise it was zero. The particle's weight was then scaled by its modulation factor.

2.4. Phase space modulation testing

Images were acquired for the following field sizes: 3x3, 10x10, and 15x15 cm^2. After modulating each particle's weight, the modulated phase space was written and then used as input for DOSXYZnrc [15] or VMC++[16] to calculate the dose delivered to a water phantom with 0.5x0.5x0.5 cm^3 voxel dimensions. Calculation of the absolute dose from modulated fluences was performed using an adaptation of the method previously developed by our group [13] for IMRT dose calculations. For each field size, 5 billion particles were requested in the MC runs. The percent depth dose curves and inplane dose profiles at several depths were compared with benchmarked dose distributions from complete MC runs as calculated by the VIMC system [17, 18].

2.5. Implementation to VIMC software

This method of the dose reconstruction has been implemented as an option to VIMC software [17, 18] enabling its user-friendly use for conformal or IMRT treatments (including RapidArc™) on any scanned phantom or patient CT dataset. Ten IMRT (and one RapidArc™) patient plans have been calculated using this option within the VIMC system and results were compared to direct MC modelling.

3. Results

3.1. Deconvolved images

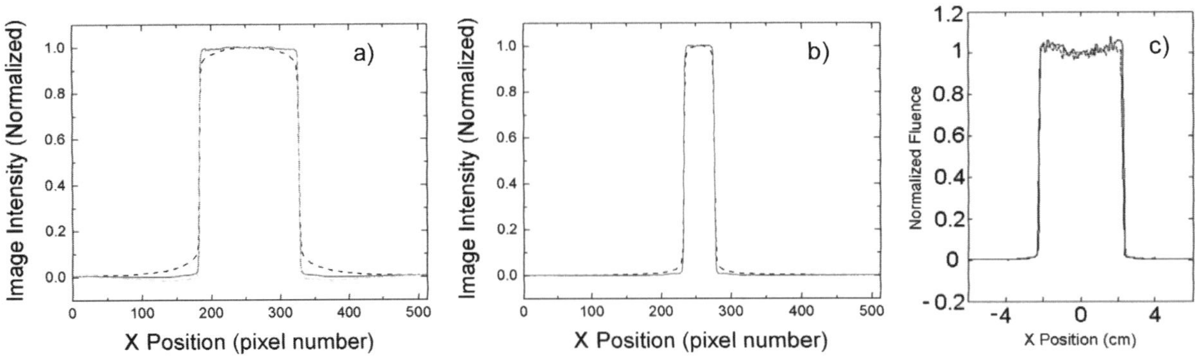

Figure 1. Profiles through (a) the portal image of 10x10 cm^2, (b) the portal image of 3x3 cm^2 fields: raw image shown as dashed black, image deconvolved with derived scatter kernel - solid red, and image deconvolved with less accurate kernel - dash-dot green; (c) PS-B fluence profile calculated from portal image (solid blue) and by BEAMnrc (dashed blue).

Journal of Physics: Conference Series **250** (2010) 012023 doi:10.1088/1742-6596/250/1/012023

Figures 1 (a, b) show normalized profiles for 10x10 and 3x3 cm^2 field portal images before and after deconvolving it with two different scatter kernels. The kernel parameters which produced the best fluences as well as the closest match in the dose distributions are $a_1 = 0.992$, $a_2 = 7.993 \times 10^{-4}$, $a_3 = 3.397 \times 10^{-5}$, $b_1 = 25.0\ cm^{-1}$, $b_2 = 2.0\ cm^{-1}$, and $b_3 = 0.27\ cm^{-1}$. Figure 1 (c) shows a comparison between PS-B fluences; one derived from a deconvolved 10x10 cm^2 field portal image using above kernel parameters and the other from a complete BEAM simulation. It is clear that the scatter is essentially completely removed from the image and the measured profile is well reproduced.

3.2. Tests of phase space modulation

Figure 2 shows the resultant dose profiles at different depths in the phantom after deconvolving images of 3x3, 10x10, and 15x15 cm^2 fields. In the percent depth dose (PDD) profiles the average differences between the profiles beyond D_{max}, are (1.0±0.7) %, (0.4±0.4) %, and (0.5±0.4) % for the 3x3, 10x10, and 15x15 cm^2 fields, respectively. Mean differences in the lateral profiles are within 1%.

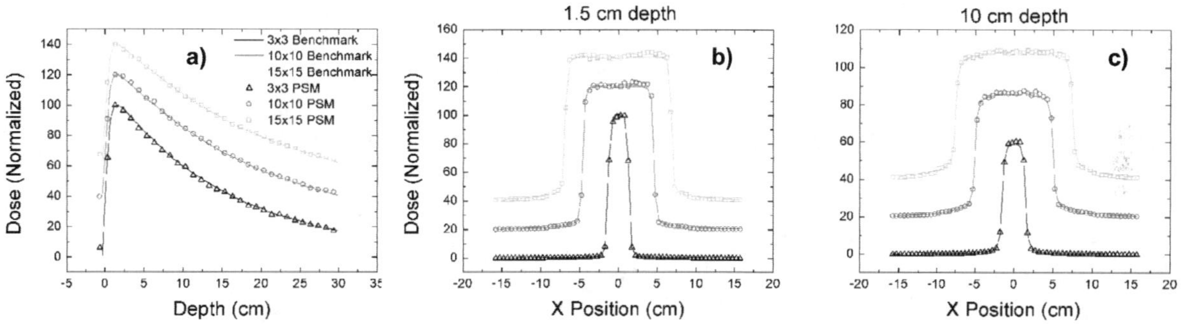

Figure 2. Percentage Depth Dose (PDD) profiles (a) and in-plane profiles at depths of 1.5 cm (b) and 10 cm (c), for 3x3, 10x10 and 15x15 cm^2 fields. All distributions are normalized by their PDD value at a depth of 1.5cm and then offset from each other for visual clarity. The offsets are 0% for 3x3, 20% for 10x10, and 40% for 15x15 cm^2 for all the plots.

3.3. 3D dose calculations for the patient plans.

IMRT dose distributions have been calculated for 10 patient plans, including one RapidArc plan. The difference in the mean dose in the outlined structures of interest was within 1% (0.04±0.9%). Considering that these distributions were calculated with 0.5% statistical uncertainty, these results are within calculation error.

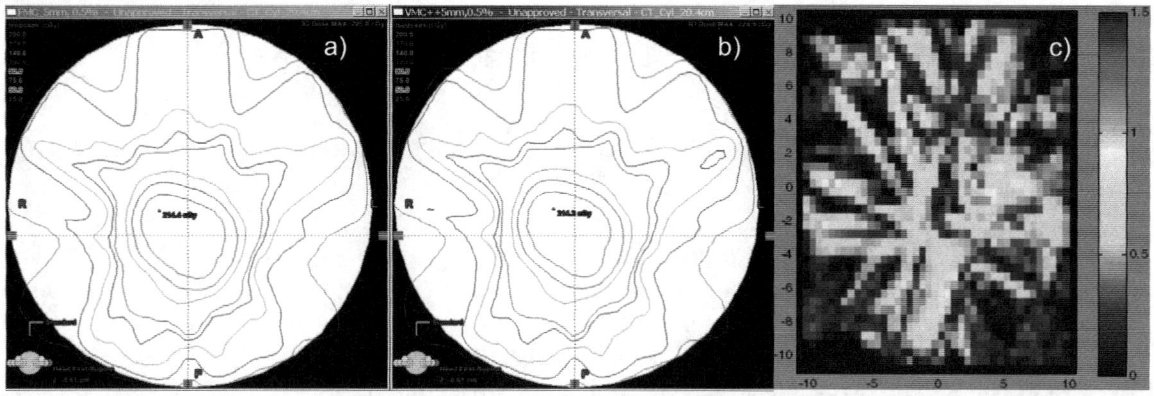

Figure 3. Isodose distributions calculated for a patient IMRT plan on cylindrical QA phantom: a) calculated by PSM method, b) calculated by direct MC modelling, c) gamma map for 2%/2 mm agreement. Notice that color scale on the difference map range from 0 to 1.5, though most voxels on this slice are within 0 – 1 range.

Figure 3 shows the dose distributions for a patient IMRT plan calculated by VIMC using the PSM method (Fig. 3a) as compared to direct simulation using VMC++ code (Fig. 3b). The dose distributions are in absolute units of Gy and have not been normalized to each other. Gamma analysis showed that 95 % of the pixels in the irradiated region passed 2 %/2 mm agreement criteria.

4. Discussion and conclusions

After testing the Phase Space Modulation method against direct MC simulations, it is clear that it works very well for reproducing known particle fluences and dose distributions modelled by MC. The modulated PS-B fluence and the benchmark fluence for the 10x10 cm^2 field in Fig. 1c show excellent agreement and the dose profiles in Fig. 2 corroborate the result by showing the final reconstructed dose distributions in close agreement with the benchmarked distributions. The method was further confirmed by calculations of IMRT plans (Fig 3). In these calculations we expect a close agreement of the dose reconstructed by PSM method with that from direct MC simulations; however we also expect some differences which are due to practicalities of the dose delivery. These include small differences in gantry positions relative to those planned, imperfections in the EPID positions relative to those expected. The dose differences seen in the Fig 3c have the "stripe" structure indicating that the main source of these differences is probably misalignment of the gantry and imager angles in ideal (modelled by MC) distributions as compared to the reconstructed ones.

5. Acknowledgements

The authors would like to acknowledge Will Ansbacher for helpful discussions on portal image dosimetry and for providing portal images from RapidArc delivery. We also acknowledge Mohammad Mohammadi of the Royal Adelaide Hospital for sharing his gamma analysis function.

References

[1] E. E. Grein, R. Lee, and K. Luchka, *Med. Phys.* 29 (2002) 2262.
[2] P. B. Greer and C. C. Popescu, *Med. Phys.* 30 (2003) 1618.
[3] W. Ansbacher, *Med. Phys.* 33 (2006) 3369.
[4] C. Kirkby and R. Sloboda, *Med. Phys.* 32 (2005) 1115.
[5] W. D. Renner, K. Norton, and T. Holmes, *J. Appl. Clin. Med. Phys.* 6 (2005) 22.
[6] W. J. C. van Elmpt, S. M. J. J. G. Nijsten, R. F. H. Schiffeleers, A. L. A. J. Dekker, B. J. Mijnheer, P. Lambin, A. W. H. Minken, W. Luo, J. Li, J. R. A. Price, L. Chen, J. Yang, J. Fan, Z. Chen, S. McNeeley, X. Xu, and C.-M. Ma, (2006)
[7] B. Warkentin, S. Steciw, S. Rathee, and B. G. Fallone, *Med. Phys.* 30 (2003) 3143.
[8] S. Steciw, B. Warkentin, S. Rathee, and B. G. Fallone, *Med. Phys.* 32 (2005) 600.
[9] G. Cranmer-Sargison, W. A. Beckham, and I. A. Popescu, *Phys. Med. Biol.* 49 (2004) 1557.
[10] I. M. Gagné and S. F. Zavgorodni, *J. Appl. Clin. Med. Phys.* 8 (2007) 33.
[11] M. Fippel, *Medical Physics* 26 (1999) 1466.
[12] C. Locke and S. Zavgorodni, *Australas. Phys. Eng. Sci. Med.* 31 (2008) 290.
[13] I. A. Popescu, C. P. Shaw, S. F. Zavgorodni, and W. A. Beckham, *Phys. Med. Biol.* 50 (2005) 3375.
[14] D. W. O. Rogers, B. R. Walters, and I. Kawrakow, in *NRC Report*, National Research Council of Canada, Ottawa, 2006.
[15] B. R. B. Walters, I. Kawrakow, and D. W. O. Rogers, in *NRC Report*, National Research Council of Canada, Ottawa, 2006.
[16] I. Kawrakow, in *Advanced Monte Carlo for Radiation Physics, Particle Transport Simulation and Applications: Proceedings of the Monte Carlo 2000 Meeting Lisbon* (A. Kling, F. Barao, M. Nakagawa, L. Távora, and P. Vaz, eds.), Springer, Berlin, 2001, p. 229.
[17] K. Bush, R. Townson, and S. Zavgorodni, *Phys. Med. Biol.* (2008) N359.
[18] S. Zavgorodni, K. Bush, C. Locke, and W. Beckham, *Radiother. Oncol.* 84 (2007) Supplement 1.

Journal of Physics: Conference Series **250** (2010) 012024 doi:10.1088/1742-6596/250/1/012024

Preliminary investigations with a photodiode-based fan-beam optical CT scanner

W. G. Campbell[a,b], A. Jirasek[a], and D. Wells[b]

[a]Dept. of Physics and Astronomy, University of Victoria, Victoria, BC, Canada

[b]BC Cancer Agency, Vancouver Island Centre, Victoria, BC, Canada

wgcampbe@uvic.ca

Abstract. An updated version of a prototype fan-beam optical computed tomography scanner is presented. The scanner uses a 0.83 mm diameter 543 nm HeNe laser beam with a 60° line-generating lens for fan creation. A pair of linear polarizers is used for light intensity control. A set of five 64-element photodiode arrays forms the detector arc. Extended dynamic range is achieved by adjusting light intensity and photodiode integration time. Two detector collimators (one single-slot, one multi-hole) are employed for comparative purposes. Highly attenuating solutions of scatter-based and absorption-based agents are examined using both types of collimation. Comparison of the results provides useful information about the persistent presence of scatter in optical CT.

1. Introduction

For more than a decade, the effects of radiation on the optical densities of polymer gels have been investigated [1]. While initial gel dosimeters utilized scatter as a mechanism for light attenuation, polyurethanes utilizing absorption for light attenuation were later composed [2]. In order to promote the use of 3D dosimeters, a relatively small number of researchers worldwide continue development in optical CT systems.

Currently, two commercial systems are available. The first—the OCTOPUS™ from MGS Research Inc. (Madison, CT, USA)—uses a pencil-beam scanning geometry. Although scan times using pencil-beam geometry are long, scatter contamination is avoided by only illuminating a small volume of the dosimeter at a time and by detector collimation. The second commercial system—the Vista™ from Modus Medical Devices Inc. (London, ON, Canada)—uses a cone-beam geometry. With a cone-beam geometry, scan times are vastly improved by acquiring full projections for multiple slices simultaneously. However, this simultaneous illumination can lead to multiple scattering, which may compromise the use of scatter-based dosimeters. Cupping artefacts are observed with the Vista™ scanner when examining polymer-based dosimeters that have been treated with uniform high doses [3]. A similar effect has also been observed in highly attenuating absorption-based phantoms [4]. The existence of similar phenomena for both scatter-based and absorption-based samples brings the cause of the cupping artefact into question.

This work presents the updated design of a fan-beam optical CT scanner that is currently under development [5]. Preliminary results examining highly attenuating scatter-based solutions and absorption-based solutions are presented.

Journal of Physics: Conference Series **250** (2010) 012024 doi:10.1088/1742-6596/250/1/012024

2. Materials and methods

2.1. Fan-beam optical CT scanner

A schematic and photograph of our system are shown in Figure 1. The light source is a 543 nm HeNe laser with 0.83 mm beam diameter and 2 mW output (Research Electro Optics, Inc.; Boulder, CO, USA). Light intensity control is performed using a pair of linear polarizers (Edmund Optics; Barrington, NJ, USA). Tests show that the polarization sensitivity of our detectors is minimal. Still, to minimize effects seen due to beam polarization, the second polarizer remains stationary while the first (closest to laser) is adjusted. Fan creation is achieved using a 60° line-generating lens (Edmunds Optics). A scanning medium is housed in a semi-circular tank designed with concentric walls so that light enters and exits the medium with normal incidence. Phantoms are contained in 1L cylindrical PET flasks (Modus Medical Devices Inc.). Flasks are suspended overhead and moved using a rotational stage and a translational stage, both of which are instructed via a Universal Motion Controller (Newport; Irvine, CA, USA). Light is detected by five 64-element photodiode arrays (Hamamatsu; Hamamatsu City, Japan). Detector elements have 0.8 mm pitch, 0.7 mm active width and 0.8 mm active height. Detectors are mounted concentrically onto one of two collimators. The first, a single-slot, provides no scatter-rejection collimation (slot h: 0.8 mm, d: 12-14 mm). The second, a multi-hole collimator, has holes cut for each individual element (w: 0.7 mm, h: 0.8 mm, d: 15 mm).

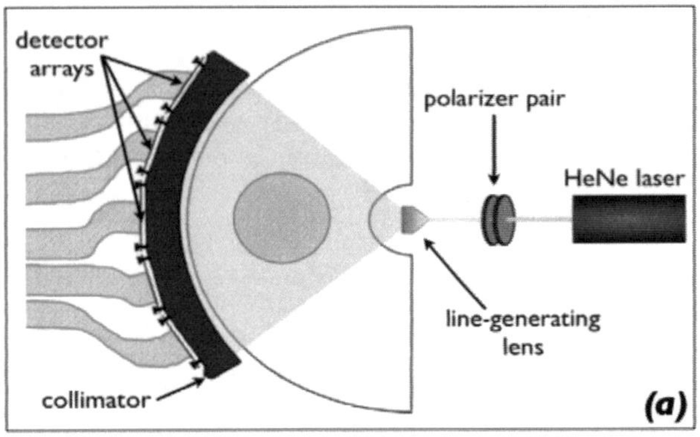

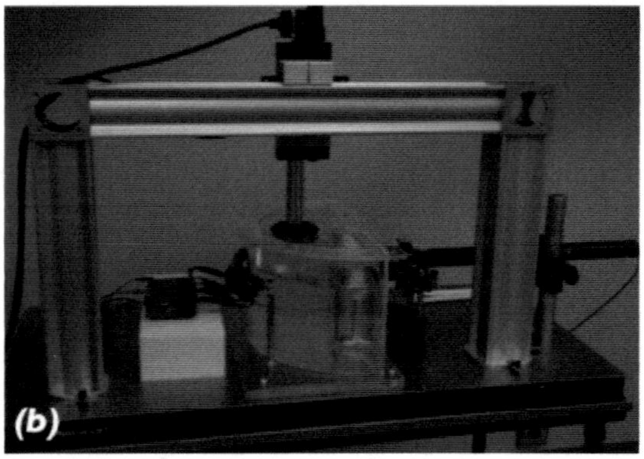

Figure 1: (a) Schematic of current setup, and (b) a photograph.

Data acquisition and collection was performed via a custom-built circuit board and five daughter boards, which were developed at the University of Victoria. Operating at the fastest speeds specified by our detectors (2 MHz clock rate), the circuit board stores scan data to onboard memory. Current research does not focus on minimizing acquisition time. At the moment, data collected from elements near array edges are compromised, which leads to ring artefacts at consistent radii in our reconstructed images (see Figure 2a). This issue has a hardware-based solution that requires modifying the current circuit board. Nevertheless, the system is capable of acquiring 6500 full-profile views in ~35 seconds before onboard memory fills and data must be output to the PC (upload time ~100 seconds).

Uniform scatter-based phantoms of diluted Duramax B-1000 polymer (Rohm and Haas; Philadelphia, PA, USA) were examined using water as the scanning medium. In order to investigate scanning protocols and their effects on image noise, a disc region of interest was selected to lie just outside of the inner rings and within the outer rings (N.B. the outermost rings in Figure 2a are the walls of the flask). The ROI sampled 46192 pixels, and relative noise (σ/μ) was used to quantify image quality. Full 360° sinograms (acquired at 1° increments) of water-filled flasks were acquired for I_o before scatterer was added through a porthole in the upper portion of the flask. Adding scatterer through the porthole allowed the same slice to be examined for I and I_o sinograms. Absorbance profiles were calculated respectively for each projection angle ($A = -\log_{10}[I/I_o]$). Data collection and image reconstruction were orchestrated using MATLAB (The Mathworks™; Natick, MA, USA).

Journal of Physics: Conference Series **250** (2010) 012024 doi:10.1088/1742-6596/250/1/012024

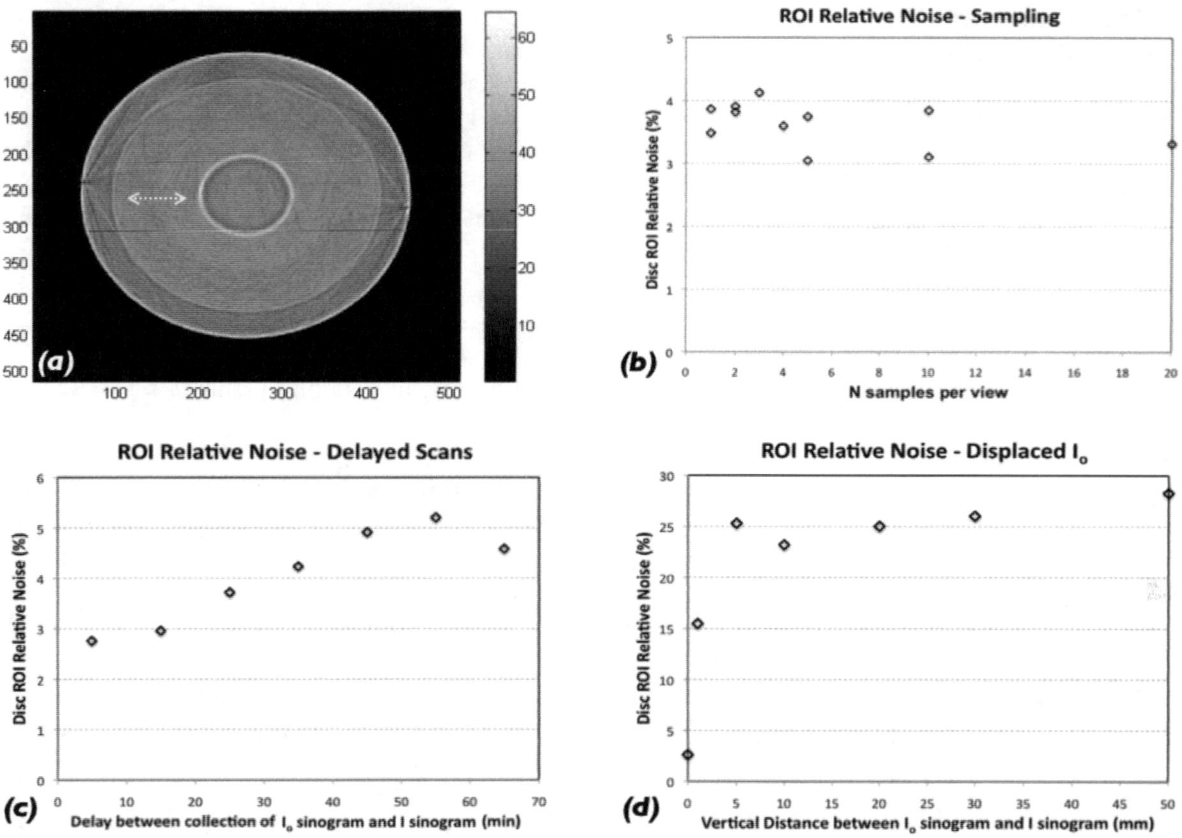

Figure 2: (a) Unprocessed image examining a dilution of Duramax B-1000 polymer *(ROI radii indicated)*, (b) relative noise acquiring multiple samples per projection, (c) relative noise when delaying scans, and (d) relative noise comparing different slices of the flask.

The stability of both our detectors and our light source is demonstrated in Figure 2b. Multiple samples were acquired and averaged for each projection angle. Image noise showed no significant gain achieved by obtaining multiple samples. Examining the length of time between acquisitions of the I_o and I sinograms, Figure 2c indicates that although noise is at its lowest with the shortest delay, there are only minor penalties for lengthening the time between acquisitions. The largest source of noise in our imaging investigations came in comparing different slices of the flask. Multiple slices of a water-filled flask were scanned to serve as I_o sinograms for a final scan. As can be seen in Figure 2d, images suffer greatly when comparing different surfaces of the flask. We attribute these differences to surface imperfections, and while they vary dramatically from slice to slice, their variations are consistent. This indicates that registering the surface of the flask may provide huge reductions in image noise. Future work aims to address this task.

2.2. Extending dynamic range

In order to interrogate samples of high absorbance ($A > 1.25$), the dynamic range of the scanner must be extended. Using the light intensity control allowed by our polarizer pair, a method of DR extension is implemented which is similar to methods used by other investigators [4, 6]. For transmission profiles containing 'dark' readings, light intensity is increased to obtain valid data, which is then normalized to fill in for missing values. The characterization curve of our polarizer pair is used for normalization. Shifting to longer photodiode integration times can further extend our DR. In this case, integration times are used to normalize the data obtained. The linearity of each method was tested before being used for these investigations. By employing a combination of the two techniques, absorbance values slightly greater than 4 are measurable.

2.3. Examining highly attenuating solutions

Concentrated solutions of scatter-based and absorption-based agents (Duramax B-1000 polymer and blue food colouring, respectively) were prepared. In order to evaluate the absorbances of these solutions, 9 dilutions of each solution were prepared. Absorbance values of these dilutions were obtained using a Cary 500 UV-VIS-NIR Spectrophotometer (Varian, Inc.; Palo Alto, CA, USA) with cuvettes of 1 mm pathlength. By considering the absorbance for a given concentration over 1 mm pathlength, theoretical absorbance values for longer pathlengths were calculated. A flask 95 mm in diameter with ~1 mm thick walls was used to contain samples of increasing concentration. Precise concentrations for each solution were scanned using the single-slot collimator. These tests were then repeated using the multi-hole collimator. For comparison between measured and theoretical absorbance values, the center three pixels of the measured absorbance sinograms were averaged (3 pixels × 360 projections) and compared against theoretical values for a 93 mm pathlength.

3. Results and discussion

Absorbances up to ~2 were accurately measured for both solution types using both collimators (see Figure 3a,b). For the highest concentration of dye, measurements using the single-slot collimator understated absorbance values (A_{th} = 4.09, A_{mSS} = 3.71), while measurements using the multi-hole collimator maintained linearity (A_{mMH} = 4.09). For scattering solutions, similar results were observed for the highest concentration that provided readings for both collimators (A_{th} = 3.84, A_{mSS} = 3.32, A_{mMH} = 3.91). Two higher concentrations provided measurements with the single-slot collimator, although these values were severely nonlinear (A_{th} = 7.55, 14.54; A_{mSS} = 3.72, 3.87).

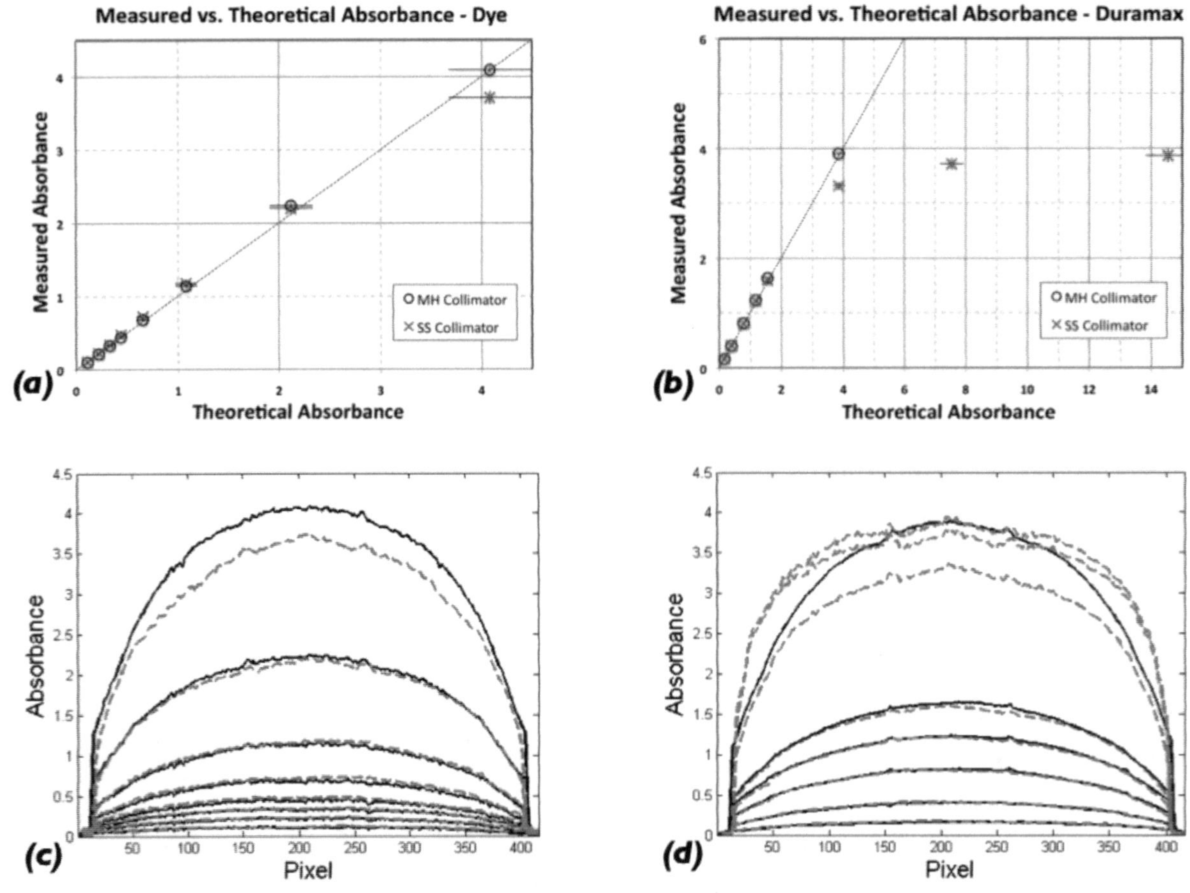

Figure 3: Measured vs theoretical absorbances for (a) dye, and (b) scatterer *(blue dotted line indicates expected values)*. Absorbance sinogram profile comparisons for (c) dye, and (d) scatterer. Multi-hole collimator results are plotted in solid-black, single-slot collimator results in dashed-red.

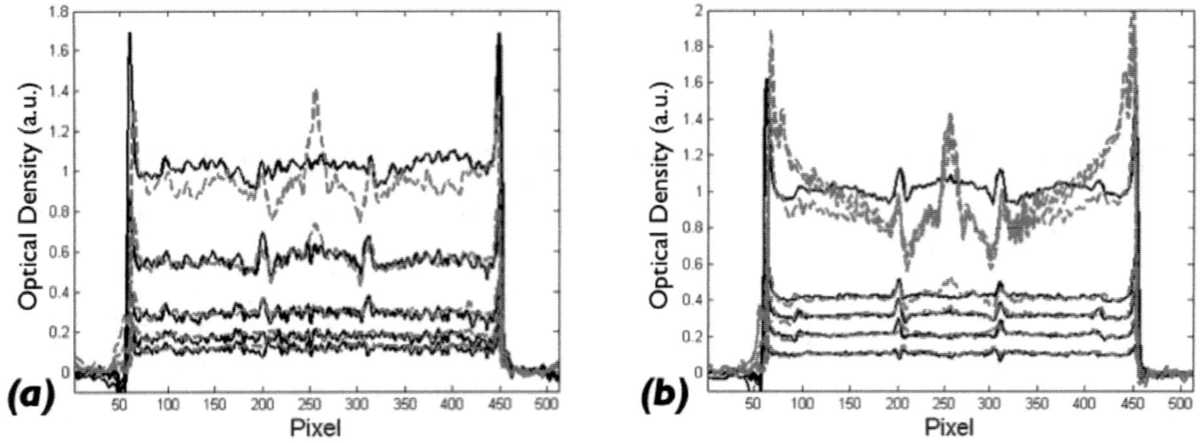

Figure 4: Profiles through reconstructed images for (a) dye, and (b) scatterer. Multi-hole collimator results are plotted in solid-black, single-slot collimator results in dashed-red.

Absorbance profiles are shown in Figure 3 and profiles through reconstructed images are shown in Figure 4. Flattening of absorbance profiles is especially apparent with the highest two concentrations of scatterer that were examined with the single-slot collimator (Figure 3d). Cupping of reconstructed images is also apparent for these concentrations (Figure 4b). The magnitude of ringing caused by array-edge data corruption is evident in the image profiles. An additional "spiking" artefact is seen in reconstructions of high absorbances with the single-slot collimator. The cause of this effect is currently unclear, though it may be due to the redistribution of light reflected off of the elements of our detectors. The multi-hole collimator limits detection of such stray light.

4. Conclusions

The current version of a fan-beam optical CT scanner has been presented. System stability has been discussed and our most recent imaging issues have been indicated. Scanning of highly attenuating phantoms using two types of collimation has provided useful information regarding the effects of scatter in optical CT. Results indicate that the understatement of absorbances for highly attenuating scatter-based and absorption-based phantoms is due to signal contamination by scattered light.

Acknowledgements

Funding from the Canadian Institutes of Health Research and the University of Victoria supported portions of this work.

References

[1] Gore, JC, Ranade, M, Maryañski, MJ and Schulz, RJ. Radiation dose distributions in three dimensions from tomographic optical density scanning of polymer gels: I. Development of an optical scanner. *Phys. Med. Biol.*, **41**, 2695-2704 (1996).

[2] Adamovics, J and Maryañski, MJ. A new approach to radiochromic three-dimensional dosimetry-polyurethane. *J. Phys Conf. Ser.*, **3**, 172-175 (2004).

[3] DeJean, P, Senden, R, McAuley, K, Rogers, M, and Schreiner, LJ. Initial experience with a commercial cone beam optical CT unit for polymer gel dosimetry I: Optical dosimetry issues. *J. Phys Conf. Ser.*, **56**, 179-182 (2006).

[4] Al-Nowais, S and Doran, SJ. CCD-based optical CT scanning of highly attenuating phantoms. *J. Phys Conf. Ser.*, **164**, 012023 (2009).

[5] Jirasek, A, Rudko, D, and Wells, D. A prototype fan-beam optical CT scanner for polymer gel dosimetry. *J. Phys Conf. Ser.*, **164**, 012025 (2009).

[6] Krstajić, N and Doran, SJ. Characterization of a parallel-beam CCD optical-CT apparatus for 3D radiation dosimetry. *Phys. Med. Biol.*, **52**, 3693-3713 (2007).

Journal of Physics: Conference Series **250** (2010) 012025 doi:10.1088/1742-6596/250/1/012025

Development of a new laser-line and CCD based optical-CT scanner for the readout of 3D radiation dosimeters

A E Papadakis[1], T G Maris[2], G Zacharakis[3], J Ripoll[3], C Varveris[4] and J Damilakis[2]

[1] Department of Medical Physics, University Hospital of Heraklion, Staurakia, 71110, Crete, Greece

[2] Department of Medical Physics, University of Crete, Medical school, Staurakia, 71110, Crete, Greece

[3] Institute of Electronic Structure and Laser, Foundation for Research and Technology Hellas (FORTH), Voutes, 71110, Crete, Greece

[4] Department of Radiation Therapy, University of Crete, Medical school, Staurakia, 71110, Crete, Greece

E-mail: apapadak@edu.med.uoc.gr

Abstract. We present initial results on the comparison of the dose readout from a three dimensional polymer gel dosimeter using two different optical-CT systems; i) a common wide field and a wide area detector optical-CT system and ii) a new 'laser-line' and wide area detector based optical-CT system. The findings presented herein highlight the advantage of the laser based over the wide field optical-CT concept for the readout of scattering 3D dosimeters. Moreover, the new 'laser-line' based optical-CT system overcomes the disadvantage of the long acquisition times required by the existing laser-based instruments.

1. Introduction

Optical computed tomography (optical-CT) has been successfully applied for the read out of three dimensional radiation therapy dosimeters. Two different optical configuration concepts have been applied. The first employs a laser point beam and a single detector, which scan the dosimeter point by point to create two dimensional maps (2D) of the light attenuation through the dosimeter [1,2]. Although this configuration constitutes the most accurate method with the measurements minimally affected by scattered light, it is slow requiring approximately 7 min of data acquisition per slice. The second optical design set up is based on a combination of a wide field light illumination source and a wide area detector [3-5]. In this configuration 2D angular image projections are acquired thereby generating multiple slices of the dosimeter simultaneously. The latter allows for very fast volumetric data acquisition times in the order of 5-6 minutes. However, the dose measurements provided by these scanners may suffer from scatter contamination owing to the wide area of detection.

The purpose of this study was to report initial results on the comparison of the dose readout derived from a wide illumination source and a wide area detector optical-CT system to the dose readout derived from a new 'laser-line' based optical-CT system.

2. Materials and methods

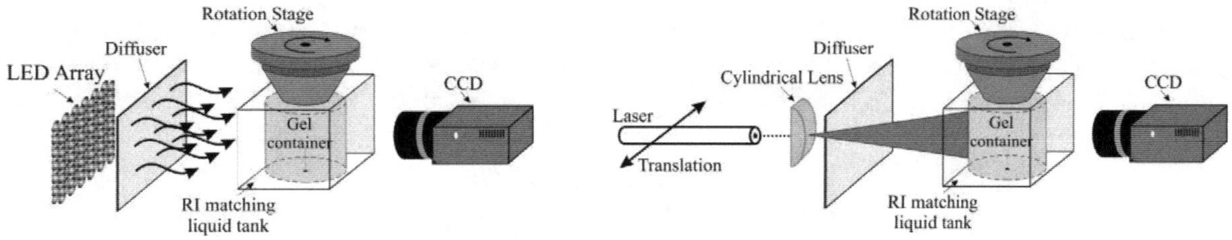

Figure 1. A schematic diagram of the wide illumination source and wide area detector optical-CT system (a) and a new 'laser-line' based optical-CT system (b).

2.1. *Design and set up of the optical-CT systems*

A schematic drawing of the two optical-CT systems used in this study is shown in Figure. 1. A detailed description of the first system shown in Figure 1a has been reported elsewhere [3]. The main difference between the two systems lies in the set up of the illumination geometry. In the second optical-CT system the illumination is provided through a combination of laser beam (633 nm) and a cylindrical lens (Figure 1b). This set up creates a laser line beam, which illuminates the dosimeter vertically through a single plane. This plane is parallel to the dosimeter's axis of rotation. The laser and the cylindrical lens are mounted on a translational stage, which can travel laterally on a direction perpendicular to the axis of rotation. After each lateral scan the dosimeter is rotated by 1 deg and the lateral scan is repeated. The total time for the acquisition of the raw data (180 deg revolution, 1 lateral scan per deg) required for the 3D reconstruction was less than 10 min. The angular sequence of the measured light intensities from a row of 640 detector pixels forms a sinogram that reconstructs the axial slice using standard filtered back projection algorithm (inverse Radon transform) [6].

2.2. *Polymer gel dosimeter*

A normoxic N-Vinylpyrrolidone based polymer gel dosimeter with tetrakis (hydroxymethyl) phosphorium chloride (VIPET) was employed in this study [7,8]. The VIPET gel material was poured into cylindrical, 2 cm diameter, glass vials. The vials were irradiated to known doses up to 48 Gy using a 6 MV linear accelerator. The above samples were independently scanned with the two optical-CT apparatus to generate attenuation coefficient μ (mm^{-1}) versus dose calibration data.

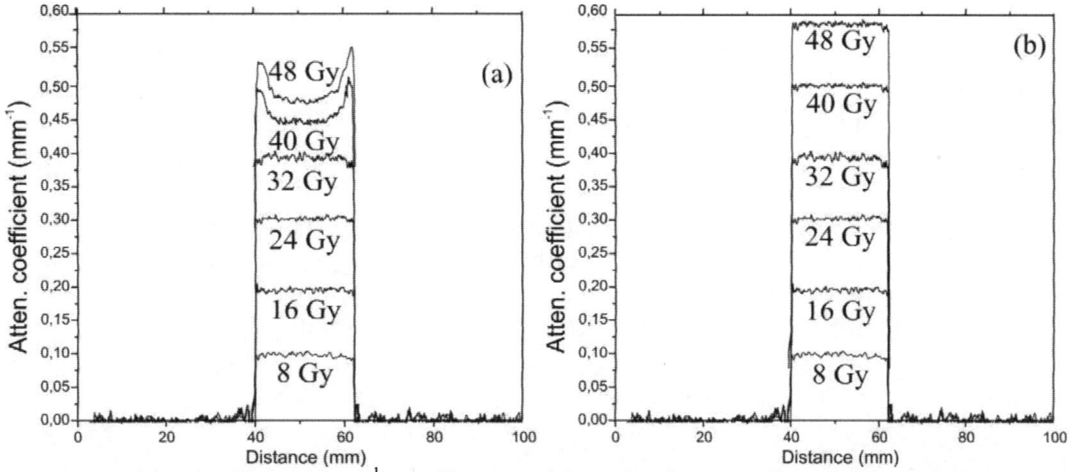

Figure 2. Attenuation coefficient (mm^{-1}) profile curves along the diameter of the dosimeters at various doses derived from the wide field illumination (a) and 'laser-line' (b) optical-CT system.

3. Results and Discussion

Figure 2 shows the 'μ' profiles taken along the diameter of the reconstructed axial slices of the calibration samples at different doses. The 'μ' profiles obtained from the wide illumination beam optical-CT system (Figure 2a) show a dishing effect at doses above 32 Gy. This is attributed to the increased scattering that is taking place within the optically denser scattering medium at high doses. This effect is not observed in the corresponding profiles derived from the 'laser-line' based system (Figure 2b). Thus the dynamic range of the laser-line beam optical-CT system is significantly wider than that of the common wide illumination beam optical-CT system. This finding confirms that the laser illumination based optical-CT systems are minimally affected by scattered light. A significant advantage of the 'laser-line' optical-CT system presented in this study compared to existing laser-based systems is the significantly reduced scanning time. This is facilitated by the 'laser-line' illumination concept. Raw projection data are acquired from all axial slice locations simultaneously without the need to translate the laser beam, which is required by the existing laser based optical-CT systems to generate 3D datasets.

Further experiments are needed to investigate the performance characteristics of the new 'laser-line' optical-CT system for the readout of 3D dosimeters.

4. References

[1] Lopatiuk-Tirpak O, Langen K M, Meeks S L, Kupelian P A, Zeidan O A and Maryanski M J 2008 Performance evaluation of an improved optical computed tomography polymer gel dosimeter system for 3D dose verification of static and dynamic phantom deliveries Med. Phys. **35** 3847-59

[2] Sakhalkar H S, Adamovics J, Ibbott G and Oldham M 2009 A comprehensive evaluation of the PRESAGE/optical-CT 3D dosimetry system Med. Phys. **36** 71-82

[3] Papadakis A E, Zacharakis G, Maris T G, Ripoll J and Damilakis J 2010 A new optical-CT apparatus for three-dimensional radiotherapy dosimetry: Is free space scanning feasible IEEE Trans. Med. Imag. **29** 1204-1212

[4] Krstajic N and Doran S 2007 Characterization of a parrallel beam CCD optical-CT apparatus for 3D radiation dosimetry Phys. Med. Biol. **52** 3693-3713

[5] Babic S, McNiven A, Battista J and Jordan K 2009 Three-dimensional dosimetry of small megavoltage radiation fields using radiochromic gels and optical CT scanning Phys. Med. Biol. **54**, 2463–2481

[6] Kak A and Slaney M 1987 Principles of computerized tomographic imaging IEEE Press, Engineering in Medicine and Biology

[7] Papadakis A E, Maris T G, Zacharopoulou F, Pappas E, Zacharakis G and Damilakis J 2007 An evaluation of the dosimetric performance characteristics of N-vinylpyrrolidone based polymer gels Phys. Med. Biol. **52** 5069-5083

[8] Pappas E, Maris T G, Zacharopoulou F, Papadakis A, Manolopoulos S, Green S and Wojnecki C 2008 Small SRS photon beam profile dosimetry performed using a pin point air ion chamber, a diamond detector, a novel silicon diode array (DOSI), and polymer gel dosimetry. Analysis and intercomparison Med. Phys. **35** 4640-4648

Journal of Physics: Conference Series **250** (2010) 012026 doi:10.1088/1742-6596/250/1/012026

Dosimetry and Quality Assurance for Static and Rotational Fields

Thomas Rockwell Mackie
Department of Medical Physics, University of Wisconsin, Madison, WI, USA

trmackie@wisc.edu

Abstract. Dosimetry calibration and quality assurance has difficulty keeping up with the rapid changes in the capability of radiation therapy delivery methods. The development of IMRT illustrated the needs for accurate modeling of small beams and the periphery of beams which overlap in the treatment field. Rotational delivery stresses quality assurance methods designed for a few treatment fields. Even the fundamental basis for radiation delivery, beam calibration, has been dramatically altered.

1. Introduction

A working party was formed by the International Atomic Energy Agency (IAEA) with cooperation of the AAPM Therapy Physics Committee to examine the issues of dosimetry of small fields and systems for which careful dosimetry techniques must be applied and which cannot be calibrated according to calibration standards like the IAEA TRS-398 or the AAPM TG-51 protocols. This committee documented the need for fine resolution when determining the dose in small fields to overcome partial volume effects which can lead to a severe under-prediction of dose. There is also a need to calibrate commercially available beams which cannot open their field sizes to the required 10 cm x 10 cm as stipulated in official protocols. In addition even those delivery systems which can be calibrated are often utilized, for example for rotational IMRT which is often composed of large numbers of small and highly irregular fields.

Daily image guidance and in-vivo imaging has also revealed that the patient is not static but changes from moment-to-moment due to breathing, day-to-day due to setup variability and organ filling and week-to-week due to weight loss and tumor shrinkage. With on-line delivery verification and dose reconstruction it is now possible to determine the dose during the course of radiotherapy. This reveals that the dose varies by amounts which are far greater than accuracy thresholds for calibration and patient-specific quality assurance protocols. The relevant dose to the patient can be determined by deformably registering the individual snap-shots of the dose distribution back to the original treatment planning patient representation.

2. Conclusions

The common factor in the recent dosimetry paradigm shift is that the fourth-dimension (time) must be explicitly included. Newer dosimetry quality assurance systems are now able to determine the dose in 4D. Temporal resolution will become as important as spatial resolution as delivery gets faster. These systems are not yet sufficient. They will also need to be set in motion or even deformed in order to provide quality assurance for 4D dosimetry calculations that are becoming available. In order to remain relevant, the name of this conference will even have to be altered in the future to reflect the change to 4D dosimetry.

Journal of Physics: Conference Series **250** (2010) 012027 doi:10.1088/1742-6596/250/1/012027

Novel Approaches to Improve the Combined Geometric and Dosimetric Performance of Existing IG-IMRT Systems

D.A. Jaffray, M. Sharpe, A. Rink, D. Letourneau

Princess Margaret Hospital, Toronto, Canada

David.Jaffray@rmp.uhn.on.ca

Abstract. The development of image-guidance technologies offer increased precision and accuracy in the placement of radiation therapy dose distributions in the human body. Kilovoltage and megavoltage systems now offer sub-mm accuracy and precision in localizing the treatment and planned isocenters. While this represents a significant improvement in our ability to target lesions, the performance of these systems in terms of their precise and accurate placement of field edges, including individual multi-leaves, is not performing at the same level. The development of dosimeters with small, imageable regions of response will allow these new systems to be evaluated with a much higher level of precision and accuracy. Furthermore, the development of completely automated systems that exploit the robotic nature of the imaging and delivery subsystems should permit the true performance of current IG-IMRT systems to be realized, quantified, and monitored. This includes adjustments to the treatment machines operating parameters, as well as, refinements of the planning system's beam model. Specifically, investigations into novel approaches of employing these detectors, automating their use to maximize system performance, and, the development of novel detectors for compatibility in MR-based systems will be discussed.

Journal of Physics: Conference Series **250** (2010) 012028 doi:10.1088/1742-6596/250/1/012028

Effective Management of FXG Gel Dosimetry

T Olding[1], J Darko[2] and L J Schreiner[1,2]

[1]Department of Physics, Queen's University, Kingston, Ontario, Canada, K7L3N6
[2]Cancer Centre of South Eastern Ontario at Kingston General Hospital, 25 King Street West, Kingston, ON, Canada, K7L5P9

E-Mail: Tim.Olding@krcc.on.ca

Abstract. The details of a calibration basis for the Fricke-xylenol orange-gelatin (FXG) gel dosimeter combined with the fast, easily accessible readout tool of cone beam optical computed tomography (CT) are described in this report. With proper controls in place, the results from a test intensity modulated radiation therapy (IMRT) treatment plan evaluation indicate that greater than 95% Low's gamma function agreement between plan and gel-measured dose using 3% dose and 3 mm distance-to-agreement criteria is achievable.

1. Introduction

This paper discusses some of the important parameters for ensuring the reliability of cone beam optical CT-based FXG gel dosimetry. Results are presented that establish the intra-batch reproducibility, dose calibration linearity, and megavoltage range energy independence of the dosimeter using a Vista™ cone beam scanner (Modus Medical Devices Inc, London, ON, Canada) for optical readout [1]. Time and temperature control criteria are set to manage the dosimetry. A dosimeter modification is then described that reduces stray light perturbation in the optical readout of a standard-sized FXG gel-filled 1 L polyethylene terephthalate (PETE) jar dosimeter and improves the overall three dimensional (3D) dose measurement.

2. Experiment

2.1. Dosimeter Preparation & Imaging

FXG gel dosimeters were prepared according to the recipe outlined by Babic *et al* (2008). After preparation, the heated formulations were poured into standard-sized 1 L PETE jars and allowed to set overnight in a refrigerator prior to irradiation. Dosimeter imaging was performed using the Vista scanner set to a camera lens aperture of f5 and 590 nm LED illumination. Reference scans were completed on the dosimeters prior to irradiation at the highest shutter exposure time possible without inducing camera pixel saturation (and the lowest gain setting) in approximately 4 minutes. After irradiation, data scans were acquired at the same camera settings as the reference scans. Optical CT attenuation images were reconstructed to an isotropic voxel resolution of 0.5 mm, unless otherwise specified. The remaining details of scanner operation are specified elsewhere [1]. Temperature measurements were taken using a digital temperature probe (TM99A-NA Digital Thermometer, Nuclear Associates, Carle Place, NY, USA). Line profiles and regions of interest in the reconstructed optical CT images of the FXG gel dosimeters and gelatin phantoms were evaluated using Microview 2.1.2 (GE Healthcare, United Kingdom) or MatLab (Mathworks, Newark, NJ, USA) software.

Journal of Physics: Conference Series **250** (2010) 012028 doi:10.1088/1742-6596/250/1/012028

2.2. FXG Gel Calibration

Gel irradiations were completed on the day after manufacture using a Varian Clinac 21iX linear accelerator (Varian Medical Systems, Palo Alto, CA, USA), or a Gulmay superficial x-ray therapy system (Gulmay Medical Inc, Buford, GA, USA). Depth fiducials were marked on the FXG-gel filled 1 L PETE jars in permanent marker, serving as reference points for registration of depth dose to attenuation in the reconstructed optical CT image. Calibration 6 x 6 cm photon and electron beams were delivered to the top surface of the gel-filled PETE jar with the lid off (typically to a maximum absorbed dose of 180-210 cGy) using the Clinac irradiator. A 200 cGy, 80 kVp delivery was performed using the Gulmay unit, employing a 6 cm cone set to the gel surface.

Calibration photon beam central axis depth dose data were obtained from 0.13 cm^3 ionization chamber-water tank measurements (CC13/TNC ionization chamber, Scanditronix-Wellhofer North America, Bartlett, TN, USA). Calibration electron beam central axis depth dose data were acquired from diode-water tank measurements (DEB002-3727, DEB022-3608, Scanditronix-Wellhofer North America). The dose data were considered to be accurate to within an uncertainty of $\pm$ 0.5% based on standard deviation of the measurement. The irradiator room temperature was not specified, but the gel dosimeters were kept in a covered holding tank filled with water maintained to 20.0-21.0 $\pm$ 0.2°C until the point of irradiation.

Figure 1a shows raw FXG gel calibration data for a range of beam energies, corrected for auto-oxidation during the time between the reference and data scans [2]. Optical CT scans of the gels were acquired at 30 $\pm$ 0.2 minutes post-irradiation in the temperature range of 20.8-21.2 $\pm$ 0.1°C. Each attenuation data point was obtained as the mean value of a 5 mm diameter single slice circular region-of-interest on the central depth dose axis of the electron (or photon) beam in the optical CT volume image. Electron beam calibrations for two dosimeters prepared from the same gel batch and irradiated with a 200 cGy, 6 x 6 cm, 20 MeV electron beam at two different dose rates are also shown in figure 1a (inset). At each depth, the mean optical attenuation values from these two dosimeters were observed to be linear and in agreement within error over the full dose-to-attenuation calibration range. Figure 1b shows the agreement between Wellhofer ion chamber-measured and linear dose-to-attenuation fit-calibrated FXG gel depth dose curves.

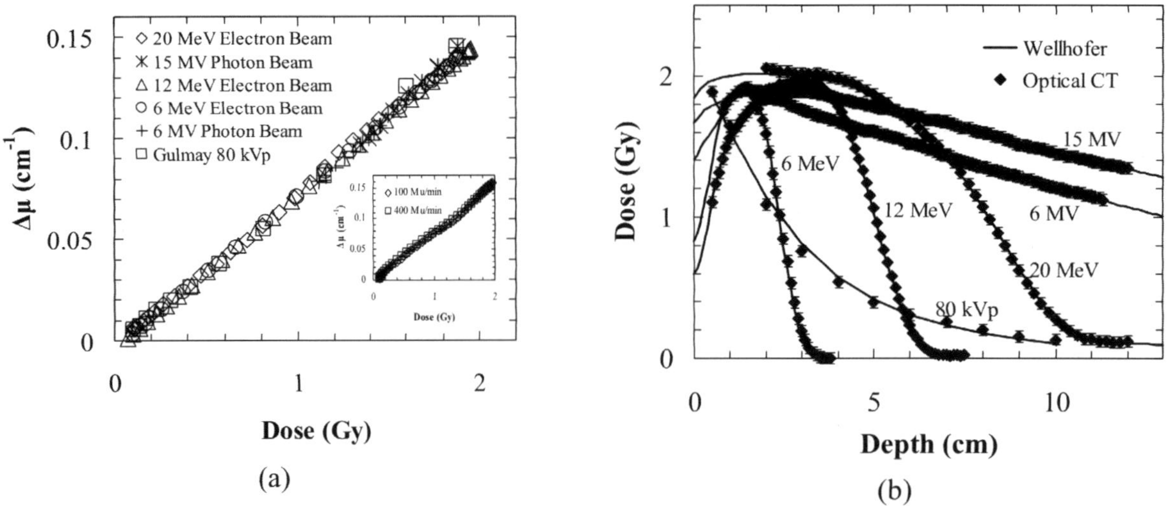

Figure 1: (a) Calibration data corrected for auto-oxidation, showing the reproducibility and energy independence of the FXG gel dosimeter. The inset plot shows the 20MeVelectron beam calibration of two dosimeters from same gel batch, delivered at different dose rates. (b) Depth dose curves of the different treatment beams from (i) the Wellhofer measurements in a water tank, and (ii) the calibrated FXG gel dosimeter (assuming a linear calibration relationship between absorbed dose and optical attenuation). For the purpose of clarity, only the line through the Wellhofer depth dose data is shown in this plot.

Journal of Physics: Conference Series **250** (2010) 012028 doi:10.1088/1742-6596/250/1/012028

2.3. Time and Temperature Controls

To examine time-based changes in the dosimetry, a FXG gel-filled 1 L PETE jar was irradiated to a surface dose of 200 cGy on the day after manufacture using the Gulmay unit with a 6 cm cone insert, then imaged over a period of 24 hours. The dosimeter was kept in a covered water bath held at 20.1 ± 0.2°C between scans and the scanner matching tank fluid was held at 20.1 ± 0.2°C during each scan. The central axis depth attenuation curves adjusted for auto-oxidation of the dosimeter [2] clearly show dose development over this time period, at a rate of approximately 2.8-3.0% per hour (figure 2a). Another time-sensitive factor seen in the dosimeter is the loss of spatial dose integrity over time due to diffusion of the ferric ion-xylenol orange dye complex [3]. For example, at the representative depth of 1 cm, the normalized attenuation changes in magnitude approximately at a rate of 0.5% every 10 minutes at the most rapidly changing point in the profile through the diameter of the jar image.

To quantify the influence of dosimeter temperature at the time of optical scanning, a FXG gel was pre-irradiation imaged at 22.0 ± 0.1°C, irradiated with a 6 x 6 cm, 20 MeV electron beam to a dose of 200 cGy at a reference depth of 3 cm, then post-irradiation imaged at different dosimeter temperatures (1 mm reconstruction cubic voxel size, see figure 2b). Note that the dosimeter was refrigerated for 3 weeks prior to post-irradiation scanning to ensure reaction completion (i.e. dosimeter development was minimal at this point). Individual scans were acquired on the Vista scanner, and auto-oxidation effects were taken into account [2]. Approximate measurements of the dosimeter temperature were acquired prior to each scan by inserting the temperature probe 1 cm into the FXG gel. The temperature of the matching tank solution was held to 22 ± 1°C during the post-irradiation scans.

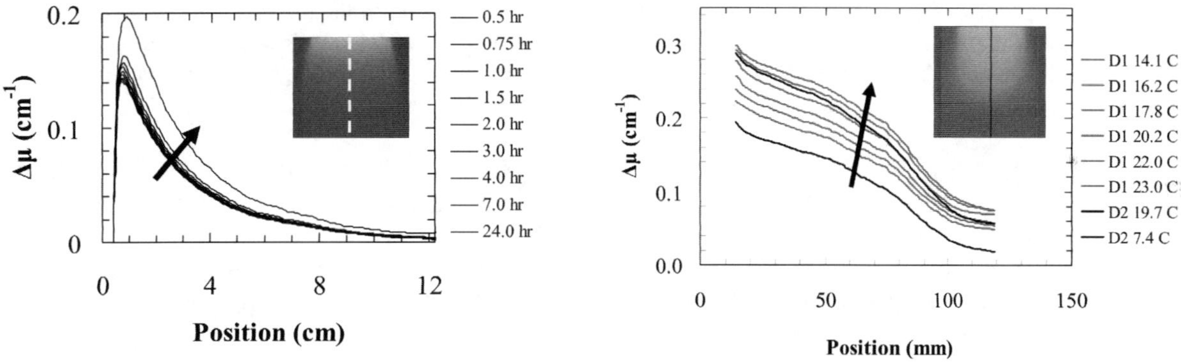

(a) (b)

Figure 2: (a) Attenuation (depth dose) development of an 80 kVp, 200 cGy Gulmay irradiation delivered to an FXG gel dosimeter. The arrow shows increasing time. (b) Central axis attenuation for a FXG gel dosimeter irradiated with a 200 cGy, 20 MeV electron beam. Two sets of scans were obtained for this dosimeter on subsequent days (D1 in gray, D2 in black), with increasing temperature shown by the arrow direction.

2.4. Dosimeter Modification

A number of sources of stray light in the optical measurement system perturb the accuracy of cone beam optical CT-based FXG gel dosimetry. One of the predominant stray light perturbations is angled scatter from the gel-forming matrix. In an effort to better manage this effect, several alternative gel-forming materials were investigated and one (a 300 Bloom bovine bone gelatin from Eastman Gelatine Corp, Peabody, MA, USA) was selected for further evaluation. Figure 3a shows the reconstructed profiles across two 5 wt% gelatin-in-water-filled 1 L PETE jars imaged in the Vista scanner (against a reference scan of the matching tank solution). One of the jars was prepared using the standard 300 bloom porcine skin gelatin and the other from the alternative bovine bone gelatin. Figure 3b shows the molecular weight distributions of these two gelatin materials determined by particle size exclusion chromatography (analysis completed by a third party through Eastman Gelatine Corp). The time at which the signal is observed is correlated against the behavior of the known

Journal of Physics: Conference Series **250** (2010) 012028 doi:10.1088/1742-6596/250/1/012028

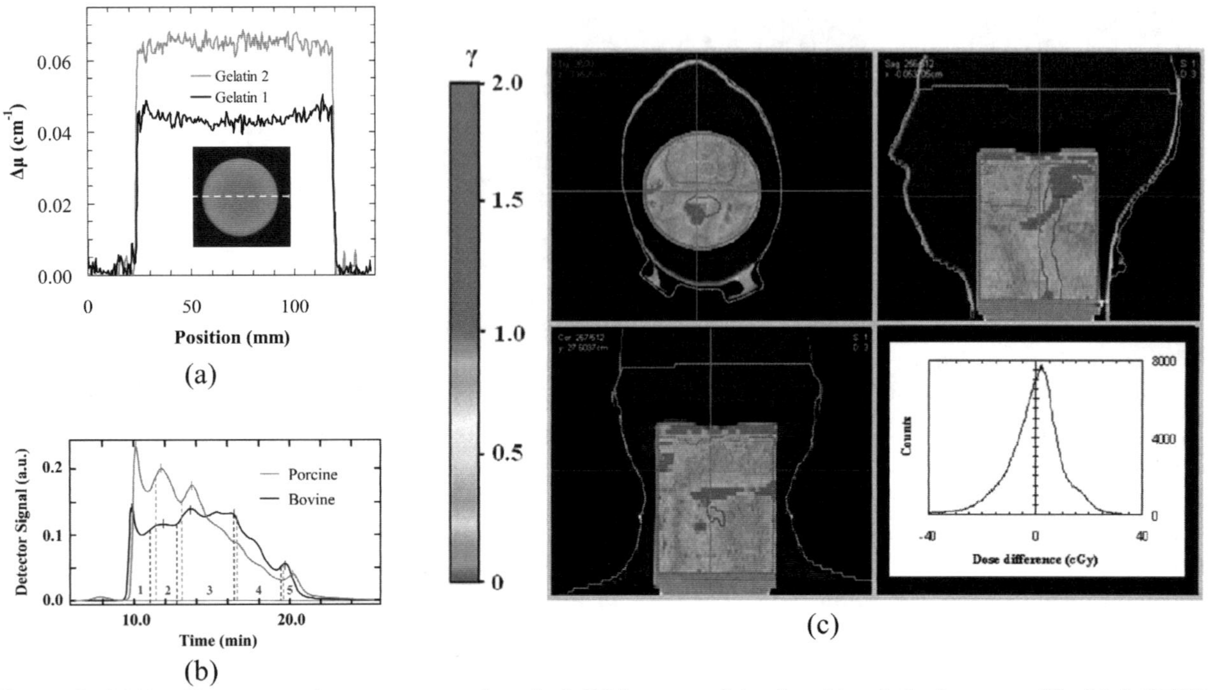

Figure 3: (a) Profiles across the reconstructed optical CT images of the 5 wt% gelatin-in water-filled 1 L PETE jars, prepared using standard 300 bloom porcine skin gelatin (Gelatin 1, black line) and 300 bloom bovine bone gelatin (gelatin 2, grey line), and imaged on the Vista scanner. (b) Size exclusion chromatograms of the two gelatin materials. (c) A 3%, 3mm Low's gamma function analysis of a 200 cGy, 6 MV IMRT treatment fraction delivered to a bovine gelatine-based FXG-gel dosimeter in-phantom. (c, bottom right inset) A dose difference histogram of the delivery, comparing treatment plan and gel-measured dose.

standards to infer the molecular weight from the hydrodynamic volume, which is related to effective particle size. Regions 1-5 in figure 3b correspond to the molecular weight ranges of >250 kilodaltons (kDa), 250-150 kDa, 150-50 kDa, 50-20 kDa, and 20-4 kDa respectively.

An IMRT study was then completed using a wax-filled reproduction of the head-and-neck from a Kyoto SBU-4 Rando anthropomorphic phantom (Capintec, Ramsey, NJ, USA) prepared at the Cancer Centre of Southeastern Ontario with a custom cavity designed to fit the standard 1 L PETE jar gel dosimeter. Two 1 L PETE containers were filled from a single batch of bovine gelatin-based FXG gel. One container was irradiated with a well characterized 12 MeV electron beam using the Varian Clinac 21iX linear accelerator and used for dose-to-attenuation calibration. The other container was inserted into the wax-filled Rando reproduction as the IMRT measurement dosimeter. A 6 field, 6MV, 200 cGy IMRT treatment fraction was then planned and delivered (using the same accelerator) to the phantom. The IMRT evaluation is not described in detail here, but is outlined more fully in a companion paper found in these proceedings. Figure 3c presents the results of a 3D Low's gamma function [4] voxel-by-voxel comparison of gel-measured dose against the reference treatment plan dose for the IMRT delivery, employing 3% dose difference and 3 mm distance-to-agreement criteria. Slightly more than 95% of the voxels within the central 8 cm diameter and 10 cm high preferred evaluation volume of the FXG dosimeter [1] satisfy the 3%, 3 mm criteria between the plan dose and gel-measured IMRT dose. The inset plot in the lower corner of figure 3c shows the dose difference histogram comparing treatment plan dose and gel-measured dose for this delivery.

3. Discussion & Conclusions

Referring to figure 1a, most of the raw dose-to-attenuation calibration data are seen to visually overlap in the plot. The fact that the dose-to-attenuation calibration data presented in the inset of figure 1a agree within error gives good indication that a measurement dosimeter can be well-calibrated by a second FXG gel-filled 1 L PETE jar dosimeter prepared from the same gel batch. In addition to indicating good reproducibility, the overlapping calibration data for the different beam energies shown in figure 1a also suggest that the FXG gel dosimeter is approximately energy independent into the kilo-voltage energy range. As a standard practice then, a dose-to-attenuation linear fit obtained from a well-characterized photon or electron beam delivered to a FXG gel dosimeter can be used to calibrate a second dosimeter prepared from the same gel batch and irradiated with different beam energy.

The results presented in figure 2a suggest that close attention to the time between dosimeter irradiation and scanning is important, as the attenuation is still changing throughout the dosimeter at a rate of ~0.5% every 10 minutes past the 30 minute mark, consistent with Kelly *et al* (1998). In order to reduce the uncertainties introduced by diffusion, it is desirable to minimize the time between irradiation and imaging of the dosimeter. However, the rate of dose development before the 30 minute mark is higher [3], which increases intra-scan variation in attenuation. A good balance between these two parameters is found by choosing a standard wait time between dosimeter irradiation and scanner readout of 30 minutes, measured to within ± 30 seconds. The importance of the fast optical cone beam scan time of 4-5 min is highlighted through these results.

Looking at figure 2b, the depth dose data indicate significant temperature dependence of the optical attenuation. This thermo-chromic behavior is believed to be primarily sourced in d-orbital spin state crossover of the ferric ion in the ferric ion-xylenol orange dye complex with increase in temperature [5]. The increase in attenuation with temperature (from the day 1 scan data) was observed to be approximately 4% per degree Celsius, suggesting that the dosimeter and scan bath temperatures should be matched and controlled to within the practical measurement limit of $\pm 0.1^{o}$C.

Substitution of the standard use 300 Bloom porcine skin gelatin with a high optical clarity 300 Bloom bovine bone gelatin in the FXG gel dosimeter formulation has the effect of reducing one of the main sources of stray light perturbation in the optical measurement system. The stray light-derived cupping artefact normally observed when using porcine gelatin (figure 3a) is reduced after bovine bone gelatin substitution (and other measurements not presented in this work are consistent with this conclusion). This is likely due to the difference in the effective particle size distributions of the two gelatin materials (figure 3b), which affects the optical attenuation (dose) readout. The overall result is then to reduce the stray light-induced shift (figure 3c inset plot) in the dosimetry normally observed with the use of the standard porcine gelatine [1,2], and to improve the FXG gel measurement of dose (figure 3c) to the target level of 95% voxel agreement using 3%, 3mm Low's gamma function criteria.

Acknowledgements
Research funding for this work has been provided by the Canadian Institutes of Health Research (CIHR) and the Cancer Centre of Southeastern Ontario.

References

[1] Olding T, Holmes O and Schreiner L J 2010 Cone beam optical computed tomography for gel dosimetry I: scanner characterization *Phy. Med. Biol.* **55** 2819-40
[2] Babic S, Battista J and Jordan K 2008 Three-dimensional dose verification for intensity-modulated radiation therapy in the radiological physics centre head-and-neck phantom using optical computed tomography scans of ferrous xylenol-orange gel dosimeters *Int. J. Radiat. Oncol. Biol. Phys.* **70** 1281-91
[3] Kelly B G, Jordan K J and Battista J 1998 Optical CT reconstruction of 3D dose distributions using the ferrous- benzoic-xylenol (FBX) gel dosimeter *Med. Phys.* **25** 1741-50
[4] Low D A and Dempsey J F 2003 Evaluation of the gamma dose distribution comparison method *Med. Phys.* **30** 2455-64
[5] Wulfsberg G 2000 *Inorganic Chemistry* (Sausalito: University Science Books) p 368

Journal of Physics: Conference Series **250** (2010) 012029 doi:10.1088/1742-6596/250/1/012029

Ionization chamber array for patient specific VMAT, Tomotherapy and IMRT QA

Sotiri Stathakis
Department of Radiation Oncology, Cancer Therapy and Research Center at the University of Texas Health Science Center, San Antonio TX 78229, USA.
Stathakis@uthscsa.edu

Abstract. The evaluation between measured and calculated dose is essential in the patient specific quality assurance procedures for intensity modulated radiation therapy. The high complexity of volumetric arc radiotherapy, Tomotherpay and intensity modulated radiation therapy deliveries attributed to the dynamic and synchronization requirements of such techniques require new methods and potentially new tools for the quality assurance of such techniques. Studies evaluating the dosimetric performance of EDR2 film and a 2D ionization chamber array quality assurance device have been performed in our institution. Our results showed that differences between the detector systems are small. The respective gamma index histograms showed that when 3% dose difference and 3mm distance to agreement are used, more than 90% of the evaluated points were within the tolerance criteria

1. Introduction

Volumetric modulated arc therapy (VMAT) is a novel extension of conventional intensity-modulated radiotherapy (cIMRT), in which an optimized three-dimensional dose distribution may be delivered in a single gantry rotation. VMAT is the predecessor to RapidArc (Varian Medical System). Leaf positions and segment weights (control points) are directly optimized, and a variable dose rate is used. RapidArc incorporates capabilities such as variable dose-rate, variable gantry speed, and accurate and fast dynamic multileaf collimators (DMLC), to optimize dose conformality, delivery efficiency, accuracy and reliability. Tests have been designed that evaluate RapidArc performance in a stepwise manner. The accuracy of DMLC position during gantry rotation, the ability to vary and control the dose-rate and gantry speed and, the combined use of variable DMLC speed and dose-rate can be independently tested. Similar to IMRT with static gantry angles and set dose rate, the RapidArc deliveries require patient specific QA. The evaluation between measured and calculated dose is essential in the patient specific quality assurance procedures for intensity modulated radiation therapy. The high complexity of volumetric arc radiotherapy delivery attributed to the dynamic and synchronization requirements of such delivery require new methods and potentially new tools for the quality assurance of such techniques. Studies evaluating the dosimetric performance of EDR2 film and commercially available quality assurance devices have been performed in our institution. Among the commercially available 2D array detector devices, based on either diode or ionization chamber arrays[1-3] show excellent characteristics in terms of linearity, repeatability, and independence from dose rate effects while at the same time have limited spatial resolution, due to the discrete although parse number of detectors available when compared to film. (figure1) Poppe et al.,[3] analyzed the influence of those parameters on accuracy of IMRT dose measurements and concluded that dose variations in realistic IMRT dose distributions contain very little, if any, spatial frequency components

Journal of Physics: Conference Series **250** (2010) 012029 doi:10.1088/1742-6596/250/1/012029

above 0.1 mm^{-1}. Spezi et al.,[4]show that the PTW Seven29 array can detect leaf positional errors down to 1mm and also spatial resolution has been discussed by others for I'matrixx[5] and Delta[4] array[6-7]. Since, our goal of the study is not to characterize or evaluate the different detector arrays which can be been found elsewhere.

2. Materials and Methods

2.1. PTW 2D Array and Octavius phantom

The PTW seven29 2D-Array consists of 729 vented plane-parallel ionization chambers with a $0.6g/cm^2$ graphite wall arranged in a 27 x 27 matrix covering an area of 27 cm x 27 cm. Each single chamber is air-filled with a cross section of 5mm x 5mm and height of 5mm. The chambers are separated from each other by 5mm. The distance between the centers of adjacent chambers is 10mm. The 2D array surrounding material is made up of polymethyl methycrylate (PMMA). The measuring system consists of the chamber array itself, which also accommodates part of the electronic devices, the array interface, and a data acquisition board for the personal computer. A dedicated phantom for the QA of rotational treatments focusing primarily on the use of the Seven29 2D ion chamber array, called Octavius was used during measurements. Octavius is made of polystyrene (physical density 1.04 g/cm3, relative electron density 1.00), and is 32cm wide and has a length of 32 cm. A 30 x 30 x 2.2 cm^3 central cavity allows the user to insert the 2D ion chamber array into the phantom. The position of the cavity is such that when the 2D array is inserted, the plane through the middle of the ion chambers goes through the center of the phantom. The measurement ranges for PTW Seven29 as specified by manufacturer are 200 mGy – 1000 Gy and 500mGy min^{-1} to 8 Gy min^{-1}. The 2D array is calibrated for absolute dosimetry in a Co^{60} photon beam at the PTW secondary standard dosimetry laboratory. An on-site cross calibration factor correcting for the quality of the beam can be measured and used by the detector acquisition software. In this work the detector array was used in absolute dose measuring mode and dose values were corrected for daily variation of linac output.

2.2. Dose Verification

2D detector arrays have two limitations of their geometrical resolution, the size of the single detector and the centre-to-centre distance between the detectors. In the case of the PTW Seven29 the first limitation has been dealt with by adjusting the dose verification routine to the measurement process. The measured values are compared to the convolution product of the calculated dose distribution and the single-detector lateral response function, and the width of this response function is sufficiently small to enable the detection of 1 mm MLC misalignments. The second limitation has been dealt with by taking two frames of each dose distribution with a step width of 5 mm in between, so that a sampling frequency of 0.2 mm^{-1} is achieved in the direction of the shift. This may be generalized by applying the 5 mm shift in two dimensions. Applying the Nyquist theorem to the measured values of realistic dose distributions, it has been shown[8] that the sampling frequency of 0.2 mm^{-1} is appropriate. Due to the optimized MLC-adapted geometry of the PTW Seven29, the required 5 mm shift between two frames of the same dose distribution can be easily and precisely achieved.

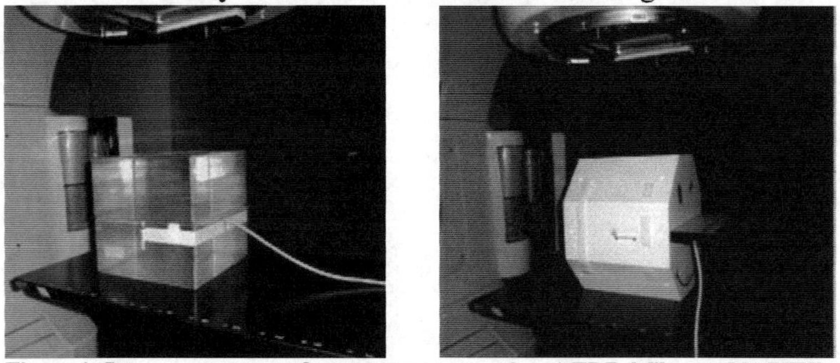

Figure 1. Setup arrangement for measurements using a) EDR 2 film, and b) PTW Seven29 Array

3. Results - CTRC experience with PTW Seven29

Our results showed that differences between the film and the PTW Seven29 measurements of IMRT, RapidArc and Tomotehrapy patient specific QAs are small (Table 1). The respective gamma index histograms showed that when 3% dose difference and 3mm distance to agreement are used, more than 90% of the evaluated points were within the tolerance criteria. The agreement as reflected through gamma index passing rates was higher on average for the Seven29 2D array compared to the film measurements.

Accurate dose validation is a key component in the successful delivery of patient treatments. The PTW Seven29 with Octavius is accurate with the clinical criterion of 3% and 3 mm and is also accurate with more stringent criteria, 2% and 2 mm, for clinically used field widths and pitches. With more obscure settings of 1.06 cm FW and pitch of 2, higher discrepancies were observed using more stringent criteria. The Seven29/Octavius and the film measurements are comparable in performance based on average gamma passing percents. Based on these measurements however, the Seven29/Octavius gives better results when using the clinical passing criteria of 3%, 3 mm as well as the more stringent 2% and 2 mm when comparing QA plans with helical TomoTherapy.

Table 1. Comparison of Gamma index values between film measurements and PTW seven29 2D Array

		RapidArc			IMRT			Tomotherapy	
PLAN NO	SITE	**Film**	**PTW Seven29**	SITE	**Film**	**PTW Seven29**	SITE	**Film**	**PTW Seven29**
1	Brain	91.69	94.55	Brain	98.24	96	Pancreas	94.6	100.00
2	Brain	99.34	98.9	H&N	98.21	96.96	Prostate	99.7	99.90
3	H&N	94.38	95.79	Liver	97.32	96.19	Anus	98.4	100.00
4	Liver	99.99	99.78	Lung	93.62	97.47	Prostate	96.4	99.70
5	Lung	99.63	98.7	Prostate	99.55	97.49	Prostate	98.9	99.90
6	Lung	99.87	99.95	Prostate	94.92	98.13	Prostate	95.1	100.00
7	Prostate	94.95	96.05	Prostate	96.72	96.77	Pelvis	98.2	99.90
8	Prostate	98.79	97.94	Prostate	98.47	97.74	Pelvis	95.1	97.70
9	Prostate	94.24	96.78	Prostate	97.94	97.88	Prostate	96.6	99.30
10	Prostate	93.98	96.2	Prostate	99.63	99.96	Prostate	91.2	96.60
11	Prostate	98.37	97.81				Prostate	97.5	100.00
12	Prostate	99.11	98.42				Liver	92	97.60
13	Prostate	92.93	99.45				Larynx	95	96.20
14	Prostate	99.23	99.78				Spine	94.3	96.80
15	Spine	96.76	98.72				Hypopharynx	97.5	93.40
	Average	**96.88**	**97.92**	Average	**97.46**	**97.46**	Average	**96.03**	**98.47**

4. Conclusions

Although rotational radiotherapy treatments are increasingly used, the developed technology is relatively new and requires careful monitoring and verification. For the verification of these treatment methods, the Seven29 2D ion chamber array provides an overall accuracy comparable to that of single ion chamber measurements when it is used in combination with the Octavius phantom. This phantom contains a compensation cavity to rectify the different collection efficiency when the array is irradiated from the rear. This QA method facilitates the pretreatment verification process by providing on-line absolute 2D dose information.

The availability of 2D arrays for patient specific quality assurance has provided another established method of accurately validating precise, measured dose distributions. The ability of current 2D array technology to provide stable and reproducible results as shown in multiple published literature, validates the future role of 2D arrays as a robust dosimeter. Current 2D ionization array systems such as the PTW Seven29, eliminate the need of film and ion chamber-based dose validation and further shorten the time associated with dose validation due to their ability of instantaneously reading, processing and analyzing the data.

References

[1] **F. B. Buonamici, A. Compagnucci, L. Marrazzo, S. Russo and M. Bucciolini, "An intercomparison between film dosimetry and diode matrix for IMRT quality assurance," Med Phys 34, 1372-1379 (2007).**

[2] **J. L. Bedford, Y. K. Lee, P. Wai, C. P. South and A. P. Warrington, "Evaluation of the Delta4 phantom for IMRT and VMAT verification," Phys Med Biol 54, N167-176 (2009).**

[3] **B. Poppe, A. Blechschmidt, A. Djouguela, R. Kollhoff, A. Rubach, K. C. Willborn and D. Harder, "Two-dimensional ionization chamber arrays for IMRT plan verification," Med Phys 33, 1005-1015 (2006).**

[4] **E. Spezi, A. L. Angelini, F. Romani and A. Ferri, "Characterization of a 2D ion chamber array for the verification of radiotherapy treatments," Phys Med Biol 50, 3361-3373 (2005).**

[5] **J. Herzen, M. Todorovic, F. Cremers, V. Platz, D. Albers, A. Bartels and R. Schmidt, "Dosimetric evaluation of a 2D pixel ionization chamber for implementation in clinical routine," Phys Med Biol 52, 1197-1208 (2007).**

[6] **V. Feygelman, D. Opp, K. Javedan, A. J. Saini and G. Zhang, "Evaluation of a 3D diode array dosimeter for helical tomotherapy delivery QA," Med Dosim (2010).**

[7] **V. Feygelman, K. Forster, D. Opp and G. Nilsson, "Evaluation of a biplanar diode array dosimeter for quality assurance of step-and-shoot IMRT," J Appl Clin Med Phys 10, 3080 (2009).**

[8] **B. Poppe, A. Djouguela, A. Blechschmidt, K. Willborn, A. Ruhmann and D. Harder, "Spatial resolution of 2D ionization chamber arrays for IMRT dose verification: single-detector size and sampling step width," Phys Med Biol 52, 2921-2935 (2007).**

Journal of Physics: Conference Series **250** (2010) 012030 doi:10.1088/1742-6596/250/1/012030

Development of a radiochromic ferric oligomer hydrogel

Kevin Jordan and Masaya Sekimoto

London Regional Cancer Program, London Health Sciences Centre, University of Western Ontario, Dept. of Medical Biophysics

Kevin.jordan@lhsc.on.ca

Abstract. Ferrous gelatin hydrogels were prepared by using sulphuric acid concentrations lower than required to maintain radiation induced ferric ions fully hydrated. The ferric hydroxyl species that are produced following irradiation exhibit a radiochromic response that can be probed with blue light. The dose distribution shapes were stable in time, indicating no long term diffusion. An over response to dose gradients was observed both in one centimeter cuvette samples and litre volumes probed with optical cone beam CT. This ferrous hydrogel may represent a model system for studying iron radiochemistry in biological systems.

1. Introduction

The radiochromic conversion of acidic aqueous solutions of ferrous ions to ferric ions ($[Fe(H_2O)_6]^{+3}$) in the pH range 0.5 to 2 is known as the Fricke reaction. The extent of the reaction is measured by recording the absorbance change in the ultraviolet spectral region at 304 nm. By performing this reaction in hydrogels, convection mixing is eliminated and spatially resolution of the dose distribution is maintained. The corresponding 3D dose distributions were initially recorded in ferrous agarose gels with magnetic resonance imaging [1] because examination of ferrous agarose hydrogels at 304 nm is problematic. The agarose gels have sufficient scatter to make them appear translucent with visible light. The scatter is much great for shorter wavelength UV light. The ferrous gelatin gels appear transparent with visible light. However, the gelatin gels still absorb and scatter ultraviolet light and causes ferrous oxidation through photochemistry. The photochemistry becomes another variable for quantitative measurements. These effects make optical CT scanning of Fricke type gelatin hydrogel impractical at 304 nm for quantitative measurements over pathlengths greater than a few mm. However, dose responses in ferrous gels can and have been studied at 304 nm for 1 cm cuvettes [2].

In less acidic solutions ferrous ions are readily oxidized and the product ferric ions are hydrolyzed and form complexes that easily precipitate from water. Dinuclear and multinuclear species form resulting in red-brown colloidal gels. Hydroxo species such as $[Fe(OH)(H_2O)_5]^{+2}$ are yellow due an ultraviolet charge transfer band which tails into the visible spectrum [3]. These species undergo photochemistry when illuminated and can result in the production of OH radicals. For this reason, aqueous iron photochemistry is under investigation as a potential mechanism for the inexpensive decomposition of pollutants in industrial effluent [4]. Iron chemistry is important in biological reactions and is considered to be base of several medical conditions. It is expected that basic iron chemistry may be effectively studied in gelatin hydrogels as model radiobiological systems. This preliminary study examines ferrous gelatin hydrogel radiochromic response at higher pHs where the anticipated distribution of species may have absorption spectra that can be optically probed with visible blue light. These multinuclear species are expected to have very low diffusion coefficients due to their higher molecular mass behaving like small polymers in hydrogel matrix. However, extrapolations from known pH dependence at higher acid concentrations predict these ferrous gels will have much lower radiation sensitivity and not useful as 3D dosimeters. This study is interested in the chemistry that is initiated by the radiation not the measurement of absorbed dose.

2. Methods

Ferrous gelatin hydrogels were prepared by adding ferrous ammonium sulphate (1 mM) to 4% gelatin solutions containing sulphuric acid at 30 C. The acid concentration was varied from 0.1 to 50 mM. Dose response and diffusion were studied with samples in 1 cm pathlength, disposable, poly(methyl methacrylate) cuvettes. One litre samples for optical CT scanning were poured into polyethylene terephthalate jars, supplied by Modus Medical Devices. Absorption measurements were performed with a Perkin and Elmer model 139, UV-VIS absorption spectrometer. Optical CT scanning was conducted with a commercial optical cone beam scanner, model Vista10 from Modus Medical Devices, modified for this experiment. The LED light box was removed and replaced with an actinic fluorescent lamp which was filtered with dark blue cobalt glass. The filtered emission spectrum was a narrow peak at 420 nm. The narrow bandpass filter at the CCD camera lens was also removed for this scan. The gel was optical CT scanned before and after axial, irradiation to 120 Gy at depth of maximum dose (d_{max}) with a 12 MeV electron beam, (field size 6 x 6 cm, SSD 100 cm, dose rate 400 cGy min^{-1} at d_{max}). The 512 image sets were reconstructed using software provided by with the Vista10 scanner. For the diffusion study, transmission images of half irradiated cuvettes (200 Gy) at 420 nm were recorded with an in-house optical cone beam scanner system at 12 pixels per mm resolution at several post irradiation times. The reference image was of air.

3. Results

The radiation sensitivities of the gel formulations were a strong function of pH. At wavelengths less than 350 nm, the response increased with lower pH. In contrast, at 420 nm, the sensitivity increased as the sulphuric acid concentration decreased from 50 to 0.1 mM. The absorption spectrum of irradiated gels peaked at less than 340 nm and tailed past 450 nm. In figure 1, transmission profiles of a cuvette that had been half irradiated to 200 Gy is shown. The profiles at 0.5 and 3.5 hours post irradiation are shown. Note the later profile was offset in order to demonstrate the stability of the profile shape. The offset was necessary due to baseline drifts due to auto oxidation and changing lamp output. Transmissions were referenced to air. Note the over response to absorbed dose in the vicinity of the dose gradient. Specifically, the sharp dip in transmission near 17 mm position. Since the profiles are identical all radiochromic reactions are complete in less than 0.5 hours and there is no diffusion of the radiochromic products.

A transverse slice from the 3D reconstruction corresponding to the depth of maximum dose is shown in figure 2. The post irradiation scan required 8 minutes and was initiated 10 minutes post irradiation. Note the enhancement in radiochromic response at the edges of the electron beam. This effect is consistent with the x-ray half sample irradiation of figure 1.

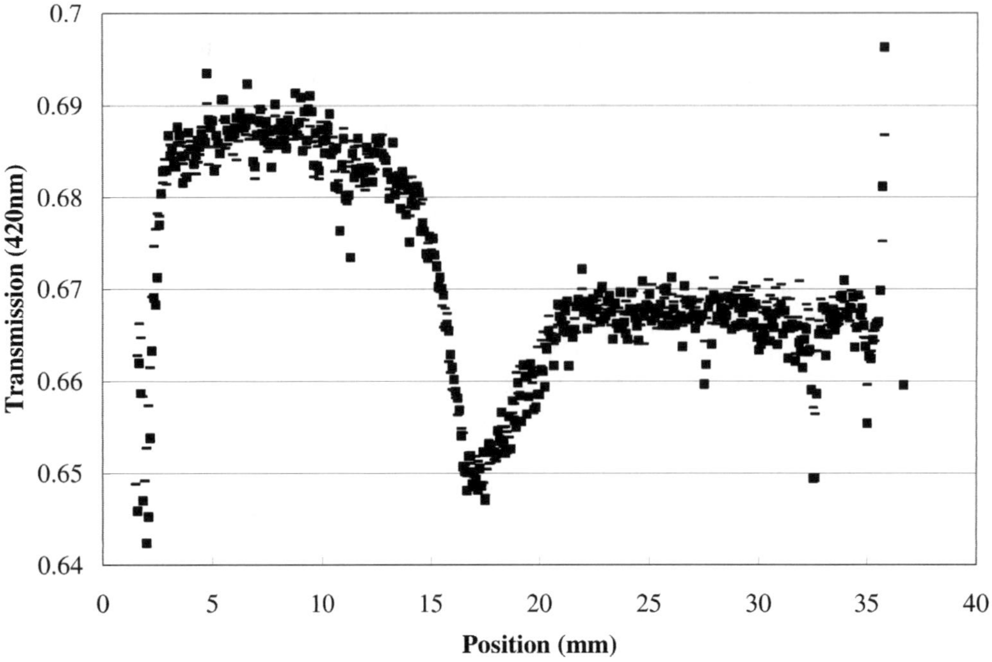

Figure 1. Transmission profiles for "ferric polymer" hydrogel with air reference recorded at 0.5(-) and 3.5(■) hours. Right half of gel cuvette (~17 to 35 mm) irradiated to 200 Gray. Note over response at dose gradient.

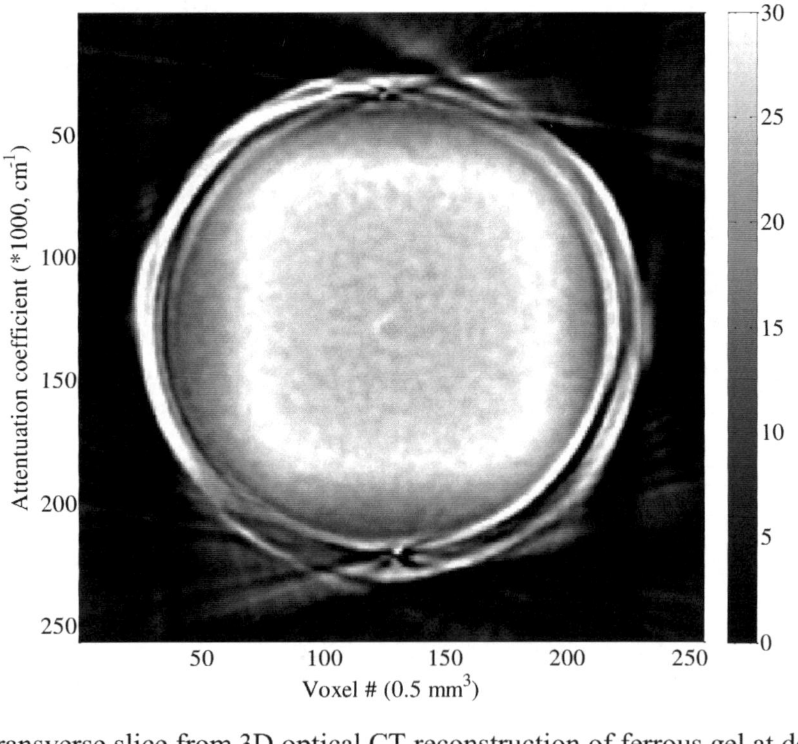

Figure2. Transverse slice from 3D optical CT reconstruction of ferrous gel at depth of maximum dose, 12 MeV, 120 Gy, 6 x 6 cm field size. Note radiochromic over response at edge of beam.

4. Summary

Ferrous hydrogels prepared with lower concentrations of sulphuric acid than required to maintain ferric oxidation species as single iron molecules do exhibit a small radiochromic response that can be optically probed with blue light. The reaction is rapid and complete in less than 10 minutes. The shape of the radiochromic 3D distribution is fixed, demonstrating that diffusion the species absorbing blue light is not measurable on the time scale of three hours. A marked over response to dose gradients was observed for both x-ray and electron beams. The nature of the over response requires further investigation but it appears that ferrous ions in the low dose region, diffuse to multinuclear species growing in the gradient regions. The results demonstrate that this material is inappropriate for conventional radiation dosimetry but may be a model system for chemical dosimetry within biological systems.

Acknowledgements: Funding was from Ontario Research and Development Challenge Fund: Ontario Consortium for Image-guided Therapy and Surgery and London Regional Cancer Program Small Grants fund.

References

[1] Gore JC, Kang YS and Schulz RJ, 1984 Measurement of radiation dose distributions by nuclear magnetic resonance (NMR) imaging, Phys Med Biol **29**, 1189-97

[2] Hill B, Bäck S, Lepage M, Simpson J, Healy B and Baldock C, 2002 Investigation and analysis of ferrous sulfate polyvinyl alcohol(PVA) gel dosimeter, Phys Med Biol **47**:4233-46

[3] Cotton FA and Wilkinson G, 1980 Advanced inorganic chemistry: A comprehensive text 4th ed, Wiley Interscience

[4] Horvath CO and Huszank R, 2003 Degradation of surfactants by hydroxyl radicals photogenerated from hydroxoiron (III) complexes, Photochem. Photobiol. Sci., **2**, 960-966

Journal of Physics: Conference Series **250** (2010) 012031 doi:10.1088/1742-6596/250/1/012031

Non-diffusing radiochromic micelle gel

Kevin Jordan and Masaya Sekimoto

London Regional Cancer Program, London Health Sciences Centre and Department of Medical Biophysics, University of Western Ontario, London, ON, Canada

Kevin.jordan@lhsc.on.ca

Abstract. The addition of Laponite, a synthetic clay nanoparticle material to radiochromic leuco Malachite Green micelle hydrogel eliminates diffusion of the cationic dye by electrostatic binding. The clay nanoparticles also increased dose sensitivity ten-fold relative to the parent gel formulation. This material is a suitable 3D water equivalent dosimeter with optical CT readout.

1. Introduction

Transparent, radiochromic hydrogels are ideal candidates for tissue or water-equivalent radiation dosimetry with optical readout. Materials with minimal scatter can be efficiently scanned by both broad-beam and single-ray optical CT scanner geometries. Also, low scatter materials are preferred for scanning larger volumes, such as cylinders up to 30 cm in diameter. Radiochromic micelle gelatin hydrogels have been reported for both leuco malachite green (LMG) [1] and leuco crystal violet [2]. The respective diffusion coefficients of these gels are 3 and 25 times lower than the reference ferrous xylenol orange gel (FX). While these gels are promising for quantitative 3D dosimetry, the current formulations have 30 and 10 times lower dose sensitivities relative to typical FX gel formulations.

Clays are known to provide binding surfaces for solvents and solutes. In particular, clays are under investigation as low cost materials for trapping dyes in effluent from industrial textile production. Laponite is a synthetic nanoparticle clay which has several chemical properties which may be useful for improving the performance of radiochromic hydrogels. Laponite, binds cationic dyes through an electrostatic interaction [3], and binds to collagen in gelatin hydrodgel [4]. At higher concentrations it directly forms transparent, colourless hydrogels. This preliminary investigation reports that introducing a small amount of Laponite into LMG micelle gel eliminates the diffusion of malachite green formed after irradiation of the hydrogel.

2. Methods

The gel formulation was: 4% gelatin, 0.1% Laponite, 15 mM trichloroacetic acid, 7 mM Triton X-100 and 0.3 mM LMG. The gelatin (porcine, 300 bloom, Sigma) was slowly added to water (at 22 C, deionized Millipore) and then stirred in a water bath at 40 C. Similarly, the Laponite RD (sample from Southern Clay Products, Tx USA) was sprinkled on water at 22 C and stirred to form a clear colourless solution. An aqueous solution containing, Triton X-100 (Sigma-Aldrich reagent grade), trichloroacetic acid (Mallinckrodt, Cat No. UN1839) and LMG (Aldrich reagent grade) was stirred at 22 C to form a transparent, pale green solution. The Laponite solution was added slowly to the warm gelatin solution and mixing continued for two hours at 40 C. After lowering the gelatin solution to

Journal of Physics: Conference Series **250** (2010) 012031 doi:10.1088/1742-6596/250/1/012031

30°C the surfactant solution was slowly added while stirring and poured into 1 cm plastic cuvettes. The samples were stored at 4°C for gelation. Samples were irradiated under conditions of full scatter with 4 or 6 MV x-ray beams from medical linear accelerators at dose rates near 4 Gy min^{-1}. For dose response measurements, transmission at 633 nm was measured between irradiations, with about 10 minutes between points. For diffusion studies, half of a cuvette was irradiated to 5 Gy and a sequence of transmission images at 590 nm were recorded at 22 C.

3. Results

The dose sensitivity of this gel increased with concentration of LMG, Laponite and gelatin. Sensitivity was also related to concentrations of surfactant and ACl_3. Increasing Laponite concentration from 0.1 to 0.2% by mass increased both scatter and dose sensitivity. In figure 1, transmission profiles recorded 0.5 and 3.5 hours post irradiation are shown. Except, for an offset due to auto-oxidation the two profiles are essentially the same. This demonstrates that the MG^+ dye that is formed is immobilized when Laponite is present. In figure 2, the dose response for this formulation is presented. The dose sensitivity is around 25% of typical FX gel formulations. All formulations investigated demonstrated a linear doses response with development times less than one minute and no fractionation effects.

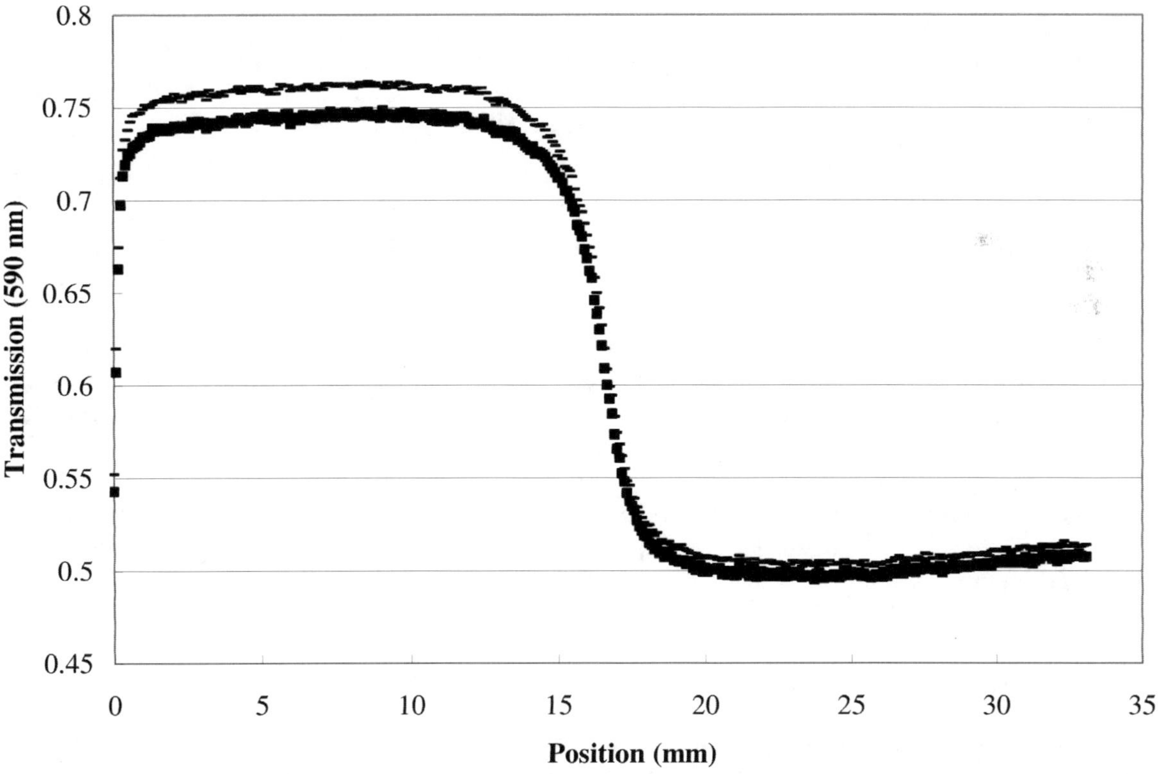

Figure 1. Transmission profiles for LMG-micelle-Laponite gel in half irradiated cuvette at 0.5 (-) and 3.5 (■) hours post irradiation, air reference.

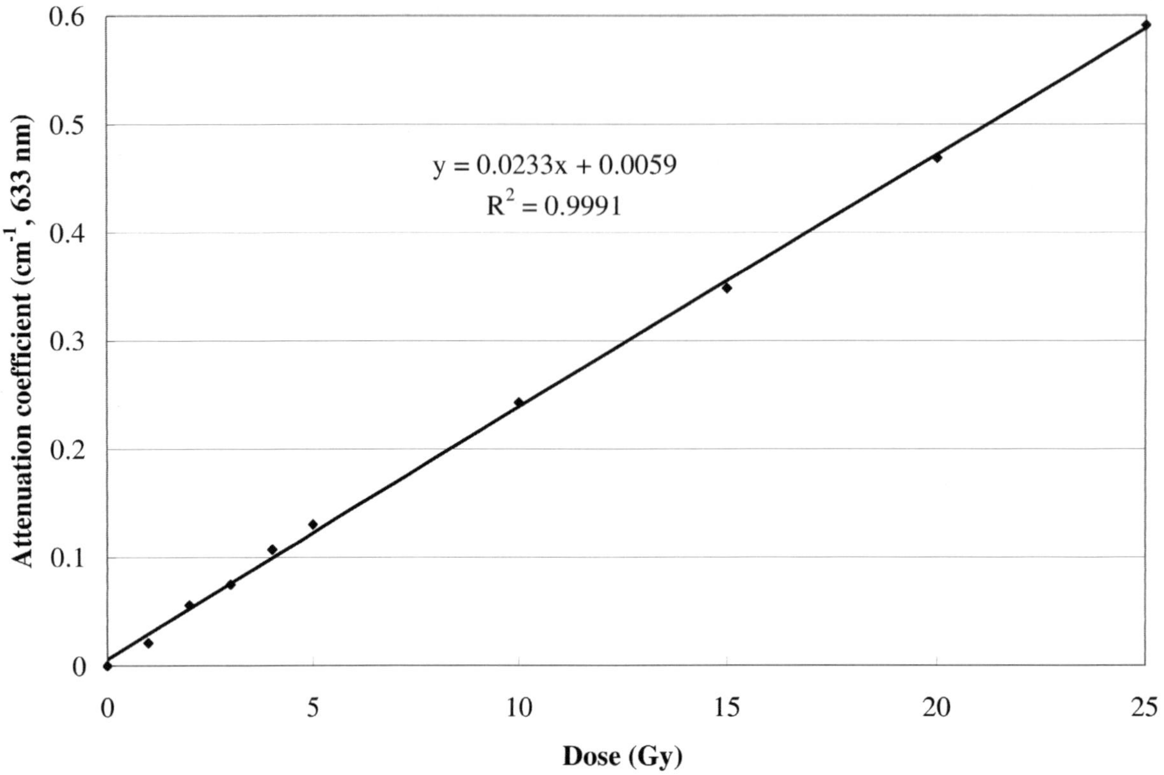

Figure 2. Radiochromic dose response of LMG-micelle gel with 0.1% Laponite.

4. Summary

Laponite RD clay nanoparticles when added to radiochromic LMG micelle gels eliminate diffusion and increase the dose sensitivity roughly ten fold. The Laponite also causes a moderate increase of optical scatter in the gelatin gels. Laponite may be a general chemical approach to elimination of diffusion in radiochromic gels that contain cationic dyes.

Acknowledgements: Funding was from Ontario Research and Development Challenge Fund: Ontario Consortium for Image-guided Therapy and Surgery and London Regional Cancer Program Small Grants fund.

References

[1] Jordan K and Avvakumov N, 2009 Radiochromic leuco dye micelle hydrogels: I. Initial investigation, Phys Med Biol **54**:6773-89
[2] Babic S, Battista J and Jordan K, 2009 Radiochromic leuco dye micelle hydrogels: II. Low diffusion rate leuco crystal violet gel, Phys Med Biol **54**:6791-6808
[3] Zhang L-M, Zhou Y-J and Wang Y, 2006 Novel hydrogel composite for the removal of water-soluble cationic dye, J Chem Tech Biotech **81**:799-804
[4] Pawar N and Bohidar H B, 2009 Surface selective binding of nanoclay particles to polyampholyte protein chains, J Chem Phys **131**:045103-1-10

Journal of Physics: Conference Series **250** (2010) 012032 doi:10.1088/1742-6596/250/1/012032

Feasibility of quantitative PET/CT dosimetry for proton therapy using polymer gels

O A Zeidan[1], W C Hsi[2], O Lopatiuk-Tirpak[1], S I Sriprisan[1], S L Meeks[1], P A Kupelian[1], Z Li[3] and J R Palta

[1]M. D. Anderson Cancer Centre Orlando, Orlando, FL

[2]ProCure, Bloomigton, IN

[3]University of Florida Proton Therapy Institute, Jacksonville, FL

Email: lenatirpak@gmail.com

Abstract. A feasibility study of proton beam PET/CT off-line quantitative dosimetry using polymer gels is presented. A newly developed proton-sensitive polymer gel dosimeter (BANG®3-Pro2) is used as a dosimeter and a tissue-equivalent phantom medium for this study. We explore a new approach to correlating measured proton 3-dimensional (3D) dose distributions directly to measured positron emission from in the gel medium using PET/CT imaging. A large cylindrical volume (2.2 Litres) of the gel was irradiated with a clinical modulated proton beam using irregular-shaped aperture geometry. The gel was imaged in a nearby PET/CT unit immediately (<3 min) after irradiation. Dose distribution in the gel was generated using an optical tomography scanning system. Direct 3D spatial comparison of dose and positron emission distributions was then performed. Profiles along the beam path show that the distal fall-off of the dose is nearly 2 cm deeper than the activity profile which is comparable to previous studies with plastic phantoms and Monte Carlo simulations of activity distributions. Planar PET and dose distributions at depth and perpendicular to beam axis show a strong one-to-one spatial correlation. This phantom study demonstrates that the gel medium could be potentially useful for quantifying various physical factors that can influence the PET activity range verification method in patients.

1. Introduction

Several studies have demonstrated that off-line PET imaging of proton beams in phantoms may potentially be a useful technique for proton therapy dosimetry [1]. Widespread clinical implementation of this technique is still challenging due to underlying physical factors that significantly affect the accuracy of this method [2]. Some of these factors include patient intra-fraction motions, variations of tissue compositions, blood perfusion, and setup uncertainties. In order to quantify PET and dose distributions in tissue-equivalent materials, direct measurement of dose and PET activity in the same phantom medium is needed. We have recently shown that a proton-sensitive formulation of the BANG3 polymer gels, BANG3-Pro2, is capable of reproducing ion chamber dose profiles in water with little or no observed quenching in the Bragg peak [3]. The gel dosimeter has very similar elemental composition by weight to that of human tissue. Therefore, the BANG3-Pro2 dosimeter is an ideal medium for systematic studies of PET and dose spatial correlations for proton beam dose and range verifications. This off-line dosimetry technique has some distinct advantages compared to previous off-line phantom verification studies. First, the homogenous gel medium acts as both as a tissue-equivalent phantom and a dosimeter. Second, the gel medium provides a large-volume dosimeter which allows for accurate and direct correlation of PET activity and dose in 3D for a wide range of beam configurations. In this study we describe our experimental approach for this unique verification method using a clinical spread-out Bragg peak (SOBP) beam and present a quantitative description of the correlation between measured PET emissions and dose distributions in 3D.

2. Materials and Methods

A large volume (2.2 Liters) of the BANG3-Pro2 gel was used for this work. The gel medium is tissue equivalent and has a physical density of 1.08 g/cm^3. The dosimetry system is composed of the gel dosimeter and a laser-based optical computed tomography scanner, OCTOPUS^{TM}-IQ, (MGS Research, Inc) with sub-millimeter scanning resolution capability. The proton-sensitive gel is provided by the manufacturer (MGS Research, Inc) in a clear plastic cylindrical container, measuring approximately 12.7 cm in diameter and 17.5 cm in height.

The gel dosimeter was irradiated at the University of Florida Proton Therapy Institute (UFPTI, Jacksonville, FL) using a modulated clinical proton beam with passive double-scattering technique. The gels were immersed under water in a Wellhofer scanning tank (IBA Dosimetry America) with the long axis of the gel cylinder coinciding with the beam axis. The water depth above the gel was adjusted so as to capture most of the 15.1 cm range modulated (5.0 cm) beam profile including the distal end. An irregular-shaped aperture was chosen for this study to evaluate 3D PET activity in the gel from complex beam aperture geometries. An overview of the experimental setup is shown in **figure 1**. The gel was irradiated for a total of 3.37 min (beam on time) to a maximum dose of 600 cGy. The dose delivered to the gel was chosen so as to maximize the induced PET activity while maintaining the ability to read the dose within the linear region of the dose response curve of the dosimeter system.

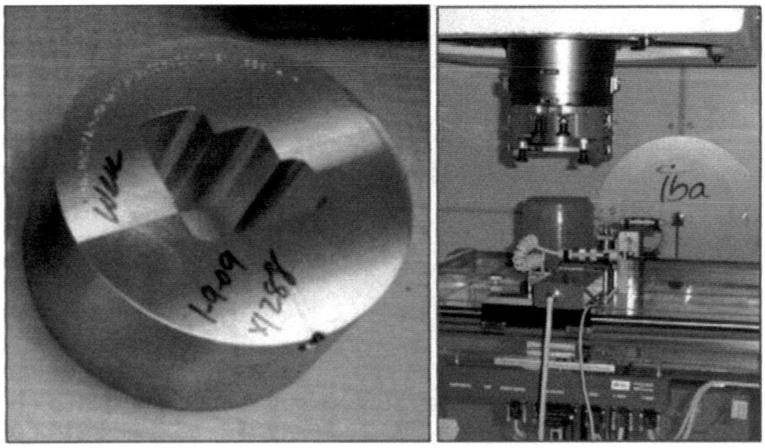

Fig 1(Left) The brass aperture used in this study. Experimental setup for gel proton beam irradiation using the Wellhofer water tank (Right). The gel is aligned vertically along the beam axis in a special mount that allows the gel to be immersed at variable water depths.

Immediately post exposure, the gel dosimeter was quickly transferred to a nearby PET/CT unit for image acquisition. The time between beam off and start of PET acquisition was 2.57 minutes. The PET/CT imaging was performed using the Philips Gemini PET/CT located on-site at UFPTI. The CT scans were obtained with 1.0 mm slice thickness and 1.172×1.172 mm^2 pixel spacing and a field of view of 60 cm. This CT study was then later used to perform CT-based phantom attenuation correction (CTAC) of the proton-irradiated gel. The PET imaging following the CT scan was them performed for 60 minutes. This acquisition time was sufficient to collect most of the induced activity within the gel volume since the total decay event counts drop by at least two orders of magnitude over this time period [4].

Journal of Physics: Conference Series **250** (2010) 012032 doi:10.1088/1742-6596/250/1/012032

At the elemental level, the gel is composed mainly of Oxygen (^{16}O), Carbon (^{12}C), Hydrogen (^{1}H), and Nitrogen (^{14}N) with w/w fractions of 0.730, 0.144, 0.101, and 0.024, respectively. Both ^{16}O and ^{12}C have have comparable β^+ reactions cross sections over therapeutic proton beam energies [4]. The half life of ^{15}O isotope is 2 minutes, while the half life for ^{11}C isotope is about 20.3 minutes. Therefore, in order to maximize the PET signal statistics from ^{15}O, the time between beam off and start of PET image acquisition needs to be minimized to be within 10 minutes after irradiation. Axial, sagittal, and coronal images if the gel's CTAC PET image are shown in **figure 2.** The PET-defined beam path is shown clearly in the coronal and axial planes, while the sagittal plane reflects the aperture geometry of the incident proton beam. The PET emission distributions from the co-registered PET/CT volume image data set was visualized and analyzed in the Eclipse^{TM} (Varian Medical Systems) treatment planning system. The geometrical correlation between the PET and gel dose data was possible using the geometrical landmarks of the gel container external dimensions and the locations fiducial BBs on the container surface.

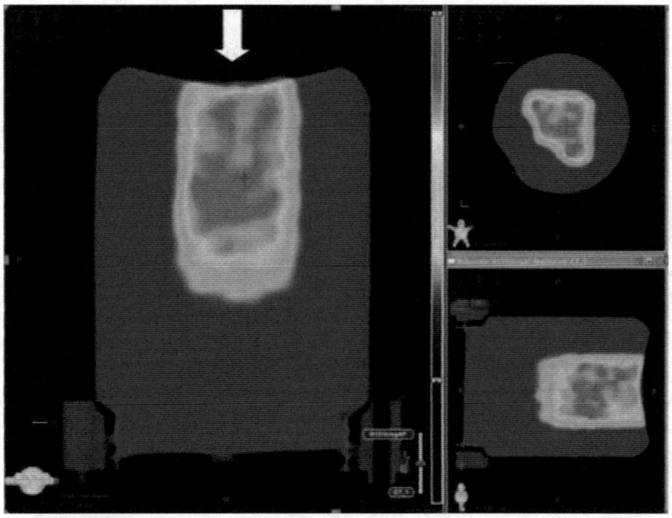

Fig 2 Co-registered PET/CT image set showing positron emission distribution in the gel from an SOBP beam delivery in the coronal, sagittal, and axial views. A cross sections view (top right) reflects the geometrical shape of the irregular aperture in the axial plane. The arrow points to the direction of beam entrance.

3. Results and Discussion

A comparison of a 2D dose distribution from gel and the corresponding PET signal distribution is shown in **figure 3.** The PET 2D distribution reflects the raw positron emission counts captured by the PET scanner. The activity map reflects the geometrical shape of the aperture shown in **figure 1** and coincides with dose distribution measured in the gel as shown in **figure 3** (left). The two maps are overlaid onto a 1.0 cm voxel grid to facilitate 2D spatial comparison. The 50% isodose line and the 50% isoemission contour were found to coincide everywhere on the grid to better than half a voxel or 2.5 mm. This is one of the unique aspects of the gel as a volumetric dosimeter. Unlike other types of dosimeters, the gel medium offers the capability to perform direct and accurate spatial correlations between PET and dose distributions within the volume of the gel itself.

The most dosimetrically relevant result from this study is the ability to perform relatively accurate spatial correlation between the distal end fall-off of the PET depth activity profile and the SOBP distal end fall off to the modulated beam along the beam axis as shown in **figure 4**. This range verification method is the most clinically relevant feature of this dosimetry technique. In-vivo dose verification plays an important role in treatment QA as beam distal end variations at depth could result in an 'overshoot' or 'undershoot' of the distal portion of the target volume [2]. Both profiles were normalized to their maximum raw values to yield the relative profiles in **figure 4**. The gel data was not acquired over the entire physical size of the gel cylinder due to the nature of the mechanical setup of

Journal of Physics: Conference Series **250** (2010) 012032 doi:10.1088/1742-6596/250/1/012032

the gel in the optical scanner scanning tank. However, the collected PET activity extends over the entire volume of the gel dosimeter, and hence, the reported activity extends beyond the dose level at which the gel dose is reported as shown in **figure 4**. An examination of the profile shapes with depth shows that the induced activity profile shape does not necessarily follow the shape of the dose profile with depth. Other studies report similar observations for both clinical modulated and pristine beams [1]. The depth trend in the positron emission profile is mainly due to cross-section energy dependence of various isotope activation reactions which yield the range of induced activity to be always smaller than the maximal penetration depth of the proton dose profile. We found that the distal 50% end of the proton dose profile to be approximately 20 mm deeper than the 50% distal end of the activity profile. This spatial offset is comparable to other studies done in other materials with similar elemental compositions and similar beam energies for SOBP beams [1]. It should be emphasized that the accuracy of the reported lateral and depth localizations of the PET and dose distributions as reported in this study are primarily limited by the spatial resolution of the PET scanner.

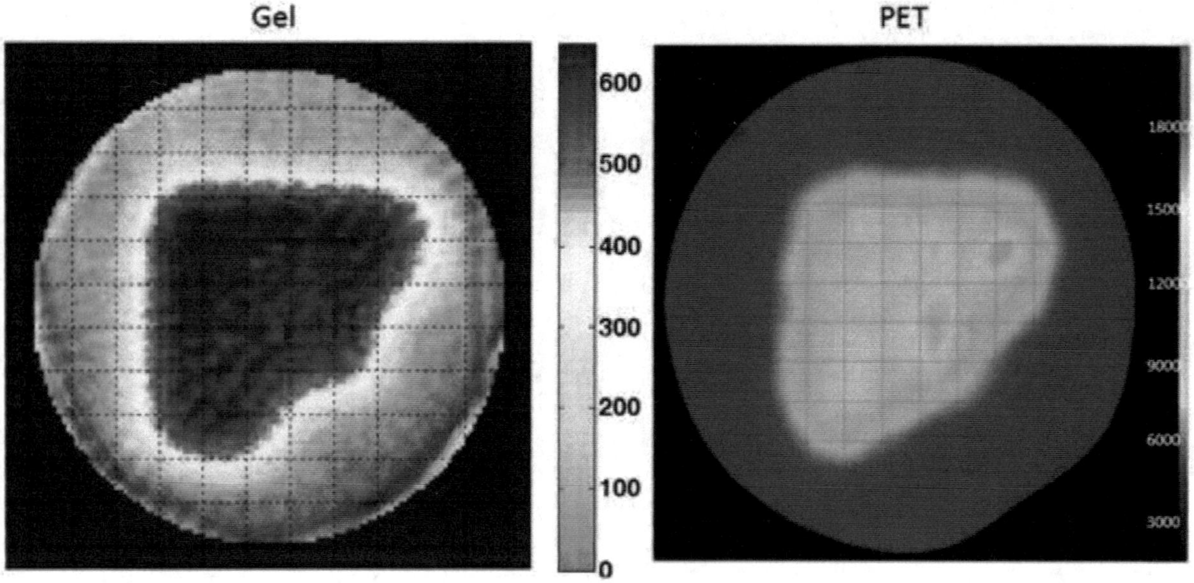

Fig 3 A 2D comparison of dose (left) and PET activity (right) at the same depth in gel. The color bars represent dose range (cGy) on the dose map and raw count range on the PET map. The distribution is generated at a depth of 13 cm in the gel (center of the SOBP plateau).

Future *in vivo* measurements can be performed to quantify the effect of the physical factors that may influence the correlation between PET activity distributions and dose using the methodology outlined in this paper. There are several attractions about using the gel system for in-depth quantitative phantom studies on PET *in vivo* dosimetry. First, the gel dosimeter can be available in a variety of volumes (smaller or larger than the size used in this study which allows for verification studies over a wide range of beam configurations. Second, the gel dosimeter can be implanted with various types of heterogeneous objects in the beam path or upstream. This allows for evaluating the effect of dose perturbations downstream on PET activity distributions. Third, the gel dosimeter can be coupled with motion phantoms to study motion dosimetry and its effects pm PET emissions from clinical proton beam. This feasibility has already been demonstrated using the BANG3 gel with photon beams.[9] Understanding and quantifying these physical factors using the methodology outlined in this work may potentially increase our confidence and understanding of PET/CT *in vivo* dosimetry, its advantages and limitations for patient site-specific treatment QA.

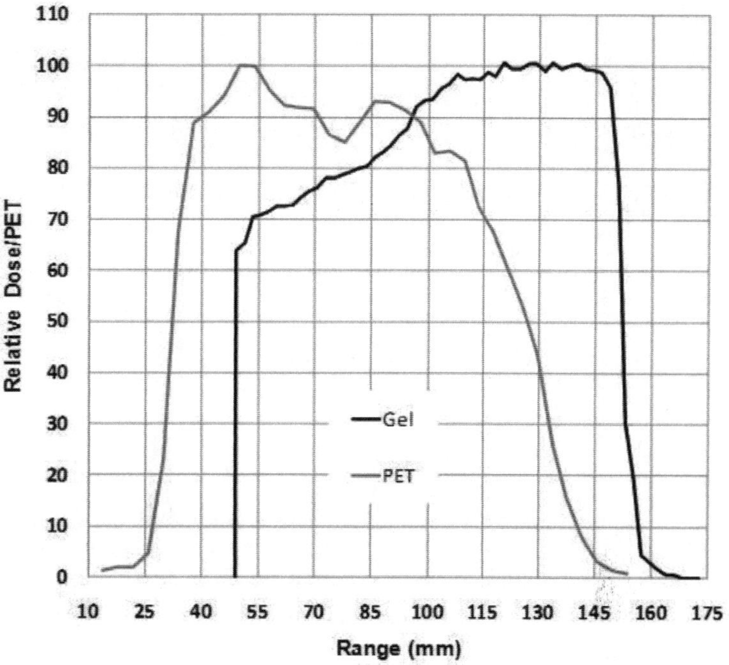

Fig 4 Overlay of depth profiles for PET activity and depth dose in gel. The distance between the 50% distal end of the PET distribution to the 50% distal dose profile is nearly 20 mm.

4. Conclusions

In this work we demonstrate a unique approach to off-line proton dosimetry and depth range verification by direct comparison of measured proton dose distributions in the gel dosimeter volume to the proton-activated positron distributions in the gel. Three-dimensional spatial correlation was possible using PET/CT imaging of the positron activated gel volume immediately post proton beam irradiation. Our results for a clinical SOBP agree well with similar studies on the distal end location of the PET emission profile with respect to the distal end location of the dose depth profile. Using this novel direct comparison methodology, correlations of distal edge doses based on PET activity distributions from clinical proton beams can potentially be quantified under various delivery conditions. This study demonstrates that polymer gels could potentially be useful in assessing various physical factors that could affect PET activity range verification method and its use as a viable patient treatment QA tool.

References

[1] Parodi K, Paganetti H, Cascio E, Flanz J B, Bonab A A, Alpert N M, Lohmann K and Bortfeld T 2007 *Med. Phys.* **34** 419
[2] Knopf A, Parodi K, Paganetti H, Cascio E, Bonab A and Bortfeld T 2008 *Phys. Med. Biol.* **53** 4137
[3] Zeidan O A, Sriprisan S I, Lopatiuk-Tirpak O, Kupelian P A, Meeks S L, Hsi W C, Li Z, Palta J R and Maryanski M J 2010 *Med. Phys.* **37** 2145
[4] Litzenberg D W, Roberts D A, Lee M Y, Pham K, Vander Molen A M, Ronningen R and Becchetti F D 1999 *Med. Phys.* **26** 992

Journal of Physics: Conference Series **250** (2010) 012033 doi:10.1088/1742-6596/250/1/012033

Initial study of 3D dose verification of multi-field proton therapy treatment along match line with polymer gel detectors

S. Avery, C. Freeman, K. Shahnazi

Department of Radiation Oncology, University of Pennsylvania Medical Center, Philadelphia,

avery@uphs.upenn.edu

Abstract. This paper is intended as an initial study for quality assurance benefits from polymer gel detectors to proton therapy treatments. Several gel types were explored for stability from batch to batch. The depth dose distributions in the gels were examined with regard to dose dependences and compared to baseline measurements. The results indicate polymer gel detectors may be able to verify dose in three dimensions along match line for proton therapy treatments

1. Introduction

The purpose of our study is to examine the dose due to match lines using polymer gel detectors for quality assurance of multi-field proton treatment plans. An advantage of gels is that you have the ability to verify dose in three dimensions. The uses of polymer gel detectors as a tool for radiation dosimetry have been well investigated in the literature [1]. Some of the advantages of gel dosimeter include potential applications of high linear energy transfer (LET) particles and proton therapy [2-4]. In the study we generate a calibration curve to interpret our experimental results on the dose response of the gel from the irradiation of protons.

2. Materials and Methods

We prepared MAGIC (methacrylic and ascorbic acid in gelatin initiated by copper) and MAGAT (methacrylic acid gelatine and tetrakis (hydroxymethyl) phosphonium chlorided) type gels for this study. The gels are prepared at standard temperature and pressure conditions (22°C, 760 mmHg). The Magic gels are composed of 9% gelatin by weight (300 bloom, (Aldrich; Milwaukee, WI), ascorbic acid (2×10^{-3} M, Mallinkrodt; Paris, KY), $CuSO_4\cdot5H_2O$ (8×10^{-5} M, Aldrich; Milwaukee WI) hydroquinone (1.8×10^{-2} M, Sigma; St Louis, MO), and HPLC grade distilled water. The MAGAT Gel was made by mixing gelatin by weight (300 bloom, Aldrich; Milwaukee, WI), HPLC grade distilled water (Sigma Aldrich, and Tetrakis (hydroxmethyl) phosphonium chloride, 80 % sodium in water (Sigma Aldrich). (Table 1)

The MAGIC gels are made by mixing 40 g of gelatine and 320 ml of water in a 500 ml flask. The gelatine is then heated to 55 deg C until the gelatine has completely dissolved and is clear. Then the gel is then cooled to 37 deg C before the following mixtures are added to the gelatine: Hydroquinone (1 gm/ 24 ml H_2O), Ascorbic Acid (0.176/ 25 ml H_2O), and Copper solution (0.01g / 15 ml H_2O). Once all components are added the solution continues to mix until completely clear (approx 2-3 min). The final component, 50 ml of Methacrylic Acid, is added to the mixture and mixes

Journal of Physics: Conference Series **250** (2010) 012033 doi:10.1088/1742-6596/250/1/012033

until homogeneous. The gel is then poured into 20 ml scintillation vials and kept in a closed box to reduce light exposure. The process to make the MAGAT gels is similar to the MAGIC process with different reagents added. We need to allow the gels to congeal at least six hours before irradiating.

Reagent	Gelatin(300 Bloom)	Methacrylic Acid	Ascorbic Acid	CuSo4:5H20	Hydroquinone	Distilled Water	Tetrakis
MAGIC	x	x	x	x	x	x	
MAGAT	x	x					x

Table 1: Composition of MAGIC gels and MAGAT gels.

Calibration curves (Figures 1 and 1a) were repeatedly created to determine the stability of the gel processing method. All gels were irradiated at room temperature inside a small water tank in the centre of a 10 x 10 field of view at 100 SSD at 2 cm depth. Due to the availability of the machines the MAGIC and MAGAT gels were irradiated with 6 MV photons for calibration using a 2300 iX (Varian) standard linac.

Using proton beams our study, all gels were irradiated at 2 cm depth with beam line option B5 (R15M10) [proton range of 15 cm, 10 cm modulation] and field size of 10x10 cm^2 at the Roberts Proton Therapy Center which is housed in the department of Radiation Oncology at the University of Pennsylvania Medical Center.

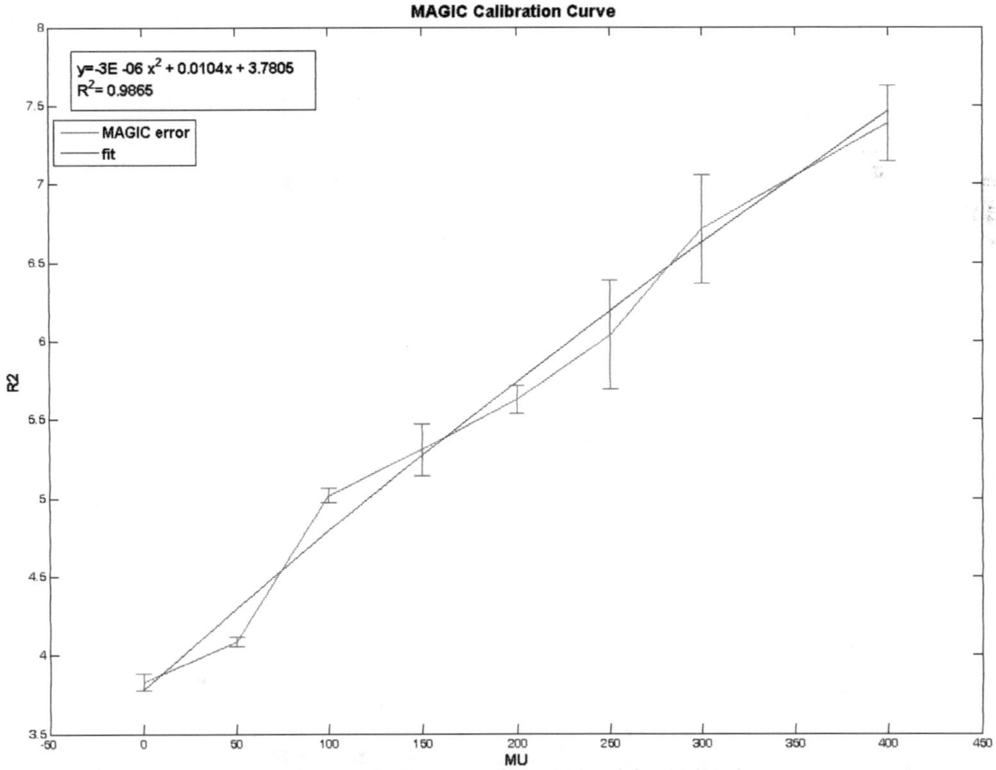

Figure 1. Calibration of MAGIC gel for 6 MV photons

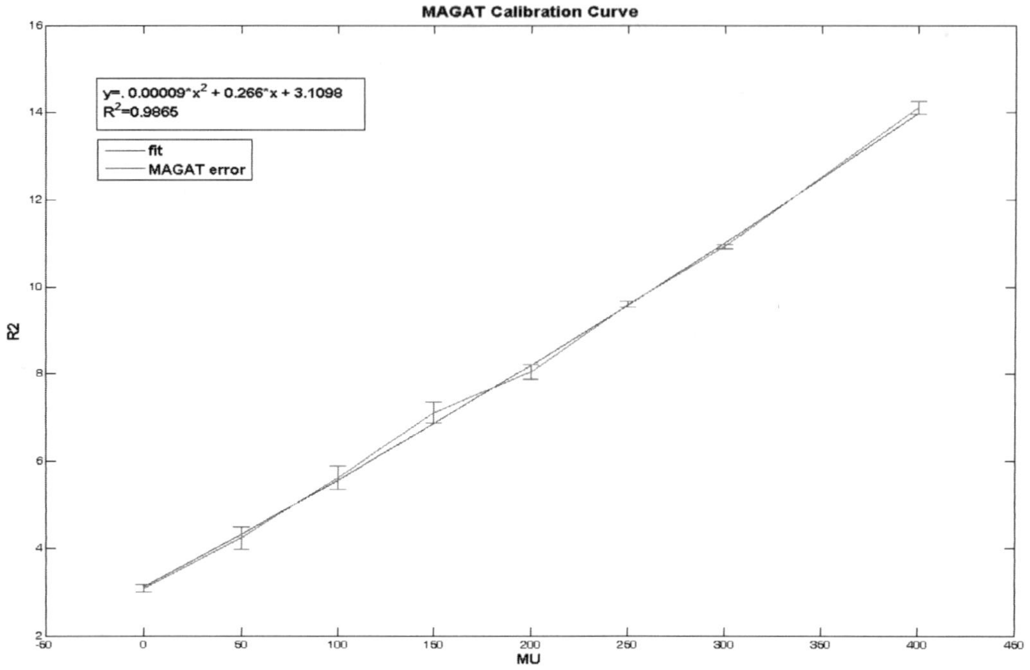

Figure 1a. Calibration of MAGAT gel for 6 MV photons

We irradiated a 500ml flask with MAGAT with multiple fields from the proton center. Figure 5 is the Monte Carlo calculated SOBP, in water, generated with the GEANT4 simulation toolkit [5]. The arrow indicates where on the SOBP we took our measurements.

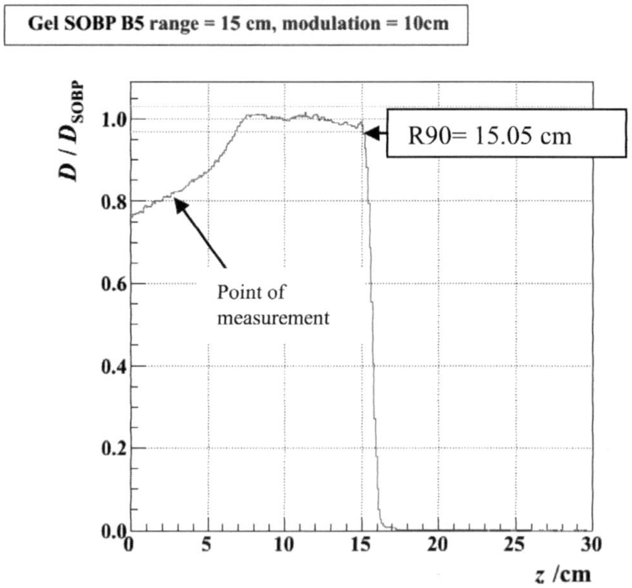

Figure 2: Location on SOBP where gels were irradiated

MR scans were performed using Siemens Esperee Magnetrom 1.5 Tesla MRI for dose read-out. We placed the gels within a Magnetrom Head Coil (Syngo MR B15) that has an 8 channel head array. The multi-spin pulse sequence parameters include a TR=8660 ms, slice thickness=2mm, 16 echoes, 26 slices and TE=22.4 ms. We used a dedicated MATLAB program to do pixel-by-pixel calculation of the R2 (=1/T2) values.

3. Results and Discussion

We determined MAGAT was the most practical gel to use based on its consistency and ease of preparation. Figure 2 shows the calibration vials with one slice from the MR scan. From the analysis of the scans using our MATLAB code we were able to calculate a calibration curve (figure 3) from the proton irradiation. For the protons the dose was given with beam line option B5 (R15M10) and dose was delivered from 50 MU to 311 MU. This equation will allow for dose to be calculated for future dose plans. We plan to use the gels to ensure dose to plans like the ones seen below in figure 5 and 6. it is important to ensure dose coverage by using a second check(gels) and ensure the dose is being delivered as planned.

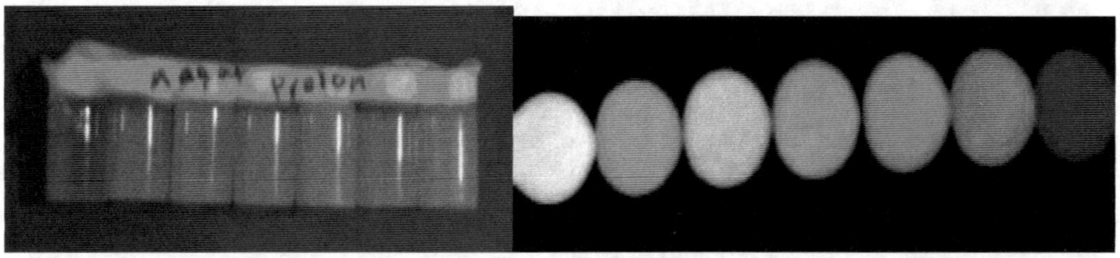

Figure 3: Picture of MAGAT proton calibration vials and corresponding MR images

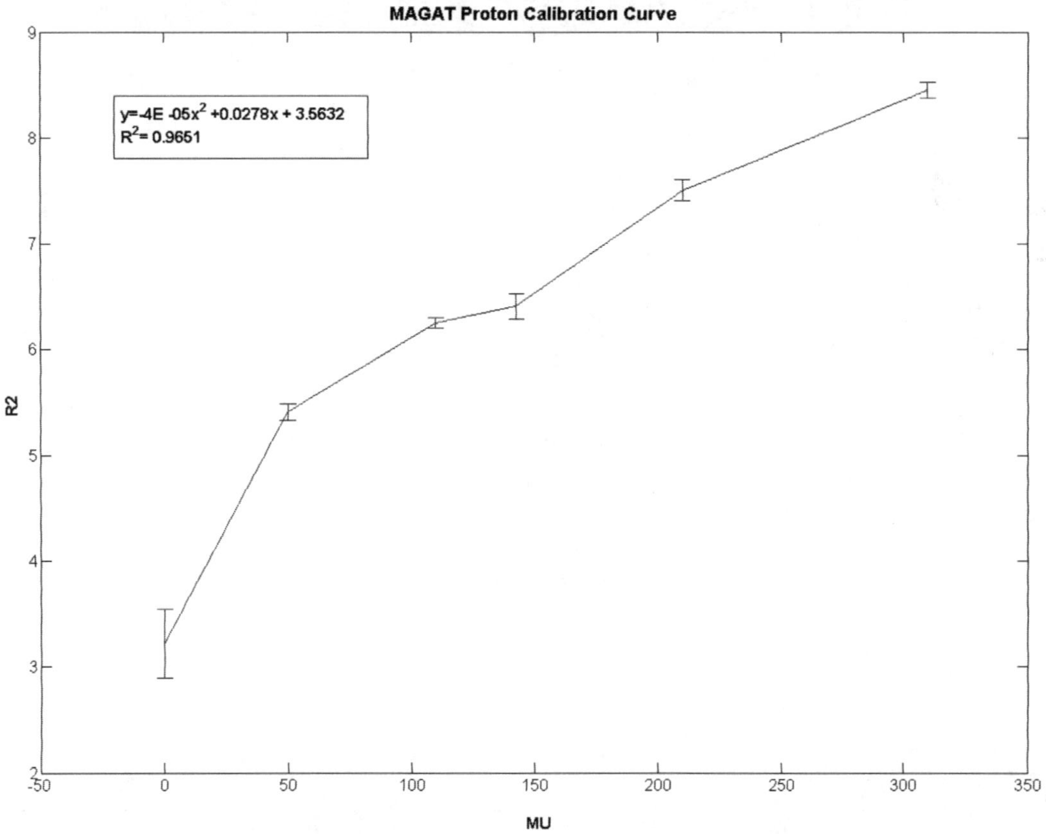

Figure 4: MAGAT gel - Proton Calibration Curve

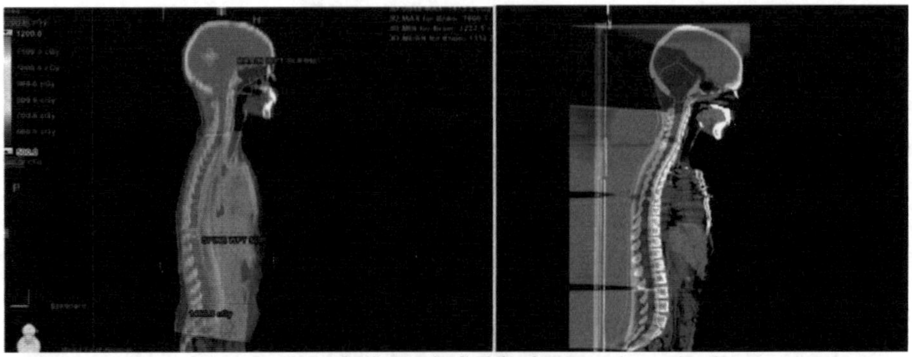

Figure 5. Left: Match fields for cranio-spinal

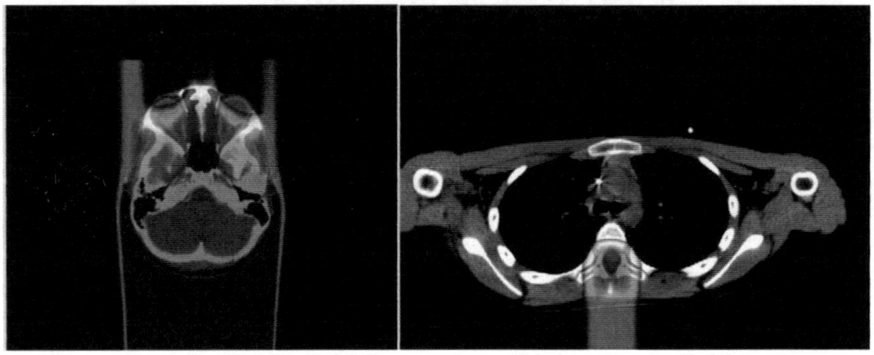

Figure 6. Left: Single field for cranium, Right: Single field for lung

4. Conclusions

In this initial study, we developed a consistent method for preparing MAGAT gel and developing a multi-spin pulse sequence for creating R2 maps. The results indicate that the gels have the potential of measuring the dose response from proton beams. Further work is needed in this study to examine the characteristics of the dose response before we use them clinically; especially in match line for proton or photon beam. The hot or cold spots should fall within the calibration curve.

References

[1] C Baldock *et al* 2010 *Phys. Med. Biol.* **55** R1
[2] Heufelder J, Stiefel S, Pfaender M et al. *Med. Phys.* 2003; **30** 1235-40
[3] Jirasek A, Duzenli C. *Med. Phys.* 2002; **29** 569-77
[4] Ramm U, Weber U, Bock M, et al. *Phys. Med. Biol.* 2007; **52** 2719-28
[5] S. Agostinelli, et al, *Nuclear Instruments and Methods in Physics Research;* A 506 (2003) 250-303

Journal of Physics: Conference Series **250** (2010) 012034 doi:10.1088/1742-6596/250/1/012034

An investigation of the response of the radiochromic dosimeter PRESAGE^{TM} to irradiation by 62 MeV protons

Shamsa Al Nowais[1], Andrzej Kacperek[2], John N H Brunt[2], John Adamovics[3], Andrew Nisbet[1,4] and Simon J Doran[1,5]

[1] Department of Physics, University of Surrey, Guildford, Surrey, UK

[2] Douglas Cyclotron, Clatterbridge Centre for Oncology NHS Trust, Wirral, UK

[3] Heuris Pharma, Skillman, NJ, USA

[4] Royal Surrey County Hospital, Guildford, Surrey, UK

[5] CRUK and EPSRC Cancer Imaging Centre, Institute of Cancer Research, Sutton, Surrey, UK

Simon.Doran@icr.ac.uk

Abstract. Measurements of the 62 MeV proton beam at the Clatterbridge Centre for Oncology using the radiosensitive plastic PRESAGE^{TM} have previously shown a dependence of the dosimeter sensitivity (dose-response slope) on the linear energy transfer (LET) of the ionising particles. This work focuses on a possible explanation in terms of track structure theory (TST). Experimental measurements of highly irradiated PRESAGE^{TM} samples established the D_{37} parameter of the theory to be of the order of 1000 Gy. Initial attempts at applying the theory showed good agreement of the theoretical and experimental values of relative effectiveness, but more work is needed to verify the model and understand its different parameters.

1. Introduction

Preliminary results — previously reported in abstract form [1-2] — have shown that optical CT of PRESAGE^{TM} dosimeters is an extremely suitable method for obtaining high resolution isotropic 3-D maps of dose from a proton beam. However, that initial work raised concerns that the dosimeter *under-responded* significantly in the high-LET region of the Bragg peak.

Previous authors have established a theoretical framework known as *track structure theory*, based on a part-theoretical / part-empirical model initially introduced by Katz in the late 1960s [3-6]. The work was later corrected by Zhang *et al.* [7] and a semi-empirical formula was proposed by Waligórski *et al.* [8] to account for dosimeter under-response. The framework was applied to polymer gel dosimetry by Jirasek and Duzenli [9], where it explained a number of key features of the gel under-response. Our aim was to extend this work and assess the degree to which the theory explains the observed effects for PRESAGE^{TM}.

2. Application of track structure theory to PRESAGE^{TM}

The δ-ray theory of track structure (TST) is attractive because it is a general-purpose radiological model that requires very few explicit details of the interactions by which dose deposition causes changes in any given detector. It is based on the statistical properties (Poisson distribution) of a single- or multiple-hit interaction of absorbed radiation with the detector, and predicts the effectiveness of high-LET radiation for most physical detectors, once a few parameters obtained from low-LET irradiations are known.

Building on the target theory of Lea [10], the basic assumption is that any radiation detector consists of radiation sensitive elements, often referred to as "targets". The target size is characteristic of the individual detector, which, in the case PRESAGE™, is the leuco dye molecule. These sensitive elements have two states "non-activated" / "activated": for PRESAGE™, this would correspond to "non-absorbing" / "absorbing" or "light green" / "dark green". The targets are converted between the two states by a radiation "hit" (or in more complex models multiple hits). They are embedded in a more or less passive matrix acting as an energy transfer medium. We further assume that the targets are bathed with a locally uniform field of secondary electrons following x- or γ- irradiation and that the *probability of a hit* can be calculated from standard Poisson statistics. The resultant response of the detector as a whole is obtained by (i) calculating the radial dose distribution around the path of the penetrating ion, which leads to an inhomogeneous distribution of radiation effects; (ii) integrating over the whole irradiated volume to find the average response of the medium; (iii) calculating the response after a specified fluence of particles.

The reason that high-LET radiation leads to an under-response in chemical detection systems is that the dose immediately surrounding the particle track is extremely high (~10^5 Gy at a distance of 1 nm from the path of a proton). Some of this dose is "wasted", as all the targets in the vicinity are activated by a much smaller dose. By contrast for the low-LET radiation, the given dose is "spread out" over a larger number of elements.

The relative effectiveness, *RE*, is the ratio of sensitivities of the dosimeter for the two radiations being compared [11], in this case, x-rays (k_γ, low-LET) and protons (k_p, high-LET). To apply track structure theory, two detector parameters are required: (a) D_{37} is the experimentally-measured dose at which e^{-1} (~37%) of the elements remain inactivated when the sample is irradiated with x-rays. It can be shown that $k_\gamma=1/D_{37}$. (b) a_0 is the size of the sensitive element and can be estimated either based on known physical characteristics of the detector (e.g., the diameter of the leuco-dye molecule) or from track structure theory itself. Using these inputs, *RE* can be calculated theoretically according to:

$$RE = \frac{k_p}{k_\gamma} = D_{37}\frac{\sigma}{LET} \qquad (1)$$

where *LET* is the average energy deposited per unit path length in a unit density material, as obtained from the collision stopping power of the moving ion, which is equal to LET_∞, and σ is the relevant interaction cross-section, calculated using track structure theory (see [9] for details). *RE* can also be measured experimentally at different points along the depth-dose curve (i.e., at different LET values), using the data from [1]. The purpose of this study is therefore to compare the theoretical and empirical values of *RE*, to determine whether track structure theory is applicable for the case of PRESAGE™. It is hoped that a better understanding of the PRESAGE™ response to protons will allow an improved dosimeter to be created that shows less of an under-response in the high-LET regions.

3. Materials and methods

3.1 Determination of $k_\gamma = 1/D_{37}$

As is well known, PRESAGE™ has a very wide dynamic range and previous experiments [12] had shown that irradiation of standard cuvettes with a 10 mm path length led to samples that were too dark to be measured using standard techniques. Furthermore, it was not feasible to use a standard hospital linac to irradiate samples to the high levels (several thousand Gy) required to determine saturation.

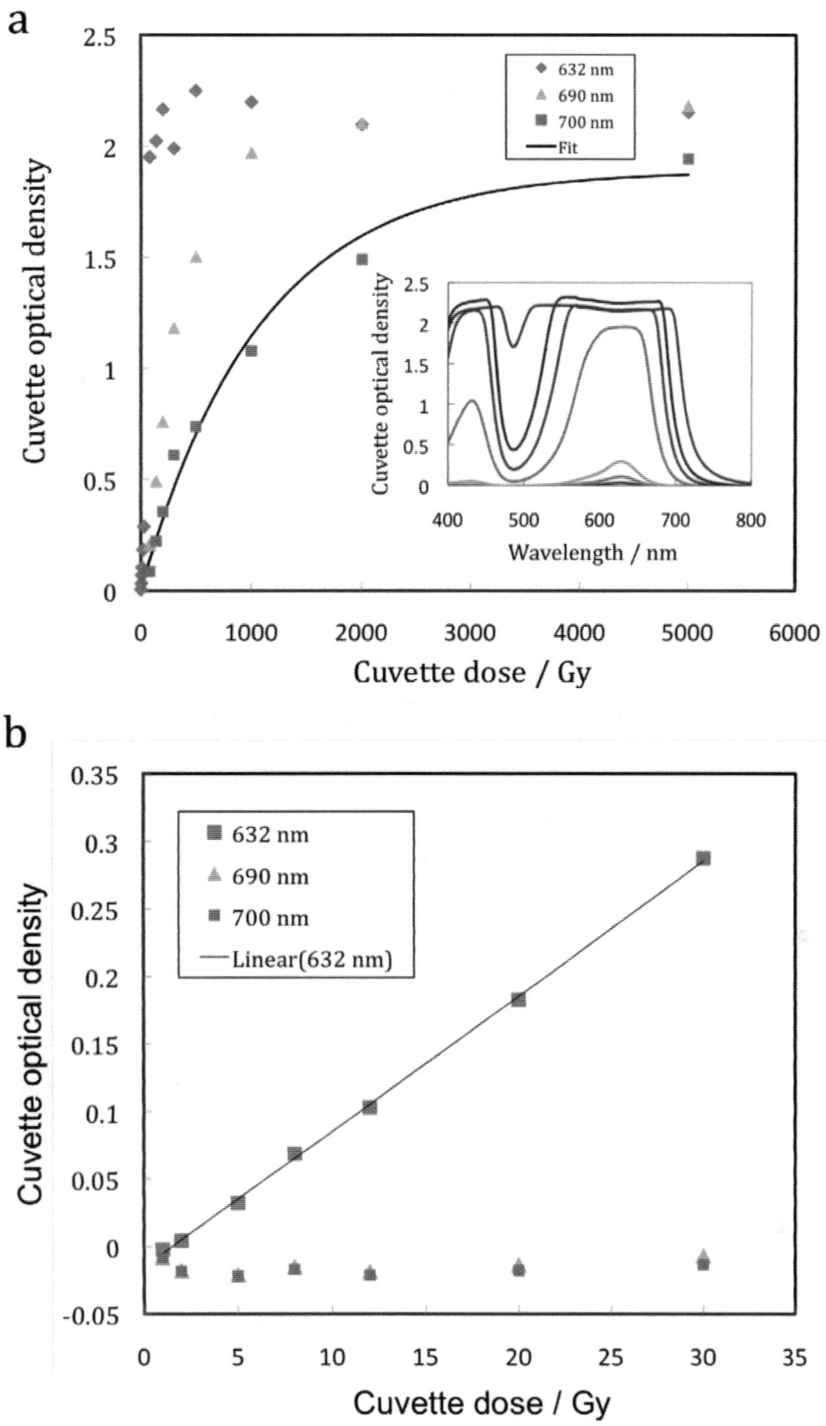

Figure 1: (a) Relationship between dose and optical density for the 15 cuvettes irradiated at ESRF, as measured at three different wavelengths. Inset is the complete spectrum for a representative set of cuvettes. Note how the intense coloration of the PRESAGE™ causes the spectrophotometer to saturate at the normal measurement wavelength of 632 nm. Using the data for 700 nm, we are able to fit a curve to the ideal single-hit detector model. (b) Enlarged version of the low-dose behaviour showing an excellent linear relationship between dose and OD_{632}, but with OD_{690} and OD_{700} too insensitive to dose to make any effective measurement at low dose. (Note: it was difficult to make reliable estimates of the experimental error and so error bars have not been included for this figure.)

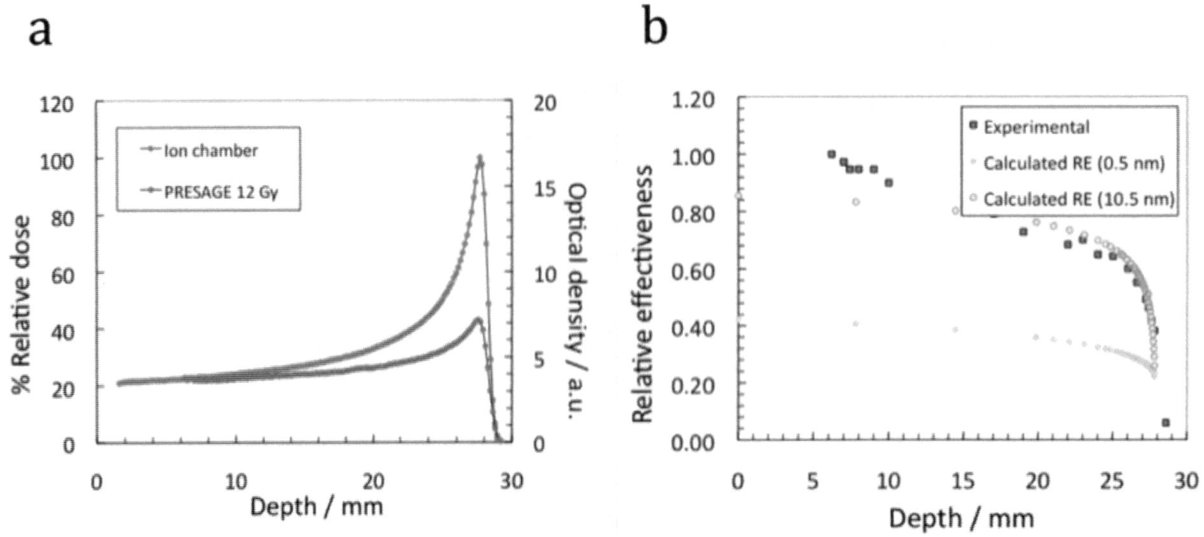

Figure 2: (a) Data used to calculate the "experimental" value of relative effectiveness; (b) comparison of experimental and theoretical values of relative effectiveness for two different target sensitive element sizes a_0=0.5 nm and a_0=10.5 nm.

Thus, ultra-micro UV cuvettes (Eppendorf UVette®) with 2 mm optical path-length were filled with PRESAGE™ ($C_{270}H_{460}N_{19}O_{68}Br$, z_{eff} ~ 9.6, 1.08 g/cm^3) and irradiated at the ID17 Microbeam Radiation Therapy (MRT) beamline at the European Synchrotron Radiation Facility (ESRF) in Grenoble, France. Operating in filtered broad-beam mode, with a spectrum ranging from ~35 to 350 keV, the dose delivery rate by the beam can be up to 16 kGy s^{-1}. Each cuvette was placed individually on a stand and irradiated, and the sequence yielding the following nominal doses over the 15 cuvettes: 1, 2, 5, 8, 12, 20, 30, 50, 80, 140, 200, 300, 500, 2000 and 5000 Gy. Two unirradiated control samples were measured to establish a base-line. The optical absorbance spectra of the samples were measured using a CARY 5000 UV-Vis-NIR spectrophotometer (Varian, Inc.). Despite the low optical path length, the highly irradiated samples were still almost completely opaque at the normal measurement wavelength of 632 nm, with OD_{632} >> 2, and could not be measured by our apparatus. To obtain a value of D_{37}, we assumed that the x-ray dose-response of PRESAGE™ measured at 700 nm would scale linearly with that at 632 nm, whilst leaving the optical absorbance still within the dynamic range of the spectrophotometer. Future measurements on thin PRESAGE™ films might remove the need for this.

3.2 Determination of a_0

For a physico-chemical detector the approximate size of the sensitive element a_0 may simply be the size of the sensitive molecule. For leuco malachite green, this is approximately 0.65 nm, from tabulated values [13]. However, it is also possible to derive a value of "effective radius" from target theory itself, which suggests that a_0 may be ~10 nm, larger than the physical size of the molecule.

3.3 Determination of relative effectiveness

The LET for protons in PRESAGE™ in MeVcm^2/mg was calculated using an Energy-LET-Range converter software application [14], by entering the elemental target densities weighted by their relative stoichiometry. A MathCad spreadsheet was used to implement the track structure theory calculations. The experimental value of *RE* was estimated by making the initial approximation that the absolute efficiencies of the protons at the entrance channel and x-rays are similar (both are considered to be "low-LET" radiations). The ratio of the optical and ion chamber results was then found for all depths and then multiplied by a calibration value obtained from the entrance dose. Although the ion chamber measurements were made in a water phantom, depths have been converted to equivalent depths in PRESAGE™ simply by dividing the ratio of densities.

4. Results

The results of the spectrophotometer analysis are displayed in Figure 1. Applying a standard Poisson model to the single-hit target, yields OD $\propto 1 - \exp(-D/D_{37})$ and a fit to this model gave a value of 1070 Gy for the D_{37} parameter. Previously reported results [1-2] for the 62 MeV proton beam are repeated for convenience in Figure 2(a), where the Bragg curve imaged with our CCD-based optical CT scanner is compared with relative dose measurements made with a PTW/Markus (parallel-plate) ionization chamber. The corresponding relative effectiveness values are shown in Figure 2(b). Notice that the size of the targets is a free parameter in the theory and clearly has a big influence on the results. Further verification of the correctness of our model is necessary.

5. Conclusions

Track structure theory provides a good basis for an explanation of the variation with proton LET of the dose-response of PRESAGE™. The under-response at the Bragg peak is attributed to the large doses deposited by delta rays close to the track of a proton, which causes detector elements to saturate. The theory does not specify the nature of the sensitive element, though the size of the element is an important parameter for the model. The experimental and theoretical values of relative effectiveness of PRESAGE™ show similar trends. However, more work is needed to understand the model fully.

6. References

[1] Doran, S.J., et al., True-3D scans using PRESAGE™ and Optical-CT: A case study in proton therapy J Phys Conf Ser, 2006. **56** p. 231–234.
[2] Al-Nowais, S., et al., A preliminary analysis of LET effects in the dosimetry of proton beams using PRESAGE ™ and optical CT. Applied Radiation and Isotopes, 2009. **67**(3): p. 415-418.
[3] Butts, J.J. and R. Katz, Theory of RBE for heavy ion bombardment of dry enzymes and viruses. Radiation Research, 1967. **30**(4).
[4] Katz, R., Track structure theory in radiobiology and in radiation detection. Nuclear Track Detection, 1978. **2**(1): p. 1-28.
[5] Katz, R., S.C. Sharma, and M. Homayoonfar, The structure of particle tracks, in *Topics in radiation dosimetry*, F.H. Attix, Editor. 1972, Academic press: New York. p. 317-383.
[6] Kobetich, E.J. and R. Katz, Width of heavy ion tracks in emulsion. Bulletin of the American Physical Society, 1968. **13**(1): p. 130-&.
[7] Zhang, C.X., D.E. Dunn, and R. Katz, Radial-distribution of dose and cross-sections for the inactivation of dry enzymes and viruses. Radiation Protection Dosimetry, 1985. **13**(1-4): p. 215-218.
[8] Waligorski, M.P.R., R.N. Hamm, and R. Katz, The radial-distribution of dose around the path of a heavy-ion in liquid water. Nuclear Tracks and Radiation Measurements, 1986. **11**(6): p. 309-319.
[9] Jirasek, A. and C. Duzenli, Relative effectiveness of polyacrylamide gel dosimeters applied to proton beams: Fourier transform Raman observations and track structure calculations. Medical Physics, 2002. **29**(4): p. 569-577.
[10] Lea, D.E., *Actions of radiations on living cells*. Second ed. 1962, New York: Cambridge University Press.
[11] Hansen, J.W., Experimental investigation of the suitability of the track structure theory in describing the relative effectiveness of High-LET irradiation of physical radiation detectors. 1984, Riso National Laboratory, DK-4000 Roskilde, Denmark.
[12] Al-Nowais, S., et al., An attempt to determine the saturation dose for PRESAGE™. J Phys. Conf. Ser., 2009. **164**: p. 012043.
[13] Koltuniewicz, A.B. and E. Drioli, *Membranes in Clean Technologies: Theory and Practice*. Vol. 1. 2008: John Wiley and Sons Ltd.
[14] Zajic, V., Energy vs. LET vs. Range calculator, htttp://tvdg10.phy.bnl.gov/LETCalc.html.

Journal of Physics: Conference Series **250** (2010) 012035 doi:10.1088/1742-6596/250/1/012035

Determination of the depth dose distribution of proton beam using PRESAGE ™ dosimeter

L Zhao [1, 2], I J Das[1, 2], Q Zhao[1], A Thomas [3], J Adamovics [4] and M Oldman [3]

[1]Midwest Proton Radiotherapy Institute, Bloomington, IN, USA 47404

[2]Department of Radiation Oncology, School of Medicine, Indiana University, Indianapolis, IN,USA 46202

[3]Department of Radiation Oncology, Duke University Medical Center, Durham, NC, USA 27710

[4]Heuris Pharma LLC, 412 Sunset Rd., Skillman, NJ, USA 08558

Email: lzhao@mpri.org

Abstract. PRESAGE™ dosimeter dosimeter has been proved useful for 3D dosimetry in conventional photon therapy and IMRT [1-5]. Our objective is to examine the use of PRESAGE™ dosimeter for verification of depth dose distribution in proton beam therapy. Three PRESAGE™ samples were irradiated with a 79 MeV un-modulated proton beam. Percent depth dose profile measured from the PRESAGE™ dosimeter is compared with data obtained in a water phantom using a parallel plate Advanced Markus chamber. The Bragg-peak position determined from the PRESAGE™ is within 2 mm compared to measurements in water. PRESAGE™ shows a highly linear response to proton dose. However, PRESAGE™ also reveals an underdosage around the Bragg peak position due to LET effects. Depth scaling factor and quenching correction factor need further investigation. Our initial result shows that PRESAGE™ has promising dosimetric characteristics that could be suitable for proton beam dosimetry.

1. Introduction

Proton beam therapy is changing the landscape of cancer patient care with promising results in many sites due to its unique physical characteristics. Proton beam commissioning is labor intensive and requires validation from several dosimetric systems. Depth dose measurements are performed for beam commissioning and patient specific field measurements in proton therapy. There is a need for a simple and convenient method for verifying depth dose profiles in proton beam therapy, especially for treating small tumor volumes at shallow depths. Ion chambers are most reliable [6], but film [7], MOSFET [8] and optically stimulated luminescent detector (OSLD) [9] have been advocated. A recent study shows that BANG3 polymer gel is capable of reproducing ion chamber dose data for modulated and un-modulated Bragg peak beams with different clinical beam energies [10]. Using PRESAGE™ and optical CT scanner for proton beam dosimetry was initially studied by Doran et al [11]. In this work, we have now extended these studies to include a comparison between PRESAGE™ dosimeter and ion chamber.

Journal of Physics: Conference Series **250** (2010) 012035 doi:10.1088/1742-6596/250/1/012035

2. Methods and Materials

Three 8 cm diameter cylindrical shaped PRESAGE™ samples with a length of 6 cm were irradiated at 300 MU, 500 MU and 1000 MU respectively using un-modulated proton beams from the passive scattering beam delivery system at the Midwest Proton Radiotherapy Institute in Bloomington, Indiana. The beam energy was chosen in a way so that the proton beam stopped in the samples. The long axis of the dosimeter was placed along the beam central axis. A circular 10 cm in diameter aperture was used. To avoid light sensitivity, dosimeters were wrapped and kept in light tight plastic during exposure and shipment for optical readout measurement. For comparison, an advanced Markus parallel-plate ionization chamber (PTW-Freiburg GmbH, Freiburg, Germany) was placed in a Wellhofer scanning WP700 water tank for depth dose measurements (data at depth less than 2 cm was discarded due to front wall thickness offset). The PRESAGE™ dosimeter was scanned with 1 mm increment over the entire length by an optical-CT scanning system [1] to determine the percent depth dose distribution.

3. Results and Discussions

Figure 1 shows a comparison of measured data from PRESAGE™ samples showing signal attenuation coefficients along the central axis indicating radiation dose response. Figure 2 shows a highly linear dose response of PRESAGE™ . The comparison of percent depth dose curves after normalizing data to the depth 2 cm is shown in Figure 3. PRESAGE™ showed an under-response in the Bragg peak compared to ion chamber. The peak-to-plateau dose ratio (the ratio of the dose at the peak of the Bragg peak to that at near-zero depth) obtained from PRESAGE™ samples was 2.34 which was about 13% less than the peak-to-plateau dose ratio registered from ionization chamber measurement in water (2.69). This linear energy transfer (LET) quenching effect is similar to what have been reported from Gafchromic EBT film dosimetry in proton beam [7] and BANG gel dosimetry in proton beam [12].

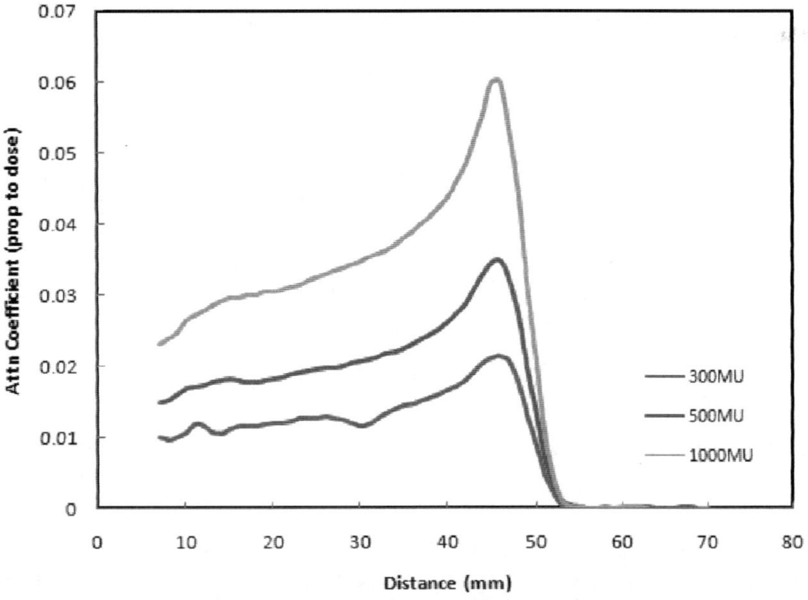

Figure 1 Comparison of measured attenuation coefficients along the central axis of PRESAGE™ samples indicating radiation dose response.

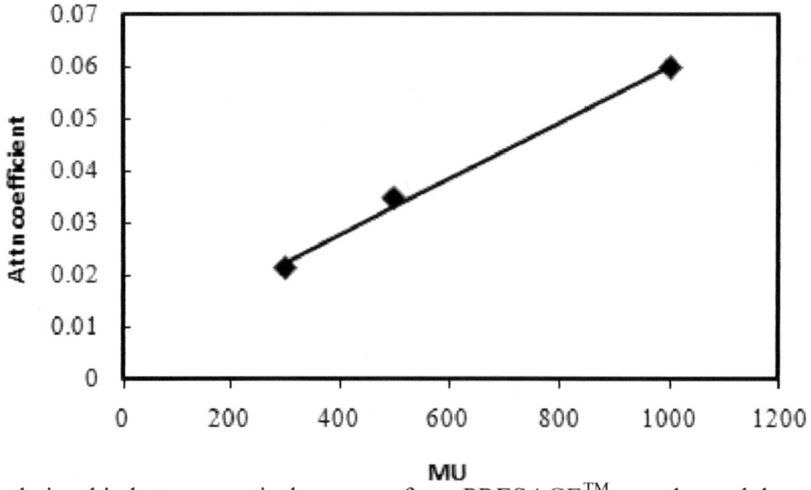

Figure 2 The relationship between optical response from PRESAGE™ samples and dose monitor units.

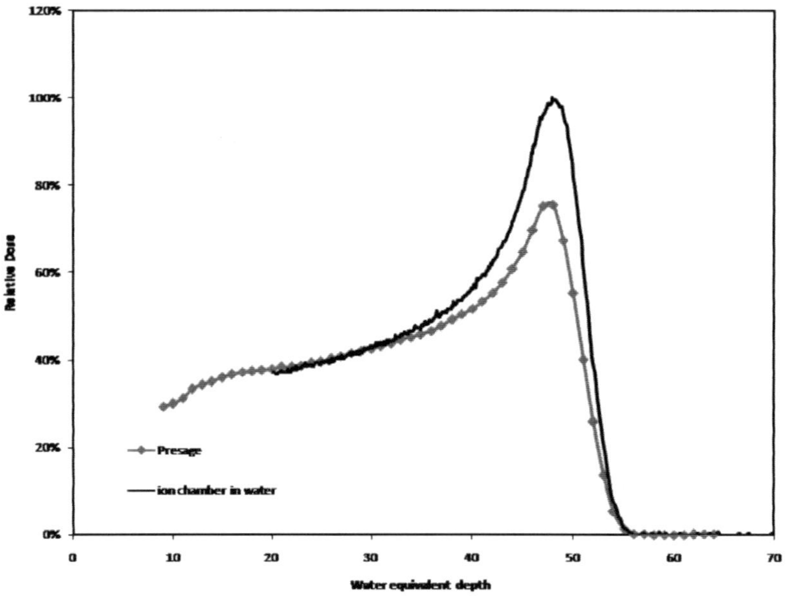

Figure 3 Percent depth dose curves obtained from PRESAGE™ and ionization chamber measurements in water.

4. Conclusions

PRESAGE™ dosimeter has unique characteristics that could be utilized for proton beam therapy. Our measurements in proton therapy are promising and show potential for clinical use especially for small fields. However, to achieve a clinically-acceptable level of accuracy, depth scaling factor as a function of proton energy and quenching correction factor will be subjected to further investigation.

References

[1] Oldham M, Guo P, Gluckman G and Adamovics J 2006 IMRT verification using a radiochromic/optical-CT dosimetry system J. Phys. 56 221-4

[2] Sakhalkar H, Sterling D, Adamovics J, Ibbott G and Oldham M 2009 Investigation of the feasibility of relative 3D dosimetry in the Radiologic Physics Center Head and Neck IMRT phantom using presage/optical-CT Med. Phys. 36 3371-7

[3] Sakhalkar H S, Adamovics J, Ibbott G and Oldham M 2009 A comprehensive evaluation of the PRESAGE/optical- CT 3D dosimetry system Med. Phys. 36 71-82

[4] Guo P, Adamovics J and Oldham M 2006 A practical three-dimensional dosimetry system for radiation therapy Med. Phys. 33 3962-72

[5] Clift C, Thomas A , Adamovics J, Chang Z, Das I and Oldham M 2010 Toward acquiring comprehensive radiosurgery field commissioning data using the PRESAGE/ optical-CT 3D dosimetry system Phys. Med. Biol. 55 1279-93

[6] ICRU Report 78 2007 Prescribing, Recording, and Reporting. Proton-Beam Therapy International Commission on Radiation Units and Measurements Bethesda Imprint Bethesda, MD 7 No 2

[7] Zhao L and Das I 2010 Gafchromic EBT film dosimetry in proton beams Phys. Med. Biol. 55 N291-301

[8] Wolanski M, Cheng C, Gautam A S and Das I 2010 Dosimetric characteristics of the OneDose MOSFET detector for in-vivo dosimetry in proton beam therapy (PBT) Particle Therapy Co-Operative Group 48 meeting (Germany)

[9] Reft C 2009 The energy dependence and dose response of a commercial optically stimulated luminescent detector for kilovoltage photon, megavoltage photon, and electron, proton, and carbon beams Med. Phys. 36 1690-99

[10] Zeidan O A, Sriprisan S I,Lopatiuk-Tirpak O,Kupelian P A, Meeks S L, His W C, Li Z, Palta J R and Maryanski M J 2010 Dosimetric evaluation of a novel polymer gel dosimeter for proton therapy Med. Phys. 37 2145-52

[11] Doran S, Al-Nowais S, Krstaji N, Adamovics J, Kacperek A and Brunt J 2006 True-3D scans using PRESAGE™ and Optical-CT: A case study in proton therapy J. Phys.: Conf. Ser. 56 231

[12] Heufelder J, Stiefel S, Pfaender M, Ludemann L, Grebe G and Heese J 2003 Use of BANG polymer gel for dose measurements in a 68 MeV proton beam Med. Phys. 30 1235-40

An evaluation of Genipin gel as a water equivalent dosimeter for megavoltage electron beams and kilovoltage x-ray beams

Tina Gorjiara[1], Robin Hill[1, 2], Zdenka Kuncic[1], Stephen Bosi[1, 3] and Clive Baldock[1]

[1] Institute of Medical Physics, School of Physics, University of Sydney, NSW 2006, Australia

[2] Department of Radiation Oncology, Royal Prince Alfred Hospital, Sydney, NSW 2050, Australia

[3] Department of Radiation Oncology, Prince of Wales Hospital, Randwick, NSW 2031, Australia

clive.baldock@sydney.edu.au

Abstract. Genipin gel is a radiochromic gel with the potential to be used as a three dimensional (3D) dosimeter. An ideal dosimeter should present radiologically water equivalent properties. In this work, we have evaluated the water equivalency of genipin gel by calculating its radiological properties, such as mass and electron density, effective atomic number, fractional interaction probabilities, mass energy absorption coefficient and mass stopping powers as well as depth doses for kilovoltage x-ray and megavoltage electron beams. Based on the results of this study, we conclude that genipin gel is a water equivalent dosimeter.

1. Introduction

Genipin, extracted from the fruits of *Gardenia jasminoides Ellis*, has been used widely in herbal medicine [1]. Genipin can form molecular cross links with gelatin's proteins. The result is a light to dark blue radiochromic gel with the property that its colour bleaches after irradiation. By quantifying the change in colour using imaging modalities such as optical CT, it is possible to determine the delivered radiation dose [2, 3]. Radiochromic genipin gel has a fast bleaching process which produces a dose image with sufficient stability and sensitivity to be used as a 3D dosimeter [4].

Water equivalency is one of the characteristics of an ideal dosimeter and can be assessed by determination of mass density and electron density; effective atomic number; mass energy absorption coefficient; mass stopping power and relative dosimetry measurements such as depth dose [5-7].

Water equivalency of PRESAGE and several polymer gel dosimeters have been investigated previously [5, 8, 9]. In this work we have evaluated the water equivalency and radiological properties of genipin gel by calculating a number of parameters such as: electron density, effective atomic number, photon interaction probabilities, mass energy absorption coefficient and mass stopping power. We also determined depth doses in each of the dosimeters using Monte Carlo calculations for 6 and 12 MeV electron beam and 50 and 280 kVp x-ray beams.

2. Material and Methods

The genipin gel was made using a previously described methodology [2]. The mass density of the genipin gel was measured using a helium multi-pycnometer (Quantachrome instruments, Florida,

USA). Once the mass density was known, the elemental composition and electron density were calculated. Effective atomic number was calculated using Mayneord formula [10].

The probability of photon attenuation by photoelectric absorption, Compton scatter and pair production is defined by their fractional interaction cross sections. The cross sectional data were obtained from the NIST XCOM database over the energy range 1 keV - 20 MeV [11]. The NIST x-ray attenuation database was also used to obtain elemental energy absorption coefficients. The energy absorption coefficients of genipin and water were calculated using the mixture rule [12]. The mass stopping powers were similarly calculated using the NIST ESTAR database [13].

Relative depth dose calculations for the genipin gel dosimeter and water were performed using the EGSnrc/BEAMnrc Monte Carlo package (Version 4, NRC, Ottawa, Canada) [14]. The 6 and 12 MeV electron beams were produced by a Varian 21iXs linear accelerator (Varian Medical Systems, city, state, USA). The beams were modeled using BEAMnrc with a field size of 10×10 cm^2 and the electron and photon energy cutoff parameters, ECUT and PCUT, were set to 0.561 MeV and 0.01 MeV, respectively. For the electron beam simulation, the incident beam was a parallel electron beam with 0.5 cm radius and 6.7 MeV kinetic energy for 6 MeV. For 12 MeV electron beam, incident beam had 0.6 cm radius and 13.2 MeV kinetic energy. The kilovoltage x-ray beams were generated by a Pantak Therapax DXT300 kilovoltage x-ray therapy unit with a tube potential varying from 50 to 280 kVp [15].

We calculated depth dose curves in the genipin gel and water using the DOSXYZnrc user code and the phase space files generated by BEAMnrc for the electron and x-ray beams. For the electron beams, a $60\times60\times30$ cm^3 phantom was defined which had a 0.25 cm voxel size in the buildup region. For the kilovoltage x-ray beams, a $20\times20\times10$ cm^3 phantom was defined in DOSXYZnrc with a total of 20 voxels of 0.5 cm thickness in the depth direction.

3. Results

Table 1 shows the calculated elemental composition, mass density, electron density and effective atomic number of the genipin gel dosimeter and water. The effective atomic number of genipin gel is 1.4% higher than water while its mass density is almost equal to that of water.

The calculated fractional interaction probabilities for the genipin gel and water over the energy range 1 keV - 20 MeV are plotted in Figure 1. The maximum difference in the photoelectric absorption and Compton scattering fractional interaction probabilities of the genipin gel compared to water are less than 0.1%.

Table 1- Elemental composition by fractional weight (w_k), mass and electron densities and effective atomic number for genipin gel and water.

Material	w_H	w_C	w_N	w_O	w_S	Mass density	Electron density ($\times10^{23}$ e.cm^{-3})	Effective atomic number
Genipin gel	0.1105	0.01522	0.005216	0.8696	0.003108	1.001	3.352	7.518
Water	0.1119	-	-	0.8881	-	1.000	3.343	7.417

Journal of Physics: Conference Series **250** (2010) 012036 doi:10.1088/1742-6596/250/1/012036

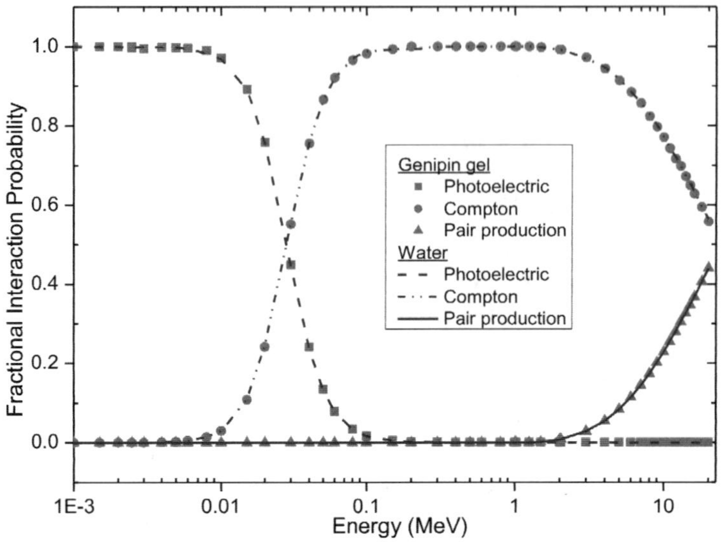

Figure1. Fractional interaction probabilities for the genipin gel dosimeter and water.

The ratio of the mass energy absorption coefficients of the genipin gel and water are presented in Figure 2 for the photon energy range 10 keV - 20 MeV. In this plot, a peak is noticeable from 10 – 100 keV which is less than 3% higher than water. This can be attributed to the small amount of the high atomic number components.

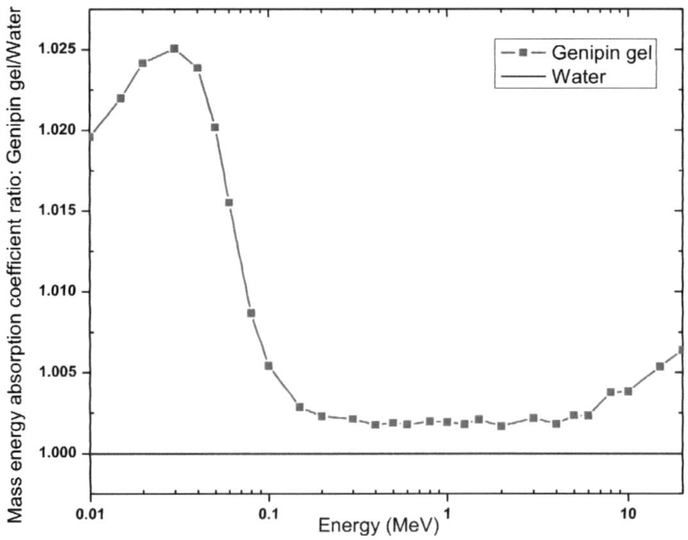

Figure 2. The mass energy absorption ratio for genipin gel relative to water.

Figure 3 shows plots of the ratios of the collisional, radiative and total stopping powers for genipin gel to that of water. It is evident that the stopping power of genipin gel has a maximum difference of 0.3% compared with the value of water.

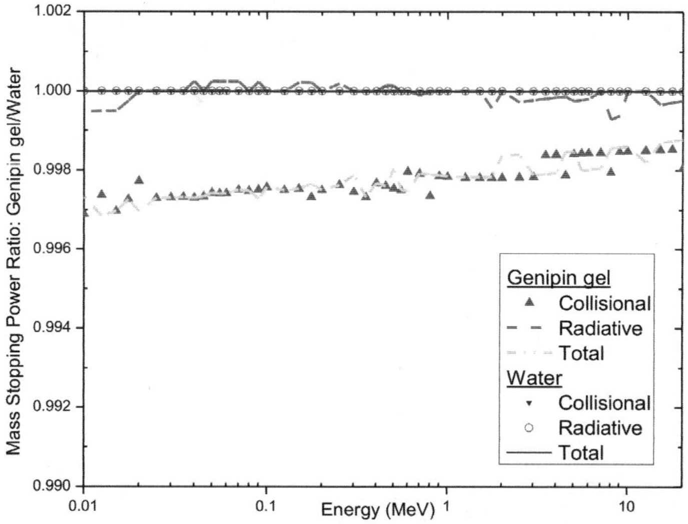

Figure 3. The collisional, radiative and total mass stopping power ratios for the genipin gel to water.

Figure 4 shows the Monte Carlo calculated depth doses for the genipin gel and water using 6 and 12 MeV electron beams. The relative dose differences with the value of water are up to 2% for the genipin gel dosimeter at a radiological depth of 10 cm.

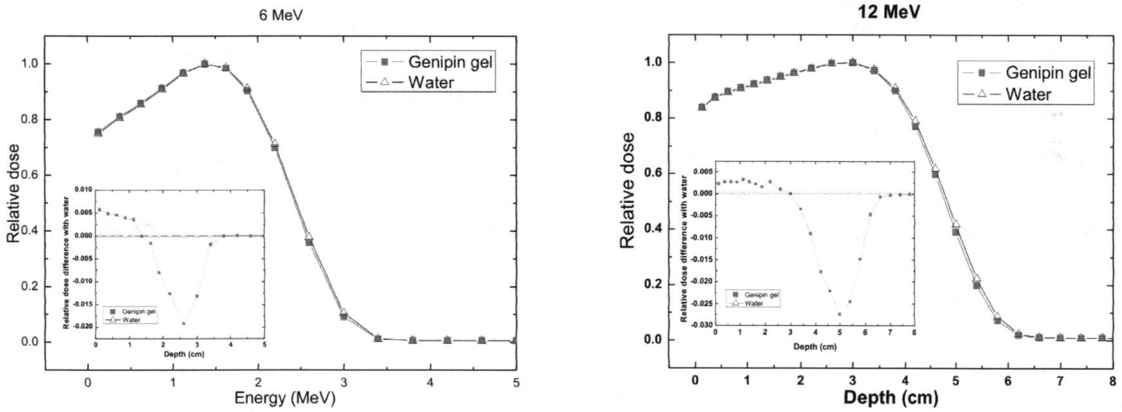

Figure 4. Monte Carlo depth dose curves for the 6 and 12 MeV electron beam for the genipin gel dosimeter and water.

Figure 5 shows the kilovoltage Monte Carlo calculated depth dose curves for the genipin gel and water. For 50 kVp, the relative dose difference with the value of water is less than 0.5% while for 280 kVp, the relative dose differences with the value of water is negligible.

4. Conclusion

Our results demonstrate that genipin gel can be considered a radiologically water equivalent dosimeter. In comparison to PRESAGE and polymer gel dosimeters such as MAGAT, MAGIC and PAGAT, genipin gel shows more water equivalent radiological properties as well as dose response [5, 8].

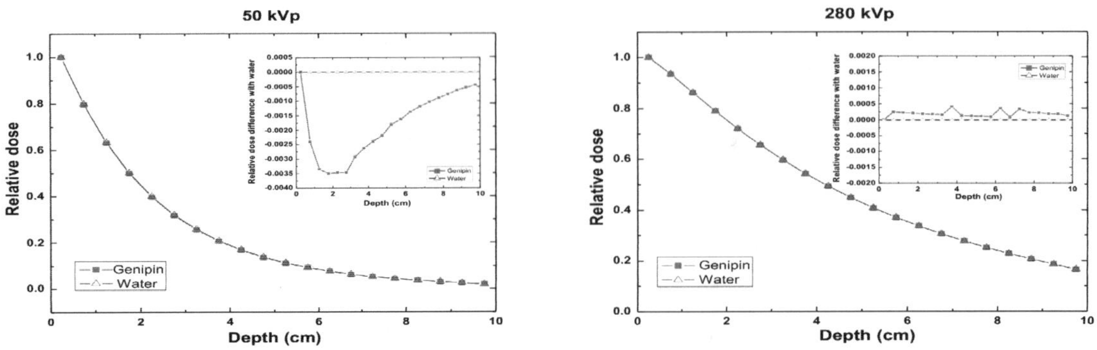

Figure 5. Monte Carlo calculated relative depth dose curves for 50 kVp and 280 kVp x-ray beams for the genipin gel and water.

References

[1] Yao, C.H., et al., Preparation of networks of gelatin and genipin as degradable biomaterials. Materials Chemistry and Physics, 2004. 83(2-3): p. 204-208.
[2] Jordan, K., Optical CT scanning of cross-linked radiochromic gel without cylinder wall. Journal of Physics: Conference Series, 2009. 164: p. 012029.
[3] Cho, S.J., et al., A study on the characteristic of genipin-gelatin gel as a radiochoromic dosimeter, in The Ninth Asia Oceania Congress of Medical Physics. 2009: Thailand.
[4] Jordan, K. Zero diffusion radiochromic genipin-gelatin dosimeter for three dimensional radiation dosimetry. in Conference Proceeding COMP annual meeting. 2008. Quebec.
[5] Brown, S., et al., Radiological properties of the PRESAGE and PAGAT polymer dosimeters. Applied Radiation and Isotopes, 2008. 66(12): p. 1970-1974.
[6] Midgley, S.M., Measurements of the X-ray linear attenuation coefficient for low atomic number materials at energies 32-66 and 140 keV. Radiation Physics and Chemistry, 2005. 72(4): p. 525-535.
[7] Baldock, C., et al., Polymer gel dosimetry. Physics in Medicine and Biology, 2010. 55(5): p. R1-R63.
[8] Venning, A.J., et al., Radiological properties of normoxic polymer gel dosimeters. Medical Physics, 2005. 32(4): p. 1047-1053.
[9] Keall, P. and C. Baldock, A theoretical study of the radiological properties and water equivalence of Fricke and polymer gels used for radiation dosimetry. Australasian physical & engineering sciences in medicine 1999. 22: p. 85-91.
[10] Khan, F.M., The physics of radiation therapy. 4th ed, ed. J. Pine. 2010, Philadelphia: Lippincott Williams & Wilkins.
[11] Berger, M.J., et al., XCOM: Photon Cross Sections Database. NIST Standard Reference Database 8 (XGAM). 2009, National Institute of Standards and Technology: Gaithersburg
[12] Hubbell, J.H. and S.M. Seltzer, Tables of X-Ray Mass Attenuation Coefficients and Mass Energy-Absorption Coefficients 2004, National Institute of Standards and Technology: Gaithersburg.
[13] Berger, M.J., et al., ESTAR, PSTAR, and ASTAR: Computer Programs for Calculating Stopping-Power and Range Tables for Electrons, Protons, and Helium Ions 2005, National Institute of Standards and Technology: Gaithersburg.
[14] Rogers, D.W.O., Fifty years of Monte Carlo simulations for medical physics. Physics in Medicine and Biology, 2006. 51(13): p. R287-R301.
[15] Kim, J., et al., An investigation of backscatter factors for kilovoltage x-rays: a comparison between Monte Carlo simulations and Gafchromic EBT film measurements. Physics in Medicine and Biology, 2010. 55(3): p. 783-797.

Journal of Physics: Conference Series **250** (2010) 012037 doi:10.1088/1742-6596/250/1/012037

MAGIC-*f* gel dosimetry for clinical electron beam

T A Pianoschi[1], M Alva[1], M Santanna[2], O Baffa[1] and P Nicolucci[1]

[1]Department of Physics and Mathematics, University of São Paulo, Av. Bandeirantes 3900, CEP:14010-901- Ribeirão Preto – SP – Brazil.

[2]Hospital of Cancer at Barretos (HCB), Av. Antenor Duarte Villela – 1331, CEP: 14784- 400, Barretos – SP –Brazil.

E-mail: thatianealves@yahoo.com.br

Abstract. Electron beam therapy is well-known for its characteristic of low penetration and high dose rate. The input the dosimetric parameters like percentage depth dose (PDD) and beam profiles is required for treatment planning systems. Polymeric gel is a valuable tool for dose measurements, specially the MAGIC-*f* gel dosimeter have been shown as a feasible dosimeter for use in radiotherapy due to its characteristics good spatial resolution and the ability to construct tridimensional dose distributions. The aim of this work is to study the MAGIC-*f* response for an electron beam of 6MeV using different doses and dose rates and also the determination of dosimetric parameters, like PDD and beam profile, for this clinical beam. From the obtained results it can be infered that MAGIC-*f* polymeric gel can be used as a complementary dosimetric tool for electron beams due to its characteristics of high spatial resolution and tridimensional dose distributions.

1. Introduction

Electron beams are widely used in radiotherapy with advantages in the irradiation of near-surface targets and in the steep falloff of the dose with depth, reducing the absorbed dose in healthy tissues [1]. Because of its characteristics the clinical application of this radiation beam requires precise planning and dosimetric measurements. Useful dosimetric parameters in radiotherapy treatment planning systems, like PDD and beam profile, are determined at the maximum, practical and therapeutic ranges as the dose at depths of 100%, 90% and 50% doses [2-3].

Polymeric gel dosimeters have been studied for some applications in radiotherapy due to its characteristics of high spatial resolution and the determinations of tridimensional dose distributions in treatment volumes [4]. Standard dosimeters do not present the tridimensional (3D) property, because their responses are in one-dimensional or two-dimensional, like an ionization chamber and film, respectively. Tridimensional dosimetry is important in treatments that have high dose gradients like conformal radiotherapy (IMRT) and electron beams [5]. Especially to electron beams the sensitivity, spatial resolution, dose, energy and dose rate dependence, as well as other dosimetric parameters, are still to be investigated for this dosimeter and there is no consolidated data in literature [6].

This work proposes the response evaluation of the MAGIC-*f* dosimeter for a 6 MeV electron beam. Measurements of dose, dose rate and dosimetric parameters like PDD and beam profile were measured with this dosimeter. The gel dosimeter responses were compared with clinical data.

Journal of Physics: Conference Series **250** (2010) 012037 doi:10.1088/1742-6596/250/1/012037

2. Materials and methods

2.1. MAGIC-*f fabrication*

The samples of MAGIC-*f* polymeric gel were prepared following the literature [7] and were used to fill three phantoms: glass tubes (Vacutainer ®), 6 cm length and 0.5 cm radio, for the study of the gel response with dose and dose rate; glass tubes of 20 cm length and 1 cm radio, for PDD determinations and PPMA cubes of 10 x 5 x 5 cm^3 for beam profiles.

2.2. Irradiation

The irradiation was performed in a Varian 2100 acelerator with electron beams of 6 MeV, in a 50 x 50 x 50 cm^3 PPMA phantom, filled with water, as shown in figure 1a. Figure 1b shows the phantoms after irradiation with doses of 0.5 Gy to 20 Gy. The irradiations for the dose rate study were performed in the range of 80 to 400 UM/min with 2 Gy. For the analysis of the dosimetric parameters, like PDD and beam profile, the samples were irradiated with 2 Gy with 400 UM/min. The figure 2 shows the irradiated phantoms for the determination of the PDD and beam profile.

(a)

(b)

Figure 1. Calibration of MAGIC*f:* (a) irradiation geometry, (b) visualization after irradiation.

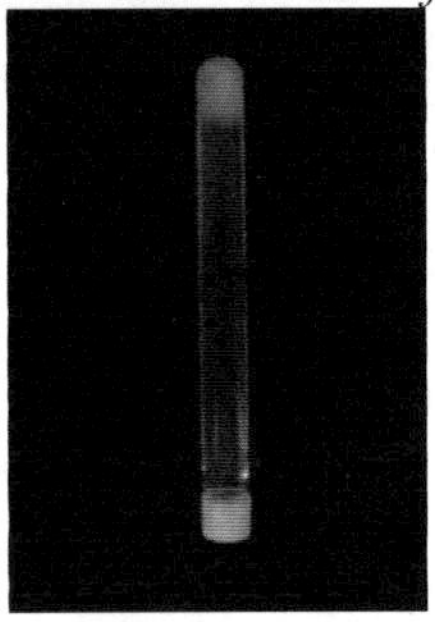
(a)

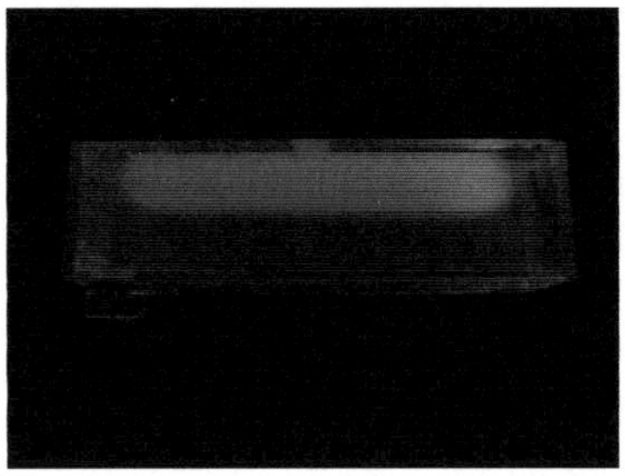
(b)

Figure 2. Irradiated phantoms with 6 MeV: (a) PDP, (b) Beam profile.

2.3. *Relaxation measurements*

The relaxation time measurements were performed in a 3 T Philips RMN tomography using a head coil. The RM images were acquired with a multi spin-eco sequence of 16 ecos, TE 15ms, TR 4000ms, resolutions of 250 μm.

3. Results and discussion

The results obtained for the response evaluation of dosimeter were compared with the clinical dosimetric parameters. The results of the linearity and dose rate response are shown in figure 3.

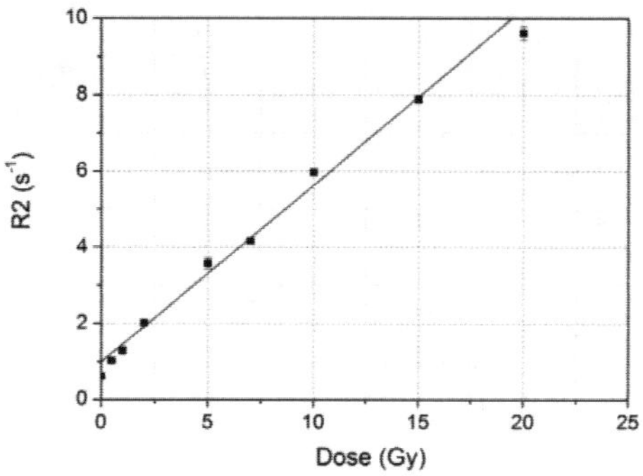

(a)

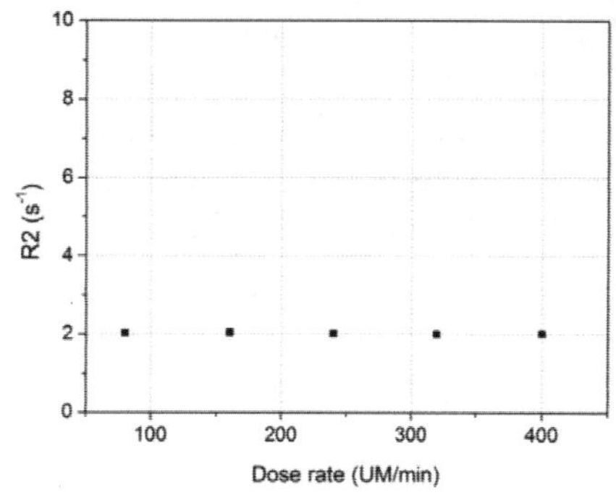

(b)

Figure 3. MAGIC-*f* response for 6 MeV: (a) dose linearitiy, (b) dose rate.

Gel response versus dose showed a linear behavior, with correlation coefficient, r^2, approximately of 0.996 and the maximum standard deviation from the average readings was 0.14 s^{-1}. A maximum difference of 1.7% was found for the R2 values, when a dose rate of the 400 UM/min (clinical standard dose rate) is compared to the other dose rates (80 – 320 UM/min) with the maximum standard deviation from the average readings was 0.04 s^{-1}.

The PDD curves and beam profile, at the dose maximum depth, were obtained with the MAGIC-*f* gel and are shown in figure 4.

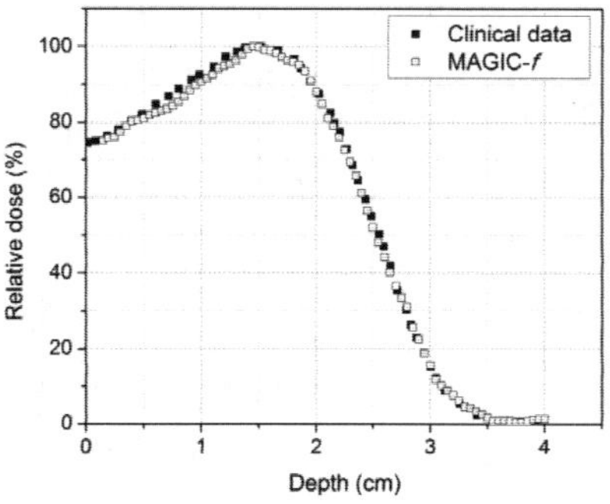

(a)

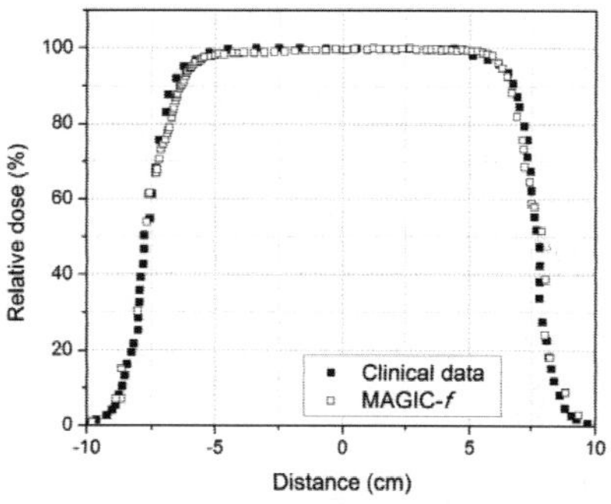

(b)

Figure 4. Dosimetric parameters for 6 MeV: (a) PDP; (b) beam profile.

The PDD and beam profile curves obtained with MAGIC-*f* were compared with clinical data showing differences lower than 1.3% and 2.5%, respectively. Therefore, in concordance to the results shown in figure 4, MAGIC-*f* can determine dosimetric parameters for electron beams.

4. Conclusions

MAGIC-*f* gel showed dose linearity and the dose rate dependency was lower than 1.2%. The dosimetric parameters for electron beams, like percentage depth dose and beam profile, show similar results as the clinical data used as reference for this study. Although, these resutls are preliminar for 3D dosimetric can be infered that MAGIC-*f* polymeric gel can be used to measure these dosimetric parameters and can be used like a complementary dosimetric tool for electron beams, presenting the important characteristic of high spatial resolution and allowing the determination of 3D dose distributions in electron beam therapy.

Acknowledgments

Authors gratetfully acknowledge Hospital do Câncer de Barretos, Hospital das Clínicas de Ribeirão Preto and J.L. Aziani for the tecnical support and CAPES for the financial support.

References

[1] Faddegon B, Sawkey D, O'shea T, McEwen M and Ross C 2009 *Med. Phys.* **36** 4578 – 91

[2] IAEA 2001 *Absorbed dose determination in external beam radiotherapy: na international code of practice for dosimetry based on standards of absorbed dose to water* IAEA-TRS-398 (Viena: International Atomic Energy Agency)

[3] IAEA 2003 *Review of radiation oncology Physics: A handbook for teachers and students* (Viena: International Atomic Energy Agency)

[4] Trapp J V, Partridge M, Hansen V N, Childs P, Bedford J, Warrington A P, Leach M O and Webb S 2004 *Phys. Med. Biol.* **49** 1625 – 35

[5] Wong C J, Ackerly T, He C, Patterson W, Powell C E, Ho A, Qiao G, Solomon D H, Meder R and Geso M 2007 *App. Radiat. Istopes.* **65** 1160 – 64

[6] Jin L, Ma C M, Elddib A, Price R A, Chen L, Wang L, Chi Z, Xu Q, Sherif M and Li J M 2008 *Phys. Med. Biol.***53** 6009 – 25

[7] Fernandes J P, Pastorello B F, de Araújo D B and Baffa O 2008 *Med. Phys.* **53** N53-N58

Journal of Physics: Conference Series **250** (2010) 012038 doi:10.1088/1742-6596/250/1/012038

A Liquid Scintillator System for Dosimetry of Photon and Proton Beams

Beddar S

Department of Radiation Physics, Unit 94, The University of Texas M. D. Anderson Cancer Center, 1515 Holcombe Boulevard, Houston, Texas 77030

abeddar@mdanderson.org

Abstract. We have developed a 3D system based on liquid scintillator (LS) for the dosimetry of photon and proton therapy. We have validated the LS detector system for fast and accurate quality assurance of IMRT and proton therapy fields. Further improvements are required to optimize the quantitative analysis of the light output provided by the system in photon beams. We have also demonstrated its usefulness for protons as it can determine the position and the range of proton beams. This system has also been shown to be capable of fast, sub-millimeter spatial localization of proton spots delivered in a 3D volume and could be used for quality assurance of IMPT. Further developments are on-going to measure beam intensities in 3D.

1. Introduction

The LS detector system consist of a 20x20x20 cm3 tank of LS surrounded by a light tight enclosure viewed by a CCD camera placed 80 cm from the center of the tank to decrease the effects of stray radiation. The LS (BC-531, Saint Gobains) has a physical density of 0.87 g/cm^3 and emits light in the blue region of the spectrum when irradiated. The LS tank is centered on the isocenter, and the radiation beam is directed perpendicular to the tank. During irradiation, the CCD camera acquires images of the light distribution in either rapid or continuous imaging mode. A deconvolution algorithm is applied to the images to reduce the effects of scattered light. The measured light distribution is compared to the dose distribution measured by ionization chambers

2. Methods

2.1. Photon Beams

In experiments measuring x-ray fields from a conventional linac, the measured light signal showed good agreement with the predicted light image based on the delivered dose. Differences occurred due to the light scattering and light reflection. An integration time of 150 ms was adequate to provide a resolution of 1 cGy at a dose rate of 400 cGy/min. Frame-by-frame acquisition demonstrated the advantage of reduced noise over a single long-time acquisition.

Figure 1a below depicts the experimental setup of our detector system shown on the right (a 6-MV photon radiation is delivered by a Varian Clinac 21EX, with a gantry rotating around the detector). A sketch of the components in this detector is shown in Figure 1b. It consists of a radiation sensitive compartment filled with a liquid scintillator, a light tight housing and a compartment for the CCD camera. The optical components are perpendicular to the beam direction to minimize noise from stray radiation.

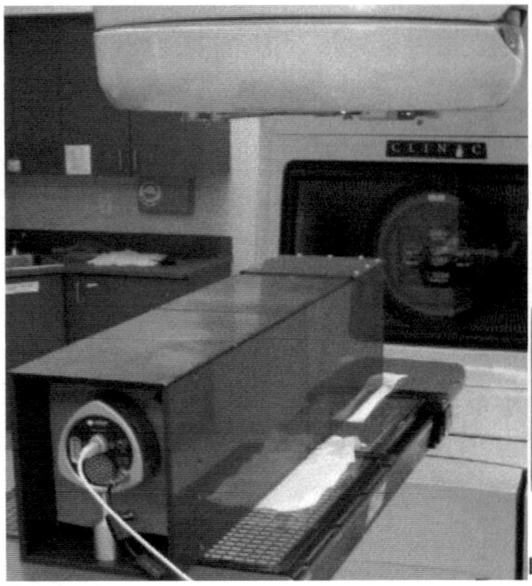

Figure 1a

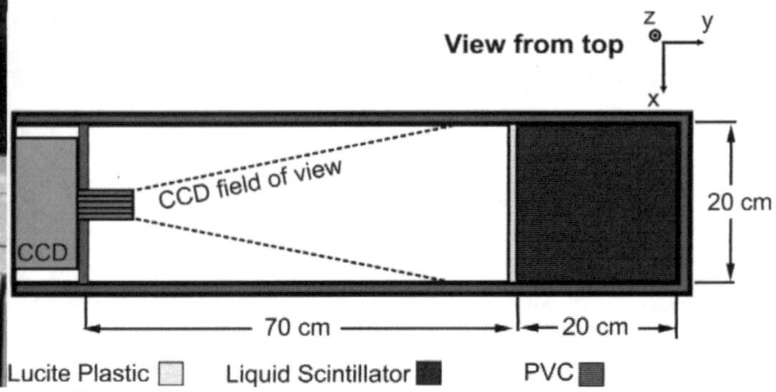

Figure 1b

Figure 2 shows a comparison between the measured light and the predicted light from the projected dose. Both images are similar. However, the data within the measured light image (top) are noisier than within the forward projected convoluted dose (bottom). This is because of the stray radiation from the treatment head. Shielding the camera would improve the quality of image.

Further work is required to optimize the blurring filter kernel and to automate the entire workflow for this IMRT QA method using such a detector system.

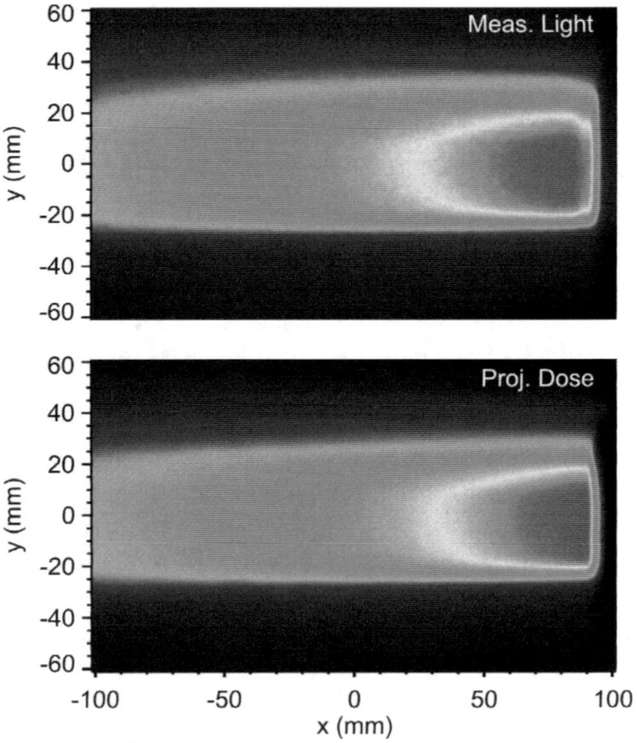

Figure 2

2.2 Proton Beams

In experiments on passive scattered proton beams, quenching was observed throughout the SOBP, resulting in a decrease in signal with depth. However, for a given depth, the light signal increased linearly with the number of MUs delivered. The depth of passive scattered proton beams can be determined within 1 mm.

In experiments on scanned proton beams, the lateral position of a proton pencil beam could be determined within 0.6 mm, and the depth of pencil beams could be determined within 0.4 mm. Rapid imaging of scanned proton beams (Figure 3) shows promise for providing beam positioning information in real time as the spot scanning fields are being delivered. Monte Carlo calculations show that it is possible to calculate quenching correction factors for individual pencil beams.

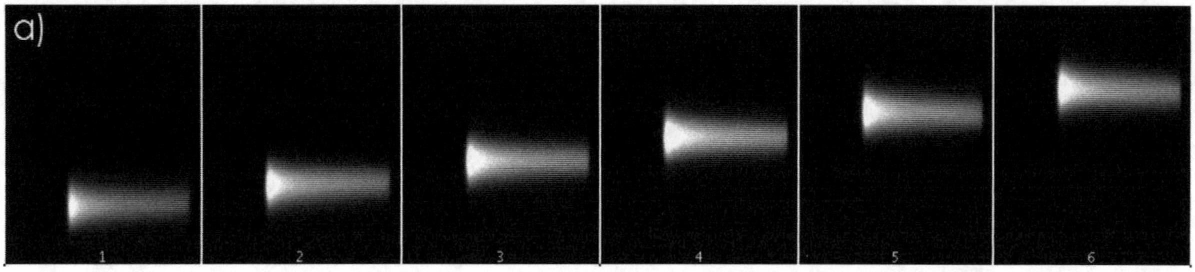

Figure 3: A sequence of 6 frames acquired while the proton beam was swept vertically (x-direction).

Beam energy (MeV)	Range shifter (cm of water)	Nominal range (cm)	Measured range (cm)
121.2	0	10.50	10.52
121.2	1	9.50	9.54
161.6	0	17.60	17.61
161.6	1	16.60	16.62
161.6	5	12.60	12.54

The above table shows a comparison between the measured and the nominal proton range measured with the LS system for two beam energies with various range shifters.

Figure 4. Measurements of dose distributions produced by a 120 MeV passively scattered proton beam with a field size of 2x2 cm^2 (to approximate a pencil beam from a passive scattering system). The nominal beam penetration in water was 5 cm.
Scintillation light profile with the LS detector system compared to depth dose measurement with an ion chamber. Quenching effects within the LS decrease the light signal in the proton Bragg peak. However, Monte Carlo calculations can provide correction factors to correct for the quenching artifacts. (All the curves were normalized at 2.0 cm)

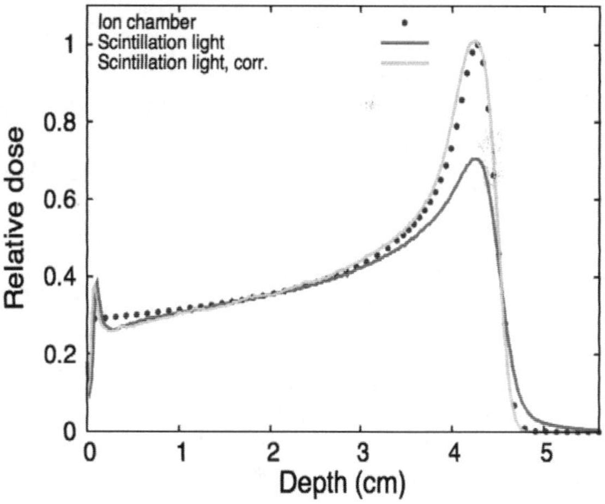

Figure 4

3. Conclusions

A liquid scintillation (LS) detector system has been developed for verification of 3D dose distributions. We demonstrate the utility of such system for fast and accurate quality assurance of IMRT and proton therapy fields.

References

[1] Beddar S, Archambault L, Sahoo N, Poenisch F, Chen GT, Gillin MT, and Mohan R. Exploration of the potential of liquid scintillators for real-time 3D dosimetry of intensity modulated proton beams. Med Phys 36(5):1736-1743, 2009.

[2] Pönisch F, Archambault L, Briere TM, Sahoo N, Mohan R, Beddar S, and Gillin MT. Liquid scintillator for 2D dosimetry for high-energy photon beams. Med Phys 36(5):1478-1485, 2009.

Dosimetry of a New P-32 Ophthalmic Applicator

[1,2] Chang Heon Choi, [3]Hyon Soo Han, [3]Kwang-Jae Son, [3]Ul Jae Park, [1,2]Sung-Joon Ye

[1]Department of Radiation Oncology, Seoul National University College of Medicine, Seoul Korea 110-744

[2]Institute of Radiation Medicine, Medical Research Center Seoul National University College of Medicine, Seoul Korea 110-744

[3]Hanaro Applications Research, Korea Atomic Energy Research Institute, Daejeon Korea 305-353

sye@snu.ac.kr

Abstract. **Purpose**: A ^{90}Sr/Y applicator has been used as a β-source for postoperative irradiation after pterygium excision. As an alternative to ^{90}Sr/Y irradiation, we proposed treatments with ^{32}P. This study aims to provide the dosimetry for this new applicator. **Method and Materials**: A ^{32}P ophthalmic applicator, of which the design and materials were optimally designed by Monte Carlo simulations, was fabricated by the Korea Atomic Energy Research Institute. The absorbed dose at the front surface of the cylindrical ^{32}P applicator was measured by using an extrapolation ionization chamber. Radiochromic film (RCF) was used to measure depth dose distributions and dose profiles at various depths. A micro-MOSFET detector was used for depth dose measurements. **Results**: The absorbed dose rates to the reference point (i.e., at the centre on the front surface of ^{32}P applicator) were 0.238 ± 0.012 cGy/s for an extrapolation ionization chamber, 0.280 ± 0.001 cGy/s for RCF, and 0.257 ± 0.020 cGy/s for MOSFET. The axial depth dose rate was reduced into approximately 1/10 as ^{32}P betas penetrate every 2 mm depth. Measured data in depths of 1 mm to 3.5 mm agreed with Monte Carlo data. Due to non-uniform absorption of ^{32}P into an absorbent disk, the dose at the centre of trans-axial plane was 2%-4% less than the peak dose around the periphery. We confirmed no leakage of ^{32}P activities and negligible exposure rate around the hand grip of the applicator. **Conclusions**: The ^{32}P applicator can deliver uniform therapeutic doses to the surface of the conjunctiva, while sparing the lens better than ^{90}Sr/Y applicators. The doses at any points from the ^{32}P applicator can be calculated by using these measured data sets. The safety of ^{32}P applicator was confirmed. However, prior to the clinical application of every new applicator, safety, dose uniformity, and absorbed dose rate at the reference point should be carefully evaluated by the method developed in this study.

1. Introduction

^{90}Sr/Y applicator has been widely used as a pure β-source for postoperative irradiation after pterygium excision. As an alternative to ^{90}Sr/Y irradiation, we proposed treatments with pure betas of a ^{32}P-source. We developed new ^{32}P ophthalmic applicators. The dosimetric characteristics of ^{32}P were calculated by using Monte Carlo simulations and shown to be advantageous over ^{90}Sr/Y irradiation. In

Journal of Physics: Conference Series **250** (2010) 012039 doi:10.1088/1742-6596/250/1/012039

order to optimize the proper design and materials, Monte Carlo simulations were performed. The applicators were manufactured by the Hanaro reactor group of KAERI (Korea Atomic Energy Research Institute), according to this proposed design. Prior to the clinical application of new ophthalmic applicators, three dosimetric characteristics (absolute dose rate, central axis depth doses, and trans-axial dose profiles) were determined. Dosimetric characteristics were determined using these dosimeters (extrapolation ionization chamber, radiochomic film, and semiconductor detector). Finally, the measurement depth dose and trans-axial dose profile were compared with Monte Carlo calculations. Based on this comparison, the feasibility for ^{32}P ophthalmic applicators was evaluated.

2. Methods and Materials

High-energy betas of ^{90}Sr/Y (maximum 2.27 MeV from ^{90}Y) can deliver therapeutic doses to the sub-conjunctival tendon within 1 mm from the applicator surface. The average energy of emerging betas from a ^{90}Sr/Y applicator is about 0.9 MeV. Furthermore, its long half-life (28.8 yrs) requires additional precautions for its production, storage, and disposal. Even though ^{32}P has a lower maximum energy (1.71 MeV) than ^{90}Sr/Y, its beta spectrum has a peak near 0.7 MeV and an average energy comparable to that of ^{90}Sr/Y. ^{32}P can deliver doses to the target lesion with similar or better effectiveness than ^{90}Sr/Y. ^{32}P is an isotope that can be produced relatively easily by fast or thermal neutron activation in a nuclear reactor. Its short half-life (14.3 days) makes it a less radiohazardous material.

To optimize the structure and materials of an applicator, dose distributions for a variety of designs were evaluated the Monte Carlo method. Based on this study, an optimal ^{32}P applicator was fabricated in KAERI. The component of applicators is shown in Fig. 1.

The radiochromic film (RCF, GAFCHROMIC EBT2, ISP Corporation, USA) is used to provide the dose distribution of ^{32}P applicator in water (or tissues). A flatbed scanner (Expression 10000XL, Epson, USA) used on this study is a high-quality scanner. We evaluated the film using an in-house program. The films were calibrated against 15 MV photon beam of a clinical linac. RCFs were used to measure depth-dose data and trans-axial dose profiles in joint phantoms. An extrapolation chamber (Bohm extrapolation chamber, PTW, Germany) is considered the standard device for the dosimetry of beta particle sources, The applicator was located on 5 cm distance from the detector surface. For this study, Micro-MOSFET dosimeters (Mobile MOSFET, Bestmedical Canada) were selected because of its high spatial resolution. To measure the surface dose, MOSFET was placed on the top of solid water phantom. The central axis depth dose was measured at the surface, 1 mm, 2 mm, 3 mm, and 4 mm depths.

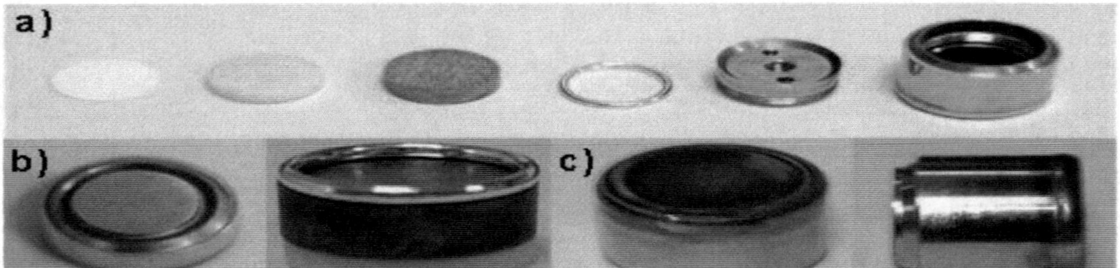

Fig 1 A prototype of ^{32}P applicator and parts a) various components (isotope, sealing, and protector, etc.), b) a part of grip, c) eye-contact surface.

3. Results

The absorbed dose rates to the reference point (center of the circular surface) in water were 0.238 ± 0.012 cGy/s for an extrapolation ionization chamber, 0.280 ± 0.001 cGy/s for radiochromic films (RCF), and 0.257 ± 0.020 cGy/s for MOSFET dosimeters(Table 1). The axial depth doses measured by RCF and MOSFET decreased very rapidly (more than exponentially) with increasing depths (Fig

2). The dose rate was reduced into approximately 1/10 as ^{32}P betas penetrate every 2 mm depth. Being considered very high dose gradient and measurement setup uncertainties, measured data sets in depths of 1 mm to 3.5 mm agreed with Monte Carlo data. Due to non-uniform absorption of ^{32}P to an absorbent disk, the dose at the center of trans-axial plane was 2%-4% less than the peak dose around the periphery (Fig 3). But it provided better trans-axial dose uniformity than Monte Carlo data that assumed a uniform absorption of ^{32}P. We confirmed no leakage of ^{32}P activities in the applicator and negligible exposure rate around the hand grip (close to the background level).

Table 1. Absorbed dose rate (cGy/s) to water at the reference point measured by three different dosimeters.

Depth (mm)	Extrapolation chamber	Radiochromic film	MOSFET
0 (surface)	0.238± 0.012	0.280 ± 0.001	0.257±0.020

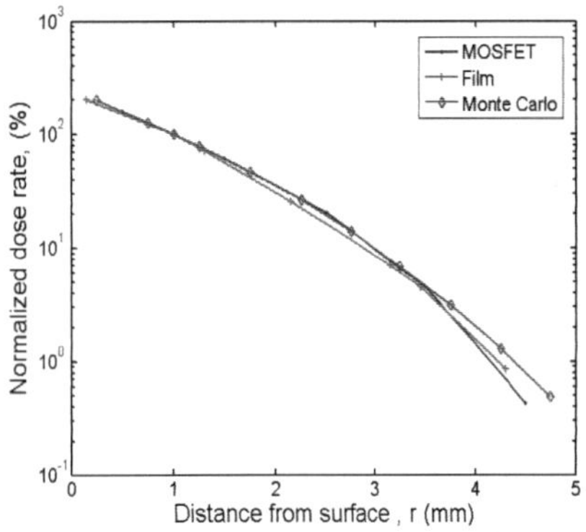

Fig 2. Measured and calculated (MCNP) depth dose distributions for the ^{32}P planar source.

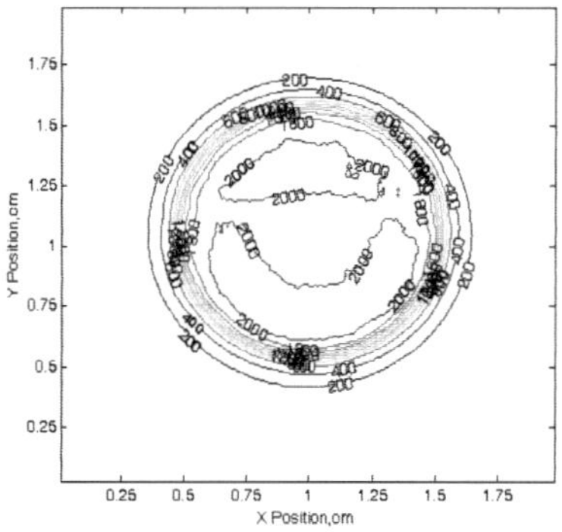

Fig 3. Relative dose distributions for the ^{32}P source obtained with radiochromic film.

4. Discussion & Conclusion

The clinical doses at any points from the ^{32}P applicator can be calculated from these measured data sets. The ^{32}P applicator can deliver uniform therapeutic doses to the surface of the conjunctiva for postoperative irradiation of pterygium, while sparing the lens better than $^{90}Sr/Y$ applicators. The safety of the ^{32}P applicator was confirmed. However, prior to the clinical application of every new source, the integrity, dose uniformity, and absorbed dose rate at the reference point should be evaluated by the methods described in this study. We successfully developed all dosimetry procedures necessary for the clinical implementation of the ^{32}P ophthalmic applicator.

References

[1] M. Jurgenliemk-Schulz, L. J. Hartman, J. M. Roesink, R. J. Tersteeg, I. van Der Tweel, H. B. Kal, M. P. Mourits, and H. K. Wyrdeman, "Prevention of pterygium recurrence by postoperative single-dose betairradiation: A prospective randomized clinical double-blind trial," Int. J.Radiat. Oncol., Biol., Phys. 59, 1138–147 (2004).

[2] Y. K. Park, S.-J. Ye, I. H. Kim,W.R.Wee, M. K. Kim, H.S.Han, K.-J. Son, and U. J. Park, "Potential use of P-32 ophthalmic applicator: Monte Carlo simulations for design and dosimetry," Med. Phys. 35, 1854-1858 (2008).

[3] C. G. Soares, S. Vynckier, H. Ja rvinen, W. G. Cross, P. Sipila, D. Flu hs, B. Schaeken, F. A. Mourtada, G. A. Bass and T. T. Williams, "Dosimetry of beta-ray ophthalmic applicators: Comparison of different measurement methods," Med. Phys. 28(7) , 1373-1384 (2001).

Investigation on Tissue Equivalent Normoxic Polymer Gel Dosimeter using In-house Laser CT scanning system

D. Senthil Kumar and E. James Jebaseelan Samuel

Photonics Division, School of Advanced Sciences, VIT University, Vellore, India.

E-mail: kalaimeenasen@gmail.com

Abstract. Optical Computed Tomography has wide applications in the treatment of cancer. In continuation of this, an in-house Laser CT scanner has been built for "3D gel dosimetry". The Laser CT (LCT) scanner plays a major for Gel dosimeter or phantom readout and in clinical radiation therapy as a 3-Dimensional Radiation Dosimetry. A gel dosimeter which absorbs dose in a tissue-equivalent manner and allows the measurement of spatial distribution of the deposited dose is required. The normoxic PAGAT (Polyacrylamide Gelatin and Tetrakis) gel is used as a dosimeter for this analysis. When laser passes through this gel phantom, absorption and scattering takes place and combined to attenuation. The optical attenuation coefficient of the laser can be obtained by measuring its intensity after passing through the gel by means of a sensor. Reconstruction using Mat Lab algorithm provides 3D dose distribution.

1. Introduction

Now days Stereotactic Radiotherapy, 3D Conformal Radiotherapy and Intensity Modulated Radiotherapy are the modern and common radiation delivery technique intents accurate delivery of radiation dose. These techniques are based on increased complexity of radiation delivery methods and therefore require verification of dose distribution by a precise dosimetric method. Gel dosimeters are capable of recording dose distributions in 3 dimension manner. At present, magnetic resonance imaging has been most widely used for the evaluation of absorbed dose distributions in gel dosimeters. However, due to the cost, problem of accessibility, optical CT scanning of gel has been used by several investigators as an exact and substitute to Magnetic Resonance Imaging [1-4]. In this work we exhibit the construction of an in-house laser CT scanning system and some initial investigations using it with the normoxic PAGAT Gel Dosimeter.

2. Materials and Method

2.1. Gel Preparation

The gel used for this study was the normoxic PAG Gel named as PAGAT, which had been previously evaluated with X-ray CT and Magnetic Resonance Imaging system [5, 6]. The normoxic ''PAGAT'' gel described elsewhere [6-8] was prepared under normal atmospheric conditions. The gel preparation was carried out as follows: 5% gelatin was added to the magnetic stirred glass flask containing 89% of triple distilled water and allowed to swell for 10–15 min. Then the gelatin solution was heated to the temperature of 50 ^{0}C for complete dissolution of the gelatin. After obtaining the clear gelatin solution 3.5% BIS and 3.5% Acrylamide were added and stirred continuously until the mixture was dissolved homogeneously. The antioxidant 10 mM THP was added to the solution under stirring just before transferring the gel into Pol yethylene Teraphthalate cylindrical recipients of s ize 10 c m diameter and 13 cm long. All the Poly ethylene Teraphthalate cy lindrical recipients containing the PAGAT gel wer e sealed properly to avoid the further diffusion of ox ygen and were kept in a refrigerator for one full night for gelling. All chemicals were from Sigma Aldrich.

2.2. Gel Irradiation

To evaluate the dose response of pol ymer gel to gamma radiations, the poly mer gel dosimeters were irradiated aft er the complete gellation. The c enter o f the phantom w as aligne d with the cr osshairs of the treatment field. The poly mer gel dosimeters were irradiated with a direct square field size of 5 x 5 cm^2 for ga mma rays using Co-60 source and the do se is 12 Gy . The source-to-surface distance (SSD) was 100 cm . One dosimeter was kept unirradiated for background subtraction. The irradiated gel dosimeters were left in a refrigerator for one day before imaging to allow the complete post irradiation polymerization.

2.3. Laser CT scanning system

The Laser C T scanner wa s constructed on the principl e of a first generation X-ray CT is shown in figure 1. This scanner consisted of an aquarium , a rotating table, a red colour di ode laser (650 nm) as light source, a detector (Photodiode). The aquarium of size 30 x 30 x 30 cm^3 was fabricated using an

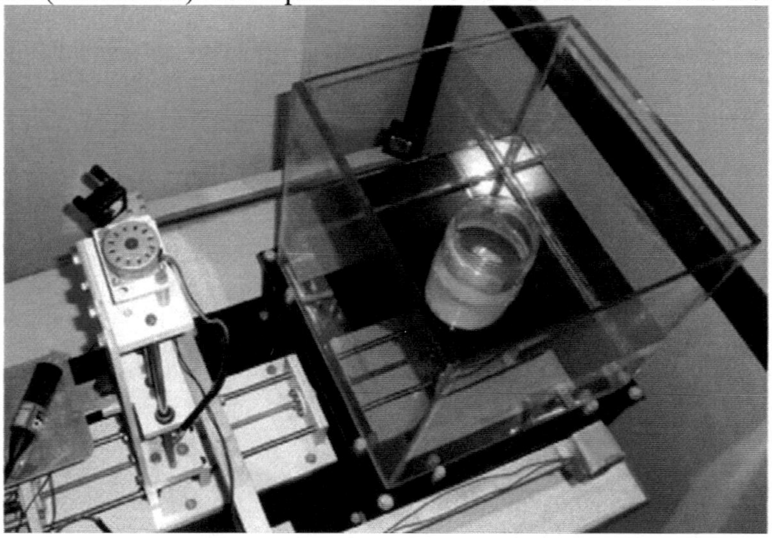

Figure 1. In-house fabricated Laser CT scanner.

imported Perspex with wall thickness of 1 mm on the laser t ransmission faces. Using oil seal technique, the rotating table was fixed in the aquari um and a bipolar stepper motor was connected for rotation. This rotating table could be r otated at an angle of 1 °. The linear motion of the source and the detector was achieved by the horizontal rails connected with a ball screw which was driven by another bipolar stepper motor. Figure 2 explains the schema tic function of the Laser CT scanner. A sensor has to measure th e intensity of a laser beam after passi ng through the sample (in y direction). This laser beam has to scan the sample in ‘x’ and ‘z’ direction. The cylinder-shaped sample has to be rotated (θ) around its s ymmetry axis which is parallel to the ‘z’ axis. One challenge is the refraction of the laser beam when passing through the bottle walls which are not at right angle to the beam line. This effect can be m inimized by keeping the gel vessel within an aquarium which is fill ed with a liquid of the same refractive index as the gel, realiz ed by a water and gly cerol mixture. For all three movements it has to be considered, whether the sample or the laser beam should move. The movements are executed by stepper motors driven by stepper motor drivers, which are co ntrolled by the softwar e using a Dat a Acquisition Card (DAq card). The analog inputs were converted into digital outputs by connecting this DAq card as an interface between the computer and the Laser CT scanner.

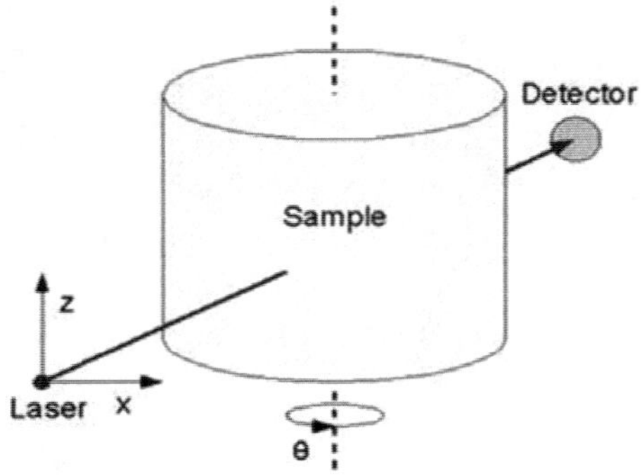

Figure 2. Schematic sketch of Laser CT scanner's functionality.

2.4. Scanner Control and Image Reconstruction

The software will control the scanner, acquire and store the data and perform the necessary calculations for data procession and transformation. The "Image Reconstruction Software" (IRS) converts the scanner data into 3-dimensional image data. The purpose of this software is to apply an "Inverse Radon Transformation" to the scanner raw data. Matlab readily provides the "Filtered Back-Projection algorithm", an efficient algorithm for the Inverse Radon Transformation. Thus the IRS consists of only few lines of Matlab script, reading a TIFF formatted file with the scanner raw data, applying the filtered back-projection algorithm and writing the result into another TIFF formatted file.

3. Results and Discussion

Figure 3 & figure 4 represents the top view of pre & post irradiated gel phantoms. The transparency

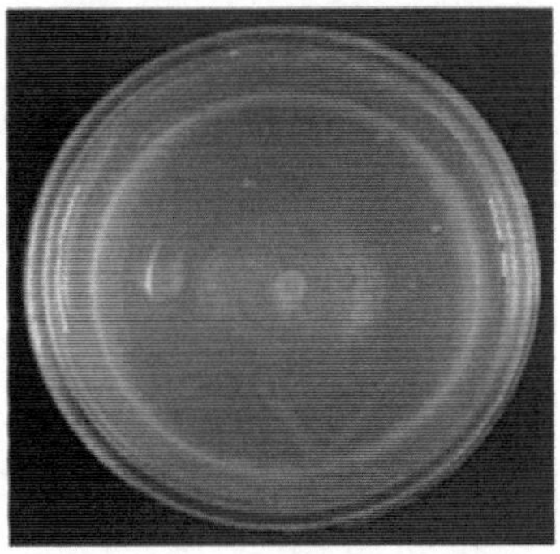

Figure 3. Top view of Pre irradiated Gel Phantom.

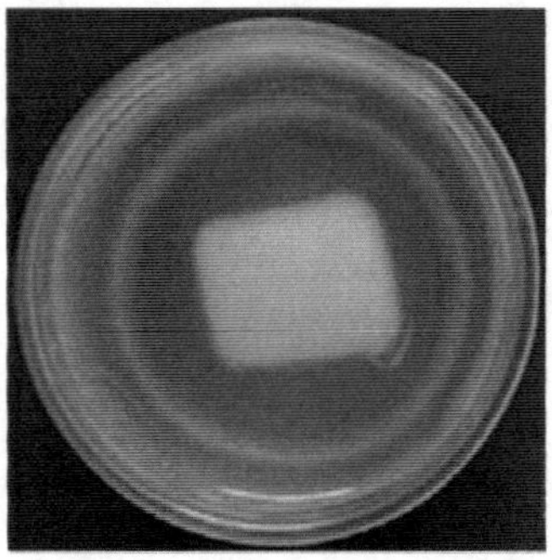

Figure 4. Top view Post irradiated Gel Phantom.

variation clearly differentiates the pre & post irradiated phantoms. The pre-irradiation image was subtracted from the post-irradiation scan and the subtracted image were analysed. The reconstruction for a single slice has been done using the Filtered Back Projection algorithm implemented in Mat lab software. Data analysis like getting Dose profile for the reconstructed slices and subtraction of post irradiated images from pre irradiated images was performed with in-house code developed in Mat Lab. Figure 5 represents the image of the pre irradiated gel slice where, a line is drawn on the pre and post

Figure 5. 2D Image of CT Reconstructed Slice of Pre irradiated gel.

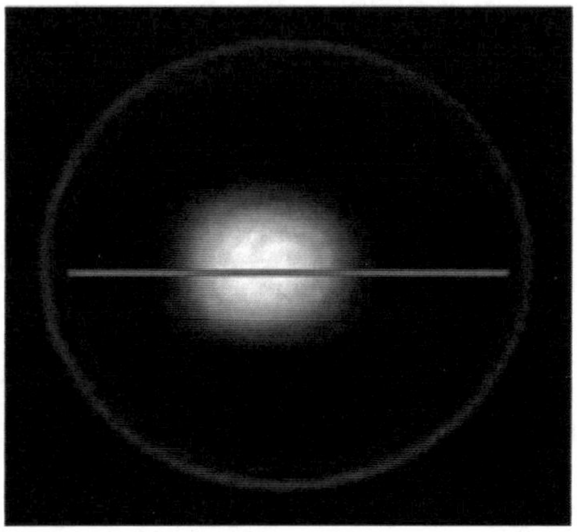

Figure 6. 2D Subtracted Image of CT Reconstructed Slices of Post irradiated gel.

irradiated gel slice to get their corresponding dose profile. Figure 6 represents the subtracted slice of the post irradiated gel phantom. Figure 7 represents the dose profile for the pre irradiated gel slice with distance across the gel dosimeter. Figure 8 represents the dose profile at 4cm depth for a post irradiated gel slice with distance across the gel.

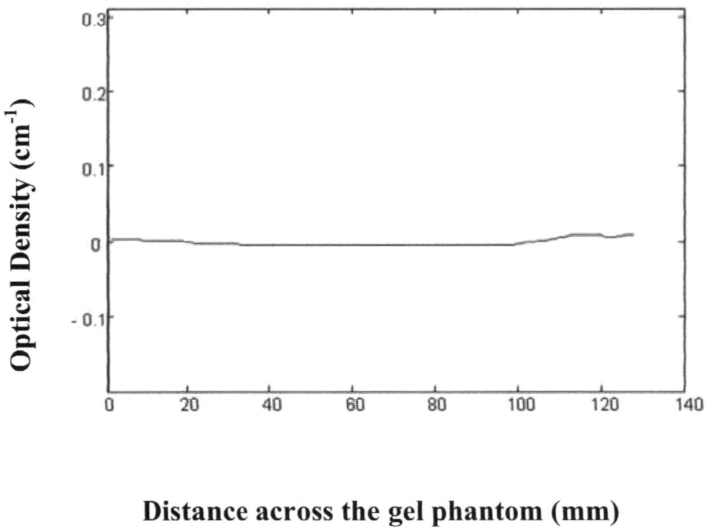

Figure 7. Dose Profile for Pre irradiated gel slice (Distance across the gel phantom vs Optical Density).

In figure 7 and figure 8, the distance across the gel phantom has been plotted in the x axis and optical density has been plotted in the y axis.

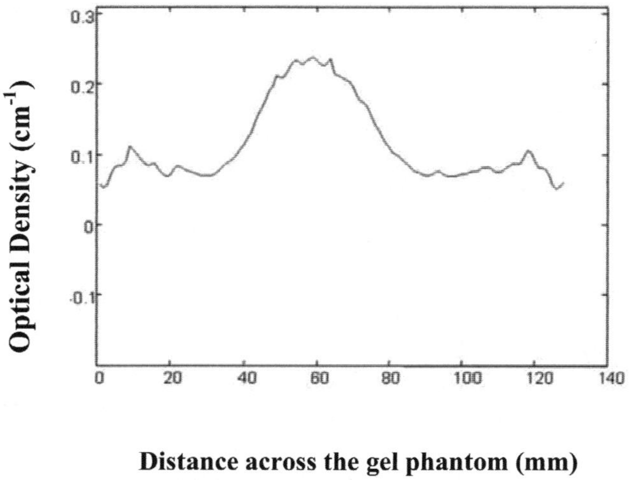

Figure 8. Dose Profile for Post irradiated gel slice at 4 cm depth (Distance across the gel phantom vs Optical Density).

4. Conclusions

A Laser CT scanner for reading a tissue equivalent gel was constructed indi genously and its initial performance for reading an irradiated gel dosimeter was investigated. The inte nsity measurement and the motions of the scanner were performed with th e sample. The maxi mum resolution obtained using this scanner was 0.5 mm. The Reconstructed Im age for the irradi ated slic e and its data analysis was performed with in-house code developed in Mat Lab. From the results, we conclude that the scanner is suitable for reading normoxic gels for clinical dosimetry.

Acknowledgements

The authors are grateful to the AER B, Government of In dia for their fi nancial supp ort and t he management of VIT University for their continuous encouragement. The authors also acknowledge the guidance pro vided by Pro f. B. Paul Ravindran, Ch ristian Medical College & Hospitals, Vellore and Mr. P. Sellakumar, Bangalore Institute of Oncology in performing this work.

References

[1] Oldham M, Siewerdsen J H, Shett y A, Jaffra y D A 2001 High resolution gel-dosim etry by optical-CT and MR scanning *Med. Phys.* **28** 1436-1445

[2] Oldham M, Siewerdsen J H, Kumar S, Wong J, Jaffray D A 2003 Optical-CT gel-dosimetry I: Basic investigations *Med. Phys.* **30** 623-634

[3] Oldham M, Kim L 2004 Optical-CT gel-dosim etry II: Optical artifacts and geom etrical distortion *Med. Phys.* **31** 1093-1104

[4] Xu Y, Wuu C S, Mary anski M J 2004 Perform ance of a co mmercial optical CT scanner and polymer gel dosimeters for 3-D dose verification *Med. Phys.* **31** 3024-3033

[5] Brindha S, Venning A J, Hill B, Baldock C 2004 Experimental study of attenuation properties of normoxic polymer gel dosimeters *Phys. Med. Biol.* **49** N353-N361

[6] Venning A J, Hill B, Brindha S, Healy B J, Baldock C 2005 Investigation of PAGAT Polymer gel dosimeter using magnetic resonance imaging *Phys. Med.Biol.* **50** 3875-3888

[7] Hill B, Venning A, Baldock C 2005 The dose re sponse of normoxic pol ymer gel dosim eters measured using X-ray CT *Brit. J. Radiol.* **78** 623-630

[8] Brindha S, Paul B R, Baldock C 2006 Optim ization of imaging protocol of an X-Ray CT scanner for evaluation of normoxic polymer gel dosimeters *J. Med. Phys.* **31** 72-77

Journal of Physics: Conference Series **250** (2010) 012041 doi:10.1088/1742-6596/250/1/012041

Cone Beam Optical CT Investigation on Tissue Equivalent Normoxic Polymer Gel Dosimeter

D. Senthil Kumar and E. James Jebaseelan Samuel

Photonics Division, School of Advanced Sciences, VIT University, Vellore, India.

E-mail: kalaimeenasen@gmail.com

Abstract. A potential method has emerged in the form of water-equivalent "3D gel dosimetry" using optical computed-tomography (optical-CT) which enables accurate, high resolution, 3D measurement of dose distributions associated with modern radiation treatments.. Optical Cone Beam CT (CBCT) scanner plays a major role for Gel dosimeter readout and clinical radiation therapy as 3-Dimensional Radiation Dosimetry. The normoxic PAGAT (Polyacrylamide Gelatin and Tetrakis) gel is used as a dosimeter for this cone beam CT analysis due to its tissue equivalent behaviour. Applying a uniform background subtraction of open field intensity resulted in cone beam CT reconstructed attenuation coefficient for a PAGAT Gel Dosimeter.

1. Introduction

Modern radiotherapy techniques such as stereotactic radiotherapy (SRT) stereotactic radiosurgery (SRS), three dimensional conformal radiotherapy (3DCRT) and intensity modulated radiation therapy (IMRT) aim to deliver a high dose to the tumour while sparing the surrounding normal healthy tissues. As a result of these complicated treatment techniques there is a need for a 3- dimensional (3D) dose verification system. However, currently available dosimeters such as ion chambers, diodes, thermo luminescent dosimeters and films are limited to point (or) planar measurement. Multiple measurements are required for obtaining the 3-dimensional dose distribution using the above dosimeters. Gel dosimetry attempts to meet the requirements of 3D radiation dose distribution. Gel dosimetry is tissue equivalent [1] and it acts as a gel phantom as well as gel dosimeter so there is no need for dose perturbation correction. Optical Computed Tomography provides an inexpensive and easy to use alternative imaging modality [1-5]. However, optical computed tomography approaches introduce novel challenges because of the inherent complications from the index of refraction matching at interfaces, multiple scattering in polymer systems [3, 4, and 6]. Improvements in radiotherapy treatments based on an increased complexity of the treatment approach require accurate determination of the absorbed dose. Computer calculations have frequently been used to estimate three-dimensional dose distributions in complex geometries and therefore it becomes important to validate these by accurate three-dimensional measurements. As the dose is delivered from multiple directions, the phantom must represent a three dimensional body [5]. Due to its flexibility, tissue equivalence, ability to act as both the detector and phantom and provide spatial distribution of absorbed dose, gel dosimetry offers the potential to provide the desired dosimetric environment for complex radiotherapy delivery techniques. To date, MR scanning, which is expensive and can be susceptible to subtle sources of error [3] has been utilized as the main readout system for gel dosimetry. Gore *et al* [2] proposed optical CT scanning of PAG gels, an approach analogous to first-generation x-ray CT with the x-ray source replaced by a visible laser and the x-ray detector replaced with a light-sensitive photodiode. Kelly [4] demonstrated this technique for ferrous benzoic xylenol (FBX) gel dosimeter by applying Cone Beam CT imaging principle used in x-ray imaging to optical CT using CCD array detector. In this paper we present some initial investigations of the use of a Vista Cone Beam Optical CT scanner (Modus Medical Devices Inc., London, Canada) in relation with normoxic PAGAT Gel Dosimeter. This approach of recording transmission images of the gel is the inverse of x-ray cone beam CT if you consider only the rays, which contribute to image formation [7].

2. Materials and Method

2.1. Gel Preparation

The conventional polymer gel dosimeter used in this study was the normoxic ''PAGAT'' gel described elsewhere (Venning et al., Hill et al and Brindha et al.,) [8-10] and was prepared under normal atmospheric conditions. The PAGAT polymer gel consisted of 3.5% (w/w) N, N-methylene-bis-acrylamide (BIS), 3.5% (w/w) acrylamide, 5% (w/w) gelatin, 89% triple distilled water and 10mM tetrakis hydroxymethyl phosphonium chloride (THP) as antioxidant. The gel preparation was carried out as follows: 5% gelatin was added to the magnetic stirred glass flask containing 89% of triple distilled water and allowed to swell for 10–15 min. Then the gelatin solution was heated to the temperature of 50 °C for complete dissolution of the gelatin. After obtaining the clear gelatin solution 3.5% BIS and 3.5% Acrylamide were added and stirred continuously until the mixture was dissolved homogeneously. The antioxidant 10 mM THP was added to the solution under stirring just before transferring the gel into the Polyethylene Teraphthalate cylindrical jars of size 10 cm diameter and 13 cm long. All the polyethylene teraphthalate cylindrical jars containing the PAGAT gel were sealed properly to avoid the further diffusion of oxygen and were kept in a refrigerator for one full night for gelling.

2.2. Gel Irradiation

To evaluate the dose response of polymer gel to gamma radiations, the polymer gel dosimeters were irradiated after the complete gellation. The center of the phantom was aligned with the crosshairs of the treatment field. The polymer gel dosimeters were irradiated with a direct square field size of 5 x 5 cm^2 for gamma rays using Co-60 source with a dose of 12 Gy. The source-to-surface distance (SSD) was 100 cm. One gel dosimeter was kept unirradiated for background subtraction. The irradiated gel dosimeters were left in a refrigerator for one day before imaging to allow the complete post irradiation polymerization.

2.3. Cone Beam Optical CT scanner

CT images were obtained with the Vista Optical Cone Beam CT scanner. The scanner uses 630 or 590 nm LED array to provide illumination and the projection data are acquired with a 1024 x 768 pixels, 10-bit monochrome CCD camera. The acquisition and reconstruction time using the Feldkamp back projection algorithm for the above mentioned gel dosimeter takes around 10 minutes. For all runs, reference scans were taken immediately before irradiation and dosimetry scans were performed 24 hours after irradiation to avoid temporal instabilities as the radiation-induced polymerization proceeded. The data shown in this work were taken at 590 nm (yellow light) and reconstructed images had a voxel resolution of 0.5 mm. Data analysis was performed with in-house code developed in Mat Lab (Math works, Natwick, USA).

The post irradiated gel phantom or dosimeter can be kept in a place for imaging which consists of a uniform LED light source of 590 nm that could provide parallel rays, A Perspex made aquarium of size 15 cm^3 of wall thickness of 5 mm on the light transmission regions with a stepper motor holder positioned upwards to rotate the gel phantom to 360 degrees, and a CCD camera an imaging device to image the gel phantom are the three main parts required for the construction of an optical cone beam CT scanner. The gel phantom was rotated using a stepper connected at the top of the phantom holder.

The motion of the stepper motor was controlled with the help of a Personal Computer. The Vista Cone Beam Optical CT scanner used for this work is shown in figure 1.

Figure 1. Vista Optical Cone Beam CT Scanner.

3. Results and Discussion

Due to irradiation, polymerization occurs on the gel dosimeters. Hence the monomers presented in the gel are converted in to polymers. The pre-irradiation images were subtracted from the post-irradiation scans and subtracted images were analysed. The pre and post irradiation image data set of 340 frames in 360^0 of 1024 x 768 pixels were loaded for reconstruction. The reconstruction has been done using

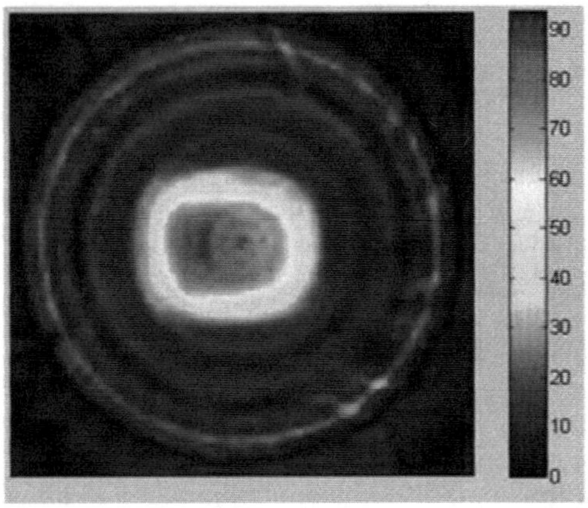

Figure 2. 2D Subtracted Image with colour bar of CT Reconstructed Slices of Post irradiated gel at depth of 3cm.

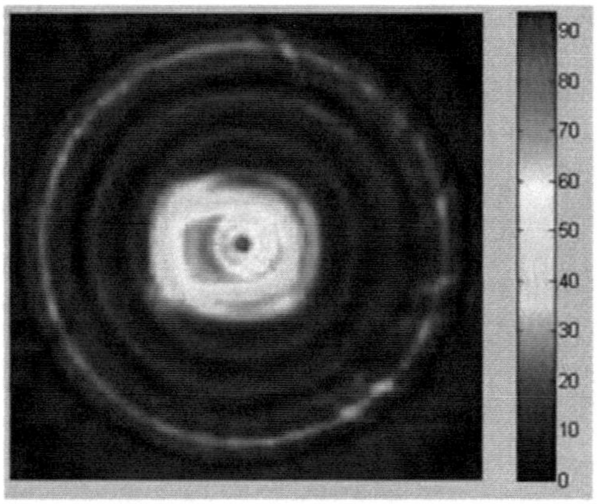

Figure 3. 2D Subtracted Image with colour bar of CT Reconstructed Slices of Post irradiated gel at depth of 6cm.

the Feldkamp filtered back projection algorithm implemented in Vista Scan software and slice by slice subtracted image of the post irradiated slices has been obtained in 3 cm and 6 cm depths shown in figure 2 and figure 3. Data analysis like getting Dose profile in the reconstructed slices and subtraction of post irradiated images from pre irradiated images was performed with in-house code developed in Mat Lab (Math works, Natwick, USA). Figure 4 represents the dose profile at 3cm depth of the post irradiated gel phantom whereas figure 5 represents the dose profile at 6cm depth. In the figures

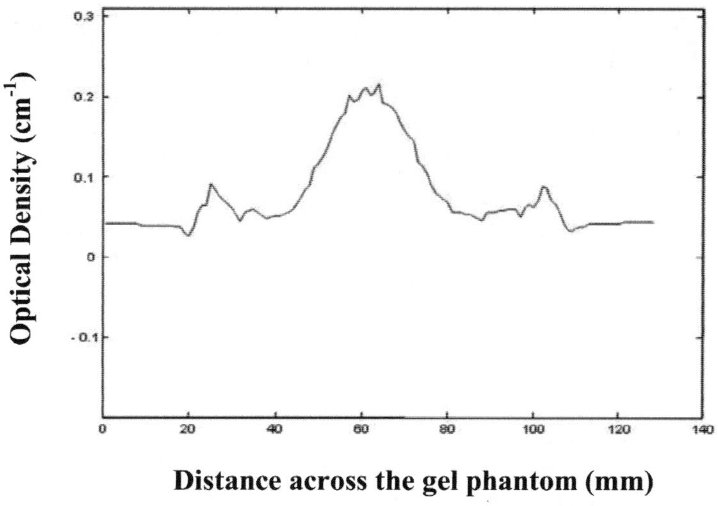

Figure 4. Dose Profile for Post irradiated gel slice at 3cm depth.

mentioned above, the distance across the gel phantom has been plotted in the x axis and the optical density (OD) has been plotted in the y axis.

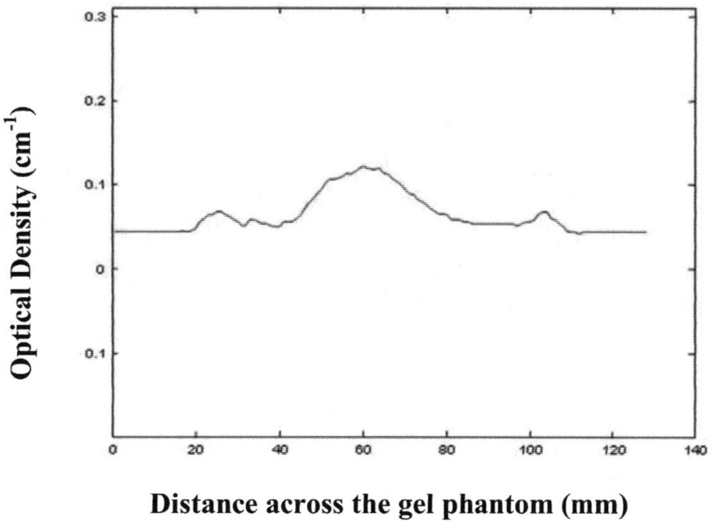

Figure 5. Dose Profile for Post irradiated gel slice at 6cm depth.

4. Conclusions

The experience with Vista Optical Cone Beam CT scanner of dual LED sources and a CCD camera has been performed with the tissue equivalent normoxic PAGAT Gel Phantom. The profile plots for the reconstructed slices after subtracting with the pre irradiated slice were acquired using in-house code developed in Matlab. The CT reconstructed images for the corresponding slices suggested that it is one of the cheapest and easiest methods for mapping gel dose distribution. Corresponding Iso dose distribution for the post irradiated gel phantom can be obtained by further enhancing the Matlab software. Investigating the feasibility of using Optical CT Gel Dosimetry for the verification of IMRT, SRS, SRT and comparing this Vista Optical CT results with pre-existing 3D commercially available planning software will be the future studies.

Acknowledgements
The authors are grateful to the DRDO, Govt. of India for their financial support and the management of VIT University for their perpetual support and encouragement. The authors acknowledge the guidance and encouragement provided by Mr. John Miller, Jennifer Dietrich from Modus Medical Devices, Inc. Canada, Dr. Kevin Jordan, Medical Physicist, University of Western Ontario, Canada, Dr. B. Paul Ravindran, Professor of Radiation Physics, Department of Radiation Oncology, Christian Medical College & Hospitals, Vellore, Mr. P. Sellakumar, Medical Physicist, Bangalore Institute of Oncology, Bangalore and Dr. R. Srinivasan from VIT University, Vellore.

References

[1] Maryanski M J, Zastavker Y Z, Gore J C 1996 Radiation dose distributions in three dimensions from tomographic optical density scanning of polymer gels: II Optical properties of the BANG polymer gel *Phys Med Biol.* **41** 2705–2717
[2] Gore J C, Ranade M, Maryanski M J, Schulz R J 1996 Radiation dose distributions in three dimensions from tomographic optical density scanning of polymer gels: I Development of an optical scanner *Phys Med Biol.* **41** 2695–2704
[3] Oldham M, Siewerdsen J H, Kumar S, Wong J, Jaffray D A 2003 Optical-CT gel-dosimetry I: Basic investigations *Med Phys.* **30** 623–634
[4] Kelly R G, Jordan K J, Battista J J 1998 Optical CT reconstruction of 3 D dose distributions using the ferrous-benzoic-xylenol (FBX) gel dosimeter *Med Phys.* **25** 1741–1750
[5] Kron T 2001 Radiation therapy requirements: what do we expect from gel-dosimetry? *Proc 2nd Int. Conf. Radiotherapy Gel Dosimetry.* 2–9
[6] Doran S J, Koerkamp K K, Bero M A, Jenneson P, Morton E J, Gilboy W B 2001 A CCD based optical CT scanner for high-resolution 3D imaging of radiation dose distributions: equipment specifications, optical simulations and preliminary results *Phys Med Biol.* **46** 3191-3213
[7] Jordan K J, Pajak T M, Pionteck C, Battista J J 2001 Optical cone beam tomography with low pressure sodium light *Proc 2nd Int Conf Radiotherapy Gel Dosimetry.* 172–174
[8] Venning A, Hill B, Brindha S, Healy B J, Baldock C 2005 Investigation of PAGAT Polymer gel dosimeter using magnetic resonance imaging *Phys Med Biol.* **50** 3875-3888
[9] Hill B, Venning A, Baldock C 2005 The dose response of normoxic polymer gel dosimeters measured using X-ray CT *Brit J Radiol.* **78** 623-630
[10] Brindha S, Paul B R, Baldock C 2006 Optimization of imaging protocol of an X-Ray CT scanner for evaluation of normoxic polymer gel dosimeters *J Med. Phys.* **31** 72-77

Journal of Physics: Conference Series **250** (2010) 012042 doi:10.1088/1742-6596/250/1/012042

Preliminary Dose Calibration Results Using a TENOMAG Polymer Gel Dosimeter and Optical CT (VISTA^{TM})

Suk Lee[1], Jang Bo Shim[1], Kyoung Hwan Chang[1], Yuanjie, Cao[1], Jaeyoun Yi[2], Jinwoo Park[2], Sam Ju Cho[3], Sang Hoon Lee[4], HyunDo Huh[5], Kwang Hwan Cho[6], Chul Kee Min[7], Dae Sik Yang[1], Young Je Park[1], Won Seob Yoon[1], Chul Yong Kim[1]

[1] Department of Radiation Oncology, College of Medicine, Korea University, Seoul, Korea
[2] INFINITT Healthcare, Seoul, Korea
[3] Department of Radiation Oncology, Eulji University, Seongnam, Korea
[4] Department of Radiation Oncology, Cheil General Hospital & Women's Healthcare Center, Kwandong University College of Medicine, Seoul, Korea,
[5] Department of Radiation Oncology, Inha University Hospital, Incheon, Korea
[6] Department of Radiation Oncology, Soonchunhyang University Boocheon Hospital, Boocheon, Korea
[7] Department of Radiation Oncology, Konyang University Hospital, Daejeon, Korea

sukmp@korea.ac.kr

Abstract. We present a preliminary result of dose calibration for 3D dose evaluation using gel dosimetry in radiotherapy. In this work, we measured the percentage depth dose (PDD) and dose linearity using a normoxic polymer gel dosimeter (TENOMAG) and a commercial cone-beam optical CT scanner (VISTA^{TM}, Modus Medical Devices, Inc., London, ON, Canada). We confirmed that the PDD measured for 1 and 2 Gy had a build up region at 0.4 cm depth. However, the PDD measured for 3 and 4 Gy had not a build up region. In all cases, the PDD has not greatly decreased with depth above 4 cm depth. Moreover we confirmed that the optical density proportionally increased with dose at 1, 1.5 and 2 cm depth, while the response has not greatly increased with dose for other depths. In this study, we confirmed that we will be able to perform the dose calibration using optical density obtained from a TENOMAG gel data scanned with an optical CT scanner.

1. Introduction

A study on dose calibration is still lacking in 3D gel dosimetry [1-3]. We present a preliminary result of dose calibration for 3D dose evaluation using gel dosimetry in radiotherapy. In this work, we measured the percentage depth dose (PDD) and dose linearity using a normoxic polymer gel dosimeter (TENOMAG) and a commercial cone-beam optical CT scanner (VISTA^{TM}, Modus Medical Devices, Inc., London, ON, Canada) [4, 5].

2. Materials and method

We have been calibrated the geometry and rotation of the optical CT scanner using a general calibration method developed by Modus Medical Devices. As shown in figure 1, we set up the laser alignment the water phantom and gel dosimeter in order to obtain the CT scan images of gel dosimeter. After CT scanning, we divided gel dosimeter into four equal parts to deliver the 1 ~ 4 Gy in dosimeter. We used a 10 MV photon beam (iX, Varian Medical System, Palo Alto, CA) for dose calibration. After irradiation, the dosimeter was scanned in the VISTA™ scanner. The scanned data were reconstructed using VistaRecon software (Feldkamp reconstruction algorithm with Hamming filter) and we measured a dose linearity and PDD using dose distribution of optical density.

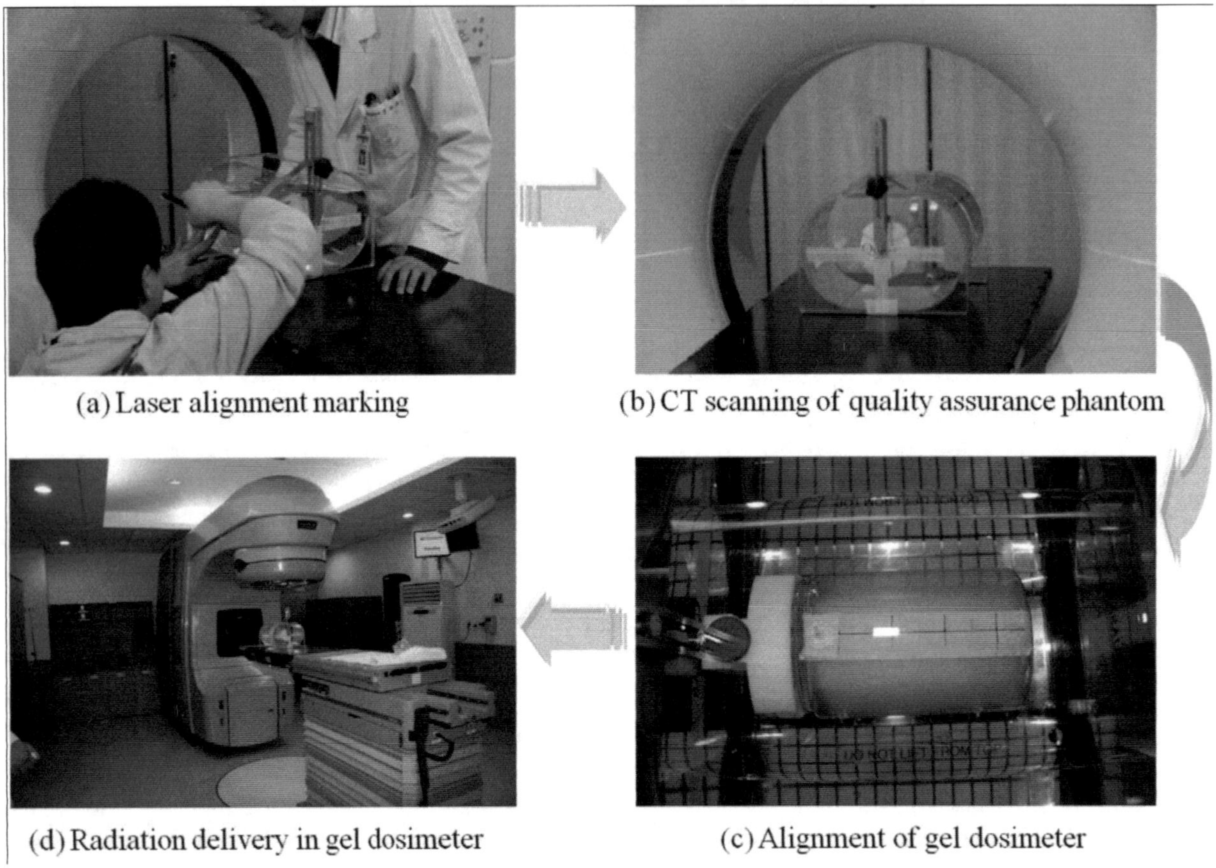

Figure 1. It shows that schematic diagram of experimental process of dose calibration: (a) Laser alignment marking; (b) CT scanning of quality assurance phantom; (c) Alignment of gel dosimeter; (d) Radiation delivery in gel dosimeter.

3. Results and discussion

3.1. Percentage depth dose (PDD)

Figure 2 (b) shows the relationship between the percentage depth dose and optical density for different doses. We confirmed that the PDD measured for 1 and 2 Gy had a build up region at 0.4 cm depth. However, the PDD measured for 3 and 4 Gy had not a build up region. In all cases, the PDD has not greatly decreased with depth above 4 cm depth.

3.2. Dose linearity

Figure 2 (c) shows the relationship between the dose and optical density for various depths. We confirmed that the optical density proportionally increased with dose at 1, 1.5 and 2 cm depth, while the response has not greatly increased with dose for other depths.

Journal of Physics: Conference Series **250** (2010) 012042 doi:10.1088/1742-6596/250/1/012042

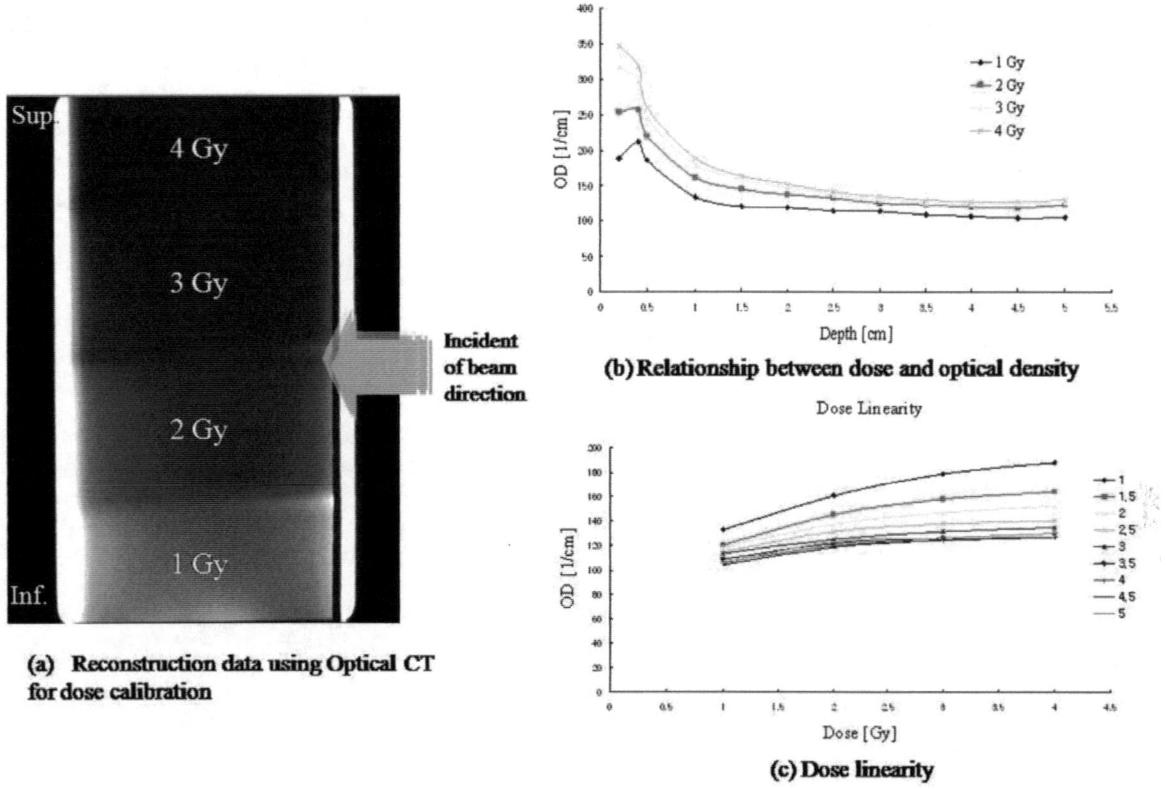

Figure 2. It shows that the percentage depth dose (PDD) and dose linearity using a normoxic polymer gel dosimeter and a commercial cone-beam optical CT scanner: (a) After irradiation, reconstructed data; (b) Relationship between and dose and optical density; (c) Results of dose linearity.

4. Conclusion

In this study, we confirmed that we will be able to perform the dose calibration using optical density obtained from a TENOMAG gel data scanned with an optical CT scanner.

5. Acknowledgements:

The work was supported by a Korea University Grant. The authors thank Dr. John Miller from the Modus Medical Devices, Inc., London, ON, Canada.

References

[1] Oldham M, Guo P, Gluckman G and Adamovics J. 2006 IMRT verification using a radiochromic/optical-CT dosimetry system *J. Physics.: Conf. Ser.:* **56** 221-224

[2] Sakhalkar H S, Adamovics J, Ibbot G and Oldham M 2009 A comprehensive evaluation of the PRESAGE/optical-CT 3D dosimetry system *Med. Phys.* **36** 71-82

[3] S Babic, J Battista and Jordan K 2008 Three-dimensional dose verification for intensity-modulated radiation therapy in the radiological physics centre head-and-neck phantom using optical computed tomography scans of ferrous xylenol-orange gel dosimeters *Int J. Radiat.*

Oncol. Biol. Phys. **70** 1281-1291
[4] SJ Cho, DO Shin and Suk Lee et al. 2007 Development of a novel normoxic polymer gel dosimeter (TENOMAG) Journal of Korean Physical Society. **52** 1798-1805
[5] SJ Cho, Su Jung Shim and Suk Lee et al. 2010 Dose Response Characteristics in Polymer Gel for the Composition Ratio Journal of Korean Physical Society. **56** 147-152

Journal of Physics: Conference Series **250** (2010) 012043 doi:10.1088/1742-6596/250/1/012043

Review of recent advances in radiochromic materials for 3D dosimetry

Kevin Jordan

London Regional Cancer Program, London Health Sciences Centre
Department of Medical Biophysics, University of Western Ontario, London, ON, Canada

Kevin.jordan@lhsc.on.ca

Abstract. Recent papers concerning radiochromic films, plastics and hydrogels for 3D dosimetry are summarized. The utility of Presage™, a radiochromic plastic, with optical CT readout was demonstrated for the following applications: motion and gated treatment delivery, commissioning of small fields for radiosurgery, ^{192}Ir high dose rate brachytherapy source commissioning and as a 3D insert for IMRT credentialing tests with Radiological Physics Centre (RPC) phantoms. Preliminary performance for characterizing microbeams from a synchrotron with optic projection tomography readout demonstrated resolution of an 83 micron diameter beam. Hydrogel chemistries based on nonionic micelles for leuco malachite green and leuco crystal violet demonstrated that low diffusion gels can be designed by choosing product dyes that are poorly soluble and water and tend to remain in the micelles. Turnbull blue chemistry has been successfully adapted to form a non-difffusing gel as well. The performance of ferrous xylenol orange hydrogel layers doped with boron to form neutron dosimeters demonstrated another practical application. Polymerization hydrogels are alternate materials that can be read with optical CT scanners. High dose gradient applications in brachytherapy with 90Sr/90Y sources and proton dosimetry are presented for comparison.

1. Introduction

This review provides a summary of works published since the 5th International Conference on Radiotherapy Gel Dosimetry, DOSGEL 2008. The material draws primarily on search results within the PUBMED and Web of Science databases using the keywords; 3D, three-dimensional, dosimetry and radiochromic. Polymerization gels have been treated as a separate topic since many gel dosimetry researchers have been focused on these materials and chemistry and readout reviews were presented earlier in this conference. However, several polymer gel papers that involve optical CT readout have been included here as well for convenient comparison. Radiochromic materials are generally read by optical transmission techniques. In particular optical computed tomography (CT) scanners are the primary technology for 3D scanning. There are several scanner geometries and they are broadly classed as single ray – single detector or broad beam (fan beam, cone beam and collimated beam) – array detectors. The single ray - single detector geometries are least sensitive to scatter in contrast to high acceptance angle cone beam systems being the most sensitive. For this reason, the quality of the reconstructed 3D dose distribution is intimately connected to both the optical performance of the dosimeter and readout scanner. Advances in scanner design, data processing, algorithms for reconstruction and materials development provide steady improvements in quantitative tissue/water equivalent 3D dosimetry. Fast scanners make 3D dosimetry more practical to perform in a clinical setting and allow gels such as ferrous-xylenol orange to accurately record high dose gradients generated with megavoltage radiation beams. Materials research to develop higher performance materials concentrates on: low diffusion chemistries, rapid dose response, adequate sensitivity for the typical two Gray radiotherapy treatments and low scatter in order to achieve the highest spatial resolution and dynamic range.

There are currently three materials for radiochromic 3D dosimetry; film (EBT2, International Specialty Products), gels and plastics (Presage™, Heuris Pharma LLC). By default EBT2 film is becoming the film standard for radiotherapy QA measurements and a dedicated session is included in this conference. In this review, there is overlap with many of the other presentations because most of the current polymer gel formulations can be read in 3D with optical CT, magnetic resonance imaging (MRI) and x-ray CT. The issue of scatter contamination in transmission profiles is generally addresses by choosing scanner geometries with low acceptance angles such as single ray- single detector or fan-beam with linear array detectors. However, substantial progress in scatter rejection for cone beam CT provides stimulus for continued scanner development of all geometries. As well, the optical scanners can easily be adapted for film densitometry providing added value.

In order to compare different dosimetry systems for a specific application there are a few basic parameters required in order to determine which materials and readout scanners are adequate. Irradiation criteria will include: dose range, dose rates, energy of radiation beam, spatial resolution and volume of interest. The dose sensitivity, which is expressed as the change in optical linear attenuation coefficient per Gray (units cm^{-1} Gy^{-1}) can quickly eliminate many materials if the maximum dose is around 2 Gy. A very useful metric is the dose resolution of a system [1]. It provides a measure of the minimum dose that can be measured with a specified uncertainty for a given spatial resolution. For example, a comparison at a fixed spatial resolution such as 1x1x1 mm voxel size quickly ranks the relative performance of complete systems. However, it does not identify the limiting factors which may be related to the dosimeter or the readout mechanism. Temperature needs to be controlled in order to conduct reproducible chemical reactions and the thermal history of the material may also be a factor. Most materials that have sufficient dose sensitivity are likely also sensitive to light and heat. Drifts in absorption coefficients are generally due to auto-oxidation rates. Ferrous xylenol orange gels continue to darken whereas materials based on leucodyes may darken or bleach depending on the specific chemistry. The rate at which the radiochromic reaction proceeds can have a negative effect. For example, if diffusion of radiation induced chromophores is an issue then the reaction needs to be fast so that 3D scanning can be completed prior to degradation of the dose distribution gradients. Fast reactions are very useful from a practical perspective as well. Some systems now allow for 3D scanning and reconstruction to be completed in less than 30 minutes post irradiation. Dose response data is important for development of radiation sensitive materials. Common issues include: linearity, low dose response, dose rate and fractionation effects. Often there are measureable differences with sample volume that make usage of cuvettes inadequate for calibration of larger volumes. Both the scatter and absorption coefficients are important for estimation of material performance. For radiochromic reactions, transparent and initially colourless materials are preferred in order to maximize dynamic range and spatial resolution. Material composition is required for calculation of electron density and mass density for scaling to water or tissue equivalent dimensions.

A recent review by Doran has provided a concise history of optical CT and dosimeters for 3D dosimetry [2].

2. Dosimeter developments and applications

The following papers contain a mix of dosimeter material results as well as instrumental innovations that were required to attain these signal to noise ratios. The applications papers also contain information related to improvements in materials, instruments and data analysis.

2.1 Radiochromic film

Silver based film is being rapidly phased out due to increasing costs of the dwindling commercial supply. The major film manufacturers have evolved in response to the near universal adaptation of solid state array detectors for image detection. Film dosimetry in practice is becoming limited to the set of radiochromic films available from International Specialty Products. The particular product for megavoltage radiotherapy dosimetry is referred to as EBT2. There are many recent publications concerning evaluation of this product. A detailed uncertainty analysis for radiochromic film scanned with a flatbed document scanner, demonstrated an uncertainty of 0.4% is achievable [3]. As the quality of 3D dose distributions improves, analogous uncertainty calculations will be required when comparing to equivalent planes with film measurements. EBT film is no longer commercially available and the early batches of is successor EBT2 have demonstrated some variability in energy response below 100 keV [4]. EBT film has been the standard 2D dosimeter for comparison with selected planes of 3D dosimeter results. This implies that researchers using this material need to be proficient in its use in order to attain quality data for comparisons.
A report of stacking 8 EBT film strips to record near surface doses, at depths ranging from 0.012 to 0.18 cm, reported agreement with parallel plate ion chamber and thin TLD measurements [5].

2.2 Radiochromic plastic: Presage™

A very promising material for practical 3D dosimetry is the radiochromic plastic, Presage™. Leucodyes and halogenated hydrocarbons are dissolved in polyurethane to form a robust dosimeter. The material does not exhibit diffusion and can be formulated to cover any dose range of relevance to radiotherapy. It is anticipated that versions of this material will become one of the common 3D dosimeters for routine clinical usage. Most results are reported with leucomalachite green (LMG) formulations and readout near 633 nm. This wavelength is very convenient since EBT films also have their main absorption peak at 639 nm. This allows optical CT scanners optimized for Presage™ to also read EBT film. Since early Presage™ formulations contained halocarbons initiator content in the range of 5 to 20% by mass the low energy response is expected to be larger than for water. Also the carbon to oxygen ratio is much greater than in hydrogels leading to an effective atomic number higher than for water. These effects were quantified in a Monte Carlo calculation comparing PAGAT polymer gel and a specific formulation of Presage™ [6]. The calculations reported the differences in depth doses for 6 and 15 MV photon beams relative to water for both dosimeters. Results demonstrated that PAGAT was nearly water equivalent and the Presage relative depth dose in increased up to 7% at 30 cm depth. Recent formulations are reported to be more water equivalent. Other researchers have prepared Presage™ based on published papers and patent and have obtained similar results. The sensitivity and long term stability as a function of LMG and initiator content was recently reported by Mostaar et al [7].

Sakhalkar et al have reported on the performance of Presage™ for the specific application of 3D dosimetry within the IMRT head and neck phantom from the Radiological Physics Centre (RPC) [8-10]. The dosimetry insert contains TLD's and orthogonal EBT films. Substituting a Presage cylinder for the insert allowed for 3D dosimetry within the pocket. Multislice scanning with a commercial laser CT scanner (OCTOPUS 5x) revealed better agreement with EBT film results than with the treatment planning calculation. A set of four samples were used to demonstrate interdosimeter reproducibility to within 2% for dose difference and 2 mm distance to agreement relative to the average response. This particular formulation demonstrated a postirradiation increase in absorption that scaled uniformly throughout the dose distribution. The successful implementation of Presage™ into the RPC credentialing process will provide high profile application of 3D dosimetry.

The use of Presage™ to verify gated radiotherapy delivery has been reported by Brady et al [11]. This dosimetry problem has been recognized as difficult for point dosimeters and 3D dosimeters have provided convincing data for the utility of 3D dosimeters. A CCD camera with telecentric lens was used to capture the transmission images. Improvements to this broad beam optical CT scanner are also reported to increase signal to noise levels 3-fold over previous reports.

Full 3D dosimeters have been used for small field dosimetry. They have the advantage of being both the phantom and detector, which simplifies the irradiation process. Because Presage™ has no reported diffusion it is anticipated to perform well for commissioning small fields for radiosurgery. Clift et al presented depth doses, profiles and output factors for Presage scanned with CCD camera with telecentric lens and laser CT (OCTOPUS 5x) which agreed with EBT film and previous commissioning data measurements [12]. The results demonstrate the efficiency of 3D dosimetry for commissioning small fields.

Proton dosimetry is yet another dosimetry application that involves high dose gradients. Because of the high LET of the protons near the end of their range, recombination of radicals has made quantitative chemical dosimetry inadequate for this particular problem. An under response has generally been observed with hydrogels (exception BANG3-Pro gel, see below) and previous Presage™ formulations. This issue has been re-examined and an under-response of 20% was observed for the Bragg peak [13]. Microbeam irradiation with synchrotrons has generated exciting radiobiology results that are key for understanding the mechanisms of grid therapy. These beams are typically less than 100 microns in diameter and require much higher spatial resolution than is required for IMRT type distributions. Only dosimeter materials that do not exhibit diffusion are relevant for measuring these high gradients. Early results for Presage™ with a high resolution optical projection tomography system revealed dose lines less than 80 micron widths could be resolved [14]. Brachytherapy is another common dosimetry problem in which high dose gradients are present. The utility of Presage™ to accurately record the distribution from a high dose rate ^{192}Ir source were presented by Wai et al [15].

2.3 Micelle hydrogels

The addition of surfactants above a threshold concentration allow micelles to form. In hydrogels, the interior of the micelles provides an nonpolar chemical environment in which nonpolar reactants can dissolve. This innovation allows the opportunity to design radiation sensitive reactions based on nonpolar chemistry within hydrogels [16]. These reactions could be polymerization or radiochromic reactions. The initial results have focused on radiochromic systems. Non-ionic surfactants such as Triton X100 with LMG and trichloroacetic acid form transparent gelatin hydrogels. Following a 5 Gy preirradiation, these gels demonstrate a linear response. The linearity is quantified by obtaining a direct correlation between a 12 MeV electron beam depth dose and the corresponding attenuation coefficient as determined by optical CT. This initial formulation had a dose sensitivity of 0.004 cm^{-1} Gy^{-1} which is approximately 3% of typical FX formulations. It also had a diffusion coefficient approximately 3 times lower than FX gels. While this gel demonstrated feasibility of micelles for developing new radiation sensitive gels it has limited value for clinical dosimetry based on its low sensitivity. A subsequent study with leuco crystal violet (LCV) provided a much improved material. The sensitivity was doubled and the diffusion was 10 times less than the LMG version [17]. The LCV gel demonstrated a linear response with no requirement for preirradiation. These gels are easy to prepare and provide a convenient alternative to FX gels provided higher doses can be delivered. One application where they can be a valid alternative to FX gels is for commissioning of small fields [18].

2.4 Radiochromic Turnbull Blue gel

An alternative radiochromic chemistry is based on the production of Turnbull blue $(Fe(II)Fe(III)CN_6)^+$. A new hydrogel based on phytagel and ferric cyanide chemistry had been reported [19]. This material has an absorption peak around 690 nm, very low diffusion rate and dose sensitivity around 0.005 cm^{-1} Gy^{-1}. This sensitivity is similar to the LMG micelle gel but with the advantage of no diffusion. Samples demonstrate sensitivity to ambient lighting and a measureable auto-oxidation rate which is non uncommon for radiation sensitive materials.

2.5 Ferrous xylenol orange gels

Ferrous xylenol orange gelatin gel continues to be a useful material in spite of the diffusion rate. In cases where irradiation times are short, gradients modest (typical IMRT type gradients, <10% per mm) and scan times are short these gels provide adequate results. Care is required concerning thermal history. Depending on wavelength of readout a preirradiation dose may be required in order to establish a subsequent linear dose response. The recent innovations have concentrated on the addition of boron to the gels in order to perform neutron dosimetry [20-23]. Calculations reporting the water equivalence of FX and several formulations of polymerization gels have also been reported [24,25].

2.6 Polymer hydrogels

A considerable fraction of 3D hydrogel dosimetry research has concentrated on polymerization gels in which monomers undergo free radical induced polymerization. The literature has been summarized in a review article authored by many of the pioneers in this area [26]. The commercial versions are referred to as BANG™ gels. Several formulations are available, optimized for specific dosimetry problems. The utility of BANG gel with readout by high resolution optical CT has been reported for high gradient applications [27,28]. Verification of gated treatment delivery has also been reported [29]. In the case of proton dosimetry, a specific formulation has successfully recorded the Bragg peak, demonstrating equivalence to ion chamber depth dose curves [30].

3 Optical CT scanner design

Incremental improvements to scanner designs for laser CT and broad beam systems with CCD cameras have been included in previous references related to application studies. Another geometry exploring the performance of free space scanning has been reported. Elimination of the aquarium and refractive index matching liquid would simplify scanning if sufficient projection data could be obtained. The work compares polymer gel dose distributions with the prototype optical scanner and MRI reporting similar results for specific geometries [31]. Further work may be required to determine if this approach is useful for scanning complete sample volumes.

4 Summary

Radiochromic materials development continues to be an active area of research. The development of several strategies to produce low or zero diffusion materials allows many high dose gradient applications to be effectively measured with radiochromic systems providing an alternative to polymerization gels. Optical CT continues to improve as experimenters address issues with specific geometries and explore alternate geometries. Several scanner geometries now report full 3D data scans in less than 10 minutes which minimizes the impact of diffusion for IMRT type gradients.

References

[1] Baldock C, Lepage M, Back S A J, Murry P J , Jayasekera P M, Porter D and Kron T 2001 Dose resolution in radiotherapy polymer gel dosimetry: effect of echo spacing in MRI pulse sequence *Phys. Med. Biol.* **46** 449-60

[2] Doran S, 2009 The history and principles of chemical dosimetry for 3-D radiation fields: Gels, polymers and plastics, *Appl. Radiat. Isot.* **67** 393-8

[3] Bouchard H, Lacroix F, Beaudoin G, Carrier J-F and Kawrakow I, 2009 On the characterization and uncertainty analysis of radiochromic film dosimetry, *Med. Phys.* **36** 1931-46

[4] Sutherland JGH and Rogers DWO, 2010 Monte Carlo calculated absorbed-dose dependence of EBT and EBT2 film, *Med. Phys.* **37** 1110-16

[5] Chiu-Tsao ST, Chan MF, 2009 Photon beam dosimetry in the superficial buildup region using radiochromic EBT film stack, *Med. Phys.* **36** 2074-83

[6] Brown S, Venning A, De Deene Y, Vial P, Oliver L, Adamovics J, Baldock C, 2008 Radiological properties of the PRESAGE and PAGAT polymer dosimeters, *Appl. Radiat. Isot.* **66** 1970-4

[7] Mostaar A, Hashemi B, Zahmatkesh MH, Aghamiri SM, Mahdavi SR, 2010A basic dosimetric study of PRESAGE: the effect of different amounts of fabricating components on the sensitivity and stability of the dosimeter, *Phys. Med. Biol.* **55** 903-12

[8] Sakhalkar HS, Adamovics J, Ibbott G, Oldham M, 2009 A comprehensive evaluation of the PRESAGE/optical-CT 3D dosimetry system, *Med. Phys.* **36** 71-82

[9] Sakhalkar H, Sterling D, Adamovics J, Ibbott G, Oldham M, 2009 Investigation of the feasibility of relative 3D dosimetry in the Radiological Physics Center Head and Neck IMRT phantom using Presage / optical CT, *Med. Phys.* **36** 3371-77

[10] Sakhalkar HS, Adamovics J, Ibbott G, Oldham M, 2009 An Investigation into the Robustness of Optical-CT Dosimetry of a Radiochromic Dosimeter Compatible with the RPC Head-and-Neck Phantom, *J. Phys.* **164** 12059

[11] Brady SL, Brown WE, Clift CG, Yoo S, Oldham M, 2010 Investigation into the feasibility of using PRESAGE/optical-CT dosimetry for the verification of gating treatments, *Phys. Med. Biol.* **55** 2187-201

[12] Clift C, Thomas A, Adamovics J, Chang Z, Das I, Oldham M, 2010 Toward acquiring comprehensive radiosurgery field commissioning data using the PRESAGE/optical-CT 3D dosimetry system, *Phys. Med. Biol.* **55** 1279-93

[13] Al-Nowais S, Doran S, Kacperek A, Krstajic N, Adamovics J, Bradley D, 2009 A preliminary analysis of LET effects in the dosimetry of proton beams using PRESAGE and optical CT, *Appl. Radiat. Iso*t. **67** 415-8

[14] Doran SJ, Brochard T, Adamovics J, Krstajic N, Bräuer-Krisch E 2010 An investigation of the potential of optical computed tomography for imaging of synchrotron-generated x-rays at high spatial resolution, *Phys. Med. Biol.* **55** 1531-47

[15] Wai P, Adamovics J, Krstajic N, Ismail A, Nisbet A, Doran S, 2009 Dosimetry of the microSelectron-HDR Ir-192 source using PRESAGE and optical CT, *Appl. Radiat. Isot.* **67** 419-22

[16] Jordan K, Avvakumov N, 2009 Radiochromic leuco dye micelle hydrogels: I. Initial investigation, *Phys. Med. Biol.* **54** 6773-89

[17] Babic S, Battista J, Jordan K, 2009 Radiochromic leuco dye micelle hydrogels: II. Low diffusion rate leuco crystal violet gel, *Phys. Med. Biol.* **54** 6791-808

[18] Babic S, McNiven A, Battista J, Jordan K, 2009 Three-dimensional dosimetry of small megavoltage radiation fields using radiochromic gels and optical CT scanning, *Phys. Med. Biol.* **54** 2463-81

[19] Solc J, Spevácek V, 2009 New radiochromic gel for 3D dosimetry based on Turnbull blue: basic properties, *Phys. Med. Biol.* **54** 5095-107

[20] Mangueira TF, Silva CF, Coelho PR, Campos LL, 2010 Gamma/neutron dose evaluation using Fricke gel and alanine gel dosimeters to be applied in boron neutron capture therapy, *Appl. Radiat. Isot.* **68** 791-4

[21] Carrara M, Fallai C, Gambarini G, Negri A, 2010 Fricke gel-layer dosimetry in high dose-rate brachytherapy, *Appl. Radiat. Isot.* **68** 722-5

[22] Bartesaghi G, Burian J, Gambarini G, Marek M, Negri A, Viererbl L, 2009 Evaluation of all dose components in the LVR-15 reactor epithermal neutron beam using Fricke gel dosimeter layers, *Appl. Radiat. Isot.* **67** S199-201

[23] Gambarini G, Vanossi E, Bartesaghi G, Carrara M, Mariani M, Negri A, Burian J, Viererbl L, Klupak V, Rejchrt J, 2009 Dose imaging in a thorax phantom with lung-equivalent volume at the epithermal neutron beam of LVR-15 reactor, *Appl. Radiat. Isot.* **67** S214-7

[24] Taylor ML, Franich RD, Trapp JV, Johnston PN, 2009 Electron interaction with gel dosimeters: effective atomic numbers for collisional, radiative and total interaction processes, *Radiat. Res.* **171** 123-6

[25] Taylor ML, Franich RD, Trapp JV, Johnston PN, 2008 The effective atomic number of dosimetric gels, *Australas Phys Eng Sci Med* **31** 131-8

[26] Baldock C, De Deene Y, Doran S, Ibbott G, Jirasek A, Lepage M, McAuley KB, Oldham M, Schreiner LJ, 2010 Polymer gel dosimetry, *Phys Med Biol* **55** R1-63

[27] Massillon-Jl G, Minniti R, Mitch MG, Maryanski MJ, Soares CG, 2009 The use of gel dosimetry to measure the 3D dose distribution of a 90Sr/90Y intravascular brachytherapy seed, *Phys Med Biol* **54** 1661-72

[28] Massillon-JL G, Minniti R, Soares CG, Maryanski MJ, Robertson S, 2010 Characteristics of a new polymer gel for high-dose gradient dosimetry using a micro optical CT scanner, *Appl Radiat Isot* **68** 144-54

[29] Lopatiuk-Tirpak O, Langen KM, Meeks SL, Kupelian PA, Zeidan OA, Maryanski MJ, 2008 Performance evaluation of an improved optical computed tomography polymer gel dosimeter system for 3D dose verification of static and dynamic phantom deliveries, *Med Phys* **35** 3847-59

[30] Zeidan OA, Sriprisan SI, Lopatiuk-Tirpak O, Kupelian PA, Meeks SL, His WC, Li Z, Palta JR and Maryanski MJ, 2010 Dosimetric evaluation of a novel polymer gel dosimeter for proton therapy, *Med Phys* **37** 2145-52

[31] Papadakis A, Zacharakis G, Maris T, Ripoll J, Damilakis J, 2010 A new optical-CT apparatus for three-dimensional radiotherapy dosimetry: Is free space scanning feasible? *IEEE Trans Med Imaging* Mar 18. [Epub ahead of print]

How to perform dosimetry with Optical CT

Cheng-Shie Wuu and Y Xu

Department of Radiation Oncology, Columbia University, New York, NY, USA.

csw6@columbia.edu

Abstract. Both polymer gels and PRESAGE radiochromic solid dosimeter, in conjunction with optical CT scanning system, have been employed to measure 3-D dose distribution. The 3-D dose maps obtained from these systems can provide a useful tool for dose verification on complex treatments such as IMRT, radiosurgery, and RapidArc. These complex treatments present high dose gradient regions in the boundaries between the target and the surrounding critical organs. Dose accuracy in these areas can be critical, and may affect the treatment. There is a pressing need for a dosimeter that allows for accurate determination of 3-D dose distribution with high spatial resolution. In this paper, performance of polymer gels and PRESAGE dosimeter with optical CT scanning is reviewed and evaluated in terms of their sensitivity calibration, irradiation, optimization of scanning procedures, precision, and accuracy. Clinical applications of optical-CT dosimetry are presented.

1. Introduction

It has long been recognized that 3-D dosimetric information is important in clinical situations where high dose gradients exist such as IMRT and stereotactic radiosurgery. Both polymer gels and PRESAGE radiochromic plastic dosimeter, in conjunction with optical CT scanner, have been employed to record the radiation dose distribution in three-dimensions [1-3]. When irradiated, Polymer gels become opaque through polymerization and PRESAGE presents colour change. Their optical densities increase in response to absorbed dose. The 3-D dose information can be reconstructed based on the readout of optical computerized tomography (CT) of polymer gels and PRESAGE. In 1996, Gore et al [1] demonstrated the potential of this optical-CT imaging technique for polymer gel dosimeters, although the concept of this technique can be traced further back. The history and principles of optical CT for scanning 3-D radiation dosimeters were reviewed in details previously by Doran [4-5]. The fundamental science for polymer gels has recently been thoroughly reviewed by Baldock *et al.* [6]. This article will discuss the practical aspects of optical scanners with polymer gels and PRESAGE in implementing a 3-D dose distribution of clinical applications.

2. Optical-CT systems

The collected data for the optical-CT technique are optical projections acquired either by scanning a laser across the dosimeter sample and detected by a photodiode detector, or by passing a broad light field through the sample and collected by CCD or CMOS area detectors. The principles of optical computed tomography are similar to that for the x-ray CT. Detailed information and the mathematical procedure of filtered back-projection used in the 3-D reconstruction of dose image can be found in the reference [5]. Besides introduction of various optical-CT systems, the main objective of this article is to review and discuss the operation and procedures as to how to perform clinical 3-D dosimetry with optical CT systems.

2.1. First-generation optical-CT scanner

The concept of original optical-CT scanner, designed by Gore *et al*, is illustrated in Figure 1(a). We describe here the basic running principle and the configuration of a commercial product, OCTOPUS 5x (MGS Research, Inc.), of this invention. Briefly, a single He-Ne laser beam (633-nm wavelength) is guided by a stepper motor and a series of mirrors and lenses to scan across a rectangular water tank containing the gel/PRESAGE sample repeatedly with a fixed field of view. The light transmitted through the tank is collected by a photodiode during each cycle of translational motions as single projection data. After each full translational scan, the sample mounted on a turntable is rotated by a small angular increment and followed by the linear scan. In earlier version of OCTOPUS scanner, the turntable rotates 360 degrees to collect projection data for each planar image. OCTOPUS x5 has improved imaging time to approximate 8-9 min per slice by reducing the rotation to 180 degrees, since each scan is duplicated in 360-degree scan. In addition, the manufacturer discontinued the use of the reference diode, as shown in Figure 1(a), for higher sampling rate. The laser drift and output variation are taken into account in the driver software. The tank is otherwise filled with a mixture of water, glycerol, and a blue dye.

First-generation optical-CT systems have been demonstrated to be reproducible and accurate, and can be employed, in conjunction with polymer gels and PRESAGE radiochromic plastic dosimeter, to measure 3-D dose distribution with good quality [7-11]. They are also most efficient at minimizing the artefacts caused by scattered light. However, the first-generation systems have not been routinely used in the clinical application, partially because of its slow scanning speed.

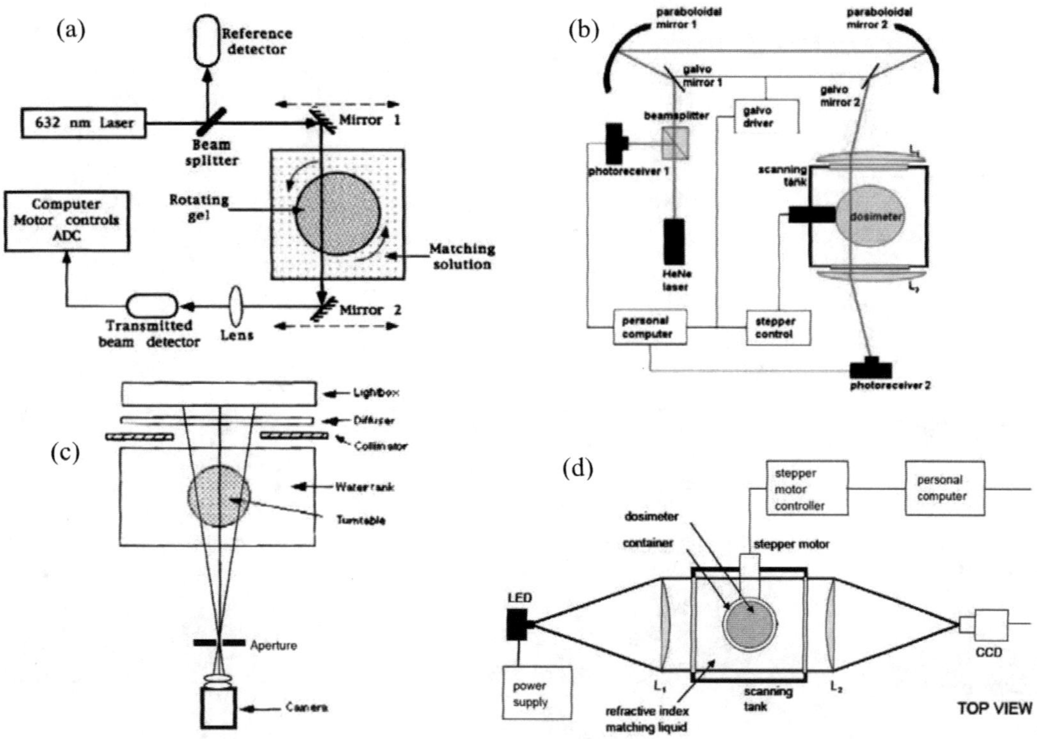

Fig. 1. Schematic diagram of the different types of optical CT scanners (diagram reproduced from reference [6]) : (a) first-generation laser system (Gore et al, reference [1]); (b) fast laser scanner (Krstajic *et al,* reference [14]); (c) cone-beam CCD scanner (Wolodzko *et al,* reference [15]); (d) parallel-beam CCD scanner (Krstajic *et al*, reference [17])

2.2. Fast optical-CT scanners

In the design of fast optical-CT scanner, several methods for increasing the speed of the scanning laser beam have been proposed. A number of groups have introduced the approach of a rotating mirror, in which the laser was reflected by the rotating mirror instead of the mechanical translation [12-13]. Another approach is the use of a pair of galvanometer-controlled mirrors proposed by Krstajic and Doran [14], as shown in Figure 1(b). A faster alternative is the use of a broad beam with CCD or CMOS area detectors. A cone-beam configuration was proposed by Wolodzko et al [15] (Figure 1(c)). A parallel-beam configuration was proposed by Doran et al [16], and telecentric optics was added to minimize sensitivity to scattered light in a refined design [17]. Similar to x-ray CT, the presence of scattered light is a well known source of artifact in the 3-D dose reconstruction using optical CT scanning of polymer gels. Both Xu et al [18] and Islam et al [19] reported a cross-shaped artifact that appeared in the optical-CT dose images, especially in a high-dose irradiation where the signal-to-noise ratio is relatively small. It should be noted that artifact becomes noticeable when scattering is a significant portion of the signal. Successful 3-D dose verifications of complex IMRT have been reported using first-generation optical-CT systems [7-8, 10]. While scatter correction is still the main obstacle for broad-beam with CCD systems to become a successful 3-D dosimeter, the first-generation optical-CT scanner can be used as a "gold standard" [18].

3. 3-D radiation dosimetry by optical CT scanning of polymer gels/ PRESAGE

3-D dose distributions of various radiotherapy techniques have been obtained by optical-CT scanning of polymer gels on the basis of light-scattering [7-8]; as well as optical scanning of PRESAGE dosimeter based on light-absorbance [9-10]. In this paper, the complete process of 3-D dose determination using the optical CT approach, with either polymer gel or PRESAGE, will be reviewed.

3.1. Polymer gels/PRESAGE sensitivity calibration

The characterization of both polymer gels and PRESAGE radiochromic solid dosimeter have been well studied, including diffusion effect, temperature effect, dose rate dependence and energy dependence [3, 19-21]. It was concluded that, for both polymer gel and PRESAGE dosimeters, neither photon energy nor dose rate has an effect on the dose response for all the megavoltage photon and electron beams tested. One important step in the determination of dose distribution using optical-CT based polymer gels/PRESAGE dosimeter is the sensitivity calibration. To achieve this goal, three calibration methods have recently been used to study the optical dose response of BANG®3 polymer gels and compare their performances [20]. As shown in Fig.1, the methods include a) measurements across a series of small glass-vials dosimeters irradiated to graded doses; b) optical-CT scanning of a single Barex cylindrical gel phantom irradiated to graded doses with small photon fields; c) percent depth dose comparison between optical-CT scans and beam data for a single 16 MeV electron field.

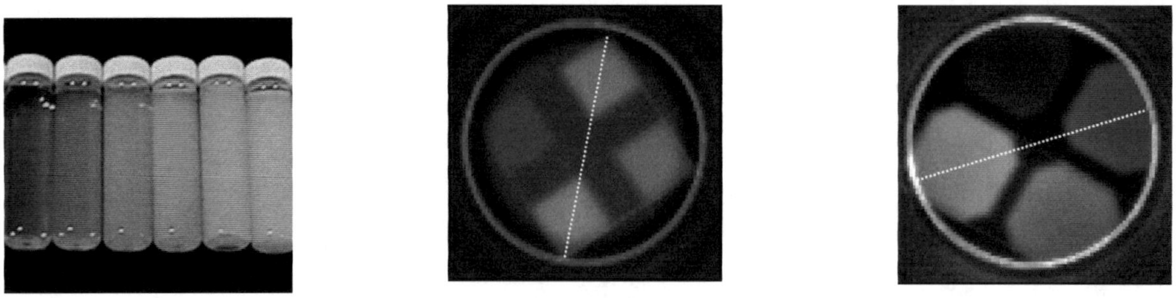

(a) (b) (c)

Fig. 2. Optical dose response calibration using three different methods described above: (a) glass vials of BANG polymer gels irradiated to graded doses; and reconstructed images from irradiations of (b) four 4 cm x 4 cm photon fields; c) four 6 cm x 6 cm electron fields. Reproduced from reference [20].

In that study, the optical response of the BANG®3 gels was found to be linear and energy-independent within the uncertainty of the experimental method used. The dose response curves obtained from the six Barex gel phantoms agree within 2.2%, indicating a good reproducibility of the gel dose response within phantoms of the same geometry. The dose sensitivity of the BANG®3 gels obtained from the small glass vial experiment was substantially different from that of the large Barex phantoms, owing possibly to the difference in temperature inside the two types of phantoms during gel formation and irradiation [21], and possible oxygen contamination of the glass vial walls. In addition, it was observed that the readouts were sensitive to the positioning of the laser path relative to the glass vials. The maximum difference among the 6 readings taken for one glass vial was found to be 15%. The ratio between the standard deviation and the mean value for the group of 6 readings taken from a glass vial could be as much as 6%. It was suggested that cautions need to be taken with the dose sensitivity calibration method using small glass vials irradiated to graded doses in optical-CT of BANG®3 gels. The large variation in laser path through a glass vial may be reduced if an optical cuvette with flat surface is used instead [3]. Based on the energy independence and the dose response reproducibility established for the BANG®3-Barex gel phantoms, the PDD matching method using electron depth dose curve from ion chamber measurement as a benchmark is demonstrated to be a convenient and relatively accurate way for dose response calibration in optical-CT scanning of BANG®3 gel dosimeters. A 16 MeV electron field irradiation with a field size of 6 cm or larger is appropriate for this purpose. The calibration method using multiple small-field irradiations is in general more labor intensive and less accurate than the PDD matching approach in that it can't provide optical density values for all desired doses.

PRESAGE™ is a solid 3-D dosimetry material consisting of an optically clear polyurethane matrix, containing a leuco dye that exhibits a radiochromic response when exposed to ionizing radiation. It has the advantages of insensitivity to oxygen, radiation induced absorption contrast rather than scattering contrast, and a solid plastic texture that can be molded to a variety of shapes and sizes without an external container. Guo et al [3] have investigated the basic characteristics of PRESAGE dosimeter, including optical response linearity, dose rate dependence, reproducibility, stability, special changes in absorption, and temperature effect. In that study, sensitivity calibration of the PRESAGE was performed with PRESAGE cuvettes, as shown in Fig. 3(A).

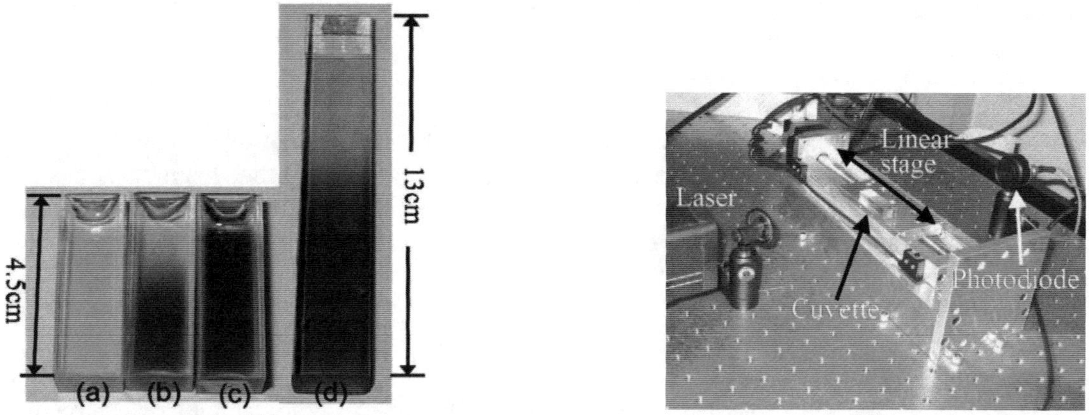

(A) (B)

Fig. 3. (A) Pictures of various cuvettes, (a) nonirradiated cuvette. (b)irradiated cuvette by 6 MeV electrons. (c) irradiated cuvette vy 6 MV photons. (d) A PRESAGE™ column without an external container, irradiated by 16 MeV electrons. (B) Picture of the linear optical scanner. A He-Ne laser is the light source. Reproduced from reference [3].

The linear optical scanner used to monitor optical density variations along the long axis of the PRESAGE samples is shown in Fig. 3(B). The optical response sensitivity was determined from the plot of normalized depth OD changes along the long PRESAGE columns for a 16 MeV electron and 6 MV photon irradiations plotted against the calculated PDD. Good agreement between the normalized depth OD change and the PDD was observed. The OD change versus the absorbed dose along the long PRESAGE column, irradiated by the 16 MeV electrons, was used to determine the sensitivity from a linear fit. The linear fit showed a high linearity between OD change and absorbed dose, with RMSE < 1%. However, "over response" at the low dose region, for both 6 MeV and 16 MeV electrons, was observed (at tails of the PDD curves). This over response phenomenon was also observed at the dose region below 2 Gy in the linear fit of dose range spanning from 1 Gy to 12 Gy. While the reason for this is not fully clear, it is always prudent to perform the sensitivity calibration covering only the full dose range that is used in the specific clinical application (e.g. 0 to 3 Gy for a fractionated external beam treatment; 2 to 20 Gy for a stereotactic radiosurgery).

At the present time, there is not enough data to show if polymer gels and/or PRESAGE dosimeter can be consistently used for absolute dosimetry. More studies are needed in terms of the uniformity, stability, and consistency of polymer gels/PRESAGE; as well as the optimization of sensitivity calibration, dosimeter irradiation, and optical scanning procedures.

3.2. Irradiation of dosimeters

In the optical CT scanning of gel/PRESAGE dosimeters, the net optical density varies in a much wider range than it does in the 2-D scan of the same plan, because of the longer and varying path lengths of the laser beam through the dosimeter. The usable dynamic range of a scanner is limited by several factors, including the signal-to-noise ratio and the electronic and mechanic noises in the scanner. Therefore, gel/PRESAGE sensitivity and the dose delivered should be optimised such that the optical density increments along all possible paths during scanning of an irradiated dosimeter never exceed a certain maximum. As such, readout uncertainty can be minimized and reconstruction artefacts can be avoided. On the other hand, the sensitivity of the dosimeter should be high enough to maximize the dynamic rage of the scanning system, while a high signal-to-noise ratio and thus high dosimetric accuracy can be maintained. Xu *et al* developed a method for determining the minimum transmitted signal along all scan paths through a gel of known dose response irradiated with a planned dose distribution [23]. This calculation allows either the dose to be set for the optimal transmission range of the dosimeter or to prepare a dosimeter with an optimal sensitivity for the delivered dose. In that study, optical CT experiments were performed to determine the range of transmitted signal that would allow a reconstruction accuracy of less than 4%. It was concluded that, given the optical CT scanning system (OCTOPUS, MGS Research, Inc), optical densities from 0.5 to 2.5 met the 4% criteria. The optimal dose delivery for each dosimeter may vary depending on the sensitivity and the dosimeter size. It is suggested that, for clinical dose verification such as IMRT, the dosimeter sensitivity be adjusted to its optimum such that verification can be performed at clinical dose range.

Temperature effects on both polymer gels and PRESAGE have been reported [3, 20]. The polymer gels/PRESAGE can have a different optical response to irradiation at a different temperature. The temperature sensitivity of the PRESAGE dosimeter around room temperature is about 1.7% OD variation per degree. Uniform temperature throughout the dosimeter volumes during calibration and phantom irradiation is essential in order to achieve an accurate dose measurement. Special cautions should be taken when the phantom size is significantly different from the samples (e.g. cuvettes) used for sensitivity calibration. In general, it is adequate to keep the dosimeters under the irradiation environment for 24 hrs before the experiment.

3.3. Scanning process

The majority of optical-CT dosimetry of polymer gels/PRESAGE, especially those related to clinical dose verification, was obtained from the first-generation laser scanning systems. The most significant sources of artifacts in optical CT scanning of polymer gels/PRESAGE result from reflection and refraction of light at the surface of the dosimeter. One key step to minimize the artifact in dose reconstruction is to match the refractive index of the tank liquid (water, glycerol, and a blue dye) with the refractive index of the polymer gels. The addition of a dye to the matching fluid is to match the light attenuation between the gel and the liquid. This allows for maximum use of the dynamic range of the light detection system. Depending on the differences in refractive index between the matching fluid, the gels, and gel container wall, an edge artifact was observed in the outer 1-2 cm of the gel container [7]. Optical CT scanning with refractive index matching liquids currently involves open systems for placement of the gel samples. Jordan has a thorough discussion of matching liquids that are used in current scanners [25]. Experimenters should consider: toxicity, safety (flammable), vapor pressure, viscosity, chemical stability, color stability, disposal and corrosive nature of solutions when choosing suitable materials for matching liquid. The matching procedure can be lengthy and time-consuming. Vandecasteele and De Deene have developed an automated procedure [26], based on a dry and wet off-axis refraction measurement of the gel dosimeter used. Based on the determined refractive index of the gel and the fluid, the required glycerol concentration can be determined from all the experimental data. An automated pump and mixing system was developed for concentration adjustment. For PRESAGE/optical CT scanner, the matching fluid currently used is a mixture of octyl salicylate and methoxy octyl cinnamate [10]. With a better refractive index matching of the fluid, the use of a more attenuating fluid, and an improved PRESAGE formulation, the edge artifact can be reduced to 3mm in the outer region of the dosimeter.

The only commercially available first-generation optical CT laser scanner is the OCTOPUS™, manufactured by MGS Research Inc. In the OCTOPUS™ series of optical CT scanners, the data acquisition process is controlled by a windows executable program written with TESTPOINT (CEC Inc., Billenca, MA). The scanning field of view, the pixel size and the number of pixels per projection are defined by the users. The data acquisition rate and the velocities of the 3 stepper motors are optimized by the control program. After the data acquisition process is completed, the obtained laser signals from the data acquisition board are imported into an image reconstruction program written in MATLAB. The image reconstruction program reconstructs the optical density distributions slice by slice using the filtered back projection method and a set of filters. The total time for obtaining a full set of 3D images of a dosimeter, ranging from 4 to 8 hours, depends on the dimension of the dosimeter, pixel size, and the version of the scanner.

The reconstructed 3D optical density map from the image reconstruction program represents the three dimensional relative dose distribution within the dosimeter scanned. Sensitivity calibration procedures can be performed to obtain the absolute dose distribution delivered to the dosimeter. The viewing and the preliminary analysis of the reconstructed 3D relative or absolute dose distributions can be done using most of the standardized imaging software, ImageJ (http://rsb.info.nih.gov/ij), for example. Detailed analysis of the reconstructed images from the gel measurement, including the comparison of the 3D dose distribution with those from the planning system or other measurement methods, can be done using in-house developed computer software. In addition, evaluation tools for 3D dose comparison can be incorporated in the computer program.

3.4. Artifacts from light scattering

The dose response of optical-CT based polymer gel dosimeter arises from light scattering at the radiation-induced polymer micro-particles in the gel. Similar to the scattered radiation in x-ray CT, the scattered light can be a potential source of artifacts in the optical CT dose reconstruction, as reported by Xu *et al* and Islam *et al* [7, 27]. As shown in Fig. 4, a cross-shaped artifact that appeared in optical-CT reconstructed dose images of a high-dose square-field irradiation. The distortions of the images in Figs. 4(b) and 4(c) can be attributed primarily to the significant contribution from scattered photons at high ODI. In polymer gels, optical attenuation is caused by the Rayleigh/Mie scattering of light on polymer particles. Therefore, a collection of some of the scattered photons at the detector's aperture is inevitable but normally does not cause any detectable error in the reconstructed image. With increasing optical density increments, the number of scattered photons produced increases whereas the intensity of the transmitted primary beam decreases (low signal-to-noise ratio). The fraction of the scattered light in the signals collected will increase and eventually become a significant source of error. The cross artifact was confirmed to result from light scattering in optical-CT Monte Carlo simulations by Oldham *et al* [28]. The impact caused by light scattering can be minimized to less than 4% uncertainty, if the optical density is between 0.5 and 2.5 as described in section 3.2 above.

PRESAGE dosimeter exhibits optical contrast through light absorbance rather than light scattering, and has an advantage for the optical CT scanning technique since the scattering artifact displayed above is not an issue.

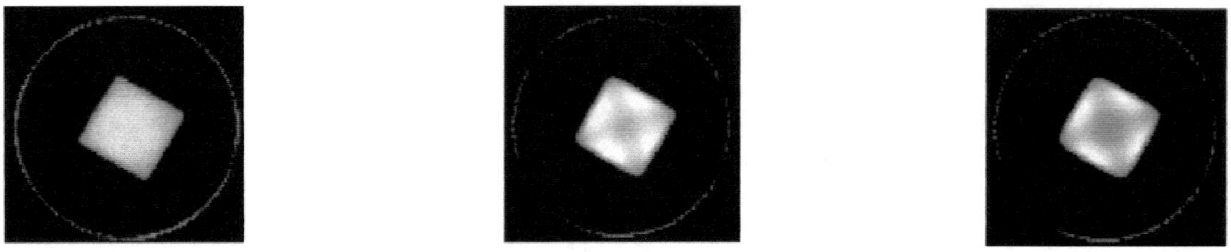

(a) (b) (c)

Fig. 4 Reconstructed dose images for 3 axial slices at depth 6, 4, and 3 cm, from a 6 cm x 6 cm single 6 MV field with 100 cm SSD. The maximum optical density increments for these planes (left to right) are 2.48, 2.8 and 3.1, respectively. Reproduced from reference [7].

4. Dose verification in clinical applications

Both polymer gels and PRESAGE dosimeter, in conjunction with the optical CT system OCTOPUS, have been employed to verify complex 3D dose distribution, including IMRT [8,10] and stereotactic radiosurgery [29]. Fig. 5 shows comparisons of IMRT dose distributions in three planes, obtained from gel (blue), EDR2 film (green), and the Helios IMRT treatment planning system (red). The results show good agreement among all three techniques at selected planes. PRESAGE/optical-CT scanner was also applied to verify a complex IMRT dose distribution [10]. A highly modulated IMRT plan was delivered to a cylindrical PRESAGE dosimeter, and the dose distribution was readout using a commercial optical CT scanner OCTOPUS. As shown in Fig. 6, comparisons were made with GAFCHROMIC EBT film measurements, and the calculated dose distribution from ECLIPSE treatment planning system. Excellent agreements were observed.

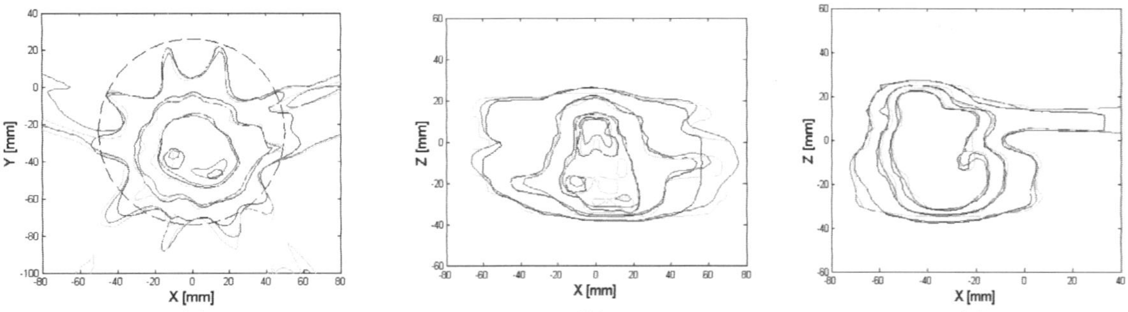

(a) (b) (c)

Fig. 5. Comparison of dose distributions at (a) axial central plane (40%, 60%, 100%, and 115%), (b) coronal central plane (40%, 60%, 100%, and 115%), and (c) sagittal central plane (30%, 70%, and 100%). Dose distributions were obtained from treatment planning calculation (red), polymer gel measurement (blue), and EDR2 measurements (green).

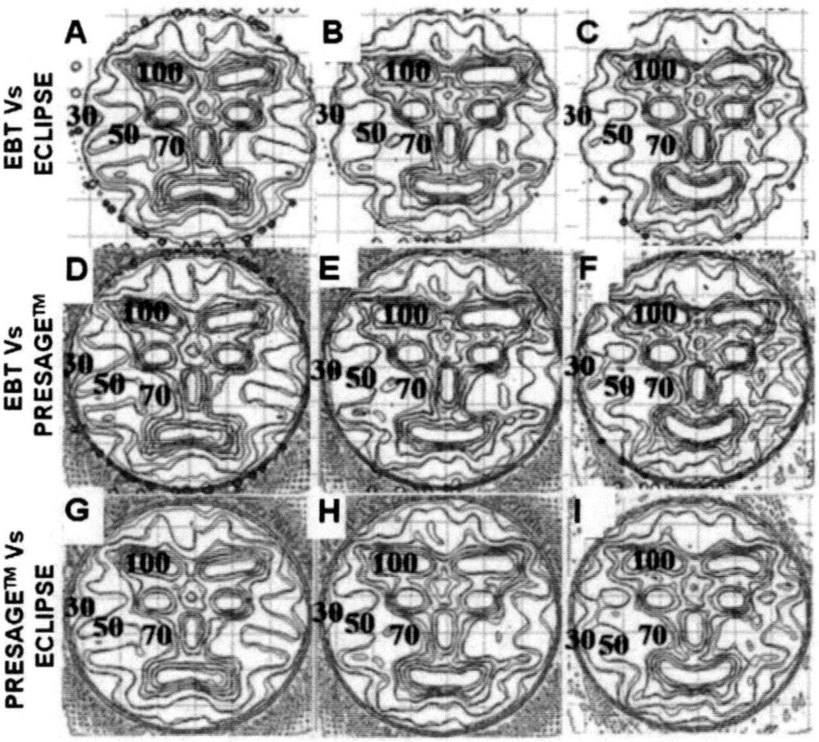

Fig. 6 Isodose overlay plots of the 100%, 90%, 80%, 70%, 50%, and 30% isodose lines for the three selected slices between PRESAGE™, ECLIPSE®, and EBT film dose distributions. The first row [(a)-(c)] compares isodose lines between EBT (blue) and ECLIPSE® (red). The second row [(d)-(f)] compares isodose lines between EBT (blue) and PRESAGE™ doses (red). The last row [(g)-(i)] compares isodose lines between ECLIPSE® (blue) and PRESAGE™ (red) doses. Reproduced from reference [10].

Radiosurgery, either cranial or extracranial, is a procedure that precisely delivers large radiation doses to tumors and other relevant anatomical targets in a single fraction or in a small number of fractions (typically up to five). Therefore, the success of stereotactic radiosugery requires precise targeting and rapid dose falloff in surrounding normal tissues. We have employed both polymer gel and PRESAGE dosimeters to measure the 3-D dose distribution of an IMRT radiosurgery plan for a 2.5 cc brain tumor. Figure 7 shows dose distribution comparison at the central axial plane between measured (polymer gels and PRESAGE dosimeter) and calculated dose distributions. Both measured and calculated dose distributions are in reasonable agreement. However, the isodose lines from both gel and PRESAGE measurements show more variation than those from the planning calculation. These discrepancies may be partly attributed to the fact that the calculation grid for the planning system is 2.5 mm yet the resolution of measurements is 1 mm, and partly attributed to the impurities and air bubbles in the phantoms. It appears that the accuracy and precision of the 3-D dosimetry technique, with optical CT scanning of polymer gel/PRESAGE, requires further improvement for detailed 3-D dose verification.

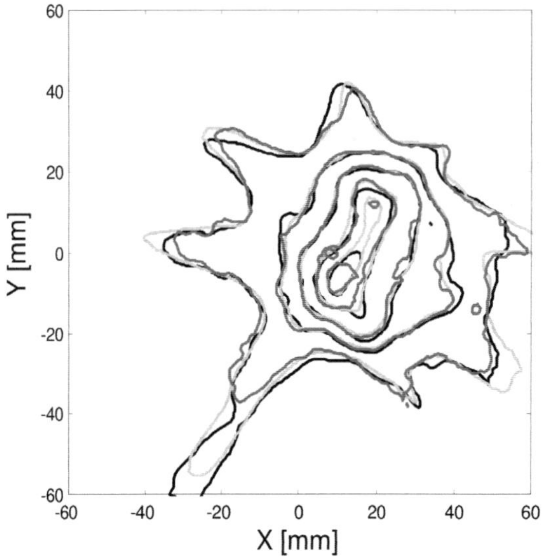

Figure 7. Comparison of the isodose lines (110, 100, 90, 80, 50, and 30 percent) from the Eclipse planning system (green lines) and measured dose distributions (Gel and PRESAGE) for the central axial slice. Gel: red lines; PRESAGE: black lines.

5. Conclusion

With the increasing use of complex treatments such as IMRT, stereotactic radiosurgery, and RapidArc, there is an ergent need for a dosimeter that allows for accurate 3-D dose distribution measurement. Both polymer gels and PRESAGE solid dosimeter, with an optical CT scanning system, have been used to verify complex 3D dose distributions from IMRT and radiosurgery plans. The results demonstrate that an accurate 3D dose distribution with high resolution can be implemented. Future development should be focused on improvement of the scanning speed, as well as the consistency and uniformity of the dosimeters.

References

[1] Gore, J.C., M. Ranade, and M.J. Maryanski 1996 Radiation dose distributions in three dimensions from tomographic optical density scanning of polymer gels: I. Development of an optical scanner. Phys. Med. Biol. **41**(12) 2695-2704.

[2] Maryanski, M.J., Zastavker, Y.Z., Gore, J.C. 1996 Radiation dose distributions in three dimensions from tomographic optical density scanning of polymer gels: II. Optical properties of the BANG polymer gel. Phys. Med. Biol. **41** 2705-2717.

[3] Guo, P.Y., J.A. Adamovics, and M. Oldham, 2006 Characterization of a new radiochromic threedimensional dosimeter. Med. Phys. **33**(5) 1338-1345.

[4] Doran, S. and N. Krstajic, 2006 The history and principles of optical computed tomography for scanning 3-D radiation dosimeters. J Phys. Conf. Ser. **56** 45-57.

[5] Doran, S. 2009. The history and principles of optical computed tomography for scanning 3-D radiation dosimeters: 2008 update. J Phys. Conf. Ser. **164** 012020.

[6] Baldock C. et al. 2010 Polymer gel dosimetry. Phys. Med. Biol. **55**: R1-R63.

[7] Xu, Y., C.S. Wuu, and M.J. Maryanski 2003. Performance of laser CT scanning of polymer gel as a tool for three dimensional dose verification. Med. Phys. **30**(6) 1426-1426

[8] Wuu, C.S. and Y. Xu, 2006 Three-dimensional dose verification for intensity modulated radiation therapy using optical CT based polymer gel dosimetry. Med. Phys. **33**(5) 1412-1419.

[9] Guo, P.Y., J. Adamovics, and M. Oldham 2006 A practical three-dimensional dosimetry System for radiation therapy. Med. Phys. **33**(10) 3962-3972.

[10] Oldham, M., et al. 2008 An investigation of the accuracy of an IMRT dose distribution using

two and three-dimensional dosimetry techniques. Med. Phys. **35**(5)2072-2080.
[11] Sakhalkar H.S., Admovics J., Ibbott, J., Oldham M. A 2009 comprehensive evaluation of the PRESAGE/optical-CT 3D dosimetry system. Med. Phys. **36** 71-82.
[12] Wuu, C.S., et al. 2003 Dosimetry study of Re-188 liquid balloon for intravascular Brachytherapy using polymer gel dosimeters and laser-beam optical CT scanner. Med. Phys. **30**(2) 132-137.
[13] Van Doorn, T., et al. 2005 A fast, high spatial resolution optical tomographic scanner for measurement of absorption in gel dosimetry. Australasian Physical & Engineering Sciences in Medicine **28**(2) 76-85.
[14] Krstajic N. and Doran, S.J. 2007 Fast laser scanning optical-CT apparatus for 3D radiation dosimetry. Phys. Med. Biol. **52:** N257-N263.
[15] Wolodzko, J.G., C. Marsden, and A. Appleby 1999 CCD imaging for optical tomography of Gel radiation dosimeters. Med. Phys. **26**(11) 2508-2513.
[16] Doran, S. and N. Krstajic 2006 The history and principles of optical computed tomography for scanning 3-D radiation dosimeters. J Phys. Conf. Ser., 2006. **56**: p. 45-57.
[17] Krstajic, N. and S.J. Doran 2006 Focusing optics of a parallel beam CCD optical tomography apparatus for 3D radiation dosimetry. Phys. Med. Biol. **51** 2055-2075.
[18] Sakhalkar, H.S. and M. Oldham 2008 Fast, high-resolution 3D dosimetry utilizing a novel optical-CT scanner incorporating tertiary telecentric collimation. Med. Phys. **35**, 101-111.
[19] M. J. Maryanski, C. Audet, and J. C. Gore 1997 Effects of cross-linking and temperature on the dose response of a BANG polymer gel dosimeter *Phys. Med. Biol.* **42**, 303-311.
[20] Xu, Y., Wuu, C.S., and Maryanski, M.J. 2010 Sensitivity calibration procedures in optical-CT scanning of BANG 3 polymer gel dosimeters. *Med. Phys.* **37**(2) 861-868.
[21] Adamovics, J. and M.J. Maryanski 2006 Characterisation of PRESAGE (TM): A new 3-D radiochromic solid polymer dosemeter for ionising radiation. *Radiation Protection Dosimetry* **120**(1-4) 107-112
[22] Sedaghat M, Hubert-Tremblay V, Tremblay L, Bujold R and Lepage M 2008 Volume-dependant internal temperature increase within polymer gel dosimeters during irradiation Proceedings of the 5th international conference on radiotherapy gel dosimetry, Crete, Greece, 2008. 109-112.
[23] Xu Y, Wuu C.S., and Maryanski M. 2003 Determining optimal gel sensitivity in optical CT scanning of gel dosimeters *Med. Phys.* **30** 2257–2263
[24] Salomons G J, Park Y S, McAuley K B and Schreiner L J 2002 Temperature increases associated with polymerization of irradiated PAG dosimeters *Phys. Med. Biol.* **47** 1435-1448.
[25] Jordan K. Advances in optical CT scanning for gel dosimetry 2004 J. Phys.: Conf. Ser. **3** 115.
[26] Vandecasteele, J. And De Deene Y. 2008 Optimization of a fast optical CT scanner for nPAG gel Dosimetry *J. Phys. Conf. Ser.* **164** 012024
[27] Islam K T S, Dempsey J F, Ranade M, Maryanski M J, Low D A 2003 Initial evaluation of Commercial optical CT-based 3D gel dosimeter *Med. Phys.* **30** 2159-2168
[28] Oldham M 2004 Optical-CT scanning of polymer gels *J. Phys.: Conf. Ser.* **3** 122-135

Achieving accurate radiochromic optical-CT imaging when using a polychromatic light source

A Thomas*, M Pierquet, M Oldham

Dept of Radiation Oncology, Duke University Medical Center, Durham, NC, USA

*Corresponding author: ast5@duke.edu

Abstract. Optical-CT performed with a broad spectrum light source can lead to inaccurate reconstructed attenuation coefficients (and hence dose) due to 'spectral warping' as the beam passes through the dosimeter. Some wavelengths will be attenuated more strongly than others depending on the absorption spectrum of the radiochromic dosimeter. A simulation was run to characterize the error introduced by the spectrum warping phenomena. Simulations of a typical dosimeter and delivered dose (6cm diameter, 2 Gy irradiation) showed reconstructed attenuation coefficients can be in error by >12% when compared to those obtained from a monochromatic scan. A method to correct for these errors is presented and preliminary data suggests that with the correction, polychromatic imaging can yield imaging results equal in accuracy to those of monochromatic imaging. The advantage is that polychromatic imaging may be less sensitive to prominent schlerring artefacts that are often observed in telecentric optical-CT scanning systems with tight bandwidth filters applied.

1. Introduction

An accompanying abstract introduces the DLOS scanner, a novel large field-of-view telecentric optical-CT scanner for fast 3D radicohromic dosimetry. The light source employed in this system is a red LED, and the spectrum, although narrow (~ 25*nm* FWHM), can introduce errors into the measured dose through the phenomena of 'spectral warping'. 3D radiochromic dosimetric materials have been designed for use with extremely narrow bandwidth light sources typical to that of a laser (< .1*nm*) corresponding to the materials peak absorption wavelength upon irradiation. The absorption spectrum [1, 2] of the dosimetric materials used in 3D radiochromic dosimetry can vary drastically over the spectrum of an LED light source, thus introducing differences in readouts of a laser system to that of an LED system. Spectral warping arises due to the preferential absorption of photons at the dosimeters peak absorption wavelength leaving a higher ratio of photons at both higher and lower wavelengths as light is transmitted through the dosimeter. This process leaves an artificially high number of photons to be detected for each projection leading to errors in the sinogram and reconstruction; a process similar to spectral hardening or beam hardening in X-ray imaging, although here the spectrum is not getting harder. This area has been addressed extensively in X-Ray imaging [3-9] although does not translate directly to dosimetry. One other common difference between area and laser scanning imaging can be light scatter [10]; the aim of this work, however, was to characterize the magnitude of the spectral warping effect and develop a method to overcome it.

2. Methods

All the measurements and corrections made within are specific to a PRESAGE Radiochromic dosimeter and an in-house area based CCD telecentric lens optical-CT scanning system with a 3*W* Red LED light source. Section 2.1 describes the process of collecting spectral information used in determining the magnitude of the

Journal of Physics: Conference Series **250** (2010) 012045 doi:10.1088/1742-6596/250/1/012045

spectral warping. In order to better understand the spectral warping, section 2.2 details the signal received at the detector for a traditional laser source and the LED source used in the DLOS system. Also included in this section is a method for making corrections to LED projection data to mimic what the laser projection would return. These sections are followed by a simulation of a uniformly irradiated dosimeter measured by a laser source and an LED source to show the errors introduced, and the implementation of the correction on a two field plan delivered to a 16*cm* dosimeter.

2.1. Spectral Information

The 1st step in obtaining a correction for an LED light source system is to understand the spectrum. Spectral information was required for the unattenuated LED, the dosimeters absorption spectrum and the detectors spectral response. The LED spectrum was measured with an Ocean Optics HR 2000 spectrometer and can be seen in figure 1. This information determined the range of wavelengths to consider when looking for the absorption response characteristics of the radiochromic dosimeter. In order to determine the spectral response of the dosimeter, three 1*cm* thick cuvettes were irradiated to .5, 2 and 4*Gy* and the change in optical density (ΔOD) over the LED spectrum (580 - 690*nm*) was measured with a Genesys 20 spectrophotometer from Thermo Scientific. The spectral response of the camera was obtained by using the manufacturer's documentation on the chips quantum efficiency.

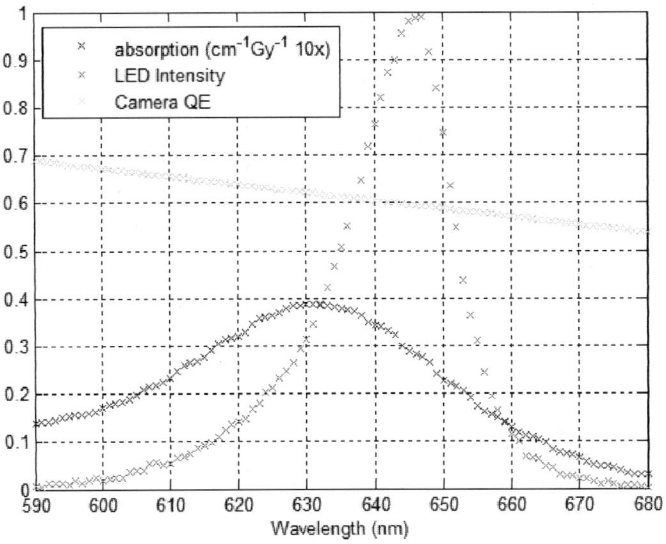

Figure 1 – LED, quantum efficiency and dosimeter absorption spectrum. It can be seen from analysis of the plot that the LED spectrum will be preferentially distinguished at 633*nm* and less so at higher and lower energies causing a spectrum deformation giving an artificially high signal collected as the photons at higher and lower wavelengths are not being attenuated as strongly.

2.2. A Correction: Laser v LED projections

Once the spectral information has been gathered one can begin to develop a correction for the LED light source. Next, an understanding of the signal collected for the laser v the LED is needed. With the dosimeters absorption spectrum, the LED's power spectrum and the cameras quantum efficiency spectrum from the manufacturer, calculations can be made to determine the signal collected from the different light sources by the following equation:

$$S_{LED} = \frac{\sum_{\lambda=590nm}^{\lambda=680nm} I_\lambda \cdot \eta_\lambda \cdot e^{-\mu_\lambda \cdot D \cdot d}}{\sum_{\lambda=590nm}^{\lambda=680nm} I_\lambda \cdot \eta_\lambda} \qquad \text{(eqn 1)}$$

Where S_{LED} is the signal at the detector for an image projection through the setup in figure 1. I_λ is the intensity of the light source for wavelength λ. η_λ is the quantum efficiency of the detector for λ. μ_λ is the attenuation coefficient for λ. d is the pathlength travelled through the dosimeter and D is the dose deposited. Similarly, the laser's signal can be written as:

$$S_{633} = e^{-\mu_{633} \cdot D \cdot d} \qquad \text{(eqn 2)}$$

Where S_{633} is the detected signal and μ_{633} is the attenuation coefficient of the dosimeter at 633*nm*.

Calculations were made with the above data for different levels of attenuation (ie $D \cdot d$) assuming a uniform attenuation along the pathlength. Comparisons could then be made to relate the ideal 633*nm* signal to that of the LED spectrum. Knowing this relationship allows for a correction factor to be applied to each point on the sinogram, thus taking the raw data of an LED spectrum and converting it to what would have been measured by a tighter HeNe light source. The specific correction factor can be written as:

$$CF_{attn} = \left(\frac{S_{633}}{S_{LED}} \right)_{attn} \qquad \text{(eqn 3)}$$

Where CF_{attn} is the attenuation specific correction factor for a pathlength with an average dose, *D*, and a pathlength, *d*. Inspection of the plots in figure 1 indicate the LED spectrum will be preferentially absorbed at 633*nm* allowing the spectrum to broaden. This phenomena is similar to beam hardening with X-rays, but instead of the beam becoming more energetic and monoenergetic, it appears much more uniform and multichromatic.

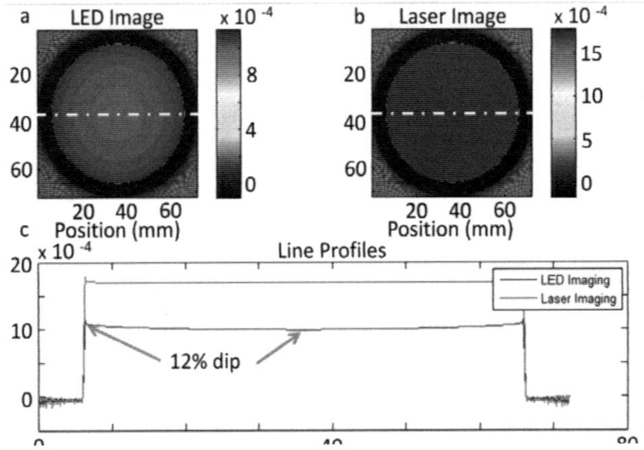

Figure 2 – Simulated image of what a reconstruction of a 6cm dosimeter irradiated uniformly to 2*Gy* would appear with the data taken from (a) LED or (b) HeNe laser light source. (c) Line profiles of the two simulated images showing a 12% dip in the image acquired with a typical LED light source and the difference in reconstruction values which would prevent any absolute dosimetry from taking place.

2.3. Spectral warping effects via simulation

To show the effect of spectral warping in practice a Matlab simulation was run mimicking the response of a uniformly circularly irradiated 6*cm* diameter dosimeter to 2*Gy*. Two sinogram data sets were created, both based on imaging the *ΔOD*, one mimicking the data acquired from an LED light source and the other from a laser light source. The sinograms consisted of 180 projections over 180 degrees with .1*mm* pixels. The images generated from the sinograms represented attenuation coefficients per pixel for the specific wavelengths in the simulated LED and laser scans. These data were forward projected and summed over the appropriate wavelengths to produce projection data. The reconstructions were then compared for differences.

2.4. Application of correction

A 16*cm* diameter dosimeter was irradiated with two orthogonal square fields to a dose of 2*Gy* and later imaged in an in house LED light source system (DLOS) with and without a 1*nm* bandwidth laser line filter and in the MGS commercial scanner with a HeNe laser light source. 360 projections were taken 1 degree apart with 25 images taken at each projection angle for the LED based images and 360 projections over 360 degrees were also done with the laser based scanner. The projection data was processed and placed into respective sinograms. The correction factor was applied to each point on the unfiltered sinogram. The sinograms were reconstructed and the resulting images compared for differences.

3. Results & Discussion

3.1. Simulation results

The simulation gave results very similar to what one would expect. Figure 2c shows the dramatic difference returned in attenuation coefficient values. Differences of greater than 70% for just a 2*Gy* irradiation over a 6*cm* diameter were seen for the LED and laser sources. Any calibration attempts for moving to absolute dosimetry would need to consider these discrepancies if using a broad band light source such as an LED.

Figure 2a & 2c also exhibit a common imaging artefact present in optical-CT imaging, a depression in the center of the dosimeter. In this case the depression was 12% below the edges of the dosimeter. While there have been several causes of such depressions reported, "spectral warping", the preferential distinction of photons at 633*nm* one of them to consider. Figures 2b &2c show how utilizing only one wavelength in the reconstruction can return results on par with the designed expected results of the dosimeter & scanner system as a whole. The reconstruction returns values that have the expected shape and value such that the potential for calibration exists in shifting to absolute dosimetry.

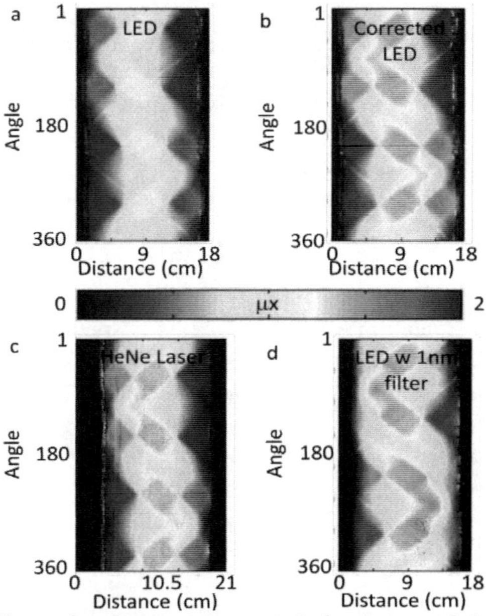

Figure 3 – Prescan corrected sinograms of the two field plan from (a) an LED light source (b) the corrected LED light source (c) a HeNe laser light source and (d) an LED light source with a 1*nm* FWHM centered at 633*nm*. The values in (b), (c) and (d) are all consistent while those in (a) are significantly lower.

3.2. Application results

Similar to the simulated results, the images reconstructed with the LED based system have the general shape of the reconstruction, but are not consistent with the laser based scanners in terms of accuracy and contrast. Figure 3 shows 4 sinograms where the post irradiation sinogram was divided pixel by pixel by the pre irradiation sinogram collected from (a) an uncorrected LED system, (b) a corrected LED system, (c) a HeNe laser system, and (d) a LED system with a 1*nm* filter centered at 633*nm* in the optical chain.

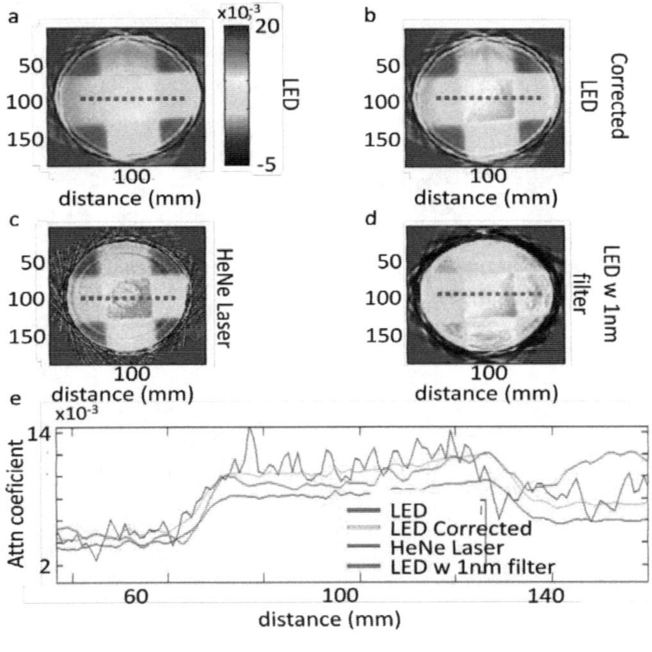

Figure 4 - Reconstructed images of the two field plan taken from (a) an LED, (b) corrected LED, (c) HeNe laser and (d) an LED with a 1*nm* FWHM filter. (e) Line profiles of the images corresponding to the dashed red line in the four images. The corrected LED and the HeNe laser show the same values throughout the profile displaying the effectiveness of the LED correction.

Comparisons of these images show very striking resemblances in the Corrected LED sinogram and the HeNe sinogram. Figure 4 displays the correlating image reconstructions in a, b, c & d with a line profile of each in e. Again there is good correlation between the corrected LED image and the HeNe image with the major difference being the HeNe image has more noise associated with it although the values are very similar. Looking at the $1nm$ FWHM bandwidth LED image it is evident that the filter makes the system much more sensitive to any schlering bands present in the dosimeters during the curing process.

4. Conclusion

This work looked at the magnitude of error introduced by 'spectral warping' when performing optical-CT scans utilizing a polychromatic LED light source, and described a correction method to overcome it. This correction has the ability to eliminate discrepancies previously seen between laser scanning imaging and LED light source imaging. The correction is highly dependent upon accurate input data for the detectors quantum efficiency, the LED spectrum, and the dosimeters spectral absorption response to radiation over the spectrum of the LED. Use of these corrections should lead to more accurate dosimetry with results comparable to the laser scanning optical-CT scanning systems.

Acknowledgements: Supported by NIH R01 CA100835

References

[1] Kelly, R.G., K.J. Jordan, and J.J. Battista, Optical CT reconstruction of 3D dose distributions using the ferrous-benzoic-xylenol (FBX) gel dosimeter. Med Phys, 1998. **25**(9): p. 1741-50.
[2] Adamovics, J. and M.J. Maryanski, Characterisation of PRESAGE: A new 3-D radiochromic solid polymer dosemeter for ionising radiation. Radiat Prot Dosimetry, 2006. **120**(1-4): p. 107-12.
[3] Ruegsegger, P., et al., Standardization of computed tomography images by means of a material-selective beam hardening correction. J Comput Assist Tomogr, 1978. **2**(2): p. 184-8.
[4] Peng, G.H., et al., [Hardening correction model of energy spectrum for X-ray TICT in testing composites workpiece]. Guang Pu Xue Yu Guang Pu Fen Xi, 2007. **27**(4): p. 823-6.
[5] Joseph, P.M. and R.D. Spital, A method for correcting bone induced artifacts in computed tomography scanners. J Comput Assist Tomogr, 1978. **2**(1): p. 100-8.
[6] Idris, A.E. and J.A. Fessler, Segmentation-free statistical image reconstruction for polyenergetic x-ray computed tomography with experimental validation. Phys Med Biol, 2003. **48**(15): p. 2453-77.
[7] Hopkins, F., et al., Analytical corrections for beam-hardening and object scatter in volumetric computed tomography systems, in 16th World Conference of Nondestructive Testing. 2004: Montreal, Canada.
[8] Elbakri, I.A. and J.A. Fessler, Statistical image reconstruction for polyenergetic X-ray computed tomography. IEEE Trans Med Imaging, 2002. **21**(2): p. 89-99.
[9] De Man, B., et al., An iterative maximum-likelihood polychromatic algorithm for CT. IEEE Trans Med Imaging, 2001. **20**(10): p. 999-1008.
[10] Jordan, K. and J. Battista, Linearity and imageuniformity of the Vista™ optical cone beam scanner, in 4th International Conference on Radiotherapy Gel Dosimetry. 2006: Sherbrooke, Canada.

Creation of sophisticated test objects for quality assurance of optical computed tomography scanners

A. T. Abdul Rahman[1], Elke Bräuer-Krisch[2], Thierry Brochard[2], John Adamovics[3], Steve Clowes[1], David Bradley[1], Simon Doran[1,4]

[1]Department of Physics, University of Surrey, Guildford, Surrey, UK

[2]European Synchrotron Radiation Facility, Grenoble, France

[3]Rider University, Lawrenceville, NJ, USA

[4]CRUK and EPSRC Cancer Imaging Centre, Institute of Cancer Research, Sutton, UK

Simon.Doran@icr.ac.uk

Abstract. Optical computed tomography (CT) shows great potential for radiation therapy dose verification in 3D. However, an effective quality assurance regime for the various scanners currently available still remains to be developed. We show how the favourable properties of the PRESAGE™ radiochromic polymer may be exploited to create highly sophisticated QA phantoms. Five 60 mm-diameter cylindrical PRESAGE™ samples were irradiated using the x-ray microbeam radiation therapy facility on the ID17 biomedical beamline at the European Synchrotron Radiation Facility. Samples were then imaged on the University of Surrey parallel-beam optical CT scanner and were designed to allow a variety of tests to be performed, including linearity, MTF (three independent measurements) and an assessment of geometric distortion. A small sample of these results is presented. It is clear that, although the method produces extremely high quality test objects, it is not practical on a routine basis, because of its reliance of a highly specialised radiation source. Hence, we investigated a second possibility. Two PRESAGE™ samples were illuminated with ultraviolet light of wavelength 365 nm, using cheap masks created by laser-printing patterns onto overhead projector acetate sheets. There was good correlation between optical density (OD) measured by the CT scanner and the expected UV "dose" delivered. The results are highly encouraging and a proposal is made for a scanner test regime based on calibrated and well characterised PRESAGE™ samples.

1. Introduction

The technique of optical computer tomography (CT) [1-2] has shown great potential for the mapping of radiation doses in three dimensions, as well as in 3-D biological imaging [3]. However, it is clear that careful attention to quality assurance (QA) is necessary if the long term aim of the technique is to be able to provide any form of "gold standard" for 3-D dose mapping. In other fields of medical imaging, particularly diagnostic scanning, QA is now routine, with prescribed imaging protocols and specialist companies providing QA phantoms. Whilst a number of academic studies e.g., [4-6] have previously used phantoms to characterise properties of optical CT scanners such as linearity and geometric distortion, the degree to which sophisticated test objects can be manufactured mechanically

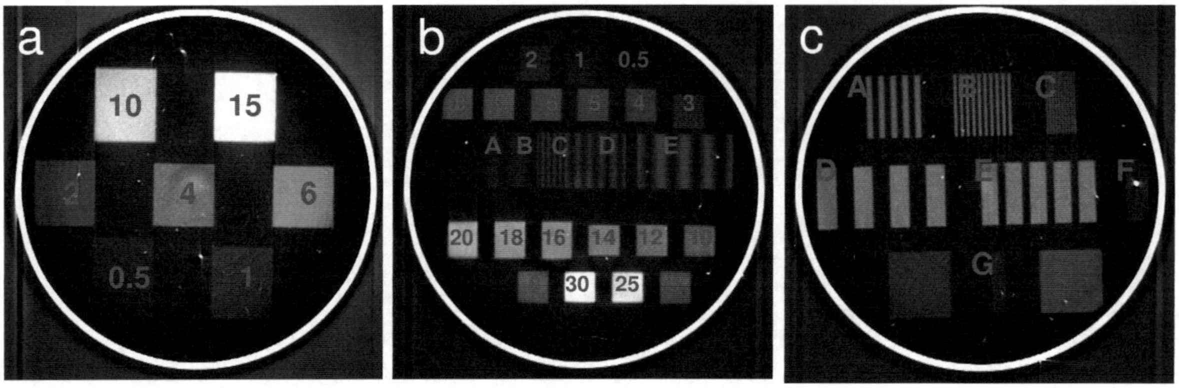

Figure 1: Reconstructed slices from three PRESAGE™ phantoms irradiated at ESRF: (a) Sample 1: 7 10 × 10 mm^2 squares 0.5 to 15 Gy; (b) Sample 2: 19 5 × 5 mm^2 squares 0.5 to 30 Gy and 5 sinusoidal patterns: (A) 3 lp/mm, (B) 2 lp/mm, (C) 1 lp/mm, (D) 0.5 lp/mm, (E) 0.25 lp/mm; (c) Sample 3: resolution test pattern resolution test pattern of different line widths, each depositing 6 Gy: (A) 1 mm, (B) 0.5 mm, (C) 0.25 mm, (D) 3 mm, (E) 2 mm, (F) 0.1 mm, (G) 0.05 mm. Two 10 × 10 mm^2 squares were irradiated with 2 and 4 Gy respectively for calibration purposes.

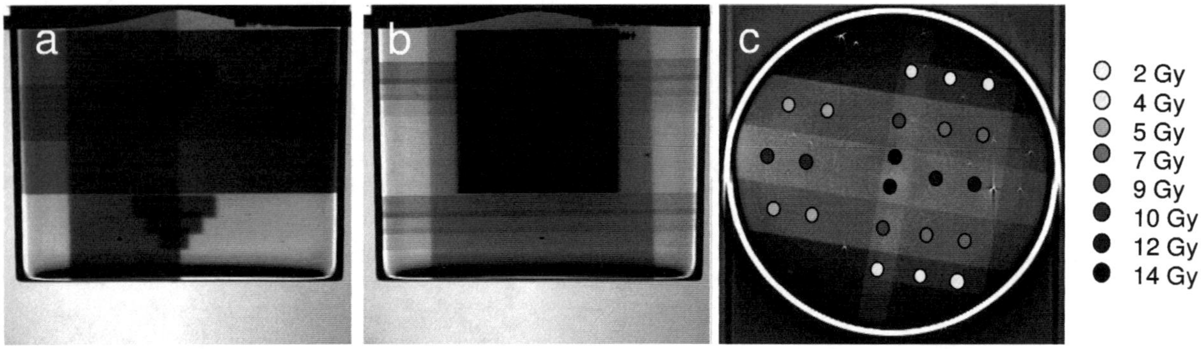

Figure 2: Sample 4: (a, b) projection images from two angles; (c) single slice from the reconstructed 3-D image with markers corresponding to regions analysed in Fig. 5d.

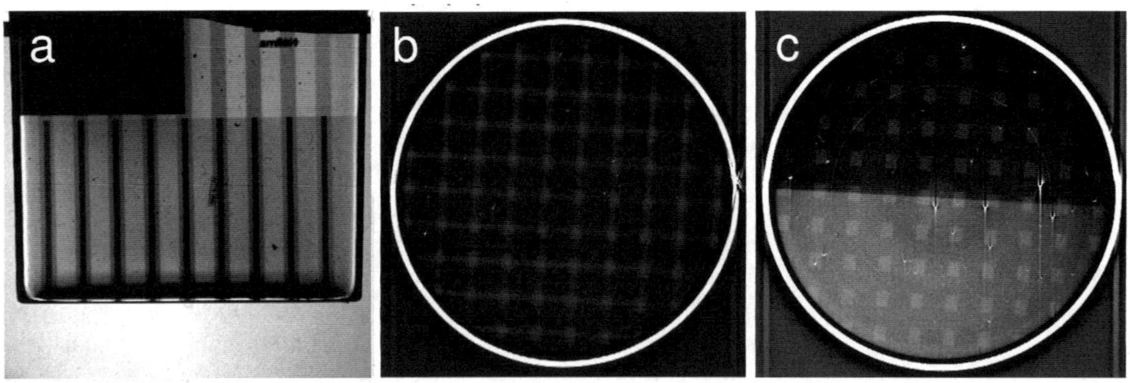

Figure 3: Sample 5: Multipurpose geometric distortion mapping phantom: (a) typical projection image (b,c) reconstructed transverse slices at two different levels.

is limited. Nowhere is this demonstrated more graphically than in the quantitative validation of optical CT microscopy. The type of India-ink-in-gel phantom created by Sharpe *et al.* [3] from would not be suitable for verifying optical density values or assessing image distortion, but to construct something better mechanically would be an enormous technical challenge.

The radiosensitive polymer PRESAGE™ has a number of favourable properties, the most important for this work being mechanical strength (a solid rather than a gel), stability over the temperature range routinely encountered in clinical and research practice, and long-term stability of the dose pattern created by irradiation. Although the optical density values are not completely fixed, changes over time are predictable and can be calibrated. In this abstract we demonstrate how these properties allow the creation of robust samples of great sophistication via complex irradiations. We then suggest how, given appropriate calibrations, standard samples can be created at little cost by using UV irradiation. Finally, we propose a test framework for comparing different optical CT scanners, linked to a spectrophotometric standard.

2. Materials and Methods

2.1. Phantoms irradiated using synchrotron x-rays

Dose mapping was carried out using PRESAGE™ dosimeters in the form of cylinders of diameter 60 mm and height 60 mm, and calibration measurements were carried out with small PRESAGE™ samples supplied in standard $10 \times 10 \times 45$ mm^3 optical cuvettes. (Similar experiments, reported elsewhere, were also performed using PRESAGE microscopy samples with a range of diameters 20 mm and below.) Irradiations were carried out at the European Synchrotron Radiation Facility (ESRF) in Grenoble on the ID17 biomedical beamline. This beamline uses a "wiggler" source to produce an intense, highly collimated synchrotron x-ray beam from the 6 GeV circulating electrons. The x-ray spectrum covers the range 50 to 350 keV with a mean energy of 107 keV and peak at 83 keV [7]. The irradiations were performed using a "dose-painting" technique, by translating the samples through the beam to provide various patterns mentioned.

Five phantoms were created to allow the investigation of linearity, spatial resolution and MTF, and geometric fidelity of the scanner. In addition, the samples demonstrate the dose-integration properties of the dosimeter and allow the measurement of depth-dose curves. **Sample 1** (Figure 1a) was composed of seven 10×10 mm^2 squares covering a dose range of 0.5 – 15 Gy. **Sample 2** (Figure 1b) was inspired by the standard USAF optical test target (Edmund Optics NT54-803), which combines a linearity / absolute optical density test with an MTF test. The linearity test increased the number of squares to 19, each 5×5 mm^2, extending the dose range to 0.5 Gy – 30 Gy, whilst also allowing us to test the in-plane reproducibility of measurements by creating four squares each with dose 8 Gy. The MTF test was made from 5 regions of sinusoidally-varying dose with different periods, deposited using a macro to automate the ID17 goniometer and shutter. **Sample 3** uses a different method to test MTF: it contains nine slit patterns, creating near-ideal line-pairs with spacings from 0.14 lp/mm up to 2.75 lp/mm. This sample illustrates graphically the enormous advantage of the synchrotron irradiation over a standard clinical linac, where field penumbrae make it impossible to create well-defined areas of uniform dose at this scale without the construction of an intricate collimator [8]. **Sample 4** (Figure 2) is a pattern of multiple irradiated fields, each with simple geometry, allowing us to test our assumption that the PRESAGE™ dosimeter is a perfect integrator. It also contains a number of "knife edge" structures that can be used for a further MTF analysis. Finally, **Sample 5** (Figure 3) is a regular grid of 2×2 mm^2 square dots. The absolute location of these is known and hence we can determine very precisely whether the measured images contain any distortion. In our view this is a superior method to the needle phantom used by [4-5], because the dots are not opaque and so image artefacts are not created. The sample is further irradiated from the side. By such means, it is possible to create alternating layers that allow distortion in the z-direction to be calculated, too.

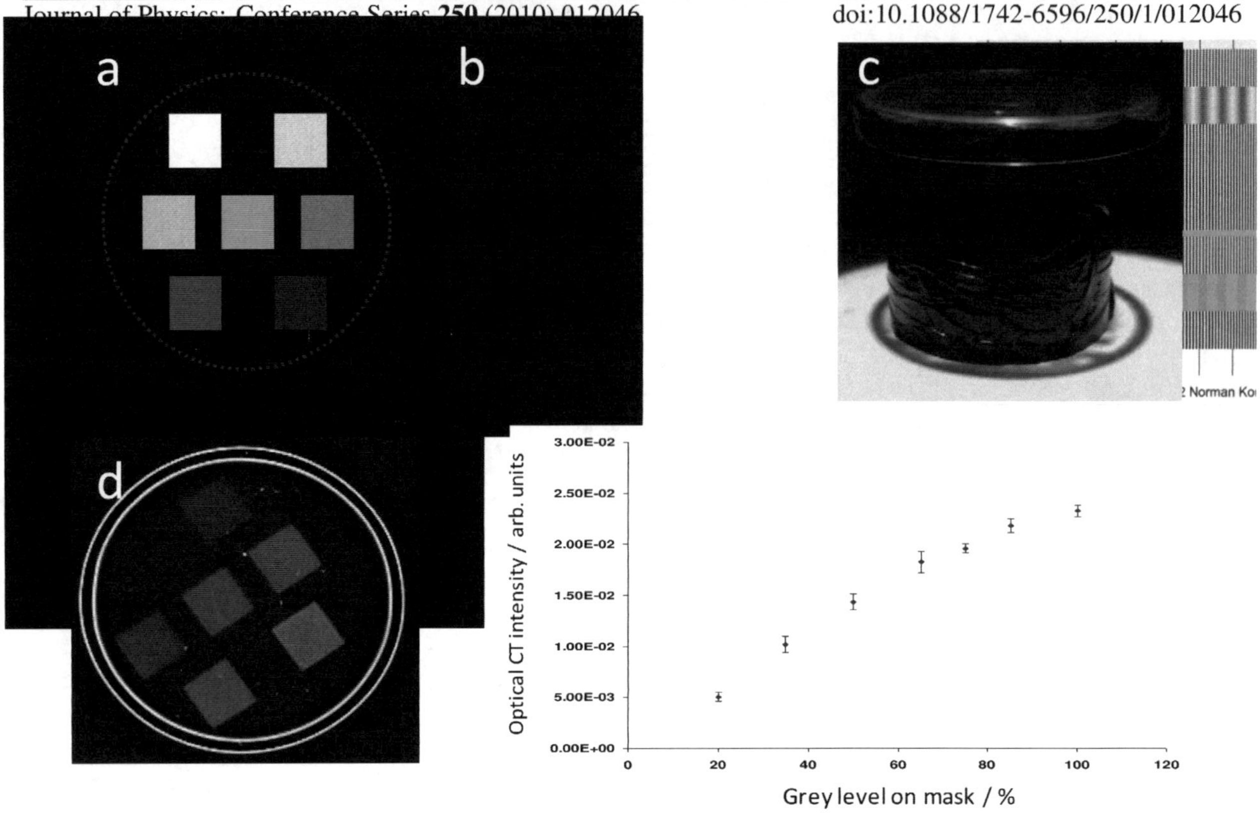

Figure 4: The UV irradiation setup. (a) Test pattern of seven squares (10×10 mm^2) of different transparency levels; (b) resolution test pattern (5 lp/mm to 12 lp/mm); (c) PRESAGE^{TM} during irradiation was covered with black plastic except at its surface where the mask was located; (d) optical CT image of UV phantom created using mask in (a); (e) relationship between grey level in mask definition file to optical CT output, illustrating that a good calibration should be possible.

2.2. Phantoms irradiated using ultraviolet light

A severe criticism of the above work might be that one could not possibly base any realistic QA regime on samples that required a synchrotron for their creation. As an alternative, we sought to make use of the known UV sensitivity of PRESAGE™. This has previously been regarded as a problem and samples are routinely wrapped in black plastic to avoid unwanted exposure during experiments, transit and storage. To our knowledge, there have been no detailed studies of the dose-response of PRESAGE^{TM} to UV.

A computer drawing package was used to create a pattern of seven squares, resembling Sample 1, having different grey levels between 100% (white) and 20%. This was laser-printed onto an overhead projector acetate sheet to form a UV mask. A second mask was created from part of a resolution test pattern (5 lp/mm to 12 lp/mm) [9], allowing us to investigate the spatial resolution achievable using a UV radiation. Masks were placed on top of cylindrical PRESAGE^{TM} samples and irradiated for 20 minutes from the top with a uniform beam of 365 nm UV light of intensity 9.95 mW/cm^2 using the lamp of a mask aligner.

2.3. Imaging methodology

All samples were imaged using the University of Surrey parallel-beam scanner. For these scans, 800 projections of matrix size 512×512 pixels were acquired and reconstructed into a stack of 512×512 axial images using filtered back projection using in-house software written in IDL.

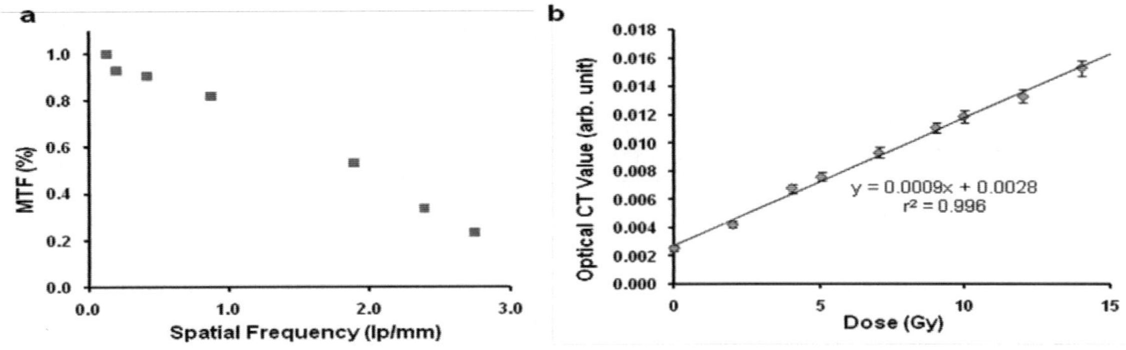

Figure 5: Selection of results from the various scanner characterizations performed: (a) MTF calculated from Sample 3; (b) Dose integration test performed using the pattern in Sample 4.

3. Results and Discussion

Optical CT images of the various samples are shown in Figures 1−4. A variety of tests was performed on the 3-D images, of which (for reasons of space) just a small selection of results is shown in Figure 5. The samples are suitable for performing linearity, MTF, geometrical distortion and signal-to-noise analysis, together with measuring the depth dose in both x-ray and UV samples (the latter being, to our knowledge, a novel result). Further samples demonstrated the artefacts resulting from scanner (but not dosimeter) saturation and could make quantitative measurements of this effect to compare with theory.

We propose that an appropriate calibration regime would consist of the manufacture of a number of such samples from a single batch of PRESAGE^{TM}, together with a set of cuvettes (one per sample). The cuvettes (calibrated using a spectrophotometer) would accompany samples, experiencing the same temperature history. Phantoms would be imaged at regular intervals and exchanged between sites.

Acknowledgement: The authors gratefully acknowledge funding for experiment MI-949 from ESRF.

References

[1] Tarte, B.J. and T. van Doorn. Optical scanning of ferrous sulfate gels for radiotherapy treatment dosimetry. in *APSEM/BECON Australasian Conference on Physical Science and Engineering in Medicine / Biomedical Engineering Conference* 1993.

[2] Gore, J.C., et al., Radiation dose distributions in three dimensions from tomographic optical density scanning of polymer gels: I. Development of an optical scanner. Phys Med Biol 1996. **41**: p. 2695-2704.

[3] Sharpe, J., et al., Optical Projection Tomography as a tool for 3D microscopy and gene expression studies. Science, 2002. **296**(5567): p. 541-545.

[4] Oldham, M. and L. Kim, Optical-CT gel-dosimetry II: Optical artifacts and geometrical distortion. Med Phys, 2004. **31**(5): p. 1093-1104.

[5] Olding, T., O. Holmes, and L.J. Schreiner, Cone beam optical computed tomography for gel dosimetry I: scanner characterization. Phys Med Biol, 2010. **55**: p. 2819-2840.

[6] Guo, P. and et al., Quality assurance in 3D dosimetry by optical-CT. Journal of Physics: Conference Series, 2006. **56**(1): p. 191.

[7] Doran, S.J., et al., An investigation of the potential of optical computed tomography for imaging of synchrotron generated x-rays at high spatial resolution. Phys Med Biol, 2010. **55**: p. 1531-1547.

[8] Bayreder, C., et al., The spatial resolution in dosimetry with normoxic polymer-gels investigated with the dose modulation transfer approach. Med Phys, 2008. **35**(5): p. 1756-1769.

[9] Koren, N. Lens test chart. 2003 [access date 2010 9 May 2010]; Available from: http://www.normankoren.com/Tutorials/MTF5.html#newchart.

Journal of Physics: Conference Series **250** (2010) 012047 doi:10.1088/1742-6596/250/1/012047

An investigation into a new re-useable 3D radiochromic dosimetry material, Presage^{REU}

Michael Pierquet, Andrew Thomas, John Adamovics[1], Mark Oldham

Duke University Medical Center, Durham, NC
[1] Heuris Pharma LLC, Lawrenceville, NJ

Corresponding author email: mark.oldham@duke.edu

Purpose: To investigate the dosimetric properties of a new Presage formulation which exhibits a reversible color change on exposure to radiation. Presage^{REU} offers the intriguing possibility of the first re-useable 3D dosimetry material. **Method and Materials:** Small volumes of Presage^{REU} in 1x1x5cm optical cuvettes were irradiated and re-irradiated under a variety of conditions and times to investigate a range of properties including re-usability, dose-rate dependence, dose sensitivity, temporal response, energy sensitivity, and temperature dependence. **Results:** The radiation induced change in optical density (OD) was found to be linear with dose after initial and subsequent irradiations. After the first irradiation OD was observed to clear in ~2 weeks when stored at room temperature. 3 subsequent irradiations of the same cuvettes showed a very similar strong OD response, although there was a significant increase between this response and that achieve at initial irradiation. **Conclusion:** The Presage^{REU} formulation shows strong potential as the first re-useable 3D dosimetry material. When dosimeters are stored at room temperature (~22°C) clearing can occur in 2-3 weeks.

1. Introduction

PRESAGE was first introduced as a material for 3D dosimetry by Adamovics and Maryanski 2006 [1]. Attractive properties included a radiochromic response, facilitating accurate optical-CT imaging, a lack of requirement for an external container, and a robustness to exposure to the laboratory and clinical environment (except for a sensitivity to exposure to UV light). Since then, several formulations have been characterized, and significant variations in temporal responses have been observed [2,3,4]. In most of the formulations we have studied, the optical-density (OD) response is observed to gradually increase with time after irradiation. Recently new formulations were discovered where the reverse has been observed, and the OD change gradually fades or 'clears', until the dosimeter returns to approximately its original state. This new formulation has been termed Presage^{REU}, and has an estimated $Z_{eff} = 8.1$ and density of 1.07g/cm^3, with CT number 130 ± 20 HU. The exact mechanism controlling the rate of reversibility is still under investigation, but is believed to clear due to a new LMG (lueco malichite green) derivative synthesized at Heuris Pharma, LLC (Skillman, NJ). In this work we investigate the key properties of the Presage^{REU} formulation as well as its suitability for repeat irradiations.

2. Materials and Methods

2.1 Cuvette studies

The reusable formulation was initially characterized using small volume studies where optical cuvettes filled with Presage^{REU} were irradiated in a reference geometry to known doses. A solid water setup (Fig. 1) with bolus was used as the main arrangement. A cuvette-sized depression was removed from a 1.5 cm thick sheet of tissue equivalent bolus enabling cuvette positioning with center at 5.5 cm depth. Cuvettes were irradiated in a 10x10 cm field on the surface of the phantom (SSD=100cm), with default dose rate of 400 MU/min.

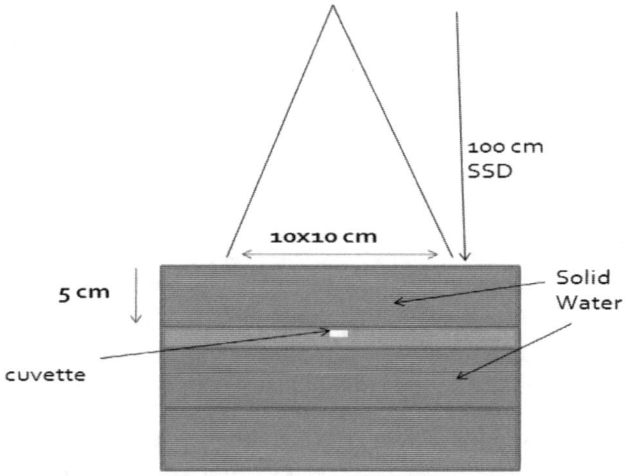

Fig. 1. Irradiation geometry for the cuvette studies. Optical cuvettes of 1x1x5cm were used. Reproducible dose was delivered by embedding the cuvettes in bolus material

The initial OD dose response of Presage^{REU} was determined by irradiating 6 cuvettes to doses of 0.5, 1, 3, 6, 10, and 15 Gy. The dosimeters were then stored at room temperature for at least 2 weeks to allow clearing. Re-useability was investigated by re-irradiating the same cuvettes with the same doses as the initial irradiation. The dosimeters were allowed to clear again, prior to subsequent irradiations. This cycle was repeated 4 times. After each irradiation the radiation induced OD change was immediately recorded. Each cuvette received the same dose on subsequent irradiations (e.g. the cuvette that initially received 1Gy, also received 1Gy on subsequent irradiations). Optical density (OD) measurements of the Presage^{REU} -filled cuvettes were made pre and post irradiation using the Thermo Scientific Genesys 20 spectrophotometer. Prior experiments determined the optimal scanning wavelength (i.e. where the maximum radiation induced OD change occurred) was 632 nm, which is consistent with previous PRESAGE studies [2]. All further OD measurements were therefore made at this wavelength. The key quantity determined from irradiated cuvettes is ΔOD i.e. the radiation induced change in OD.

2.2 Large volume studies

A preliminary study was initiated to investigate whether the OD clearing phenomena observed in the small volume cuvettes also occurred in larger volumes. To study this, a 10cm diameter Presage^{REU} cylinder was irradiated with a simple 2-field 6X plan to 4Gy at isocenter. The dosimeter was stored in the dark at room temperature and scanned at several time-points after irradiation and over the first 45 days.

3. Results and Discussion

3.1 Small volume cuvette studies

The immediate sensitivity (radiation induced OD change) of the 6 Presage^{REU} cuvettes exposed to multiple irradiations (doses of 0.5, 1, 3, 6, 10, and 15 Gy) is shown in figure 2. The first irradiation (June 2009) showed a linear relationship between OD and absorbed dose, with a slope of 0.0206cm^{-1}/Gy. 336 hours after irradiation, the OD of each cuvette had reduced significantly and remained stable relative to the 0 Gy control. The sensitivity of the same cuvettes to the second irradiation was very surprisingly observed to increase by a factor of ~2 when compared with the first irradiation. For the next 3 re-irradiations the same sensitivity was observed as on the second irradiation. The last re-irradiation in March, 2010 showed a slight decrease in dose sensitivity compared to the 3 immediately prior irradiations. Error bars are omitted from the figures due to the magnitude of the errors on the data being smaller than the size of the marker representing each data point.

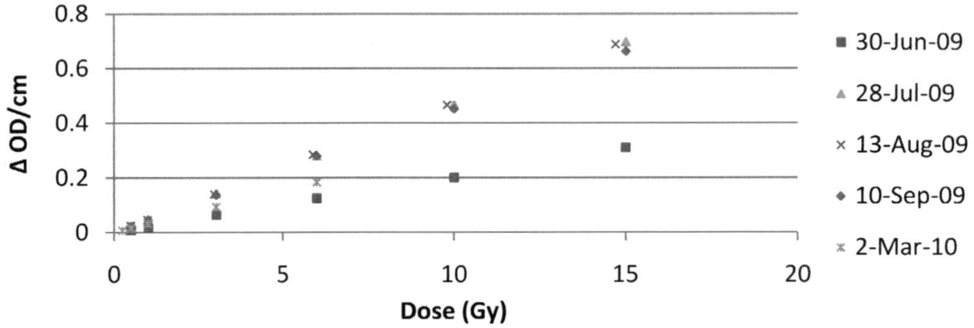

Figure 2. The same 6 Presage^{REU} cuvettes were irradiated to varying doses (.5-15Gy) on 5 occasions. The dose sensitivity on the first irradiation was surprisingly observed to be about half that of the other 4 irradiations.

We speculate that the depressed initial OD response of Presage^{REU} could be due to an aging phenomenon, where the sensitivity of Presage^{REU} gradually increases over the first few weeks after manufacture as the dosimeter gradually completes the curing process. Some support for this interpretation came from another experiment involving a different batch of the same formulation. 3 cuvettes were irradiated after storage for a month at room temperature (~22°C), where full curing had presumably occurred. The initial dose sensitivity was now found to be 0.0432Gy^{-1}, similar to the higher sensitivity reported in figure 2. The rate of decay of the OD in the 6 cuvettes after the first irradiation is shown in figure 3.

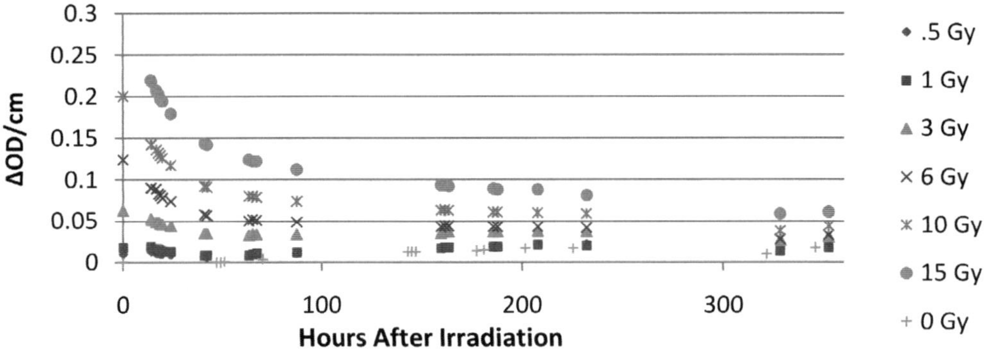

Figure 3. Optical clearing of small volumes of Presage^{REU} stored at room temperature.

3.2 Large volume study

Figure 4 shows the decay of the signal in the dosimeter over a 42 day period. Images in (a) and (b) are prescan corrected reconstruction images of the dosimeter windowed identically. Inspection of the figure shows the radiation induced attenuation is clearly fading over time.

4. Conclusions

A new type of Presage is shown to have potential for re-useability. This represents a notable milestone in the history of 3D dosimetry. Our preliminary results from irradiating small volumes of Presage have found that 5 re-irradiations are feasible, with the background returning to near baseline after each. The data suggests that many more irradiations would also be feasible. Several observations warrant further study, in particular: the rate of clearing after each irradiation (may be slightly increased), the sensitivity with time after curing and repeat irradiations, and any volume effect (effect of size of dosimeter on re-useability and sensitivity). A full characterization of Presage^{REU} is currently being conducted.

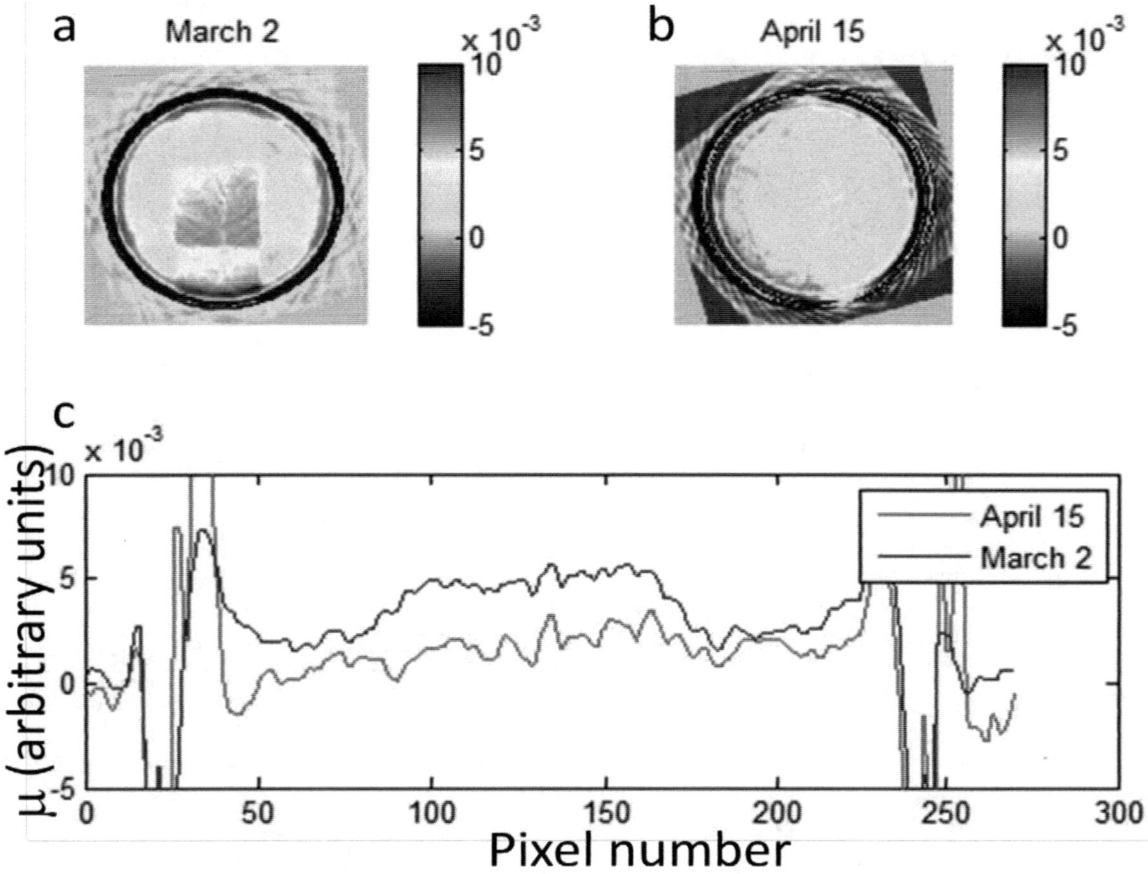

Figure 4. Optical clearing of a large volume dosimeter over a 43 day period. Reconstruction images (a & b) at the same angle windowed identically to show the decay in the attenuation coefficient over time. (c) Line profiles of (a) and (b).

Acknowledgements: Supported by NIH R01CA100835.

References

[1] Adamovics J, Maryanski M J 2006 Characterisation of PRESAGE: A new 3-D radiochromic solid polymer dosemeter for ionising radiation *Radiation Protection Dosimetry* **120**(1-4) 107-112
[2] Guo P Y, Adamovics J, Oldham M 2006 Characterization of a new radiochromic three-dimensional dosimeter *Med. Phys* **33** 1338-1345
[3] Sakhalkar H S, Adamovics J, Ibbott G, Oldham M 2009 A comprehensive evaluation of the PRESAGE/optical-CT 3D dosimetry system *Medical Physics* **36** 71-82
[4] Clift C, Thomas A et al. 2010 Toward acquiring comprehensive radiosurgery field commissioning data using the PRESAGE/optical-CT 3D dosimetry system *Physics in Medicine and Biology* **55** 1279-1293
[5] Sakhalkar H S, Oldham M 2007 Fast, high-resolution 3D dosimetry utilizing a novel optical-CT scanner incorporating tertiary telecentric collimation *Med. Phys* **35** 101-110

Journal of Physics: Conference Series **250** (2010) 012048 doi:10.1088/1742-6596/250/1/012048

Effects of adding glycerol and sucrose to ferrous xylenol orange hydrogel

Kevin Jordan and Masaya Sekimoto

London Regional Cancer Program, London Health Sciences Centre, University of Western Ontario, Dept. of Medical Biophysics

Kevin.jordan@lhsc.on.ca

Abstract. Glycerol and sucrose were substituted up to 40% by mass for water in ferrous xylenol gelatin hydrogel (FX). Both materials increased the refractive index of the aqueous component of the gels and lowered the optical scatter coefficient. Diffusion of the FX products was reduced 3-fold at 40% substitution levels. The radiation response was more stable with glycerol.

1. Introduction

Gelatin hydrogels such as ferrous xylenol orange (FX) that contain around 4% gelatin by mass are among the most transparent radiochromic materials available for 3D dosimetry with optical CT readout. Because the materials are relatively inexpensive, large volume dosimeters could be prepared with this material. Specifically, samples up to 30 cm in diameter may be useful for specialized dosimetry problems requiring complete sampling of large volumes. Optical cone beam CT can be modified in order to scan large volumes [1]. But with a diffuse source this geometry is more prone to stray light than other geometries such as single ray – single detector laser CT scanner geometries. Dosimeters that exhibit the least amount of scatter are expected to result in the greatest spatial resolution and most accurate reconstructed attenuation coefficients. It is anticipated that radiosensitive gels with higher scatter will have inferior optical performance when scanned with cone beam optical CT systems [2].

One approach to stray light reduction in transmission images through this hydrogel is to add chemicals that increase the effective refractive index of the aqueous component of the gel [3]. The effect of adding glycerol to FX gelatin hydrogel was reported earlier [4]. In this study sucrose and glycerol were compared for scatter reduction in FX gelatin hydrogels. Other dosimeter parameters examined were: dose sensitivity, auto-oxidation and diffusion rates.

2. Methods

Reference FX gelatin hydrogrels contained: 4% by mass gelatin (porcine, type A, 300 Bloom), 0.3 mM ferrous ions, 0.05 mM xylenol orange and 50 mM sulphuric acid. Sucrose or glycerol was substituted for fractions of the water during gel preparation. They were added after the gelatin was dissolved in water at 40C. All materials except the sucrose were from Sigma-Aldrich. The sucrose was purchased as coarse granulated cane sugar locally. The glycerol was pre-treated to scavenge oxidizing species by adding zinc metal pellets and stirring for several minutes. Samples were prepared with dopant concentrations ranging up to 40% by mass. Above this level the materials were very viscous and difficult to pour as optically uniform samples. Gelatin hydrogels doped with 50 mM H_2SO_4, 30% glycerol or 40% sucrose were placed in 10 cm pathlength custom plastic cuvettes for scatter measurements.

Attenuation spectra were recorded from 400 to 700 nm with a Perkin Elmer model 139 UV-VIS spectrometer. Dose response measurements were performed on single one cm pathlength PPMA cuvettes with sequential irradiations. Transmission measurements were performed at 30 to 32 minutes postirradiation. A Varian 600C medical accelerator (4 MV photon beam) was used for irradiations. The diffusion experiment included, irradiating half of a 1 cm cuvette sample to 5 Gy and recording sequential transmission images over several hours at 590 nm with an in-house optical cone beam scanner. The spatial resolution was 12 pixels per mm. Profiles from the sequence of images were fit to calculate a linear diffusion coefficient [5].

3. Results and discussion

Increasing the hydrogel gelatin content from 4% to 8% effectively doubled the scatter coefficient at all visible wavelengths. Approximately a 30% reduction in scatter was obtained by substituting 30% glycerol or 40% sucrose for water in the hydrogels.

The FX gels became stiffer, were darker orange and had increased auto-oxidation rates as a result of increased doping levels. The profiles at 0.5 and 4.5 hours post irradiation for gels with 40% sucrose or glycerol are shown in figure 1 with offsets to correct for auto-oxidation. Note that the radiochromic response of the sucrose sample continued to increase after 0.5 hours but the glycerol gel response was stable. This longer development time with sucrose is problematic for reproducible measurements.

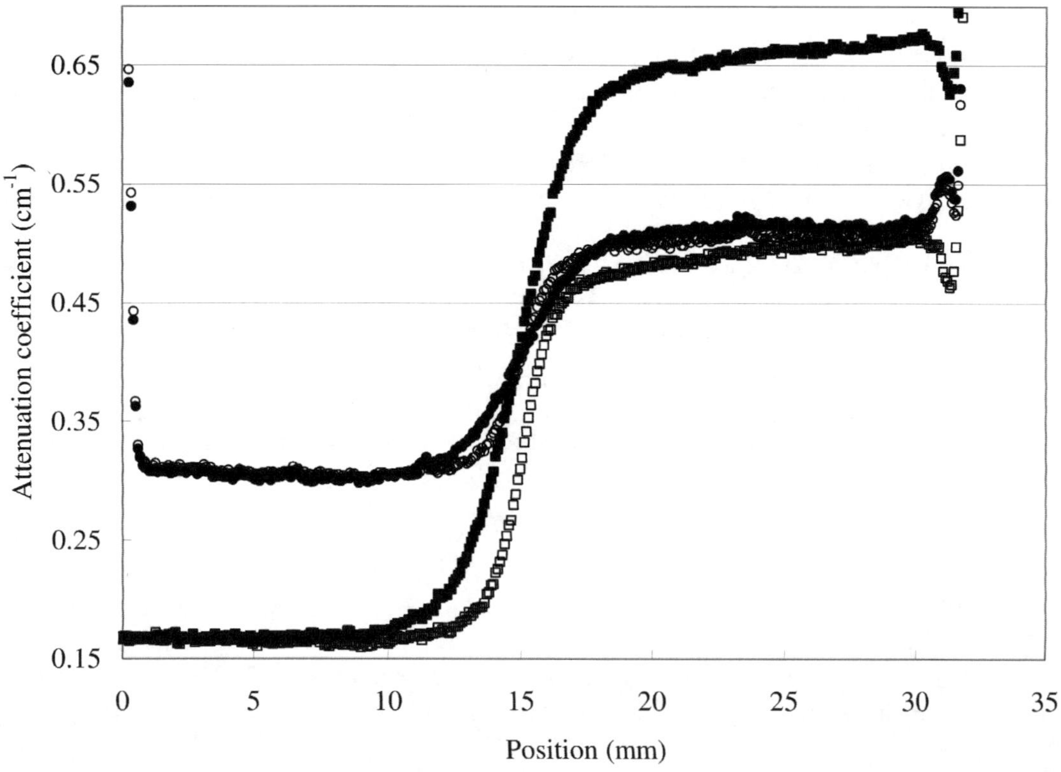

Figure 1. Post irradiation attenuation profiles for FX gels: 40% sucrose (0.5(□), 4.5(■) hours), 40% glycerol (0.5(○), 4.5(●) hours). Right half received 5 Gy. See text concerning long development time with sucrose.

Profiles were recorded from 0.5 to 5 hours after irradiation in 0.5 hour intervals. The fitted data resulted in measured diffusion coefficients of 0.145 and 0.175 $mm^2 hr^{-1}$ for 40% sucrose and glycerol gels respectively. These values are three times lower than the 0% reference FX gel. Note that earlier work with 10% gelatin gels would have predicted negligible reduction in diffusion rates [6]. Also, glycerol has been reported to increase dose response at low concentrations and decrease response at high levels in Fricke solutions [7]. Sucrose follows a similar trend of increasing dose sensitivity at low concentrations [8] and decreasing sensitivity at higher levels in FX gelatin hydrogels, this work.

4. Summary

The addition of either sucrose or glycerol lowered scatter in FX gels. However, minimizing gelatin content was a more significant parameter. Both additives lowered diffusion to a similar degree. However, the post irradiation signal from the sucrose gel samples continued to increase for several hours. Sucrose is optically active and may introduce nuisance, polarization artifacts for optical scanning [9]. Since, glycerol lowered both scatter and diffusion while stiffening the gel it should be considered as useful method to improve optical performance when scanning larger gel volumes and for experiments that require a lower diffusion rate. It is recommended that 4% gelatin and 30% glycerol be used when minimal scatter FX gel is required.

Acknowledgements: Funding was from Ontario Research and Development Challenge Fund: Ontario Consortium for Image-guided Therapy and Surgery and London Regional Cancer Program Small Grants fund. Initial studies of glycerol in FX gel was presented at World Congress on Medical Physics and Biomedical Engineering 2000 and funded by National Cancer Institute of Canada.

References

[1] Jordan K and Battista J, 2006 Small, medium and large optical cone beam CT, 4th International Conference on Radiotherapy Gel Dosimetry(DOSGEL 2006), J of Phys: Conference Series **56**: 214-216

[2] Bosi S G, Brown S, Sarabipour S, De Deene Y and Baldock C 2009 Modelling optical scattering artefacts for varying pathlength in a gel dosimeter phantom Phys Med Biol **54** 275-83

[3] Franklin J and Wang Z Y, 2002 Refractive index matching: A general method for enhancing the optical clarity of a hydrogel matrix, Chem Mater **14**: 4487-89

[4] Jordan K, Coelho L and Chu K, 2000 Reduction of Optical Scatter and Diffusion in Radiochromic Gel Containing Glycerol , proceedings of World Congress on Medical Physics and Biomedical Engineering

[5] Babic S, Battista J and Jordan K, 2009 Radiochromic leuco dye micelle hydrogels: II. Low diffusion rate leuco crystal violet gel", Phys Med Biol **54**:6791-6808

[6] Nixon J R, Georgakopoulos P P and Carless J E, 1967 Diffusion from gelatin-glycerol-water gels, J Pharm Pharmac **19**:246-252

[7] Lamprecht I, Umlauf C and Stein W, 1968 On the influence of various concentrations of glycerol on the yield in an iron(II)-sulfate dosimeter (Fricke dosimeter), Z Naturforsch **23 b**:320-325

[8] Hill B, Back S A J, Lepage M, Simpson J, Healy B and Baldock C, 2002 Investigation and analysis of ferrous sulfate polyvinyl alcohol (PVA) gel dosimeter, Phys Med Biol **47**: 4233-46

[9] Wood M F G, Guo X and Vitkin I A, 2007 Polarized light propagation in multiply scattering media exhibiting both linear birefringence and optical activity: Monte Carlo model and experimental methodology, J of Biomed Optics **12**: 014029

Journal of Physics: Conference Series **250** (2010) 012049 doi:10.1088/1742-6596/250/1/012049

Diode Arrays and QA of Advanced Techniques

Alonso N. Gutiérrez, Ph.D., Oscar Calvo, M.Sc.

Department of Radiation Oncology, Cancer Therapy and Research Center at the University of Texas Health Science Center, San Antonio TX 78229, USA.

gutierrezan@uthscsa.edu

Abstract. Dosimetric verification of delivery for intensity-modulated radiotherapy (IMRT) treatment plans is critical to ensure accurate and safe patient treatments. Commonly, a point dose measurement using a calibrated ion chamber as well as a planar dose measurement using film was traditionally implemented for dosimetric quality assurance (QA) of treatment plans. However, new products have become commercially available in which both an absolute and coarse planar dose measurements can be acquired simultaneously by the use of an array of detectors—either ion chamber- or diode-based. Currently, two devices, MapCHECK and Delta^4, utilize diode technology for planar dose measurements with Delta^4 implementing an orthogonal biplanar arrangement versus the common singular plane. Both devices have been thoroughly clinically characterized with more published experience in the literature available for the MapCHECK due to the novelty of Delta^4. In this review, an overview of basic diode dosimetry and both diode array systems is presented with an emphasis on our research and clinical experience of the Delta^4.

1. Introduction

The delivery verification of patient specific intensity-modulated radiotherapy (IMRT) plans is critical in order to ensure accurate and safe delivery of precise patient treatment plans. While multiple means exist to accomplish such verification, common clinical practice has established a need for an absolute point dose measurement along with a method of measuring a planar dose distribution as standard. Traditionally, a calibrated ion chamber along with radiochromic or radiographic film has been used to validate treatment deliveries. However, new electronic arrays of ion chambers and diode detectors have become commercially available and have facilitated and shortened the time required for delivery verification due to their ease of setup and instantaneous absolute dose readout.

Two array devices currently on the market employ diodes as means of radiation measurement. Diode detectors have a role in radiation dosimetry due to their high sensitivity, high spatial resolution, and quick response time when compared to ionization chambers. Their small detector size (<1 mm) makes them ideal for measuring complex IMRT planar dose distributions with minimal volume averaging effect. Even though diodes exhibit excellent reproducibility and good mechanical stability, diodes do show slight dose-rate, temperature, direction and energy dependence—all of which must be accurately characterized prior to use as an absolute dosimeter.

The basic structure in a silicon diode is a p-n junction. This junction is typically manufactured by lightly doping a pure silicon substrate to establish either an n-type or p-type diode. The diode is termed n-type when the substrate is doped with phosphorous and p-type when the substrate is doped with boron. A surface region is then heavily doped with impurities of the opposite

Journal of Physics: Conference Series **250** (2010) 012049 doi:10.1088/1742-6596/250/1/012049

type to form a p-n junction. By doing so, a region, known as a depletion region, with steep charge gradient and strong electric field is formed at the p-n junction. Because of this strong electric field, charge collection is possible without the need of an external bias. The incident ionizing radiation generates electron-hole pairs throughout the diode. The electrons on the p side and holes on the n side diffuse toward the p-n junction and are swept across the junction by the strong electric field. The total current measured is a sum of the radiation induced current and the leakage current of the diode and is proportional to the dose. [1]

2. Diode Arrays

2.1. MapCHECK

The MapCHECK diode array (SunNuclear Corp. Melbourne FL) contains 445 n-type solid-state diode detectors. The inner 221 detectors cover the central 10.0 x 10.0cm^2 area and are arranged in a zig-zag pattern so that the diagonal spacing between any given two detectors is 0.707cm. The remaining 224 detectors are arranged in a similar pattern with double the diagonal spacing between detectors, namely 1.414cm. A total area of 22.0 x 22.0cm^2 is covered by 445 diodes of 0.8 x 0.8mm^2 active area.

The response of the MapCHECK has been thoroughly characterized and evaluated. Jursinic and Nelms [2] and Létourneau *et al.* [3] examined the linearity and temperature dependence of the MapCHECK detectors and found that the diode response was linear within the range of the radiation dose delivered—up to 2.95Gy. All diodes were calibrated to within 1.0% response of each other with excellent reproducibility—standard deviation of 0.15%. A temperature dependence of 0.57%/°C was noted and was necessary to correct for absolute dosimetric measurement.

A clinical evaluation and comparison of the MapCHECK unit to an ion chamber array performed by Li *et al.* shows that no field size or SSD dependence was observed within the range of the field sizes and SSDs used in the study at both 6 MV and 18MV photon energies. The MapCHECK showed negligible errors (<1%) when measuring doses of more than ~8cGy, but exhibited errors of ~3% when measuring doses on the order of 1cGy. [4] Moreover, the MapCHECK showed a less than 1% discrepancy and more than 90% of the detectors within 0.5% response variation. Excellent gamma passing rates were obtained when compared with the planar dose distributions from the treatment planning system for IMRT fields.

2.2. Delta^4

The Delta^4 system (ScandiDos AB, Uppsala, Sweden) consists of 1069 p-type silicone diodes arranged on two orthogonal arrays placed in a 22.0cm diameter cylindrical polymethyl methacrylate (PMMA) phantom. The arrays lie orthogonal with respect to each other however are rotated with respect to the treatment couch—essentially forming an "X." The diode active area is 1.0mm in diameter and the diodes are placed on a varying square grid with 0.5cm spacing for the central 6.0 x 6.0cm^2 region and 1.0cm spacing in the remaining 20.0 x 20.0cm^2 measurement area.

The Delta^4 system is a bit more advanced than the MapCHECK in terms of radiation acquisition, for it measures the dose per pulse as oppose to a continuous sampling. This is accomplished by synchronizing the sampling of the Delta^4 to the linear accelerator through a coaxial cable attached to the synch output. Doing so, the measurements in the electrometer are only acquired shortly before, during, and after the radiation pulse providing excellent dose resolution. Additionally, corrections are applied to the raw diode readings to account for rotational sensitivity, depth, and field size dependence since the beam parameters are known from the treatment planning system.

In a recent study, Feygelman *et al.* evaluated the Delta^4 and its performance in conjunction with the helical tomotherapy system for patient specific delivery QA. [5] It was shown that the short-term reproducibility of the Delta^4 was 0.1%, dose linearity was <0.1%, dose-rate dependence was 0.4% and absolute calibration accuracy was 0.4%. Additionally, the Delta^4 response to rotational

Journal of Physics: Conference Series **250** (2010) 012049 doi:10.1088/1742-6596/250/1/012049

irradiation changed by no more than 0.2%. Their clinical validation of nine IMRT plans showed a mean γ(3%/3mm) passing rate of 97.0 ± 2.7%.

2.2.1. 3D Dose Model Validation

The Delta^4 system uses the calculated dose distribution from the treatment planning system to generate a semi-measured 3D dose volume. The 3D semi-measured dose calculation methodology utilizes the response detected for a given radiation ray and scales the calculated dose along the path based on the response of the two diode detectors. Due to the detector geometry of the phantom, any given ray will intersect both detector planes.

In order to validate the semi-measured 3D dose calculation algorithm, measurements were performed using a TomoTherapy™ HiArt™ and a calibrated pinpoint PTW N31006 chamber with a sensitive volume of 0.016cc for point dose measurements. A section of the Delta^4 PMMA slab was modified to accept the ion chamber. Eight (n=8) quality assurance plans were created and scaled to span a range of doses from 20% to 100% (20% steps) and 200% of the prescribed dose from the tomotherapy TPS. At each delivery, a point dose measurement was recorded. Point dose measurements and semi-measured point doses were compared against the calculated point doses of tomotherapy.

Good agreement was noted among all measurements. Tomotherapy delivery results showed a mean point dose difference between the tomotherapy TPS and ion chamber of 1.3% and a mean point dose difference between the tomotherapy TPS and semi-measured dose of 2.1%. The Delta^4 semi-measured point dose showed good agreement (<1.0%) with the pinpoint chamber for the majority of measurements.

2.2.2. SBRT TomoTherapy Validation

Implementation of the Delta^4 for delivery verification of coarsely hypofractionated treatments, particularly stereotactic body radiotherapy (SBRT), using helical tomotherapy overcomes the current necessity of scaling the delivery quality assurance (DQA) procedure in order to ensure that the maximum dose delivered is within the dynamic range of the film being utilized. By using the Delta^4, scaling errors, possibly due to MLC leaf latency, would be minimized or eliminated, and the un-scaled treatment plan would be verified.

Five patients (n=5) with single or multiple small lesions in either the lung or liver were treated with SBRT techniques using the tomotherapy unit. Their five different DQA plans were delivered—3 liver patients and 2 lung patients—over the course of 4 months. Treatment plans delivered doses per fraction ranging from 10.0-20.0Gy. Corresponding DQA plans were generated and delivered on the Delta^4.

Excellent agreement was found to exist for all patients between measured and calculated dose distributions. The mean γ(3%/3mm) for all plans was 99.3% [98.7-100%]. Our preliminary results indicate that the Delta^4 is an effective and efficient method for delivery quality assurance of high dose per fraction treatment plans.

2.2.3. *UTHSCSA Delta^{4}Experience*

The Delta^4 system has been commissioned for delivery verification of all tomotherapy treatment plans delivered at the University of Texas Health Science Center San Antonio since January of 2009. During this time period, over 200 patient delivery validations have been performed. Pass criteria for the dose validations were defined as a γ index of 90% or greater. Gamma parameter tolerances were set to 3% or 3mm for detectors with measured doses between 20 to 500% of the reference dose. The reference dose was defined as the maximum measured dose. Based on all DQA measurements, a mean γ(3%/3mm) value of 96.5 ± 2.8% was computed with values ranging from 86.5 to 100%.

3. Conclusion

The availability of diode arrays for patient specific quality assurance has provided another established method of accurately validating precise, measured dose distributions. The ability of current diode technology to provide stable and reproducible results as shown in multiple published literature validates the future role of diode arrays as a robust dosimeter. Current diode array systems such as the MapCHECK and Delta^4 eliminate the need of film and ion chamber-based dose validation and further shorten the time associated with dose validation due to their ability of instantaneously reading, processing and analyzing the data.

References

[1] Zhu TC, Saini AS. Diode Dosimetry for Megavoltage Electron and Photon Beams. Clinical Dosimetry Measurements in Radiotherapy. Med Phys Publishing. 2009.

[2] Jursinic PA, Nelms BA. A 2-D diode array and analysis software for verification of intensity modulated radiation therapy delivery. *Med Phys*. 2003;70(2);199.206.

[3] Letourneau D, Gulam M Yan D, et al. Evaluation of a 2D diode array for IMRT quality assurance. *Radiother Oncol*. 2004;70(2);199-206.

[4] Li CG, Yan G, Liu C. Comparison of two commerical detector arrays for IMRT quality assurance. J Appl Clin Med Phys. 2009:10(2);62-74.

[5] Feygelman V, Opp D, Javedan K, et al. Evaluation of a 3D Diode Array Dosimeter for Helical TomoTherapy Delivery QA. *Med Dos*. 2010. (In Press)

[6] Low D, Harms WB, Mutic S, et. al. A technique for the quantitative evaluation of dose distributions. *Med Phys*. 1998;25(5):656-61.

Patient Specific Quality Assurance: Transition from IMRT to IMAT

Jennifer O'Daniel, Ph.D., Shiva Das, Ph.D., Jackie Wu, Ph.D., Fang-Fang Yin, Ph.D.

Department of Radiation Oncology, Duke University Medical Center, Durham, NC

jennifer.odaniel@duke.edu

Abstract. The purpose of this study was to test a patient-specific quality assurance (QA) protocol for intensity-modulated arc radiotherapy (IMAT), and to evaluate the use of an intensity-modulated stationary radiotherapy QA device (2D ion chamber array). Thirty-nine IMAT treatment plans for brain, spine, and prostate were analyzed using 3 methods: ion chamber (1D absolute, n=39), film (2D relative, coronal/sagittal, n=8), and 2D ion chamber array ("ICA," 2D absolute, coronal/sagittal, n=39) measurements. All measurements were compared to the treatment planning system (TPS) dose calculation with gamma analysis (3%, 3mm distance-to-agreement criteria) or absolute point dose comparison. The ICA measurements were also directly compared to film and ion chamber for validation. Absolute 1D measurements agreed well calculation (ion chamber: average deviation 1.4%, range -0.9% to 2.8%; ICA: average deviation 0.7%, range -1.8% to 2.9%). Relative 2D measurements also showed good agreement with calculation (>93% of pixels in all films passing gamma, >90% of pixels in all ICA measurements passing gamma). ICA and film relative dose results were highly similar (> 90% of pixels passing gamma in 94% of QAs). Coronal and sagittal ICA measurements were statistically indistinguishable by the paired t-test with a hypothesized mean difference of 0.2%. Ion chamber and ICA absolute dose measurements usually agreed well, but had disparities of 2-3% in 18% of plans. After validating the new IMAT implementation with ion chamber, film, and ICA, we reduced our QA from 5 (ion chamber, film, and ICA) to 2 measurements (ion chamber and single ICA) per plan. The ICA (Matrixx®, IBA Dosimetry) was validated in relative analysis mode, but ion chamber measurements are recommended for absolute dose comparison.

1. Introduction

With the recent clinical introduction of intensity-modulated arc therapy (IMAT), new patient-specific quality assurance (QA) procedures must be developed. Techniques can be adapted from patient-specific QA of static intensity-modulated radiotherapy (IMRT). However, the additional uncertainties due to the new delivery method, specifically varying gantry angle and dose rate, must be investigated. Film and single ion chambers are the "gold standard" for QA, and can be used to both validate IMAT implementation and the use of new QA measurement tools, such as a 2D ion chamber array ((Matrixx®, IBA Dosimetry, Bartlett, TN). Previous work has reported on novel 3D dosimetry techniques for the QA of IMAT treatment plans [1], on initial QA results for a small set of IMAT treatment plans using one measurement technique [2], on Monte Carlo simulations of IMAT treatment plans [3], and on an overview of implementing RapidArc at the University of Alabama at Birmingham [4]. However, a comprehensive patient-specific QA protocol based on results from a large number of

Journal of Physics: Conference Series **250** (2010) 012050 doi:10.1088/1742-6596/250/1/012050

patients, multiple treatment sites, and using multiple measurement devices for cross-validation has not yet been presented. Additionally, while the 2D ion chamber array has been validated for IMRT QA [5-7], it has not been validated for IMAT QA with a large study. In this work we have explored an effective and efficient end-to-end QA protocol for IMAT and evaluated the suitability of an IMRT QA device (2D ion chamber array) for IMAT QA.

2. Materials and Methods

We extensively analyzed 39 IMAT treatment plan QAs involving multiple treatment sites (brain, spine, and prostate), and QA devices (cylindrical ion chamber (n=39), film (n=8), and ion chamber array ("ICA", n=39)). All measurements were compared to the Eclipse v8.5 treatment planning system (TPS) dose calculations on hybrid QA plans (patient plan calculated on a CT image set of the QA device). Plans were delivered using RapidArc (Varian Medical Systems, Palo Alto, CA). For 2D analysis, gamma analysis was used with 3%, 3mm distance-to-agreement criteria and 5% of maximum dose threshold. Absolute 1D dose comparisons compared (1) the dose measured by the 4 central ion chambers in the ICA to the TPS dose calculated at those points and (2) the dose measured by the single ion chamber to TPS calculation. Measurements "passed" if the 1D dose agreed within 3% and the 2D comparison had > 90% of pixels passing the gamma analysis. The ion chamber and film results were also directly compared to the ICA measurements.

3. Results

Cylindrical ion chamber, film and ICA measurements each agreed well with calculation, indicating that the IMAT plans wer deliverable within the mechanical and dose constraints of the linear accelerator (Table 1, Fig. 1, Fig. 2). Gamma analysis between film and ICA showed excellent agreement (Table 1). While in general ICA passing rates were high, the ICA measurement produced a lower passing rate than expected in the presence of high dose gradients (Fig. 3). Ion chamber and ICA absolute dose measurements showed a similar trend, but had disparities of 2-3% in 18% of plans (Fig. 4). Coronal and sagittal ICA measurements were statistically indistinguishable (Fig. 5). Through cross-validation of our QA devices, we were able to reduce from the initial 3.6hr QA procedure (3 film + 2 ICA + 1 ion chamber) to a 1.6hr reduced QA procedure (2 ICA + 1 ion chamber), and finally to our current streamlined 1.2hr QA procedure (1 ICA + 1 ion chamber).

Table 1: Film vs. Eclipse and ICA; %Pixels Passing Gamma

		Film vs. Eclipse		Film vs. ICA	
	Plan #	Coronal	Sagittal	Coronal	Sagittal
Brain	1	98.8	99.0	93.1	98.4
	2	99.5	97.2	97.2	89.0
	3	99.4	97.4	96.3	97.3
Prostate bed	14	97.2	98.9	94.6	97.0
	15	99.1	99.6	95.3	96.8
Prostate	23	97.8	99.6	97.8	99.1
	24	97.6	99.5	99.5	99.4
Prostate + SV	32	93.1	98.6	93.9	99.0

Journal of Physics: Conference Series **250** (2010) 012050 doi:10.1088/1742-6596/250/1/012050

Fig. 1: Absolute 1D Dose Comparison: Measurement vs. Calculation

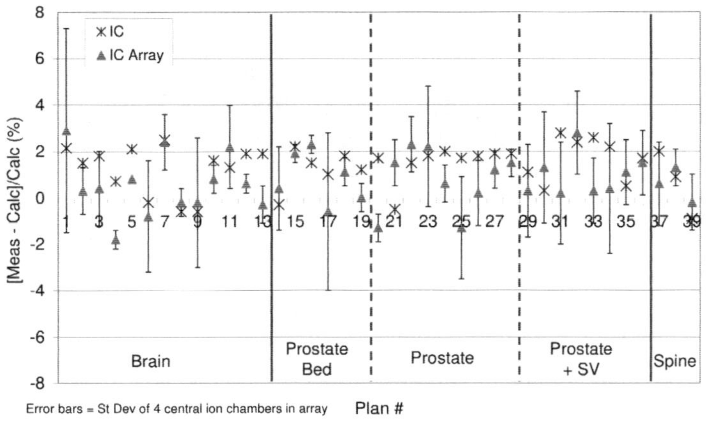

Fig. 2: Absolute 2D Dose Comparison:
%Pixels Passing 3%, 3mm Gamma Analysis

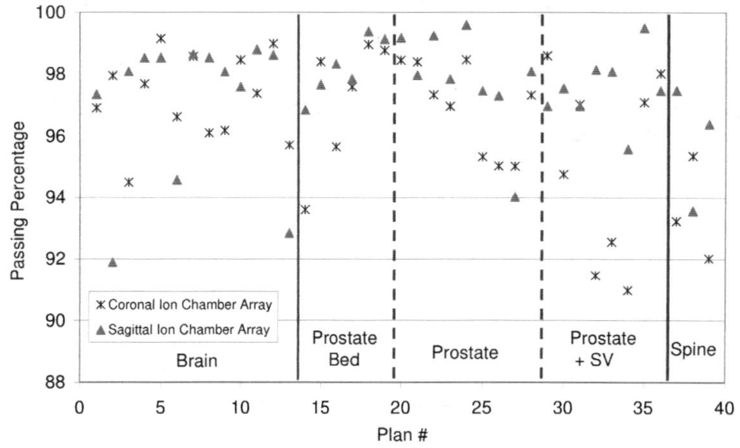

Fig. 3: ICA Dose Overestimation in the Presence of a Perpendicular High Dose Gradient

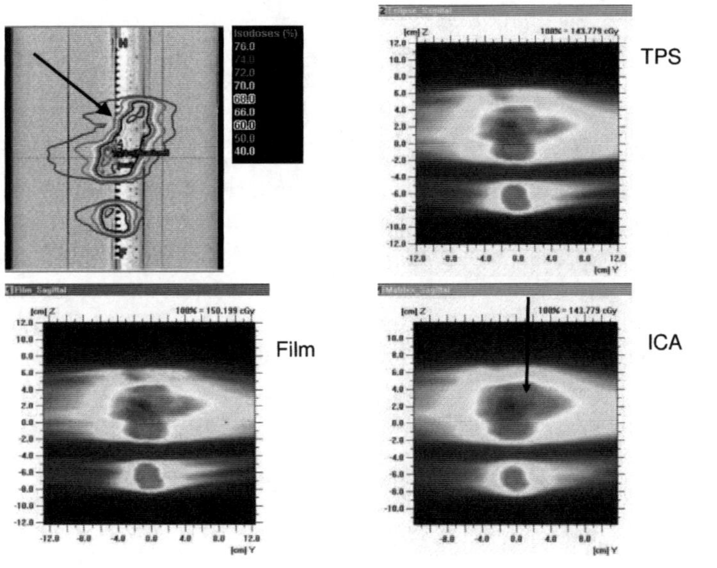

Fig. 4: Ion Chamber vs. Ion Chamber Array
[Measured - Calculation]/Calculation (%)

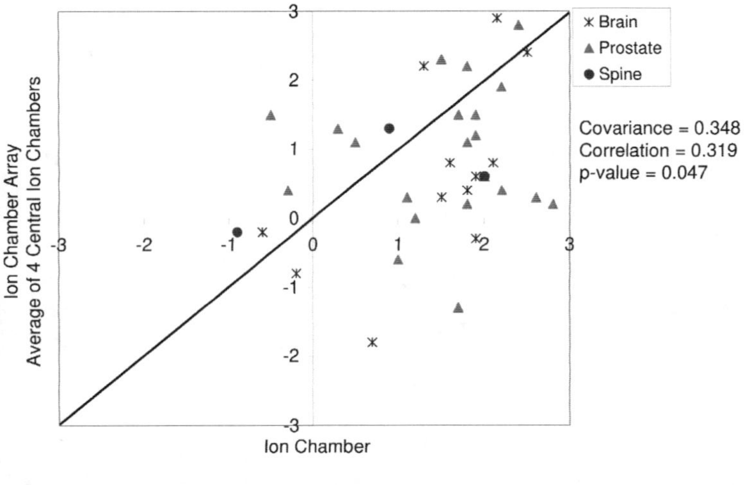

Fig. 5: Ion Chamber Array - Coronal vs. Sagittal
% Pixels Passing Gamma

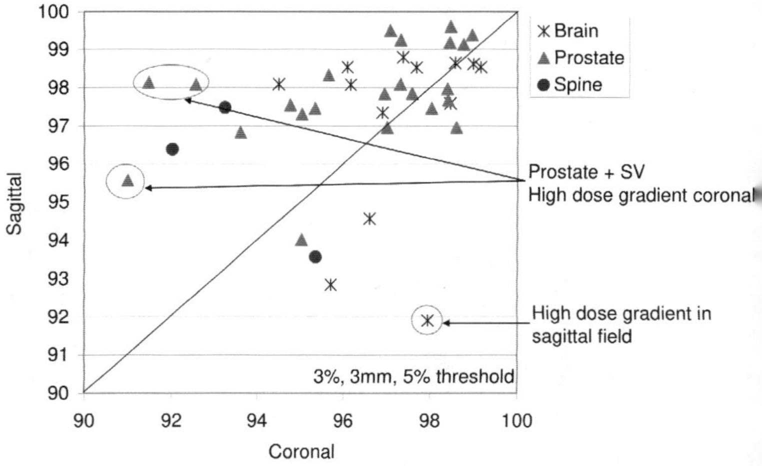

4. Conclusions

By testing the IMAT implementation for 39 patient plans with ion chamber, film, and ion chamber array, we cross-validated our QA devices and were able to maintain an effective QA protocol while reducing the amount of measurements required. Ion chamber, film, and ion chamber array results all agreed well with the calculated dose, though the ion chamber array results could be unreliable when perpendicular high-dose gradients could not be avoided. Film and ion chamber array results agreed well. Cylindrical ion chamber and ion chamber array measurements were reasonably similar, but discrepancies of up to 3% were observed. Therefore we continue to rely on the cylindrical ion chamber for absolute dose comparisons. We have subsequently reduced our IMAT QA process from 5 measurements per patient (cylindrical ion chamber, coronal ion chamber array, sagittal ion chamber array, coronal film, sagittal film) to 2 measurements per patient (cylindrical ion chamber and single-plane ion chamber array).

References

[1] Schreibmann E, Dhabaan A, Elder E, *et al.* Patient-specific quality assurance for VMAT treatment delivery. Med Phys 2009; 36: 4530-5.

[2] Korreman S, Medin J, Kjaer-Kristoffersen F. Dosimetric verification of RapidArc treatment delivery. Acta Oncol 2009; 48: 185-91.

[3] Bush K, Townson R, Zavgorodni S. Monte Carlo simulation of RapidArc radiotherapy delivery. Phys Med Biol 2008; 53: N359-70.

[4] Popple R, Fiveash J, Brezovich I, et al. Rapidarc radiation therapy: first year experience at the university of Alabama at Birmingham. Int J Radiat Oncol Biol Phys. Epub 6 March 2010.

[5] Stasi M, Giordanengo S, Cirio R, *et al.* D-IMRT verification with a 2D pixel ionization chamber: dosimetric and clinical results in head and neck cancer. Phys Med Biol 2005; 50: 4681-94.

[6] Herzen J, Todorovic M, Cremers F, *et al.* Dosimetric evaluation of a 2D pixel ionization chamber for implementation in clinical routine. Phys Med Biol 2007; 52: 1197-1208.

[7] Dobler B, Streck N, Klein E, et al. Hybrid plan verification for intensity-modulated radiation therapy (IMRT) using the 2D ionization chamber array I'mRT MatriXX-a feasibility study. Phys Med Biol 2010; 55: N39-55.

Journal of Physics: Conference Series **250** (2010) 012051 doi:10.1088/1742-6596/250/1/012051

Tumor-tracking radiotherapy of moving targets; verification using 3D polymer gel, 2D ion-chamber array and biplanar diode array

Sofie Ceberg[1], Marianne Falk[2], Per Munck af Rosenschöld[2], Herbert Cattell[7], Helen Gustafsson[1,3], Paul Keall[6], Stine S Korreman[2,4,5], Joakim Medin[2], Fredrik Nordström[1], Gitte Persson[2], Amit Sawant[6], Michelle Svatos[7], Jens Zimmerman[2] and Sven ÅJ Bäck[1]

[1]Department of Medical Radiation Physics, Lund University, Skåne University Hospital, Malmö, Sweden
[2]Department of Radiation Oncology, Rigshospitalet, Copenhagen, Denmark
[3]Department of Medical Physics, ACT Health, Canberra, Australia
[4]Niels Bohr Institute, Copenhagen University, Denmark
[5]Department of Human Oncology, University of Wisconsin-Madison, United States
[6]Department of Radiation Oncology, Stanford University, Stanford, California, USA
[7]Varian Medical Systems, Inc., Palo Alto, California, USA

sofie.ceberg@med.lu.se

Abstract. The aim of this study was to carry out a dosimetric verification of a dynamic multileaf collimator (DMLC)-based tumor-tracking delivery during respiratory-like motion. The advantage of tumor-tracking radiation delivery is the ability to allow a tighter margin around the target by continuously following and adapting the dose delivery to its motion. However, there are geometric and dosimetric uncertainties associated with beam delivery system constraints and output variations, and several investigations have to be accomplished before a clinical integration of this tracking technique. Two types of delivery were investigated in this study I) a single beam perpendicular to a target with a one dimensional motion parallel to the MLC moving direction, and II) an intensity modulated arc delivery (RapidArc®) with a target motion diagonal to the MLC moving direction. The feasibility study (I) was made using an 2D ionisation chamber array and a true 3D polymer gel. The arc delivery (II) was verified using polymer gel and a biplanar diode array. Good agreement in absorbed dose was found between delivery to a static target and to a moving target with DMLC tracking using all three detector systems. However, due to the limited spatial resolution of the 2D array a detailed comparison was not possible. The RapidArc® plan delivery was successfully verified using the biplanar diode array and true 3D polymer gel, and both detector systems could verify that the DMLC-based tumor-tracking delivery system has a very good ability to account for respiratory target motion.

1. Introduction

The desire to increase the conformity of the dose distribution in external radiotherapy has resulted in advanced treatment procedures, for instance techniques using intensity modulated beams and arcs [1-3]. The advantages of these techniques are the increased possibility to deliver a high absorbed dose to the target volume while minimizing the dose to normal tissues. However, intra-fractional tumor motion, mostly due to respiration, can be a major challenge to the ambition to deliver the desired dose distributions. An illustration of how the high dose volumes shrink due to motion is presented in figure 1, where polymer gel measurements are compared to the treatment planning system (TPS) for moving targets treated using arc therapy. Large respiratory motion is generally accounted for by increased margins, which implies an increased risk of morbidity from late toxicity. It is therefore important to find better ways to take respiratory motion into account when treating targets in the thorax region.

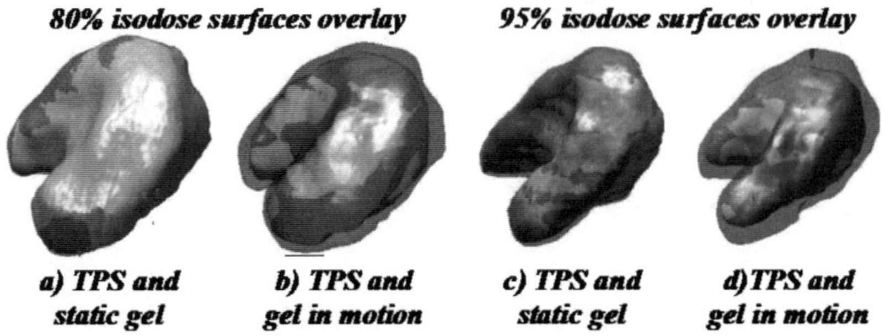

Figure 1. Result from a respiratory-motion investigation during an intensity modulated volumetric arc delivery. Each image shows an overlay of two isodose surfaces. The red is generated in the treatment planning system and the green is polymer gel measurements. The 3D images in b) and d) clarify that delivered high dose volumes shrink due to motion.

A novel promising motion-compensation strategy uses the dynamic multileaf collimator (DMLC) to continuously align and reshape the treatment machine apertures so as to following the target motion in real time [4, 5]. The advantage of this so called real-time DMLC tumor-tracking radiation delivery is the ability to allow a tighter margin around the target by continuously following, and adapting the dose delivery to its motion. This kind of treatment delivery represents a new level of complexity, and a thorough dosimetric verification is therefore highly desirable. Recently published data find that DMLC tracking together with the intensity modulated radiotherapy technique RapidArc®, is capable of improving the dose distribution delivered to a moving target [6] and improves the accuracy of RapidArc® delivery [4]. In both studies the biplanar diode array Delta^4® (ScandiDos, Inc., Sweden) was used for dose verification, and further studies using a true 3D detector would be of great interest. Using high resolution gel dosimetry [7], the absorbed dose can be obtained in the entire irradiated volume and the response is independent of the direction of the incident radiation. Previous work demonstrate the use of polymer gel dosimetry to verify dynamic radiotherapy, both under respiratory-like motion [8], as well as in intensity modulated arc therapy delivery [9]. This makes polymer gels potentially useful for tracking measurements during both conventional and arc delivery.

The aim of this study was to carry out a dosimetric verification of a DMLC-based tracking delivery using true 3D polymer gel dosimetry, a 2D ionization chamber array, and a biplanar diode array. Results are presented from a feasibility study using a single beam perpendicular to a target with a one dimensional motion parallel to MLC motion direction, and from a verification of a target with a diagonal motion to the leaf trajectory during an intensity modulated arc delivery.

2. Material and methods

2.1. The detector systems

The 2D ionization chamber array *seven29*® (PTW, Freiburg, Germany), with 729 measurement points consists of 27 x 27 cubic chambers with a center-to-center distance of 10 mm. The chamber size is 0.125 cm^3 and the detector housing material is PMMA. The *seven29*® was used for measurements during the feasibility study. The biplanar diode array Delta^4® (ScandiDos, Inc., Sweden) is a cylindrical PMMA phantom with two orthogonal detector boards. The dedicated software can interpolate measured data from the two planes to obtain a 3D matrix. The center-to-center distance is 5 mm in a central 6x6 cm^2 region of the detector and 10 mm in the rest of the 20 x 20 cm^2 measurement area in each of the planes. The Delta^4® was used for measurements during the tumor-tracking arc delivery. The 3D normoxic polyacrylamide gel (nPAG) detector system was based on 3% w/w acrylamide and 3% w/w *N,N'*-methylenebisacrylamide. Gelatine was used as the matrix substance and tetrakis(hydroxymethyl)-phosphonium chloride as an oxygen scavenger. The remaining constituent was ultra pure deionized water. The chemical information and the mixing procedure is published elsewhere [8]. For each batch, an identical un-irradiated gel phantom was used to acquire a background value and gel vials irradiated to known doses were used to assure the linearity of the gel dose response.

2.2. The tumor-tracking radiotherapy delivery

2.2.1. The feasibility study.

The real-time DMLC motion-tracking system was used together with an optical real-time positioning system (RPM™, Varian medical systems). A respiratory-like target motion was simulated mechanically. The movement was perpendicular to the beam direction and parallel to the leaf motion, with a 2 cm peak-to-peak motion extent and a 5 second period. A small circular, r =15 mm, 6 MV beam with a target dose of 2 Gy and a constant dose rate of 300 MU/min was delivered to three identical 250 ml nPAG flat-sided phantoms and the seven29®. A 20 mm thick PMMA build up was placed on top of the seven29® and 5 mm thick PMMA back scatter was placed below. The delivery was carried out in the following modes: 1) detector in motion and the tracking system disconnected (i.e. motion mode) 2) detector at rest and the tracking system disconnected and (i.e. static mode) and 3) detector in motion and the tracking system connected (i.e. tracked mode).

2.2.2. The arc delivery with diagonal target motion.

The DMLC real-time motion-tracking system was used together with the optical part of the ExacTrac system (BrainLab, Germany). The respiratory-like target motion was simulated mechanically as in 2.2.1, with 4 s period. A RapidArc® 6 MV lung plan, with a 45 degree collimator rotation and a 358-degree arc rotation was delivered using 790 MU. The plan was delivered with a varied dose rate of up to 600 MU/min to three identical 500 ml circular nPAG phantoms and to the Delta^4® system. The prescribed dose to the 4.86 cm^3 target volume was 4 Gy and the delivery was carried out in the same 3 modes as presented in 2.2.1 and in a additionally static mode with the tracking system connected. To avoid any possible contributions from discrepancy between measurements and TPS when evaluating the DMLC tracking performance, the measurements during motion were evaluated to the static measurement for the respective detector system. Further, a more strict gamma criterion of 2%/1mm was used within the volumes enclosed by the 50% and 90% iso-dose surface.

2.3. Gel-readout and analysis

Magnetic resonance imaging (MRI) of the gel was used to evaluate the absorbed dose response. The images were acquired using a 1.5 T MRI unit (Siemens Symphony, Siemens Medical Systems) and a 32-echo multi spin echo sequence with an inter-echo spacing of 25 ms. In-house developed software was used for T2 calculations [10], and MATLAB 7.4.0 was used for image processing and 3D rendering. The R2 data of the irradiated gel phantoms was converted to relative absorbed dose using background subtraction and normalization in a region of homogenous dose. An in-house developed 3D gamma evaluation program was used to compare all the gel measurement. Both the *seven29*® and Delta^4® detector system uses its own analysis software. However the 3D gamma evaluation method in Delta^4® is based on the same theory as the one developed in-house [11].

In the feasibility study the raw unfiltered gel data are presented and for the arc delivery study, the raw data was smoothed with a 3x3x3 box-filter and resampled from 1x1x3 mm^3 to 1x1x1 mm^3 voxel size to enable 3D gamma evaluation.

3. Result and discussion

3.1. The feasibility study

As expected, the motion introduced a significant dose-blurring when the tracking system was disconnected (figure 2a). The sharp MLC defined field edge was clearly visible in the static measurement (figure 2b). When tracking was applied, the dose-blurring due to motion was well compensated for (figure 2c). The distance between the 50% absorbed dose contour of the static and tracked gel-measurements was less than 1 mm for 96% of the points along the contour (max 1.4 mm). The absorbed dose in planes through the depth of dose maximum in the static and tracked gels was also compared. A 91% pass rate was reached for a gamma criterion of 3%/3mm for a region where the absorbed dose was greater than 80%. For doses >20% the pass rate was 82% where most of the dose points failing the gamma criterion occurred in the penumbra region, possibly due to motion induced dose-smearing effects, which were not adequately compensated. Overall, the *seven29®* measurements agreed well with the gel measurements (figure 3). However, due to the limited spatial resolution of the 2D array a detailed comparison was not possible. The feasibility study showed that high resolution 3D polymer gel dosimetry has a potential to perform a dosimetric verification of a tumor-tracking delivery.

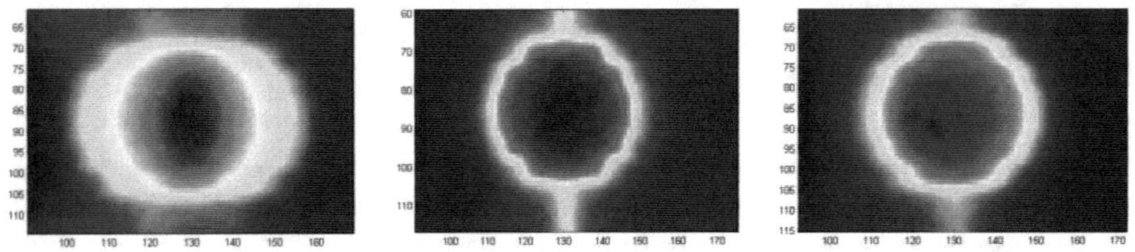

Figure 2a) Gel in motion and the tracking system disconnected. ***b) Gel at rest and the tracking system disconnected*** ***c) Gel in motion and the tracking system connected***

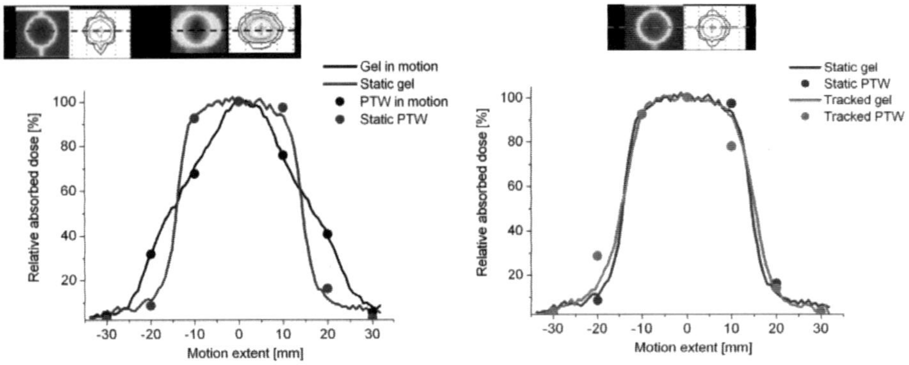

Figure 3

To the left, profiles through the gel and seven29® from the static and non-tracked measurements.

To the right, profiles through the gel and seven29® from the static (same blue profiles as to the left) and tracked measurements.

3.2. The arc delivery with diagonal target motion

Good agreement was found when comparing TPS and the static gel measurements, as well as TPS and the static Delta^4® measurements. The 3D gamma analysis was carried out using the criteria 3%/3mm and pass rates above 97% were obtained for both detector systems when comparing to TPS. The 2%/1mm gamma evaluation between measured volumes in motion without MLC tracking vs. static measurements, within volumes enclosed by the 50% and 90% iso-dose surface, resulted in pass rates between 25%-37% (table 1) which suggests a significant motion induced dose-blurring when the tracking system was disconnected. The gamma

Table 1. 3D gamma evaluation results using criterion 2%/1mm

Detector system	Delta^4®	Gel
Measurement *vs.* reference	pass rate [%] for $\gamma<1$	
Within the 90% iso-dose surface		
Tracked *vs.* static target	70.6	96.8
Moving *vs.* static target	35.3	37.1
Within the 50% iso-dose surface		
Tracked *vs.* static target	87.2	82.1
Moving *vs.* static target	36.2	25.2

evaluation comparing tracked and static measurements resulted in pass rates between 70%-97%. The highest pass rate were obtained with the gel measurements in the volume enclosed by the 90% isodose level. A relatively low pass rate, 70%, was obtained with Delta⁴® in the same volume, and further studies will be carried out to investigate the reason to this result. Nevertheless, overall the motion was well compensated for when the tracking system was connected (Figure 4 and 5).

4. Conclusion

The ability of a DMLC-tracking system to account for target motion during radiotherapy was investigated by carrying out measurements using polymer gel, *seven29*® and Delta⁴®. Both a feasibility experiment and an intensity modulated volumetric arc delivery were investigated. Good agreement in absorbed dose was found between the static and tracked measurements.

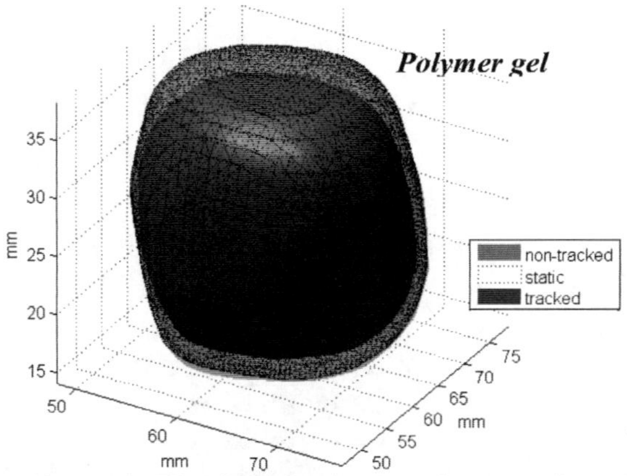

Figure 4. The 90% isodose surface overlay of the three gel measurements. The wire framed volume represents the static gel and the transparent blue represents the tracked gel. The red non-tracked volume visualizes the reduction of the 90% isodose volume.

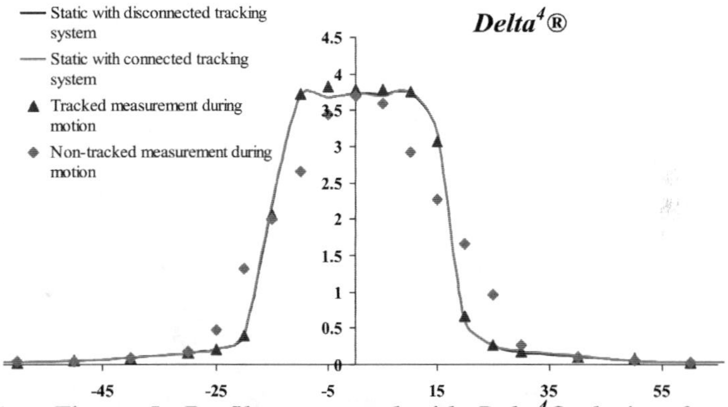

Figure 5. Profiles measured with Delta⁴® during four conditions; the detector at rest and the tracking system disconnected, the detector at rest and the tracking system connected, the detector in motion and the tracking system connected and the detector in motion and the tracking system disconnected.

References

[1] T. R. Mackie et al., Tomotherapy: a new concept for the delivery of dynamic conformal radiotherapy. Med Phys 20 (1993) 1709.
[2] K. Otto. Volumetric modulated arc therapy: IMRT in a single gantry arc. Med Phys 35 (2008) 310.
[3] C. X. Yu. Intensity-modulated arc therapy with dynamic multileaf collimation: an alternative to tomotherapy. Phys Med Biol 40 (1995) 1435.
[4] M. Falk et al. Real-time dynamic MLC tracking for inversely optimized arc radiotherapy. Radiother Oncol 94 (2010) 218.
[5] R. McMahon et al. A real-time dynamic-MLC control algorithm for delivering IMRT to targets undergoing 2D rigid motion in the beam's eye view. Med Phys 35 (2008) 3875.
[6] J. Zimmerman et al. DMLC motion tracking of moving targets for intensity modulated arc therapy treatment: a feasibility study. Acta Oncol 48 (2009) 245.
[7] C. Baldock et al. TOPICAL REVIEW Polymer gel dosimetry. Phys Med Biol 55 (2010) R1.
[8] S. Ceberg et al. Verification of dynamic radiotherapy: the potential for 3D dosimetry under respiratory-like motion using polymer gel. Phys Med Biol 53 (2008) N387.
[9] S. Ceberg et al. RapidArc treatment verification using polymer gel dosimetry. Journal of Physics: Conf Ser 164 (2009) 012052.
[10] A. Karlsson et al. Image Processing Software for 3D Dose Evaluation in Gel Dosimetry. World Congress on Medical Physics and Biomedical Engineering, 2003.
[11] D. A. Low et al. A technique for the quantitative evaluation of dose distributions. Med Phys 25 (1998) 656.

Journal of Physics: Conference Series **250** (2010) 012052 doi:10.1088/1742-6596/250/1/012052

3D Analysis of Intensity-Modulated Radiation Therapy Quality Assurance Measurement using a 2D Diode Array

M. Lawrence[1], T. Cullip[1], L. Potter[1], J. Lian[1], S. Wang[1], Z. Xu[1], K. Burkhardt[1], S. Chang[1]

Department of Radiation Oncology, University of North Carolina, Chapel Hill, NC, United States of America

E-mail: Michael_Lawrence@med.unc.edu

Abstract. Intensity-modulated radiation therapy (IMRT) quality assurance (QA) is often performed using a 2D device and compares measured and computed fluence maps to determine if a field passes or fails certain dose and position criteria. The effects of a measured deviation to the 3D patient spatial dosimetry and dose volume histogram (DVH) are largely unknown because they cannot be analyzed using commercial 2D array IMRT QA systems. We report an in-house treatment planning system (TPS) PLanUNC based 3D IMRT QA analysis approach that has been used in our institution for the past ten years when 2D fluence map IMRT QA failed. In this approach the measured 2D fluence maps are imported back to PLanUNC and used to re-compute 3D patient dosimetry including DVHs. The 2D fluence map IMRT QA criteria is that the measured dose for 95% of the detectors is within 5% of the planned dose, and that the distance-to-agreement be within 4mm (5%/4mm). 22 IMRT plans that had at least one field fail initial QA using MapCHECK 2 are examined using our 3D QA approach. The DVH analysis shows that 19/22 plans that failed initial QA were within 2% of the planned target and critical structure DVHs. 3/22 IMRT plans were found to have DVH difference greater than 2%. The 3D analysis of 2D IMRT QA result shows that when a fluence map QA fails for a single field, often it is clinically insignificant in terms of patient 3D dosimetry

1. Introduction

Pre-treatment IMRT QA measurement is necessary to insure patient safety and quality delivery of a radiotherapy treatment plan. Multiple commercial IMRT QA devices exist, and for step-and-shoot IMRT a 2D array of detectors is commonly used to measure radiation dose maps for all fields while keeping the gantry at 0 degrees [1, 2, 3, 4]. The acquired radiation dose maps are then compared to the dose maps created by the treatment planning software. The commercial IMRT QA product reveals if a field fails or passes a set comparison dose and distance-to-agreement criteria [5]. Currently, the ability to perform 3D dosimetric analysis and comparison is lacking, however commercial vendors are working to import 3D analysis [6]. Further, if an IMRT field fails, there are limited tools to determine the cause of the failure and to determine the effect of the failed field on the overall delivered dose. In this work, we present a method for in depth IMRT QA analysis that incorporates in-house software into our treatment planning system to analyze possible causes of QA failures and the 3D dosimetric result of 2D measured radiation dose maps.

Journal of Physics: Conference Series **250** (2010) 012052 doi:10.1088/1742-6596/250/1/012052

2. Methods and Materials

IMRT QA

Currently we use the MapCHECK 2 for QA of step-and-shoot IMRT QA plans. The MapCHECK 2 has 1527 diode detectors spaced 7mm apart across a 26cm X 32cm 2D detector array. IMRT field dose maps are created by the treatment planning system and imported into the MapCHECK 2 analysis software. The MapCHECK 2 is placed on the treatment table and all IMRT fields are measured with the gantry at 0 degrees. The measured and planned radiation dose maps are compared, as shown in figure 1. Field passing criteria is set to 95% of detector points must be within 5% of the planned dose and have a 4mm distance-to-agreement, (5%/4mm). When an IMRT field fails QA, all of the patient's measured IMRT intensity maps are imported to our treatment planning software PLanUNC to analyze the impact of the measured deviation in fluence maps on 3D patient dosimetry.

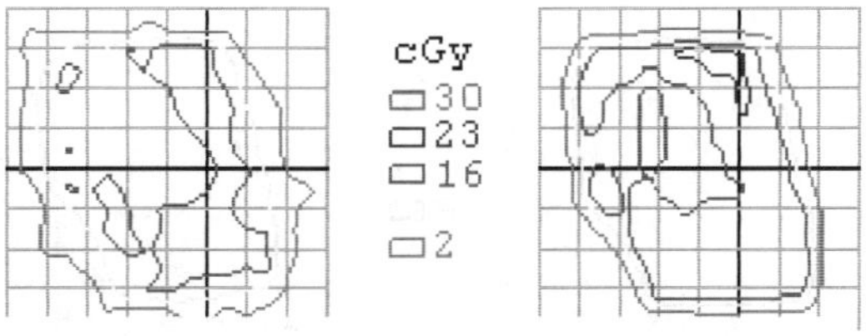

Figure1: MapCHECK2 comparison of planned (right) and measured (left) IMRT dose maps.

3D dose analysis of 2D IMRT QA Failure

When an IMRT field fails the 2D IMRT QA, the measured 2D dose map and the diode detector positions are exported to PLanUNC for 3D dose analysis. The MUs and/or the MLC leaf position of the initial treatment plan segments are adjusted in PLanUNC so that the resulting fluence map matches the IMRT QA measurement. This process creates a treatment place based on the IMRT QA. Then 3D dosimetry is compared between the original treatment plan and the IMRT QA based plan, as shown in figure 2. The IMRT plan is generally approved by the patient's physician for clinical use if the deviation in corresponding target and critical structure DVHs are within 2%.

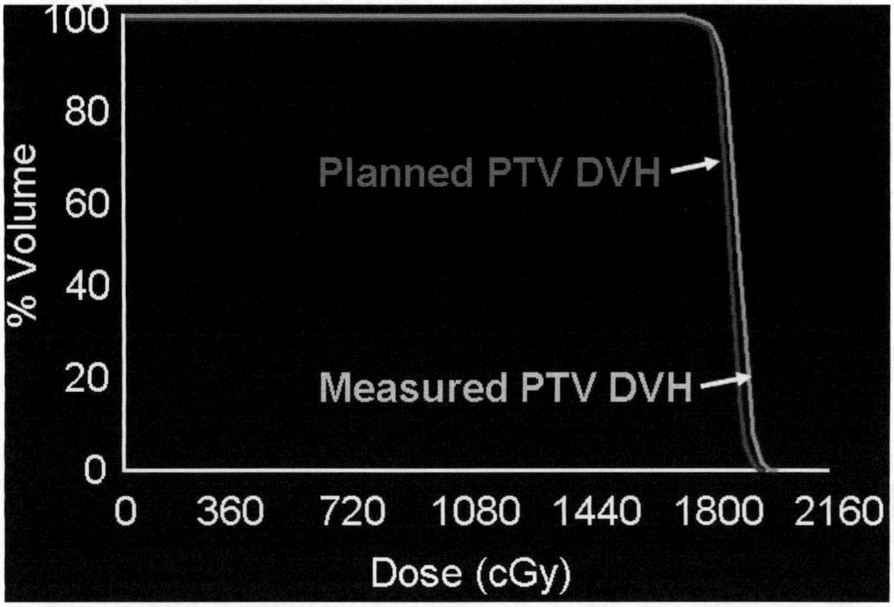

Figure 2: Comparison of planned (red) and measured (green) DVHs for the tumor volume.

Journal of Physics: Conference Series **250** (2010) 012052 doi:10.1088/1742-6596/250/1/012052

Results

We report the analysis of 22 IMRT QA measurements that failed the conventional 2D IMRT QA. Of the 22 failures, 19 were found to have treatment target volume DVH within 2% of the treatment plan. Of the 22 IMRT QA failures, 3 were found to have measured target DVHs differ the planned DVHs by more than 2%. All data is shown below in table 1.

Patient	# Fields Failed (Total Fields)	Lowest % Points Pass	DVH % Difference
1	5 (8)	75.7	1.0
2	8 (9)	80.9	3.0
3	5 (8)	71.5	3.0
4	4 (8)	75.6	3.0
5	2 (9)	93.3	1.0
6	1 (7)	89.5	2.0
7	2 (7)	57.5	1.0
8	8 (9)	83.4	1.0
9	3 (7)	91.8	1.0
10	2 (7)	91.5	1.0
11	1 (9)	94.2	1.0
12	8 (9)	88.7	1.0
13	9 (9)	79.4	1.0
14	1 (8)	84.5	1.0
15	3 (7)	78.6	1.0
16	3 (8)	70.9	1.0
17	6 (7)	80.8	2.0
18	2 (9)	70.2	1.0
19	1 (7)	90.6	1.0
20	4 (5)	84.5	2.0
21	1 (9)	85.8	1.0
22	7 (8)	82.1	1.0

Table 1: Analysis data for the 22 patients analyzed. The number of failed fields, and total treatment fields is given. The lowest percent of detector points passed reveals the lowest passing rate for the failed fields. The DVH percent difference represents the difference between the measured and planned target and critical structure DVHs.

Conclusions

We have demonstrated that an in-house 3D analysis of conventional 2D IMRT QA is a useful tool to evaluate the clinical impact of the IMRT QA failure. We showed that only 3/22 plans that failed this criteria (5%/4mm) has an impact of more than 2% in treatment plan target and normal structure DVHs. Therefore use of a more strict passing criteria in 2D fluence map IMRT QA may not result in tangible clinical benefit.

Journal of Physics: Conference Series **250** (2010) 012052 doi:10.1088/1742-6596/250/1/012052

References

[1] Nielsen M, Wilson J, McCabe R, Khan F, MacPherson M. Validation of VMAT delivery using a commercial 2D diode array. *Med Phys* 2009;36:4309.

[2] Cheong K, Kang S, Lee M, Kim S, Park S, Hwang T, Kim K, Oh D, Bae H. Evaluation of delivered monitor unit accuracy of gated step-and-shoot IMRT using a two-dimensional detector array. *Med Phys* 2010;37:1146-1151.

[3] Geurts M, Gonzalez J, Serrano-Ojeda P. Longitudinal study using a diode phantom for helical tomotherapy IMRT QA. *Med Phys* 2009;36:4977-4983

[4] Yan G, Lu B, Kozelka J, Liu C, Li J. 2009 Calibration of a novel four-dimensional diode array. *Med Phys* 2009;37:108-115.

[5] Kruse J. On the insensitivity of single field planar dosimetry to IMRT inaccuracies. *Med Phys* 2010;37:2516-2524.

[6] Lawrence M, Potter L, Cullip T, Xu Z, Deschesne K, Chang S. Comparison of a 2D and 3D array of diodes for IMRT QA. *Med Phys* 2009;36:2570.

Journal of Physics: Conference Series **250** (2010) 012053 doi:10.1088/1742-6596/250/1/012053

Study of dosimetric water equivalency of PRESAGE® for megavoltage and kilovoltage x-ray beams

Tina Gorjiara[1], Robin Hill[1, 2], Jung-Ha Kim[1], Zdenka Kuncic[1], John Adamovics[3], and Clive Baldock[1]

[1] Institute of Medical Physics, School of Physics, University of Sydney, NSW 2006, Australia

[2] Department of Radiation Oncology, Royal Prince Alfred Hospital, Sydney, NSW 2050, Australia

[3] Department of Chemistry and Biology, Rider University, Lawrenceville, NJ 08648, USA

clive.baldock@sydney.edu.au

Abstract. PRESAGE is a dosimeter that is suitable for 3D dosimetry. To be used as an ideal dosimeter, however, it should present radiologically water equivalent properties. In this work, we have investigated the radiological properties of three different PRESAGE® formulations. The radiological water equivalence was assessed by comparing the photon cross sections and radiation dosimetry properties of the three different PRESAGE® formulations with the corresponding values for water. Relative depth doses were calculated using Monte Carlo methods for 75, 125, 180 and 280 kVp and 6 MV x-ray beams. Based on the results of this study, the PRESAGE® formulations with lower halogen content are more dosimetrically water equivalent.

1. Introduction

The PRESAGE® dosimeter is composed of radiochromic components (leuco dyes) and halogen-containing free radical initiators that undergo a colour change after irradiating with ionizing radiation. As PRESAGE® is a three dimensional (3D) dosimeter, it has clinical application in modern radiation treatment techniques [1, 2].

To be used in radiation dosimetry, a 3D dosimeter should ideally be radiologically water equivalent [3]. The radiological properties of a material can be assessed by comparing the following parameters to the corresponding values for water: effective atomic number (Z_{eff}); mass energy absorption coefficient; mass stopping power; mass density (ρ); and electron density (ρ_e)[4-6].

The radiological properties of the original PRESAGE® formulation have been investigated previously [6]. It was found that the PRESAGE® formulation had an effective atomic number (Z_{eff}) that is 17% greater than that of water. Two subsequent PRESAGE® formulations have since been developed, with lower halogen content to potentially improve water equivalency over a larger energy range. In this work we have investigated and compared the radiological dosimetric properties and the water equivalency of the three PRESAGE® formulations. We calculated photon cross sections and mass energy absorption coefficients. We also determined depth doses in each of the dosimeters using Monte Carlo calculations for 6 MV and 75, 125, 180 and 280 kVp x-ray beams.

Journal of Physics: Conference Series **250** (2010) 012053 doi:10.1088/1742-6596/250/1/012053

2. Materials and Methods

The elemental composition of each PRESAGE® formulation was calculated using their molecular formula and are summarized in Table 1. We henceforth refer to the lower halogen content PRESAGE® formulations as LHP1 and LHP2, and we refer to the original PRESAGE® formulation as OP. All three PRESAGE® formulations have been provided by the same manufacturer (Heuris Pharma, New Jersey, USA).

Table 1: Effective atomic number, elemental composition, and fractional weight (w_k) for materials of interest.

Material	Z_{eff}	Formula	w_H	w_C	w_N	w_O	w_S	w_{Cl}	w_{Br}
Lower halogen PRESAGE® 1 (LHP1)	7.688	$C_{1758}H_{3000}N_{121}O_{442}S_4Cl_{30}Br_1$	0.08850	0.6178	0.04960	0.2069	0.003800	0.03112	0.002300
Lower halogen PRESAGE® 2 (LHP2)	7.740	$C_{889}H_{1541}N_{60}O_{228}S_{10}Cl_1Br_1$	0.09050	0.6224	0.04899	0.2126	0.01870	0.002100	0.004700
Original PRESAGE® (OP)	8.652	$C_{481}H_{842}N_{30}O_{129}Cl_9Br$	0.08920	0.6074	0.04460	0.2172	-	0.03340	0.008400
Water	7.417	H_2O	0.1119	-	-	0.8881		-	-

Photon cross sections for photoelectric absorption, Compton scatter and pair production were obtained from the NIST XCOM database over the energy range 10 keV - 20 MeV [7]. The elemental data of mass energy absorption coefficients were taken from the NIST X-ray attenuation database and were calculated for each PRESAGE® formulation using the mixture rule and elemental compositions, and were then compared to those for water.

Depth dose calculations for the PRESAGE® dosimeters were performed using the EGSnrc/BEAMnrc Monte Carlo package (Version 4, Research Council of Canada, Ottawa, Canada) [8]. The phase space files for each of the x-ray beams were generated using the BEAMnrc user code. In all cases, the vendor geometric and material specifications were used in the modeling. The 6 MV x-ray beam was produced by a Varian 21iXs linear accelerator (Varian Medical Systems, city, state, USA). The beam was modeled with a field size of 10×10 cm^2 and the electron and photon energy cutoff parameters, ECUT and PCUT, were set to 0.7 MeV and 0.01 MeV, respectively. The kilovoltage x-ray beams were generated by a Pantak Therapax DXT300 kilovoltage x-ray therapy unit with a tube potential varying from 50 to 280 kVp using a previously published methodology [9]. For the kilovoltage Monte Carlo calculations, ECUT and PCUT were set to 0.521 MeV and 0.001 MeV, respectively.

We calculated depth dose curves in the three PRESAGE® formulations and water using the DOSRZnrc user code and phase space files generated by BEAMnrc for the 6 MV x-ray beam. Within DOSRZnrc, each phantom was defined as a cylinder. In the buildup region, the first 8 voxels were defined with 0.25 cm thickness and the next 16 voxels with 0.5 cm thickness [6]. For the kilovoltage x-ray beams, we calculated depth doses in a similar manner but using the DOSXYZnrc user code. Each phantom was defined in DOSXYZnrc with a total of 20 voxels of 0.5 cm thickness along the central axis.

For the comparison of the phantom materials, we determined the density thickness which is defined as the physical depth (cm) × density (g cm^{-2}) [10].

3. Results

The calculated photon cross sections for the three different PRESAGE® formulations and water over the energy range 10 keV - 20 MeV are plotted in Figure 1. As indicated in Figure 1a, LHP1 and LHP2 exhibit significantly more water equivalent photoelectric absorption cross sections than OP. This may be attributed to a reduction in the proportion of bromine in the new PRESAGE® formulation compared to OP. LHP2 has the most water equivalent Compton scattering cross section. However, the pair production cross section of OP is closer to the value of water. This can be attributed to the pair production cross section being approximately proportional to Z^2/A at these energies and the highest fractional weighting is for oxygen in water, whereas it is for carbon in the PRESAGE® formulations.

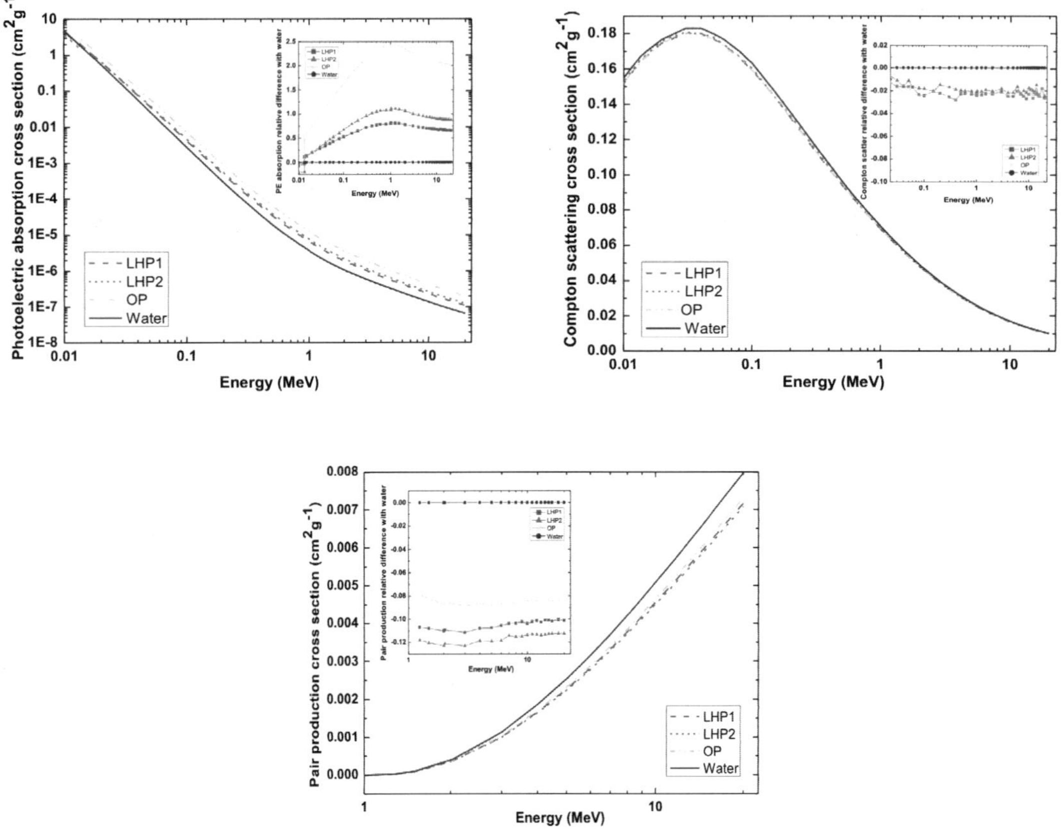

Figure1. Photon cross sections for (a) photoelectric absorption (b) Compton scattering and (c) pair production for the three PRESAGE® formulations and for water.

The ratio of the mass energy absorption coefficients of the PRESAGE® formulations and water are presented in Figure 2 for the photon energy range 10 keV - 20 MeV. A rapid increase in the relative mass energy absorption coefficient is noticeable over energies 10 - 40 keV, followed by a rapid decline towards 100 keV. At the peak, the mass energy absorption coefficient of OP is more than 75% higher than water, while the mass energy absorption coefficients of LHP1 and LHP2 are approximately only 20% higher than water.

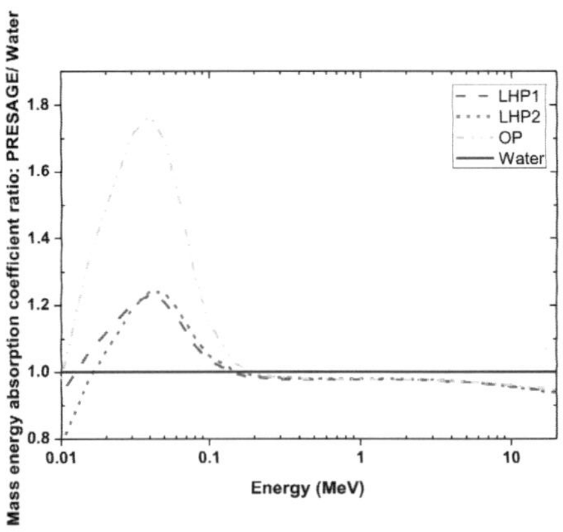

Figure 2. The mass energy absorption ratio for LHP1, LHP2 and OP relative to water.

Figure 3 shows the Monte Carlo calculated depth doses for the three PRESAGE® formulations and water using a 6 MV x-ray beam. The depth doses are scaled according to the density thickness. The dose differences relative to the value of water are up to 2 % for the lower halogen containing PRESAGE® formulations and up to 5% for OP at a radiological depth of 10 g cm^{-2}.

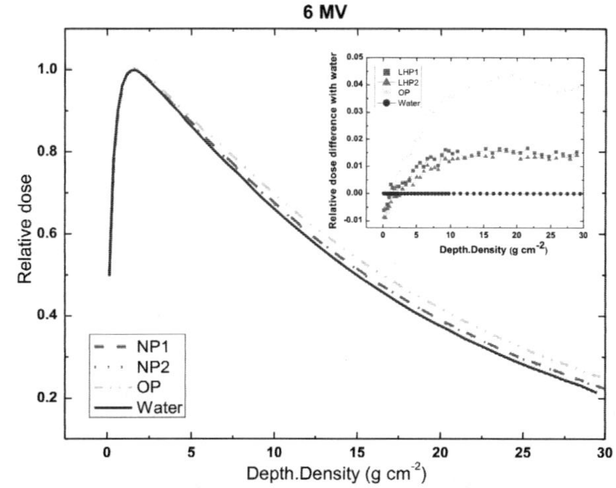

Figure 3. Monte Carlo relative density corrected depth dose curves for a 6 MV x-ray beam for the three PRESAGE® formulations and for water.

Figure 4 shows the kilovoltage Monte Carlo calculated dose to density corrected depth for the PRESAGE® formulations and water. The depth dose curves suggest that the new PRESAGE® formulations LHP1 and LHP2 are more water equivalent than OP in terms of dose response. As the x-ray beam energy increases, the differences between the relative doses decreases, but the depth dose curves for the PRESAGE® formulations then begin to exhibit less attenuation than water for energies above 200 kVp. The changeover in relative depth dose curves from 180 kVp to 280 kVp can be attributed to the rapid decrease in the mass energy absorption coefficient at energies approaching 100 keV (see Figure 2), which is close to the mean energy of these beams.

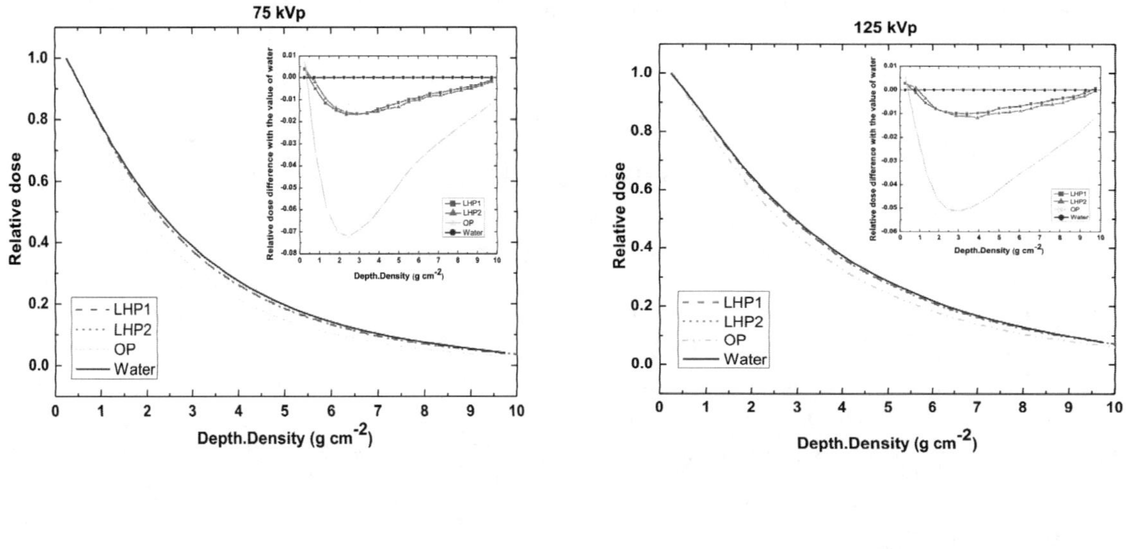

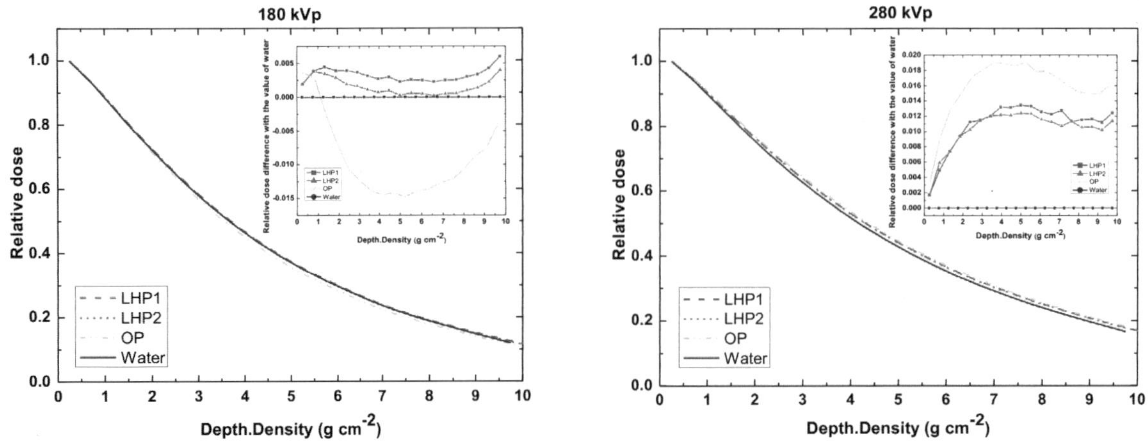

Figure 4. Monte Carlo calculated density corrected relative dose curves for 75 kVp, 125 kVp, 180 kVp and 280 kVp photon beams for the three PRESAGE® formulations and water.

4. Conclusion

These results indicate that the PRESAGE® formulations with lower halogen contents show more radiologically water equivalent properties than the original PRESAGE® formulation. The dose response of the lower halogen content PRESAGE® is comparable to the typical dose response of polymer gel dosimeters.

References

[1] Adamovics, J. and M.J. Maryanski, Characterisation of PRESAGE (TM): A new 3-D radiochromic solid polymer dosemeter for ionising radiation. Radiation Protection Dosimetry, 2006. **120**: p. 107-112.

[2] Adamovics, J. and M.J. Maryanski, A new approach to radiochromic three-dimensional dosimetry-polyurethane. Journal of Physics: Conference Series, 2004. **3**: p. 172-175.

[3] Baldock, C., et al., Polymer gel dosimetry. Physics in Medicine and Biology, 2010. **55**(5): p. R1-R63.

[4] Keall, P. and C. Baldock, A theoretical study of the radiological properties and water equivalence of Fricke and polymer gels used for radiation dosimetry. Australasian physical & engineering sciences in medicine 1999. **22**: p. 85-91.

[5] Venning, A.J., et al., Radiological properties of normoxic polymer gel dosimeters. Medical Physics, 2005. **32**(4): p. 1047-1053.

[6] Brown, S., et al., Radiological properties of the PRESAGE and PAGAT polymer dosimeters. Applied Radiation and Isotopes, 2008. **66**(12): p. 1970-1974.

[7] Berger, M.J., et al., XCOM: Photon Cross Sections Database. NIST Standard Reference Database 8 (XGAM). 2009, National Institute of Standards and Technology: Gaithersburg

[8] Rogers, D.W.O., Fifty years of Monte Carlo simulations for medical physics. Physics in Medicine and Biology, 2006. **51**(13): p. R287-R301.

[9] Hill, R., et al., An evaluation of ionization chambers for the relative dosimetry of kilovoltage x-ray beams. Medical Physics, 2009. **36**(9): p. 3971-3981.

[10] Khan, F.M., The physics of radiation therapy. 4th ed, ed. J. Pine. 2010, Philadelphia: Lippincott Williams & Wilkins.

Journal of Physics: Conference Series **250** (2010) 012054 doi:10.1088/1742-6596/250/1/012054

Commissioning of Brachytherapy TPS Using a 2D-Array of Ion Chambers

Mammo Yewondwossen and Jim Meng

Department of Radiation Oncology, Dalhousie University, Halifax, Nova Scotia, Canada

mammo.yewondwossen@cdha.nshealth.ca

Abstract. Purpose/Objective(s): Unlike external beam treatment planning system (TPS) commissioning, a brachytherapy TPS requires little input, no modeling with most brachytherapy TPS calculations are based on the AAPM TG-43 formalism. The dose distribution using the AAPM TG-43 dose formalism is usually compared with the calculations using the Sievert summation, Monte Carlo simulation or dose distributions measured by with GAFCHROMIC film. These methods have shown an agreement of within 5% compared to the AAPM TG-43 dose formalism. The purpose of this study is to report our experience of using a 2D-array of ion chambers (MatriXX Evolution, IBA Dosimetry) for dosimetric verification of conformal CT-based high dose rate (HDR) brachytherapy.Material/Methods: After benchmarking the new TPS against the old TPS used in the clinical for several years and comparing with MATLAB calculated dose distribution, the dose calculation accuracy of TPS systems was investigated by measuring dose distributions experimentally using MatriXX Evolution. The phantom used for this technique consists of multiple catheters, the IBA MatriXX detector and a slab of RW3 to provide full scattering conditions. The TPS dose distribution was calculated on the CT scan of this phantom. The measured and TPS calculated distributions were compared in IBA Dosimetry OmniPro-I'mRT software. Results: The average absolute dose difference over the ROI was 1.67% and the gamma agreement index computed for a distance to agreement of 3 mm and a dose difference of 3% showed agreement for 98.7% of all pixels, with gamma ≤ 1. Conclusion: We have found that MatriXX 2D dosimetric technique provides a fast and accurate way to validate a brachytherapy TPS for both commissioning and quality assurance.

1. Introduction

High-dose-rate (HDR) brachytherapy has proven to be a highly successful radiation treatment in the management of different types of cancers. The brachytherapy Treatment Planning System (TPS) is an important component of radiation treatment process. Commissioning and quality assurance of a commercially available brachytherapy TPS is essential to ensure accurate dose delivery. External beam TPS commissioning requires measurement and input of the beam characteristics into the TPS followed by some kind of modeling or fitting. Unlike external beam TPS, brachytherapy TPS requires little input and no modeling. The radiation source types used in HDR brachytherapy TPS are well characterized and documented in scientific literature. The common algorithm employed by brachytherapy TPSs is the AAPM TG-43 approach [1], greatly simplifying the effort required in commissioning.

One of the challenges encountered during implementation of a new brachytherapy TPS into a clinical environment is verifying the dose calculation accuracy. The complex and variable nature of the treatment planning process makes it essential to verify delivered absolute dose distributions, especially during commissioning and quality assurance. The dose distribution generated by the TPS using the AAPM TG-43 dose formalism is usually compared with the calculations using the Sievert summation, Monte Carlo simulation or dose distributions measured by with Gafchromic film. Gafchromic film is an excellent tool for dosimetry but is a costly consumable, requires careful calibration and specific methods for digitization with sufficient signal-to-noise characteristics. Recently, two-dimensional ionization chamber arrays have become increasingly popular for intensity-modulated radiation therapy (IMRT) plan verification. The MatriXX Evolution (IBA Dosimetry, Germany), an array of 1020 ionization chambers[2,3] has been developed and used widely for this purpose. The aim of this study is to evaluate the applicability of MatriXX Evolution to brachytherapy TPS plan verification.

2. Materials and methods

A kilovoltage X-ray therapy unit (Gulmay D3300, Gulmay Ltd., Chertsey, UK) was used to calibrate the matrix detector, according to the procedure detailed in the Radiation Therapy Committee Task Group 61 of American Association of Physics in Medicine (AAPM TG-61 protocol) [4]. Before patient specific dose verification can be carried out, absolute calibration of the detector response (quantified by the factor k_{user}) must be determined experimentally by the user. This factor converts the charge collected by the internal electrometer to the dose deposited in the detector plane at a given calibration depth and field size [2]. Absolute calibration of the detector was performed by irradiating the detectors with a normally incident 300 kV field, to a known dose using a circular 5 cm diameter open-ended applicator at 30 cm focus-to-surface distance.

The new Oncentra Brachy version 3.2 TPS was commissioned by: 1) benchmarking test plans against the Nucletron Plato system, v 14.3.7 which was used in the clinical routine for several years in our department, and 2) comparing the dose distribution generated by the TPS with MATLAB calculated (using calculation algorithms of AAPM TG-43) dose distribution. We then evaluated the TPS experimentally using MatriXX Evolution. To generate a brachytherapy treatment plan for the experimental conditions, Nucletron Freiburg Flap Applicator Set was used. The Freiburg Flap Applicator Set is a flexible mesh style surface mold made of a flexible silicone rubber with 36 channels for flexible implant tubes with a separation of 10 mm and a channel length of 24 cm. The Freiburg Flap maintains a fixed distance of 5 mm from the applicator to skin. This apparatus was positioned to cover the detector area of the matrix detector (Figure 1, left). Slabs of Solid Water were added above and below the matrix detector and Freiburg Flap to provide full scattering conditions. A Cone Beam CT (CBCT) scan for the treatment planning was obtained with 2.5 mm slice thickness.

Based on the CBCT data of this phantom, a treatment plan were calculated using the TPS (Nucletron Oncentra Brachy version 3.2.) The plan consisted of 3 active channels to mimic the Fletcher GYN applicator set (Figure 1, right) and was set to deliver 2 Gy at 2cm lateral to tandem axis. The matrix detector was irradiated according to this plan, which was exported directly to the console from the TPS. The data acquisition on the matrix detector was done in the movie mode at the sampling frequency of 1 sample per second.

The planned dose matrix in the detector plane was exported from the planning system for comparison with measurement using the Gamma index method. A gamma value greater than 1 was considered as outside tolerance range, using dose difference and distance criteria of 3% and 3 mm, respectively.

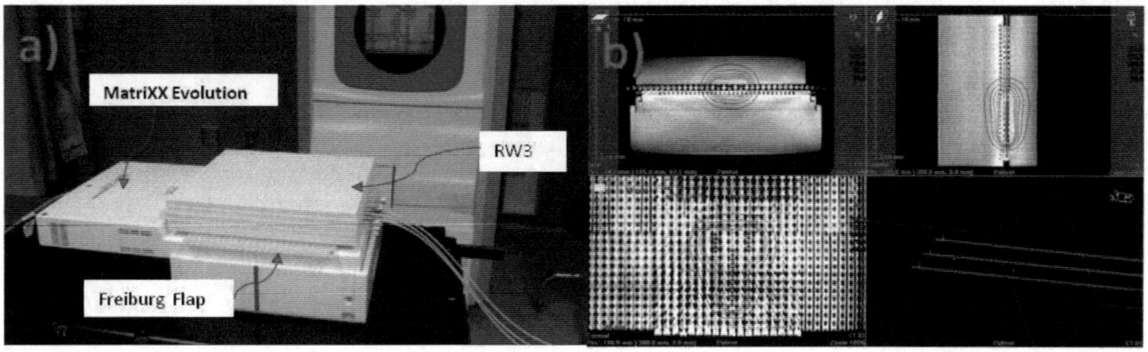

Figure 1. a) The experimental set-up b) Dose distribution of a plan mimicking a GYN treatment.

3. Results and discussion

Figures 2 show the comparison between the measured and planned planar dose distributions. The ROI shown in the figure includes 10% of the delivered dose. The average absolute dose difference over the ROI was 1.67% and minimum and maximum being 0 and 11.35%, respectively. The result shows agreement for 98.7% of all pixels, with gamma ≤ 1. These preliminary results demonstrate that the matrix detector can be used for rapid and convenient verification of brachytherapy plans with similar dose tolerances as for IMRT plan verification.

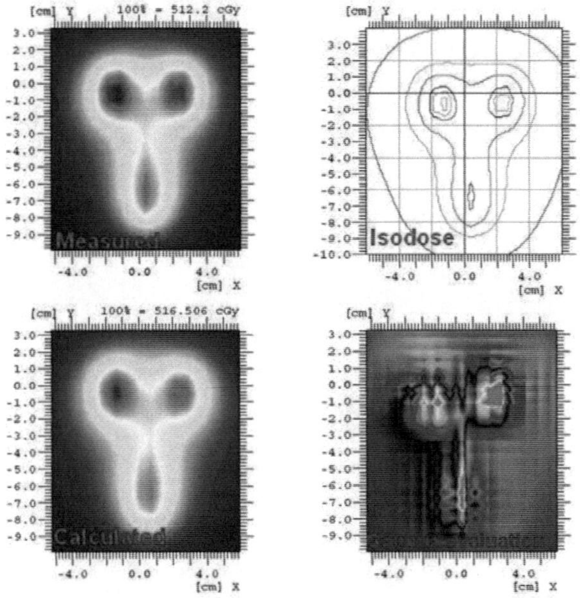

Figure 2. Comparison calculated and the measured dose distribution.

Acknowledgements: We are grateful to Dr. James Robar and Alma Harlow for their support during this work.

4. References

[1] Nath, R., Anderson, L.L., Luxton, G., Weaver, K.A., Williamson, J.F., Meigooni, A.S., Dosimetry of interstitial brachytherapy sources: Recommendations of the AAPM Radiation Therapy Committee Task Group No. 43, Med. Phys. 22 (1995) 209–234.

[2] Herzen J., Todorovic, M., Cremers, F., Platz, V., Albers, D., Bartels, A., Schmidt, R., Dosimetric evaluation of a 2D pixel ionization chamber for implementation in clinical routine Phys. Med. Biol. 52 (2007) 1197–208

[3] Lessard E, Pouliot J. Inverse planning anatomy-based dose optimization for HDR-brachytherapy of the prostate using fast simulated annealing algorithm and dedicated objective function. Med Phys 2001; 28:773e779.

[4] Ma, C. M., Coffey, C. W., DeWerd, L. A., Nath, R. , Liu, C. , Seltzer, S. M., and Seuntjens, J., AAPM protocol for 40 - 300 kV x-ray beam dosimetry in radiotherapy and radiobiology: Report of Task Group 61, Med. Phys. 28 (2001), 868 - 893

Journal of Physics: Conference Series **250** (2010) 012055 doi:10.1088/1742-6596/250/1/012055

3D dosimetry fundamentals: gels and plastics

M. Lepage[1] and K. Jordan[2]

1- Centre d'imagerie moléculaire de Sherbrooke, Université de Sherbrooke, 3001 12e Avenue Nord, Sherbrooke, QC, Canada

2- London Regional Cancer Program, London Health Sciences Centre and Department of Medical Biophysics, University of Western Ontario, London, ON, Canada

Martin.Lepage@USherbrooke.ca and Kevin.Jordan@lhsc.on.ca

Abstract. Many different materials have been developed for 3D radiation dosimetry since the Fricke gel dosimeter was first proposed in 1984. This paper is intended as an entry point into these materials where we provide an overview of the basic principles for the most explored materials. References to appropriate sources are provided such that the reader interested in more details can quickly find relevant information.

1. Introduction

The current use of hydrogels to measure 3D radiation dose distributions can be traced to the work of Gore *et al.* with a ferrous agarose system and magnetic resonance imaging (MRI) [1]. Radiochromic gels for chemical dosimetry of ionizing radiation were reported in 1951, but their 3D potential could not be exploited until digital computers and optical computed tomography were developed [2]. The key features of radiosensitive hydrogels for 3D dosimetry are their water or tissue equivalent response and their inherent 3D nature. The previous Dosgel proceedings contain a series of excellent reviews including several that focus on the chemistry if radiation sensitive hydrogels [3-7]. These reviews are the best starting point for studying gel dosimetry.

Two types of hydrogels have dominated the research: gels that contain monomers that undergo radical initiated polymerization due to absorbed dose and radiochromic hydrogels. The polymer gel dosimeters can be probed in 3D with MRI, ultrasound, x-ray and optical computed tomography (CT). While the radiochromic gels are usually read with optical CT to generate a full 3D dose distribution. Radiochromic plastics (Presage™) will be covered by other presenters at this meeting. Based on previous reports each of these materials has specific features that make them preferred for certain applications and it is anticipated that all will contribute to 3D dosimetry. Currently Presage™ (Heuris Pharma LLC, USA) and BANG (MGS Research, USA) are available for purchase and all other formulations require preparation by the users. Other considerations for material development for clinical usage include: ease of preparation and disposal, reproducibility of results, toxicity and raw material costs.

2. Water radiolysis

The water content of radiosensitive hydrogels ranges from 99% for ferrous agarose [1] and genipin gels to 80% for certain polymerization gel formulations [5]. The residual mass is predominantly organic materials such as gelatin and monomers. Dose deposition due to ionizing radiation is referred to as water radiolysis. The text by Spinks and Woods [8] is a particularly important reference for radiation chemistry in general and the review by Appleby [3] at Dosgel'99 is a concise summary relative to hydrogels. In the case of megavoltage radiotherapy the majority of dose is deposited by collisions of high energy electrons with water molecules that initiate a cascade of chemical reactions. The relative concentrations of water radiolysis products are primarily determined by the presence of dissolved oxygen and the sample's pH. The most important products are H_2O_2 and the radicals e^-_{aq}, H^+, and $^\bullet OH$ in the absence of oxygen [9, 10]. Hydrogels prepared near room temperature and exposed to the air will contain approximately 1 mM of oxygen. In that case, new radicals such as $^\bullet O_2^-$ and $HO_2^\bullet$ are formed while e^-_{aq}, H^+ are absent [9]. The water can be subdivided into a bound fraction and a free or bulk fraction. Bound and bulk water have thoroughly studied with NMR and in the specific case of hydrogels the review by Audet is recommended [11]. A direct link between the measured NMR characteristics of polymer gel dosimeters and theory was subsequently proposed [12]. Characterizing chemical reactions in the mixed bound-free water environment of hydrogels will provide insight for chemical gel dosimetry and model biological systems. For example, glycerol a common additive to gelatin hydrogels, stabilizes the collagen structure by enhancing the degree of bound water [13] resulting in stronger gels.

3. Hydrogels

The most common polymer gelling agents investigated for 3D dosimetry in hydrogels have been agarose (polysaccharide), gelatin (protein) and polyvinyl alcohol. Agarose gels are prepared at temperatures near 70°C. The elevated temperature has been problematic for many of the radiation chemistries that were investigated. For example, heating reduces the amount of dissolved oxygen and increases auto-oxidation rates [14]. This can lead to chemically nonuniform gels with unnecessarily elevated background levels for optical studies. In the case of ferrous-xylenol orange gels, the thermally increased autooxidation of ferrous ions during gel preparation produces dark gels before irradiation. Agarose gels are also translucent which minimizes their usefulness for 3D optical scanning. Gelatin is currently the most common gelling agent for preparing radiosensitive hydrogels. Hydrogel samples, with gelatin concentrations around 5% by mass, appear transparent and exhibit transmission levels in excess of 60% for 10 cm pathlengths will yellow or red light. If required, even transmission levels approaching 85% can be achieved by mechanical filtering of the gelatin solution or by the addition of refractive index matching materials such as sucrose or glycerol. Reducing scatter in gel translates into higher spatial resolution and greater dynamic range of optical measurements. For MRI and x-ray CT readout are likely insensitive to the inhomogeneities that cause optical scatter. The most common type of gelatin for hydrogels has been porcine, acid cured with a Bloom strength of 300. This gelatin type forms the stiffest gels. Depending on the formulation, gelatin gels typically melt between 28 and 34°C. In practice, if the gels are being handled in order to perform an irradiation experiment or transport to a readout instrument, temperatures exceeding 23°C may influence the gel uniformity. Crosslinking can provide radiosensitive hydrogels with melting points exceeding 60°C [15, 16]. However, crosslinking generally increases optical scatter and transverse magnetic relaxation rate as well. Gelatin can act either as a radiation sensitizer or desensitizer, depending on the radiochemistry under investigation [4, 6, 11, 16]. Polyvinyl alcohol (PVA) has also been investigated as a gelling agent. The initial interest in this synthetic polymer was due to the high purity, chemical simplicity, relative chemical inertness and low diffusion coefficients for ferrous xylenol orange gels. The high viscosity of PVA gels was problematic for preparation of large volume samples. Air bubbles became easily trapped when handling and became sources of nonuniformities, especially for optical readout. The scatter coefficients were also higher than analogous gelatin gels and continued to increase as the samples aged. For these reasons PVA gels have not been used for 3D dosimetry, but are quite helpful for investigating radiochemistries in hydrogels. Agarose, gelatin and PVA have all been sensitizers for ferrous and ferrous xylenol orange gels [17].

Journal of Physics: Conference Series **250** (2010) 012055 doi:10.1088/1742-6596/250/1/012055

While gelatin has been useful, higher performance hydrogels with better optical and mechanical properties could be developed. For example, deformable gels would provide a unique solution to a relevant clinical issue, dose warping.

4. Fundamental principles of 3D dosimeters

4.1. Ferrous gel chemistry

Ferrous gels are based on adding gelling agents to the ferrous chemical dosimeter that is often referred to as the Fricke dosimeter. Gore et al demonstrated that 3D dose distributions could be recorded with ferrous gels and readout with MRI [1]. The chemistry of these gels was reviewed at the first gel dosimetry meeting, Dosgel'99 by [3, 18]. The dose sensitivity, tissue equivalence and 3D nature of these gels generated adequate interest in this specialty to generate this sequence of meetings dedicated to 3D dosimetry. However, materials with greater sensitivity and lower diffusion are required for routine clinical use. This need has generated many innovations in gel chemistry and readout technologies. These gels will continue to be important for fundamental radiochemistry studies but have been surpassed by many other radiosensitive gels.

4.2. Ferrous xylenol orange gel chemistry

The addition of xylenol orange to the Fricke solution allowed sensitive and reproducible detection of the ferric ions generated by absorbed dose with visible light. A thorough investigation of this innovation was conducted by Gupta and Nilekani [19]. Appleby first introduced xylenol orange into ferrous agarose gels [3]. However, due to optical scatter from the agarose hydrogel, effective optical scanning was limited to 2D slabs of gel [20]. The substitution of gelatin for agarose resulted in an effective transparent radiochromic hydrogel for 3D optical CT scanning [21]. FX gels have been the most widely studied radiochromic systems to date due to their ease of preparation and adequate dose sensitivity. However, subtle variations in dose responses have been reported between and within research groups. It had been surmised that the main variation was chemical purity of xylenol orange from different manufacturers. But, detailed dose fractionation studies concluded that several Fe(III)-xylenol orange species existed in equilibrium and that sampling at different wavelengths gave different dose responses [22]. FX gels continue to be investigated and are likely to be a standard for comparison of new gel chemistries. FX gels with lower diffusion rates and higher melting points are required.

$$\mathrm{Fe(III)_1\text{-}X_2}\ (480\ \mathrm{nm}) \leftrightarrow \mathrm{Fe(III)_1\text{-}X_1}\ (540\ \mathrm{nm}) \leftrightarrow \mathrm{Fe(III)_2\text{-}X_1}\ (590\ \mathrm{nm}) \quad (1)$$

4.3. Radiochromic micelle gels

Examining reviews of chemical dosimeters reveals that many are not water based [8, 23]. One approach to designing radiosensitive hydrogels is to form hydrid materials containing nonpolar nanodomains. The simplest approach involves identifying surfactants and reactants that are compatible with the formation of transparent hydrogels. This approach is general and is an active area of research on an industrial scale as society migrates from hydrocarbon solvents to "green chemical syntheses". The initial reports have demonstrated that insoluble leuco dyes can be incorporated into gelatin hydrogels with nonionic surfactants [16, 24]. Diffusion measurements revealed that the more water soluble, cationic dye Malachite Green had a higher diffusion coefficient. This observation suggested that either uncharged dyes or ionic surfactants may form dye-micelle pairs leading to radiochromic gels with very low diffusion rates (Figure 1).

Journal of Physics: Conference Series **250** (2010) 012055 doi:10.1088/1742-6596/250/1/012055

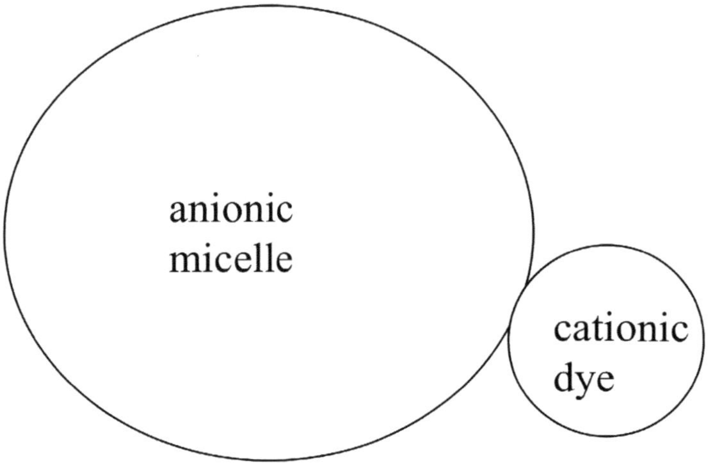

Figure 1. Micelle dye ion pair, model system for low diffusion radiochromic hydrogel.

4.4. Radiochromic crosslinked gels

Genipin crosslinks gelatin to form a blue hydrogel. At low pH these gels are radiosensitive and the blue crosslinked gels fade with absorbed dose. This is another chemical approach to developing non-diffusing radiochromic hydrogels. The crosslinking provides stable gels with lower concentrations of gelatin (Figure 2). For example, 2% gelatin crosslinked genipin gels are stable at room temperature. The crosslinking also allows preparation of free standing gels that can be immersed in an acidic solution. These gels can be scanned optically, without interference of vessel walls and the gel response is uniform throughout allowing the potential for surface dosimetry as well [25]. The chemical simplicity of this system and the small amount of organic materials added, make this gel a candidate for reference water equivalent dosimetry. It may prove especially useful for 3D dosimetry at lower energies, for example with 100 kVp x-ray beams.

4.5. Radiochromic plastics

A promising material for routine clinical 3D dosimetry is the radiochromic plastic, Presage™ [26]. The host plastic is polyurethane and the radiochromic conversion of the colourless leucomalachite green to the dye malachite green is initiated by halogenated hydrocarbon free radicals. The system is robust and available for purchase, eliminating the need for a preparation laboratory. There has been a continuous improvement in optical quality and tissue equivalence from subsequent formulations (results to be presented at this meeting). High dose sensitivity, linearity and lack of diffusion are characteristics of this material.

4.6. Polymer gel dosimeters

Polymer gel dosimeters have been proposed in 1993 as materials that can integrate a radiation dose distribution in three dimensions [27]. Since their inception, many different variations on the gel composition have been tested and used. Fundamental processes influencing the radiation response of these materials have been unveiled, yielding a thorough understanding of the factors affecting the radiation dose sensitivity and stability, both spatial and temporal [12, 28-32]. In 2001, an antioxidant was added to the gel composition, which scavenged oxygen molecules present in the gel during manufacture under normal atmosphere [33]. A full understanding of the effect of antioxidants in gels is starting to emerge [33-35].

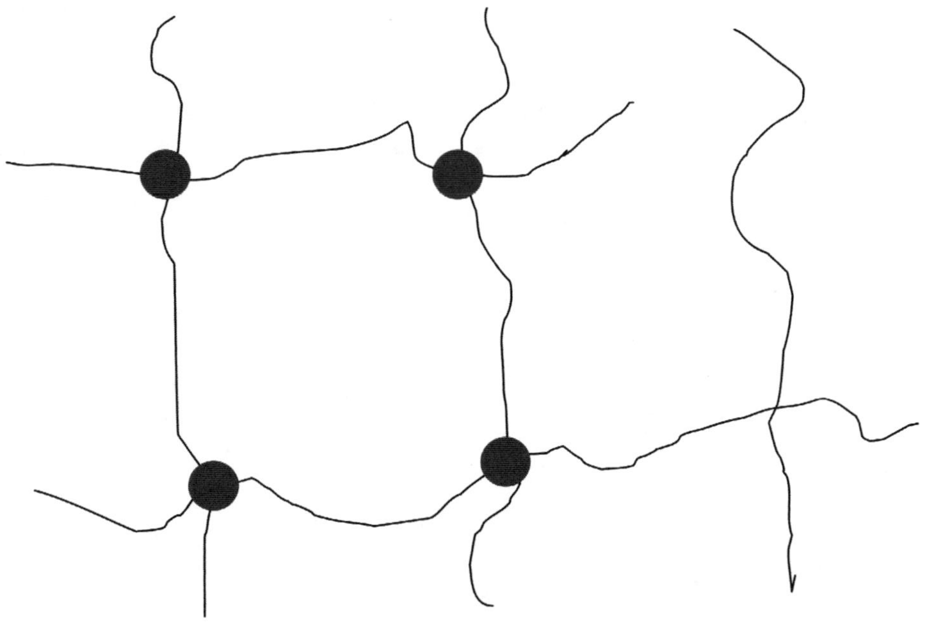

Figure 2. Schematic of radiochromic crosslinked gel matrix. Blue circles represent sites of crosslinking by genipin.

The main constituents of polymer gel dosimeters are water, monomers and a gelling agent. Traditionally, gelatin is used as the gelling agent and forms a three dimensional matrix into which the monomers are dispersed. Monomers are relatively small molecules that are characterized by a carbon-carbon double bond that can easily enter in a chemical reaction. Under irradiation, water radiolysis generates free radicals and molecular species; the former are highly reactive chemical species that can react with any gel constituent.

4.6.1. *Polymerization process.* Figure 3 provides a very simplified (and certainly non-exhaustive) schematic of the different chemical reactions that take place in a polymer gel dosimeter. When a water free radical reacts with a monomer, a process which is called "initiation", a monomer radical is formed, which can react with another monomer, forming a polymer radical, and so on in a process called propagation. This is a crucial step where monomers react together, forming a polymer. Presence of this newly formed polymer modifies the physical properties of the dosimeter and different techniques can be used to probe these changes. A calculation based on reasonable assumptions suggested the number of monomer units in a polymer chain was roughly 10^4 [12]. Termination of the polymerization reaction is also an important step since an absence of termination would result in full polymerization of the monomers after an insignificant radiation dose. Two polymer radicals can react together, a polymer radical can react with gelatin or with a water free radical; any of these will terminate the polymerization reaction. Oxygen is a very reactive chemical species that will rapidly react with water free radicals or with monomer radicals, efficiently inhibiting or terminating the polymerization reaction (figure 3). Preparing the dosimeter gel in an oxygen-free environment (*i.e.*, a nitrogen-purged glove-box) can almost totally expel oxygen from the solution, at the expense of a more labour-intensive manufacturing process.

The polymerization process is rather fast in principle; that is when all constituents have easy access to other constituents. Certain monomers are hydrophobic (difficult to solubilize in water, for example

N,N'-methylene-bisacrylamide), which means their polymerization will yield a hydrophobic polymer. Access of other monomers becomes restricted and, as a consequence, the polymerization slowly propagates for hours.

4.6.2. *Role of gelatin.* So what is the role of gelatin in all this? Its main function is to provide a three dimensional matrix into which the polymer formed cannot diffuse, thus preserving the spatial integrity of the radiation dose distribution. However, the porosity of a gelatin matrix is such that small molecules [36], and perhaps also monomers, can almost freely diffuse. A direct consequence is that the radiation dose profile from a half-blocked field differs slightly from the expected "step" irradiation profile: a polymerization "overshoot" is observed at the edge of the field [37, 38]. This was interpreted as diffusion of monomers from the unirradiated region to the irradiated region and later confirmed by computer simulations [39]. However, gelatin is also known for its role as "scavenger" of water free radicals in acrylamide-based dosimeters [40]. As a result, higher concentration of gelatin leads to a decreased sensitivity since less water free radicals become available for reaction with

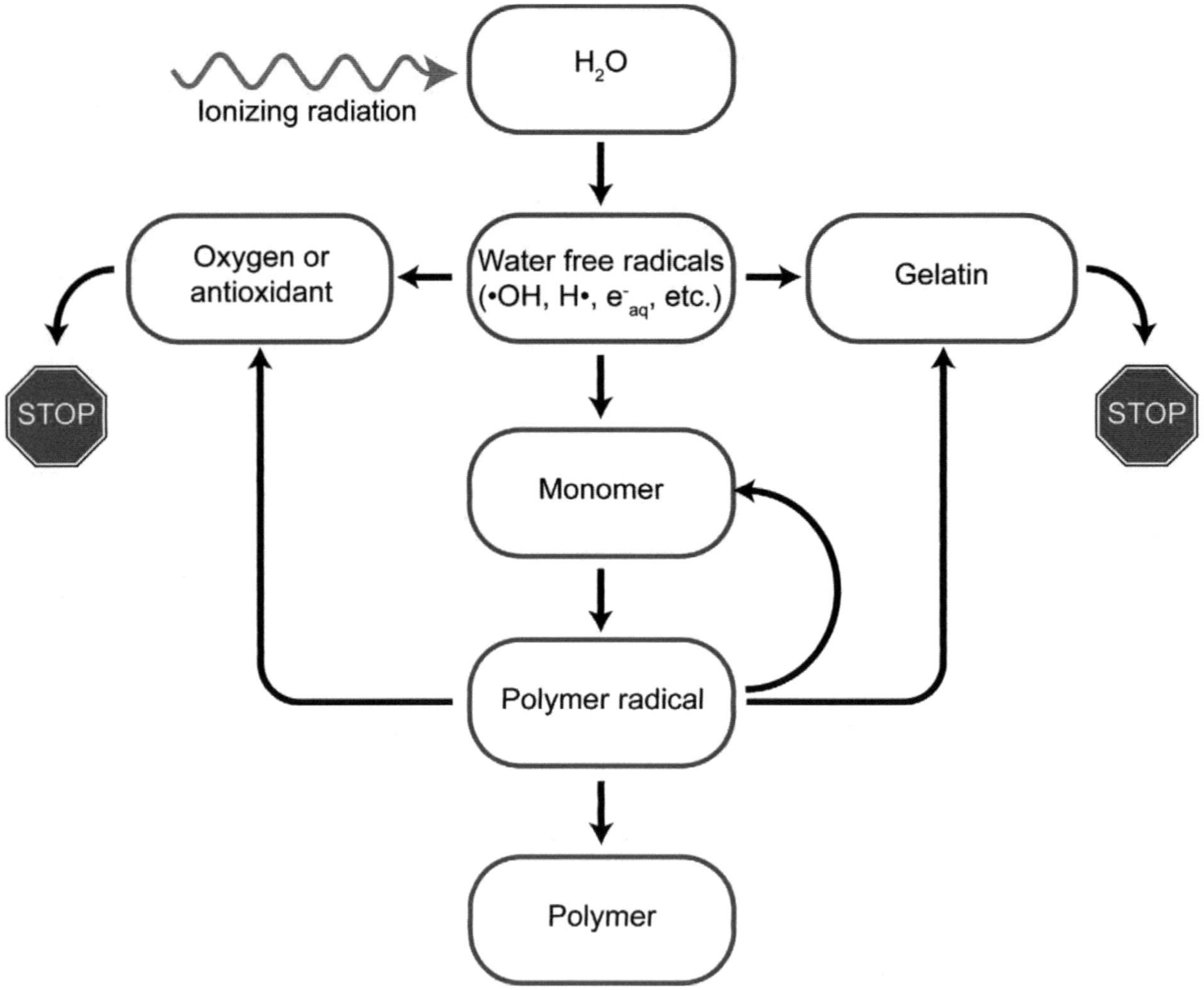

Figure 3. Schematic representation of the interaction between different chemicals in a polymer gel dosimeter, more specifically for an acrylamide-based dosimeter in a deoxygenated environment.

monomers to initiate the polymerization process. Increasing gelatin concentration lead to a decreased dose sensitivity in Fricke-gelatin dosimeters [41].

5. Conclusion

Many different materials are available for 3D dosimetry. Their usefulness is maximised when their composition and response is matched with the capabilities of a reading device [42]. A clear message is that there is not a single material that is universally optimized for every application and for every measurement device; materials and devices must be selected together for a given application.

References

[1] Gore JC, Kang YS and Schulz RJ, 1984 Measurement of radiation dose distributions by nuclear magnetic resonance (NMR) imaging, Phys Med Biol 29: 1189-97
[2] Day MJ and Stein G, 1951 Chemical dosimetry of ionizing radiations, Nucleonics 8: 34-51
[3] Appleby, A 1999 The radiation chemical basis of gel dosimetry *Proceedings of the 1St International Workshop on Radiation Therapy Gel Dosimetry* (Lexington, KY, USA: Canadian Organization of Medical Physicists)
[4] McAuley KB 2006 Fundamentals of polymer gel gosimeters, J Phys Conf Ser 56: 35-44
[5] Jirasek A, McAuley KB and Lepage M, 2009 How does the chemistry of polymer gel dosimeters affect their performance? J Phys Conf Ser 164: 012003
[6] Schreiner LJ 2004 Review of Fricke gel dosimeters, J Phys Conf Ser 3: 9-21
[7] Jirasek A 2006 Experimental investigations of polymer gel dosimeters, J Phys Conf Ser 56: 23-34
[8] Spinks, J W Tand Woods, R J 1990 *An Introduction to Radiation Chemistry* (John Wiley & Sons)
[9] von Sonntag, C 2006 *Free-Radical-Induced DNA Damage and its Repair: A Chemical Perspective* (Berlin: Springer)
[10] Ferradini C and Jay-Gerin J-, 2000 The effect of pH on water radiolysis: A still open question - A minireview, Res Chem Intermed 26: 549-65
[11] Audet, C 1999 The NMR relaxometry of radiotherapy gel dosimeters *Proceedings of the 1St International Workshop on Radiation Therapy Gel Dosimetry* (Lexington, KY, USA: Canadian Organization of Medical Physicists)
[12] Lepage M, Whittaker AK, Rintoul L and Baldock C, 2001 13C-NMR, 1H-NMR and FT-Raman study of the radiation-induced modifications in radiation dosimetry polymer gels, J Appl Polym Sci 79: 1572-81
[13] Penkova R, Goshev I, Gorinstein S and Nedkov P, 1999 Stabilizing effect of glycerol on collagen type I isolated from different species, Food Chem 66: 483-7
[14] Healy BJ, Zahmatkesh MH, Nitschke KN and Baldock C, 2003 Effect of saccharide additives on response of ferrous-agarose-xylenol orange radiotherapy gel dosimeters, Med Phys 30: 2282-91
[15] Fernandes JP, Pastorello BF, De Araujo DB and Baffa O, 2008 Formaldehyde increases MAGIC gel dosimeter melting point and sensitivity, Phys Med Biol 53: N53-8
[16] Jordan K and Avvakumov N, 2009 Radiochromic leuco dye micelle hydrogels: I. Initial investigation, Phys Med Biol 54: 6773-89
[17] Hill B, Bäck SA, Lepage M, Simpson J, Healy B and Baldock C, 2002 Investigation and analysis of ferrous sulfate polyvinyl alcohol (PVA) gel dosimeter, Phys Med Biol 47: 4233-46
[18] Bäck, S Å Jand Olsson, L E 1999 Ferrous gels (FeGel) and MRI for absorbed dose measurements; 15 years of development *Proceedings of the 1St International Workshop on Radiation Therapy Gel Dosimetry* (Lexington, KY, USA: Canadian Organization of Medical Physicists)
[19] Gupta BL and Nilekani SR, 1998 Ferrous ion oxidations by (·)H, (·)OH and H2O2 in aerated FBX dosimetry system, Radiat Phys Chem 53: 643-50
[20] Tarte BJ, Jardine PA and Vandoorn T, 1996 Laser-scanned agarose gel sections for radiation field mapping, Int J Radiat Oncol Biol Phys 36: 175-9
[21] Kelly RG, Jordan KJ and Battista JJ, 1998 Optical CT reconstruction of 3D dose distributions

using the ferrous- benzoic-xylenol (FBX) gel dosimeter, Med Phys 25: 1741-50

[22] Babic S, Battista J and Jordan K, 2008 An apparent threshold dose response in ferrous xylenol-orange gel dosimeters when scanned with a yellow light source, Phys Med Biol 53: 1637-50

[23] Holm, N Wand Berry, R J ed 1970 *Manual on Radiation Dosimetry* (Marcel Dekker Inc.)

[24] Babic S, Battista J and Jordan K, 2009 Radiochromic leuco dye micelle hydrogels: II. Low diffusion rate leuco crystal violet gel, Phys Med Biol 54: 6791-808

[25] Jordan K 2009 Optical CT scanning of cross-linked radiochromic gel without cylinder wall, J Phys Conf Ser 164: 012029

[26] Adamovics J and Maryanski MJ, 2006 Characterisation of PRESAGE™: A new 3-D radiochromic solid polymer dosemeter for ionising radiation, Radiat Prot Dosimet 120: 107-12

[27] Maryañski MJ, Gore JC, Kennan R and Schulz RJ, 1993 NMR relaxation enhancement in gels polymerized and cross-linked by ionizing radiation: a new approach to 3D dosimetry by MRI, Magn Reson Imaging 11: 253-8

[28] Baldock C, Rintoul L, Keevil SF, Pope JM and George GA, 1998 Fourier transform Raman spectroscopy of polyacrylamide gels (PAGs) for radiation dosimetry, Phys Med Biol 43: 3617-27

[29] Jirasek AI, Duzenli C, Audet C and Eldridge J, 2001 Characterization of monomer/crosslinker consumption and polymer formation observed in FT Raman spectra of irradiated polyacrylamide gels, Phys Med Biol 46: 151-65

[30] De Deene Y, Hanselaer P, De Wagter C, Achten E and De Neve W, 2000 An investigation of the chemical stability of a monomer/polymer gel dosimeter, Phys Med Biol 45: 859-78

[31] Lepage, M, Whittaker, A K and Baldock, C 1999 A 13C and 1H NMR study of acrylamide gels used in radiation dosimetry *Annual Meeting of the American Association of Physicists in Medicine* (Nashville, TN, USAP: AAPM)

[32] Lepage M, Whittaker AK, Rintoul L, Bäck SÅJ and Baldock C, 2001 Modeling of post-irradiation events in polymer gel dosimeters, Phys Med Biol 46: 2827-39

[33] Fong PM, Keil DC, Does MD and Gore JC, 2001 Polymer gels for magnetic resonance imaging of radiation dose distributions at normal room atmosphere, Phys Med Biol 46: 3105-13

[34] De Deene Y, Hurley C, Venning A, Vergote K, Mather M, Healy BJ and Baldock C, 2002 A basic study of some normoxic polymer gel dosimeters, Phys Med Biol 47: 3441-63

[35] Fong, P M, Keil, D C and Gore, J C 2001 Radiation dosimetry using MRI of normoxic polymer gels, Proc Intl Soc Mag Reson Med (Glasgow, UK : International Society for Magnetic Resonance in Medicine) pp 901

[36] Baldock C, Harris PJ, Piercy AR and Healy B, 2001 Experimental determination of the diffusion coefficient in two-dimensions in ferrous sulphate gels using the finite element method. Australas Phys Eng Sci Med 24: 19-30

[37] De Deene Y, Reynaert N and De Wagter C, 2001 On the accuracy of monomer-polymer gel dosimetry in the proximity of a high-dose-rate 192Ir source, Phys Med Biol 46: 2801-25

[38] Maryañski MJ, Schulz RJ, Ibbott GS, Gatenby JC, Xie J, Horton D and Gore JC, 1994 Magnetic resonance imaging of radiation dose distributions using a polymer-gel dosimeter, Phys Med Biol 39: 1437-55

[39] Fuxman AM, McAuley KB and Schreiner LJ, 2005 Modelling of polyacrylamide gel dosimeters with spatially non-uniform radiation dose distributions, Chem Eng Sci 60: 1277-93

[40] Lepage M, Whittaker AK, Rintoul L, Bäck SÅJ and Baldock C, 2001 The relationship between chemical processes and transverse relaxation times in polymer gel dosimeters, Phys Med Biol 46: 1061-74

[41] Audet C and Schreiner LJ, 1997 Multiple-site fast exchange model for spin-lattice relaxation in the Fricke-gelatin dosimeter, Med Phys 24: 201-9

[42] Baldock C, Lepage M, Bäck SÅJ, Murry PJ, Porter D and Kron T, 2001 Dose resolution in radiotherapy polymer gel dosimetry: effect of echo spacing in MRI pulse sequence, Phys Med Biol 46: 449-60

Journal of Physics: Conference Series **250** (2010) 012056 doi:10.1088/1742-6596/250/1/012056

Small Field: dosimetry in electron disequilibrium region

Timothy C. Zhu

Department of Radiation Oncology, University of Pennsylvania, Philadelphia, PA 19104

tzhu@mail.med.upenn.edu

Abstract. Small fields are more commonly used for radiation therapy because of the development of IMRT, stereotactic radiosurgery, and other special equipments such as Cyberknife and Tomotherapy. The dosimetry in the sub-centimeter field can result in substantial uncertainties because of the presence of electron disequilibrium due to the large dose gradients in the field. It is further complicated by the introduction of various radiation detectors, which usually perturb the conditions of disequilibrium. Hence additional corrections are required to maintain the dosimetric accuracy previously achieved for standard radiation dosimetry. A review of small field dosimetry provides some insights into the methods to characterize the detector convolution kernel and other methods to characterize detector perturbation effect.

1. Introduction

Various types of radiation detectors are used to measure relative dose distributions for therapeutic electron and photon beams. Most detectable signals from a radiation detector are correlated to the energy imparted $\delta\varepsilon$ to the detector [1]. This, in turn, is related to the absorbed dose in the detector, D_{det}. Table I lists types of radiation detectors, their signals, and their relationship to the dose in the sensitive volume of the detectors. The meaning of the dosimetric quantities can be found in Refs. [2-7]. In this review we will focus on the first three types of detectors, especially ionization chambers and diodes, since they are the most commonly used in acquiring beam data.

Table 1. Example of radiation detectors

Type	Measured Quantities	Relation to D_{det}
Ionization chambers	Charges (Q)	$D_{det} = (W/e)_{air} \cdot Q/m \cdot \beta$
Si Diode detectors	Charges (Q)	$D_{det} = (W/e)_{Si} \cdot Q/m \cdot \beta$
TLD	Light intensity (I)	$D_{det} \propto \int I dt$
Fricke dosimeter	Fe^{3+} content (ΔA)	$D_{det} = \Delta A/\rho_t G_t$
Film	Optical density (OD)	$D_{det} \propto OD$
Scintillation dosimeter	Light intensity (I)	$D_{det} \propto I(\lambda)$
Gel dosimeter	Optical density change (ΔOD)	$D_{det} = \Delta OD/\rho l(\varepsilon_m)_t G_t$
Calorimetry	Temperature (DT)	$D_{det} = c_p \cdot DT$

When such a measuring device is introduced into a radiation field, it tends to disturb the field. For small field dosimetry, one major concern is the spatial resolution of the detector since the signal correlates to the integral (or mean) dose in the detector and we are interested in the dose in the

Journal of Physics: Conference Series **250** (2010) 012056 doi:10.1088/1742-6596/250/1/012056

medium at one point. This correction can be done using the concept of detector convolution kernel. This is somewhat different from conditions for absolute dosimetry in large fields (> 4 cm) where the electron fluence distribution over the detector tends to be uniform (although not always possible)!

Small field dosimetry is becoming more important in radiation.[8] This review will try to be consistent with recommendation of the proposed AAPM TG155 as much as possible and focus on photon beam small field dosimetry. At the time of publication, the final report of TG155 is not yet available.

2. Perturbation factors and detector convolution kernel

The perturbation factors commonly used for absolute dosimetry can be expanded to electron disequilibrium regions using the concept of detector convolution kernel. Excellent reviews for absolute dosimetry can be found in that of Nahum [6, 9], Alm Carlsson [1], and Svensson and Brahme [10]. However, there are few papers on the perturbation factors under electron disequilibrium conditions.

If the detector is small compared to the range of the charged particles (i.e. it behaves as a Bragg cavity), one can write [9]

$$D_{med}(P) = \overline{D}_{\det} \cdot s_{med,\det} \cdot p\,, \qquad (1)$$

where p is the perturbation of the electron fluence, $s_{med,det}$ is the mass collision stopping power ratio between medium and the detector material. Note that the dose to the medium has been written as $D_{med}(P)$ to indicate that the value required is at a specific position P, which can be shifted from the center of the detector. $\overline{D}_{\det}$ is the mean dose in the detector. This expression is valid under electron disequilibrium, where $\overline{D}_{\det} \neq \overline{K}_{\det}$. Nahum [9] pointed out that the cause of p is the perturbation of the secondary electron fluence in the medium:

$$p = \frac{\int_{\Delta} (\Phi_E)^P_{med} (L_\Delta / \rho)_{\det} dE + [(\Phi(\Delta)^P_{med} (S(\Delta)/\rho)_{\det} \cdot \Delta]}{\int_{\Delta} (\Phi_E)_{\det} (L_\Delta / \rho)_{\det} dE + [(\Phi(\Delta)_{\det} (S(\Delta)/\rho)_{\det} \cdot \Delta]}\,, \qquad (2)$$

where Φ_E is the electron fluence and L_Δ/ρ is the restricted stopping power ratio, all other quantities have their usual meaning [4]. For absolute dosimetry (under electron equilibrium) this perturbation factor is usually very small (a few percent) so that one can separate it into multiplication of several independent perturbation factors. This separation is not usually possible in electron disequilibrium regions where the perturbation can be much larger. One of the most common separations is the wall correction factor and displacement factor [11, 12]. The displacement (or replacement) factor is further separated into electron fluence correction and gradient correction [11]. Bjärngard and Kase [13] argue that this latter separation is non-physical because it violates Fano's theorem [14]. MC simulation provides a general solution to quantify p for a specific dosimeter for small fields [15].

Supposing that a detector is introduced into a medium under a radiation field, the dose disturbance of the detector can be generally expressed as

$$\overline{D}_{\det} = \iiint g(\mathbf{r}-\mathbf{r}') \cdot D_{med}(\mathbf{r}') d\mathbf{r}' \qquad (3)$$

where $g(\mathbf{r})$ is the detector convolution kernel, $\overline{D}_{\det}$ is the mean dose determined by the detector, and D_{med} is the dose in the medium. Compared to Eq. (1), $g(\mathbf{r})$ describes the additional dose average effect of an extended detector volume. g must have an finite range reflecting the size of the detector and the range of secondary electrons. In principle, g can include all the detector disturbance, which cannot be separated into independent effects like wall correction and displacement correction. We will not explore here how to determine g theoretically since it relates doses (or secondary electron fluences) with and without the detector and is thus extremely complicated. One could, however, try to determine g experimentally for a specific radiation quality. As a first order of approximation, Eq. (3) can be simply considered as an average of the dose distribution in the medium with a weighting function, g, that depends on the detector perturbation. The convolution formalism assumes that g has the same shape everywhere within a given radiation quality.

The concept of detector convolution kernel is more useful for correcting profile measurement in regions with steep gradients because g can be independently measured. Extensive literature exists on the experimental determination of g [16-19]. All measurement techniques take advantage of the ability to modify the lateral fluence distribution of the radiation, thus D_{med}. Figure 1 shows the shape of the detector convolution kernel for two thimble ionization chambers (*Wellhöfer* IC-10 and IC-3) and a p-type Scanditronix diode (2.5-mm diameter). These measurements were made using a 6-MV step x-ray field and the two-dimensional detector response kernels are assumed to be cylindrically symmetric. Zhu *et al* find that the following function fits the data best:[20]

$$g(r) = \frac{3}{2\pi R^2}\sqrt{1 - r^2/R^2} \quad \text{for ionization chambers} \qquad (4)$$

$$g(r) = \frac{1}{\pi w^2} e^{-r^2/w^2} \quad \text{for diode}$$

The kernels are normalized using $2\pi\int g(r)rdr = 1$. Dawson *et al* found a substantial correlation between the effective radius, R, and beam energy, which tends to increase with increasing photon energy [21].

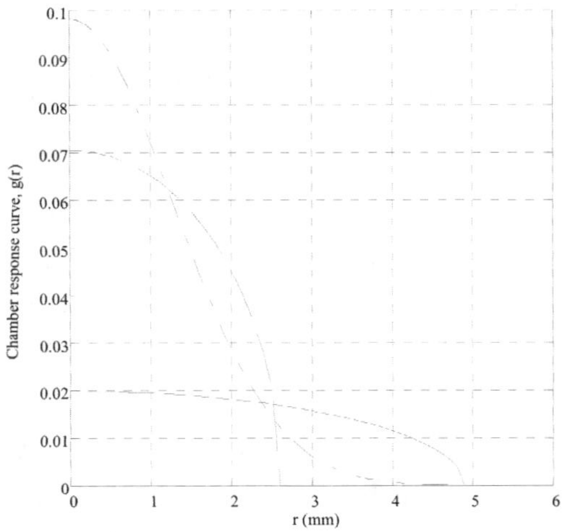

Fig. 1. Detector convolution kernel $g(r)$ measured for *Wellhöfer* 0.1 cc IC-10 (solid) and 0.03 cc IC-3 (dashed) ion chambers, and a Scanditronix p-type diode (dash-dotted).

A general formalism [22] for small field absolute dosimetry uses Eq. 1 in a similar way as that used for a broad beam using AAPM TG-51 [12] or IAEA TRS-398 [23], but is modified to use a machine specific reference field (if 10x10 cm^2 field is not available) [24]. The ratio of doses between the clinical field to the reference field at the reference depth is termed *output ratio in water. Output ratio in water* can be separated into *output ratio in air* and *phantom scatter factor*.[25] Studies have shown a difference between the true output ratios (e.g., MC generated) and those measured using a specific dosimeter [22, 26-32]. Limited studies quantify this field size dependent correction factor for various dosimeters by MC simulation.[8] Similarly TMR/ OAR for small fields can deviate from those measured using a dosimeter if the detector size is too large.[8, 33] These differences can be explained by the detector convolution method (Eq. 3), which account for the effects caused by the lack of CPE and the penumbra effect due to obscuring the x-ray source.

3. Depth dose measurements

3.1 Buildup region

The golden standard for dose measurements in the buildup region is the extrapolation parallel-plate chamber. However, these measurements are time consuming. Silicon diodes with small active volume in the radiation direction has been found to have little perturbation effect [34, 35]. Without special correction, a thimble-type ionization chamber overestimates the dose in the buildup region because of its relatively large volume.

A parallel-plate chamber with fixed plate separation can also be used to measure the depth dose in the buildup region. However, correction is needed for the fixed plate separation. Velkley *et al* studied the relationship between the correction factor and plate separation for an extrapolation parallel-plate chamber at different depths and photon energies. They found that the parallel-plate chamber with fixed plate separation tends to overestimate dose in the buildup region [36]. They were the first to introduce a correction factor ξ in the form of [36]

$$P'(d) = P(d) - \xi(E, d/d_{\max}) \times l, \tag{5}$$

where l is plate separation, d is depth to front surface of chamber, P' is corrected percentage buildup, and P is the percentage buildup obtained with chamber with plate separation l. ξ decreases with increasing photon energy and increasing depth for a fixed photon energy. For their particular extrapolation chamber design, Velkley *et al* found ξ to be 4.5%/mm, 3%/mm, 1.6%/mm, and 1%/mm for ^{60}Co, 4MV, 8MV, and 25MV, respectively, on the surface.

Nilsson and Montelius [37] performed one of the most extensive studies of the fluence perturbation of parallel-plate ionization chamber under electron disequilibrium conditions and concluded that it is not possible to correct the perturbation effect in one parallel-plate chamber with fixed plate separation with correction factors obtained with extrapolation chambers of other dimensions. Monte Carlo simulation for parallel-plate chamber perturbation effect has been performed but only for depths beyond d_{max} [38]. Correction factors ξ for specific chambers have been experimentally determined by Mellenberg [39]for the Markus chamber (PTW 30-329), Gerbi [40] for the Attix chamber, and Gerbi and Khan [41] for a range of commercially available fixed-separation parallel-plate chambers (Capintec PS-033, the PTW 30-329 Markus chamber, and Memorial chambers with circular or rectangular electrodes). Gerbi and Khan [41] concluded that the Velkley correction (Eq. (5)) produces acceptable results (within 2%) for all chambers but the Marcus chamber, because of its small guard, is still on average 11%, 6%, and 3.7% too high after correction for ^{60}Co, 6MV, and 10MV photon energies, respectively. Rawlison *et al* reviewed Velkley correction factors for commercially available parallel-plate chambers and pointed out that Velkley's formula needs to be modified to include influence of the chamber geometry and density of wall material [42]. They provided an improved formula:

$$P'(d) = P(d) - c(E) \cdot (l/w) \cdot \rho^{0.8} e^{-4.0d/d_{\max}} \tag{6}$$

where w is the inner diameter of wall, ρ is density of wall material (g/cm^3), $c(E)$ = 43%, 27%, 15% for Co-60, 6MV, 18MV, respectively. All other parameters are the same as Eq. (5). This modified equation works well even with the Markus chamber. Gerbi and Kahn [43] also examined chamber response in the buildup region for obliquely incident photon beams and found that the overresponse also changes with incident angle. They recommend using an extrapolation chamber or a well-guarded fixed-separation plane-parallel ionization chamber with a plate separation of 1 mm or less. Another important correction factor for parallel-plate ionization chamber is the polarity effect, which can be easily corrected by averaging the charges for both polarities. Several references discuss its origin [44-47] Gerbi and Kahn [48] and Aget and Rosenwald [45] give the polarity effect for several commercially available chambers for photon and electron beams. Diode is found to be a good detector for PDD measurement due to its small detector size.[35]

3.2 TMR/PDD measurements beyond buildup region

The major difference between PDD for small field and that for conventional field size is the effect caused by the change field profile vs. depth. As a result, the measured PDD may be different from the true PDD due to the increasing field width vs. increasing depth for a fixed SSD (Eq. 1). To remedy that, a simpler method is to measure TMR directly with the detector centered on the isocenter.[49] The variation of stopping power ratio vs. depth is found to be small for small fields.[8, 50, 51]

4. Small field profile measurements

Artificial broadening caused by the detector size is illustrated in Fig. 2, taken from Ref. [16]. Dawson *et al* studied the relation between penumbra width and ionization chamber inner diameter for a series of specially designed chambers with inside diameter varying between 0.3 to 1.4 cm [21]. They concluded that the P20-80 or P10-90 penumbral degradation associated with the inside diameter of the chamber can be corrected using a standard slope of 0.5 cm/cm inside diameter, regardless of the chamber type and radiation quality. Metcalfe *et al* [52]compared penumbra measured with ionization chambers, TLD, Scanditronix diode, and film and concluded that the diode (2.5 mm diameter) gives result consistent with film and very closely (within 0.2 mm) models the response of a point detector. TLD (1×1×6 mm) gives better result. We have tested a new 0.03 cm^3 ionization chamber (Wellhöfer IC-3 with 4-mm inner diameter) for penumbra measurement and found the result agrees with diode to within 0.5 mm.

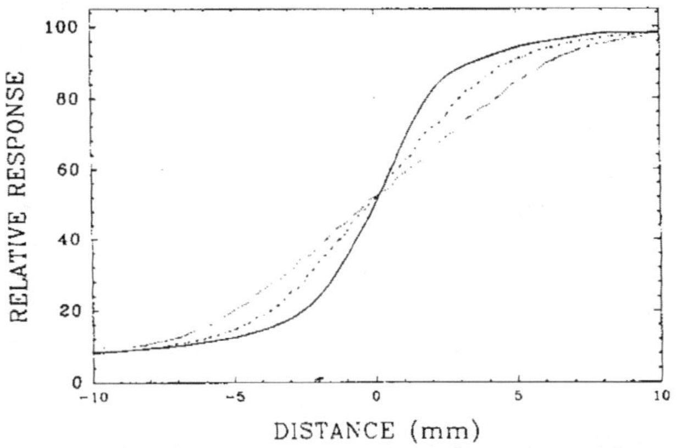

Fig. 2. Penumbra for a 6 MV x-ray beam obtained with 9.5 mm (dotted) and 5.8 mm (dashed) inner diameter ionization chamber. Full curve is measured with film. Taken from Ref. [16].

Existing studies have proven definitively that the detector convolution kernel can explain the broadening of the measured penumbra due to the detector size [16-19, 21, 53]. It is possible to extrapolate the true penumbra using the detector convolution kernel. However, the current deconvolution algorithm is very susceptible to noise and requires manual tuning to eliminate the noise effect [18]. This problem should be solvable if both the penumbra and detector convolution kernel are expressed as analytical functions. Several studies give various analytical expressions for the penumbra [52, 54] and the detector convolution kernel (Eq. (4)).

5. Small field output measurements

For output characterization of stereotactic cones, detector convolution kernel can still be used, one has to consider the effect in two-dimensions. Because the penumbra for small field may cut into the central-axis, the output can also decrease as a result. These effects can be expressed using detector convolution kernel as [20]

$$F(x,y) = \iint g(x-x', y-y') \cdot f(x',y')dx'dy' , \qquad (7)$$

where *F(x,y)* is the measured profile, *f(x,y)* is the true profile, and *g* is the detector convolution kernel. On the central-axis, the reduction of output factor, *CF*, can be expressed as [20]

$$CF = F(0,0) = \iint g(x',y') \cdot f(x',y')dx'dy' . \qquad (8)$$

Figure 3 compares calculated and measured correction factors for ionization chambers with detector convolution kernel described in Fig. 1. Here we have assumed that both *g* and *f* are cylindrically symmetric, i.e., they are a function of $\sqrt{x^2+y^2}$. Deconvolution for Eq. (7) with cylindrical symmetry can be performed using inverse Abel transform. [55]

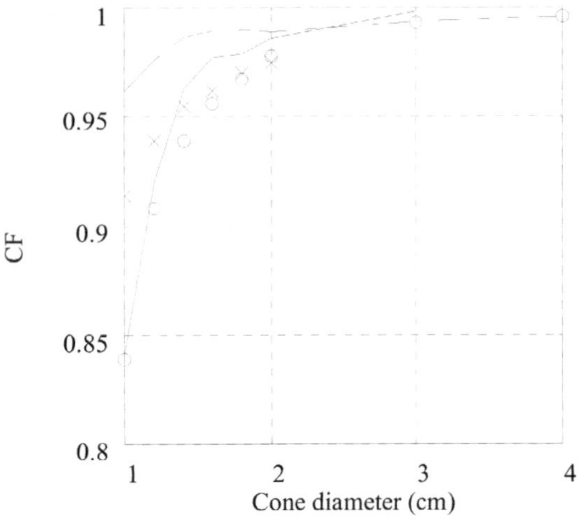

Fig. 3: Comparison between theoretical and measured *CF* for chambers. Lines (solid for 0.1 cc chamber, dashed for 0.03 cc chamber) are calculation using Eq. (9). Symbols (circle for 0.1 cc chamber and cross for 0.03 cc chamber) are measured *CF*, determined by the ratio of output measured by ionization chamber to that measured by diode.

Several studies have been performed to compare the output factors measured for the stereotactic cones by various detectors [8, 49, 56-58]. Table 2 lists the results for relative output factors for a 6MV x-rays from a Clinac 2500 accelerator [49]. Serago *et al* [49]concluded that small diode and TLD seem to be the appropriate choices of radiation detectors to determine relative dose for small field sizes. Bjärngard *et al* performed Monte-Carlo simulation and found the diode detector breakdown for cone diameter less than 0.5-cm [56].

One of key procedure to develop the correction factor for small field is an reliable algorithm to perform deconvolution. Several recent studies have provide effective method to perform such task while achieve reasonable uncertainties.[59, 60]

Table 2: Relative output factors for 6 MV x-rays measured with ionization chamber (IC), diode, TLD, and film

Cha	mber /detector						
	IC	IC	IC	IC	Diode	TLD	Film
Field size	0.6 cm^3	0.125 cm^3	Markus	Soft x-ray	2 mm	3 × 3 mm	XV-2
10×10 cm	1.00	1.00	1.00	1.00	1.00	1.00	1.00
Diameter							
8.0 mm					0.76	0.73	0.74
11.0 mm			0.74	0.72	0.82	0.80	0.80
14.0 mm		0.75	0.83	0.80	0.85	0.85	0.85
17.5 mm		0.82	0.87	0.85	0.88	0.87	0.88
21.0 mm	0.76	0.87	0.90	0.89	0.89	0.88	0.90
24.0 mm	0.84	0.90	0.91	0.91	0.90	0.90	0.91
26.5 mm	0.89	0.92	0.92	0.92	0.91	0.90	0.92
31.5 mm	0.92	0.92	0.93	0.93	0.91	0.91	0.92
36.5 mm	0.93	0.93	0.93	0.93	0.92	0.92	0.92

6. Concluding Remarks

The perturbation factor *p* for detectors has only been vigorously studied for electron beams and for photon beams under the condition of transient charged particle equilibrium, not suitable for small field dosimetry. These studies are mostly done for ionization chambers. More importantly, these conditions restrict the electron fluence distribution to be uniform across the detector. The detector convolution kernel expanded this to regions where electron fluence distribution is not uniform. However, substantial works remains to understand the beam quality dependence of the detector convolution kernel and more theoretical work such as Monte-Carlo simulation is needed to understand the origin of dosimetry uncertainties.

7. Acknowledgment

The author wishes to thank Drs. Indra Das, Alan Nahum, and Bengt E Bjärngard for valuable advises on small field dosimetry.

References

[1] G. Alm Carlsson, "Theoretical Basis for Dosimetry," in *The dosimetry of ionizing radiation.* vol. I, K. R. Kase, B. E. Bjarngard, and F. H. Attix, Eds. Orlando, FL: Academic Press, 1985, pp. 1-75.

[2] ICRU, "Radiation Dosimetry: Electron beams with energies between 1 and 50 MeV," International Commission on Radiation Units and Measurements, Bethesda, MD 1984.

[3] A. Dutreix and A. Bridier, *Dosimetry for external beams of photon and electron radiation* vol. I. Orlando, FL: Academic Press, 1985.

[4] H. E. Johns and J. R. Cunningham, *The Physics of Radiology*, 4th ed. Springfield, IL: Charles C Thomas Publisher, 1983.

[5] L. J. Schreiner and T. Olding, "Gel Dosimetry," in *Clinical Dosimetry Measurements in Radiotherapy*, D. W. O. Rogers and J. Cygler, Eds. Medison, Ws: Medical Physics Publishing Corp, 2009, pp. 979-1025.

[6] A. E. Nahum, "Cavity Theory, Stopping-Power Ratios, Correction Factors," in *Clinical Dosimetry Measurements in Radiotherapy*, D. W. O. Rogers and J. Cygler, Eds. Medison, Ws: Medical Physics Publishing Corp, 2009, pp. 91-136.

[7] S. Beddar and T. M. Briere, "Plastic Scintillation Detectors," in *Clinical Dosimetry Measurements in Radiotherapy*, D. W. O. Rogers and J. Cygler, Eds. Medison, Ws: Medical Physics Publishing Corp, 2009, pp. 1059-1081.
[8] I. J. Das, G. X. Ding, and A. Ahnesjo, "Small fields: nonequilibrium radiation dosimetry," *Med Phys,* vol. 35, pp. 206-15, Jan 2008.
[9] A. E. Nahum, "Perturbation Effects in dosimetry: Part I. Kilovoltage x-rays and electrons," *Phys Med Biol,* vol. 41, pp. 1531-1580, 1996.
[10] H. Sevnsson and A. Brahme, *Recent Advances in Electron and Photon Dosimetry*. New York: Plenum, 1986.
[11] A. TG21, "A protocol for the determination of absorbed dose from high-energy photon and electron beams," *Med Phys,* vol. 10, pp. 741-71, Nov-Dec 1983.
[12] P. R. Almond, P. J. Biggs, B. M. Coursey, W. F. Hanson, M. S. Huq, R. Nath, and D. W. Rogers, "AAPM's TG-51 protocol for clinical reference dosimetry of high-energy photon and electron beams," *Med Phys,* vol. 26, pp. 1847-70, Sep 1999.
[13] B. E. Bjarngard and K. R. Kase, "Replacement correction factors for photon and electron dose measurement," *Med Phys,* vol. 12, pp. 785-7, 1985.
[14] U. Fano, "Note on the Bragg-Gary Cavity Principle for Measuring Energy Dissipation," *Radiat. Res.,* vol. 1, pp. 237-40, 1954.
[15] F. Crop, N. Reynaert, G. Pittomvils, L. Paelinck, C. De Wagter, L. Vakaet, and H. Thierens, "The influence of small field sizes, penumbra, spot size and measurement depth on perturbation factors for microionization chambers," *Phys Med Biol,* vol. 54, pp. 2951-69, May 7 2009.
[16] C. H. Sibata, H. C. Mota, A. S. Beddar, P. D. Higgins, and K. H. Shin, "Influence of detector size in photon beam profile measurements," *Phys Med Biol,* vol. 36, pp. 621-31, 1991.
[17] K.-S. Chang, F.-F. Yin, and K.-W. Nie, "The effect of detector size to the broadening of the penumbra - a computer simulated study," *Med Phys,* vol. 22, pp. 1663-66, 1996.
[18] P. Charland, E. El-Khatib, and J. Wolters, "The use of deconvolution and total least squares in recovering a radiation detector line spread function," *Med Phys,* vol. 23, p. 1407, 1998.
[19] F. Garcia-Vicente, J. M. Delgado, and C. Peraza, "Experimental determination of the convolution kernel for the study of the spatial response of a detector," *Med Phys,* vol. 25, pp. 202-207, 1998.
[20] T. C. Zhu, C. Liu, F. Bova, and J. R. Palta, "Convolution based correction to ionization chamber measured output for small fields," *Med Phys,* vol. 23, p. 1166, 1996.
[21] D. J. Dawson, J. M. Harper, and A. C. Akinradewo, "Analysis of physical parameters associated with the measurement of high-energy x-ray penumbra," *Med Phys,* vol. 11, pp. 491-497, 1984.
[22] R. Alfonso, P. Andreo, R. Capote, M. S. Huq, W. Kilby, P. Kjall, T. R. Mackie, H. Palmans, K. Rosser, J. Seuntjens, W. Ullrich, and S. Vatnitsky, "A new formalism for reference dosimetry of small and nonstandard fields," *Med Phys,* vol. 35, pp. 5179-86, Nov 2008.
[23] P. Andreo, D. T. Burns, K. Hohlfeld, M. S. Huq, T. Kanai, F. Laitano, V. G. Smyth, and S. Ynckier, "Absorbed Dose Determination in External Beam Radiotherapy: An International Code of Practice for Dosimetry Based on Standards of Absorbed Dose to Water," *IAEA Technical Report Series No. 398,* 2000.
[24] S. Dieterich, C. Cavedon, and E. W. YWilcox, "Dosimetry of Small Photon Beams Used for Stereotactic Radiosurgery/Radiotherapy," in *Clinical Dosimetry Measurements in Radiotherapy*, D. W. O. Rogers and J. Cygler, Eds. Medison, Ws: Medical Physics Publishing Corp, 2009, pp. 637-667.
[25] T. C. Zhu, A. Ahnesjo, K. L. Lam, X. A. Li, C. M. Ma, J. R. Palta, M. B. Sharpe, B. Thomadsen, and R. C. Tailor, "Report of AAPM Therapy Physics Committee Task Group 74: in-air output ratio, Sc, for megavoltage photon beams," *Med Phys,* vol. 36, pp. 5261-5291, 2009.

[26] J. Fan, K. Paskalev, L. Wang, L. Jin, J. Li, A. Eldeeb, and C. Ma, "Determination of output factors for stereotactic radiosurgery beams," *Med Phys,* vol. 36, pp. 5292-300, Nov 2009.
[27] J. D. Fontenot, W. D. Newhauser, C. Bloch, R. A. White, U. Titt, and G. Starkschall, "Determination of output factors for small proton therapy fields," *Med Phys,* vol. 34, pp. 489-98, Feb 2007.
[28] C. McKerracher and D. I. Thwaites, "Head scatter factors for small MV photon fields. Part I: a comparison of phantom types and methodologies," *Radiother Oncol,* vol. 85, pp. 277-85, Nov 2007.
[29] C. McKerracher and D. I. Thwaites, "Head scatter factors for small MV photon fields. Part II: the effects of source size and detector," *Radiother Oncol,* vol. 85, pp. 286-91, Nov 2007.
[30] C. McKerracher and D. I. Thwaites, "Phantom scatter factors for small MV photon fields," *Radiother Oncol,* vol. 86, pp. 272-5, Feb 2008.
[31] A. J. Scott, A. E. Nahum, and J. D. Fenwick, "Monte Carlo modeling of small photon fields: quantifying the impact of focal spot size on source occlusion and output factors, and exploring miniphantom design for small-field measurements," *Med Phys,* vol. 36, pp. 3132-44, Jul 2009.
[32] O. A. Sauer and J. Wilbert, "Measurement of output factors for small photon beams," *Med Phys,* vol. 34, pp. 1983-8, Jun 2007.
[33] E. Pappas, T. G. Maris, F. Zacharopoulou, A. Papadakis, S. Manolopoulos, S. Green, and C. Wojnecki, "Small SRS photon field profile dosimetry performed using a PinPoint air ion chamber, a diamond detector, a novel silicon-diode array (DOSI), and polymer gel dosimetry. Analysis and intercomparison," *Med Phys,* vol. 35, pp. 4640-8, Oct 2008.
[34] G. Rikner, "Silicon diodes as detector in relative dosimetry of photon, electron, and proton radiation fields." vol. Ph.D. Uppsala: Uppsala University, 1983.
[35] T. C. Zhu and A. S. Saini, "Diode Dosimetry for Megavoltage Electron and Photon Beams," in *Clinical Dosimetry Measurements in Radiotherapy*, D. W. O. Rogers and J. Cygler, Eds. Medison, Ws: Medical Physics Publishing Corp, 2009, pp. 913-939.
[36] D. E. Velkley, D. J. Manson, J. A. Purdy, and G. D. Oliver Jr., "Build-up region of megavoltage photon radiation sources," *Med Phys,* vol. 2, pp. 14-19, 1975.
[37] B. Nilsson and A. Montelius, "Fluence perturbation in photon beams under nonequilibrium conditions," *Med Phys,* vol. 13, pp. 191-195, 1985.
[38] D. W. O. Rogers, "Calibration of parallel-plate chambers: Resolution of several problems by using Monte Carlo calculations," *Med Phys,* vol. 19, pp. 889-899, 1992.
[39] D. E. Mellenberg Jr, "Determination of build-up region over-response corrections for a Markus-type chamber," *Med Phys,* vol. 17, pp. 1041-1044, 1990.
[40] B. J. Gerbi, "The response characteristics of a newly designed plane-parallel ionization chamber in high-energy photon and electron beams," *Med Phys,* vol. 20, pp. 1411-1415, 1993.
[41] B. J. Gerbi and F. M. Kahn, "Measurement of dose in the buildup region using fixed-separation plane-parallel ionization chambers," *Med Phys,* vol. 17, pp. 17-26, 1990.
[42] J. A. Rawlinson, D. Arlen, and D. Newcombe, "Design of parallel plate ion chambers for buildup measurements in megavoltage photon beams," *Med Phys,* vol. 19, pp. 641-648, 1992.
[43] B. J. Gerbi and F. M. Kahn, "Plane-paralle ionization chamber response in the buildup region of obliquely incident photon beams," *Med Phys,* vol. 24, pp. 873-878, 1997.
[44] G. Wickman and T. Holmstrom, "Polarity effect in plane-parallel ionization chambers using a dielectric liquid as ionization medium," *Med Phys,* vol. 19, pp. 637-640, 1992.
[45] H. Aget and J. C. Rosenwald, "Polarity effect for various ionization chambers with multiple irradiation conditions in electron beams," *Med Phys,* vol. 18, pp. 67-72, 1991.
[46] D. M. Galbraith, J. A. Rawlinson, and P. Munro, "Dose errors to charge storage in electron irradiated plastic phantoms," *Med Phys,* vol. 11, pp. 197-203, 1984.
[47] L. O. Mattsson and H. Svensson, "Charge build-up effects in insulating phantom materials," *Acta Radiol Oncol,* vol. 23, pp. 393-399, 1984.

[48] B. J. Gerbi and F. M. Kahn, "The polarity effect for commercially available plane-parallel ionization chambers," *Med Phys,* vol. 14, pp. 210-215, 1987.

[49] C. F. Serago, P. V. Houdek, G. H. Hartmann, D. S. Saini, M. E. Serago, and A. Kaydee, "Tissue maximum ratio (and other parameters) of small circular 4,6,10,15, and 24 x-ray beams for radiosurgery," *Phys Med Biol,* vol. 37, pp. 1943-1956, 1992.

[50] F. Sanchez-Doblado, P. Andreo, R. Capote, A. Leal, M. Perucha, R. Arrans, L. Nunez, E. Mainegra, J. I. Lagares, and E. Carrasco, "Ionization chamber dosimetry of small photon fields: a Monte Carlo study on stopping-power ratios for radiosurgery and IMRT beams," *Phys Med Biol,* vol. 48, p. 2081=2099, 2003.

[51] K. Eklund and A. Ahnesjo, "Fast modelling of spectra and stopping-power ratios using differentiated fluence pencil kernels," *Phys Med Biol,* vol. 53, pp. 4231-47, Aug 21 2008.

[52] P. Metcalfe, T. Kron, A. Elliot, and T. Wong, "Dosimetry of 6 MV x-ray beam penumbra," *Med Phys,* vol. 20, pp. 1439-1442, 1993.

[53] P. D. Higgins, C. H. Sibata, L. Siskind, and J. W. Sohn, "Deconvolution of detector size effect for small field measurement," *Phys Med Biol,* vol. 36, pp. 621-31, 1995.

[54] T. Kron, A. Elliott, and P. Metcalfe, "The penumbra of a 6-MV x-ray beam as measured by thermoluminescent dosimetry and evaluated using an inverse square root function," *Med Phys,* vol. 20, pp. 1429-38, Sep-Oct 1993.

[55] H. Treuer, R. Boessecke, W. Schlegel, G. H. Hartmann, R. P. Muller, and V. Sturm, "The source-density function: determination from measured lateral dose distributions and use for convolution dosimetry," *Phys Med Biol,* vol. 37, pp. 1895-1909, 1992.

[56] B. E. Bjarngard, J. S. Tsai, and R. K. Rice, "Dose on the central axis of narrow 6-MV x-ray beams," *Med Phys,* vol. 17, pp. 794-799, 1990.

[57] M. J. Engler and G. L. Jones, "Small-beam calibration by 0.6- and 0.2- cm3 ionization chambers," *Med Phys,* vol. 11, pp. 822-26, 1984.

[58] C. Clift, A. Thomas, J. Adamovics, Z. Chang, I. Das, and M. Oldham, "Toward acquiring comprehensive radiosurgery field commissioning data using the PRESAGE Optical-CT 3D dosimetry system," *Phys Med Biol,* vol. 55, pp. 1279-1293, 2010.

[59] A. Djouguela, D. Harder, R. Kollhoff, S. Foschepoth, W. Kunth, A. Ruhmann, K. Willborn, and B. Poppe, "Fourier deconvolution reveals the role of the Lorentz function as the convolution kernel of narrow photon beams," *Phys Med Biol,* vol. 54, pp. 2807-27, May 7 2009.

[60] G. Bednarz, M. Saiful Huq, and U. F. Rosenow, "Deconvolution of detector size effect for output factor measurement for narrow Gamma Knife radiosurgery beams," *Phys Med Biol,* vol. 47, pp. 3643-9, Oct 21 2002.

SRT and SBRT: Current practices for QA dosimetry and 3D

S H Benedict[1], J Cai[2], B Libby[1], M Lovelock[3], D Schlesinger[1], K Sheng[1], and W Yang[1]

[1] University of Virginia Department of Radiation Oncology, Charlottesville, VA 22908
[2] Duke University Department of Radiation Oncology, Chapel Hill, NC 27710
[3] Memorial Sloane Kettering, NY, NY 10065

Email : SHB4X@hscmail.mcc.virginia.edu

Abstract. The major feature that separates stereotactic radiation therapy (cranial SRT) and stereotactic body radiation therapy (SBRT) from conventional radiation treatment is the delivery of large doses in a few fractions which results in a high biological effective dose (BED). In order to minimize the normal tissue toxicity, quality assurance of the conformation of high doses to the target and rapid fall off doses away from the target is critical. The practice of SRT and SBRT therefore requires a high-level of confidence in the accuracy of the entire treatment delivery process. *In SRT and SBRT confidence in this accuracy is accomplished by the integration of modern imaging, simulation, treatment planning and delivery technologies into all phases of the treatment process; from treatment simulation and planning and continuing throughout beam delivery.* In this report some of the findings of Task group 101 of the AAPM will be presented which outlines the best-practice guidelines for SBRT. The task group report includes a review of the literature to identify reported clinical findings and expected outcomes for this treatment modality. Information in this task group is provided for establishing an SBRT program, including protocols, equipment, resources, and QA procedures.

1. Introduction

Over 4000 publications spanning several decades have affirmed the clinical usefulness of stereotactic radiosurgery (SRS) in the treatment of benign and malignant lesions as well as functional disorders. The radiobiological rationale for SBRT is similar to that for SRS; delivering a few fractions of large dose in relatively short overall treatment time results in a more potent biological effect. The clinical outcomes of SBRT for both primary and metastatic diseases compare favorably with surgery with minimal adverse effects. In addition, the limited number of treatment fractions makes SBRT more convenient for the patient, and a potentially more cost-effective treatment modality than traditional radiation therapy.

For both SRS and SBRT patient-specific quality assurance is a critical component in the overall quality control of radiation treatment as it provides assurance that the delivered dose will accurately match the planned dose distribution. In particular, it is important to verify that the treatment device is physically capable of delivering the planned dose distribution and that patient setup in the treatment room matches as closely as possible the patient setup at time of simulation.

Patient-specific QA becomes even more critical for Stereotactic Body Radiation Therapy (SBRT). SBRT differs from traditional radiation therapy in that it features highly conformal, hypofractionated dose delivery. Current SBRT protocols generally involve 3-5 treatments with a dose of 6-22 Gy per fraction to sites such as the spine, liver, and lung[3, 6, 7, 9, 10, 28, 33]. The desired biological effect is achieved both by fractionation and perhaps more importantly by the differential dose delivered to targeted and normal tissue; the goal is to minimize the volume of normal tissue exposed to a high dose of radiation. Therefore in SBRT the traditional gross tumor volume (GTV) and clinical tumor volume (CTV) as described in ICRU 62[12] are often treated interchangeably[29, 35, 36]. SBRT is also highly dependent on image guidance for localization and repositioning. The requirements of large doses and highly accurate targeting in SBRT mean that special attention needs to be paid to all aspects of the treatment for each patient, including immobilization, localization, pre-treatment dose verification and review of on-board imaging by the physician.

2. Problems associated with dosimetry of small/narrow field geometry

SRT, SBRT and IMRT routinely use small fields and beamlets of less than 10 mm in diameter in order to achieve the desired, highly-focused and precisely modulated dose distribution. Measurement of small photon beams is complicated by the loss of lateral electronic equilibrium, volume averaging, detector-interface artifacts, collimator effects, and detector position-orientation effects. Due to the small dimensions and steep dose gradients of photon beams used in SRS/SRT and IMRT, an appropriate dosimeter with a spatial resolution of approximately 1 mm or better (stereotactic detectors) is required to measure the basic dosimetry data, *e.g.*, the total scatter factor (SF, or relative output factor), tissue-maximum ratio (TMR), and off-axis ratios (OARs). Even with stereotactic detectors, careful detector-phantom setup, and detailed dose corrections, one might still find more than 10% discrepancies among the measurements of very small fields (< 10 mm in diameter). MLC-shaped fields have more geometry and dosimetry uncertainties than those of the circular cones. Li *et al*. demonstrate that large errors are often caused by a small setup error or measuring point displacement from the central ray of the beam[18]. For small MLC fields, the collimator leaf-edge effect is almost independent of the depth but is closely related to the field size and type of MLC. The volume effect becomes significant when the detector diameter is comparable to the half size of the small fields.

For the profile (off-axis ratio) measurement of the small photon beams, Higgins *et al*. demonstrated a simple approach to unfolding the chamber size artifact from measured small-beam profiles using typical cylindrical chambers by de-convolving the detector-response artifact from each point in the profiles. T*he maximum inner diameter of a detector should be less than half the FWHM of the smallest beam measured in order for the deconvolution of the detector-size effect to work properly.*

3. Comprehensive QA Program for SRS and SBRT

Patient-specific QA procedures for SRS/SBRT should be developed as an integrated part of a comprehensive ongoing QA program in the clinic. Therefore, before implementing an SBRT program, the clinic first needs to determine which system(s) will be used and develop QA procedures to match. SBRT-enabled systems often have specialized equipment such as immobilization systems, localization systems, and on-board imaging systems which are not always found in the clinic. In other cases, the entire system is specialized for SBRT (e.g. the Accuray Cyberknife). For example, Table 1 summarizes the stereotactic localization and image guidance strategies used by commercially-available systems. These specialized components require detailed and specialized QA procedures, over and beyond the general guidelines for external beam radiotherapy as specified in the AAPM Reports of TG 40, 142, and 45[16, 23, 15].

Table 1. Comparison of typical characteristics of 3D/IMRT radiotherapy and SBRT

Characteristic	3D/IMRT	SBRT
Dose / Fraction	1.8 – 3 Gy	6 – 30 Gy
# Fractions	10 – 30	1-5
Target definition	CTV / PTV (gross disease + clinical extension): tumor may not have a sharp boundary.	GTV / CTV / ITV/ PTV (well-defined tumors: GTV=CTV)
Margin	Centimeters	Millimeters
Physics / dosimetry monitoring	Indirect	Direct
Required setup accuracy	TG40, TG142	TG40, TG142
Primary imaging modalities used for treatment planning	CT	Multi-modality: CT/MR/PET-CT
Redundancy in geometric verification	No	Yes
Maintenance of high spatial targeting accuracy for the entire treatment	Moderately enforced (moderate patient position control and monitoring)	Strictly enforced (sufficient immobilization and high frequency position monitoring through integrated image guidance)
Need for respiratory motion management	Moderate – must be at least considered	Highest
Staff training	Highest	Highest + special SBRT training
Technology implementation	Highest	Highest
Radiobiological understanding	Moderately well understood	Poorly understood
Interaction with systemic therapies	Yes	Yes

Table 2 summarizes recommendations for annual, monthly, and daily QA activities for SBRT clinics which enable verification of overall device accuracy.

Table 2. Achievable accuracies reported in the literature categorized by body site and immobilization / repositioning device

Author / year	**Site**	**Immobilization / Repositioning**	**Reported accuracy**
Lax-1994[17]	Abdomen	Wood frame / stereotactic coordinates on box to skin marks	3.7mm Lat 5.7mm Long
Hamilton-1995[8]	Spine	Screw fixation of spinous processes to box	2mm

Murphy-1997[20]	Spine	Frameless / Implanted fiducial markers with real time imaging and tracking	1.6mm radial
Lohr-1999[19]	Spine	Body cast with stereotactic coordinates	≤ 3.6mm mean vector
Yenice – 2003[37]	Spine	Custom stereotactic frame and in-room CT guidance	1.5mm system accuracy, 2-3mm positioning accuracy
Chang-2004[2]	Spine	MI™ BodyFix with Stereotactic Frame / linac / CT on rails with 6D robotic couch	1 mm system accuracy
Tokuuye-1997[30]	Liver	Prone position Jaw and arm straps	5mm
Nakagawa-2000[22]	Thoracic	MVCT on linac	Not reported
Wulf-2000[34]	Lung, Liver	Elekta™ body frame	3.3mm lat 4.4 mm long.
Fuss-2004[4]	Lung, liver	MI™ BodyFix	Bony anatomy translation 0.4, 0.1, 1,6 mm (mean X, Y, Z); Tumor translation before image guidance 2.9, 2.5, 3.2 mm (mean X, Y, Z)
Herfarth-2001[10]	Liver	Leibinger body frame	1.8-4.4 mm
Nagata-2002[21]	Lung	Elekta™ body frame	2mm
Fukumoto-2002[3]	Lung	Elekta™ body frame	Not reported
Hara-2002[9]	Lung	Custom bed transferred to treatment unit after confirmatory scan	2mm
Hof-2003[11]	Lung	Leibinger body frame	1.8 – 4 mm
Timmerman-2003[28]	Lung	Elekta™ body frame	Approx. 5mm
Wang-2006[32]	Lung	Medical Intelligence Body Frame stereotactic coordinates / CT on rails	0.3 ± 1.8mm AP -1.8 mm±3.2mm Lat 1.5 mm ± 3.7 mm SI

As an example specifically relevant to SBRT, Table 3 lists published repositioning accuracies of various SBRT immobilization schemes. The reported accuracies vary significantly depending on the particulars of the approach, however in general errors in target localization and patient repositioning can be sorted into two categories: (1) set-up errors, which can be greatly reduced with proper localization procedures and (2) organ motion, especially motion due to the respiratory cycle, which can be highly dependent on the specific geometry and respiratory characteristics of the patient. QA procedures must therefore be developed that can address repositioning errors at the device level as well as the patient level.

Table 3. Summary of published QA recommendations for SBRT and SBRT-related techniques

Journal of Physics: Conference Series **250** (2010) 012057 doi:10.1088/1742-6596/250/1/012057

Source	**Purpose**	**Proposed Test**	**Reported Achievable Tolerance**	**Proposed Frequency**
Ryu et al, 2001[25]	End-to-end localization accuracy	Stereo x-ray/DRR fusion	1.0 to 1.2 mm root mean square	Initial commissioning and annually thereafter
Ryu et al, 2001[25]	Intrafraction targeting variability	Stereo x-ray/DRR fusion	0.2 mm average, 1.5 mm maximum	Daily (during treatment)
Verellen et al, 2003[31]	End-to-end localization accuracy	Hidden target (using stereo x-ray/DRR fusion)	0.41 ± 0.92 mm	Initial commissioning and annually thereafter
Verellen et al, 2003[31]	End-to-end localization accuracy	Hidden target (using implanted fiducials)	0.28 ± 0.36 mm	Initial commissioning and annually thereafter
Yu et al, 2004[38]	End-to-end localization accuracy	Dosimetric assessment of hidden target (using implanted fiducials)	0.68 ± 0.29 mm	Initial commissioning and annually thereafter
Sharpe et al, 2006[26]	CBCT mechanical stability	Constancy comparison to MV imaging isocenter (using hidden targets)	0.50±0.5mm	Baseline at commissioning and monthly thereafter
Galvin et al., 2008[5]	Overall positioning accuracy, including image registration (frame-based systems)	Winston-Lutz test modified to make use of the in-room imaging systems	<=2mm for multiple couch angles	Initial commissioning and monthly thereafter
Palta et al, 2008[24]	MLC accuracy	Light field, radiographic film, or EPID	<0.5mm (especially for IMRT delivery)	Annually
Solberg et al, 2008[27]	End-to-end localization accuracy	Hidden target in anthropomorphic phantom	1.10 ± 0.42 mm	Initial commissioning and annually thereafter
Jiang et al, 2008[13]	Respiratory motion tracking and gating in 4D CT	Phantoms with cyclical motion	N/A	N/A
Bissonnette et al, 2008[1]	CBCT geometric accuracy	Portal image vs. CBCT image isocenter coïncidence	±2mm	daily

Table 4 summarizes these two types of errors and lists strategies for verifying target localization, for both inter- and intra-fraction variations. Image guidance can often be used to correct for setup errors. However, organ motion is not as easily accommodated, and further strategies that include compensation for this motion must be employed.

Table 4. Types of positional errors and error management strategies for SBRT

		STRATEGY		
		Immobilization/Setup Aids	**Off-line**	**On-line**
Setup Errors	Inter-fraction	Alignment/Constraint Standard procedures Lasers/Light Field on Tattoos Thermoplast masks Tape Bite Blocks Vacu-Form molds/casts	Conventional weekly port film practice Statistical Approaches: i) Population-based thresholds. ii) Individual-based thresholds.	MV-Radiographs (conventional pre-ports) EPID MV-Radiographs On-line kV Radiographs (with/without markers) Optical Video Monitoring
	Intra-fraction	Thermoplast Body Casts Stereotactic Head Frame Stereotactic Body Frame	(see note [a])	MV Fluoroscopic kV Fluoroscopic Optical Video Monitoring Optical Reflectors/Markers
Organ Motion	Inter-fraction	Breath-hold Consistent Time-of-Day Active Breathing Control Specifications (bladder/rectum, full/empty) Patient position (prone/supine)	Off-line strategies based on repeat CT scans.	On-line Computed Tomography (CT-on-a-rail, Cone-Beam CT, Tomotherapy) Ultrasound Other imaging modalities (MRI, Ultrasound)
	Intra-fraction	Breath-hold Compression Plate Active Breathing Control	(see note [a])	Respiratory Gating Cardiac Gating MV/kV Fluoroscopy of surrogates for organ motion

[a] *Devices and procedures often serve not only to provide accurate inter-fraction alignment (setup aids) but also to constrain against intra-fraction motion (immobilization devices); hence, the distinction between inter- and intra-fraction strategies is blurred in this case.*

[b] *Although off-line correction strategies do not address intra-fraction variability directly, such strategies may provide margins which better accommodate such variability provided the inter- and intra-fraction motions are from the same distribution.*

4. Simulation and Treatment Planning

As discussed earlier, one of the characteristics of SBRT is that treatments are simulated and planned with the goal of providing the smallest possible margins to minimize the dose to normal tissue, account for motion of the tumor, and provide a repeatable set up. A variety of techniques for accounting for tumor motion within the PTV have been described and are summarized in the report of AAPM task group 76.[14] The efficacy of many of these techniques (for instance, breath-hold techniques) can vary on a patient-specific basis and therefore require patient-specific QA procedures to verify their appropriateness in any given situation.

Treatment planning may require patient-specific quality assurance procedures to ensure the dose distribution is within the appropriate dose-volume constraints for both the target and all relevant organs at risk (OAR).

The clinic should have established procedures for a second check and patient specific QA for SBRT treatments. An axial CT slice is shown for a spinal SBRT plan in Figure 1, with the PTV contoured in red.

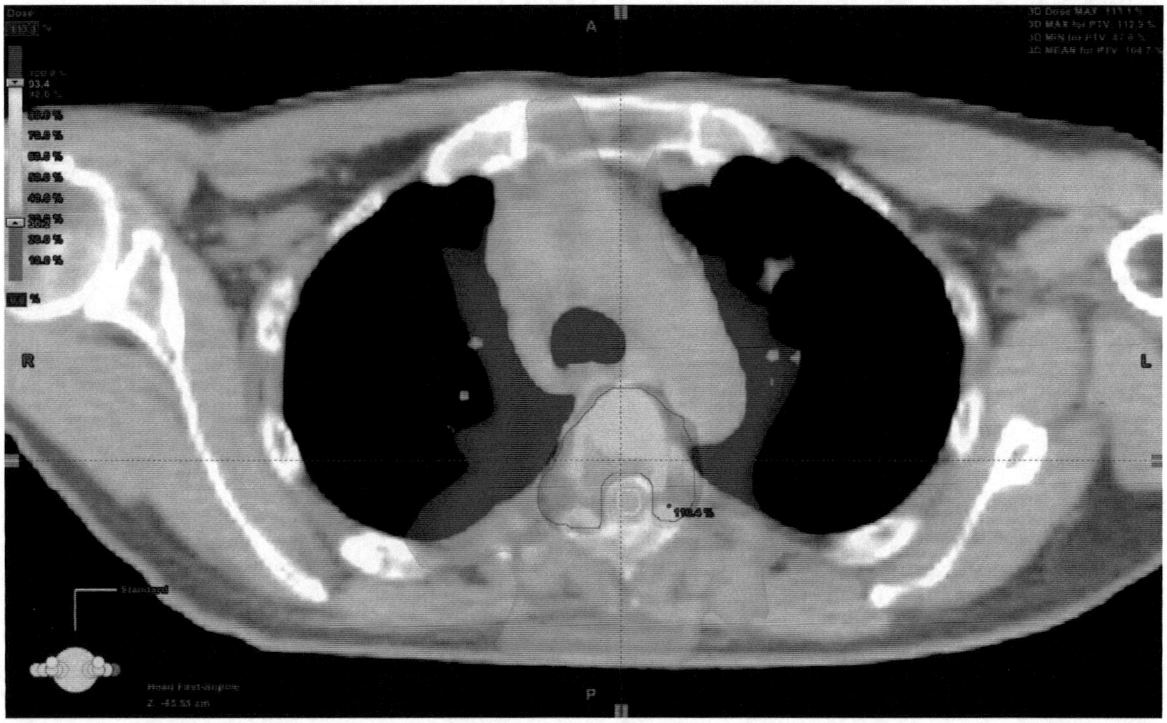

Figure 1. Axial CT slice for a spinal SBRT treatment showing the PTV (red contour) and isodose distribution

Figure 2 provides an example of strategies that could be applied to patient-specific pre-treatment dose verification. Figure 2a illustrates patient specific QA that was performed for the treatment in Figure 1 on the TomoTherapy® Hi-Art system® (TomoTherapy Inc.) for a spinal SBRT treatment. Film is used to determine the measured isodose curves for a single plane, which is then compared to the planar dose determined from the planning system, along with a point dose at the plane of the film and the gamma distribution. Additionally, the dose to a point (usually isocenter) is determined using an ionization chamber, which is then compared to the calculated dose at that point. An SBRT plan for the same patient was developed for delivery on a Varian Trilogy™ linear accelerator (Varian Medical Systems Inc.), with the QA performed using a MatriXX array (IBA Dosimetry GmbH). The isodose curves for a single plane for this treatment are shown in Figure 2b, along with the gamma distribution. The QA program used for

SBRT needs to be well understood by the planning team, including changes in the plan, such as scaling of the monitor units so that film is not saturated, that need to be performed for the specific QA technique.

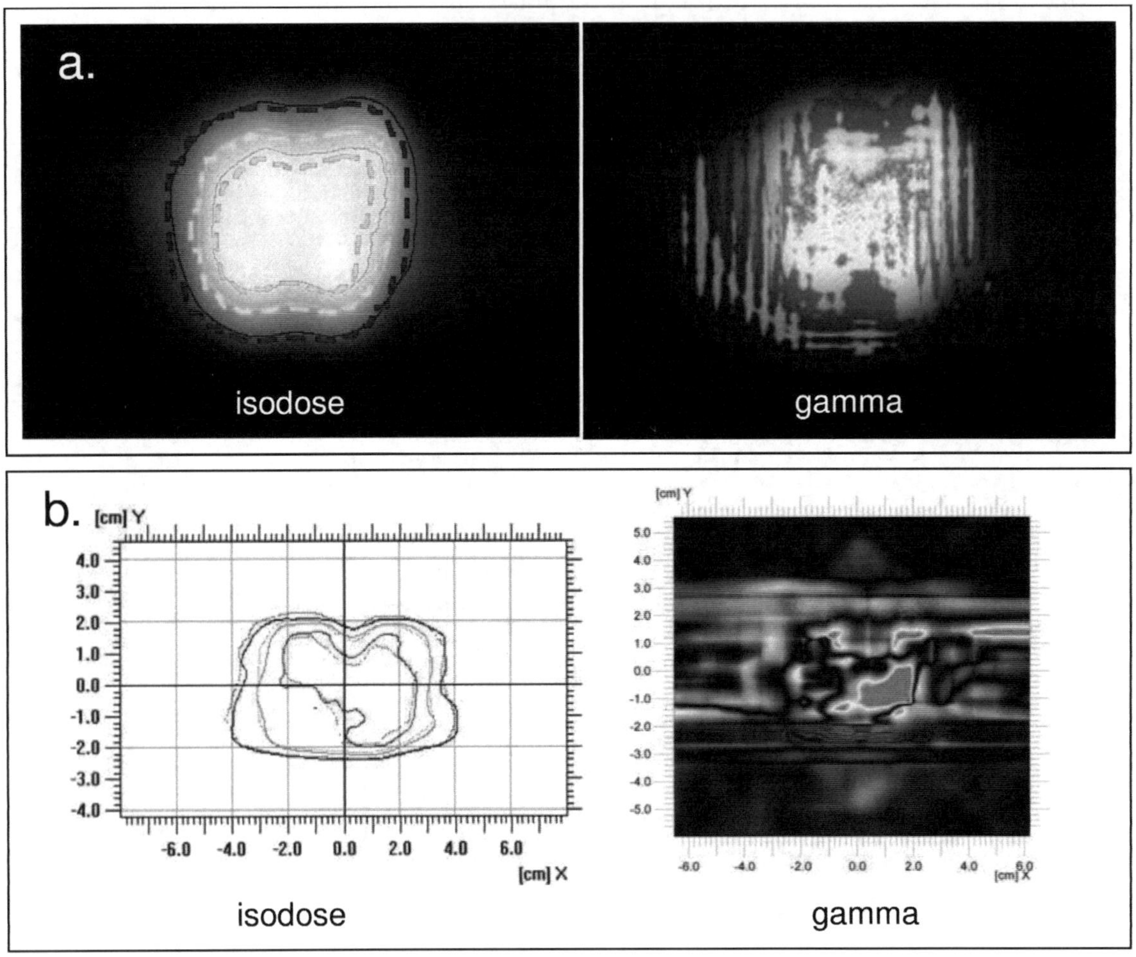

Figure 2. QA performed for the plan in Figure 1 on the a) TomoTherapy Hi-Art and b) Varian Trilogy linear accelerators.

5. Patient repositioning and treatment delivery

Current SBRT systems rely on image guidance for patient repositioning at the beginning of each fraction of treatment. Typically, simulation images are transferred to the treatment console and are co-registered with kV and/or MV images acquired with the on-board imaging from the treatment device. Offsets in the resultant co-registration are detected as setup shifts required to bring the patient into optimal setup correspondence with the simulation position. The clinic should have specific quality assurance procedures to evaluate the quality of the OBI imaging as well as the quality of the final patient setup. This may include a procedure that requires the OBI be reviewed and approved by the physician for each fraction.

6. Conclusion

Journal of Physics: Conference Series **250** (2010) 012057 doi:10.1088/1742-6596/250/1/012057

The authors have presented four tables summarizing the generalized patient specific QA issues associated with SBRT, and these include:
Table 1 Comparison of typical characteristics of 3D/IMRT radiotherapy and SBRT
Table 2: Reported accuracy of SBRT immobilization and repositioning systems.
Table 3: Summary of QA recommendations for SBRT systems
Table 4: Error-analysis strategy for set-up and organ motion.

A dosimetric analysis for a spinal SBRT is presented in two figures which demonstrates the issues that must be addressed in verification of dose planning and delivery. For further and more comprehensive discussion on patient specific QA for SBRT the readers are advised to refer to the AAPM Task Group 101.

References

[1] Bissonnette J P, Moseley D, White E, Sharpe M, Purdie T and Jaffray D A 2008 Quality assurance for the geometric accuracy of cone-beam CT guidance in radiation therapy *Int J Radiat Oncol Biol Phys* **71** S57-61

[2] Chang E L, Shiu A S, Lii M F, Rhines L D, Mendel E, Mahajan A, Weinberg J S, Mathews L A, Brown B W, Maor M H and Cox J D 2004 Phase I clinical evaluation of near-simultaneous computed tomographic image-guided stereotactic body radiotherapy for spinal metastases *Int J Radiat Oncol Biol Phys* **59** 1288-94

[3] Fukumoto S, Shirato H, Shimzu S, Ogura S, Onimaru R, Kitamura K, Yamazaki K, Miyasaka K, Nishimura M and Dosaka-Akita H 2002 Small-volume image-guided radiotherapy using hypofractionated, coplanar, and noncoplanar multiple fields for patients with inoperable Stage I nonsmall cell lung carcinomas *Cancer* **95** 1546-53

[4] Fuss M, Salter B J, Rassiah P, Cheek D, Cavanaugh S X and Herman T S 2004 Repositioning accuracy of a commercially available double-vacuum whole body immobilization system for stereotactic body radiation therapy *Technol Cancer Res Treat* **3** 59-67

[5] Galvin J M and Bednarz G 2008 Quality Assurance Procedures For Stereotactic Body Radiation Therapy *Int J Radiat Oncol Biol Phys* **71** S122-S5

[6] Gerszten P C, Burton S A, Welch W C, Brufsky A M, Lembersky B C, Ozhasoglu C and Vogel W J 2005 Single-fraction radiosurgery for the treatment of spinal breast metastases *Cancer* **104** 2244-54

[7] Gerszten P C, Ozhasoglu C, Burton S A, Vogel W J, Atkins B A, Kalnicki S and Welch W C 2004 CyberKnife frameless stereotactic radiosurgery for spinal lesions: clinical experience in 125 cases *Neurosurgery* **55** 89-98; discussion -9

[8] Hamilton A J, Lulu B A, Fosmire H, Stea B and Cassady J R 1995 Preliminary clinical experience with linear accelerator-based spinal stereotactic radiosurgery *Neurosurgery* **36** 311-9

[9] Hara R, Itami J, Kondo T, Aruga T, Abe Y, Ito M, Fuse M, Shinohara D, Nagaoka T and Kobiki T 2002 Stereotactic single high dose irradiation of lung tumors under respiratory gating *Radiother Oncol* **63** 159-63

[10] Herfarth K K, Debus J, Lohr F, Bahner M L, Rhein B, Fritz P, Hoss A, Schlegel W and Wannenmacher M F 2001 Stereotactic single-dose radiation therapy of liver tumors: results of a phase I/II trial *J Clin Oncol* **19** 164-70

[11] Hof H, Herfarth K K, Munter M, Essig M, Wannenmacher M and Debus J 2003 The use of the multislice CT for the determination of respiratory lung tumor movement in stereotactic single-dose irradiation *Strahlenther Onkol* **179** 542-7

[12] ICRU 1999 ICRU Report 62. Prescribing, recording and reporting photon beam therapy (supplement to ICRU Report 50)

[13] Jiang S B, Wolfgang J and Mageras G S 2008 Quality assurance challenges for motion-adaptive radiation therapy: gating, breath holding, and four-dimensional computed tomography *Int J Radiat Oncol Biol Phys* **71** S103-7

[14] Keall P J, Mageras G S, Balter J M, Emery R S, Forster K M, Jiang S B, Kapatoes J M, Low D A, Murphy M J, Murray B R, Ramsey C R, Van Herk M B, Vedam S S, Wong J W and Yorke E 2006 The management of respiratory motion in radiation oncology report of AAPM Task Group 76 *Med Phys* **33** 3874-900

[15] Klein E E, Hanley J, Bayouth J, Yin F F, Simon W, Dresser S, Serago C, Aguirre F, Ma L, Arjomandy B, Liu C, Sandin C and Holmes T 2009 Task Group 142 report: quality assurance of medical accelerators *Med Phys* **36** 4197-212

[16] Kutcher G J, Coia L, Gillin M, Hanson W F, Leibel S, Morton R J, Palta J R, Purdy J A, Reinstein L E, Svensson G K and et al. 1994 Comprehensive QA for radiation oncology: report of AAPM Radiation Therapy Committee Task Group 40 *Med Phys* **21** 581-618

[17] Lax I, Blomgren H, Naslund I and Svanstrom R 1994 Stereotactic radiotherapy of malignancies in the abdomen. Methodological aspects *Acta Oncol* **33** 677-83

[18] Li S, Rashid A, He S and Djajaputra D 2004 A new approach in dose measurement and error analysis for narrow photon beams (beamlets) shaped by different multileaf collimators using a small detector *Med Phys* **31** 2020-32

[19] Lohr F, Debus J, Frank C, Herfarth K, Pastyr O, Rhein B, Bahner M L, Schlegel W and Wannenmacher M 1999 Noninvasive patient fixation for extracranial stereotactic radiotherapy *Int J Radiat Oncol Biol Phys* **45** 521-7

[20] Murphy M J 1997 An automatic six-degree-of-freedom image registration algorithm for image-guided frameless stereotaxic radiosurgery *Med Phys* **24** 857-66

[21] Nagata Y, Negoro Y, Aoki T, Mizowaki T, Takayama K, Kokubo M, Araki N, Mitsumori M, Sasai K, Shibamoto Y, Koga S, Yano S and Hiraoka M 2002 Clinical outcomes of 3D conformal hypofractionated single high-dose radiotherapy for one or two lung tumors using a stereotactic body frame *Int J Radiat Oncol Biol Phys* **52** 1041-6

[22] Nakagawa K, Aoki Y, Tago M, Terahara A and Ohtomo K 2000 Megavoltage CT-assisted stereotactic radiosurgery for thoracic tumors: original research in the treatment of thoracic neoplasms *Int J Radiat Oncol Biol Phys* **48** 449-57

[23] Nath R, Biggs P J, Bova F J, Ling C C, Purdy J A, van de Geijn J and Weinhous M S 1994 AAPM code of practice for radiotherapy accelerators: report of AAPM Radiation Therapy Task Group No. 45 *Med Phys* **21** 1093-121

[24] Palta J R, Liu C and Li J G 2008 Quality assurance of intensity-modulated radiation therapy *Int J Radiat Oncol Biol Phys* **71** S108-12

[25] Ryu S I, Chang S D, Kim D H, Murphy M J, Le Q T, Martin D P and Adler J R, Jr. 2001 Image-guided hypo-fractionated stereotactic radiosurgery to spinal lesions *Neurosurgery* **49** 838-46

[26] Sharpe M B, Moseley D J, Purdie T G, Islam M, Siewerdsen J H and Jaffray D A 2006 The stability of mechanical calibration for a kV cone beam computed tomography system integrated with linear accelerator *Med Phys* **33** 136-44

[27] Solberg T D, Medin P M, Mullins J and Li S 2008 Quality assurance of immobilization and target localization systems for frameless stereotactic cranial and extracranial hypofractionated radiotherapy *Int J Radiat Oncol Biol Phys* **71** S131-5

[28] Timmerman R, Papiez L, McGarry R, Likes L, DesRosiers C, Frost S and Williams M 2003 Extracranial stereotactic radioablation: results of a phase I study in medically inoperable stage I non-small cell lung cancer *Chest* **124** 1946-55

[29] Timmerman R D 2008 An overview of hypofractionation and introduction to this issue of seminars in radiation oncology *Semin Radiat Oncol* **18** 215-22

[30] Tokuuye K, Sumi M, Ikeda H, Kagami Y, Murayama S, Nakayama H, Kawashima M and Ishii H 1997 Technical considerations for fractionated stereotactic radiotherapy of hepatocellular carcinoma *Jpn J Clin Oncol* **27** 170-3

[31] Verellen D, Soete G, Linthout N, Van Acker S, De Roover P, Vinh-Hung V, Van de Steene J and Storme G 2003 Quality assurance of a system for improved target localization and patient set-up that combines real-time infrared tracking and stereoscopic X-ray imaging *Radiother Oncol* **67** 129-41

[32] Wang L, Feigenberg S, Chen L, Pasklev K and Ma C C 2006 Benefit of three-dimensional image-guided stereotactic localization in the hypofractionated treatment of lung cancer *Int J Radiat Oncol Biol Phys* **66** 738-47

[33] Whyte R I, Crownover R, Murphy M J, Martin D P, Rice T W, DeCamp M M, Jr., Rodebaugh R, Weinhous M S and Le Q T 2003 Stereotactic radiosurgery for lung tumors: preliminary report of a phase I trial *Ann Thorac Surg* **75** 1097-101

[34] Wulf J, Hadinger U, Oppitz U, Olshausen B and Flentje M 2000 Stereotactic radiotherapy of extracranial targets: CT-simulation and accuracy of treatment in the stereotactic body frame *Radiother Oncol* **57** 225-36

[35] Wulf J, Hadinger U, Oppitz U, Thiele W and Flentje M 2003 Impact of target reproducibility on tumor dose in stereotactic radiotherapy of targets in the lung and liver *Radiother Oncol* **66** 141-50

[36] Wulf J, Hadinger U, Oppitz U, Thiele W, Ness-Dourdoumas R and Flentje M 2001 Stereotactic radiotherapy of targets in the lung and liver *Strahlenther Onkol* **177** 645-55

[37] Yenice K M, Lovelock D M, Hunt M A, Lutz W R, Fournier-Bidoz N, Hua C H, Yamada J, Bilsky M, Lee H, Pfaff K, Spirou S V and Amols H I 2003 CT image-guided intensity-modulated therapy for paraspinal tumors using stereotactic immobilization *Int J Radiat Oncol Biol Phys* **55** 583-93

[38] Yu C, Main W, Taylor D, Kuduvalli G, Apuzzo M L and Adler J R, Jr. 2004 An anthropomorphic phantom study of the accuracy of Cyberknife spinal radiosurgery *Neurosurgery* **55** 1138-49

Journal of Physics: Conference Series **250** (2010) 012058 doi:10.1088/1742-6596/250/1/012058

Dose Verification of Stereotactic Radiosurgery Treatment for Trigeminal Neuralgia with Presage 3D Dosimetry System

Z. Wang[1], A. Thomas[1], J. Newton[1], G. Ibbott[2], J. Deasy[3], M. Oldham[1]

[1] Duke University Medical Center, Durham, NC [2] MD Anderson Cancer Center, Houston, TX [3] Washington University, St. Louis, MO

Zhiheng.wang@duke.edu

Abstract. Achieving adequate verification and quality-assurance (QA) for radiosurgery treatment of trigeminal-neuralgia (TGN) is particularly challenging because of the combination of very small fields, very high doses, and complex irradiation geometries (multiple gantry and couch combinations). TGN treatments have extreme requirements for dosimetry tools and QA techniques, to ensure adequate verification. In this work we evaluate the potential of Presage/Optical-CT dosimetry system as a tool for the verification of TGN distributions in high-resolution and in 3D. A TGN treatment was planned and delivered to a Presage 3D dosimeter positioned inside the Radiological-Physics-Center (RPC) head and neck IMRT credentialing phantom. A 6-arc treatment plan was created using the iPlan system, and a maximum dose of 80Gy was delivered with a Varian Trilogy machine. The delivered dose to Presage was determined by optical-CT scanning using the **D**uke **L**arge field-of-view **O**ptical-CT **S**canner (DLOS) in 3D, with isotropic resolution of 0.7mm^3. DLOS scanning and reconstruction took about 20minutes. 3D dose comparisons were made with the planning system. Good agreement was observed between the planned and measured 3D dose distributions, and this work provides strong support for the viability of Presage/Optical-CT as a highly useful new approach for verification of this complex technique.

1. Introduction

The radiation dose for stereotactic radiosurgery (SRS) treatment of trigeminal neuralgia is usually several folds higher than typical SRS doses for brain tumors [1]. The aperture of the radiation beam for trigeminal neuralgia SRS treatment is usually very small, i.e., 4~5 mm in diameter. The dose measurement and characterization of such treatment can be challenging since conventional dose measurement devices, such as ion chambers, may exhibit volume averaging artifacts, and also it is difficult to accurately place the dosimeter at the absolute maximum dose point. Presage dosimetry system can measure the three-dimensional (3D) dose distribution with very fine spatial resolution, and therefore can potentially be an ideal tool to measure the dose for trigeminal neuralgia SRS. This study investigated the feasibility of using the Presage dosimetry system to measure the 3D dose distribution of a trigeminal SRS treatment with a Radiological Physics Center (RPC) head neck (HN) phantom. The measured dose was compared with the planned dose.

2. Materials and Methods

Journal of Physics: Conference Series **250** (2010) 012058 doi:10.1088/1742-6596/250/1/012058

2.1. Phantom setup

A cylindrical Presage dosimeter (~12cm diameter by 15cm length) was inserted into the RPC HN phantom at the position of typical trigeminal neuralgia SRS treatment. A Brainlab head ring was mounted to the phantom, as shown in figure 1.

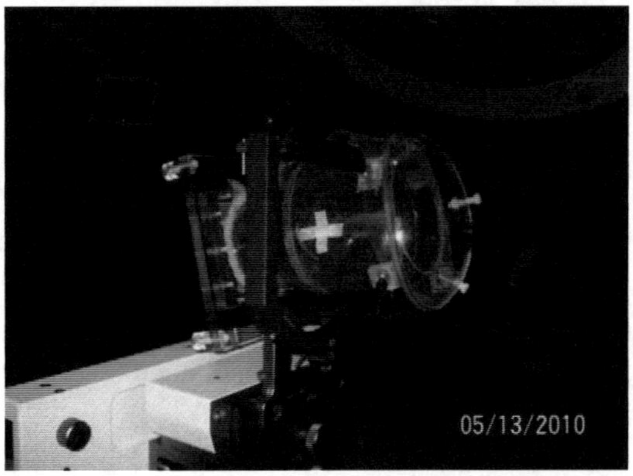

Figure 1: HN phantom mounted in a head ring.

2.2. SRS Planning

A CT scan of the phantom mounted on the head ring was performed with a Brainlab localizer. The slice thickness of the CT images was 0.625 mm. The CT images were sent to a Brainlab iPlan system for treatment planning.

The CT images were first localized with iPlan and an isocenter was placed at a typical trigeminal neuralgia SRS treatment position. 7 arcs with a total of 700 degrees were used in the treatment plan. 6 MV photons with a circular collimator of 4 mm in diameter were used for treatment. 80 Gy was prescribed to the maximum dose point. The dose calculation voxel size was set to 1 mm x 1 mm x 1 mm. The Presage dosimeter was then taken out of the phantom for pre-scanning using the Duke Large Optical-CT Scanner (DLOS) described in an accompanying abstract (Thomas et al. [2], see also Sakhalkar et al. [3]).

2.3. Treatment Delivery

A Novalis TX linear accelerator (LINAC) was used for treatment. The dosimeter was reinserted into the phantom before the treatment. Accurate and reproducible insert alignment within the phantom was achieved by marks scratched on the dosimeter, and aligned with matching marks on the RPC phantom. The phantom was then mounted to the LINAC couch as shown in Fig. 1,and localized to the isocenter position with a localization box. A set of cone beam CT (CBCT) images was then obtained and matched to the planning CT. The phantom was then automatically moved to the treatment position based on CBCT matched result. 7 arcs were then delivered at 1000 MU/minute.

2.4. Dose Measurement

The Presage dosimeter was taken out from the phantom after the treatment was completed and immediately scanned with the DLOS optical CT scanner.

The calibration curve was obtained separately with 5 calibration points determined from small cuvettes of Presage from the same batch (1x1x5cm) irradiated to varying known doses.. The linearity of dose response of Presage dosimeter was also tested up to 80 Gy. The calibration curve was applied to convert Optical-CT pixel values (corresponding to optical attenuation coefficients) to radiation doses. Planning CT doses were exported from iPlan SRS planning system and fused to the Presage scans for comparison.

3. Results

3.1. SRS Plan Isodoses

The isodoses of the SRS plan was displayed in figure 2. The monitor units (MU) numbers for the arcs were 2844, 2851, 2518, 2450, 2558, 2479, and 2306.

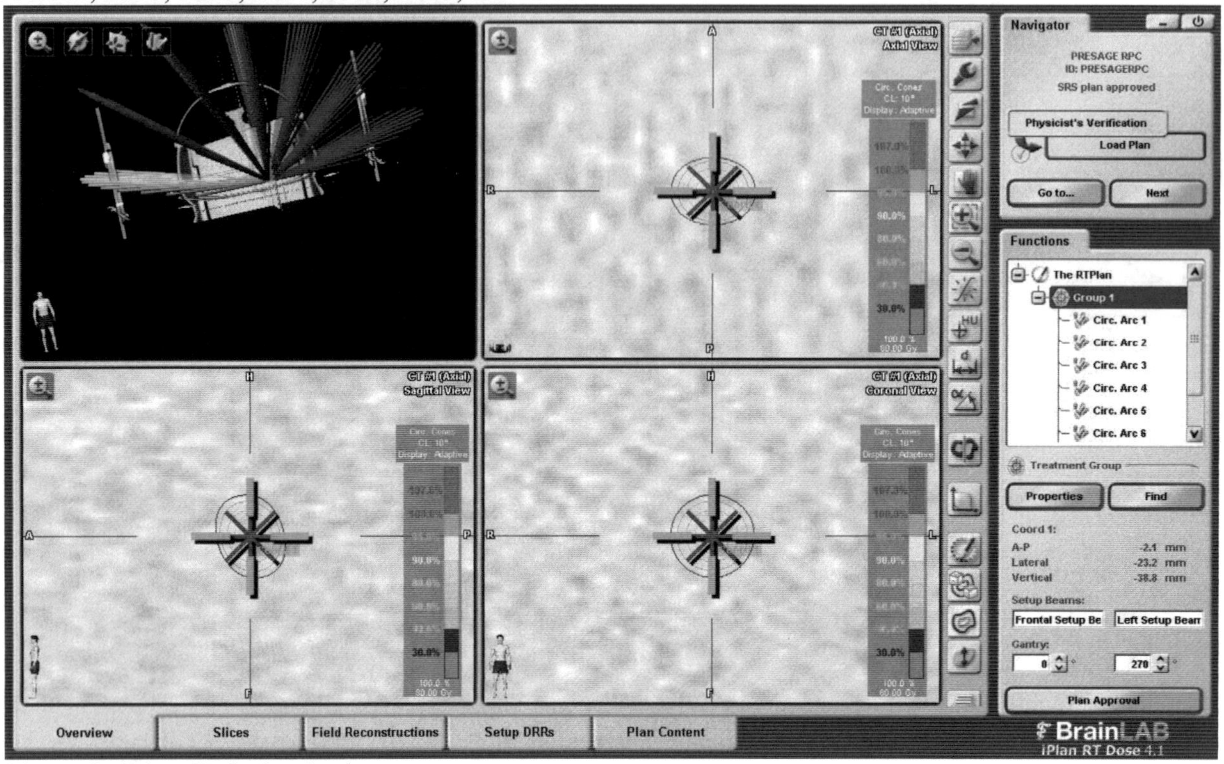

Figure 2: iPlan SRS plan isodoses

3.2. Presage Linearity Test and Calibration

The Presage test showed a linear response within the whole dose range up to the maximum dose we tested of 80Gy. The calibration curve is shown in Figure 3.

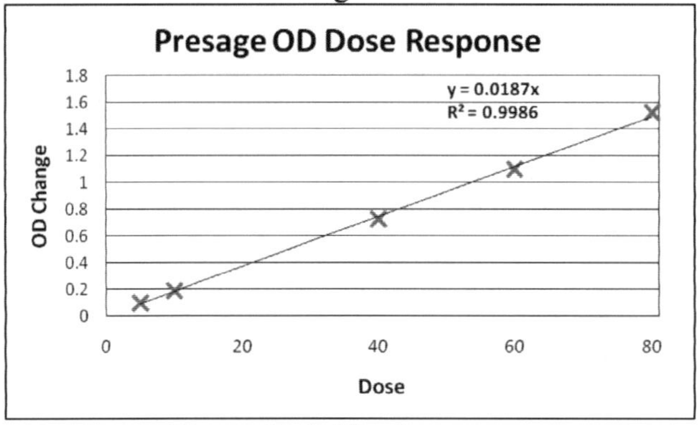

Figure 3: Calibration curve

3.3. Presage dose measurement and comparison

The calculated and measured 3D dose distributions (iPlan and Presage respectively) were imported into CERR [4], registered, and compared using overlays, and profiles (Figure 4 a-c). The dose profiles of the measured and corresponding planning doses are plotted on the right side of the pictures. The relative dose-differences are displayed below the profiles.

The dose volume histograms (DVH) were also plotted for both the measured and planned dose, as shown in Fig. 4 d). The maximum dose as measured with Presage was 85.0 cGy. It matched reasonably well with the prescribed maximum dose of 80.0 Gy.

Journal of Physics: Conference Series **250** (2010) 012058 doi:10.1088/1742-6596/250/1/012058

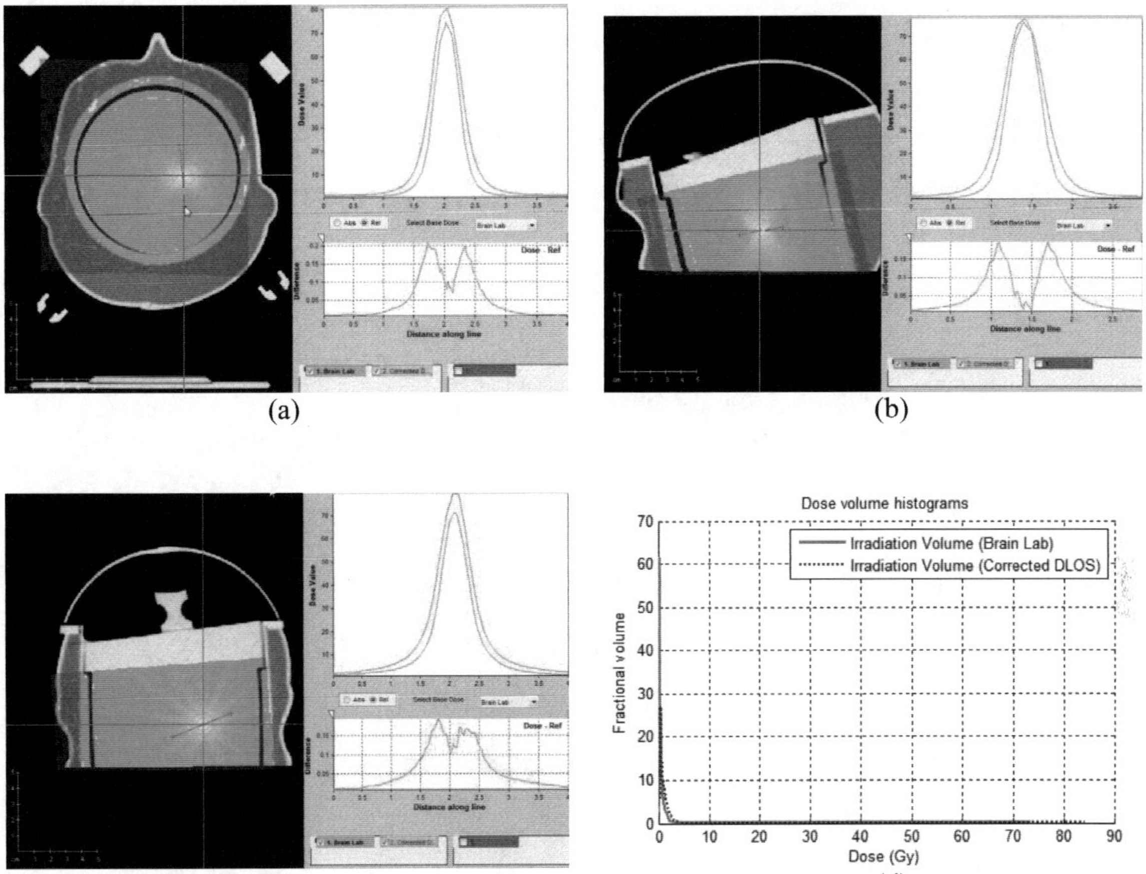

(a) (b) (c) (d)

Figure 4: Measured doses and profiles (a-c), and dose volume histogram (d).

4. Discussion and Conclusions

The maximum dose as measured by the Presage dosimeter was slightly higher than the planned maximum dose. We speculate that the higher dose reported by Presage could be closer to the actual delivered dose because of the fine spatial resolution of the Presage measurement, i.e., a reduced volume averaging effect when compared with the commissioning data used in iPlan. The profiles of the measured dose were slightly wider than the planned dose by about 1 mm. This difference may also be real, caused by the imperfect isocenter of the treatment machine. As the gantry and couch rotate between different arcs, slight variations in the isocenter position may lead to a broadening of the delivered distribution when compared to the ideal calculation in iPlan which has no allowance for gantry sag and couch drift. This feasibility study showed that Presage dosimetry system works well for the dose measurement of the trigeminal neuralgia SRS treatment.

References

[1] Z.A. Smith et al, Dedicated linear accelerator radiosurgery for the treatment of trigeminal neuralgia. J. Neurosurg. 99:511-516, 2003.

[2] A. Thomas et al, An investigation into the feasibility of a fast, large field-of-view, telecentric optical-CT scanning system for 3D radiochromic dosimetry. Abstract #

[3] H.S. Sakhalkar et al, A comprehensive evaluation of the PRESAGE/optical-CT 3D dosimetry system. Med. Phys. 36:71-82, 2009

[4] J.O. Deasy et al, CERR: A Computational Envirment for Radiotherapy Research. Med. Phys. 30:979-85, 2003.

Journal of Physics: Conference Series **250** (2010) 012059 doi:10.1088/1742-6596/250/1/012059

3D geometric gel dosimetry verification of intraprostatic fiducial guided hypofractionated radiotherapy of prostate cancer

Fredrik Nordström[1,2], Sofie Ceberg[1,2], Sacha af Wetterstedt[2], Per Nilsson[3,4], Crister Ceberg[5], Sven ÅJ Bäck[1,2]

[1]Department of Medical Radiation Physics, Lund University, Malmö, Sweden

[2]Department of Radiation Physics, Skåne University Hospital, Malmö, Sweden

[3]Department of Radiation Physics, Skåne University Hospital, Lund, Sweden

[4]Department of Radiation Sciences, Umeå University, Umeå, Sweden

[5]Department of Medical Radiation Physics, Lund University, Lund, Sweden

Fredrik.Nordstrom@med.lu.se

Abstract. This pre-study is aimed to investigate the feasibility of a normoxic polyacrylamide gel (nPAG) dosimeter with implanted gold fiducials to evaluate the geometric precision, including setup correction strategies, in the delivery of hypofractionated treatments. For this purpose a phantom consisting of three parts was constructed: (1) the patient simulating volume, providing realistic scatter conditions and weight, (2) a bottle containing the active dosimetric volume and (3) the gold fiducials and the fiducial support structure. A 6.1 Gy prostate IMRT treatment was delivered to the phantom using the sliding-window technique. The phantom was positioned prior to the treatment using the implanted fiducials and kV on-board imaging. An overlay of the 95% isosurface of the TPS calculated dose distribution and the measured dose distribution using gel showed good agreement. The clinical target volume (CTV) was well centred inside the 95% isodose surface of the measured volume. It was shown for the evaluated case that the use of on-board imaging and integrated setup correction tools could be used to compensate for a deliberately introduced offset in CTV position. The study showed that MRI based nPAG gel dosimetry can be used to verify setup correction procedures using implanted gold fiducials.

1. Introduction

Several studies have suggested that dose-escalated external radiotherapy could prolong the freedom from failure time, measured as a rising prostate-specific antigen (PSA) [1-3]. On the other hand, preliminary results from a randomized MD Anderson study reported an increase of rectal toxicity in the dose-escalation arm by a factor of two [4]. If the prostate is localized prior to treatment and by adjusting the patient position to account for the inter-fraction prostate movement, the planning target volume (PTV) could be reduced [5]. This would reduce the dose to the surrounding normal tissue, including rectum, and potentially reduce side-effects of the treatment. Image guided radiotherapy (IGRT) has been widely adopted to minimize the effects of inter-fractional motion. Intraprostatic gold fiducials are the most wide-spread approach for IGRT of the prostate in clinical practice [6]. The importance of prostate localization is undoubtedly higher when increasing the fraction dose.

A randomized multicenter phase III study (HYPO-RT-PC) of patients with intermediate risk prostate cancer has been initiated [7]. Conventional fractionation (2.0 Gy in 39 fractions) to a total absorbed dose of 78.0 Gy is compared with hypofractionation (6.1 Gy in 7 fractions) to a total absorbed dose of 42.7 Gy. All fractions are delivered with either conventional (3D-CRT) or intensity modulated radiotherapy (IMRT). The position of the prostate, as determined by the implanted gold fiducials, is verified prior to each fraction with kV/MV portal imaging or cone beam CT.

Polymer gel dosimetry can be used to measure absorbed dose distributions in a complete volume with high spatial resolution [8, 9]. The use of gel dosimetry has also been found feasible for verification of dynamic delivery [10, 11].

This pre-study aims to investigate the feasibility of a normoxic polyacrylamide gel (nPAG) dosimeter with implanted gold fiducials to evaluate the dosimetric consequences of the setup correction strategies used at the centers participating in the HYPO-RT-PC study.

2. Material and methods

2.1. Gel preparation

In this study a single batch of normoxic polyacrylamide gel (nPAG) was used, based on 3% w/w acrylamide (electrophoresis grade, ≥99%, powder, Sigma Aldrich) and 3% w/w N,N′-methylenebisacrylamide (electrophoresis grade, ≥98%, powder, Sigma Aldrich). Gelatine (300 bloom, Sigma Aldrich) was used as the matrix substance and tetrakis(hydroxymethyl)–phosphonium chloride (techn. ~80% in water, Sigma Aldrich) was used as an oxygen scavenger. The remaining constituent was ultra-pure deionized water (resistivity > 18.2 MΩ cm). The method used for gel preparation has been described elsewhere [12]. Vials containing the gel were irradiated with an absorbed dose ranging from 1 to 7 Gy in order to assure the linearity of the gel dose response for this batch of gel.

2.2. Phantom

The phantom consists of three parts: the patient simulating volume, providing realistic scatter conditions and weight, a bottle containing the active dosimetric volume and the fiducials and the fiducial support structure (figure 1). The outer part of the phantom was created using polystyrene slabs of 5 mm thickness that were glued together and modified to an oval shape to simulate the body outline in the pelvic region. A hole was drilled in the center of the polystyrene that fits an oxygen resistant glass bottle, containing the nPAG gel. The fiducials (1 mm diameter, 5 mm length) were fixated at the ends of three 5 mm thick polymethyl methacrylate (PMMA) rods attached to the lid of the bottle.

2.3. CT scanning and treatment planning

Treatment planning was performed on planar CT images of the phantom with a slice thickness of 3 mm, in accordance with the HYPO-RT-PC study protocol. Structures from a dummy patient was imported onto the phantom and a six field, 6 MV, sliding-window IMRT treatment plan was created with a fractional dose of 6.1 Gy (figure 1). The planning target volume (PTV) included the prostate (CTV) with a margin of 7 mm in all directions. All DVH constraints and technical aspects of the study protocol were fulfilled. The TPS calculated dose matrix was interpolated from 2.5 × 2.5 × 3 mm to 1 × 1 × 3 mm using cubic spline.

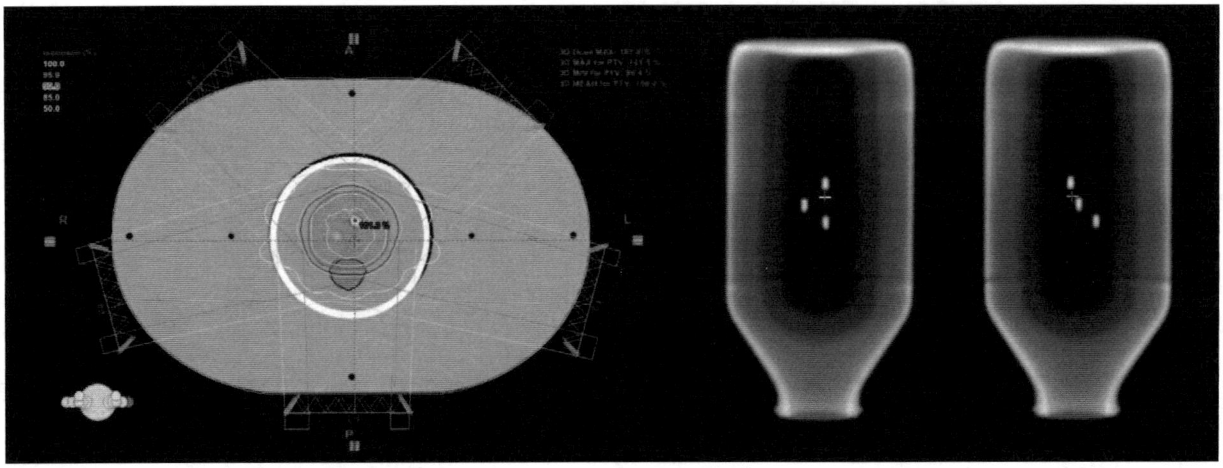

Figure 1. The six field, 6MV, sliding-window IMRT treatment plan with structures from a dummy patient (left). The fiducials are clearly visible in the digital reconstructed radiographs (DRR) generated in 0° and 270°, respectively (middle and right).

2.4. Treatment delivery

A Clinac 2100C/D (Varian medical systems) equipped with an x-ray on-board imager (OBI) was used for treatment delivery. The phantom was intentionally placed at an offset position from iso-center, with a rotation of the glass bottle around the longitudinal axis. Planar imaging at 0° and 270° using the OBI was performed and the integrated automatic correction possibilities (longitudinal, lateral and vertical couch movement) in the Aria verification system (Varian medical systems) were used to correct for the simulated misplacement. The dose-rate was 400 monitor units per minute and Portal Dosimetry (Varian medical systems) was performed prior to the gel dosimetry in order to exclude possible machine-dependent errors.

2.5. Magnetic resonance imaging

Magnetic resonance imaging (MRI) of the gels was carried out 24 hours after the irradiation, using a 1.5 T MRI unit (Siemens Medical Systems) and a circularly polarized receive-only head coil. The images were acquired using a 32-echo multi spin echo sequence with an inter-echo spacing of 25 ms and a repetition time of 4000 ms. The voxel size was 1 × 1 × 3 mm. To obtain an accurate background signal, an unirradiated gel bottle of the same dimensions was also scanned. An in-house developed software was used for calculation of the transversal relaxation rate ($R2 = 1 / T2$) [13].

3. Results and discussion

The γ pass-ratio between the measured and calculated EPID dose distribution agreed within 96 % using a 3% / 3 mm criterion, confirming that there was no machine dependent delivery errors.

The planar images obtained with the OBI at 0° and 270° were matched with the DRRs generated in the TPS using the Aria (Varian medical systems) verification system, and the resulting geometrical corrections were used to account for the present offset (figure 2, table 1).

Journal of Physics: Conference Series **250** (2010) 012059 doi:10.1088/1742-6596/250/1/012059

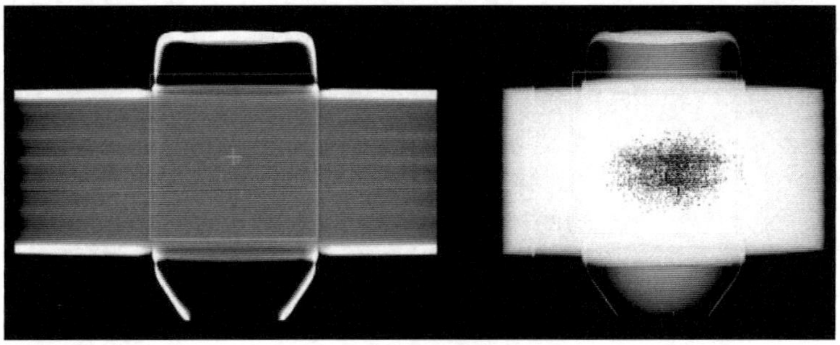

Figure 2. kV images of the phantom taken with the OBI at 0° (left) and 270° (right). The images were matched against the DRR generated in the TPS and used to correct for the offset.

Table 1. Position of the phantom relative to the iso-center determined using an external laser system (LAP Laser Applications).

	Initial position	**Position after correction**
Longitudinal	-7 mm	+2 mm
Lateral	+3 mm	-1 mm
Vertical	+5 mm	+2 mm
Rotation	6.3° CCW	6.3° CCW

The gel results were normalized to a TPS calculated dose using a region of homogenous dose in a slice without the presence of the fiducials or the fiducial support structure. Evaluation of the irradiated vials confirmed the linearity of the gel dose response for this batch of gel ($R^2 = 0.98$). An overlay of the 95% isosurface of the TPS calculated dose distribution and the measured dose distribution using gel showed good agreement (figure 3, left). The numbers of voxels inside the 95% dose level were 46586 and 45303 for the calculated and measured dose distribution, respectively. Further, the CTV volume was well centred inside the 95% isodose surface of the measured volume (figure 3, right). An increase of the R2 value was apparent in the region surrounding the fiducial support structure. This MRI read-out artefact typically corresponded to a relative dose increase of 25% in a radius of 1 cm around the structure compared to the TPS calculated dose distribution. However, this effect did not compromise the determination of the location of the 95% isodose surface as the fiducials were located centrally in the volume of interest.

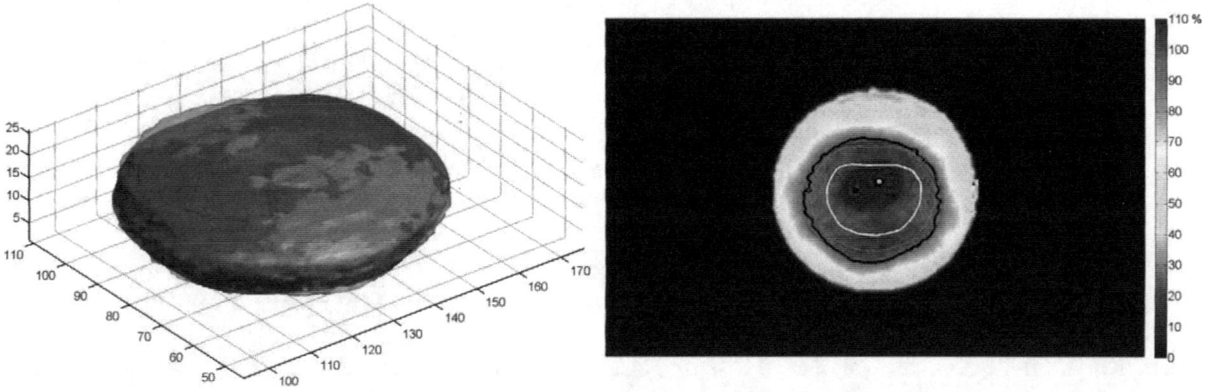

Figure 3. An overlay of the TPS calculated dose distribution (red) and the measured dose distribution (green) using alpha blending (left). Only the 95% isodose are shown. A 2D slice with both the measured and calculated 95% isodoses (black), as well as the CTV and PTV structures (white) are shown to the right.

4. Conclusions

This study showed that MRI based nPAG gel dosimetry can be used to verify setup correction procedures using implanted gold fiducials. For the case used in this pre-study, it was shown that the use of on-board imaging and integrated setup correction tools could be used to compensate for a deliberately introduced offset in CTV position. By expanding this study to include more patient cases, it should be possible to draw more general conclusions about the IGRT setup correction capabilities. This would be of value in order to determine the part of the PTV margin magnitude intended to account for inter-fractional movement of the prostate.

Acknowledgements

This study was financially supported by the Swedish Cancer Society.

References

[1] Hanks G E, Lee W R, Hanlon A L, Hunt M, Kaplan E, Epstein B E et al. 1996 Conformal technique dose escalation for prostate cancer: Biochemical evidence of improved cancer control with higher doses in patients with pretreatment prostate-specific antigen ≥10 ng/ml. *Int J Radiat Oncol Biol Phys* **35** 861-8

[2] Pollack A, Zagars G K 1997 External beam radiotherapy dose response of prostate cancer. *Int J Radiat Oncol Biol Phys* **39** 1011-8

[3] Zelefsky M J, Leibel S A, Gaudin P B, Kutcher G J, Fleshner N E, Venkatramen E S et al. 1998 Dose escalation with three-dimensional conformal radiation therapy affects the outcome in prostate cancer. *Int J Radiat Oncol Biol Phys* **41** 491-500

[4] Sorey M R, Pollack A, Zagars G, Smith L, Antolak J, Rosen I 2000 Complications from radiotherapy dose escalation in prostate cancer: preliminary results of a randomized trial. *Int J Radiat Oncol Biol Phys* **48** 635-42

[5] Balter J M, Lam K L, Sandler H M, Littles J F, Bree R L, Ten Haken R K 1995 Automated localization of the prostate at the time of treatment using implanted radiopaque markers: Technical fesibility. *Int J Radiat Oncol Biol Phys* **33** 1281-6

[6] Kupelian P A, Langen K M, Willoughby T R, Zeidan O A, Meeks S L 2008 Image-guided radiotherapy for localized prostate cancer: treating a moving target. *Semin Radiat Oncol* **18** 58-66

[7] HYPO-RT-PC Phase III study of HYPOfractionated RadioTherapy of intemediate risk localised Prostate Cancer, *http://www.controlled-trials.com/ISRCTN45905321*

[8] Vergote K, De Deene Y, Duthoy W, De Neve W, Achten E, De Wagter C 2004 Validation application of polymer gel dosimetry for the dose verification of an intensity-modulated arc therapy (IMAT) treatment. *Phys Med Biol* **49** 287-305

[9] Isbakan F, Ûlgen Y, Bilge H, Ozen Z, Agus O, Buyuksarac B 2007. Gamma Knife 3-D dose distribution near the area of tissue inhomogeneities by normoxic gel dosimetry. *Med Phys* **34** 1623-30

[10] Gustavsson H, Karlsson A, Bäck S Å J, Olsson L E 2003 MAGIC-type polymer gel for three-dimensional dosimetry: Intensity-modulated radiation therapy verification. *Med Phys* **30** 1264-70

[11] Ceberg S, Karlsson A, Gustavsson H, Wittgren L, Bäck S Å J 2008 Verification of dynamic radiotherapy: the potential for 3D dosimetry under respiratory-like motion using polymer gel. *Phys Med Biol.* **53** 387-96

[12] Karlsson A, Gustavsson H, Månsson S, McAuley K B, Bäck S Å J 2007 Dose integration characteristics in normoxic polymer gel dosimetry investigated using sequential beam irradiation. *Phys Med Biol* **52** 4697-706

[13] Karlsson A 2007 Characterization and clinical application of normoxic polymer gel in radiation therapy dosimetry. *Ph.D.-thesis, Lund University, Malmö*

IC3DDose: The 6th International Conference on 3D Radiation Dosimetry IOP Publishing
Journal of Physics: Conference Series **250** (2010) 012060 doi:10.1088/1742-6596/250/1/012060

On the use of polymer gels for assessing the total geometrical accuracy in clinical Gamma Knife radiosurgery applications

A. Moutsatsos[1], P. Karaiskos[1,2], L. Petrokokkinos[1,3], K. Zourari [1], E. Pantelis[1], L. Sakelliou[3], I. Seimenis[4], C. Constantinou[5], A. Peraticou[5] and E. Georgiou[1].

[1]Medical Physics Laboratory, Medical School, University of Athens, 75 Mikras Asias, 115 27 Athens, Greece

[2]Medical Physics and Gamma Knife Department, Hygeia Hospital, Kifisias Avenue & 4 Erythrou Stavrou, Marousi, 151 23 Athens, Greece

[3]Nuclear and Particle Physics Section, Physics Department, University of Athens, Panepistimioupolis, Ilisia, 157 71 Athens, Greece

[4]"Ayios Therissos" Medical Diagnostic Center, 92 Sp. Kyprianou Avenue, 2033, Nicosia, Cyprus

[5]Bank of Cyprus Oncology Center, 32 Acropoleos Avenue, 2006, Nicosia, Cyprus

Corresponding author's e-mail: armouts@phys.uoa.gr

Abstract. The nearly tissue equivalent MRI properties and the unique ability of registering 3D dose distributions of polymer gels were exploited to assess the total geometrical accuracy in clinical Gamma Knife applications, taking into account the combined effect of the unit's mechanical accuracy, dose delivery precision and the geometrical distortions inherent in MR images used for irradiation planning. Comparison between planned and experimental data suggests that the MR-related distortions due to susceptibility effects dominate the total clinical geometrical accuracy which was found within 1 mm. The dosimetric effect of the observed sub-millimetre uncertainties on single shot GK irradiation plans was assessed using the target percentage coverage criterion, and a considerable target dose underestimation was found.

1. Introduction

Gamma Knife (GK) stereotactic radiosurgery (SRS) is a well established treatment approach for the management of a wide variety of intracranial lesions. The efficiency of the technique is based on the high geometrical accuracy in target localization and the high precision in the delivery of a therapeutic radiation dose to the target, which facilitates restriction of the dose to surrounding critical structures. Besides the excellent mechanical accuracy of a GK unit (less than 0.5mm) [1], the total clinical accuracy of the technique may be relented due to the spatial distortion inherent in the magnetic resonance (MR) images used for target volume prescription [2].

This work presents an experimental procedure based on polymer gels used to assess the total geometrical accuracy in clinical GK applications taking into account every link in the GK treatment chain; from patient imaging and irradiation planning on clinical MR images, to patient positioning and GK dose delivery.

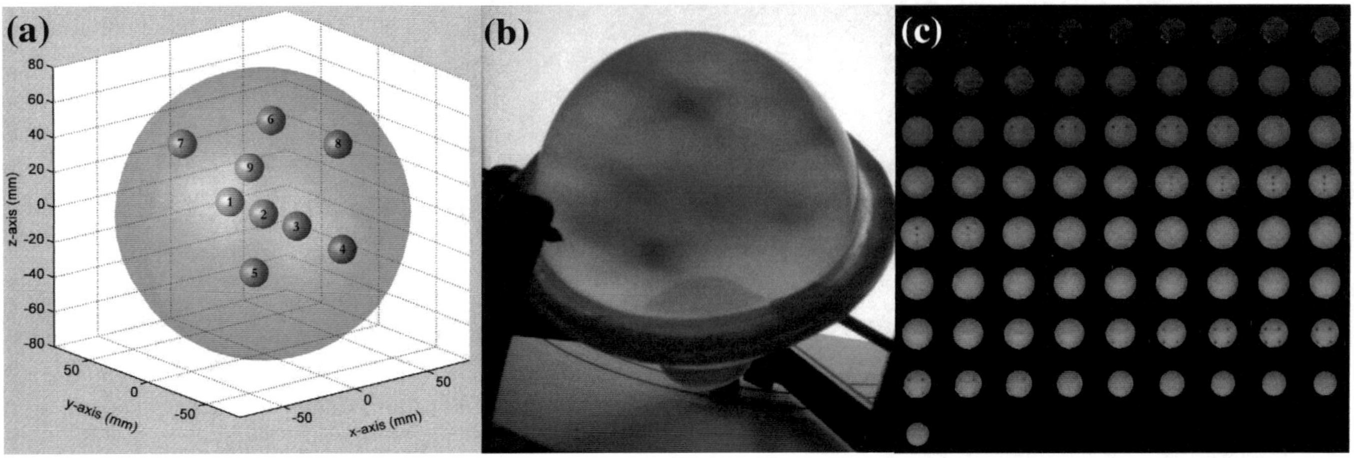

Figure 1. (a) A graphical representation of the GK irradiation plan. An identification number is assigned to each 8mm shot corresponding to its irradiation order. The origin of the coordinate system used was fixed to the planned coordinates of the second shot to facilitate presentation. **(b)** A photograph of the irradiated gel phantom mounted on the Leksell stereotactic frame. **(c)** Montage of the 73 axial T2-weighted MR images acquired from the irradiated gel phantom.

2. Materials and Methods

2.1. Polymer gel and experimental setup

The VIP gel formulation [3,4] (4% *N,N*-methylenebisacrylamide, 8% *N*-vinylpyrrolidone, 7.5% gelatin,, 0.0008% copper sulfate and 0.007% ascorbic acid diluted in hyperpure water with resistivity $>18\text{M}\Omega\cdot\text{cm}$ [3,5]) was used for the purposes of this work. The VIP elemental composition, along with its mass density (d=1.031 g/cm^3), ensure its tissue-like properties regarding MR imaging. A spherical, 16 cm diameter PMMA flask filled with gel served as a head phantom and was used to accurately reproduce every link in the GK treatment chain.

2.2. Irradiation planning and delivery

Planning was performed on pre-irradiation CT images of the gel phantom, which are known to suffer from negligible geometrical distortion. The irradiation scheme involved the delivery of nine (9) single shots using the 8mm helmet collimator. The distribution of the shots within the phantom's volume was programmed so as to cover its major part and preclude any "cross-talking" between adjacent shots. A schematic representation of the irradiation plan is given in figure 1(a). Irradiation was delivered using a GK model 4C unit equipped with an automated positioning system (APS). A maximum dose of 25 Gy was programmed to be delivered at the planned center of each shot using the GammaPlan treatment planning system (TPS). This dose level lies safely within the linear dose-response region of the VIP gel formulation, while being much higher than the lower limit of dose detection (~2.5 Gy) [6]. A photograph of the irradiated gel phantom is presented in figure 1(b).

2.3. Magnetic Resonance Imaging

The irradiated gel phantom mounted on the Leksell stereotactic frame (with the localization box in-place) was MR-scanned at 1.5T 3 days post-irradiation using a protocol commonly employed for GK irradiation planning purposes. This involved a volume selective (3D), T2-weighted, turbo spin echo (TSE) pulse sequence (160 ms echo time, 2700 ms repetition time, 145.76 kHz per pixel receiver bandwidth). 73 axial (x-y plane) partitions were reconstructed with a voxel size of $0.4\times0.4\times1.5$ mm^3. As seen in figure 1(c), the strong T2-weighting of the clinical MR sequence used results in adequate image contrast for visualizing the radiation-induced polymerization. The set of the acquired MR images was subsequently used to construct a 3D matrix of MR signal intensities.

2.4. 3D data manipulation procedure

A custom-written algorithm [7], based on standard imaging routines, was used to accurately define the spatial coordinates of the center of mass for each of the GK single shot dose distributions registered to the gel by exploiting the symmetry that the corresponding MR signal intensity distributions present in space with respect to this point.

In the 3D MR signal intensity matrix of the irradiated gel phantom, the signal iso-intensity surfaces corresponding to a delivered shot were considered as solid bodies in space, and the spatial coordinates of the center of mass of each one were calculated. The coordinates of the center of mass of each shot were then determined as the average of the spatial coordinates resulting from 8 different iso-intensity surfaces for that shot, with the standard deviation in each dimension denoting the corresponding experimental uncertainty. The 8 intensity values of the iso-signal surfaces used to determine the center of mass coordinates of each shot were chosen so as to cover adequately the linear upslope and downslope parts of the "through-shot" signal profiles in all three dimensions. The above procedure was repeated for all 9 delivered shots. In order to preclude any systematic discrepancies stemming from the performance of this tool, the planned spatial coordinates of the 9 delivered shots were determined by applying the same algorithm to the 3D dose matrix of the irradiation plan which was DICOM exported from the GammaPlan TPS. The reference coordinate system used in this work for reporting planned and experimental results for the spatial coordinates of the delivered shots coincides with that utilized by the GammaPlan TPS for exporting DICOM RT dose data (see figure 1(a), but note that in this figure the origin of the coordinate system was fixed to the spatial coordinates of the second shot).

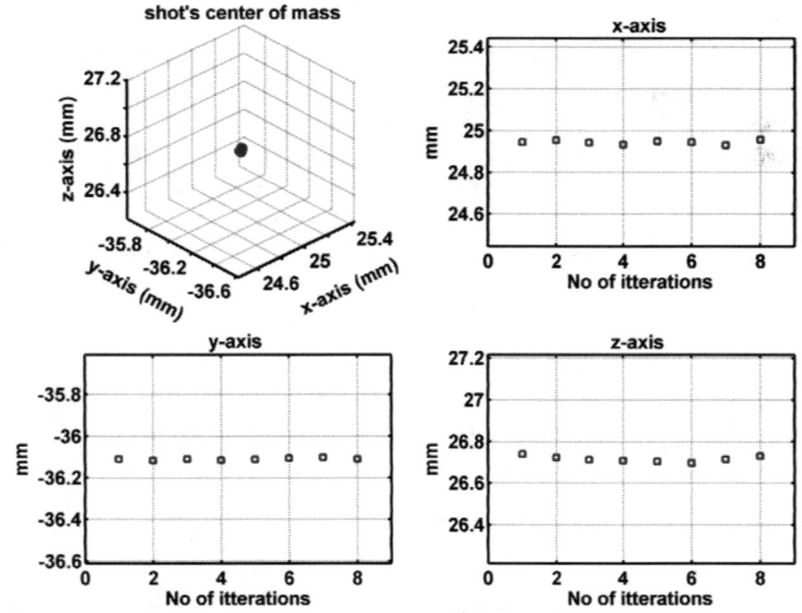

Figure 2. Polymer gel results for the center of mass coordinates of the 8th delivered shot. The x-axis label "No of iterations" corresponds to the number of the MR signal iso-intensity surfaces used.

3. Results and Discussion

Figure 2 presents experimental results for the spatial coordinates of the center (CoC) of the 8th delivered shot (see figure 1(a)). As seen, the algorithm described in the previous section locates the shot's center of mass with sub-millimeter precision. Corresponding results were obtained for the rest 8 shots.

Figure 3 presents results for the total geometrical accuracy of the GK unit and the specific clinical MR imaging protocol used in this work. For each shot, the experimentally defined CoC were subtracted from the corresponding planned values to determine the spatial components (denoted as dx, dy and dz) of the total displacement vector, which in turn were used to calculate its magnitude, dR. As clearly shown, the total displacement of each shot is dominated by its component along the x-axis dimension (a mean dx of 0.53 mm was calculated). It is also worth noting that the calculated dx values are all positive, implying that the experimentally determined shot centers are systematically shifted in x-axis (see figure 1(a)), i.e., along the frequency encoding direction in MRI acquisition. These findings are in agreement with published results, according to which the difference in the magnetic susceptibility between the fiducial markers of the Leksell localization box and the polymer gel phantom results in a geometrical shift of the MR images along the frequency encoding direction. A theoretical calculation [8] based solely on susceptibility effects yielded an image shift of 0.3 mm for the specific MR imaging parameters used in this work. This finding, along with figure 3 data, indicates that the susceptibility-induced geometrical distortion is the dominant factor leading to geometrical inaccuracies for the clinical GK application studied. Overall however, the calculated dR values suggest that the total geometrical accuracy is well within 1mm.

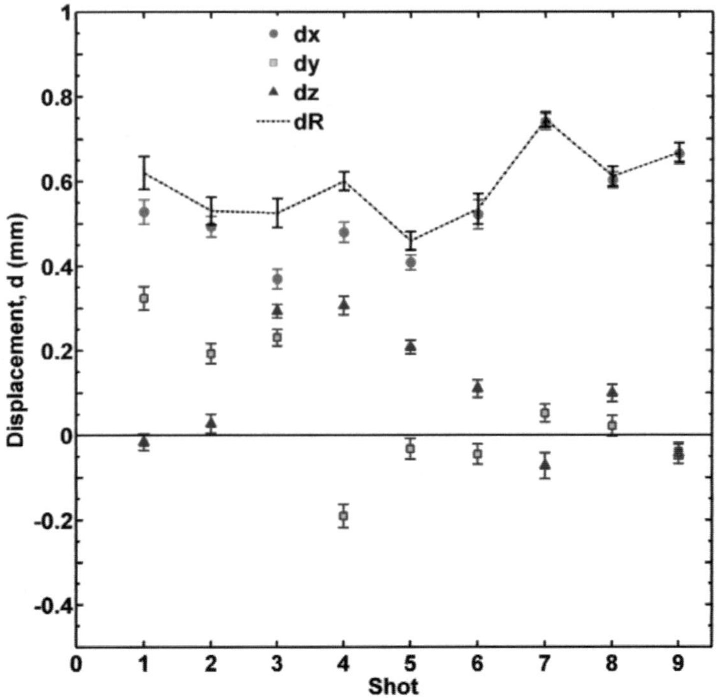

Figure 3. The total displacement, dR, calculated for each of the delivered shots along with its spatial components in each dimension (denoted with dx, dy and dz).

In an effort to assess the dosimetric effect that geometrical uncertainties of the order of the calculated dR values would induce to a clinical single shot GK application, a commonly used criterion for the evaluation of 3D irradiation plans was employed; that of percentage target coverage defined as the percentage of the target volume (TV) covered by a therapeutic iso-dose surface. Using the planned dose distribution data (see Sec. 2.2) in the GammaPlan TPS, a target volume was defined within each shot and the percentage target coverage by the 18 Gy iso-dose surface (i.e. 72% of the 25 Gy prescribed at the center of each shot) was calculated. The same calculation was then performed after "relocating" the shot in order to meet the experimentally determined spatial coordinates of its center. Results are summarized in table 1. As clearly shown, the sub-millimeter displacement found for each shot results in significant reduction in the target percentage coverage, implying that clinically the target receives a lower dose than that prescribed (with the difference being of the order of 10%).

Table 1. Percentage target coverage results calculated using the GammaPlan TPS for the planned and the experimentally defined coordinates of the center (CoC) of each shot.

Shot	Tumor Volume (TV) (in cm^3)	% coverage of TV by the 18 Gy iso-dose surface	
		Planned CoC	Experimental CoC
1	0.3414	95	87
2	0.3189	97	91
3	0.3218	96	89
4	0.3164	97	93
5	0.3148	97	93
6	0.3115	97	90
7	0.3254	97	93
8	0.3004	98	95
9	0.3176	97	92

This drastic effect stems from the steep dose gradient in all three dimensions of the GK single shot dose distributions, and is expected to soften when multiple shots are used to cover the target volume.

It should be noted that in the above analysis potential geometrical distortions in the CT images and diffusion of the polymer gel monomer, which could constitute sources of systematic uncertainties, were not taken into account since their effect in the total clinical geometrical accuracy was deemed negligible.

4. Conclusions

Results of this study show that the total geometrical accuracy in the specific clinical GK application examined is better than 1 mm. The susceptibility-induced geometrical distortion in the MR images constitutes the major contributor to accuracy degradation. Sub-millimeter geometrical uncertainties in single shot GK applications were found to have a significant effect in percentage target coverage.

Acknowledgement

This work was partly supported by the Cyprus Research Promotion Foundation and the European Regional Development Fund (Project: ΥΓΕΙΑ/ΔΥΓΕΙΑ/0308(ΒΙΕ)/12). A.M. was supported by the Greek Ministry of National Education and Religious Affairs (Project: "Irakleitos").

References

[1] Heck B, Jess-Hempen A, Kreiner H J, Schopgens H and Mack A. Accuracy and stability of positioning in radiosurgery: long-term results of the Gamma Knife system. Med Phys **34**, 1487-1495 (2007).

[2] Baldwin L N, Wachowicz K and Fallone B G. A two-step scheme for distortion rectification of magnetic resonance images. Med Phys **36**, 3917-3926 (2009).

[3] Pantelis E, Antypas C, Petrokokkinos L, Karaiskos P, Papagiannis P, Kozicki M, Georgiou E, Sakelliou L and Seimenis I. Dosimetric characterization of CyberKnife radiosurgical photon beams using polymer gels. Med Phys **35**, 2312-2320 (2008).

[4] Petrokokkinos L, Kozicki M, Pantelis E, Antypas C, Fijuth J, Karaiskos P, Sakelliou L and Seimenis I. Characterization of a new polymer gel for radiosurgery dosimetry using Magnetic Resonance Imaging. Journal of Instrumentation **4**, P06018 (2009).

[5] Kozicki M, Maras P, Rybka K, Bieganski T, Kadlubowski S and Petrokokkinos L. On the Development of the VIPAR Polymer Gel Dosimeter for Three-Dimensional Dose Measurements. Macromolecular Symposia **254**, 345-352 (2007).

[6] Petrokokkinos L, Moutsatsos A, Karaiskos P, Kouridou V, Pantelis E, Papagiannis P and Seimenis I. On the use of VIP gel dosimetry in HDR brachytherapy. Journal of Physics: Conference Series **164**, 012051 (2009).

[7] Moutsatsos A, Petrokokkinos L, Karaiskos P, Papagiannis P, Georgiou E, Dardoufas K, Sandilos P, Torrens M, Pantelis E, Kantemiris I, Sakelliou L and Seimenis I. Gamma knife output factor measurements using VIP polymer gel dosimetry. Med Phys **36**, 4277-4287 (2009).

[8] Watanabe Y, Perera G M and Mooij R B. Image distortion in MRI-based polymer gel dosimetry of gamma knife stereotactic radiosurgery systems. Med Phys **29**, 797-802 (2002).

Journal of Physics: Conference Series **250** (2010) 012061 doi:10.1088/1742-6596/250/1/012061

MAGAT gel dosimetry for its application in small field treatment techniques

N Gopishankar[1,2], S Vivekanandhan[2], S S Kale[1], G K Rath[3], S Senthil Kumaran[4], Sanjay Thulkar[5], V Subramani[3], M A Laviraj[3], R K Bisht[1], A K Mahapatra[1]

[1]Department of Neurosurgery and Gammaknife, [2]Department of Neurobiochemistry, [3]Department of Radiation Oncology, [4]Department of NMR, [5]Department of Radiodiagnosis, All India Institute of Medical Sciences, New Delhi, India.

E-mail: gshankar1974@yahoo.com

Abstract. Purpose of this work is to present the role of in-house manufactured MAGAT gel for treatment verification in small field dosimetric techniques such as Gammaknife (GK) and intensity-modulated radiation therapy (IMRT). Magnetic resonance imaging (MRI) is one of the most extensively used imaging technique for polymer gel dosimetry hence we used this method for gel evaluation. Different MR scanners and MRI sequences were used in this study for obtaining calibration plot between R2 and absorbed dose. An experimental plan was created for Gammaknife and IMRT. The prepared gel was filled in spherical glass phantom and in-house designed human head shape phantom for verification purpose. We used 8 TE values for all the imaging sequences for two reasons. Firstly it is sufficient enough to give good signal to noise ratio. Second considering the enormous scanning time involved in multiple spin echo sequence. MATLAB based in-house programs were used for R2 estimation and dose comparison. The isodose comparison with MAGAT gel showed reasonable agreement for both Gammaknife and IMRT techniques.

1. Introduction

The development of a new polymer gel named MAGIC paved the way for manufacturing gels on the bench top in laboratory. These types of polymer gels became known as normoxic gel dosimeters as presence of oxygen in the gel was removed by introducing antioxidants in the chemical composition. A typical polymer gel dosimetry involves three steps including gel fabrication, gel filling in anthromorphic phantom containers and vials for calibration, gel irradiation and finally imaging of polymerized gel. However there is no "gold standard" for 3D dosimetry. Through this work we present our methodology to verify small field techniques using in-house manufactured MAGAT gel.

2. Materials and Methods

Two different MAGAT gel preparations were undertaken. The first preparation was used for GK dosimetry and the second preparation was used for IMRT dosimetry. Three MRI scanners (Siemens Avanto (1.5 T), Siemens Sonata (1.5 T) and a newly installed Philips Achieva (3 T)) were used for this study. The gel samples scanned with Philips Achieva MRI were converted to T2 maps using the inbuilt scanner algorithm. Whereas gel samples scanned in MRI Avanto and MRI sonata were converted to dicom images and R2 calculations were made using Maximum Likelihood Estimation

Journal of Physics: Conference Series **250** (2010) 012061 doi:10.1088/1742-6596/250/1/012061

(MLE) algorithm coded in MATLAB programme. The codes have an option of subtracting background and skipping the first echo from R2 calculation. For plan verification in GK and IMRT, spherical glass container and in-house made acrylic phantom were used respectively.

Figure 1. The spherical glass phantom filled with MAGAT gel fixed in the Leksell frame for experimental irradiation in Gammaknife.

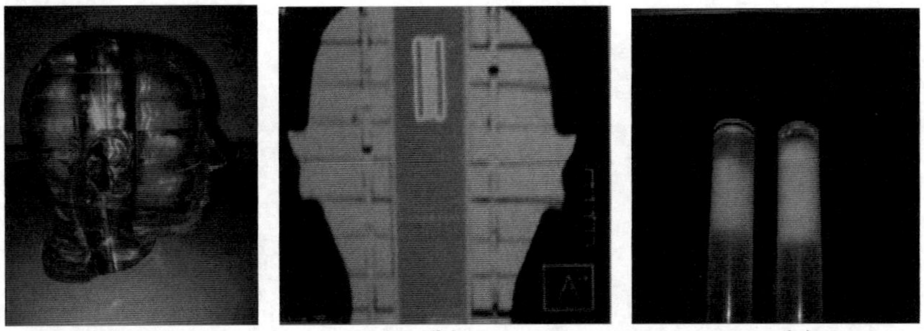

(a) (b) (c)

Figure 2 (a). Locally designed human head shaped acrylic phantom used for experimental irradiation of IMRT plan. **(b)** An experimental target created in the phantom which is doughnut shape target (Coronal View). **(c)** Two cylindrical inserts of human shaped phantom showing gel irradiation region created by seven field and nine field IMRT.

2.1. Gel Preparation

The gel preparation was done as per the composition given in Table 1. In composition B hydroquinone was added to avoid any polymerization initiation before irradiation.

Table 1. MAGAT gel composition used for various studies.

	Composition A	Composition B
Gelatin	8 %	6 %
Methacrylic acid	9 %	9 %
THPC	10 mM	10 mM
H2O	83 %	85 %
Hydroquinone		0.05 mM
Gel Volume	3000 mL	1500 mL
Dose Range	0-17 Gy	0-5 Gy

2.2. Calibration Method

Calibration of the gel for GK verification was carried out with a Cobalt unit (Theratron 780) using source to axis distance (SAD) technique. The cylindrical glass vials had a wall thickness of 1 mm, an outer diameter of 13 mm, and a length of 95 mm. The vials were placed in the water-filled tank and irradiated at a depth of 10 cm from the water surface to the middle of the test tube with a single field of 10 x 10 cm^2. To obtain calibration curves, eleven test tubes were irradiated with doses in the range of 0-17 Gy on the same day of treatment planning execution to the spherical glass phantom.

Calibration of gel for IMRT study was carried out with 11 Acrylic tubes of 15 cm length, 2 cm outer diameter, 1.5 cm inner diameter. Five vials each were used for calibration to doses from 50 cGy to 500 cGy, one as control vial. The tubes were placed in water tank and irradiated with parallel opposed 6 MV photons from CL2300CD (Varian) and SynergyS (Elekta) with field size of 16 x 16 cm^2.

2.3. Gel irradiation

For Gammaknife irradiation three shots with 8, 14 and 18 mm collimator were used for creating an experimental plan in Leksell GammaPlan (LGP 5.32 version). See figure.1. For IMRT treatment verification an experimental cylindrical target and doughnut shaped target were created in Eclipse TPS and Pinnacle TPS respectively. A seven field IMRT (Eclipse 6.5) and Nine field IMRT plan (Pinnacle Version ADAC, Superscript (3) 8.0M Philips Medical System) was created respectively.

Table 2. MRI scanning parameters used for gel scanning.

Scan date	Scanner name	MRI sequence	Gel cylinder	Gel vials	TR	No of echoes	TE	Slice thickness	NAX	FOV
18/1/2010	Philips Achieva(3T)	CPMG (TSE)	7 Field IMRT	7	5321 ms	8	15 ms	3 mm	1	256
18/1/2010	Philips Achieva(3T)	GRASE (EPI)	7 Field IMRT	7	7210 ms	8	15 ms	3 mm	1	256
23/1/2010	Magnetom Avanto (1.5T)	CPMG	7 Field IMRT	7	7210 ms	8	22 ms	3 mm	1	256
30/1/2010	Magnetom Sonata(1.5T)	CPMG	9 Field IMRT	7	7210 ms	8	22 ms	3 mm	1	256
6/2/2010	Magnetom Avanto (1.5T)	CPMG	7 Field IMRT	12	5600 ms	8	22 ms	3 mm	1	256
6/2/2010	Magnetom Avanto (1.5T)	CPMG	9 Field IMRT	7	5600 ms	8	22 ms	3 mm	1	256
7/7/2009	Magnetom Sonata(1.5T)	CPMG	Gamma knife	11	7000 ms	8	14 ms	2 mm	1	256

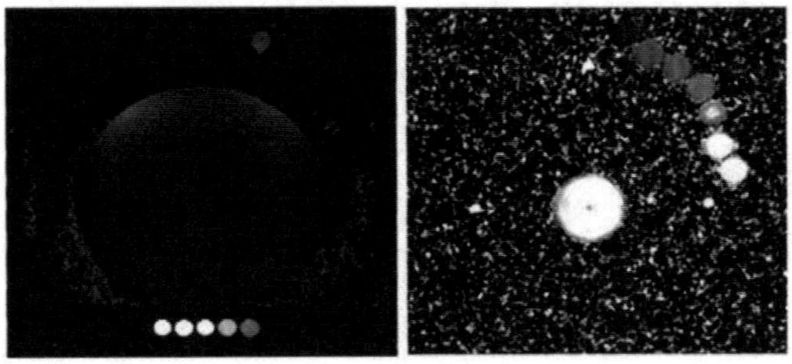
(a) (b)

Figure 3 (a) illustrates the R2 map of the spherical container with the 11 vials placed in either side. **(b)** shows the cylindrical gel insert with seven acrylic vials. Three small white spots indicate the fiducial points created using Vitamin E capsules.

3. Results and Discussion

Overall in this study we did six scans with calibration vials used for IMRT verification. We restricted our spin echo sequence to 8 as we assumed it was sufficient enough to give good SNR. Also considering the availability of scanner the imaging sequence was chosen in this fashion for all the scanning. Time period between scans in each machine was around five to seven days. The gel vials used for Gammaknife verification was scanned only once as shown in Table.2. In Philips Achieva machine the gel was first scanned within two hours after removal from refrigerator and was not allowed to stabilize to MRI room temperature deliberately. R^2 was 0.951 and 0.967 for TSE and GRASE sequence.

We used MLE for other scans as it is more reliable for noisy data than other algorithms. For the scanning done in the fifth day after the first scan and seventh day after the second scan the R^2 was 0.993 and 0.994 respectively. See figure. 4b & 4c. During this study gel vials were left in the MRI room for 4 to 5 hours before scanning. Final scan was done on the sixth day from the third scan. The gel vials were left in the scanning room for more than 10 hrs to attain room temperature. The calibration plot showed a R^2 of 0.998 which was better than its predecessor. In the calibration plot for GK, good linearity was observed between 2 to 14 Gy with R^2 value of 0.996. A second degree polynomial fit for the whole measurement for 0 to17 Gy gave R^2 of 0.996. However the correct functional form to fit calibration curves is mono or bi exponential fit [4]. In this work we have shown only one IMRT plan with nine field for treatment verification comparison. A comprehensive 3D comparison was done for Gammaknife and IMRT treatments with MAGAT gel. Figure 5 a & b show isodose comparison for Gammaknife and IMRT at one z position.

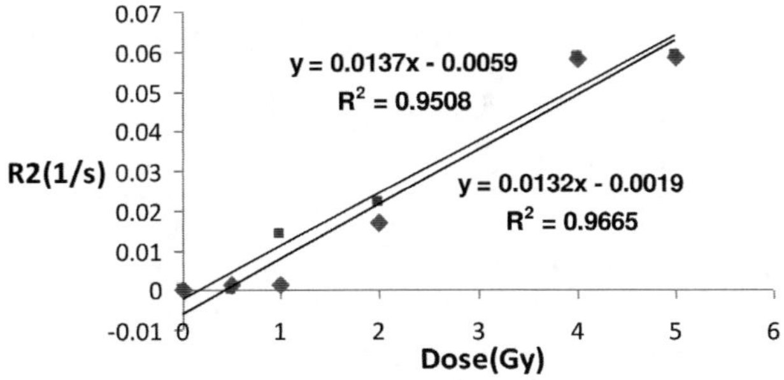

(a)

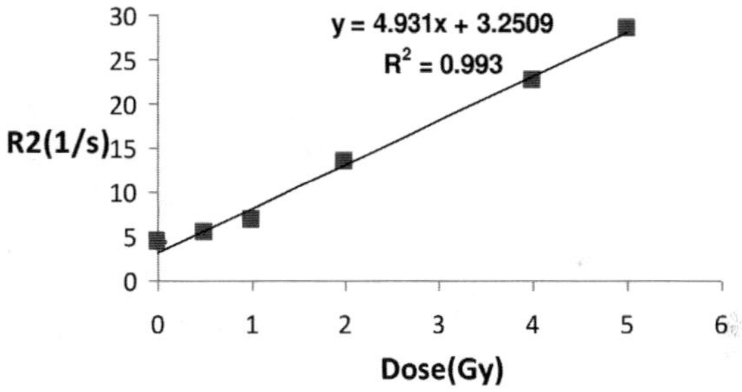

(b)

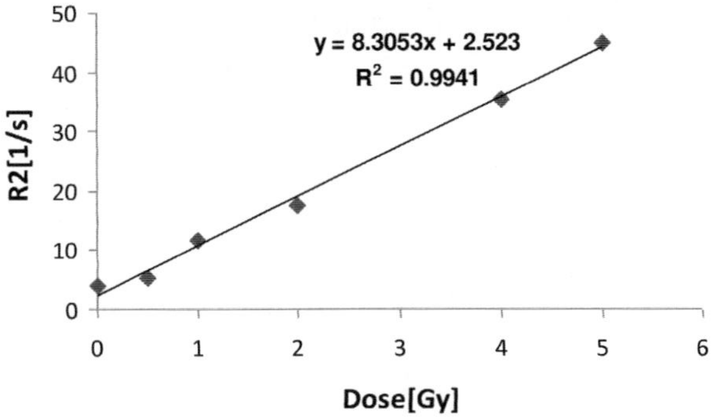

(c)

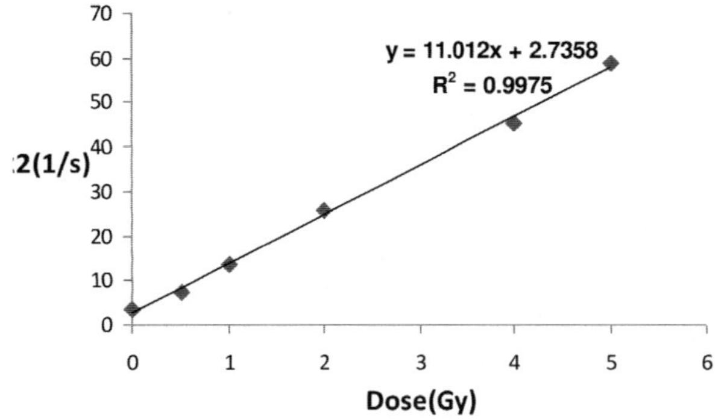

(d)

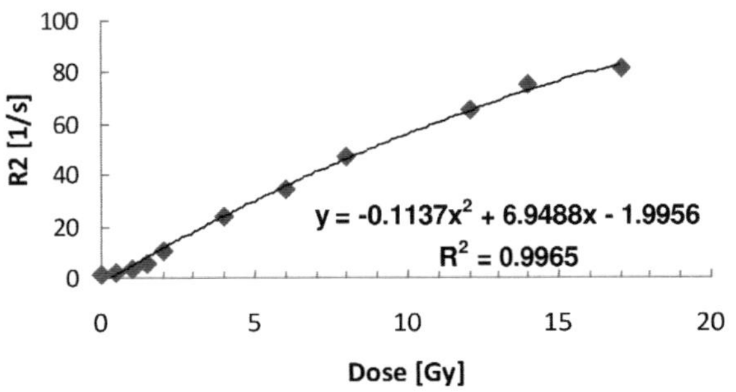

(e)

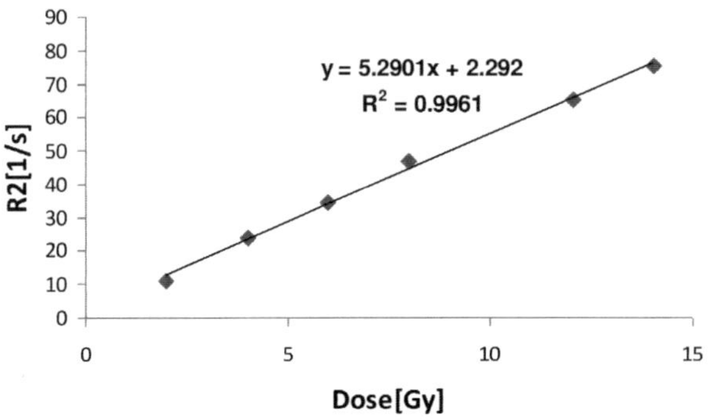

(f)

Figure 4 (a-d) shows calibration curves plotted between R2 and Dose for IMRT verification. **Figure 4 e & 4 f** show calibration plot for Gammaknife verification.

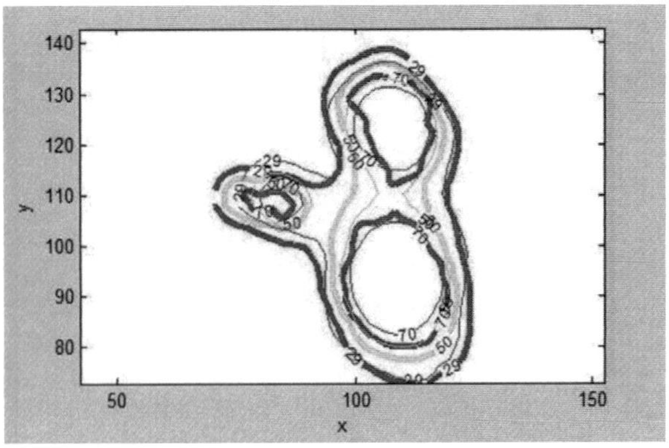

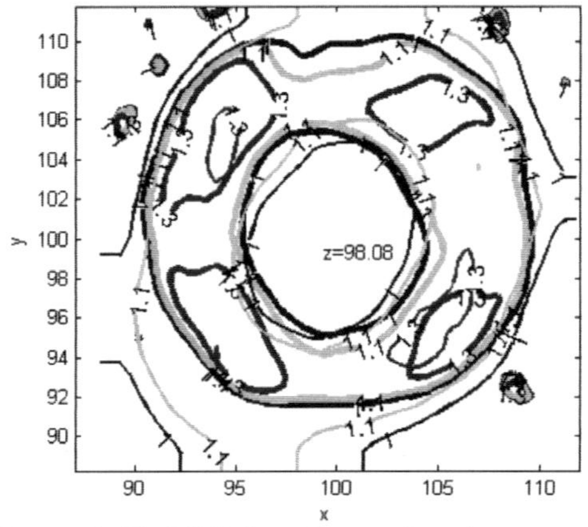

Figure 5 (a) Isodose comparison between calculated Gammaknife TPS dose (thin line) and measured MAGAT Gel dose (composition A) (thick line). **(b)** Isodose Comparison between calculated IMRT dose (Pinnacle) (thin line) and MAGAT Gel dose (composition B) (thick line).

4. Conclusion

MAGAT gel is useful for treatment verification, provided certain dosimetric limitation of gel is known before using it for dosimetry. We were able to show that measurements done with different MRI scanners gave different R2 values and hence the gel measurement is scanner dependent. Dosimetric verification with MAGAT gel for small field techniques such as Gammaknife, IMRT showed reasonable agreement.

Acknowledgment

We gratefully acknowledge AERB, India for their support through project no N-964.

References

[1] De Deene Y, Baldock C. Optimization of multiple spin-echo sequences for 3D polymer gel dosimetry, Phys.Med.Biol. 2002; 47:3117-3141.

[2] Watanabe Y, Akimitsu T, Hirokawa Y et al.: Evaluation of dose delivery accuracy of Gamma Knife by polymer gel dosimetry, J Appl Clin Med Phys. 2005;6:133-142.

[3] Watanabe Y, Kubo H. Comparison of R2 estimation algorithms for polymer gel dosimetry. Med Phys. [abstract]. 2009; 36(6):2611.

[4] Baldock C, De Deene Y, Doran S et al. Polymer Gel Dosimetry, Phys.Med.Biol. 2010;55:R1-R63.

[5] Hurley C, Venning A and Baldock C. A study of a normoxic polymer gel dosimeter comprising methacrylic acid, gelatin and tetrakis (hydroxymethyl) phosphonium chloride (MAGAT), Appl. Radiat.Isot.63: 443-56; 2005.

Journal of Physics: Conference Series **250** (2010) 012062 doi:10.1088/1742-6596/250/1/012062

Multiple slot array collimator to minimize stray light in optical cone beam CT

Kevin Jordan, Jonatan Snir and Jerry Battista

London Regional Cancer Program, London Health Sciences Centre
Department of Medical Biophysics, University of Western Ontario, London, ON, Canada
Kevin.jordan@lhsc.on.ca

Abstract. Acceptance of stray light in transmission images is a general imaging problem. A commercial optical cone beam CT scanner, Vista10 was modified to reduce stray light. Further stray light reduction was accomplished by placing vertical slot aperture arrays between the object and the diffusive light source. Comparisons of single and seven aperture arrays demonstrated that transmissions equivalent to scatter free conditions can be achieved with a multiple array collimator. Results for uniform liquids and finger gel phantoms demonstrate that small objects in a bright background are a more stringent test of stray light rejection.

1. Introduction

Optical cone beam CT uses one of the simplest data acquisition geometries. Transmission images through the radiation sensitive dosimeter material are recorded as the object is rotated within an aquarium containing a reference liquid with a similar refractive index [1]. The light source consists of a planer diffuser backlit by light emitting diodes. The geometry is inversely analogous to x-ray cone beam CT with the diffuser sheet replacing the flat-panel CCD array detector and the image forming aperture of a camera substituting for the point source. Improving scatter corrections for x-ray cone beam CT is an active area of academic and commercial research [2].

Investigations of stray light in the Modus optical cone beam CT scanner (Vista10, Modus Medical London, Canada), suggest that increasing the distance of the diffuser from the rotation axis, removing the bandpass filter at the camera lens and using a compensating light distribution can reduce the amount of stray light in transmission measurements. Aperture arrays are also efficient for measuring residual stray light that can be subtracted from raw transmission data. One particularly effective and simple approach is to employ such an array with equally spaced open and blocked areas. Multiple images are recorded and "stitched" together to form open and shadow images. The difference image represents the primary light for input to the CT image reconstruction algorithm [3]. The initial work involved one slot array at the input face of the aquarium. In this study, the implementation of multiple slot arrays is compared to assess performance of the stray light reduction strategy.

2. Methods

The Vista10 scanner was modified by removing the uniform diffuse light source and substituting a larger diffuser at 100 cm from the rotation axis, as shown in Figure 1. A filtered line source was constructed from three LED's (Luxeon III) mounted vertically to a strip heat sink, each weakly

Journal of Physics: Conference Series **250** (2010) 012062 doi:10.1088/1742-6596/250/1/012062

collimated with a lens and spectrally filtered with a 10 nm bandpass filter centred at 590 nm. The LED's backlight the diffuser sheet. Slot arrays were manually cut from radiographic film. The first aperture array had slots widths and separations of 1.0 cm. Subsequent apertures were scaled according to system magnification and adjusted to have a ratio of 60% open to 40% blocked areas. Three images were acquired with the first slot array, centred and translated ± 7 mm. The central 70% of the open and shadow areas were stitched together to form open and shadow images. A difference image is a close approximation to the primary image required for CT reconstruction. For the multi-array sets, the additional arrays were positioned according to the divergence of the modified scanner. Reference images were obtained by collapsing the optical cone beam scanner to a fan beam geometry by placing horizontal slots at the aquarium and near the diffuser plate.

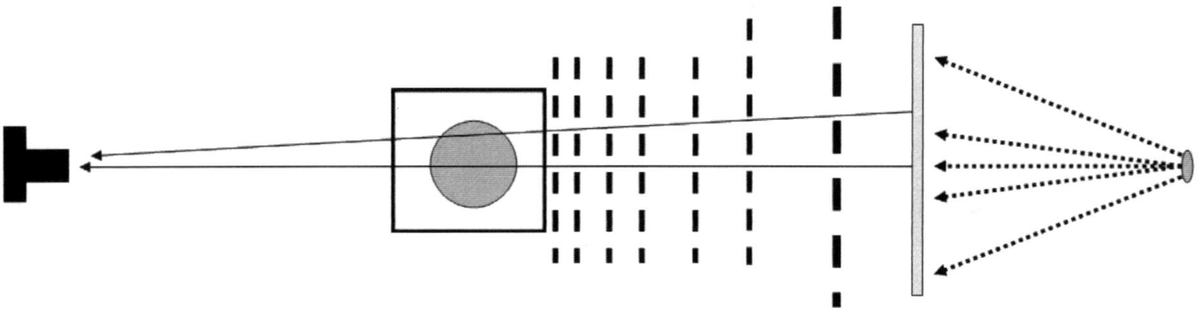

Figure 1. Top view schematic of modified Vista10 optical cone beam CT scanner. From right to left: LED array, diffuser plate, image forming rays, vertical slot arrays, aquarium, sample, CCD camera.

Test solutions were made by adding Intralipid and trace amount of carbon black nanoparticles suspended in micelles of Triton X-100 to water. The Intralipid provided a linear attenuation coefficient for scatter of approximately 0.1 cm^{-1}. This simulated twice the amount of scatter as in a 4% gelatin hydrogel. The carbon black concentration was adjusted to provide varying degrees of absorption. Hydrogen peroxide was added to eliminate microbial growth and extent shelf life of the solutions [4]. The reference solution was water. The transmission values of the solutions were independently measured in 10 cm quartz cuvettes with a He-Ne laser at 594 nm. In a similar manner, carbon black nanoparticle micelles gels were prepared as stable hydrogel finger phantoms with diameters of 2.54, 1.27 and 0.48 cm [5]. The finger phantoms represent a very difficult geometry for transmission imaging since the object is surrounded by a bright background. For these finger phantom experiments the reference sample was a uniform gel (without "fingers") from the same batch.

3. Results

The solution transmission values were chosen to extend to the lower limits that still provided accurate results. Figure 2, shows transmission images for the darkest solution recorded with zero and 7 slot arrays. The images have been adjusted to the same window and level for comparison. Note the cylinder wall is clearly seen once stray light is reduced. The transmission measurements are summarized in Table I.

Figure 2. Transmission images of carbon black micelle solution with intralipid, same window and level: top image, open geometry (zero slot arrays), bottom image (7 slot arrays). Note the cylinder wall as a shadow line in slot opening (left bottom) and as a bright line in slot shadow (right bottom) compared to open image where cylinder walls are not visible due to stray light.

	Solution 1	Solution 2	Solution 3	Solution 4
Laser (594 nm)	0.081	0.236	0.430	0.499
Open	0.075	0.233	0.422	0.487
Single slot array	0.079	0.236	0.428	0.496
Seven slot arrays	0.081	0.238	0.430	0.497

Table I. Linear attenuation coefficients (cm^{-1}) of carbon black nanoparticle micelle solutions with Intralipid scattering. Stray light was minimized using vertical slot arrays.

The linear attenuation coefficients measured for the fan beam (single horizontal slot, 1 cm height), open field, single slot array and seven slot array collimators are shown in table II. The 2.54 cm diameter finger demonstrated the expected trend; as stray light is reduced in the system, measurements converge to the fan beam result. Similar trends are seen for the smaller 1.27 and 0.48 cm fingers but the results are noisier since the change in transmission is proportionally lower.

		Finger diameter (cm)	
	2.54	1.27	0.48
Slot (reference)	0.301 / 0.306	0.286 / 0.286	0.347 / 0.347
Open	0.287 / 0.288	0.271 / 0.276	0.324 / 0.326
Single slot array	0.291 / 0.298	0.279 / 0.281	0.331 / 0.325
Seven slot array	0.300 / 0.309	0.277 / 0.283	0.357 / 0.346

Table II. Linear attenuation coefficients (cm^{-1}) of carbon black nanoparticle micelle hydrogel "finger" phantom. Paired data correspond to independent measurements before and after a 180 degree rotation of the finger phantom.

4. Summary

Stray light in optical cone beam CT can be reduced by incorporating several innovations: extending the distance from the diffuser to the object, removing the bandpass filter at the camera lens (reduces glare), customizing the light source distribution and introducing collimating aperture arrays. Fan beam geometry provides an internally consistent geometry for comparisons with cone beam stray light reduction strategies. Finally multiple slot aperture arrays can further reduce stray light and increase the quantitative dynamic range of the transmission images to yield more accurate reconstructions.

Acknowledgements: The authors Jordan and Battista held a licensing agreement with Modus Medical Devices concerning optical cone beam CT. Funding was from Ontario Research and Development Challenge Fund: Ontario Consortium for Image-guided Therapy and Surgery and London Regional Cancer Program Small Grants fund. Jennifer Dietrich from Modus Medical Devices assisted with scanner modifications.

References

[1] J. G. Wolodzko, C, Marsden and A. Appleby, 1999 CCD imaging for optical tomography of gel radiation dosimeters, Med Phys **26**: 2508-2513,

[2] Zhu L, Xie Y, Wang J and Xing L, 2009 Scatter correction for cone beam CT in radiation therapy *Med Phys* **36** 2258-68

[3] Granton P, Battista J and Jordan K, Stray Light in Optical Cone-Beam Computed Tomography: Techniques for Measurement and Supression, in preparation for Phys Med Biol

[4] Jordan K and Battista J, 2009 A stable black-refractive-index-matching liquid for optical CT scanning of hydrogels, J of Phys: Conference Series **164**: 012045

[5] Jordan K and Battista J, 2009 A transparent black non-diffusing micelle gel for optical CT performance evaluation phantoms, J of Phys: Conference Series **164**: 012046

Journal of Physics: Conference Series **250** (2010) 012063 doi:10.1088/1742-6596/250/1/012063

Light scattering in optical CT scanning of Presage dosimeters

Y Xu [1], J Adamovics [2], J C Cheeseborough [3], K S Chao [1] and C S Wuu [1]

[1] Department of Radiation Oncology, Columbia University, New York, NY 10032

[2] Department of Chemistry, Rider University, Lawrenceville, NJ 08648

[3] Department of Radiation Oncology, Weill Cornell Medical College, New York, NY10065

E-mail: yx2010@columbia.edu

Abstract. The intensity of the scattered light from the Presage dosimeters was measured using a Thorlabs PM100D optical power meter (Thorlabs Inc, Newton, NJ) with an optical sensor of 1 mm diameter sensitive area. Five Presage dosimeters were made as cylinders of 15.2 cm, 10 cm, 4 cm diameters and irradiated with 6 MV photons using a Varian Clinac 2100EX. Each dosimeter was put into the scanning tank of an OCTOPUS™ optical CT scanner (MGS Research Inc, Madison, CT) filled with a refractive index matching liquid. A laser diode was positioned at one side of the water tank to generate a stationary laser beam of 0.8 mm width. On the other side of the tank, an in-house manufactured positioning system was used to move the optical sensor in the direction perpendicular to the outgoing laser beam from the dosimeters at an increment of 1 mm. The amount of scattered photons was found to be more than 1% of the primary light signal within 2 mm from the laser beam but decreases sharply with increasing off-axis distance. The intensity of the scattered light increases with increasing light attenuations and/or absorptions in the dosimeters. The scattered light at the same off-axis distance was weaker for dosimeters of larger diameters and for larger detector-to-dosimeter distances. Methods for minimizing the effect of the light scattering in different types of optical CT scanners are discussed.

1. Introduction

Gel dosimeter is a promising 3D dose verification system that could have wide applications in routine clinical radiation therapy practice. Among the imaging methods studied for gel dosimetry so far, optical CT is the only approach that utilizes equipments designed for gel dosimeters with expected accuracy and spatial resolution. The advantage of the optical CT approach over other dose readout methods for gel dosimetry is commonly acknowledged [1-2]. It has been postulated that gel dosimetry using a bench-top optical CT scanner will be a valuable tool for patient-specific treatment dose verification, periodic quality assurance of radiation therapy units, and commissioning of new treatment techniques and machines.

In optical CT scanning of 3D dosimeters, the dose distribution recorded in a dosimeter is represented by the 3D optical density map caused by the light attenuation and/or absorption from the radiation induced micro-particles in the dosimeter. Therefore, only the portion of the incoming laser beam that *survives* the multiple Rayleigh-Mie scattering and/or light absorption in the forward direction should be used for image reconstruction [3-4]. Artifacts caused by the scattered light in different optical CT scanners were reported and discussed previously [5-8].

The purpose of this study is to give a quantitative estimation of the intensity of the scattered light in regions where light signals might be collected for image reconstruction in optical CT-based 3D dosimetry. Even though the measurements were done for the Presage dosimeters inside the scanning tank of an OCTOPUS™ research CT scanner, the results can be used as a general guideline to minimize the effect of light scattering in the design and operation of different optical CT scanners.

2. Materials and methods

The Presage dosimeters used in this study are one cylinder of 4 cm diameter, one cylinder of 10 cm diameter, and three cylinders of 15.2 cm diameter (see Fig. 1a). All the dosimeters were irradiated with 40 cm x 40 cm, 6 MV photon fields using a Varian 21EX linear accelerator. Before the irradiations, the background optical density drop (OD) across each dosimeter was measured and the monitor units for irradiating the dosimeters were estimated according to the following criteria:

1. The optical density drops across the three dosimeters of 15.2 cm diameter should vary substantially in the range from 0 to 4.
2. The optical density value for one of the 15.2 cm diameter dosimeter should be comparable to those for the dosimeters of 4 cm and 10 cm diameters.

a) b)

Figure 1: Materials used in this study: a) Presage dosimeters of 4 cm, 10 cm, and 15.2 cm diameters; b) laser beam, water tank, and the in-house manufactured positioning system.

The irradiated dosimeters were put on the turntable of an OCTOPUS™ optical CT scanner filled with a refractive index matching liquid. A laser diode was put on one side of the scanning tank to generate a stationary laser beam of 0.8 mm diameter and 633 nm wavelength. The position of each dosimeter in the tank was adjusted such that the laser beam passed through the central region of the dosimeter. The outgoing laser signal from the tank was recorded by a Thorlabs optical power meter with an optical sensor of 1 mm diameter sensitive area. The operating frequency of the power meter was set to be 635 nm. An in-house manufactured positioning system was used to move the optical sensor to the off-axis positions from -9 cm to 9 cm at an increment of 1 mm (Fig. 1b). The light signal at each off-axis position was recorded and the ratio to the intensity of the primary beam at the central position was taken.

Prior to the measurements on the irradiated dosimeters, the intensity of the light signal at each position for the tank filled with the refractive index matching liquid was recorded. All the 6 sets of measurements were done with a detector-to-tank distance of 2 cm. In order to study the dependence of the intensity of the scattered light on the detector-to-tank distance, the measurements were also repeated for one of the 15.2 diameter dosimeters with a detector-to-tank distance of 10 cm.

3. Results

Figure 2 shows the intensity of the light signal measured for the tank fill with the refractive index matching liquid only. The power of the outgoing laser beam from the tank was determined to be 1.532 milliwatts. For the ±1 mm off-axis positions, the primary laser beam was partially detected by the optical sensor and the recorded laser signal was significant. For off-axis distance ≥ 2 mm, the intensity of the scattered light was less than 1% of the primary laser intensity and dropped sharply with increasing off-axis distances.

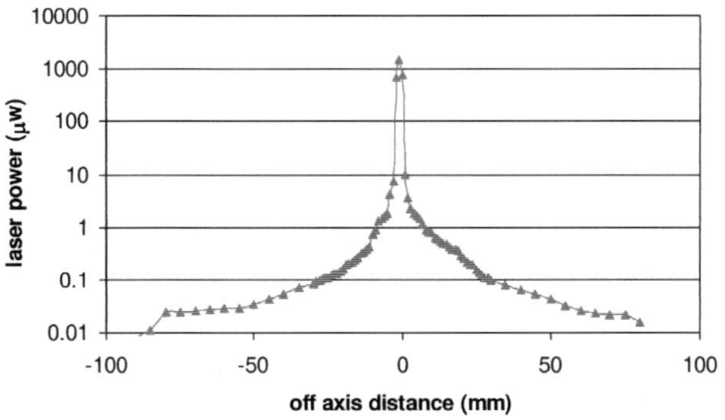

Figure 2: Distribution of the measured light intensity for the scanning tank filled with the refractive matching liquid.

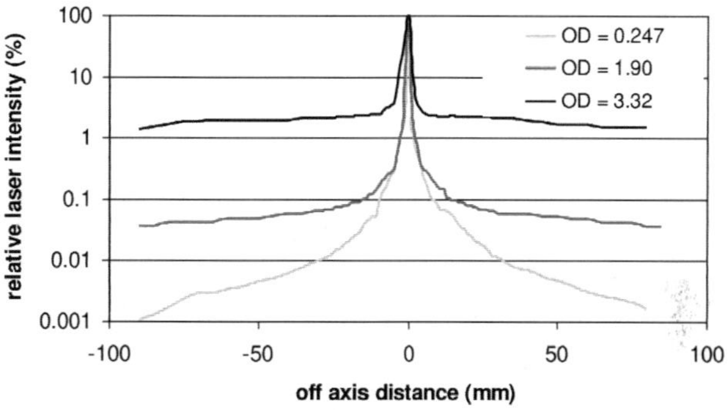

Figure 3: Comparison of the light intensity distributions for the three Presage dosimeters with 15.2 cm diameter and 0.247, 1.9 and 3.32 optical density values (OD).

Figure 3 compares the distributions of the intensity of the light signal for the three dosimeters of 15.2 cm diameter. The optical density drops across the dosimeters (defined as the logarithm of the ratio of 1.532 mw to the laser intensity at the central position) were 0.247, 1.9 and 3.32 respectively. It can be seen that the intensity of the scattered light increases with increasing light attenuation and absorption within the dosimeter. For the two dosimeters with optical density values of less than 2, the intensities of the scattered photons at more than 3 mm off-axis distances are all less than 1% of the primary beam. For the dosimeter with optical density value of 3.32, however, the intensity of the scattered light was more than 3% of the primary laser beam even for off-axis distance of 5 mm. Furthermore, the drop of the intensity of the scattered light at larger off-axis distances is slower for dosimeter of large optical density values. For the dosimeter of 3.32 optical density, the intensity of the scattered light is more than 1% of the primary laser beam even at the 9 cm off-axis position. It should be pointed out that the 3.32 optical density is beyond the valid dynamic ranges for most of the optical CT scanners [11].

Figure 4 compares the distribution of the intensity of the light signal from the dosimeter of 15.2 cm diameter with those from the dosimeters of 10 cm and 4 cm diameters. The optical density drops across the three dosimeters were 1.9, 1.83, and 1.79, respectively. For dosimeters of comparable optical density values, the intensity of the scattered light was found to decrease with increasing dosimeter diameter.

Figure 5 shows the results for the dosimeter of 15.2 cm diameter and 1.9 optical density value taken at two detector-to-tank distances (2 cm and 10 cm). The intensity of the scattered light at the same off-axis distance is weaker for larger detector-to-tank distances.

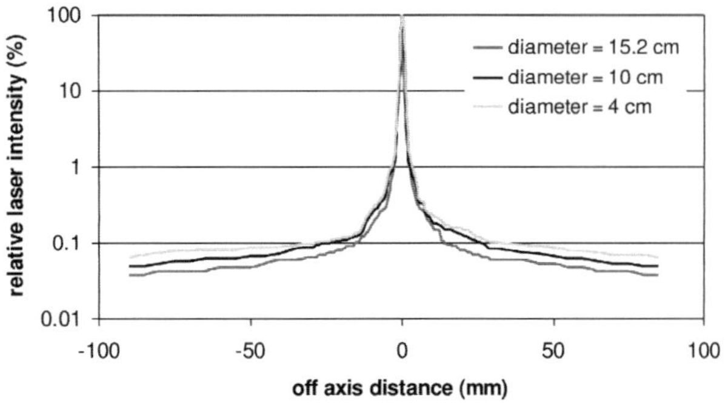

Figure 4: Comparison of the light intensity distributions for the three dosimeters of 4 cm, 10 cm, and 15.2 cm diameters and with comparable optical density values.

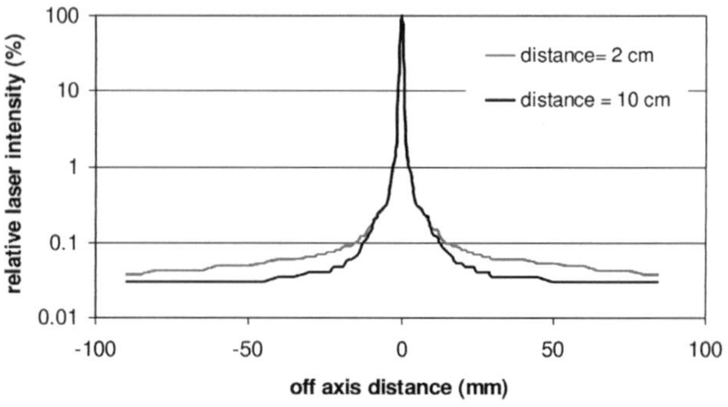

Figure 5: Light intensity distributions for a dosimeter of 15.2 cm diameter taken at detector-to-tank distances of 2 cm and 10 cm.

4. Discussion

Minimizing the effect of light scattering has been an important consideration in the design and operation of optical CT scanners. Currently, there are two groups of optical CT scanners: one is based on a single laser source coupled to a single photodiode detector and the other uses incoherent broad light source and a large area detector such as a CCD camera. The methods for handling the scattered photons in these two groups of scanners are different.

The basis for separating the scattered light from the real signal has to do with the angular distribution of the scattered photons. As is shown in Figs. 3, 4 and 5, the majority of the scattered photons are distributed within a small angle from the direction of the scanning beam inside the dosimeters. For scanners using a single laser source and a small area detector, OCTOPUS™ scanner, for example, the photodiode detector does not have to be positioned to face the scanning laser beam. Instead, the point laser source can be guided through a series of reflective mirrors (and/or lenses) to scan across the dosimeters and reach the photodiode detector. Collection of the scattered light at the detector aperture is therefore limited to photons going through multiple scattering only. Furthermore, with fine tuning of the optics and the refractive index matching, the position of the primary laser beam on the detector surface can be adjusted to be stable during the entire scanning process. The detector aperture can be reduced to a few millimeters such that the collection of the scattered photons by the detector is minimal.

When the detector has to be placed to face the light source, as is the case for the CCD camera approach, the effect of the scattered photons can be minimized in a variety of ways [9-10]. For instance, a telecentric lens can be placed in front of the CCD camera to ensure that only the incoming parallel light rays in certain directions are collected by the camera. However, considering the dimension of the camera lens, collection of the scattered photons within a small angle from the direction of the parallel light rays is inevitable. Furthermore, the scattered photons are mostly distributed in a small angle from the primary laser beam, as is shown in the present study.

An alternative for reducing the effect of light scattering is to decrease the overall optical density drop across the dosimeters scanned, at the price of a diminished overall signal-to-noise ratio. As is shown in Fig. 3, smaller attenuation of the scanning laser beam within a dosimeter can result in smaller amount of scattered photons everywhere. For dosimeters with optical density values of less than 2, the intensity of the scattered light at the position of 2 mm off-axis distance is less than 1.5%. A balance between the

Journal of Physics: Conference Series **250** (2010) 012063 doi:10.1088/1742-6596/250/1/012063

experimental uncertainties generated by the light scattering and other type of noises might be necessary in some cases.

5. Conclusions

We have performed a preliminary study on the effect of light scattering in optical CT scanning using the Presage dosimeters, the water tank of the OCTOPUS™ optical CT scanner, and an in-house manufactured light intensity measuring system. The intensity of the scattering light from the Presage dosimeters was found to decrease with increasing off-axis distance from the primary laser beam and increasing detector-to-dosimeter distance. The amount of the scattered photons was demonstrated to be smaller for dosimeters with less attenuation of the primary laser beam. For dosimeters of the same optical density value, the intensity of the scattered light is larger for smaller dosimeter diameters.

6. References

[1] Baldock C, De Deene Y, Doran S, Ibbott G, Jirasek A, Lepage M, McAuley K B, Oldham M and Schreiner L J 2010 Polymer gel dosimetry *Phys. Med. Biol.* **55**, R1-R63.

[2] Oldham M, Siewerdsen J H, Shetty A S and Jaffray D A 2001 High resolution gel-dosimetry by optical-CT and MR scanning *Med. Phys.* **28** 1436-44.

[3] Gore J C, Ranade M, Maryanski M J and Schulz R J 1996 Radiation dose distributions in three dimensions from tomographic optical density scanning of polymer gels: I. Development of an optical scanner *Phys. Med. Biol.* **41**, 2695-2704.

[4] Maryanski M J, Zastavker Y Z, and Gore J C 1996 Radiation dose distributions in three dimensions from tomographic optical density scanning of polymer gels: II. Optical properties of the BANG polymer gel *Phys. Med. Biol.* **41**, 2705-17.

[5] Islam K T S, Dempsey J F, Ranade M, Maryanski M J and Low D A 2003 Initial evaluation of commercial optical CT-based 3D gel dosimeter *Med. Phys.* **30**, 2159-68.

[6] Xu Y, Wuu C S and Maryanski M J 2003 Determining optimal gel sensitivity in optical CT scanning of polymer gels *Med. Phys.* **30**, 2257-63.

[7] Doran S J, Koerkamp K K, Bero M A, Jenneson P, Morton E J and Gilboy W B 2001 A CCD-based optical CT scanner for high-resolution 3D imaging of radiation dose distributions: equipment specifications, optical simulations and preliminary results *Phys. Med. Biol.* **46**, 3191-3213.

[8] DeJean P, Senden R, McAuley K B, Rogers M and Schreiner L J 2006 Initial experience with a commercial cone beam optical CT unit for polymer gel dosimetry II: Clinical potential *Journal of Physics: Conference Series* **56** 183–6.

[9] Krstajić N and Doran S J 2006 Focusing optics of a parallel beam CCD optical tomography apparatus for 3D radiation gel dosimetry *Phys. Med. Biol.* **51**, 2055-75.

[10] Sakhalkar H S and Oldham M 2008 High-resolution 3D dosimetry utilizing a novel optical-CT scanner incorporating tertiary telecentric collimation *Med. Phys.* **35**, 101-111

[11] Xu Y, Wuu C S and Maryanski M J 2004 Performance of optical CT scanning of polymer gels as a tool for 3D dose verification *Med. Phys.* **31** 3024-30.

ScanSim: A tool for simulating optical-CT imaging

Mark Oldham

Duke University Medical Center, Durham, NC USA

Mark.Oldham@duke.edu

Abstract. A software tool has been developed that can simulate image formation in a variety of optical-CT scanning configurations. The formalism of the simulation is introduced, including two main modes: a diverging point source mode, and a converging broad beam mode. Preliminary results are presented for scanning Presage dosimeters in both modes and immersed in refractive media of widely varying refractive index (RI), including air, water, and a fully matched medium. Pronounced differences in the edge artifacts and accuracy of reconstructed coefficients is observed. The ScanSim software is shown to be a useful tool to investigate and quantify many aspects of optical-CT image formation, including reducing dependence on matching fluids.

1. Introduction

The first optical-CT scanner for radiation dosimetry was proposed by Gore et al. [1]. This 1st generation scanning laser system has now been evaluated by several groups and found to be an accurate and reliable system with the facility to minimize artifacts from scattered and stray light [2-4]. More recently, other approaches to optical-CT scanning system have been proposed, including broad-beam scanning systems [5-9]. These systems have potential for very high spatial resolution and much faster scanning times (minutes), at the potential expense of higher susceptibility to scatter and stray light and associated artifacts. In general the scanning laser systems are used for dosimeters where the primary contrast mechanism is light scattering (e.g. polymer gels). Broad-beam systems have been explored in conditions of low-light-scatter, and dosimeters where the primary contrast mechanism is light-absorbtion (e.g. Presage, FBX gels). Despite the significant diversity in scanner design and approach, a common feature of all current systems is the requirement for the dosimeter to be immersed in a refractively matched fluid. This fluid minimizes the refraction of light incident on the curved surfaces of the dosimeter, and facilitates the measurement of line-integrals of attenuation which is critical input for reconstruction. Effective refractive matching between the dosimeter and fluid is known to be critical to reducing edge artifacts in optical-CT. Good matching also facilitates accurate dosimetry particularly in broad-beam scanning systems, which are more sensitive to this aspect. Despite the advantages to immersing the dosimeters in a refractively matched fluid-bath, this procedure is one of the most challenging, laborious and messy aspects of optical-CT scanning. The purpose of the present work was to create a software tool 'ScanSim' which could simulate the optical-CT imaging process under a wide variety of conditions both with and without a refractive matched fluid. The aim is to explore new scanning arrangements that might reduce or eliminate the need for refractively matched fluid (convenience and cost), and to reduce artefacts (improve accuracy).

2. Methods

Journal of Physics: Conference Series **250** (2010) 012064 doi:10.1088/1742-6596/250/1/012064

Two basic modes of the software were implemented (figure 1a and b) corresponding to a converging or diverging incident beam respectively. The software parameters include the refractive indices of the dosimeter and surrounding fluid, the attenuation coefficients of medium and fluid, wavelength of incident light, the radius of the cylindrical dosimeter, the divergence or convergence of the incident beam, and the position of the dosimeter. In addition rays that deflect from the optical-axis by more than a given tolerance angle (degrees) are shown as dashed, enabling visualization of the scatter rejection of a telecentric system.

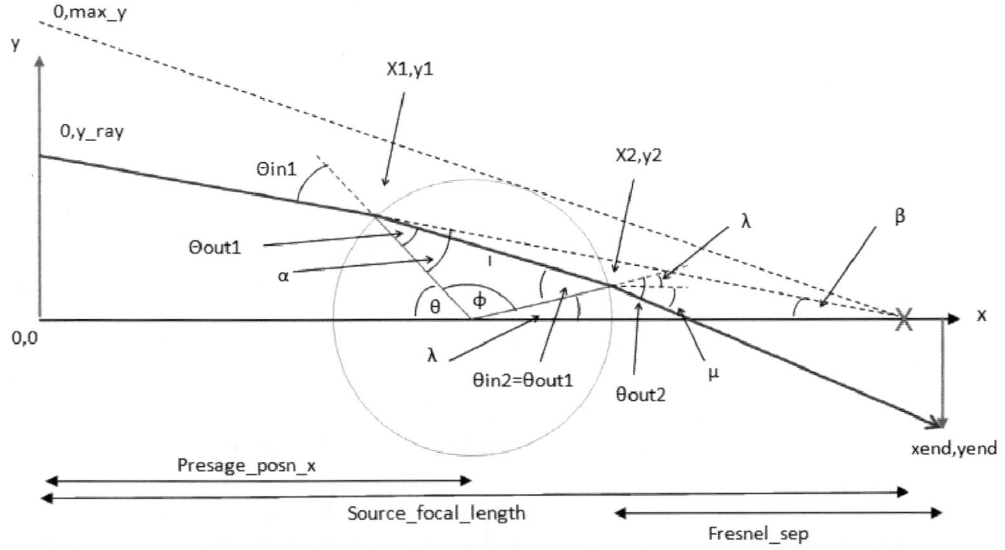

Figure 1a: Convergent broad-beam geometry. The 'ScanSim' software models the attenuation and refraction of light rays that originate at an extended source (located at x=0), and converge at the point X (large red X) in the absence of the dosimeter. The attenuation and refraction of the ray as it passes through the dosimeter (shown as the green cylinder outline) are modeled through to the imaging plane. A single example ray is shown in this figure.

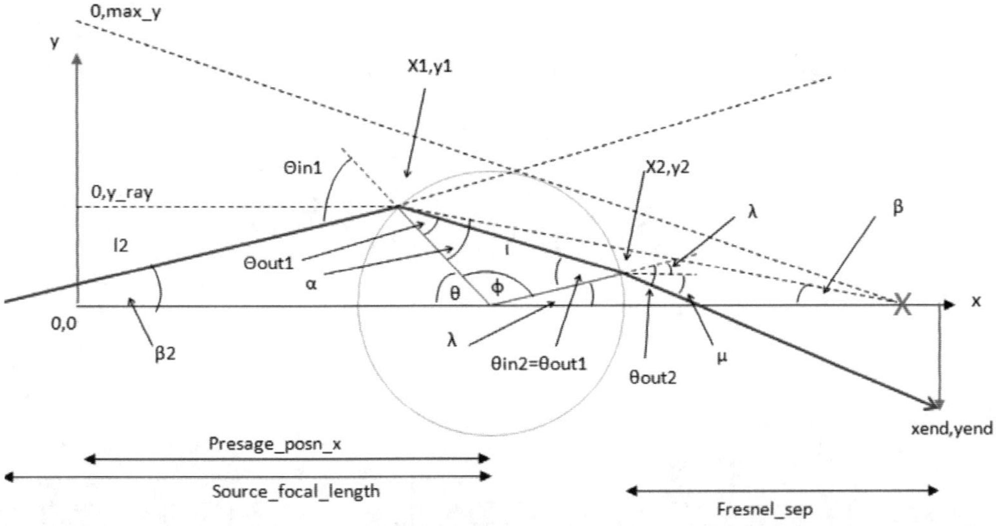

Figure 1b: Divergent broad-beam geometry. The 'ScanSim' software models the attenuation and refraction of light rays that originate at a point source (located to the left of x=0), and diverge at an angle β2. The attenuation and refraction of the ray as it passes through the dosimeter (shown as the green cylinder outline) are modeled through to the imaging plane (xend,yend). A single example ray is shown.

At present ScanSim can simulate optical-CT imaging of uniformly attenuating cylindrical non-scattering. Ray paths and associated intensity loss (caused by both attenuation within the dosimeter, and Fresnel transmission losses at interfaces) for a single projection are first simulated. Due to symmetry, all other projections corresponding to an optical-CT data set will be the same. A full acquisition data set for reconstruction is generated by re-sorting rays according to the angle of the trajectory through the dosimeter. Each ray is thus assigned to the projection in which it would be normal to the detecting plane, and thus a parallel ray reconstruction algorithm can be used to reconstruct the ensemble of generated projections. Simulations assume an ideal detector which assigns a detected pixel intensity to the corresponding line integral of attenuation. This can be achieved in the real world if the dimensions of the dosimeter are known, and the trajectory of ray-paths can be calculated a-priori, both of which are easily feasible for cylindrical dosimeters.

ScanSim simulations have been run to investigate the correlation between refraction matching, edge artefacts, and the accuracy of reconstructed attenuation coefficients. Also to investigate the potential for new in-air scanning configurations, and configurations which might use lower RI matched fluids.

3. Results and discussion

3.1 An illustrative ScanSim simulation.

In-air optical-CT imaging of a 50mm radius Presage cylinder (uniform attenuation 0.115/cm, and RI=1.501), using the divergent geometry mode (a point light source positioned 200mm away) is shown in figure 2. The user defined parameters of the simulation are shown in the left panel). The upper right figure shows ray-paths from the point source (x=0) through the Presage cylinder shown as black circle. Ray colors are artificial, and just facilitate discerning different ray-paths. The lower left panel shows both the accumulated transmitted power coefficients (green line) of each ray, and the total intensity of each ray (blue), compared to un-attenuated light. The lower middle 2 images show the sinogram data of the ensemble of generated projections, and the lower right image shows the final reconstructed image.

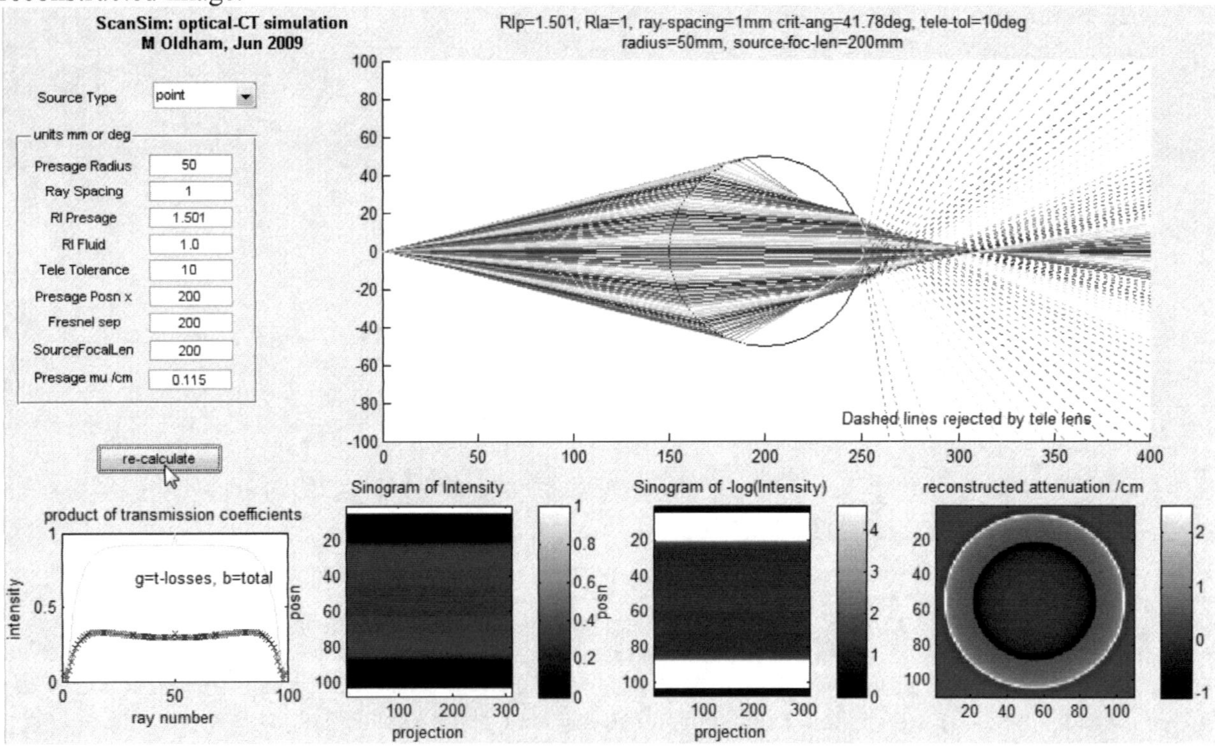

Figure 2: A ScanSim simulation in divergent mode (or point light source at x=0 upper right image). This figure illustrates the challenges of in-air scanning of Presage due to a strong lensing effect. A substantial edge artifact is observed (~20mm wide), and greatly depressed attenuation coefficients are reconstructed compared to the known input attenuation.

3.2. ScanSim investigation of the effect of variable RI matching.

A sequence of divergent beam simulations were run to investigate the correlation between refraction matching, edge artifacts, and the accuracy of reconstructed attenuation coefficients (figure 3). All parameters (except the RI of the surrounding fluid) were kept constant for all simulations, including: RI dosimeter=1.501, source-distance=400mm, ray-spacing=1mm, dosimeter-radius=50mm.

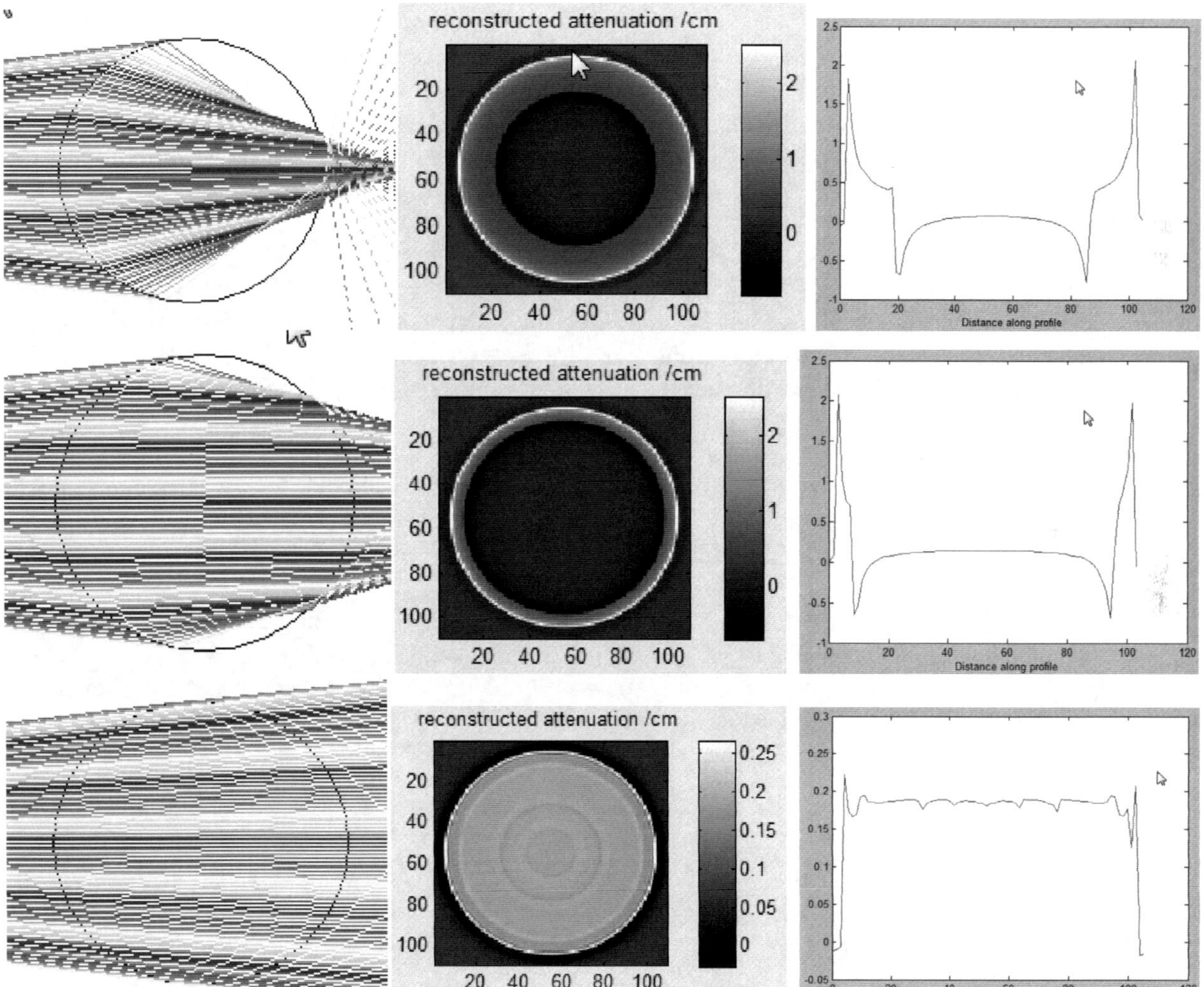

Figure 3: Results from 3 simulations. All parameters were constant except for the refractive-index of the surrounding medium, which was 1.0 (air), 1.3 (water), and 1.5 (Presage), in the upper, middle and lower rows respectively.

The simulations of figure 3 illustrate the significant effect of refractive matching on both the edge artifacts, and the accuracy of the reconstructed attenuation coefficient.

4. Conclusions

The ScanSim software is a useful tool for investigating many aspects of the mechanisms and artifacts of optical-CT imaging. Further studies are planned to explore and quantify the relationships between RI matching and edge and reconstruction artifacts, as well as to investigate new scanning configurations that may reduce the requirement for expensive and difficult-to-work with high RI fluids. Simulations assume an ideal detector which assigns a detected pixel intensity to the corresponding line integral of attenuation. This can be achieved in the real world if the dimensions of the dosimeter and ray-paths are known a-priori, as is the case with cylindrical dosimeters.

Acknowledgements: Supported by NIH R01 CA100835.

References

[1] Gore, J. C., Ranade, M., Maryanski, M. J., and Schulz, R. J., "Radiation dose distributions in three dimensions from tomographic optical density scanning of polymer gels: I. Development of an optical scanner", Phys.Med.Biol., Vol. 41, pp 2695-2704, 1996.

[2] H.S. Sakhalkar, J. Adamovics, G. Ibbott and M. Oldham, "A comprehensive evaluation of the PRESAGE/optical-CT 3D dosimetry system", Med. Phys. 36 (1), 2009.

[3] Islam KT, Dempsey JF, Ranade MK, Maryanski MJ, Low DA, Initial evaluation of commercial optical CT-based 3D gel dosimeter. Med Phys. 2003 Aug;30(8):2159-68.

[4] Xu Y, Wuu CS, Maryanski MJ., Performance of a commercial optical CT scanner and polymer gel dosimeters for 3-D dose verification. Med Phys. 2004 Nov;31(11):3024-33

[5] Baldock C, De Deene Y, Doran S, Ibbott G, Jirasek A, Lepage M, McAuley KB, Oldham M, Schreiner LJ. Polymer gel dosimetry. Phys Med Biol. 2010 Mar 7;55(5):R1-63. Epub 2010 Feb 11. Review.

[6] Oldham M, 3D dosimetry by optical-CT scanning., J Phys. 2006;56:58-71

[7] Doran SJ, Koerkamp KK, Bero MA, Jenneson P, Morton EJ, Gilboy WB. A CCD-based optical CT scanner for high-resolution 3D imaging of radiation dose distributions: equipment specifications, optical simulations and preliminary results., Phys Med Biol. 2001 Dec;46(12):3191-213.

[8] Olding T, Holmes O, Schreiner LJ., Cone beam optical computed tomography for gel dosimetry I: scanner characterization., Phys Med Biol. 2010 May 21;55(10):2819-40

[9] Clift C, Thomas A, Adamovics J, Chang Z, Das I, Oldham M. Toward acquiring comprehensive radiosurgery field commissioning data using the PRESAGE/optical-CT 3D dosimetry system., Phys Med Biol. 2010 Mar 7;55(5):1279-93

Journal of Physics: Conference Series **250** (2010) 012065 doi:10.1088/1742-6596/250/1/012065

Mathematical modelling of depth-dose response of polymer gel dosimeters

J. N. M. Chain[1], L. J. Schreiner[2,3], K. B. McAuley[1]

[1] Dept. of Chemical Engineering, Queen's University, Kingston, Canada, K7L 3N6
[2] Cancer Centre of Southeastern Ontario, Kingston, Canada, K7L 5P9
[3] Depts. of Oncology and Physics, Queen's University, Kingston, Canada, K7L 3N6

kim.mcauley@chee.queensu.ca

Abstract. A dynamic partial differential equation (PDE) model is used to simulate the effects of radiation depth-doses on polymer formation in Polyacrylamide Gel (PAG) dosimeters. Depth-doses are simulated using different types of radiation including Co^{60} gamma and 6 and 15 MV X-ray photon beams, along with 6, 9, 12, 16 and 20 MeV electron beams. Effects of monomer diffusion (edge enhancement) and temperature are studied. Diffusion results in excess polymer formation at the position of maximum dose (4% for Co^{60} γ-radiation and less for other types of radiation studied). Temperature increases on the order of 2 °C increase the mass of polymer formed by approximately 2-3%. These results provide insight for calibrating dosimeters using depth-dose information.

1. Introduction

Polymer gel dosimetry is based on the dose dependant polymerization of monomers dispersed in an irradiated gel substrate. The polymerization is well behaved when the irradiation is essentially uniform. Measurements and modelling have shown that in non-uniform conditions with large dose gradients, such as the edge of a beam, the degree of polymerization at the edge of the irradiation may be enhanced by monomer diffusing from low dose areas (e.g., Fuxman et al. [1,2], Vergote et al. [3]). This has not been characterized in more gentle dose gradients. This article uses simulations to investigate the influence of depth-doses on the behaviour of PAG dosimeters. The primary objective is to predict the amount of polymer formed and the temperature at various depths in response to Co^{60} γ-radiation, 6 and 15 MV X-ray photon beams, and 6 MeV to 20 MeV electron beams. Use of the model enables examination of monomer diffusion effects and temperature effects, which can influence the accuracy of dosimeter calibration using depth-dose information [4]. The current research focuses on 6%T 50%C PAG dosimeters containing 5% gelatin. Note that the model could also be used to simulate NIPAM-based dosimeters and PAG dosimeters with other recipes.

2. Materials and Methods

2.1. Primary Data for Depth-Doses

The percent depth-dose (PDD) data used in this study were from clinical measurements for various clinical radiation units. Clinical data under standard reference conditions (10 x 10 cm^2 field sizes at standard source to surface distance) were taken for Co^{60} gamma and 6 and 15 MV X-ray photon beams (Fig. 1a), along with 6, 9, 12, 16 and 20 MeV electron beams (Fig. 1b).

Different cubic splines were used to convert the experimental depth-dose data for the various radiation beams in Fig. 1 to a convenient functional form for use in the PDE model. Additional spline nodes were used near the maxima of the PDD curves to ensure that accurate depth-dose information was provided to the model.

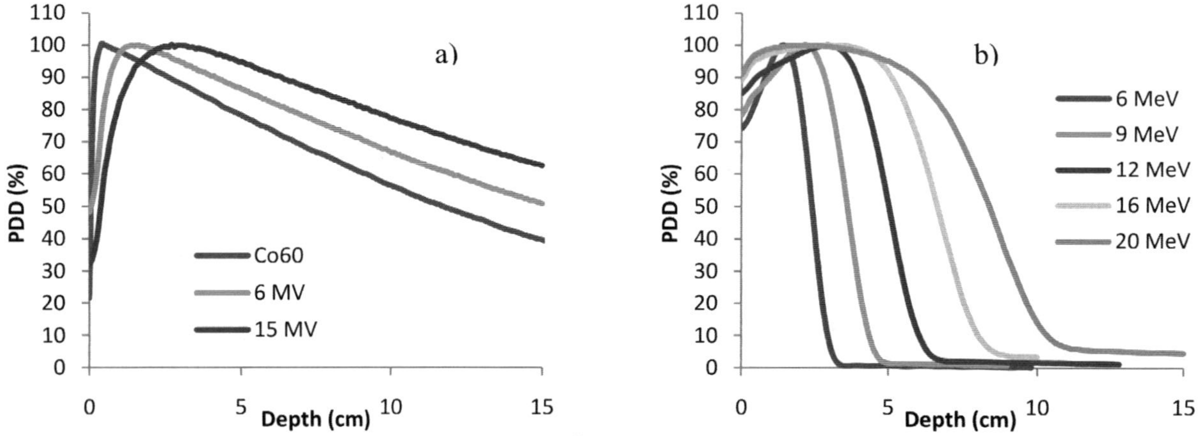

Figure 1. Percent depth-dose data by: (a) Co^{60} γ-radiation and 6 and 15 MV X-ray photon beams; (b) 6, 9, 12, 16 and 20 MeV electron beams.

2.2. Model Development
PDE material balances on chemical species in aqueous and polymer phases of PAG dosimeters were formulated by Fuxman et al. [1,2] and are extended in the current article. The revised model requires modification of the PDE material balance on primary radicals generated by water radiolysis:

$$\frac{\partial[(PR^{\cdot})]}{\partial t} = \frac{\Gamma(x)R_{PR^{\cdot}}}{\phi V_R} - r_{(PR^{\cdot})} - \frac{[(PR^{\cdot})]}{\phi}\frac{\partial\phi}{\partial t}$$

where $[(PR^{\bullet})]$ is the concentration of primary radicals, $\Gamma(x)$ identifies zones receiving or not receiving radiation, $R_{PR\bullet}$ is the rate of generation of primary radicals (dose rate times G factor), ϕ is the volume fraction occupied by the aqueous phase, V_R is the total volume of the phantom and $r_{(PR\bullet)}$ is the rate of consumption of primary radicals by subsequent reactions. Fuxman et al. [1,2] used their model to study edge enhancement at a sharp interface between irradiated and un-irradiated zones. They used $\Gamma(x)$ to distinguish locations that receive radiation ($\Gamma(x) = 1$) from locations that do not ($\Gamma(x) = 1$). When investigating the effects of PDD in the current work, $\Gamma(x)$ provides depth-dose information in the form of cubic splines. All other PDEs remain the same as those of Fuxman et al. [1,2], and boundary conditions were revised to permit heat loss through the front and back walls of the phantom shown in Fig. 2. Updated values of model parameters [5] that were fitted using experimental data [6,7], are used in the revised model.

2.3. Model Solution
Cubic polynomial equations were fitted using the spline toolbox in Matlab™. The complete system of 29 PDEs was solved using the VLUGR2 solver within Fortran [8]. Note that the PDE model can also be solved using the *pdepe* subroutine in Matlab™, but simulation times are prohibitively long. The PDEs were solved to predict the mass of polymer and the temperature vs. depth at different times. Additional simulations were performed to examine the importance of monomer diffusion and temperature effects. Simulations with diffusion turned off ($D_{M1} = 0$) were compared with simulations

that account for diffusion ($D_{M1} = 4.1 \times 10^{-6}$ cm^2/min), and simulations with heat effects turned off (T = 20°C everywhere) were compared with simulations that account for the heat released by polymerization. All simulations considered the total given dose at d_{max} to be 5 Gy; for the Co^{60} γ-radiation simulations, a dose rate was set to 2 Gy/min, for all other radiation beams the dose rate was 4 Gy/min.

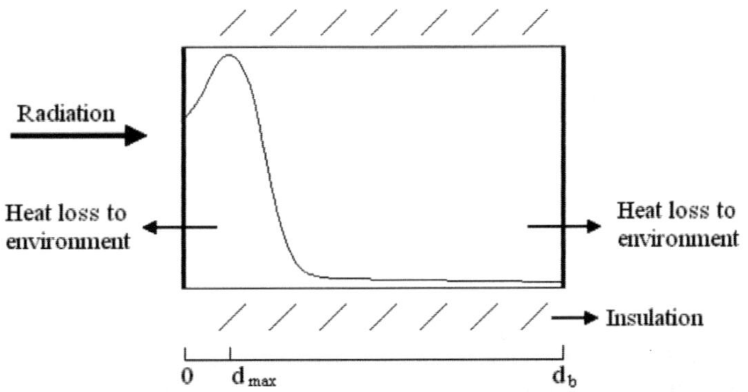

Figure 2. Schematic diagram of the irradiation of a simulated phantom with a PDD curve, where d_{max} represents the location of maximum absorbed dose, and d_b is the depth of the back wall. A value of d_b = 10 cm was used for all simulations, except for the 20 MeV electron beam, which used d_b = 15 cm because of deeper penetration.

3. Results and Discussion

Simulation results showing the amount of polymer formed 24 hours post irradiation are shown in Fig. 3b along with the corresponding depth-dose data in Fig. 3a. The amount of polymer formed for each type of radiation has a shape similar to its depth-dose curve. There are, however, some noticeable differences between the shapes of the curves, especially near the position of maximum dose. For example, the polymer formation curve for the Co^{60} γ-radiation has a more pronounced maximum than does its PDD curve. An examination of the cause of this difference is provided below.

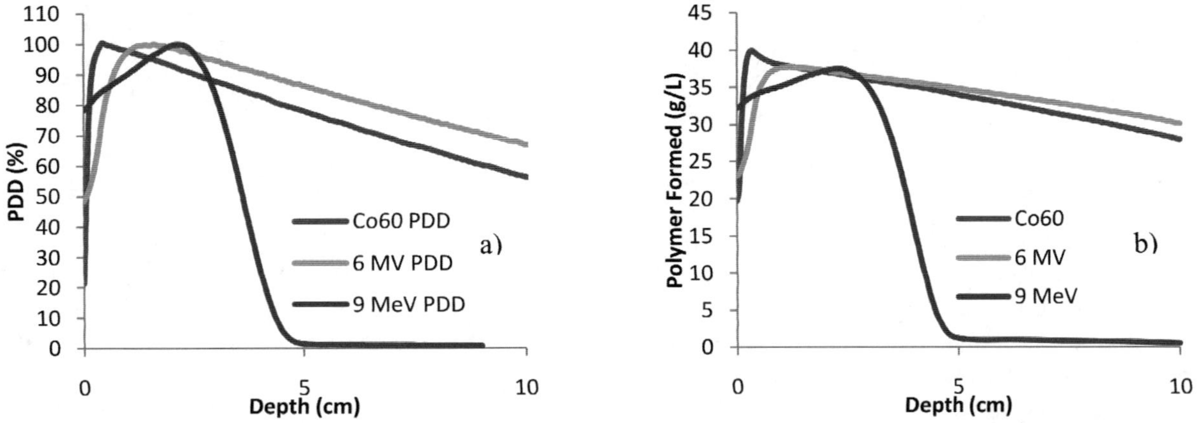

Figure 3. Comparison between (a) the depth-dose curves and (b) the amount of polymer formed 24 post irradiation for Co^{60} γ-radiation, 6 MV X-ray photon beam and 9 MeV electron beam.

3.1. Diffusion Effects

Selected simulation results comparing the model predictions with and without diffusion effects are presented in Fig. 4. Table 1 shows the percent difference in the mass of polymer formed at the location of maximum dose, with and without diffusion. Note that, when diffusion is turned on, the location of maximum dose (d_{max}) does not correspond perfectly to the location of maximum polymer formation, with this peak shifted slightly towards the adjacent area receiving less radiation where monomers are more plentiful. The simulations show that diffusion effects are more prominent near the depth of maximum dose than at greater depths where dose gradients are small and diffusion effects can hardly be noticed (see Fig. 4a). For high MV photon beams and high MeV electron beams, where the PDD curves are smoother than for Co^{60} γ-radiation, diffusion effects have only a minor influence on the amount of polymer formed (see Fig. 4b and Table 1).

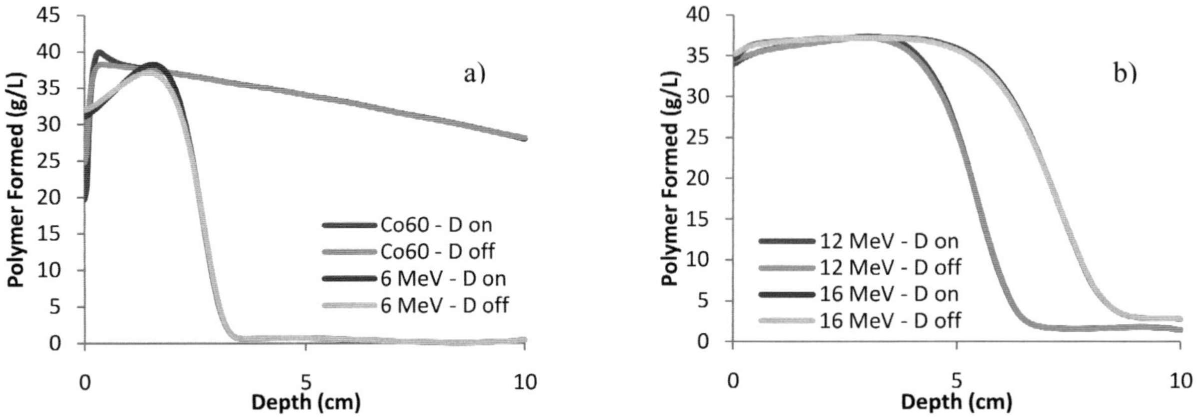

Figure 4. Comparison between normal simulations and simulations without diffusion for: a) Co^{60} γ-radiation and 6 MeV electron beam, b) 12 and 16 MeV electron beams, 24 hours post irradiation.

Table 1. Amount of polymer formed 24 hours post irradiation at the location of maximum absorbed dose for simulations with and without diffusion and with and without temperature effects.

Radiation type	Polymer formed (g/L)			Percent increase with D	Percent increase with T
	with D and T	without D	without T		
Co60 γ-radiation	39.72	38.24	39.21	3.87	1.30
6 MV X-ray photon beam	37.57	37.33	37.05	0.64	1.40
15 MV X-ray photon beam	37.5	37.38	36.64	0.32	2.34
6MeV electron beam	38.12	37.11	37.35	2.72	2.06
9 MeV electron beam	37.47	36.96	37.19	1.38	0.75
12 MeV electron beam	37.39	37.21	37.01	0.48	1.03
16 MeV electron beam	37.24	37.21	36.78	0.08	1.25
20 MeV electron beam	37.34	37.31	36.27	0.08	2.95

3.2. Temperature Effects

The temperature of the system increases as polymerization occurs, and then eventually decreases due to heat loss through the walls of the phantom. Regions where higher amounts of radiation are delivered achieve higher temperatures. Since temperature increases the rate of polymerization, more polymer tends to form in highly irradiated areas, than if the system were isothermal [7]. Simulation results showing temperature profiles and comparing the model with and without temperature effects are presented in Fig. 5b and Table 1. Note that the temperature is higher for the 20 MeV electron beam than for other types of radiation because the high dose region was more extensive (given the broad shoulder in the depth-dose curve). This results in more of the phantom volume being irradiated to a higher dose and more polymer forms. As expected, the total amount of polymer formed decreases when the temperature effects are removed from the model. Larger differences between predictions with and without temperature effects occur near the maximum dose depth than deeper in the phantoms where polymerization and heat generation rates are lower. In all simulated cases, the amount of additional polymer formed at d_{max} is less than 2.5%, except for 20 MeV which had an increase of nearly 3%, due to the higher temperature, which is shown in Fig. 5a.

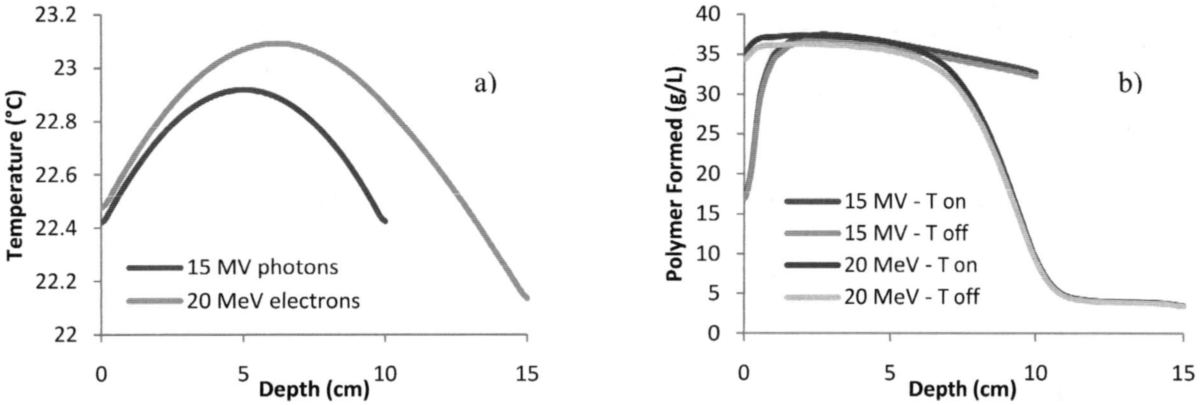

Figure 5. a) Temperature profiles in PAG dosimeters 24 hours post irradiation using a 15 MV X-ray photon beam and a 20 MeV electron beam, b) simulations with and without temperature effects.

4. Conclusions and Recommendations

A fundamental PDE model was developed to simulate depth-dose effects in PAG dosimeters. Both diffusion and temperature effects influence the amount of polymer formed at various locations. Diffusion effects are more apparent for Co^{60} γ-radiation (a 3.9% increase in polymer formed at the location of maximum dose) than for other types of radiation with flatter maxima. Temperature effects are larger when more radiation is delivered. These simulation results indicate that when calibration curves are developed using depth-dose information, more accurate results may be obtained by avoiding the depth of maximum dose, particularly for Co^{60} γ-radiation.

5. Acknowledgements

This work was funded by Mathematics of Information Technology and Complex Systems (MITACS), the Canadian Institutes for Health Research (CIHR) and the Natural Sciences and Engineering Research Council of Canada (NSERC).

References

[1] Fuxman A M, McAuley K B, Schreiner L J (2005) *Chem. Eng. Sci.* **60**, 1277
[2] Fuxman A M, McAuley K B, Schreiner L J (2003) *Macromol. Theory Simul.* **12**, 647-662
[3] Vergote K, De Deene Y, Vanden Bussche E, De Wagter C (2004) *Phys. Med. Biol.* **49**, 4507–4522
[4] Oldham M, McJury M, Baustert I B, Webb S, Leach M O (1998) *Phys. Med. Biol.* **43**, 2709-2710
[5] Koeva V I, Daneshvar S, Senden R J, Imam A H M, Schreiner L J, McAuley K B (2009) *Macromol. Theory Simul.* **18**, 495-510
[6] Babic S, Schreiner L J (2006) *Phys. Med. Biol.*, **51**, 4171
[7] Salomons G J, Park Y S, McAuley K B, Schreiner L J (2002) *Phys. Med. Biol.* **47**, 1435–1448
[8] Blom J G, Trompert R A, Verwer J G (1996) *ACM Transactions on Mathematical Software*, **22** (3), 302-328

Journal of Physics: Conference Series **250** (2010) 012066 doi:10.1088/1742-6596/250/1/012066

Mathematical modelling of response of polymer gel dosimeters to brachytherapy radiation

A.T. Nasr[1], J. N. M. Chain[1], L. J. Schreiner[2,3], K. B. McAuley[1]

[1] Chemical Engineering Dept., Queen's University, Kingston, Canada, K7L 3N6
[2] Cancer Centre of Southeastern Ontario, Kingston, Canada, K7L 5P9
[3] Depts. of Oncology and Physics, Queen's University, Kingston, Canada, K7L 3N6

kim.mcauley@chee.queensu.ca

Abstract. A dynamic partial differential equation (PDE) model is used to simulate effects of a single Ir^{192} brachytherapy seed on the amount and composition of polymer formed during polyacrylamide gel (PAG) dosimetry. Simulations are conducted for a point-source brachytherapy seed placed at the center of a 6%T 50% C anoxic PAG phantom. The seed is removed after one minute, but polymerization is simulated up to a final time of 24 hours. Simulation results indicate that changes occur in both the mass of polymer formed per unit dose and in the crosslink density as a function of the radial distance from the brachytherapy seed. For example, at a distance of 5 mm from the seed, 41 mg of polymer form per Gy of radiation absorbed (after 24 hours), whereas at a larger distance of 5 cm from the seed 75 mg of polymer form per Gy. The polymer that forms near the seed is predicted to have a higher level of crosslinking than the polymer that forms further away. These results suggest potential calibration problems that may occur during brachytherapy dosimetry using polymer gels.

1. Introduction

Models describing PAG dosimeters under spatially uniform and non-uniform external beam irradiation have been used to simulate edge enhancement and temporal instability problems in polymer gel dosimetry [1, 2]. In the current work, the model of Fuxman et al. [1] is extended to predict polymerization, crosslinking and polymer precipitation in response to brachytherapy dose delivery by a single Ir^{192} brachytherapy seed. Simulations are performed using a 6%T 50%C anoxic PAG dosimeter recipe containing 5% gelatin. Unfortunately, too little is known about reactions involving tetrakis hydroxymethyl phosphonium chloride (THPC) to account for the effects of this anti-oxidant [3, 4]. Extensions to brachytherapy dosimetry with multiple seeds are also possible, as are simulations with NIPAM-based dosimeters and simulations with oxygen contamination or anisotropic radiation delivery [5, 6]. We predict changes in the amount of polymer formed per unit dose and in polymer composition as a function of time and radial distance from the seed, providing detailed information that complements an earlier modeling study [6]. Both the mass of polymer formed and the level of bisacrylamide incorporation influence the transverse relaxation rate R_2 measured by NMR [7] and may influence the responses of other dose read-out modalities (e.g., optical and x-ray CT).

2. Materials and Models

2.1. Radial Dose Profiles

Traditionally, dose rate $\dot{D}(r)$ at a distance r from an interstitial brachytherapy source is calculated using the point source approximation [8]:

$$\dot{D}(r) = A_{app} f_{med} (\Gamma_\delta)_x \left(\frac{1}{r^2}\right) T(r) \Phi_{an} \quad (1)$$

where A_{app} is the apparent activity of the source, f_{med} is the exposure-to-dose conversion factor, $(\Gamma_\delta)_x$ is the exposure rate constant for the radionuclide in the source, $T(r)$ is the tissue attenuation factor and Φ_{an} is the anisotropy constant. Simulations were done for single point-source radiation by Ir^{192}. Values of the parameters [$A_{app} = 10\ Ci$, $f_{med} = 0.0097\ \frac{Gy}{R}$, $(\Gamma_\delta)_x = 78.17\ \frac{R\ cm^2}{Ci\ min}$, $\Phi_{an} = 1$ and

Journal of Physics: Conference Series **250** (2010) 012066 doi:10.1088/1742-6596/250/1/012066

$T(r) = 1.0128 + (5.02E-3)r - (1.18E-3)r^2 - (2.01E-5)r^3]$ were collected from several sources [8, 9]. The resulting radial dose profile in water is shown in Fig.(1, a). Note that Eq. 1 leads to an infinite dose rate at $r = 0$. To prevent numerical problems associated with very large dose rates, we used a constant value of $\dot{D}(r) = 770$ Gy/min for $r \leq 0.1$ cm and Eq. 1 for larger values of r.

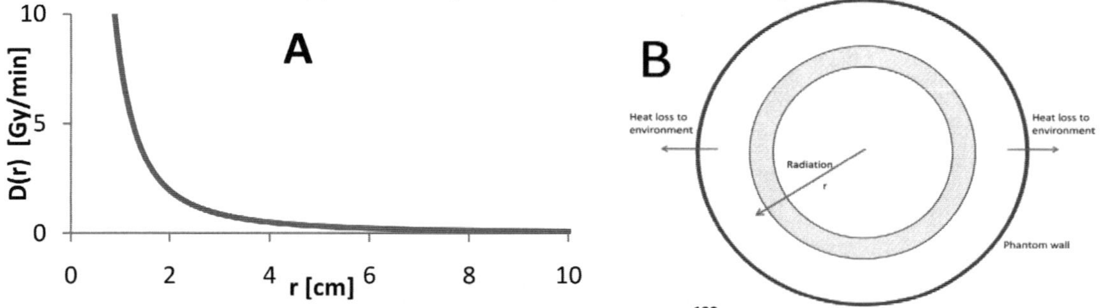

Figure 1: A) Radial dose profile for irradiation of water by a spherical Ir^{192} seed. B) Schematic diagram of the irradiation of a simulated phantom. The outer wall of the phantom was 10 cm from the brachytherapy seed located at the center.

2.2. Model Development

Partial differential equations (PDE) for material balances on chemical species in aqueous and polymer phases of PAG dosimeter developed by Fuxman et al. [1] are reformulated and extended in the current article. The revised model requires modifying the original PDE material balance on primary radicals generated by radiolysis of water:

$$\frac{\partial[(PR^{\cdot})]}{\partial t} = \frac{\Gamma(x)R_{(PR^{\cdot})}}{\phi V_R} - r_{(PR^{\cdot})} - \frac{[(PR^{\cdot})]}{\phi}\frac{\partial \phi}{\partial t} \tag{2}$$

where $\Gamma(x)$ identifies zones receiving or not receiving radiation, $R_{(PR^{\cdot})}$ is the rate of generation of primary radicals (dose rate times G factor). Fuxman et al. [1] used their model to study edge enhancement at a sharp interface between irradiated and un-irradiated zones. They used $\Gamma(x)$ to distinguish locations that receive radiation ($\Gamma = 1$) from locations that do not ($\Gamma = 0$). When investigating the effects of brachytherapy in the current work, $\Gamma(x)$ is set to one everywhere, and the dose depth rate, given by (Eq.1), is used to determine the local rate of generation of primary radicals:

$$R_{(PR^{\cdot})} = G\,\dot{D}(r) \tag{3}$$

All other PDEs that do not contain spatial derivatives remain unchanged. PDEs in Fuxman's model that contain derivatives with respect to the spatial coordinate x were reformulated in spherical coordinates to account for variation in the radial direction r. For example, the material balance on acrylamide (M1) in the aqueous phase was changed from eq. (4) to (5):

$$\frac{\partial[M_1^W]}{\partial t} = D_{M_1^W}\left(\frac{1}{\phi}\frac{\partial \phi}{\partial x}\frac{\partial[M_1^W]}{\partial x} + \frac{\partial^2[M_1^W]}{\partial x^2}\right) + r_{M_1^W} - k_m\left([M_1^W] - \frac{[M_1^P]}{\Phi_{M_1}}\right) - \frac{[M_1^W]}{\phi}\frac{\partial \phi}{\partial t} \tag{4}$$

$$\frac{\partial[M_1^W]}{\partial t} = D_{M_1^W}\left(\frac{2}{r}\frac{\partial[M_1^W]}{\partial r} + \frac{1}{\phi}\frac{\partial \phi}{\partial r}\frac{\partial[M_1^W]}{\partial r} + \frac{\partial^2[M_1^W]}{\partial r^2}\right) + r_{M_1^W} - k_m\left([M_1^W] - \frac{[M_1^P]}{\Phi_{M_1}}\right) - \frac{[M_1^W]}{\phi}\frac{\partial \phi}{\partial t} \tag{5}$$

Boundary conditions were revised to permit heat loss through the outer wall of the spherical phantom shown in Fig. 1b. Updated values of model parameters that were fitted using experimental data are used in the revised model [4, 7, 10].

3. Results and Discussion

Figure 2a shows the formation of polymer in response to the brachytherapy radiation. As expected, polymerization continues after the removal of the seed, and more polymer forms in the highly irradiated region near the seed than farther from the seed where less radiation is received. Note that a small amount of additional polymer forms between 12 and 24 hours due to long-lived radicals.

Figs. 2b and 2c show the concentration of monomers acrylamide (M1) and bisacrylamide (M2), respectively, at t=1 minute (immediately after the removal of the seed) and at longer times. As expected, a larger drop in the concentration of M2 (Fig. 2c) occurs than in the concentration of M1

(Fig. 2b) due to the higher reactivity of M2, which has two vinyl groups. As time proceeds, concentrations of monomers near the seed rise due to diffusion induced by concentration gradients.

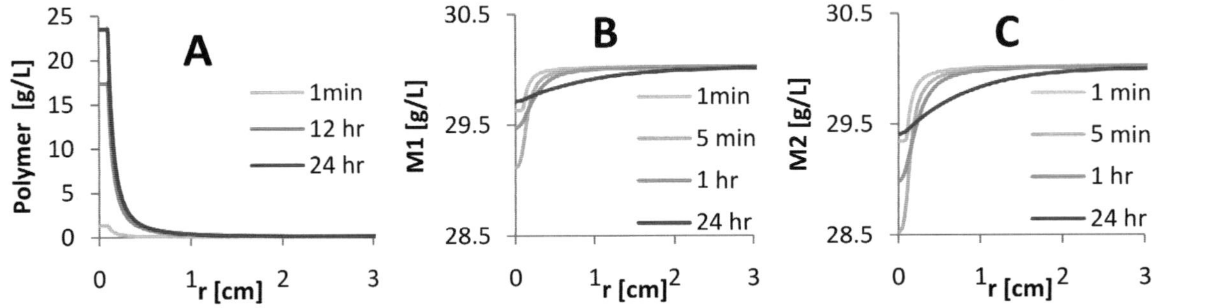

Figure2. Influence of brachytherapy radiation on a) polymer formation, b) acrylamide (M1) and c) bisacrylamide (M2) concentration gradients. Dose shown in Fig 1a was delivered for one minute before the seed was removed.

Figure 3a shows the moles of crosslinks per mass of polymer formed. Polymer formed near the seed has a larger crosslink density than the polymer formed at larger values of *r*. M2 is consumed more rapidly than M1 due to the two vinyl groups on each Bis molecule. The drop in the local Bis concentration near the seed causes Bis to diffuse toward the seed, resulting in high levels of Bis incorporation. Pendant double bonds produced by Bis incorporation are available for crosslinking.

Figure 3b shows the mass of polymer formed per unit dose of radiation delivered. Results show that the amount of polymer produced per Gy increases with *r*. Near the seed, radiation is delivered at a high dose rate causing the local radical concentration to be higher than it is at larger values of *r*. High radical concentrations result in relatively more bimolecular termination near the seed, which reduces the amount of polymer formed per radical generated. A model that accounts for the effects of THPC would likely predict less dose-rate dependence and less radial dependence on the mass of polymer formed per Gy than the current model predicts [11].

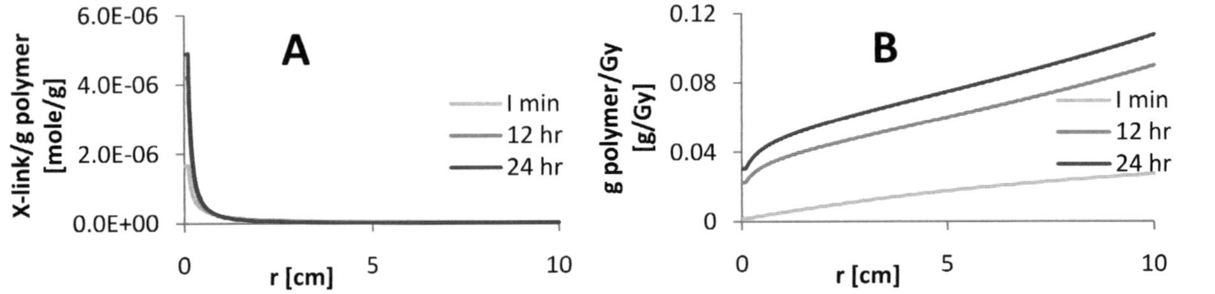

Figure 3: Influence of brachytherapy radiation on a) Crosslinks formed per mass of polymer and b) polymer formation per Gray delivered. The brachytherapy dose shown in Fig. 1a was delivered for one minute before the seed was removed.

References

[1] Fuxman A M, McAuley K B, Schreiner J (2005) *Chem. Eng. Sci.* **60**, 1277

[2] Vergote K, DeDeene Y, Vanden Bussche E, DeWagter C (2004) *Phys.Med.Biol.* **49**, 4507–4522

[3] Jirasek A, Hilts M, Shaw C and Baxter P (2006) *J. of Physics: Conference Series* **56,** 263–267

[4] Koeva V, Daneshvar S, Senden R, Imam A, McAuley K, Schreiner J (2009) *Macromol. Theory Simul.* **18**, 495-510

[5] Senden R J, De Jean P, McAuley K, Schreiner J (2006) *Phys. Med. Biol.* **51**, 3301-3314

[6] De Deene Y, Reynaert N and De Wagter C (2001) *Phys. Med. Biol.* **46** 2801-2825.

[7] Babic S, Schreiner J (2006) *Phys. Med. Biol.*, **51**, 4171

[8] Khan F M. "*The Physics of Radiation Therapy*". *Lippincott Williams & Wilkins* (1984).

[9] Peirquin B. and Marinello G. *"A Practical Manual of Brachytherapy'. Med. Phys. Publishing Corporation. Madison, Wis.* (1997).

[10] Salomons G, Park Y, McAuley K, Schreiner J (2002) *Phys. Med. Biol.* **47**, 1435–1448

[11] DeDeene Y, Vergote K, Claeys C and DeWagter C (2006) *Phys. Med. Biol.* **51** 653-675

Journal of Physics: Conference Series **250** (2010) 012067 doi:10.1088/1742-6596/250/1/012067

Monte Carlo Simulation of MAGIC-*f* gel for Radiotherapy using PENELOPE

M Alva[1], T Pianoschi[1], T Marques [1], M Santanna M[2], O Baffa[1] and P Nicolucci[1]

[1] Department of Physics and Mathematics, University of São Paulo, Av. Bandeirantes 3900 CEP:14010-901- Ribeirão Preto – SP – Brazil.

[2]Hospital of Cancer at Barretos (HCB), Av. Antenor Duarte Villela – 1331, CEP: 14784- 400, Barretos – SP –Brazil.

E-mail: alva_mirko@yahoo.com.br

Abstract. MAGIC-*f* gel has been shown as a suitable dosimeter for different techniques in radiotherapy due to its characteristics of water equivalence (effective atomic number of 7.41) and a spatial resolution better than 1mm. One way to predict the gel results is the use of simulations through PENELOPE Monte Carlo code. This simulation code was used to simulate the MAGIC-*f* gel and its use for dosimetry in conventional and conformal radiotherapy. The results obtained from the simulation were compared with experimental values. Comparisons from simulation and experimental values show mean differences of 2.88 % and 3.75% for conventional and conformal, respectively. This study shows that PENELOPE code can be simulate the components of the MAGIC-*f* gel to study and predict the gel response.

1. Introduction

Dosimetry based on polymeric gel has been widely investigated in radiotherapy since it allows the verification of dose distributions [1-2]. Due to its capability of detecting high dose gradients with high spatial resolution, MAGIC-*f* gel has been shown as an accurate tool in 3D dosimetry [3-4]. A way to study and predict MAGIC- *f* gel results is the use of computational simulation as the Monte Carlo method. From the codes more commonly used in radiotherapy PENELOPE Monte Carlo code presents the advantage of using a mixed algorithm for the simulation of photons- electrons cascades [5-7]. This code allows the “construction” of materials trough the use of the compound's chemical composition (*i.e.*, elements present and stoichiometric index, or weight fraction, of each element), mass density, mean excitation energy and energy and oscillator strength. The aim of this work was to use PENELOPE to simulate MAGIC- *f* gel dosimetric properties and to provide a valuable enhancement in the pre-constructed list of materials from PENELOPE. For validation the simulated material file (*MAGIC-f.mat*) the percentage depth doses (PDD) curves and a dose distribution were used, comparing experimental values obtained with MAGIC-*f* gel dosimeter and simulated results.

2. Materials and methods

The components of MAGIG- *f* gel (water, gelatin, methacrilic acid, copper sulfate, ascorbic acid and formaldehyde) and its characteristics, like atomic number, density and molar mass, were input into the PENELOPE 2008 code [8] to build the *MAGIC-f.mat* and *water.mat* material files and for the simulation of PDDs and the conformal treatment. The simulations used 2 x 10^9 primary particles and 0.01mm^2 pixel size.

For the measurements, MAGIC- *f* gel dosimeter was prepared following the description by Fernandes [3]. Relaxometry, weightened in T2, with a 3.0 T, NMR Philips tomography and head coil were used for the readings of the gel samples, which were perpendicularly positioned to the head coil. The images were acquired with a multi spin-echo sequence with 16 echo times, TE=20ms, TR=4000ms, matrix of 256 x 256 pixels, slices of 3mm thickness and 150 mm of FOV. The MR images were processed and analysed with a software developed in MatLab®, that produces R2 maps.

2.1 Percentage depth doses and dose distribution

Simulated PDDs were used to evaluate the *MAGIC-f.mat* constructed by the code. Simulation conditions were 10 x 10 cm^2 field size and 100 cm source-skin distance (SSD) for photon beams of 6 MV and 10 MV and phantoms of 30 x 30 x 20 cm^3 "filled" with both simulated materials.

Experimental PDDs were determined using MAGIC-*f* gel and ionization chamber (IC) of 0.6 cc for 6 MV and 10 MV beams. The measurements with the gel were performed using glass tubes of 16 cm length and 1 cm diameter filled with gel. The irradiations were carried out using PPMA cube filled with water and the glass tubes positioned in the center of the volume. A dose of 10 Gy in conventional irradiation conditions for both beams was used.

For the measurement of a coplanar dose distribution, the gel was poured into a PPMA cylindrical phantom of 10 cm diameter and 15 cm height. The irradiation of the phantom was made with 5 fields of 1 x 1 cm^2, with a Varian 2100 linear accelerator gantry angles of 0°, 45°, 90°, 135° and 180°. The target was chosen in the center of the cylinder, at 100 cm SSD. The phantom was irradiated with 10 MV beam with2 Gy per field. The conformal irradiation was simulated using PENELOPE code in the same irradiation conditions. Both distributions were also compared through the dose profile.

3. Results and Discussions

3.1 PDDs and dose distribution

Figure 1 shows the PDDs through the simulations and from experimental measurements, showing them similar behavior. A maximum difference of 1.93% and 1.88% was found for the 6 MV and 10 MV beams, respectively, when simulated PDDs with *MAGIC-f.mat* and *water.mat* are compared.

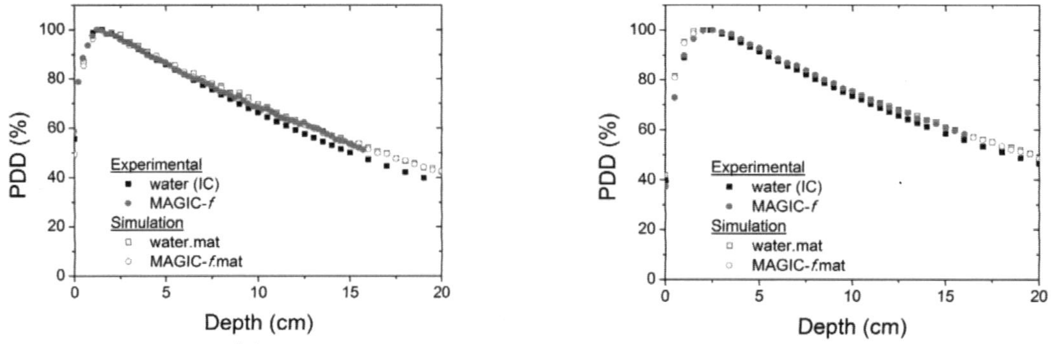

Figure 1. PDDs for water and MAGIC-*f* simulated with PENELOPE and experimental values for (a) 6 MV beam and (b) 10 MV beam.

For the 6 MV beam, a comparison between experimental gel and water (IC) shows a maximum difference of 3.20% until 10 cm depth. Beyond 10 cm the maximum difference is 5.72% in 12.5 cm. The same comparison was performed for the 10 MV beam and a maximum difference of 1.93% was found until 10 cm depth and of 2.62% at 14cm depth.

The maximum differences between experimental and simulated values are 1.46% (0.5 cm depth) and 4.3% (17cm depth) for 6 MV beam. Maximum differences of 2.14% (1 cm depth) and 3.8% (18cm depth) were found for the 10 MV beam.

The study of conformal dose distributions for 10 MV using 5 fields was performed through simulation using *MAGIC-f.mat*, shown in the figure 2.a. and experimentally using MAGIC-*f* gel dosimeter as shown in the figure 2.b.

Figure 3 shows the dose profiles obtained from both distributions.

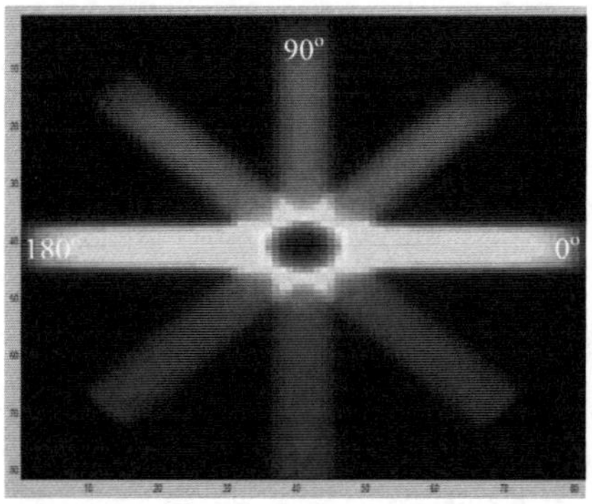

(a)

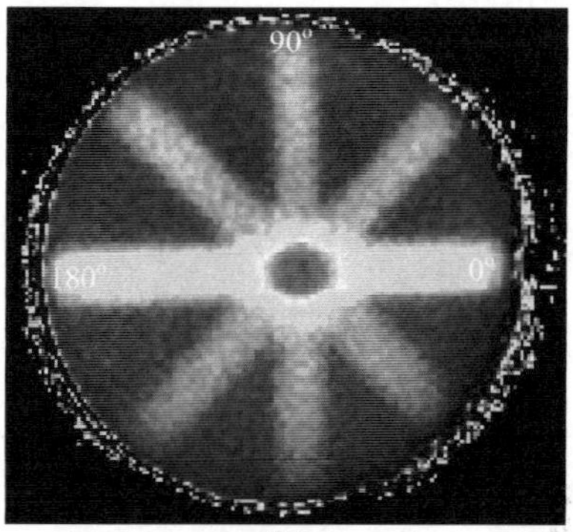

(b)

Figure 2. Dose distribution for 10 MV obtained with: (a) PENELOPE; (b) MAGIC-f dosimeter.

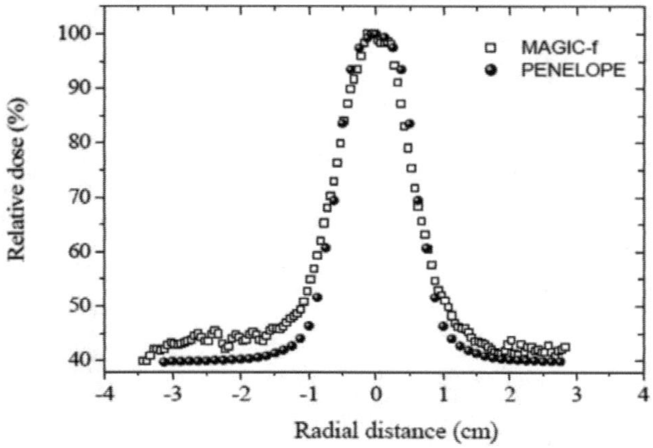

Figure 3. Dose profiles of simulated and experimental distributions.

A maximum difference of 0.5% in a radial distance of 0.55 cm inside the field (approximately half the irradiation field) was found and beyond that distance the differences between the dose distributions are less than 7%.

4. Conclusions

The simulations using PENELOPE-Monte Carlo code showed similar behaviors with the experimental values. Although discrepancies were found, specifically in the dose distribution in regions of low doses, this study shows that the code can be used to simulate the components of the MAGIC-f gel for these energies. Moreover, the *MAGIC-f.mat* file can be useful for studies of response and comparison of the gel.

Acknowledgments

This work was partially supported by CAPES. We are thankfully to the personnel from the Cancer Hospital of Barretos and the Hospital of Clinics from Ribeirão Preto. The technical support from J .L. Aziani is also appreciated.

References

[1] De Deene Y, Hurley C, Venning A, Vergote K, Mather M, Healy B J and Baldock C 2002 *Phys. Med. and Biol.* **47** 3441-63
[2] Gustavsson H, Karlsson A, Back S A J, Olsson L E, Haraldsson P, Engstrom P and Nystrom H 2003 *Med. Phys.* **30** 1264-71
[3] Fernandes J, Pastorello B, Araujo D and Baffa O 2008*Phys. Med. Biol.* **53** N53-N58
[4] Alva M, Marques T, Schwarche M, Gonçalves L R, Baffa O and Nicolucci P 2009 *IFMBE Proceedings*. **25** 248-251
[5] De Vlamynck K, Palmans H and Verhaegen F 1999 *Med. Phys.* **26** 1874-82
[6] Yamamotro T, Mizowaki T and Miyabe R 2007 *Phys. Med. Bio*. **52** 1991-2008
[7] Sempau J, Acosta E and Baró J 1997 *Nucl. Instr. and Meth. B.* **132**, 377-390
[8] Salvat F, Fernández-Varea J and Sempau J 2008 (Moulineaux, France: Nuclear Energy Agency OECD/NEA)

Journal of Physics: Conference Series **250** (2010) 012068 doi:10.1088/1742-6596/250/1/012068

High performance dosimetry calculations using adapted ray-tracing

Lancelot Perrotte, Guillaume Saupin
CEA LIST, route du panorama, 94260 Fontenay aux roses, France

E-mail: lancelot.perrotte@cea.fr, guillaume.saupin@cea.fr

Abstract. When preparing interventions on nuclear sites, it is interesting to study different scenarios, to identify the most appropriate one for the operator(s). Using virtual reality tools is a good way to simulate the potential scenarios. Thus, taking advantage of very efficient computation times can help the user studying different complex scenarios, by immediately evaluating the impact of any changes. In the field of radiation protection, people often use computation codes based on the straight line attenuation method with build-up factors. As for other approaches, geometrical computations (finding all the interactions between radiation rays and the scene objects) remain the bottleneck of the simulation. We present in this paper several optimizations used to speed up these geometrical computations, using innovative GPU ray-tracing algorithms. For instance, we manage to compute every intersectionbetween 600 000 rays and a huge 3D industrial scene in a fraction of second. Moreover, our algorithm works the same way for both static and dynamic scenes, allowing easier study of complex intervention scenarios (where everything moves: the operator(s), the shielding objects, the radiation sources).

1. Context

Preparing interventions in the nuclear field notably implies to evaluate the impact of radioactive sources on operators. In order to respect the ALARA principle, it is much relevant to find the most appropriate scenarios of intervention using simulation tools. Lots of computation codes exist to simulate the propagation of radiations, but most of them operate offline. Decreasing computation times in a noticeable way brings interactivity to the user, allowing him to interact with the scene and to intuitively study more appropriate scenarios.

Dose rate evaluation can be appraised by solving the transportation equation of Boltzmann. For radiation protection purposes, people often take advantage of simplified methods and algorithms, such as the straight line attenuation method with build-up factors [1]. Examples of radiation protection softwares using this method include MERCURAD [2] (based on the Mercure code) and VISIPLAN [3]. Besides, each volumic source is often seen as a cloud of points emitting radiations (Monte-Carlo approach of the emitting volume). The more points there are in the cloud (typically tens/hundreds of thousands of points), the more accurate is the representation of this volume.

For each generated ray, all the intersections with the objects of the scene have to be found. The final computed dose rate will depend on the depth of materials intersected by each ray. In the simulation tool we developed to perform interactive dose rate, the objects of the scene can be: analytic (sphere, box, cylinder ...), a combination of analytic objects (for instance, a pipe is an external cylinder minus an internal one), or meshed objects. The latter ones (soup of triangles) typically represent complex

objects coming from CAD tools. We are able to account for all these kinds of objects, but we will only focus on meshed objects in this paper.
The number of objects in a scene can be quite large (some thousands), leading to a scene counting millions of triangles. The number of rays coming out of all volumic sources can also be large (up to millions of rays). Thus, a very high amount of intersections must be computed for each dose rate evaluation. Moreover, all these intersections have to be sorted along each ray. Even if additional computations are necessary before getting a final result, the bottleneck of the overall simulation always remains in these ray tracing requests.

2. Relationship with traditional ray-tracing

Our concern could be thought of as a classical ray-tracing problem, but two main differences appear between our approach and traditional ray tracing. First, this approach does not require the handling of secondary (and incoherent) rays. Indeed, each rays coming from sources converge to the measure point. Hence, it is as if all rays started from this same point, since the beginning and ending points of each ray can be switched without modifying the list of objects intersected along the ray. These highly coherent rays, usually referenced as primary rays in the ray-tracing community, allow major optimizations.
The second main difference is that each intersection along each ray has to be stored. This requirement, useless in usual ray tracing, is very demanding. Moreover, these intersections also have to be sorted, in order to compute depths of materials intersected along each ray. Most of the time, in graphical raytracing applications, only the first intersection along each ray is computed.

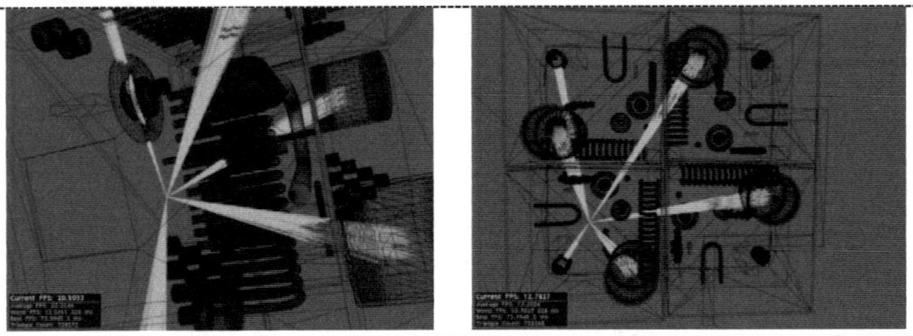

Figure 1. nuclearCase scene (738K triangles) from two different points of view. 600 000 rays are drawn, for 6 millions of intersections sorted, in 400 ms on a NVIDIA GTX 295. Rays are coloured in white and become red when crossing an object.

3. Choice of the structure of acceleration

There also exist major similarities between our case and the traditional ray-tracing. In both cases, it would be totally inefficient to test each ray against each triangle of the scene. It is then appreciable to use acceleration structures. Building an acceleration structure consists in pre-sorting the triangles, so that, for each ray, only a very few triangles will have to be effectively tested. Different types of acceleration structures can be used, depending on the use case ([4], [5], [6]). Hunt defined a grid in perspective space [7] that turns out to be the most appropriate structure for primary rays. In our case, it consists in a regular tiling of the space (in fact the *perspective space*) around the point where the dose rate has to be evaluated (see Figure 2). It can be seen as a regular tiling of the image produced by a look at the scene through a camera whose point of view is the measure point. To view all of the space, six of these cameras, *e.g.* six grids, have to be constructed.

Journal of Physics: Conference Series **250** (2010) 012068 doi:10.1088/1742-6596/250/1/012068

The perspective grid is particularly well-suited to our purpose: since each ray starts from the same point, each ray belongs to one unique cell. Therefore, the list of triangles to be checked for potential intersections is directly available, saving up significant computation time (see Figure 2).
It is the acceleration structure we chose to perform very efficient dose rate computations. Even though it is necessary to rebuild this structure regularly for static and dynamic scenes (each time the measure point moves), the construction algorithm we designed makes it possible to build it very quickly. Therefore, what seemed to be a drawback turns out to be an asset: dynamic scenes can be computed in no time thanks to this grid.

4. Results and discussion

4.1. Implementation

We bench our algorithm on an NVIDIA GeForce GTX 295 (2*896 Mo), coupled with a 4 cores, Intel(R) Xeon(R) X5550 @ 2.67GHz. After transferring the data between the CPU and the GPU, all the computations are done on the GPU. To check the results produced by our algorithm, we used traditional scenes of the ray-tracing community: Erw6, Fairy Forest, Conference and Sully.

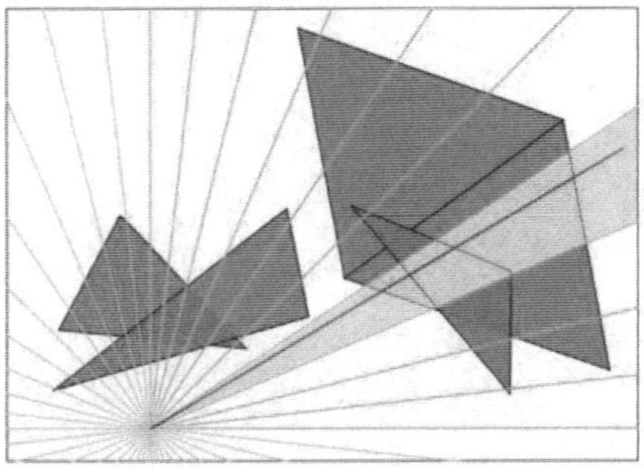

Figure 1. Example of a scene of 5 triangles viewed from above.

The lines in green represent the perspective grid; those in orange belong to the other grids (six grids to represent the whole space). The ray in red falls in the green cell of the grid. This cell directly references the two triangles to test for this ray.

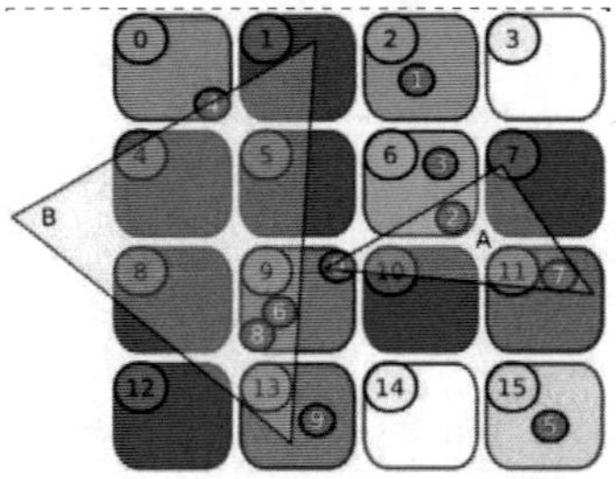

Figure 2. : Example of 2 triangles A and B in a perspective view.

The associated perspective grid has a resolution of 4*4; indices of the cells are indicated in the upper-left corner of each cell. Red circles numbered from 0 to 9 are the rays starting from the eye.

4.2. Experimentations on a nuclear-like scene

Then, we designed a study-case (called nuclearCase) illustrating the kind of scene (in terms of complexity) potentially involved in a nuclear installation. We defined 6 groups of 100 000 points, representing the six volumic radiation sources of the nuclearCase scene. All rays thrown in the present study case start from one of the 600 000 emitting points, and end at the same point (representing the position of a virtual operator, where the dose rate is evaluated).

Journal of Physics: Conference Series **250** (2010) 012068 doi:10.1088/1742-6596/250/1/012068

Table 1. Simulation times (including the building of the structure) required to measure the dose rate at a point inside various scenes. BVH is a standard acceleration structure used in traditional ray-tracing. perspGrid is the perspective grid we used. Times are given in milliseconds.

	BVH CPU 1 core	PerspGrid CPU 1 core	PerspGrid CPU 8 cores	PerspGrid GPU
Erw (804 trgs)	810	561	121	71
Conference (283K trgs)	3578	2308	427	152
Fairy (174K trgs)	2864	1653	276	112
Sully (804K trgs)	5169	3266	550	340
NuclearCase(800K trgs)	6145	3920	605	395

These rays and the overall scene are seen through an external camera. In the air, each ray is coloured white, and turns red when intersecting an object (see Figure 1).

Using our new algorithm and implementation, we compute and sort about 6 millions intersections by simulation step, in less than 400ms (see Table 1).

5. Next steps

We currently investigate some complementary optimizations, such as:

- Taking advantage of boxes bounding sources to restrict the amount of triangles potentially intersected by a group of rays related to the same source
- Dispatching the computations on several GPUs. A first check with the NVidia GTX295 card (exposing two graphical processing units) brought promising results: between 250ms and 350ms of computation, depending on the way we split all the computations on each GPU.

6. Conclusion

We briefly presented in this paper the methods and acceleration structures we use to quickly perform the geometrical computations required for the dose rate computation based on the straight line attenuation method with build-up factors. We show that important speed-up can be achieved using GPU implementation of theses methods.

We plan to further improved theses algorithms by automatically adding bounding boxed around radioactive sources, hence reducing rays triangles tests.

Then, this enhanced algorithm will be integrated to the current simulation library developped at CEA LIST, and that is part of the NARVEOS [8] industrial software from AREVA/Euriware company.

References

[1] Assad A, Chiron M, Nimal J C, Diop CMand Ridoux P 1999 Proc. of the Ninth International Conference on Radiation Shielding (Japan: Atomic Energy Society of Japan) **9** 493

[2] MERCURAD http://www.canberra.com/products/13250.asp

[3] VISIPLAN http://www.visiplan.be/navigvisiplan/vud report.htm

[4] Reshetov A, Soupikov A and Hurley J *2005 SIGGRAPH '05: ACM SIGGRAPH 2005 Papers (New York,NY, USA: ACM) 1176-1185*

[5] Wald I, Boulos S and Shirley P *2007 ACM Trans. Graph.* ***26*** *6 ISSN 0730-0301*

[6] Cadet G and Lecussan B 2007 RT '07: Proceedings of the 2007 IEEE Symposium on Interactive Ray Tracing (Washington, DC, USA: IEEE Computer Society) 63-71 ISBN 978-1-4244-1629-5

[7] Hunt W and Mark W R 2008 *IEEE/EG Symposium on Interactive Ray Tracing 2008 (IEEE/EG) 3-10*
[8] NARVEOS http://www.euriware.fr/scripts/home/publigen/content/templates/show.asp?p=399&l=fr

Journal of Physics: Conference Series **250** (2010) 012069 doi:10.1088/1742-6596/250/1/012069

3D polymer gel dosimetry and Geant4 Monte Carlo characterization of novel needle based X-ray source

Y. Liu[1], E. Sozontov[2,3], V. Safronov[4], G. Gutman[2], E. Strumban[2], Q. Jiang[5], S. Li[6]

[1] Prigge Radiation Oncology Center, Modesto, CA 95350, USA
[2] Advanced X-ray Technology, Inc., Birmingham, MI 48009, USA
[3] Russian Research Center "Kurchatov Institute", Moscow 123182, Russian Federation
[4] Institute of Crystallography (Kaluga Branch) RAS, Kaluga 248640, Russian Federation
[5] Department of Neurology, Henry Ford Hospital, Detroit, MI 48202, USA
[6] Department of Radiation Oncology, Temple University Hospital, Philadelphia, PA 19140, USA

esozontov@yahoo.com

Abstract. In the recent years, there have been a few attempts to develop a low energy x-ray radiation sources alternative to conventional radioisotopes used in brachytherapy. So far, all efforts have been centered around the intent to design an interstitial miniaturized x-ray tube. Though direct irradiation of tumors looks very promising, the known insertable miniature x-ray tubes have many limitations: (a) difficulties with focusing and steering the electron beam to the target; (b)necessity to cool the target to increase x-ray production efficiency; (c)impracticability to reduce the diameter of the miniaturized x-ray tube below 4mm (the requirement to decrease the diameter of the x-ray tube and the need to have a cooling system for the target have are mutually exclusive); (c) significant limitations in changing shape and energy of the emitted radiation. The specific aim of this study is to demonstrate the feasibility of a new concept for an insertable low-energy needle x-ray device based on simulation with Geant4 Monte Carlo code and to measure the dose rate distribution for low energy (17.5 keV) x-ray radiation with the 3D polymer gel dosimetry.

1. Basic conception and general description of the X-ray system

The essence of the offered concept [1, 2] is a two-stage production of x-ray radiation as shown in figure 1. First, a primary x-ray beam is generated by a conventional x-ray tube and guided into a hollow needle through an optical collimator. Then, the collimated x-ray beam excites a metal target installed inside the needle and causes the target to emit x-ray fluorescence that is mainly defined by the characteristic lines of the target material and parameters of the incident x-ray beam. The target can be implemented in different geometries and arrangements. That makes the needle with a built-in target a miniaturized source of radiation potentially applicable to treatment of tumors. If the needle is inserted into the tumor, the produced by the target x-ray fluorescence expands through the needle walls and can be used as a treatment beam irradiating the tumor from the inside out.

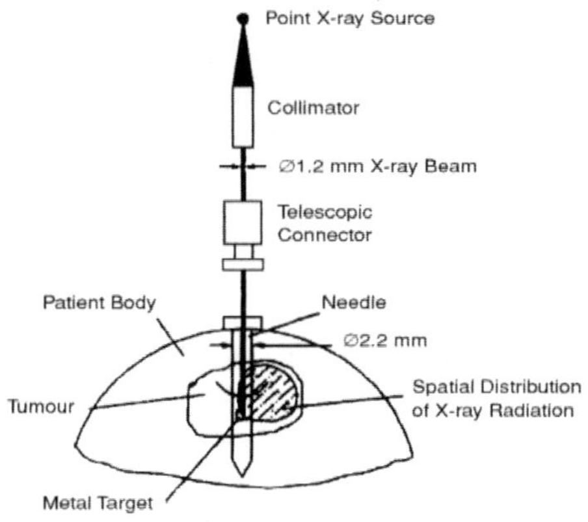

(a)

(b)

Figure 1. (a) Schematic illustration of the intensity-modulated x-ray brachytherapy system; (b) The x-ray needle based device installed in a conventional rat stereotax (Kopf Models 1760).

2. Monte Carlo simulations

The simulations were done using Geant package version 4.8.1.p02 with low energy data pack G4EMLOW version 4.0 which includes data from EPDL97 (Evaluated Photon Data Library from Livermore). This library is among the recommended by the AAPM (American Association of Physicists in Medicine) in TG-43U1 protocol [3]. Dose rate calculations were done in two simulation steps. First, a set of photon energies comprising the tube spectrum was obtained. Second, photons with these energies were directed onto a secondary target and spatial distribution of deposited dose was accumulated.

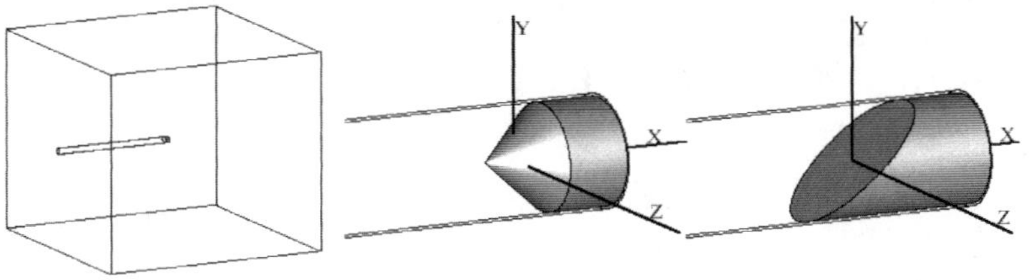

Figure 2. Simulated geometry in general (left) and secondary targets in detail with the coordinate system axes.

Four different configurations have been simulated: cone or wedge shaped secondary target made of either Molybdenum or Copper. Table 1 summarises maximal dose rates as a function of distance from needle axis within vertical longitudinal plane. Figure 3 shows constant dose rate contours within vertical longitudinal plane ($Z = 0$, see coordinate system layout at figure 2).

Table 1. Maximal dose rates (mGy/min) at given distances R from needle axis

R (mm)	**Mo-cone**	**Mo-wedge**	**Cu-cone**	**Cu-wedge**
5.25	1598	2143	2782	3452
6.75	641	1038	272	403
8.25	338	582	40	63
9.75	200	344	9.4	16
11.25	126	228	4.0	8.2
12.75	82	170	2.1	5.3
14.25	58	113	1.7	3.0
15.75	38	79	1.1	3.3
17.25	28	68	0.96	3.5
18.75	21	46	0.68	2.0
20.25	16	37	0.66	1.5

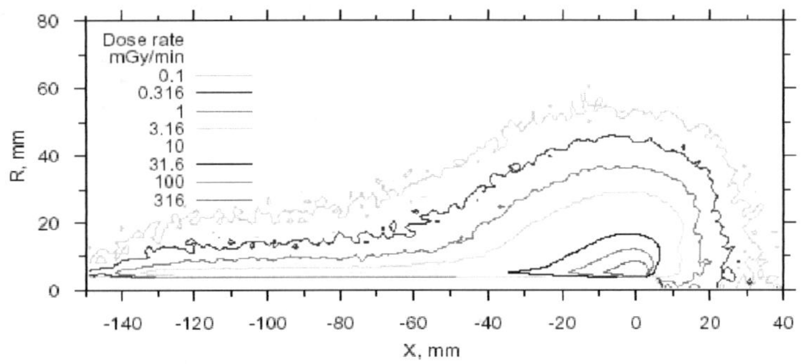

Figure 3. Full scale dose rate isolines for Molybdenum cone secondary target.

The calculated dose rate as a function of distance from the needle axis for Mo-cone secondary target was compared with experimental data. All the dose-depth profiles were measured using water-equivalent polymethyl-methacrylate (PMMA) phantoms and Harshaw LiF:Mg:Ti thermo-luminescent dosimeters (TLD-700). The experimental dose rate attenuation as a function of distance from needle axis in a PMMA phantom matched well to the Monte Carlo results as shown in figure 4.

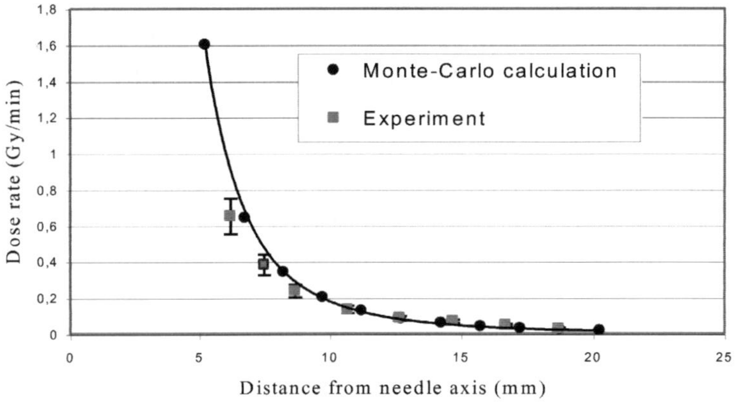

Figure 4. Dose rate as a function of distance from needle axis.

3. Polymer gel dosimetry

Gel dosimetry is a new dosimetry technique [4]. Gel dosimeters are the first and only integrating dosimeters that enable dose verification in three dimensions. Moreover, the application of a 3D dosimetry technique in clinics can give a real push to the application of advanced high-precision radiotherapy technologies.

In polymer gels, commonly known as BANG-type or PAG-type, monomers are usually dispersed in an aqueous gel matrix. The monomers undergo a polymerization reaction as a function of absorbed dose resulting in 3D polymer gel matrix. The radiation-induced formation of polymer influences Magnetic Resonance Imaging (MRI) signal relaxation properties, optical density and other physical properties that may be used to quantify absorbed radiation dose.

Gels consist of polymer network swollen in solvent. The network of flexible long-chain molecules traps the liquid medium they are immersed in. In monomer/polymer gel dosimetry, the conversion of co-monomers to polymer aggregates upon irradiation alters the mobility of surrounding water molecules. This also results in change of spin lattice relaxation rate R1 (=1/T1) and spin-spin relaxation rate R2 (=1/T2). The dose response of R2 is more pronounced than of R1 and, therefore, can be used as an effective imaging parameter.

The BANG3^{TM} polymer gel (MGS Research Inc.) has been used for both calibration and experimental measurements. Two phantom flasks filled with gel have been irradiated from within by the "treatment" radiation (the fluorescent radiation produced by the Mo pseudo target). The flask 1 was irradiated for 60 minutes by an X-ray needle with a cone-shaped pseudo target, as shown in figure 5(a). The evenly irradiated area within the gel is clearly visible, as shown in figure 5(b)

(a) **(b)**

Figure 5. The gel-filled flask irradiated by a cone-shaped pseudo-target: (a) the flask with the inserted needle, (b) irradiated area of gel-filled flask (magnified image).

After irradiation, the gel-filled flasks were scanned using a 3Tesla MRI machine. The flasks were thus imaged upright on a flat lucite sheet positioned horizontally in the head-coil so that the flasks were centered in the magnetic field. Then, images were obtained at the selected depths in the flasks. Only Transverse Relaxation T2 weighted scan data (TE echo times =120, 55, 10) were used. The Longitudinal Relaxation T1 and T2* weighted scan data didn't provide good image and, therefore, were not used. There were two options available for image slice thickness: 2mm+1mm gap or 1.5mm +0.5 mm gap. The slice thickness of 2mm+1mm gap was used for high signal to noise ratio (SNR) in the MRI images.

The R2 (R2=1/T2) map in a longitudinal slice (along the optical axis of the needle with a cone-shaped pseudo-target) was reconstructed by eigen2 software. Using the R2 map and the calibration dependence T2 vs. Dose calibration curve, the dose map was obtained by using Matlab software. The reconstructed T2 longitudinal image and the dose map of the irradiated gel (60 minutes irradiation from a cone shaped pseudo target) are shown in figure 6 (a,b). The region outlined in the small box in figure 6(a) corresponds to the area shown in figure 6(b) from the Matlab.

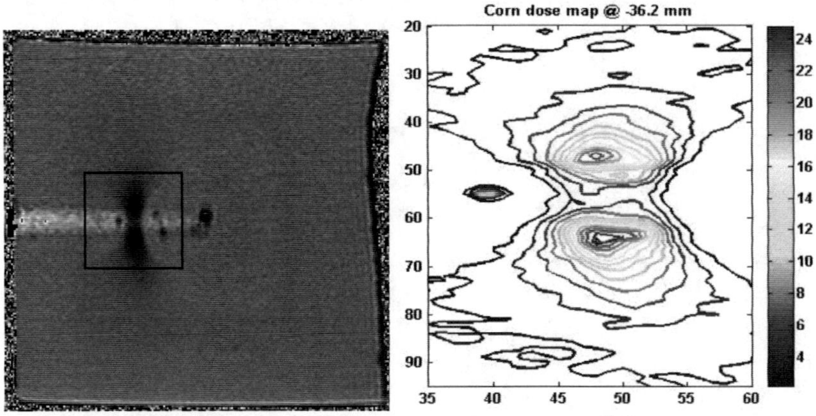

(a) **(b)**

Figure 6. The reconstructed T2 longitudinal image (a) and the dose map (b) of the irradiated gel (irradiation time 60 minutes; cone-shaped pseudo-target).

The R2 (R2=1/T2) map of a latitudinal slice (perpendicular to the optical axis of the needle with a cone-shaped pseudo-target) was also reconstructed by eigen2 software. Using the R2 map and the calibration dependence T2 vs. Dose calibration curve, the dose map was obtained by using Matlab

Journal of Physics: Conference Series **250** (2010) 012069 doi:10.1088/1742-6596/250/1/012069

software. The reconstructed T2 latitudinal image and the dose map of the irradiated gel (60 minutes irradiation from a cone shaped pseudo target) are shown in figure 7 (a,b). The region outlined in the small box in figure 7(a) corresponds to the area shown in figure 7(b).

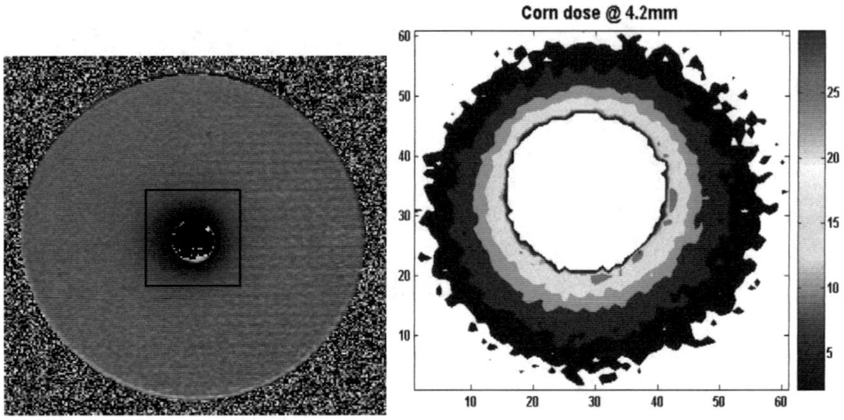

(a) **(b)**

Figure 7. The reconstructed T2 latitudinal image (a) and the dose map (b) of the irradiated gel (irradiation time 60 minutes; cone-shaped pseudo-target).

4. Conclusion

The basic concept of a novel needle x-ray system for medical applications is reported; the main principle of the system is based on a two-stage production of x-rays. We performed Monte Carlo calculations using Geant4 code and experimental measurements of dose rate distributions with polymer gel dosimeter in water equivalent phantoms for a range of x-ray intensities and various target design and materials of the fabricated x-ray needles. These studies have experimentally confirmed that the concept of this miniature needle based x-ray system is correct.

References

[1] Gutman G, Sozontov E, Strumban E, Yin F F, Lee S W and Kim J H, 2004 *Phys. Med. Biol.,* **49**, 4677-4688

[2] Gutman G, Strumban E, Sozontov E and Jenrow K, 2007 *Phys. Med. Biol.,* **52**, 1757–1770

[3] Rivard M, Coursey B, DeWerd L, Hanson W, Huq M, Ibbott G, Mitch M, Nath R and Williamson J, 2004 *Med. Phys.,* **31**, 633-674

[4] Baldock C, De Deene Y, Doran S, Ibbott G, Jirasek A, Lepage M, McAuley K B, Oldham M and Schreiner L J, 2010 *Phys. Med. Biol.,* **55**, R1–R63

Journal of Physics: Conference Series **250** (2010) 012070 doi:10.1088/1742-6596/250/1/012070

Alternative imaging modalities for polymer gel dosimetry

Andrew Jirasek

Department of Physics and Astronomy, University of Victoria, Victoria, BC, Canada, V8W 3P6

jirasek@uvic.ca

Abstract. This review summarizes recent work in the area of imaging polymer gel dosimeters using x-ray CT imaging, ultrasound, and radiation-induced changes in gel mechanical properties. In addition, recent work in the area of Raman tomographic imaging of canine bone, in conjunction with past efforts in Raman imaging of polymer gel dosimeters, raises new possibilities for new polymer gel imaging techniques.

1. Introduction

Research and development efforts in the imaging of polymer gel dosimeters has, to date, been focussed on MRI and optical CT techniques. Both of these modalities are capable of imaging irradiated polymer gel dosimeters and efforts continue in developing these modalities further. Each technique does, however, present itself with its own series of technical and implementation challenges. Hence, over the course of the history of gel dosimetry research, efforts have been made in identifying and developing alternative imaging modalities for polymer gel dosimetry. This review aims to provide an overview of these alternative techniques. The review will outline (i) a brief background of x-ray CT polymer gel dosimetry, (ii) an update of work in x-ray CT polymer gel dosimetry since the last review [1], (iii) an update to work in ultrasound and gel mechanical property imaging, and (iv) comment on prospects for Raman imaging of irradiated polymer gel dosimeters.

2. X-ray CT imaging of polymer gel dosimeters

Detailed reviews of work prior to 2008 have been reported previously [1, 2]. Hence, only a brief sketch of work prior to 2008 is given here, and is primarily aimed at providing sufficient background for the reader to understand the basics of the technique.

2.1. Density changes in irradiated polymer gel dosimeters

The underlying principle in x-ray CT polymer gel dosimetry is the fact that irradiated polymer gel dosimeters undergo a slight physical density change in proportion to the dose delivered to the gel. This density change is subsequently manifest as a change in image CT number (H, N_{CT}, see figure 1).

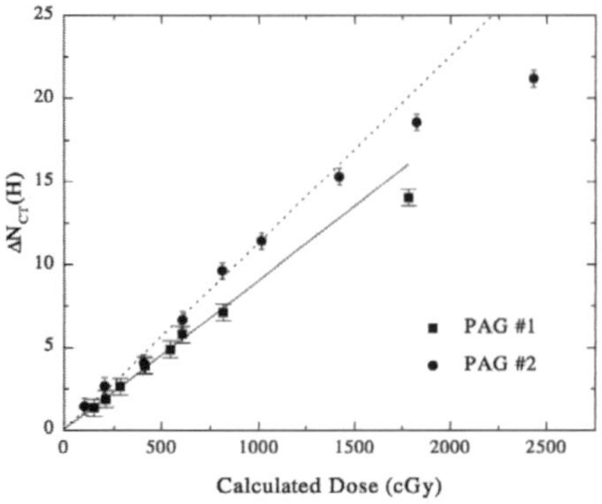

Figure 1: Change in CT number (H) for an irradiated polymacrylamide gel dosimeter (50%C, 6%T) From [3].

The first demonstrated clinical application of x-ray CT polymer gel dosimetry was undertaken by Audet *et al* [4]. Figure 2 illustrates a stereotactic irradiation of polymer gel subsequently imaged using x-ray CT. Overlaid on the image are treatment plan isodose lines.

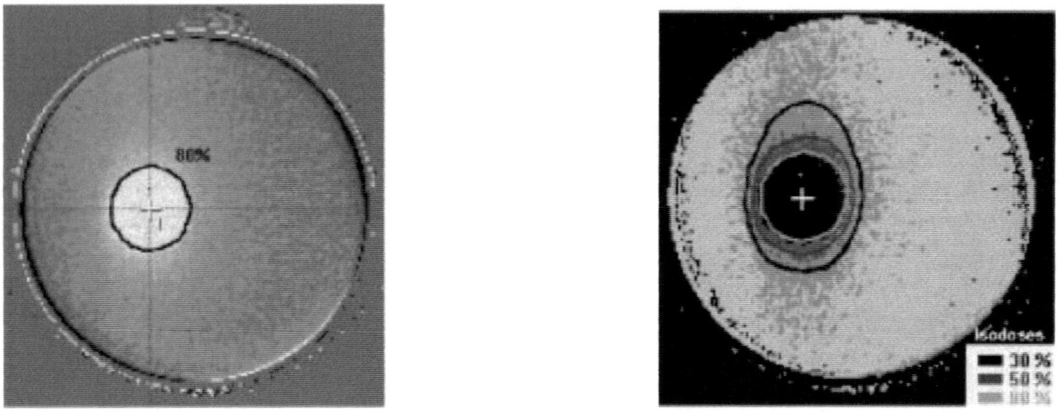

Figure 2: Stereotactic irradiation of x-ray CT-imaged polymer gel (left panel). Treatment plan isodose lines are also shown. Binned gel image with treatment plan isodose overlay shown in right panel. From [4].

2.2. Formulation studies to improve dose sensitivity

It can be seen from figures 1 and 2 that x-ray CT imaging constitutes a low contrast imaging modality, due to the minute irradiation-induced density changes with the gel. Trapp *et al* [5] undertook dose response studies of a number of polymer gel dosimeter formulations and table 1 summarizes the results of this work. From table 1 it can be seen that dose response sensitivity varies as a function of initial composition. However, even for sensitive compositions, minimum detectable dose is high, typically greater than 1 Gy. Furthermore, as shown in figure 3, dose resolution (95% confidence) remains high (several Gy) and, in fact, much higher than, for example, MRI [5]. This fact is largely responsible for the lack of clinical applications of x-ray CT polymer gel dosimetry in the literature.

Table 1: Dosimeter sensitivity when imaged with x-ray CT. From [5].

Gelling agent by % weight		Monomers by % weight							
Gelatin	Agarose	AA	Bis	Water by % weight	CT-dose sensitivity (H/Gy)	r^2	Standard error (Gy)	*P*-value	Minimum detectable dose (Gy)
2		3	3	92	0.78 ± 0.03	0.98508	0.32598	1.36E−08	1.3
3.5		3	3	90.5	0.87 ± 0.03	0.98889	0.31228	4.2E−09	2.4
5		3	3	89	0.71 ± 0.02	0.99649	0.12362	7.45E−10	1.0
6.5		3	3	87.5	0.59 ± 0.02	0.99060	0.19520	2.14E−09	3.7
8		3	3	86	0.54 ± 0.01	0.99682	0.10337	2.8E−11	2.5
5		1	1	93	0.26 ± 0.02	0.96834	0.13654	2.8E−07	2.7
5		2	2	91	0.40 ± 0.04	0.97286	0.12153	1.91E−03	1.0
5		5	5	85	1.14 ± 0.04	0.98917	0.40600	3.78E−09	1.6
5		6	6	83	1.43 ± 0.05	0.99362	0.42826	1.11E−06	1.1
	1	3	3	93	1.2 ± 0.1	0.93270	1.11288	5.77E−06	2.4
	1	4	4	91	1.3 ± 0.1	0.96552	0.58564	7.59E−05	1.5

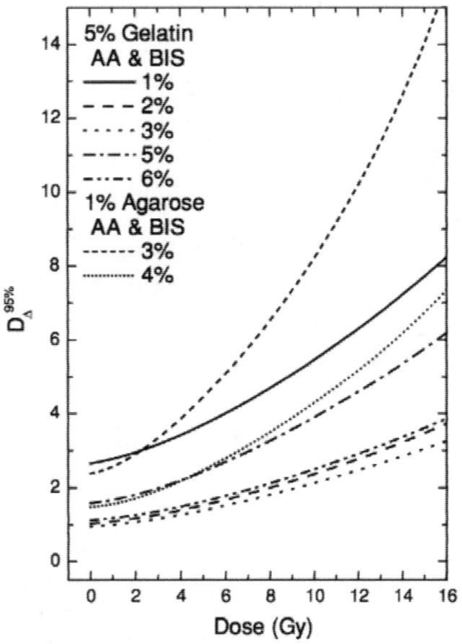

Figure 2: Dose resolution (95% confidence) for the formulations of table 1. From [5].

2.3. Image filtering

Aside from formulation studies, efforts to improve the dose resolution have been attempted through post-acquisition image filtering. Hilts and Duzenli [6] undertook a comprehensive study of a number of image filters applied to CT images of irradiated polymer gel. Table 3 illustrates the improvements in dose resolution for a range of filters and kernel sizes. While some filters look promising, it should be noted that (i) the filters can widen penumbral regions and (ii) a filters performance can vary based on the particular image in question.

Table 3: Improvements in dose resolution with image filtering. From [6].

			Low dose region		High dose region	
Filter	Mask	$\sigma_{N_{CT}}$ (H)	σ_D (cGy)	$D_\Delta^{95\%}$ (cGy)	σ_D (cGy)	$D_\Delta^{95\%}$ (cGy)
Unfiltered	...	0.93	79	219	152	421
MEAN	3×3	0.38	33	90	63	173
	5×5	0.27	23	63	44	121
	7×7	0.20	17	46	32	89
MEDIAN	3×3	0.50	42	117	81	226
	5×5	0.31	26	72	50	139
	7×7	0.09	8	21	14	40
MIDPOINT	3×3	0.49	42	117	81	224
	5×5	0.48	41	113	78	216
	7×7	0.39	33	92	64	178
ADAPTIVE	3×3	0.41	35	98	68	188
	5×5	0.26	25	70	48	134
	7×7	0.23	19	54	37	103
α-MEAN	3×3	0.41	35	97	68	188
	5×5	0.30	25	69	49	136
	7×7	0.24	20	55	39	108
SIGMA	3×3	0.63	53	147	103	285
	5×5	0.56	48	133	92	255
	7×7	0.54	46	127	88	244
SUSAN	3×3[a]	0.43	36	100	70	194

[a]Performance of the SUSAN filter is independent of mask size.

2.4. X-ray dose considerations

A further issue with implementing x-ray CT imaging for polymer gel dosimetry is the fact that CT imaging itself imparts radiation dose to the gel, potentially altering the dose information within the gel. This problem was studied by Baxter *et* al [7] who showed that, for imaging a single position within the gel and using standard imaging protocols (140kVp, 200mAs, 16 averages), CT imaging imparts <0.5Gy to the gel. Figure 4 illustrates CT dose deposite within a gel for a range of imaging methods (single position, SP; volumetric; varying slick thickness). It can be seen that care must be taken in certain scenarios, as CT dose can become a non-trivial contribution to gel dose. Ensuring that the gel is inactivated prior to imaging will alleviate any problems of CT dose within the polymer gel.

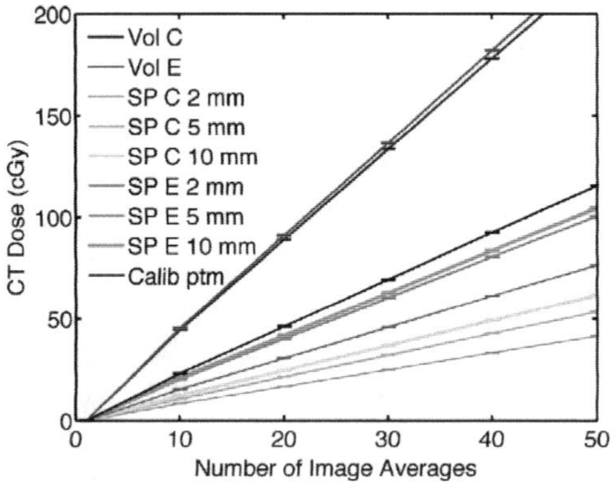

Figure 4: CT dose within a polymer gel dosimeter imaged within a single slice (SP) or volume. Different slice thicknesses are shown. From [7].

Journal of Physics: Conference Series **250** (2010) 012070 doi:10.1088/1742-6596/250/1/012070

2.5. Recent studies in x-ray CT polymer gel dosimetry

2.5.1. *Gel sensitivity:* As can be seen from prior development, aside from preliminary applications, the low contrast of x-ray CT polymer gel dosimetry has precluded researchers to undertake clinical applications using CT imaging. Rather, efforts have concentrated on improving the gel dosimeter sensitivity to radiation, with the aim of improving resulting image dose resolution. Koeva *et al* [8] undertook a series of experiments in an attempt to improve the dose sensitivity of polymer gel dosimeters. They substituted the crosslinker (bis-acrylamide) with a range of alternative crosslinkers, with the aim of reducing the primary cyclization of the bis-acrylamide crosslinker. Unfortunately, no suitable alternative crosslinkers were found in this investigation, with all tested crosslinkers giving rise to much lower gel radiation dose sensitivities. A second avenue was to incorporate a co-solvent into the manufacture process, thereby increasing the total solubility of the crosslinker. It has been previously shown [9] that increasing the total monomer/crosslinker content of the gel linearly increases the radiation dose sensitivity. However, the solubility limit of bis-acrylamide in water, the primary gel constituent, is 3%, beyond which bis-acrylamide precipitates out of solution. This presents an upper limit of dose sensitivity in traditional gel recipes. The addition of co-solvent, then, is aimed at increasing the bis-acrylamide solubility limit in the gel solution. Koeva *et al* investigated several co-solvents, with solubility limits given in table 4. Table 5 demonstrates the improvements in gel dose sensitivity, as measured with NMR, for gels with varying amounts, and types, of co-solvent. Figure 5 illustrates the improvement in gel sensitivity, as measured with x-ray CT imaging, for gels with isopropanol co-solvent. The addition of co-solvent clearly improves dose sensitivity.

Table 4: Solubility limits for bis-acrylamide for gels prepared with a range of co-solvents. From [8].

Water: cosolvent volumetric ratio	Glycerol (Gly)	Isopropanol (IPA)	N-propanol (NPA)	Sec-butanol (SBA)
10:0	3	3	3	3
9:1	**4.5**	**7**	**7**	**9**
8:2	5	8	10.5	14
7:3	**5**	**10**	**14**	**19**
6:4	5	10	17	23

Table 5: Dose sensitivities, as measured with NMR, for gels with and without co-solvent. From [8].

	Recipe				Dose sensitivity	Standard error
	%T	%C	Cosolvent	THPC (mM)	($s^{-1}\,Gy^{-1}$)	($s^{-1}\,Gy^{-1}$)
1	6	50	No cosolvent	5	0.069	0.001
2	10	50	No cosolvent	10	0.113	0.004
3	6	50	10% Gly	10	0.069	0.009
4	8	50	10% Gly	10	0.132	0.009
5	6	50	10% Iso	5	0.111	0.006
6	10	50	10% Iso	10	0.157	0.008
7	14	50	10% Iso	10	0.187	0.017
8	10	50	30% Iso	10	0.155	0.006

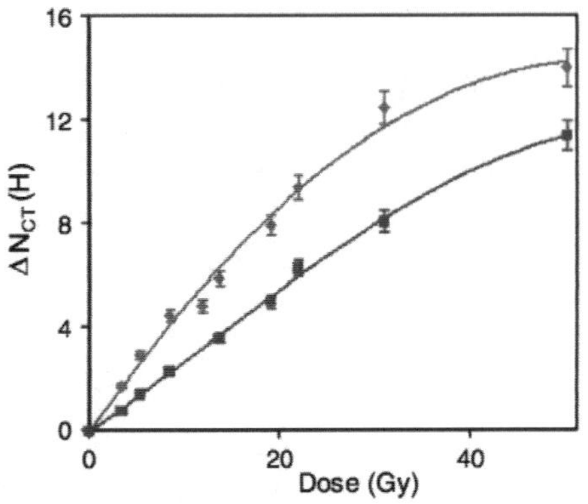

Figure 5: Improvement in dose sensitivity as measured with x-ray CT for gels with (blue trace) and without (red trace) 30% isopropanol co-solvent. From [8].

The first co-solvent for dose sensitivity enhancement to be studied in detail was glycerol [10]. It was shown that increasing glycerol concentrations within gel solutions increases irradiated gel dose sensitivity. As well, increasing total monomer/crosslinker concentration within a gel of given glycerol concentration also increases dose sensitivity, as expected (figure 6). However, increasing glycerol concentration has the effect of altering the shape of the resultant dose response curve, hence rendering calibration of dose response more difficult than in traditional dosimeters. Furthermore, the increase in dose sensitivity when using glycerol as co-solvent is not great enough to sufficiently improve the dose resolution of the system. Hence, current focus is in the investigation of isopropanol as co-solvent (see work presented at this conference).

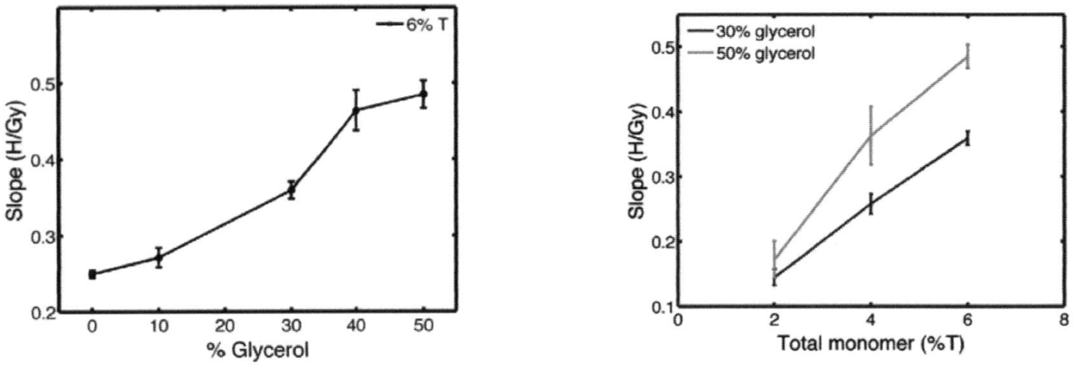

Figure 6: Improvement in dose sensitivity of irradiated polymer gels varying glycerol concentrations (left panel) and total monomer/crosslinker (right panel). From [10].

2.5.2. *Energy and modality dependence:* Recent work by Sellakumar *et al* [11] has been undertaken to study the x-ray CT imaged dose-response dependence on radiation energy and type (photon, electron). Figure 7 illustrates typical dose profiles for a photon and electron beam. Also shown are the dose response curves for PAGAT gel exposed to a range of photon and electron beam energies. Calibration curves were obtained from the measured depth dose curves. It can be seen that there is little dependence of the calibration curve on the incident beam energy. However, it is pointed out that even small differences can lead to clinically relevant errors.

Journal of Physics: Conference Series **250** (2010) 012070 doi:10.1088/1742-6596/250/1/012070

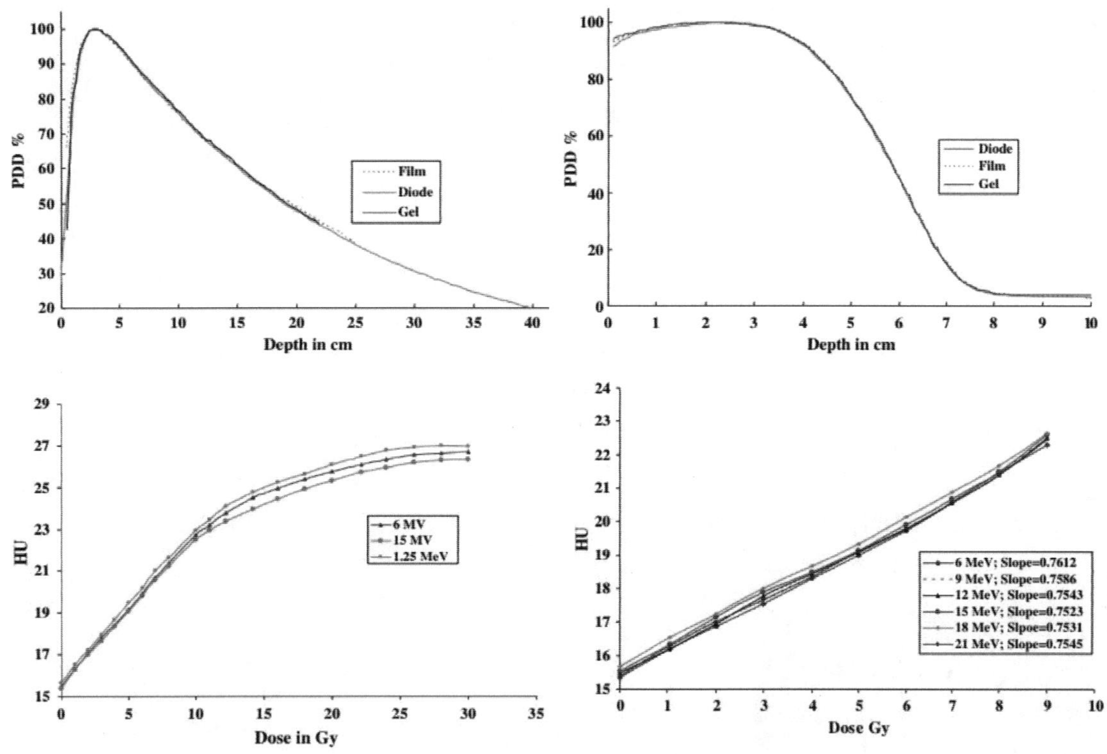

Figure 7: X-ray CT imaged PAGAT depth dose curves for a 15 MV photon beam (top left panel) and 15 MeV electron beam (top right). Also shown are constructed dose response curves for a range of incident photon (botom left panel) and electron (bottom right panel) energies. From [11].

2.6. Summary

X-ray CT offers a quick, robust, and practical imaging modality for polymer gel dosimetry. Traditionally plagued with low sensitivity and low dose resolution, efforts in the development of x-ray CT polymer gel dosimetry have centred around the improvement of the dose resolution of the system. Further studies have been undertaken on determining the incident energy-response characteristics.

3. Ultrasound and mechanical property imaging

Ultrasound imaging of polymer gel dosimeters was first proposed by Mather *et al* in 2003 [12] and is based on the change in ultrasonic properties of irradiated polymer gel. The application of the modality to polymer gel imaging, while promising, is largely in the development phase.

It has been shown that both the ultrasonic speed and attenuation within irradiated polymer gel dosimeters is dose dependent and, furthermore, highly dependent on the type of polymer gel undergoing irradiation (see figure 8) [12, 13].

Journal of Physics: Conference Series **250** (2010) 012070 doi:10.1088/1742-6596/250/1/012070

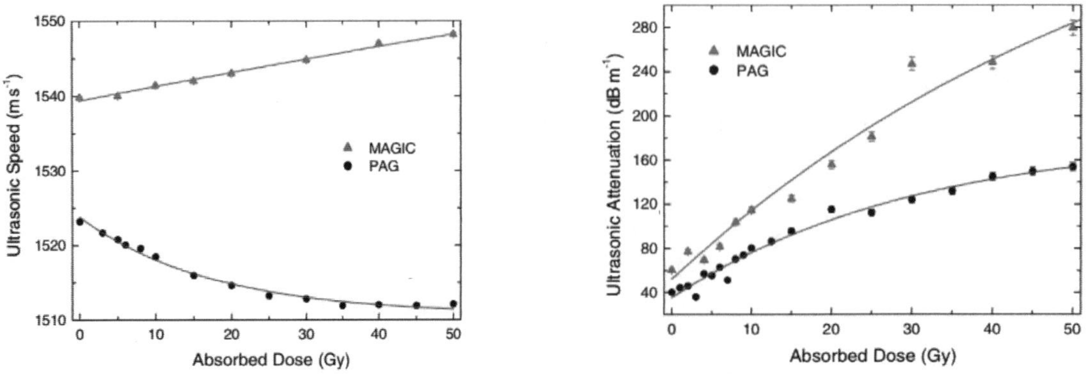

Figure 8: Ultrasonic speed (left panel) and attenuation (right panel) for irradiated PAG and MAG polymer gel dosimeters. From [12].

Mather and Baldock [14] further demonstrated the ability of time-of-flight ultrasound in imaging irradiated polymer gel dosimeters (see figure 9).

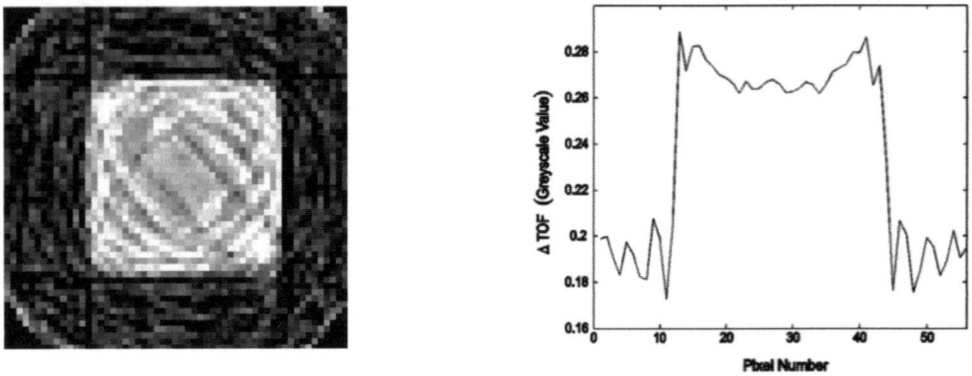

Figure 9: Ultrasound image (left panel) and profile through image (right panel) of irradiated polymer gel From [14].

More recently, Crescenti *et al* have characterized the dose-dependent variation of Young's modulus in irradiated polymer gel dosimeters (figure 10) [15]. They have been able to use elastography measurements to then image an irradiated polymer gel dosimeter (figure 11) [16]. These results are encouraging and demonstrate the possibility of mechanical-property-based imaging of irradiated polymer gel dosimeters.

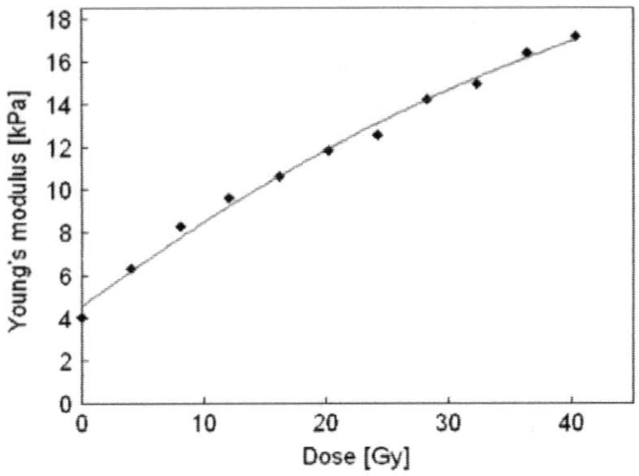

Figure 10: Variation in Youngs modulus in irradiated polymer gel dosimeters [15].

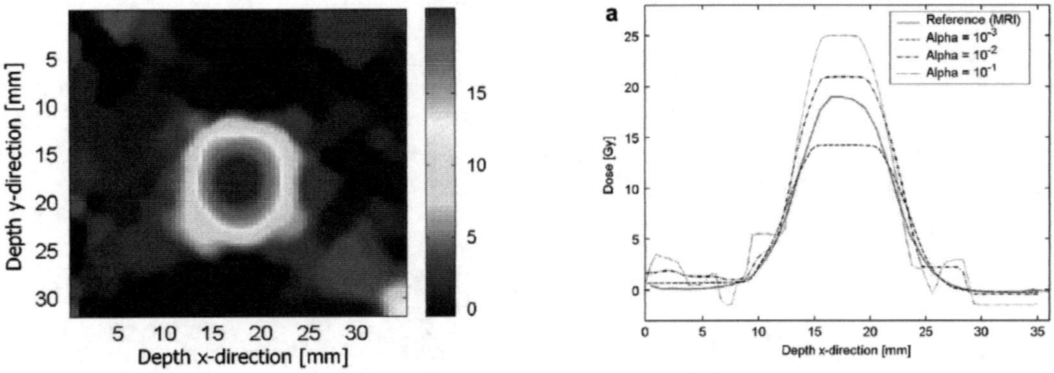

Figure 11: Elestography image (left panel) of an irradiated polymer gel dosimeter (left panel). Elastography and MRI profiles through irradiated polymer gel dosimeter (right panel). From [16].

4. Raman imaging

Raman spectroscopy, for the purposes of imaging irradiated polymer gel dosimeters, was first proposed by Rintoul *et al* [17]. They utilized the fact that Raman spectroscopy can be used to determine the dose-dependent decrease in monomer concentrations within irradiated polymer gel dosimeters. Using principle component analysis, they were able to reconstruct a depth-dose dependence within an irradiated polymer gel dosimeter (figure 12).

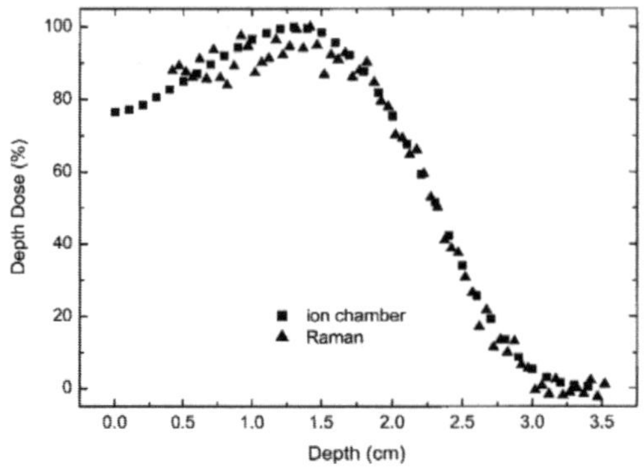

Figure 12: Reconstruction of a 6MeV depth-dose curve measured using Raman spectroscopy. Ion chamber measurements are shown for comparison. From [17].

While the results of figure 12 are not an "image" per se, this was an early demonstration of the possibility of extending Raman techniques for imaging purposes. However, recently Schulmerich *et al* [18] have demonstrated the potential if Raman microscopy for tomographic reconstructions of canine bone, as illustrated in figure 13. Bone depths varied between 2 - 4 cm. Hence, it could be envisioned that Raman microscopic imaging for polymer gel dosimetry would be potentially applicable to, for example, brachytherapy seed dose-distributions.

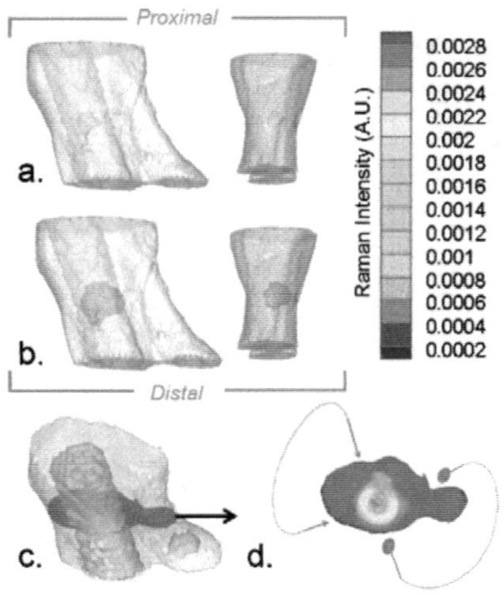

Figure 13: Raman tomographic reconstruction of canine bone From [18].

5. Conclusions

While MRI and optical CT imaging remain the primary modalities for imaging polymer gel dosimeters, alternative modalities such as x-ray CT imaging and ultrasound (and associated mechanical properties) imaging are currently under development. Both show promise in the imaging of dose distributions within irradiated polymer gel dosimeters. Finally, Raman tomographic imaging, while not tested explicitly on polymer gel dosimeters, offer an intriguing possibility for high resolution "micro"-imaging of select dose distributions.

References

[1] A Jirasek and M Hilts, An overview of polymer gel dosimetry using x-ray CT, *J. Phys. Conf. Series* **164** (2009), pp 012038–1–12.

[2] M Hilts, X-ray computed tomography imaging of polymer gel dosimeters, *J. Phys. Conf. Series* **56** (2006), 95–107.

[3] M. Hilts, C. Audet, C. Duzenli, and A. Jirasek, Polymer gel dosimetry using x-ray computed tomography: a feasibility study, *Phys Med Biol* **45** (2000), pp 2559–71.

[4] C. Audet, M. Hilts, A. Jirasek, and C. Duzenli, CT gel dosimetry technique: comparison of a planned and measured 3D stereotactic dose volume, *J. App. Clin. Med. Phys*. **3** (2002), pp 110–118.

[5] J. V. Trapp, S. A. Back, M. Lepage, G. Michael, and C. Baldock, An experimental study of the dose response of polymer gel dosimeters imaged with x-ray computed tomography, *Phys Med Biol* **46** (2001), pp 2939–51.

[6] M. Hilts and C. Duzenli, Image filtering for improved dose resolution in CT polymer gel dosimetry, *Med Phys* **31** (2004), pp 39–49.

[7] P Baxter, A Jirasek, and M Hilts, X-ray CT dose in normoxic polyacrylamide gel dosimetry, *Med. Phys*. **34** (2007), pp 1934–43.

[8] V I Koeva, T Olding, A Jirasek, L J Schreiner, and K B McAuley, Preliminary investigation of the NMR, optical and x-ray CT dose-response of polymer gel dosimeters incorporating cosolvents to improve dose sensitivity, *Phys. Med. Biol.* **54** (2009), pp 2779–90.

[9] M. Hilts, A. Jirasek, and C. Duzenli, Effects of gel composition on the radiation induced density change in PAG polymer gel dosimeters: a model and experimental investigations, *Phys Med Biol* **49** (2004), pp 2477–2490.

[10] A Jirasek, M Hilts, A Berman, and K B McAuley, Effects of glycerol co-solvent on the rate and form of polymer gel dose response, *Phys. Med. Biol*. **54** (2009), pp 907–918.

[11] P Sellakumar and E James Jebaseelan Samuel, Study on energy dependence of PAGAT polymer gel dosimeter evaluated using x-ray CT, *Radiation Measurements* **45** (2010), pp 92–97.

[12] M. L. Mather, P. H. Charles, and C. Baldock, Measurement of ultrasonic attenuation coefficient in polymer gel dosimeters, *Phys. Med. Biol.* **48** (2003), pp N269–N275.

[13] R. A Crescenti, J. C Bamber, M. Partridge, N. L Bush, and S. Webb, Characterization of the ultrasonic attenuation coefficient and its frequency dependence in a polymer gel dosimeter, *Phys. Med. Biol*. **52** (2007), pp 6747–59.

[14] M. L. Mather and C. Baldock, Ultrasound tomographic imaging of radiation dose distributions in polymer gel dosimeters: Preliminary study, *Med. Phys* **30** (2003), pp 2140–2148.

[15] R. A Crescenti, J. C Bamber, N. L Bush, and S. Webb, Characterization of dose-dependent young's modulus for a radiation-sensitive polymer gel, *Phys. Med. Biol*. **54** (2009), pp 843–857.

[16] R. A Crescenti, J. C Bamber, A. A Oberai, P.E Barbone, J. P Richter, C. Rivas, N. L Bush, and S. Webb, Quantitative ultrasonic elastography for gel dosimetry, *Ultrasound Med Biol* **36** (2010), pp 268–75.

[17] L. Rintoul, M. Lepage, and C. Baldock, Radiation dose distribution in polymer gels by Raman spectroscopy, *Appl Spectrosc.* **57** (2003), pp 51–7.

[18] M. V. Schulmerich, J. H. Cole, K. A. Dooley, M. D. Morris, J. M. Kreider, S. A Goldstein, S. Srinivasan, and B. W. Pogue, Noninvasive raman tomographic imaging of canine bone tissue, *J. Biomed. Optics* **13** (2008), pp 020506.

Journal of Physics: Conference Series **250** (2010) 012071 doi:10.1088/1742-6596/250/1/012071

Gamma Dose Distribution Evaluation Tool

Daniel A. Low, Ph.D.

Washington University, Department of Radiation Oncology, 4921 Parkview Place, St. Louis, MO 63110;

low@wustl.edu

Abstract. Quantitative comparisons of dose distributions are an integral component of a medical physicist's responsibility of assuring high quality radiation therapy dose delivery. While overlays and other displays of multiple dose distributions are useful for such evaluations, quantitative evaluations require a mathematical comparison. The dose-difference is the most straightforward method for comparing two dose distributions, but it can show large differences in steep dose gradient regions, even for relatively small misalignments. A tool, termed γ,was developed to take both dose and spatial difference into account. It does this automatically, evaluating distributions for dose difference and spatial discrepancies in regions of shallow and steep dose gradients, respectively. The tool has been used extensively in commercial dose measurement and evaluation software. This chapter describes the tool, some alternative techniques, and limitations of the tool.

1. Dose Distribution Evaluations

Dosimeters provide critically important information about the characteristics of dose distributions. They allow us to directly measure a physical phenomenon that is capable of irradiation tumors such that they are sterilized while nearby normal organs retain their function. The dose accuracy required is typically assumed to be within a few percent in order to provide the potential benefit. This is quite challenging to achieve in a complex structure such as the human body, so radiation therapy relies on sophisticated computer dose calculations to predict the dose inside the patient and allow the treatment planner to visualize the impact of selecting different beam orientations, energies, and intensities. Computer-optimized dose distributions are also used, and these have very complex dose deliveries and consequently dose distribution characteristics, such as steep conformal dose gradients that wrap around the tumors and normal organs.

The complexity of the dose delivery and the resulting distributions, and the tight requirement for accuracy, result in the need for direct measurement-based verification of the calculated dose distributions. This has led to the main purpose of this conference; the 3-dimensional measurement of dose distributions. However, having a 3-dimensional calculation and measurements are only part of the dose distribution quality assurance (QA) process. The dose distributions need to be compared to one another using methods that are quantitative, efficient, and that take maximum advantage of the limited capability for visualizing the dose distributions, which are 3-dimensional scalar fields.

The first group to publish on applying quantitative tools for dose distribution comparisons was the group of Van Dyk et al.[1] They showed that dose comparisons should consider the local dose gradient when determining the best way to compare doses. The most straightforward way to compare two dose distributions is to take the numerical difference. Presumably, if you want to know by how much the two dose distributions disagree, the difference makes it clear. However, there are practical limitations to this method. First, the dose distributions may not be precisely aligned. If one of them is measured, there is a spatial tolerance on the ability to position the detector, align the phantom to the linear accelerator, and to conduct the readout process (e.g. optical densitometry). If there is a spatial error in the measurement, the dose distribution differences that lie within steep dose gradients will be artificially enhanced by the spatial offset of the two distributions.

While a spatial offset leads to sometimes large dose differences in steep dose gradient regions, one can ask the question; why are these differences not relevant to the patient's treatment? They existed when a physicist made the dose distribution measurements, so why not consider them when evaluating the dose distribution quality? This is a valid argument, and the answer has to do with the purpose of the measurement. For most cases, the purpose of the measurement is to validate the output of the treatment planning system, or to validate the system-wide process of calculation, data transfer, and dose delivery. A reflection of the impact of measurement-induced errors is typically not desired by the person conducting or evaluating the QA results. Therefore, one of the desired features of dose distribution comparisons is to be somewhat immune to experimental error.

Van Dyk et al[1] published on the importance of applying a particular dose distribution comparison test based on the local dose gradient. In regions of shallow dose gradient, the dose difference provided the physicist with a quantitative and straightforward method for comparing two dose distributions. However, in steep dose gradient regions, the dose difference was sensitive to the small spatial offsets commonly encountered in experiments, so even if the dose distributions agreed exactly, the dose differences were very large when even a small spatial error was encountered. As an example, IMRT dose distributions often dose gradients of close to 3% mm^{-1}. Even a small position shift of one dose distribution relative to the other will cause large dose differences. In this example, a 1 mm shift will cause a 3% dose difference. Selecting a 3% dose difference criterion when comparing the two dose distributions will lead to false positive detections of regions where the calculation appears to fail to accurately model the dose distribution. Van Dyk[1] developed the concept of distance to agreement (DTA), which Harms et al[2] applied in a software tool to compare two-dimensional dose distributions.

The DTA is typically computed by measuring the closest distance between one dose distribution to where the second distribution has the same dose level. This is equivalent to determining the closest approach to a point in one distribution of the isodose line that in the other distribution that has the same dose as the point. The DTA is a non-local function, meaning that when it is evaluated at a point, the other distribution is queried at a distance. The DTA function also does not commute. While the dose difference merely changes sign when the two dose distributions are swapped, the DTA is not the same value of the roles of the two distributions are swapped. Therefore it is useful to label the two distributions as the reference and evaluated distributions. The reference distribution is the one whereby the DTA is computed point by point, while the evaluated distribution is the one queried for the closest approach of the specific isodose.

One of the nice features of the dose difference distribution is its relatively straightforward interpretation. Most physicists can glance at a dose difference distribution and see and interpret the data. Because of the impact of steep dose gradients, the dose differences may not be clinically relevant, but the meaning behind the numbers is clear. While the dose difference is overly sensitive in steep dose gradient regions, the DTA is overly sensitive in shallow dose gradient regions. As an example, if the reference dose point has dose D, lying in a shallow dose gradient region, and the evaluated dose at that point is $D+\delta D$, the distance in the evaluated distribution from the reference point that has the same dose D may be very far away. The DTA will therefore be very large, even if δD is small and has no clinical consequence. Because of the large values in shallow dose gradient regions, regions that typically dominate IMRT dose distributions, the DTA distribution is difficult, if not impossible to interpret by eye.

In steep dose gradient regions, the DTA can be interpreted as the distance between the two dose distributions. This interpretation is based on the assumption that the distance is caused primarily by a spatial offset between the two distributions. For distributions that differ by such an offset, the DTA distribution provides an effective and accurate measurement of the offset. However, as described

Journal of Physics: Conference Series **250** (2010) 012071 doi:10.1088/1742-6596/250/1/012071

above, the discrepancy may not be due to an error of the dose calculation algorithm or delivery hardware, but simply due to experimental error. While as originally described, the DTA tool provides only the magnitude of the distance; the sign of the dose difference can be added to the DTA. This provides an indication of which of the two dose distributions is greater. It can also be used to quickly determine if the discrepancy is due to a spatial offset due to one distribution being "larger" than the other.

Because the dose difference and DTA are useful in regions of shallow and steep dose gradients, their use has been combined. A composite analysis was developed by Harms et al.[2], based on the concepts of Shiu et al.[3] as applied by Cheng et al.[4] used pass-fail criteria using both the dose difference and DTA. Dose-difference and DTA test criteria are selected to determine if each point passes or fails the dose distribution comparison test. The reference points are evaluated one by one to determine their dose difference and DTA and these are compared against the criteria. At points that fail both tests, the dose distribution comparison is said to have failed, otherwise it passes. This allows either the dose difference or DTA tests to be overly sensitive, so the other test determines if the dose comparison passes or fails.

While the composite test is capable of conducting a pass-fail analysis of the dose distribution comparison, it does not indicate by how much the test is passed or failed. In other words, 3% dose difference, 3 mm DTA criteria can be exceeded by 1% and 1 mm, or 10% and 10 mm, and the result would be the same, although the clinical relevancy might differ.

In 2001, at the AAPM annual meeting, Low was explaining the dose difference, DTA, and composite tests to an employee of Computerized Medical Systems in the parking lot of one of the hotels used by the attendees. During the conversation, Dr. Low said he wished he could simultaneously test the dose difference and DTA, but the quantities, dose and distance, were not the same, so this could not be done. Instead, Dr. Low realized that if the dose difference and DTA were divided by their respective criteria, the results would be unitless, and one could evaluate both simultaneously. Because this was an analog to the composite criterion, which starts with the letter "c", the third letter of the English alphabet, he named the quantity γ, the third letter of the Greek alphabet.

2. What γ is

2.1. Definition

As in many new endeavors, there are multiple changes that take place as the ideas begin to gel. Such was the case with γ and the labeling of the two dose distributions, as well as the nomenclature used to describe the criteria and quantities. For simplicity, the nomenclature and symbology used here reflect the second γ paper by Low and Dempsey[5]. In that paper, the two dose distributions are the reference and evaluated distributions. The reference distribution is intended to be the one relative to which the other is compared. It is typically the measurement in a measurement versus calculation, or it can be a Monte Carlo calculation when calculations are compared against each other. The reference distribution is queried point by point, and γ computed individually for each. Therefore, the reference distribution can be a single point, for example an ionization chamber measurement point. The evaluated distribution needs to be a distribution, and in early γ papers, it was assumed that it had to have a relatively high spatial resolution. This is because the γ calculation can have relatively large errors in steep dose gradient regions when the spacing between evaluated distribution points is close to the DTA criterion.

The γ distribution calculation is typically conducted as a search. As in the DTA, one needs to search the local evaluated distribution. In the DTA, the value being calculated at each evaluated point is simply the Euclidean distance to the reference point. However, in the γ calculation, the distance is generalized in the dose and distance space. The generalized distance is labeled Γ. Γ is the generalized Euclidean distance in the renormalized dose and distance space, where distance and dose axes have been divided by the DTA and dose difference criteria, respectively.

Journal of Physics: Conference Series **250** (2010) 012071 doi:10.1088/1742-6596/250/1/012071

$$\Gamma(\vec{r}_e,\vec{r}_r)=\sqrt{\frac{|\vec{r}_e-\vec{r}_r|^2}{\Delta d^2}+\frac{[D_e(\vec{r}_e)-D_r(\vec{r}_r)]^2}{\Delta D^2}} \quad (1)$$

where $\vec{r}_e$ and $\vec{r}_r$ are the vector positions of the evaluated and reference points, respectively, $D_e(\vec{r}_e)$ and $D_r(\vec{r}_r)$ are the evaluated and reference doses, respectively, and Δd and ΔD are the DTA and dose difference criteria, respectively. The generalized Γ function can be computed for any pair $\vec{r}_e$ and $\vec{r}_r$, so for each reference point, there are as many values of Γ as there are evaluated points (infinite number with interpolation). The minimum value of Γ is the value of γ.

$$\gamma(\vec{r}_r)=\min\{\Gamma(\vec{r}_e,\vec{r}_r)\}\forall\{\vec{r}_e\} \quad (2)$$

Equation 2 states that γ is simply the minimum value in all of the evaluated distribution search space of Γ. While equations 1 and 2 provide the factual definition of Γ, they do not impart any intuition for what γ means and its utility.

2.2. Interpretation

The γ function is the minimum distance between two dose distributions, but in an unusual space. The distance includes not only space, but dose as well. The spatial axes are the normal ones, and there are as many as necessary to describe the evaluated dose distribution. Recall that the reference dose distribution can be a single point, so the number of dimensions is determined by the evaluated distribution. While the distance axes are the same as normal axes, they are renormalized by the DTA criterion ΔD, so they have no units. The distance corresponding to Δd is 1. There is an extra axis, which is the dose axis, and this is made possible by the renormalization of the dose axis by the dose difference criterion, ΔD. The origin is placed at the reference dose point (its position and dose), so travelling along a distance axis corresponds to moving away from the reference point and travelling along the dose axis corresponds to changing the dose difference. The distance from the origin along the dose axis is 1 at the dose difference corresponding to ΔD.

Given that the dose-difference and DTA criteria are met at the distance of 1, we define the unit circle, sphere, or hypersphere, depending on the dimensionality of the dose distributions, as the limit within which a comparison between the reference and evaluated distribution passes. Figure 1a shows a schematic of the unit circle in a one-dimensional dose distribution evaluation. The horizontal and vertical axes are the distance and dose axes, respectively. The unit circle corresponds to the region within which the hyperdistance is less than 1.

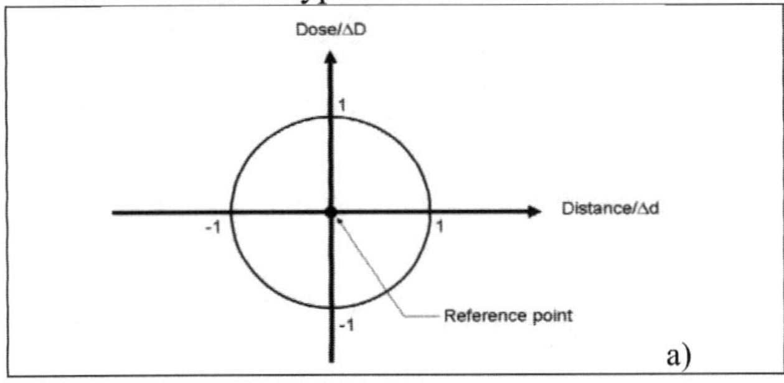

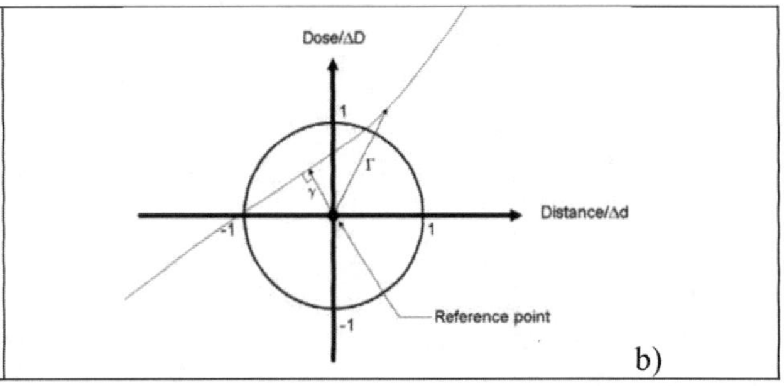

Journal of Physics: Conference Series **250** (2010) 012071 doi:10.1088/1742-6596/250/1/012071

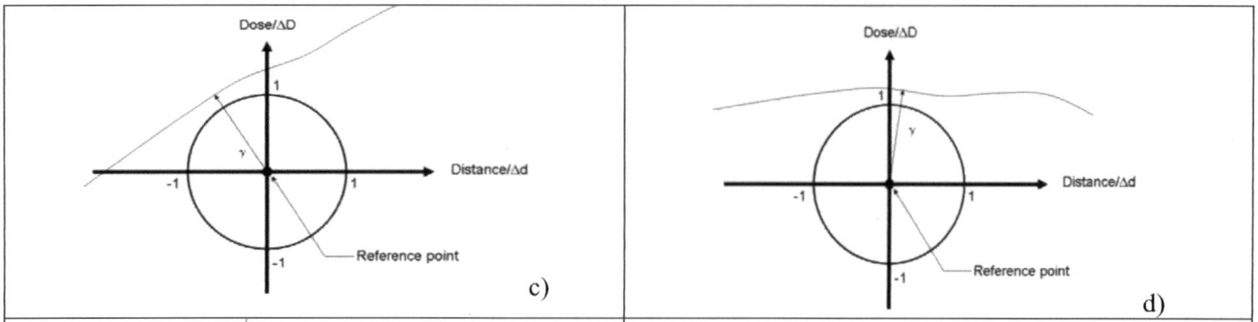

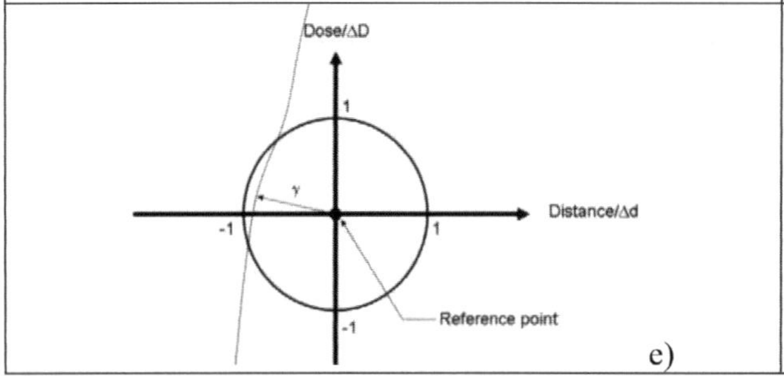

Figure 1. Schematic description of γ calculations. In each case, the thin wavy line is the evaluated distribution. a) The definition of the unit circle that defines the pass-fail γ criteria. b) A case where γ test passes. c) A case where the γ test fails. d) Shallow dose gradient case. e) Steep dose gradient case.

The evaluated distribution can be plotted in the renormalized space, and consists of a line, sheet, or hypersheet if the reference distribution is one, two, or more dimensions. In figure 1b, an example reference distribution is drawn to explain the features of the calculation. The distance from the origin to the evaluated distribution is the Γ distribution, and the minimum value is the closest that the evaluated distribution gets to the reference point. When that distance is less than (Figure 1b) or greater than (Figure 1c) 1, the evaluated distribution passed and failed the test, respectively.

The passing and failing of the γ test is very similar to the composite test of Harms et al.[2] The difference lies in the fact that, while the test does identify if the evaluated distribution passed or failed, it also identifies by how much. A value of γ=1.1 fails, but by only 10%. This corresponds to a distance of 0.3 mm, or a dose error of 0.3% for 3 mm and 3% DTA and dose difference criteria, a relatively minor infraction. On the other hand, a value of γ=3 fails by a factor of 3, corresponding to 9 mm or 9%, a large difference from the previous example.

The γ tool automatically defaults to the dose-difference and DTA tests in regions of shallow and steep dose gradients, respectively. Figures 1d and 1e show the γ evaluation under conditions of shallow and steep dose gradients, respectively. In shallow dose gradients (Figure 1d), the evaluated dose is nearly constant near the reference point, so the reference curve appears flat and the closest point lies near the dose axis. A comparison that takes place along the dose axis is essentially the dose difference, so the γ tool defaults to the dose difference test at shallow gradients. In steep dose gradients (Figure 1e), the nearest point lies in the direction of one of the spatial axes. The distance between the evaluated distribution where it crosses the spatial axis and the reference point is the DTA, so as the gradient increases, the γ direction points towards the spatial axes and automatically becomes the DTA test. The angle between the dose axis and the γ vector can be used to determine if the difference between the evaluated and reference distribution is due to the dose difference, DTA, or an intermediate reason.

An interesting question is; for real clinical data and reasonable criteria, what is the maximum angle that the γ vector makes to the dose axis? For a dose gradient of 3%/mm and DTA and dose difference criteria of 3% and 3mm, the angle the evaluated distribution makes to the dose axis is almost 72 degrees, or 18 degrees from the distance axis.

3. Other Approaches and Limitations

3.1. Other Approaches

Journal of Physics: Conference Series **250** (2010) 012071 doi:10.1088/1742-6596/250/1/012071

The original γ calculation required an exhaustive search in the evaluated dose distribution space to determine the minimum Γ value. This caused a significant amount of calculation time, especially when the evaluated dose distribution was two or three dimensional. This limitation led to a number of groups to develop methods for modifying or improving the speed of the γ calculation.

Soon after the publication of the initial γ paper, Depuydt et al. [6] identified some of the limitations of the initial report: 1) the initial γ paper did not deal with the two distributions having different spatial densities or locations, 2) the need for interpolation to reduce the artifacts caused by large spacing between evaluated points. In order to reduce the impact of these issues on the γ values, they decided to reduce the problem to whether γ was greater or less than 1 (fail or pass, respectively) rather than compute the value of γ. They employed a filter cascade process to determine which evaluated points lay within or outside the $\gamma = 1$ hypercircle. The restricted the search to regions nearby the reference point, and once finding a point for which $\gamma < 1$, they stopped the search. Barring finding points that passed, they evaluated whether a steep dose gradient caused the values of γ to be greater than 1 even though an interpolated evaluated distribution would have passed within the $\gamma = 1$ hypercircle. The intersection of the evaluated dose and the spatial axis (axes) is checked to determine if it lies within the $\gamma = 1$ hypercircle. The advantage of this approach was that it was significantly faster than the original algorithm, but it did not have the benefit of providing the user with the magnitude of γ to determine the magnitude of failure.

Moran, et al.[7] defined a dose comparison tool that examined the local dose gradient. The user selected a distance tolerance to deal with such issues as measurement spatial registration errors. The dose difference and the local dose gradients are computed. The product of the gradient and the distance tolerance indicates the maximum dose difference that could be associated with a spatial shift of the distance tolerance magnitude. The product of the gradient and the dose distance tolerance is subtracted from the dose difference magnitude to determine a gradient-adjusted difference. It is this modified dose difference that is examined.

Stock et al[8] in 2005 developed a tool for applying the γ analysis for two-dimensional dose distributions (e.g. film and a two-dimensional cross-sectional sample of a 3-dimensional dose calculation). They were the first to describe the use of the γ angle, the angle that the γ vector makes to the dose axis. Figure 2 shows an example from Stock et al.[8] of a γ distribution for a head-and-neck treatment plan. The γ distribution shows discrepancy near the posterior head, where γ values approach 2. The γ angle distribution at the same location indicates that the discrepancy lies within a region of relatively steep dose gradient and that the γ extends from approximately $\pi/4$ to $3\pi/8$. Finally, Stock et al.[8] also introduced γ histograms (also evaluated by Spezi and Lewis[9]) as a method for distilling the complex distributions into a more easily interpreted form (Figure 2c). A histogram allows the reviewer a relatively straightforward method for determining how many points are failing the γ test as well as the number of points with very large values of γ.

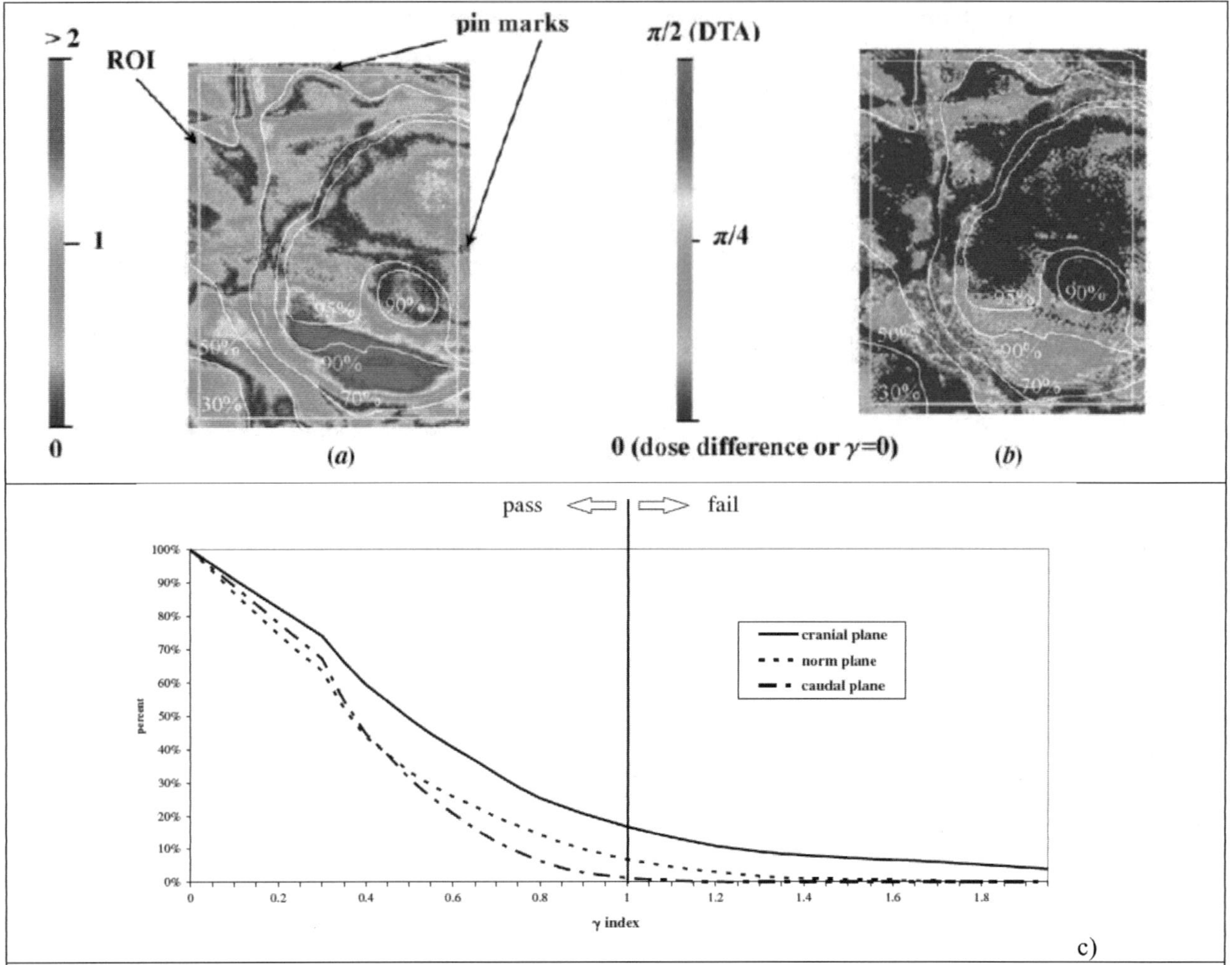

Figure 2: γ distributions for a head-and-neck quality assurance procedure. The isodose distribution is shown as white contours. a) the γ distribution. b) the γ angle distribution. c) a histogram of γ for three plane orientations. Reprinted with permission from Stock et al. [8]

Jiang et al.[10] developed a method that translated the spatial tolerance into the dose domain. They introduced the concepts of equivalent dose tolerance, maximum allowed dose difference (MADD), and the normalized dose difference (NDD). They also used dose-difference and DTA criteria (termed predetermined dose tolerance and predetermined spatial tolerance, respectively). The equivalent dose tolerance is computed by shifting the evaluated distribution until the DTA is equal to the DTA criterion. The equivalent dose tolerance is then the dose-difference between the reference point and the shifted evaluated dose distribution. If the equivalent dose tolerance (in magnitude) is less than the dose-difference, then the DTA must have been less than the DTA criterion, so only a dose-difference test is required. Jiang et al. [10] also defined an acceptance box, consisting of a rectangular region with the dose-difference and DTA criteria defining the box outlines. If the evaluated distribution intersected the box, the dose distribution passed the test. They also examined the pass-fail criteria in terms of mathematical equations. Figure 3 shows examples of the criteria graphed for the composite evaluation, γ, and the box.

The MADD defined an acceptance region in the dose-distance space such that it provided a pass/fail boundary such as shown in Figure 3 for other evaluations. The evaluated distribution is shifted until it is tangent to the MADD boundary. The MADD is the dose difference in this case, and if the dose difference is less than MADD (in magnitude), the distribution test passes. An interesting feature of the paper by Jiang et al.[10] was the appearance of γ artifacts caused by the finite evaluated dose distribution pixel spacing.

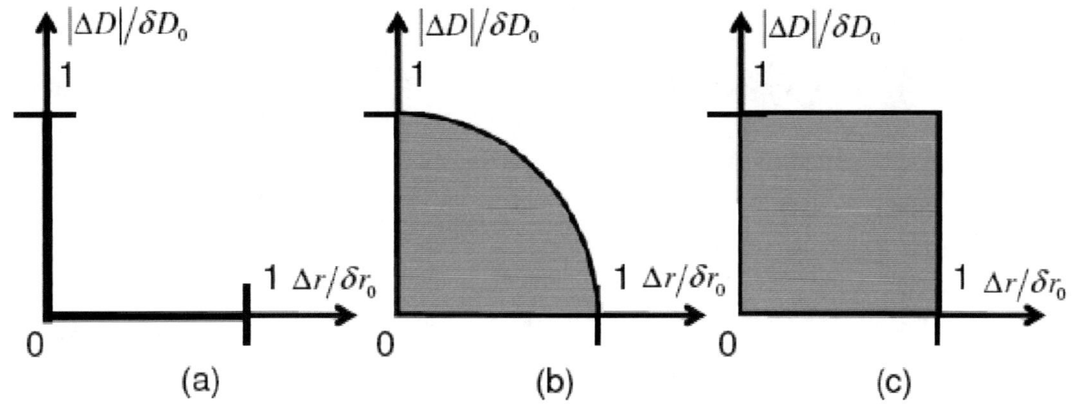

Figure 3: Acceptance criteria (one-dimensional examples) for a) the composite evaluation, b) γ, and the box method proposed by Jiang et al.[10] Figure is from Jiang et al.[10]

3.2. Calculation time and spatial resolution artifacts

Some of the concerns with the use of γ is the calculation time. When first developed, the γ distribution was calculated reference point-by-point and an exhaustive search conducted in the evaluated dose distribution for the minimum value of Γ. This did not pose a large computational burden for two-dimensional dose distributions, but was significant for three-dimensional dose distributions.

Another concern was related to γ distribution artifacts that were seen in steep dose gradient regions. Figure 4 shows an example as published by Jiang et al.[10] where two square-field dose distributions are being compared, one rotated with respect to the other. Near the field boundary, the γ distribution appears to have undulations. These are caused by the finite evaluated dose distribution spacing in steep dose gradient regions. If the γ distribution is computed only at each evaluated dose point, the actual minimum distance between the two dose distributions may be missed. This is shown in Figure 5, where the computed and interpolated values of γ are presented. When the dose distributions align such that the evaluated point sits exactly at the closest approach to the reference point, γ is accurately computed. When the dose distributions are slightly shifted, the value of γ is overestimated (Figure 5). For conditions where the dose distributions are slowly shifting (such as in Figure 4), this presents γ as a wavy distribution.

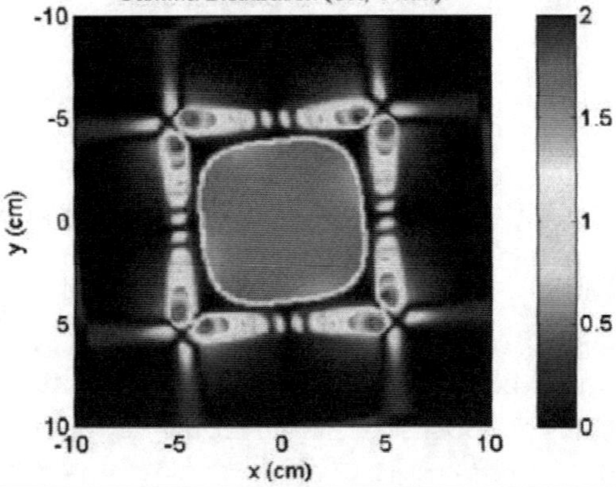

Figure 4. An example of the γ distribution from Jiang et al.[10] This example was computed using two square dose distributions, one rotated with respect to the other. For distributions with infinite spatial resolution, the distributions would not exhibit the dotted appearance at the field boundary (steep dose gradient). However, because of the behavior of γ calculations in steep dose gradient regions, the γ distribution is overestimated when the evaluated dose distribution has relatively poor spatial resolution.

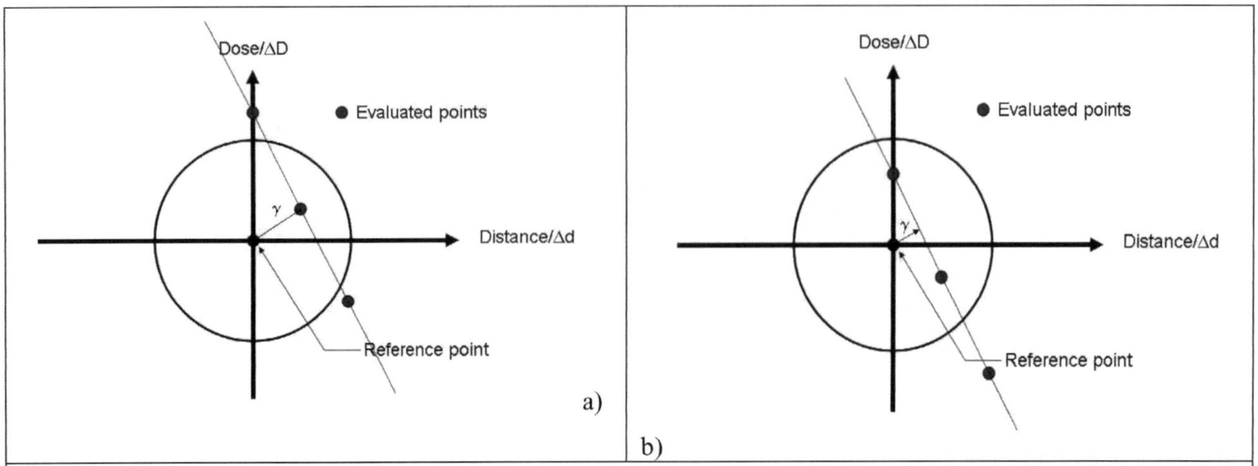

Figure 5: Example of finite dose spacing error in γ calculation errors. a) Case where interpolation is not needed, the determination of γ is accurate. b) Case where the actual value of γ is smaller than the computed value (the value as computed at the closest evaluated point).

The solution to this problem was first by interpolation, making the interpolated evaluated distribution sufficiently fine that errors in γ caused by the finite spacing were relatively small. Wendling et al.[11] dealt with the long times for computing γ accurately by limiting the search space and reducing 3D γ calculations to multiple 2D planes. Ju et al.[12] looked at γ calculations as a geometric problem. The interpolation was essentially the question of the closest approach of the linearly interpolated evaluated distribution, described as a set of line segments, to the reference position. Because this is essentially the question of the closest each line segment (interpolated evaluation distribution) gets to the reference point, Ju et al.[12] interpreted the problem geometrically.

In higher dimensions, the interpolation was more challenging. For example, a two-dimensional dose distribution would be represented by a series of four evaluation points projecting to a rectangle on the distance axes. The interpolated surface between these points is not well defined, so a fifth point, lying in the middle of the four in the spatial axis and being the mean of the four doses on the dose axis was defined and the rectangular space segmented into four triangles, each with two vertices being the evaluated points and one vertex the interpolated point. This method made the evaluated distribution into a series of triangles, which are considered simplexes, as are line segments in the one-dimensional case. The algorithm was able to rapidly evaluate the distance from the reference point to the simplexes and recursively compute the distance from the surface, edge, or points, whichever was closest. Applications of the algorithm to 3D cases showed that, not only were the artifacts caused by finite reference dose spacing gone, but the calculation was much faster, a 3D calculation requiring a couple of minutes.

3.3. Noise artifacts

One of the uses for γ was to evaluate Monte Carlo dose distributions, but these contain pixel-to-pixel noise. Because the γ calculation can be interpreted geometrically, the presence of noise would impact the closest approach determination and therefore the value of γ. This was evaluated by Low and Dempsey[5] by examining γ histograms for dose distributions with and without noise. The greatest impact of noise occurred in regions of shallow dose gradient. The impact of noise in the evaluated distribution always led to a reduction in the value of γ because the closest approach for random noise was not likely to get farther away, but closer. Figure 6 shows an example of the impact of noise on homogeneous dose distributions with differences of 6% and a dose difference criterion of 3% ($\gamma = 2$ for no noise). A histogram of γ showed that the value was 2. As pseudo-random noise was added to the reference distribution, the histogram of γ showed a reduction in the values of γ such that γ was approximately 1 for 3% noise. This implied that the impact of noise in the evaluated distribution led to a reduction in the measured value of γ that was linearly proportional to the noise (normalized by the dose-difference criterion) until the noise approached the dose difference, whereafter the values of γ were very small. A different picture appeared when noise was added to the reference distribution. In this case, because γ is computed point-by-point for the reference distribution, the impact of noise was essentially to change the dose difference at each point and the histogram was simply smeared by a width corresponding to the added noise.

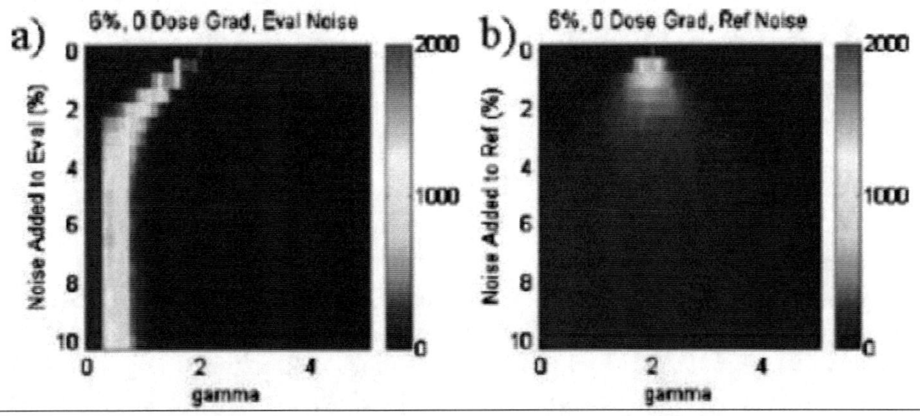

Figure 6. The impact of pseudorandom noise on γ determinations for dose distributions with no gradient and 6% dose differences. The dose-difference criterion was 3%. Shown are γ histograms stacked one per row, with the first row having 0 noise and subsequent rows having added noise. a) Noise added to the evaluated distribution, b) Noise added to the reference distribution. Figure from Low and Dempsey.[5]

These results showed that care needs to be taken when γ is used to analyze noisy data.

4. Conclusions and Future

The γ tool has proven useful in the quantitative evaluation of dose distributions. While artifacts caused by the finite evaluated dose distribution spatial resolution appeared early on, the geometric interpretation of γ allowed a more rapid and accurate computation. Most of the dose measurement and evaluation software has incorporated γ into the evaluation tool suite and the evaluation is used by most medical physicists when attempting to interpret differences between multiple dose distributions.

The original description of γ used constant criteria, often 3% and 3 mm dose-difference and DTA criteria, respectively. However, this is entirely arbitrary and should be expanded to consider clinical relevance. For example, the dose-difference at low doses could exceed 3% with no clinical consequence, or the DTA could be considered more important near a critical structure than in the periphery of the dose distribution. The criteria should be modified according to clinical needs, and ideally according to dose, and tumor and normal organ dose tolerances and spatial uncertainty. This would imply that there is a one-to-one relationship between dose and tumor and organ geometry, which is not true for measurement-based patient-specific QA. However the field is going to move towards independent, 3D calculations conducted in the patient geometry, so the γ tolerances could be spatially and dosimetrically defined.

References

[1] J. Van Dyk, R. B. Barnett, J. E. Cygler, and P. C. Shragge, "Commissioning and quality assurance of treatment planning computers," Int J Radiat Oncol Biol Phys. **26**, 261-273 (1993).

[2] W. B. Harms, Sr., D. A. Low, J. W. Wong, and J. A. Purdy, "A software tool for the quantitative evaluation of 3D dose calculation algorithms," Med Phys. **25**, 1830-1836 (1998).

[3] A. S. Shiu, S. Tung, K. R. Hogstrom, J. W. Wong, R. L. Gerber, W. B. Harms, J. A. Purdy, R. K. Ten Haken, D. L. McShan, and B. A. Fraass, "Verification data for electron beam dose algorithms," Med Phys. **19**, 623-636 (1992).

[4] A. Cheng, W. B. Harms, Sr., R. L. Gerber, J. W. Wong, and J. A. Purdy, "Systematic verification of a three-dimensional electron beam dose calculation algorithm," Med Phys. **23**, 685-693 (1996).
[5] D. A. Low and J. F. Dempsey, "Evaluation of the gamma dose distribution comparison method," Medical Physics. **30**, 2455-2464 (2003).
[6] T. Depuydt, A. Van Esch, and D. P. Huyskens, "A quantitative evaluation of IMRT dose distributions: refinement and clinical assessment of the gamma evaluation," Radiotherapy and Oncology. **62**, 309-319 (2002).
[7] J. M. Moran, J. Radawski, and B. A. Fraass, "A dose-gradient analysis tool for IMRT QA," Journal of Applied Clinical Medical Physics. **6**, 62-73 (2005).
[8] M. Stock, B. Kroupa, and D. Georg, "Interpretation and evaluation of the gamma index and the gamma index angle for the verification of IMRT hybrid plans," Physics in Medicine and Biology. **50**, 399-411 (2005).
[9] E. Spezi and D. G. Lewis, "Gamma histograms for radiotherapy plan evaluation," Radiotherapy and Oncology. **79**, 224-230 (2006).
[10] S. B. Jiang, G. C. Sharp, T. Neicu, R. I. Berbeco, S. Flampouri, and T. Bortfeld, "On dose distribution comparison," Physics in Medicine and Biology. **51**, 759-776 (2006).
[11] M. Wendling, L. J. Zijp, L. N. McDermott, E. J. Smit, J. J. Sonke, B. J. Mijnheer, and M. Van Herk, "A fast algorithm for gamma evaluation in 3D," Medical Physics. **34**, 1647-1654 (2007).
[12] T. Ju, T. Simpson, J. O. Deasy, and D. A. Low, "Geometric interpretation of the gamma dose distribution comparison technique: Interpolation-free calculation," Medical Physics. **35**, 879-887 (2008).

Journal of Physics: Conference Series **250** (2010) 012072 doi:10.1088/1742-6596/250/1/012072

Isopropanol-based polymer gel dosimeters for use with x-ray CT imaging

A. Jirasek[1], M. Hilts[2], and K. McAuley[3]

1. Dept. Physics and Astronomy, University of Victoria, BC, V8W 3P6, Canada

2. Vancouver Island Centre, BC Cancer Agency, Victoria, BC, V6R 2B6, Canada

3. Dept. Chemical Engineering, Queens University, Kingston, ON, K7L 3N6, Canada

jirasek@uvic.ca

Abstract. We report on investigations aimed at increasing the dose response sensitivity and resolution in x-ray CT imaging of polymer gel dosimeters (PGD). We incorporate isopropanol as co-solvent into the gel formulation and show that this incorporation increases dose sensitivity and dose resolution of x-ray CT imaged gel dosimeters. These gels are reproducible in response and stable post-irradiation. We apply the system to a simple 1L gel test case where 2 separate irradiations are used to generate a dose response calibration curve. A third irradiation (3-field) is then calibrated and compared to treatment planning predictions. Results indicate that isopropanol-based PGDs are promising formulations for x-ray CT gel dosimetry and that this current system outperforms previous attempts at dose reconstruction using x-ray CT imaging of PGDs.

1. Introduction

A persistent limitation of x-ray CT polymer gel dosimetry is the low dose sensitivity and resolution of the system. This limitation has precluded applications of x-ray CT polymer gel dosimetry to clinically relevant problems. Nonetheless, the system is an attractive possibility for 3D dosimetry as CT scanners are readily available in clinical settings, the system is robust and not highly sensitive to imaging temperature, and CT imaging of polymer gel dosimeters can be accomplished in a reasonably short amount of time.

Recent efforts in increasing polymer gel dosimeter radiation sensitivity have concentrated on (i) investigating alternative crosslinkers in the gel, and (ii) incorporating a co-solvent to improve the crosslinker solubility within the gel solution [1,2]. Suitable alternative crosslinkers have not, to date, been identified [1]. In terms of increasing crosslinker solubility, two co-solvents have been investigated and have produced promising results. The first, glycerol, has shown to increase resultant dose response sensitivity. However, the net increase of dose response sensitivity was shown to be limited and, furthermore, the shape of the dose response curves were shown to vary depending on the amount of co-solvent, hence making calibration of the system problematic.

We here report on using a isopropanol as co-solvent in x-ray polymer gel dosimetry. We show that isopropanol based gels (i) can significantly improve the dose sensitivity and resolution of the x-ray CT imaged gels, (ii) are stable over time, and (iii) are an attractive possibility for x-ray CT polymer gel

Journal of Physics: Conference Series **250** (2010) 012072 doi:10.1088/1742-6596/250/1/012072

dosimetry. We illustrate the application of isopropanol based x-ray CT polymer gel dosimetry to a simple 3 field irradiation.

2. Materials and Methods

2.1 Gel manufacture

Polymer gels were manufactured using 5% gelatin (Sigma-Aldrich, Oakville, Ontario, Canada), 30% isopropanol, 5mM tetrakis hydroxymethyl phosphonium chloride as antioxidant. The total amount of monomer (N-isopropylacrylamide, NIPAM, TCI Chemicals, OR, USA) and crosslinker (N,N' bis-acrylamide, Sigma-Aldrich) were varied as needed (6% - 20%T) while maintaining equal weights (i.e. 50%C) of monomer and crosslinker in each gel. The total amount of gel manufactured depended on the study (0.25 - 1L).

To begin manufacture of isopropanol based normoxic gels, water and isopropanol were heated to 30°C at which point gelatin was added. The solution was further heated to 34 °C at which point the bis-acrylamide crosslinker was added. The solution was heated and stirred to 45 °C, then cooled to 37 °C and the NIPAM was added. Once all monomer was dissolved, THPC was added. Prepared gel was transferred to 20 mL scintillation vials (Wheaton Scientific, Millville NJ, USA) or 1L jars (Modus Medical, London, Ontario, Canada).

2.2 Gel irradiation

Gels were irradiated using a Varian Clinac linear accelerator (Varian Inc, Palo Alto, CA, USA) using 6MV x-rays, a 15x20 cm^2 field size, and a dose rate of 400 cGy/min at 1.5 cm depth in water [2]. For dose response studies, gels were irradiated in a customized phantom to between 2 - 50 Gy. For the imaging application, gels were irradiated in an immobilization device with 3 separate irradiations: (i) a single PDD with 4 Gy at d_{max}, (ii) a two-field cross (~25 Gy at d_{max}), and (iii) a 3-field irradiation. The single and 2-field irradiations were used to generate a calibration curve for the polymer gel dosimeters. This calibration was then applied to the 3-field irradiation in order to convert relative response to dose. All treatment planning was performed using Eclipse (Varian Medical, Palo Alto, CA, USA).

2.3 Gel imaging

Gels were imaged using a GE HiSpeed X/i scanner (GE Medical Systems, Milwaukee WI, USA) operating at 140 kVp and 200 mAs. Image averaging (16 averages) and blank (i.e. unirradiated gel) background subtraction was utilized for each final image. For dose response studies, gel vials were placed in a customized holder, as described previously [3]. For imaging of the 1L gel phantom, a customized mask assembly was used for precise placement of irradiated and background samples.

2.4 Data processing

All images were smoothed using a 5x5 Weiner filter with 2 iterations [4].

3. Results and Discussion

3.1 Dosimeter optimization

Figure 1 illustrates the dose response curves (figure 1a) and subsequent dose resolution (figure 1b) for isopropanol-based polymer gel dosimeters with varying total monomer/crosslinker (%T) content. Also shown, for comparative purposes, is the dose response curve for an anoxic-manufactured acrylamide-based polymer gel with no co-solvent. Figure 1b illustrates that high %T gels provide dose resolutions (1σ) on the order of 0.2-0.3 H/Gy at doses up to 20 Gy. From these results, 16%T gels were chosen for all future studies.

Figure 2 illustrates the dose response reproducibility for 3 separate batches of 16%T gel (figure 2a) and imaging session reproducibility (figure 2b) 1 - 8 days post manufacture. Results indicate that isopropanol based gels are reproducible and stable in their response.

Journal of Physics: Conference Series **250** (2010) 012072 doi:10.1088/1742-6596/250/1/012072

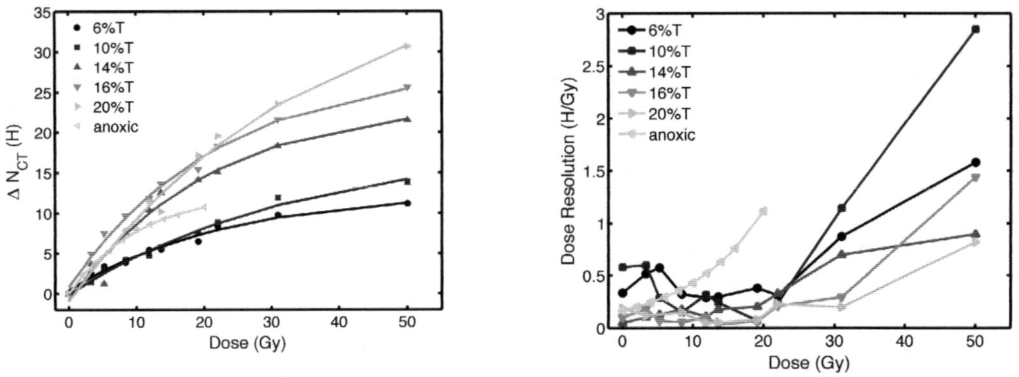

Figure 1: (Left panel) Dose response curves and (right panel) dose resolution of 30% isopropanol-based polymer gel dosimeters with varying %T. Also shown, for comparative purposes, is an anoxic (acrylamide based, 6%T) polymer gel response.

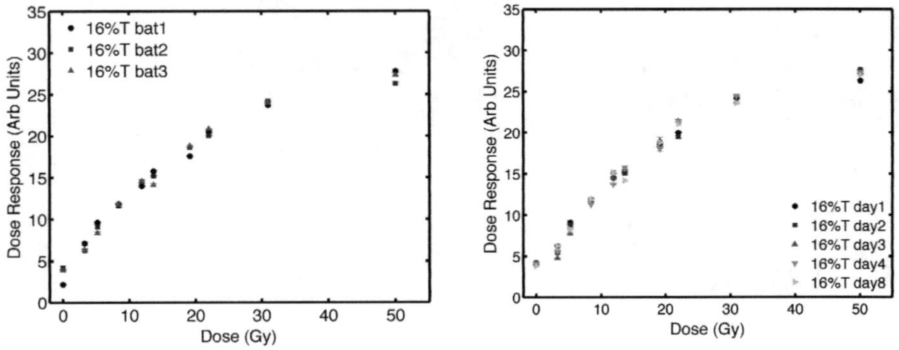

Figure 2: Dose response reproducibility for (left panel) three batches of gel and (right panel) multiple imaging sessions of a given gel (days post irradiation).

3.2 Dosimeters Imaging

Figure 3 illustrates the dose response calibration curve generated by mapping select corresponding treatment plan / gel image points in the single field and 2-field irradiations. The calibration curve of figure 3 was then applied to the single field, 2-field, and 3-field gel images in order to convert relative gel response to dose.

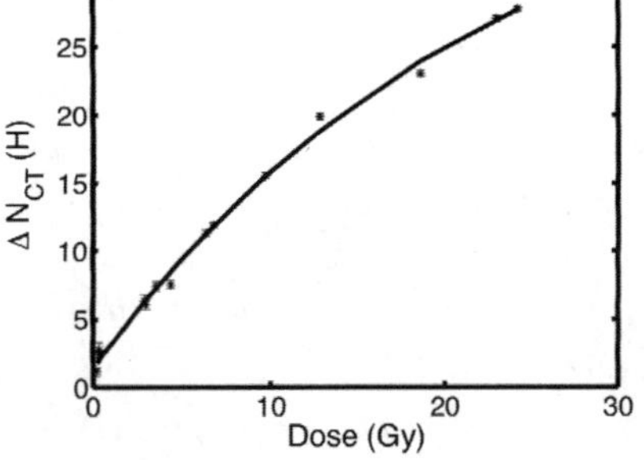

Figure 3: Calibration curve for polymer gel imaging. Curve generated by mapping select points in single field and 2-field gel images to treatment plan. (see figure 4).

Figure 4 illustrates an isodose overlay (gel image with treatment plan isodoses) and representative profiles for the single and 2-field calibration images. It can be seen that calibrated images are capable of reproducing treatment plan doses.

Figure 5(a) illustrates the gel image and treatment plan isodose overlay for the 3 field "test" irradiation. Figure 5(b,c) illustrate representative profiles for the gel and treatment plan. Figure 5(d) illustrates a dose difference map (gel image - treatment plan) for the 3-field irradiation. Results indicate that the use of a single and 2-field calibration irradiation is capable of being applied to a test irradiation. Furthermore, low-doses (<4 Gy) are accurately imaged with the CT polymer gel system, with polymer gel images capturing low-dose isodoses. This is a significant improvement over previously reported applications of x-ray CT polymer gel dosimetry [5].

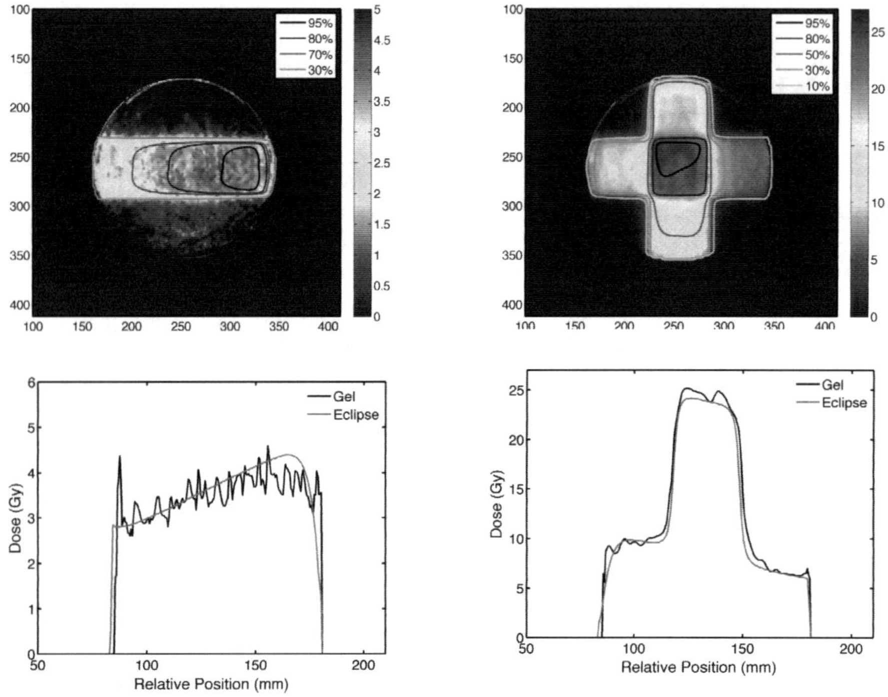

Figure 4: Gel images and treatment plan isodose overlays for PDD (4 Gy at d_{max}, top left) and 2-field (top right) calibration irradiations. Bottom left: Representative PDD profile through row 256 of PDD gel image and treatment plan. Bottom right: Representative profile through row 250 of square-field gel image and treatment plan.

4. Conclusions

These results, while preliminary, indicate that incorporating isopropanol-based polymer gel dosimeters in to x-ray CT polymer gel dosimetry increases the dose sensitivity and resolution of the system. While the dose resolution is still lower than that of, for example, MRI systems, these results indicate that x-ray CT imaging can be used to reconstruct high and low-dose information from irradiated polymer gel dosimeters. This is a significant step forward in x-ray CT polymer gel dosimetry, in that previous applications of the technique were only capable of high-dose reconstruction.

Journal of Physics: Conference Series **250** (2010) 012072 doi:10.1088/1742-6596/250/1/012072

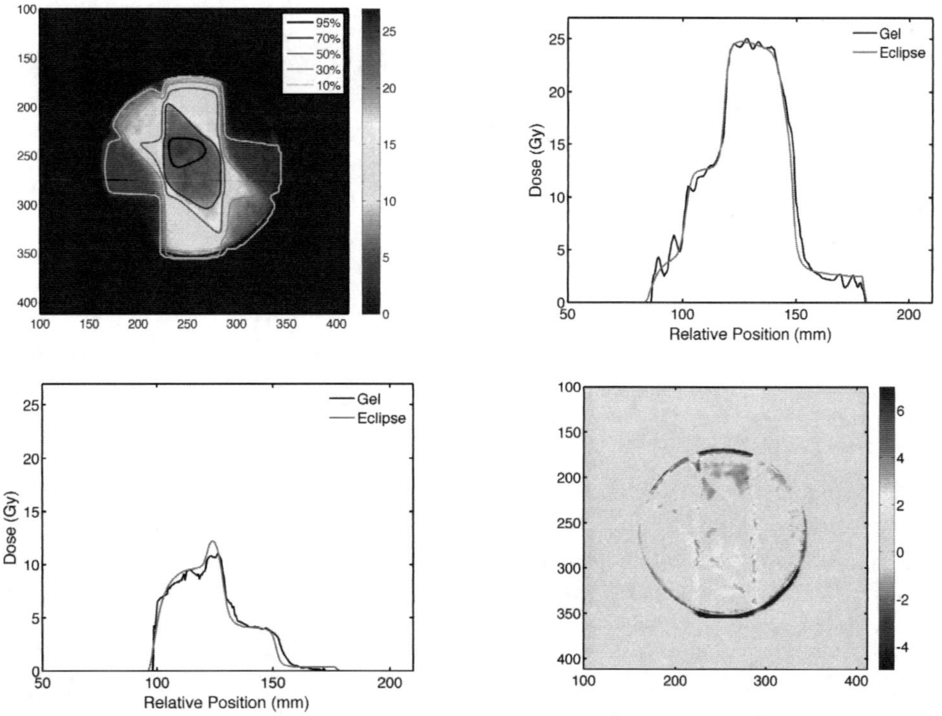

Figure 5: Top left: Gel image and treatment plan isodose overlay of 3-field irradiation. Gel image calibrated with calibration curve of figure 3. Top right: Representative profile through row 240 of gel image and treatment plan. Bottom left: Representative profile through column 200 of gel image and treatment plan. Bottom right: Dose difference map (gel image - treatment plan). Colour axis in Gy.

References

[1] V. I. Koeva, T. Olding, A. Jirasek, L. J. Schreiner, and K. B. McAuley. *Phys. Med. Biol.*, **54** 2009, pp 2779–90.

[2] A Jirasek, M Hilts, A Berman, and K B McAuley, *Phys. Med. Biol.* **54** 2009, pp 907–918.

[3] M. Hilts, A. Jirasek, and C. Duzenli, *Phys Med Biol* **49**, 2004, pp 2477–2490.

[4] M Hilts and A Jirasek, *Med. Phys.* **35**, 2008, pp1-12.

[5] C. Audet, M. Hilts, A. Jirasek, and C. Duzenli, *J. App. Clin. Med. Phys.* **3**, 2002, pp 110–118.

Polymer gel dosimetry using x-ray computed tomography: investigation of the effect of reconstruction technique

Jonathan Hindmarsh[1,2], Roger Fulton[1,3], Lyn Oliver[1,2] and Clive Baldock[1]

[1] Institute of Medical Physics, School of Physics, University of Sydney, Sydney, Australia
[2] Department of Radiation Oncology, Royal North Shore Hospital, Sydney, Australia
[3] Department of Medical Physics, Westmead Hospital, Sydney, Australia

E-mail: jhindmarsh@nsccahs.health.nsw.gov.au

Abstract. X-ray computed tomography (XCT) imaging of gel dosimeters is a promising technique that may make gel dosimetry more practical for medical physicists to use clinically. One of the issues limiting its use is the dose resolution of the system, which is determined by the system's noise and ability to resolve low contrast objects. Alternative reconstruction of CT data is shown to approximately halve the minimum detectable change in signal and increase, by up to double, the signal to noise ratio for small low contrast objects when compared with the vendor reconstruction. This has the potential to reduce the number of averages required in XCT gel dosimetry imaging protocols as well as improving dose resolution.

1. Introduction

The need for a suitable 3-D dosimetry tool for the verification of complex clinical techniques such as intensity modulated radiation therapy (IMRT), stereotactic radiation therapy (SRS and SRT) and volumetric modulated arc therapy (VMAT) has become much more evident in recent times. These techniques can produce highly conformal, localised dose distributions that need to be verified in three dimensions (3D). However the majority of methods used to verify these treatment techniques rely on one dimensional (e.g. ion chambers) or two dimensional (e.g. film) detectors [1, 2, 3]. Polymer gel dosimetry has been proposed and is now being used to clinically verify complex 3D dose distributions.

X-ray computed tomography (XCT) is a relatively new technique first proposed and demonstrated by Hilts et al in 2000 [4]. It has the potential to provide a practical solution of being more accessible and cheaper to use. Trapp [5] gave theoretical scan parameters appropriate for CT imaging of polymer gel dosimeters and the technical considerations involved in implementing XCT imaging of polymer gel dosimetry was discussed by Hilts [6].

There have been two areas of research for decreasing dose resolution and increasing sensitivity of the XCT gel dosimetry system. The first area is the use of spatial filters to remove some of the noise and improve the signal to noise ratio of the CT images by Hilts and Jirasek [7, 8]. The second area of research is investigating the change in the attenuation properties of polymer gels due to radiation exposure with the aim of increasing the Hounsfeld unit (HU) response per Gy. Trapp, et al [9, 10, 11, 12, 13, 14, 15, 16] have investigated the effect of gel composition on the radiation induced attenuation change in both normoxic and anoxic polymer gels. A summary of CT dose response for studied polymer gel dosimeters is given by Hilts [17].

The reconstruction algorithm, potential methods of maximising low contrast resolution and the sensitivity of the system have not been investigated in detail. This work compares images produced by the vendor reconstruction algorithm with images reconstructed by alternative methods.

2. Materials and Methods

2.1. Hardware

Two phantoms were used to investigate the reconstruction algorithms; the Siemens quality assurance (QA) phantom supplied with the CT scanner and a Catphan 600 phantom. The Siemens QA phantom contained a section of uniform density that was able to be extended off the end of the bed. The Catphan 600 phantom is made up of 4 modules designed to be used for quality assurance of x-ray CT scanners. The specific contents of each module are described in the Catphan 600 manual and is available from the manufacturer's website [18]. The specific modules used were:

- Module CTP404; relative electron density (ρ_e^r), slice width, circular symmetry and signal to noise ratio.
- Module CTP486; uniformity module,
- Module CTP515; sub-slice and supra-slice low contrast resolution,
- Module CTP528; high contrast spatial resolution,

The CT scanner used to acquire the raw data and vendor images was a Siemens Sensation 16 PET/CT scanner. The parameters used for scanning the phantoms were 120kV, 320mAs, 3mm slice thickness using a 16 slice helical technique with a feed/rotation of 6.6mm. The vendor reconstructed images were exported using the Dicom format while the raw projection data was exported in a Siemens proprietary format based on the Dicom format.

2.2. Software

The third party reconstruction software ran in IDL and was designed to reconstruct the raw CT transmission data contained in the Siemens format. The program reconstructed the transverse slices using either filtered back projection (FBP) with one of the filters described in [19] (analytic method) or a maximum likelihood gradient ascent algorithm, based on optimisation transfer [20, 21, 22] (iterative method). The basic reconstruction process was to (i) import projection data and extract necessary header information; (ii) extract fan beam sinogram over region to be reconstructed; (iii) convert fan beam to parallel beam sinogram (made up of 672 detectors by 1160 angles); (iv) reconstruct using analytic or iterative algorithm and (v) export

3. Results

Using the vendor reconstructions as a reference, the central slice from each of the Catphan modules and a uniform slice of the Siemens QA phantom were reconstructed. Seven sets were reconstructed using FBP, one for each of the available filters, and twelve sets were reconstructed using ML with the number of iterations set between 7 and 30 and the number of subsets set to either 20 or 40 (corresponding to 29 angles per subset and 58 angles per subset respectively). Because the sinogram was unable to be modified without affecting the resulting reconstruction quality, the images were reconstructed using all 672 detectors and 1150 angles producing reconstructions with dimensions 672×672 pixels. A representative image from each reconstruction method is displayed in figure 1 of the CTP515 low contrast module.

The reconstructed images were analysed in the way recommended in the Catphan manual [18]. The change in contrast with reducing distance between objects is known as the modulation transfer

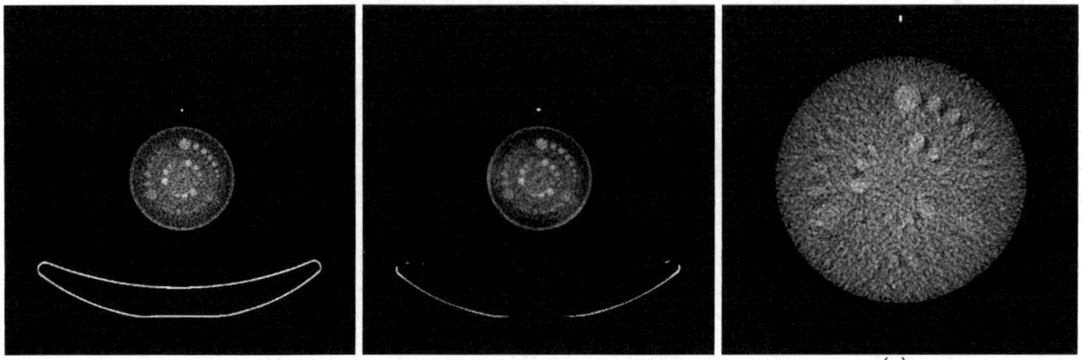

Figure 1. Reconstructions of the CTP515 module; (a) FBP using rectangular filter, (b) ML using 7 iterations and 40 subsets, (c) Siemens reconstruction using the H31s (standard) filter.

function (MTF) and was determined using the high contrast line pairs in module CTP528. Contrast is defined as

$$Contrast = \frac{ROI_{max} - ROI_{min}}{ROI_{max} + ROI_{min}} \tag{1}$$

where the region of interest (ROI) was defined by the top of the first peak and the top of the last peak. The MTF is shown in figure 2(b) which plots contrast verse line pairs per cm. Module CTP404 with its inserts of varying relative electron density ($\rho_e^r = 0.0 - 1.87$) was used to compare the response of the reconstructions to $\Delta\rho_e^r$ (figure 2(a)). The third party software produced reconstructions in units of attenuation per pixel, which needed to be converted to Hounsfeld units (HU) before being compared with the vendor reconstruction. This was performed using

$$HU = \frac{\mu_x - \mu_{H_20}}{\mu_{H_20}} \times 1000 \tag{2}$$

where μ_x is the pixel value of an unknown substance and μ_{H_2O} is the pixel value of water $\rho_e^r = 1.0$.

The low contrast module CTP515 (figure 1) contained six sets of objects with nominal contrast $0.3 - 1\%$ and diameters 2-15mm. None of the third party algorithms could adequately reconstruct ($SNR > 1.0$) the 0.3% contrast objects. However the softer algorithms (such as FBP with Hamming filter and lower iteration ML) were able to adequately reconstruct the larger 0.5% contrast objects and all were able to adequately reconstruct the 1% contrast objects down to 4mm diameter. In comparison the Vendor algorithm was only able to reconstruct the 1% contrast objects down to 9mm diameter with a $SNR > 1.0$.

Uniformity for all reconstruction methods was between $1 - 2\%$ with the variation explained by the characteristics of the different algorithms. Also there was very little difference between the uniformity of the Siemens QA phantom and the Catphan.

4. Discussion

The most obvious difference between the vendor and third party reconstructions is the size of the resulting images (compare figure 1(a) and (b) with figure 1(c)). This is because the third party reconstructions results in an image of the entire CT scanner field of view. This is unlike the Vendor reconstructions where cropping boundaries can be defined so that the resulting image is a 512×512 pixel reconstruction of the selected area only.

The MTF results (figure 2(b)) showed that both the vendor and third party reconstructions have the same limiting resolution of 6 line pairs per cm. However the vendor reconstruction does have

Journal of Physics: Conference Series **250** (2010) 012073 doi:10.1088/1742-6596/250/1/012073

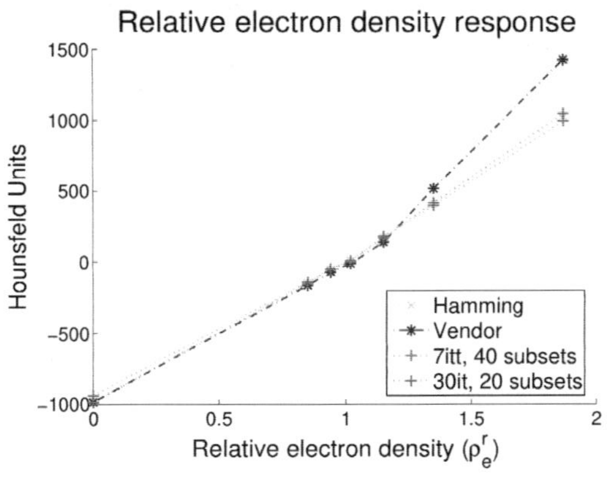

(a)

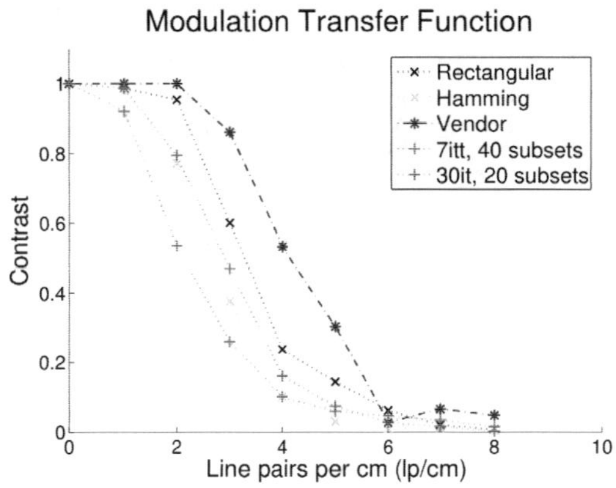

(b)

Figure 2. (a) Hounsfeld unit response to objects with different ρ_e^r. (b) Change in contrast with reducing separation

the advantage that for lp/cm >5 the contrast is greater than that produced by any of the third party reconstructions. This is a result of the expansion of the vendor reconstruction with the smaller pixels (0.42mm per pixel verse 0.75mm per pixel) able to better image the two objects. The third party reconstructions with the best MTF was the rectangular FPB reconstruction while the worst MTF was produced by the 7 iteration, 40 subset ML reconstruction. Both these results are expected since the rectangular FBP filter does not smooth the data, while the 7 iteration 40 subset ML reconstruction is only just convergent which does not produce good MTF but has benefits in low contrast situations (for example figure 1(b)).

Figure 2(a) shows the Hounsfeld unit response of the vendor and representative third party reconstructions. These are comparable along the lower half of the curve ($\rho_e^r \leq 1$) but are rather different in the upper half ($\rho_e^r > 1$) where the third party curves are more linear although two straight lines are still required to fit the data. Within the third party reconstructions the FBP filters all resulted in the same curve whereas the ML reconstructions resulted curves that differed by converged at higher iterations. With regards to the reconstruction of a uniform region, ignoring the differences between the algorithms, all the reconstruction give the same result.

The greatest impact of the third party reconstructions is in the reconstruction of the low contrast module (figure 1). This is where the third party reconstruction software has much to offer XCT gel dosimetry. Using the nominal contrast values the third party reconstructions have a minimum detectable change of $\sim 1\%$ compared with 2% for the vendor reconstruction for a single 3mm slice. At the same time the third party reconstruction is reconstructing the objects with up to double the SNR compared to the Vendor reconstruction.

The improvements above are seen best with the smooth FBP filters (for example Hann) and the just convergent ML algorithm (for example 7it with 40 subsets). Therefore using a smooth filtered FBP or just convergent ML algorithm the impact on XCT gel dosimetry is potentially two fold. The smaller minimum detectable change means that the number of images required for averaging purposes is reduced for the same endpoint. And increase SNR has the potential to cause an improvement in dose resolution.

5. Conclusion

Third party reconstruction of raw CT projection data has been shown to approximately halve the minimum detectable change in signal and increase by up to double the signal to noise ratio for small low contrast objects when compared with Vendor reconstructions. Taking these aspects into consideration, it may be possible to improve x-ray CT imaging of the irradiated gel dosimeter to such an extent that it may be possible to use it routinely in radiotherapy. The gel dosimeter would then be able to provide 3D dose evaluation of complex dose distributions created by the new complex modulated intensity x-ray techniques.

Acknowledgments

The authors thank Siemens Medical Solutions Molecular Imaging (Knoxville, TN) for providing the software used to access the raw spiral CT projection data, and Professor Johan Nuyts of Katholieke Universiteit Leuven for the third party reconstruction software used in this work.

References

[1] Geoffrey S I 2006 *Journal of Physics: Conference Series* 108
[2] Schreiner L J 2006 *Journal of Physics: Conference Series* 1
[3] Schreiner L J 2009 *Journal of Physics: Conference Series* 012001
[4] Hilts M, Audet C, Duzenli C and Jirasek A 2000 *Physics in Medicine and Biology* **45** 2559–2571
[5] Trapp J V, Michael G and Baldock C 2001b *DOSGEL 2001 Proc. 2nd Int. Conf. on Radiation Therapy Gel Dosimetry* ed Baldock C and Deene Y D (Queensland University of Technology, Brisbane, Australia)
[6] Hilts M, Jirasek A and Duzenli C 2005 *Physics in Medicine and Biology* **50** 1727–1745
[7] Hilts M and Duzenli C 2004a *Medical Physics* **31** 39–49
[8] Jirasek A, Matthews Q, Hilts M, Schulze G, Blades M W and Burner R F B 2006b *Physics in Medicine and Biology* **51** 2599–2617
[9] Trapp J V, Back S A J, Lepage M, Michael G and Baldock C 2001a *Physics in Medicine and Biology* **46** 2939–2951
[10] Trapp J V, Michael G, de Deene Y and Baldock C 2002 *Physics in Medicine and Biology* **47** 4247–4258
[11] Brindha S, Venning A J, Hill B and Baldock C 2004 *Physics in Medicine and Biology* **49** N353–N361
[12] Hilts M, Jirasek A and Duzenli C 2004b *Physics in Medicine and Biology* **49** 2477–2490
[13] Venning A J and et al 2004 *Journal of Physics: Conference Series* **3** 155
[14] Hill B, Venning A and Baldock C 2005a *British Journal of Radiology* **78** 623–630
[15] Jirasek A, Hilts M, Shaw C and Baxter P 2006a *Physics in Medicine and Biology* **51** 1891–1906
[16] Baxter P, Jirasek A and Hilts M 2007 *Medical Physics* **34** 1934–1943
[17] Hilts M 2006 *Journal of Physics: Conference Series* 95
[18] 2009 Catphan 500 and 600 manual Tech. rep. The Phantom Laboratory
[19] Buzug T M 2008 *Computed Tomography: From Photon Statistics to Modern Cone-Beam CT* (Berlin: Springer)
[20] Nuyts J, Dupont P and Mortelmans L 1996 *2nd IEEE workshop on computer intensive methods in control and signal processing: can we beat the curse of dimensionality?* ed Warwick K and Karny M (Prague, Czech Republic)
[21] Nuyts J, Man B D, Dupont P, Defrise M, Suetens P and Mortelmans L 1998 *Physics in Medicine and Biology* **43** 729–737
[22] Fessler J, Ficaro E, Clinthorne N and Lange K 1997 *IEEE Transactions on Medical Imaging* **16** 166–175

Journal of Physics: Conference Series **250** (2010) 012074 doi:10.1088/1742-6596/250/1/012074

Dose-rate dependence of PAGAT polymer gel dosimeter evaluated using X-ray CT scanner

P. Sellakumar[1], E. James Jebaseelan Samuel[2] and D. Senthil Kumar[2]

[1]HCG - Bangalore Institute of Oncology, Bangalore, Karnataka, India.
[2]Photonics Division, School of Advanced Sciences, VIT University, Vellore, Tamilnadu, India.

E-mail: psellakumar@rediffmail.com

Abstract In recent years, X-ray CT imaging of polymer gel dosimeter has been applied to direct measurement of 3D dose distributions. This work presents the dependence of PAGAT polymer gel dosimeter response to different beam dose rate using X-ray CT scanner as an evaluation tool. The normoxic PAGAT polymer gel was prepared on bench top and irradiated to different doses at different dose-rate ranging from 0.5 to 5 Gy/min. After irradiation the dosimeter samples were imaged with X-ray CT scanner using appropriate scanning protocol. The results showed that PAGAT gel dosimeter formulations investigated in this study has the smallest dose rate dependence and the change in dose sensitivity amounts to 2 % over the dose rate interval studied.

1. Introduction

In polymer gel dosimeter, the free radical induced polymerization consists of free radical generation, chain initiation, propagation and termination. The initiation of polymerization proceed through the reaction of monomer with radiation induced water radicals and the rate of generation of free radicals dependent upon the irradiation dose rate [1,2]. Higher rates of radical generation lead to higher concentration of radicals. Propagation proceeds through the formation of long polymer chain, rate of which is proportional to the radical concentration. The chain termination involves the combination of two free radicals. This behavior of dependence of propagation and termination rate on the radical concentration makes incident beam dose rate dependent polymer gel dosimeter. The dependence of the polymer gel sensitivity depends on the polymer gel constituents and concentration of gelatin [1,3-6].

Studies were carried out by Maryanski *et al* [7,8] and Novotny *et al* [9] on BANG-2 (BIS acrylamide nitrogen gelatin) gel and concluded that there is no energy and dose rate dependence of dose response for electrons in the range 2-15 MeV and the dose rate in the range 20-400 cGy/min. Ibbott *et al* [10] and Massillon *et al* [11] also reported that, the dose response of the BANG gel is independent of both dose rate and modality of the beam used. De Deene *et al* [1] studied the dose rate dependence of normoxic polymer gel dosimeters and reported that the normoxic PAG (poly acrylamide gelatin) polymer gel dosimeter has the smallest dose rate dependence while the normoxic MAG (methacrylic acid gelatin) dosimeter has the greatest dose rate dependence. Bayreder *et al* [12] stated in their study that tetrakis hydroxymethyl phosphonium chloride (THPC) based normoxic methacrylic acid polymer gel shows the insignificant dose rate dependence at the lowest dose level (D=2 Gy) and considerable dose rate effect arises for all other doses with a maximum impact at the upper limit of the linear dose range. The objective of this work is to evaluate the dependence of the PAGAT (poly acrylamide gelatin and tetrakis hydroxy methyl phosphoniun chloride) polymer gel dosimeter response to different dose rate. The response of the polymer gel dosimeter is determined by calculating the slope of the linear portion of dose-response curve.

2. Materials and Method

Three batches of PAGAT normoxic polymer gel dosimeter were prepared following the experimental procedure described elsewhere [13,14]. In summary all the dosimeters were prepared under normal atmospheric conditions from acrylamide (Sigma Aldrich), N, N^1- methylene-bis-acrylamide (BIS) (Sigma Aldrich) and gelatin (300 bloom, Sigma Aldrich). The proportions of chemical compositions (by weight) were 3.5 % BIS, 3.5 % acrylamide, 5 % gelatin and 88 % triple distilled water. During the manufacture process, gelatin was added to water at room temperature and allowed to soak for 15 minutes before heating to 50° C. While continuously stirring, acrylamide and BIS were subsequently added and stirred until complete dissolution was achieved. After the gelatin-monomer mixture was cooled down to 37° C, 10 mM of THPC was added as anti oxidant and transferred to small plastic cylindrical vials of diameter 2.5 cm and length 4 cm. All the vials were sealed with nylon lined top caps to minimize further oxygen contamination. The vials were then kept in refrigerator and left to set for 3 hours prior to irradiation.

The PAGAT gel filled vials were placed one at a time in a tissue equivalent wax phantom (20 x 22 x 22 cm^3). For the investigation of dose rate dependence of PAGAT gel dosimeter, irradiations were performed at different dose rate ranged from 0.5 to 5 Gy/min with Siemens Primus linear accelerator. First batch of gel dosimeter containing 180 vials were irradiated to the dose range of 1-30 Gy with 15 MV X-ray photon beam for each dose rate mentioned above. The second batch containing 34 vials irradiated by 6 MV X-ray beam with doses in the range of 1-30 Gy for two different dose rate of 0.5 and 2 Gy/min. The third gel batch containing 34 vials were irradiated to the dose ranging from 1-30 Gy at two different dose rate (0.74 and 1.23 Gy/min) with ^{60}Co photon beam from Alcyon-II cobalt unit. All the irradiations were performed with parallel opposed radiation beam with collimator size of 10 x 10 cm^2 so that homogeneous dose distribution was obtained across the dosimeter vials, and one gel dosimeter vial was kept unirradiated. Earlier the irradiation, absorbed dose and dose rate at the center of the dosimeter vial were measured with 0.13 cc compact chamber (Wellhofer-Scanditronix) and DOSE-1 electrometer (Wellhofer-Scanditronix) by mimicking the same set up.

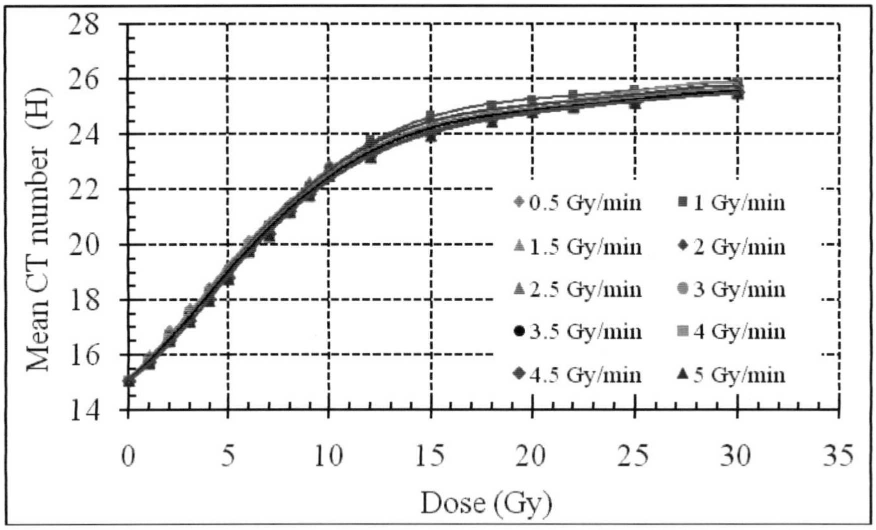

Figure 1. Dose response curves of PAGAT polymer gel dosimeters irradiated with 15 MV X-ray beam at different dose rate.

X-ray CT imaging was performed on all the dosimeters vials after 24 hours of irradiation using a Siemens Somatom Emotion CT scanner. All the dosimeter vials were kept at scanner room for about 2 hours to ensure the temperature equilibrium. The dosimeters vials were imaged with 130 kV tube voltage, 200 mA tube current, 1.5 s exposure time and slice thickness of 5 mm [15,16]. The scan was repeated 25 times and averaged together. ROI of area 10 mm^2 was drawn at the center of each vial and mean CT number ($\overline{N}_{CT}$) was extracted.

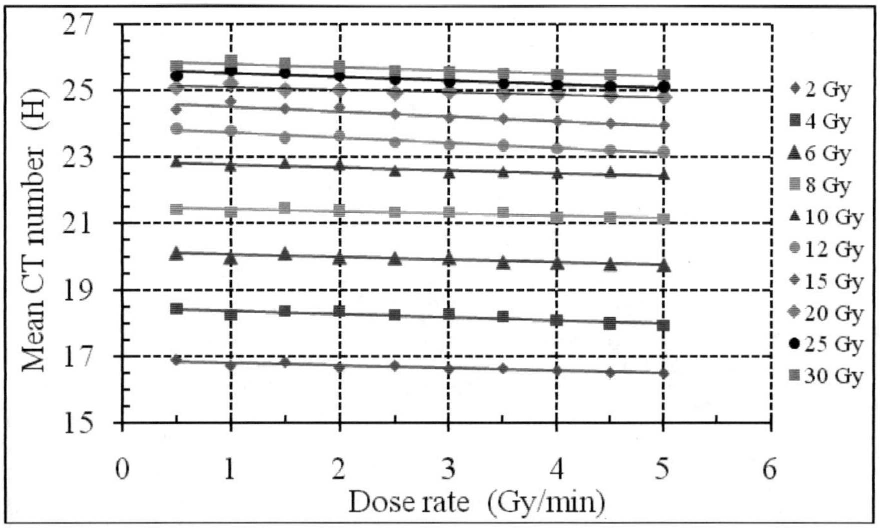

Figure 2. Mean CT number response plots of PAGAT polymer gel dosimeters as a function of dose rate irradiated with 15 MV X-ray beam to different doses.

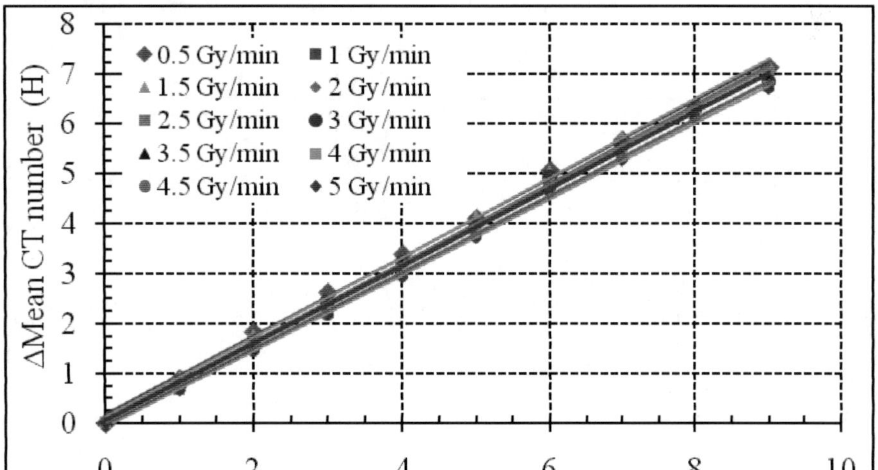

Figure 3. Linear part of dose response curves of PAGAT polymer gel dosimeters irradiated with 15 MV X-ray beam at different dose rate.

3. Results and Discussion

The response of PAGAT polymer gel dosimeter as a function of dose in the range of 0-30 Gy for different dose rate using 15 MV X-ray beam are displayed in figure 1 and figure 2. In the range 0-9 Gy dose response curve could be fitted by linear function (figure 3). Dose rate dependence of PAGAT gel dosimeters was examined by comparing the dose response determined at 2 Gy/min. The dose response curves obtained at 0.5 Gy/min and 5 Gy/min were almost identical to that obtained at 2 Gy/min. This study shows that PAGAT gel dosimeter has the smallest dose rate dependence. The change in dose sensitivity amounts to 2 % over the dose rate interval studied and the results were comparable to the previous study [17]. Figure 4 presents the dose response curves of PAGAT gel irradiated with 6 MV X-ray beam and ^{60}Co beam at different dose rates. These dose response curves are similar to that obtained with 15 MV X-rays at 2 Gy/min dose rate. Dosimeter sensitivity as a slope of linear portion (0-9 Gy) of dose response curve was also investigated as a function of dose rate for different photon beams and shown in figure 5. The dosimeter sensitivity at 0.5 Gy/min was 0.786 HGy^{-1} and at 5 Gy/min was 0.764 HGy^{-1}. The standard deviation between the dosimeter sensitivity was 0.012.

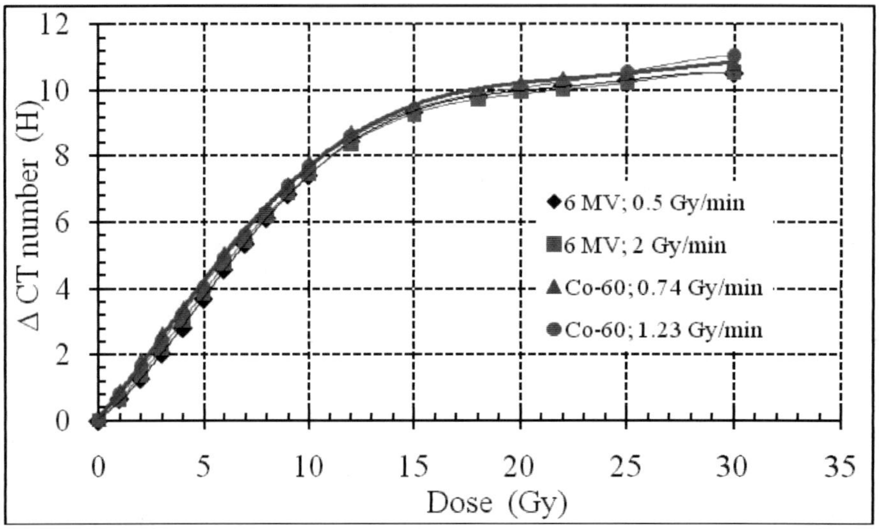

Figure 4. Dose response curves of PAGAT polymer gel dosimeters irradiated with 6 MV X-ray beam and Co-60 beam at different dose rates.

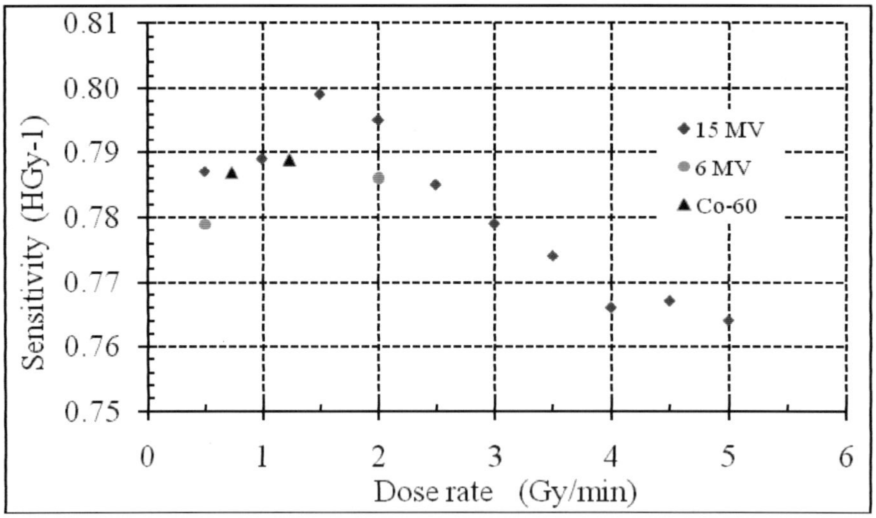

Figure 5. Sensitivity as a slope derived from the linear part of dose response curves of PAGAT polymer gel as a function of different dose rate.

Journal of Physics: Conference Series **250** (2010) 012074 doi:10.1088/1742-6596/250/1/012074

4. Conclusion

The results of this study showed that PAGAT polymer gel dosimeter formulation investigated with X-ray CT scanner exhibited trend in polymer gel sensitivity dependence on dose rate range 0.5-5 Gy/min and was clinically insignificant. But this very small dose rate dependence of PAGAT gel sensitivity might lead to potential error while using PAGAT gel for absolute dosimetry.

References

[1] De Deene Y, Vergote K, Claeys C and De Wagter C 2006 The fundamental radiation properties of normoxic polymer gel dosimeters: a comparison between a methacrylic acid based gel and acrylamide based gels *Phys. Med. Biol.* **51** 653-673

[2] Jirasek A I 2006 Experimental investigations of polymer gel dosimeters *J. Phys. Conf. Series.* **56** 23-34

[3] McAuley K B 2006 Fundamentals of polymer gel dosimeters *J. Phys. Conf. Series.* **56** 35-44

[4] Fuxman A M, McAuley K B and Schreiner L J 2005 Modeling of polyacrylamide gel dosimeters with spatially non-uniform radiation dose distributions *Chem. Eng. Sci.* **60** 1277-94

[5] Lepage M, Jayasakera P M, Back S A J and Baldock C 2001a Dose resolution optimization of polymer gel dosimeters using different monomers *Phys. Med. Biol.* **46** 2665-2680

[6] Lepage M, Whittaker A K, Rintoul L, Back S A J and Baldock C 2001b Modelling of post-irradiation events in polymer gel dosimeters *Phys. Med. Biol.* **46** 2827-2839

[7] Maryanski M J, Ibbott G S, Eastman P, Schulz R J and Gore J C 1996a Radiation therapy dosimetry using magnetic resonance imaging of polymer gels *Med. Phys.* **23** 699-705

[8] Maryanski M J, Zastavker Y Z and Gore J C 1996b Radiation dose distributions in three dimensions from tomographic optical density scanning of polymer gels: II Optical properties of the BANG polymer gel Phys. *Med. Biol.* **12** 2705-17

[9] Novotny Jr J, Spevacek V, Dvorak P, Novotny J and Cechak T 2001a Energy and dose rate dependence of BANG-2 polymer-gel dosimeter *Med. Phys.* **28** 2379-2386

[10] Ibbott G S, Maryanski M J, Eastman P, Holcomb S D, Zhang Y, Avison R G, Sanders M and Gore J C 1997 Three-dimensional visualization and measurement of conformal dose distributions using magnetic resonance imaging of BANG polymer gel dosimeters *Int. J. Radiat. Oncol.* **38** 1097-1103

[11] Massillon-JL G, Minniti R, Soares C G, Maryanski M J and Robertson S 2010 Characteristics of a new polymer gel for high-dose gradient dosimetry using a micro optical CT scanner *Appl. Radiat. Isot.* **68** 144-154

[12] Bayreder C, Georg D, Moser E and Berg A 2006 Basic investigations on the performance of a normoxic polymer gel with tetrakis-hydroxy-methyl-phosphonium chloride as an oxygen scavenger: Reproducibility, accuracy, stability and dose rate dependence *Med. Phys.* **33** 2506-2518

[13] Brindha S, Venning A J, Hill B and Baldock C 2004 Experimental study of attenuation properties of normoxic polymer gel dosimeters *Phys. Med. Biol.* **49** N353-N361

[14] Venning A J, Hill B, Brindha S, Healy B J and Baldock C 2005a Investigation of the PAGAT polymer gel dosimeter using magnetic resonance imaging *Phys. Med. Biol.* **50** 3875-3888

[15] Brindha S, Paul B R and Baldock C 2006 Optimization of the imaging protocol of an X-ray CT scanner for evaluation of normoxic polymer gel dosimeters *J. Med. Phys.* **31** 72-77

[16] Sellakumar P, Samuel E J J and Supe S S 2007 Investigation of optimal scanning protocol for X-ray computed tomography polymer gel dosimetry *Nucl. Instrum. Methods. Phys. Res. B* **264** 395-399

[17] Senden R J, De Jean P, McAuley K B and Schreiner L J 2006 Polymer gel dosimeters with reduced toxicity: a preliminary investigation of the NMR and optical dose-response using different monomers *Phys. Med. Biol.* **51** 3301-3314

Journal of Physics: Conference Series **250** (2010) 012075 doi:10.1088/1742-6596/250/1/012075

Investigation of ultrasonic properties of MAGIC gels for pulse-echo gel dosimetry

T J Atkins[1,2], V F Humphrey[2], F A Duck[1] and M A Tooley[1]

[1]Department of Medical Physics and Bioengineering, Royal United Hospital, Combe Park, Bath BA1 3NG, UK

[2]Institute of Sound and Vibration Research, University of Southampton, Southampton, SO17 1BJ, UK

Email: timothy.atkins@nhs.net

Abstract. Ultrasound has been proposed as an alternative 3D method to read out dosimetric gels. The ultrasonic properties of MAGIC gel are investigated in relation to producing a dosimetry system based on a pulse-echo technique. The sound speed, attenuation coefficient, and density of MAGIC gel were measured over a range of temperatures and doses. A non-linear relationship between ultrasonic attenuation and dose was observed, with the attenuation below 10 Gy being approximately constant. The sound speed was 1550.3 ± 1.5 m s^{-1} at 25 °C with a dose sensitivity of 0.14 ± 0.03 m s^{-1} Gy^{-1}; both properties changed with temperature and were 1535.4 ± 0.8 m s^{-1} and -0.08 ± 0.01 m s^{-1} Gy^{-1} respectively at 15 °C. The density also varied with temperature and dose, and was 1028.9 ± 0.3 kg m^{-3} with a dose sensitivity of 0.178 ± 0.004 kg m^{-3} Gy^{-1} at 26 °C. The characteristic acoustic impedance of MAGIC gel was calculated to be 1.596×10^6 kg m^2 s^{-1} at 0 Gy and 25 °C. The dose sensitivities measured indicate that changes to the characteristic acoustic impedance caused by irradiation are small, and therefore a reflector that has a similar characteristic acoustic impedance is required to maximise sensitivity of the pulse-echo technique. A suitable material has been developed and the reflection characteristics of the interface between MAGIC gel and the reflector are also temperature dependent. It is concluded that temperature management will be central to the development of any practical pulse-echo dosimetry system.

1. Introduction and background

Ultrasound has been proposed as a method of measuring the dose dependent changes induced by irradiation in polymer gels. Mather et al. [1,2] and Crescenti et al. [3] have both published data on the dose dependent changes that occur in polymer gels. Mather et al. proposed using time-of-flight tomography [2] to map radiation dose distributions, while Crescenti et al. [3] investigated the use of ultrasonic elastography. This paper records an investigation of the ultrasonic properties of a MAGIC gel as part of the development of a dosimeter based on the pulse-echo method of dose readout. The concept is to use a composite material, in which one constituent is more dose dependant than the other, resulting in scattering or reflections that are dose dependant and can be read with a conventional ultrasound scanner.

To produce an ultrasonic pulse-echo dosimetry system it is important to determine the characteristic acoustic impedance (Z) of the gel material, that is the product of the density (ρ) and speed of sound (c_0). The acoustic attenuation coefficient of the gel is also important as this will affect the received signal amplitude on which a pulse-echo dosimeter will be based. A pulse-echo readout system will require a component material that is not dose dependant; the acoustic properties for this material must also be determined.

2. Ultrasonic speed of sound and attenuation

For this study MAGIC gel, as developed by Fong el al. [4], was poured into specially constructed moulds of different thicknesses. A buffered insertion technique, with two 3.5 MHz transducers co-axially mounted in a temperature controlled water bath, was used to measure the ultrasonic speed of sound and attenuation coefficient.

2.1. Speed of sound

The speed of sound was measured at three different temperatures to determine the sensitivity of the ultrasonic properties to temperature. The measurements were performed by determining the time of flight of an ultrasound pulse through the samples with pure water being used as the reference medium. Measurements were made on three separate samples of different thicknesses (1, 2 and 3 cm) for each dose level; these were used to calculate the mean speed of sound and its standard deviation.

Figure 1 shows the variation in speed of sound with dose. The speed of sound for unirradiated gel decreased from 1550.3 ± 1.5 (1 S.D.) m s^{-1} at 25 °C to 1535.4 ± 0.8 m s^{-1} at 15 °C. The speed of sound appears to vary approximately linearly with dose over the range measured. At 25 °C the speed of sound increases with a dose sensitivity of 0.14 ± 0.03 m s^{-1} Gy^{-1}, whereas at 15 °C the speed of sound decreases with dose, with a sensitivity of -0.08 ± 0.01 m s^{-1} Gy^{-1}. The uncertainties in these slopes are quoted as 1 standard error obtained from the least-squares regression analysis.

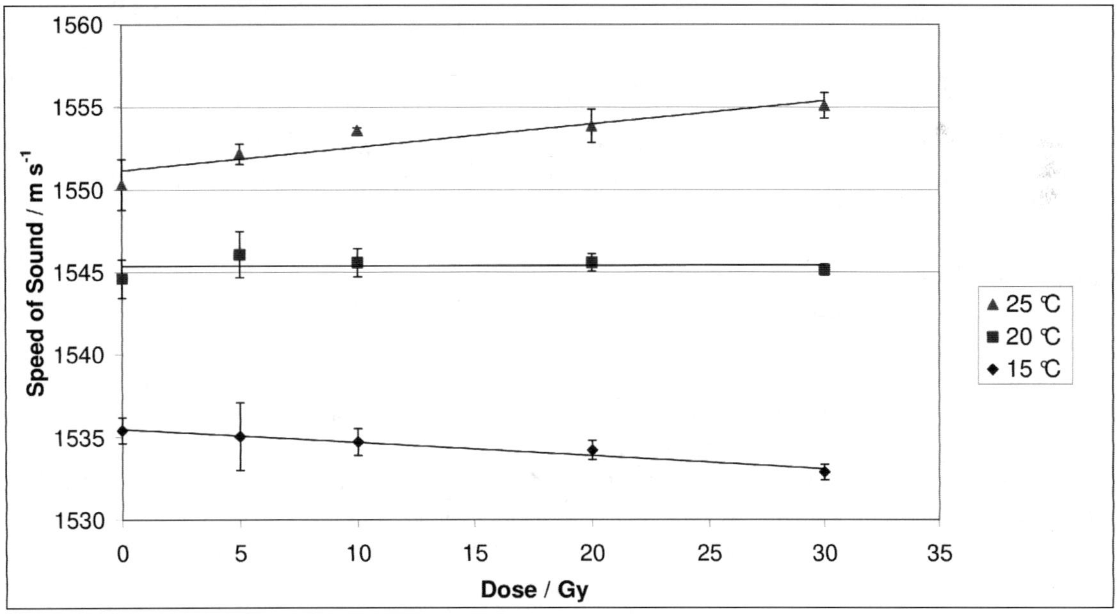

Figure 1. Variation in speed of sound of MAGIC gel with dose at three temperatures (see key). Error bars show the uncertainty (1 S.D.) calculated from the measurements for the different samples.

2.2. Attenuation

Tone burst pulses were used to measure the attenuation coefficient. A range of frequencies was selected around the centre frequency of the transducers. Attenuation was measured using a buffered insertion technique. The attenuation coefficient was calculated from the gradient of the graph of transmitted signal level as a function of thickness for three samples irradiated to the same dose. This technique removes the effect of window attenuation.

Figure 2 shows the variation of attenuation coefficient of MAGIC gels with dose at the discrete frequencies used for measurement. The measurements were made at a temperature of 25 °C. There appears to be little change in attenuation coefficient for each frequency below 10 Gy irradiated dose, whereas above 10 Gy the attenuation coefficient appears to increase with dose. The attenuation coefficient is greater for higher ultrasonic frequencies.

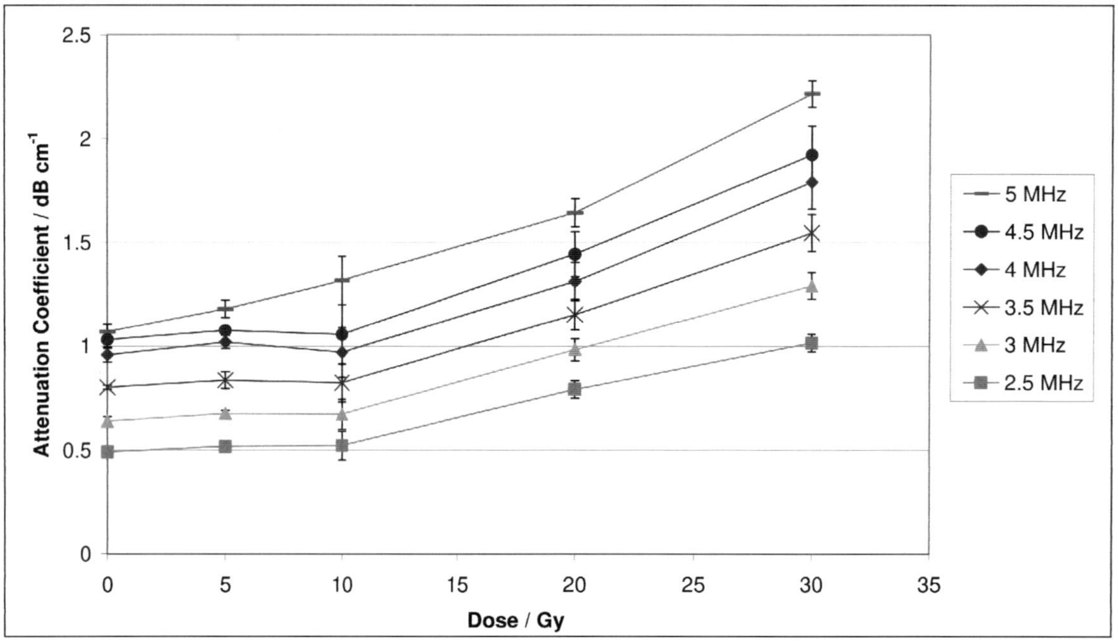

Figure 2. Attenuation coefficient measured at discrete frequencies (see key) for MAGIC gel irradiated to different doses at 25 °C. Error bars show the uncertainty (1 standard error) calculated from regression analysis.

3. Density

To determine the dose dependency of the density of MAGIC gel, some gel was poured into flexible-walled sachets. The sachets were irradiated to a range of doses from 0 to 30 Gy. Subsequently the densities of the MAGIC gel sachets were measured at 3 different temperatures. Calculation of the densities was based on measurements of the mass of the samples and the volume of fluid displaced by the samples when they were immersed in water.

Figure 3 shows the variation in density with dose. The density of unirradiated MAGIC gel increased from 1028.9 ± 0.3 kg m^{-3} at 26 °C to 1033.3 ± 0.3 kg m^{-3} at 14.5 °C. The density increases linearly with dose over the range measured with a dose sensitivity of 0.178 ± 0.004 kg m^{-3} Gy^{-1}.

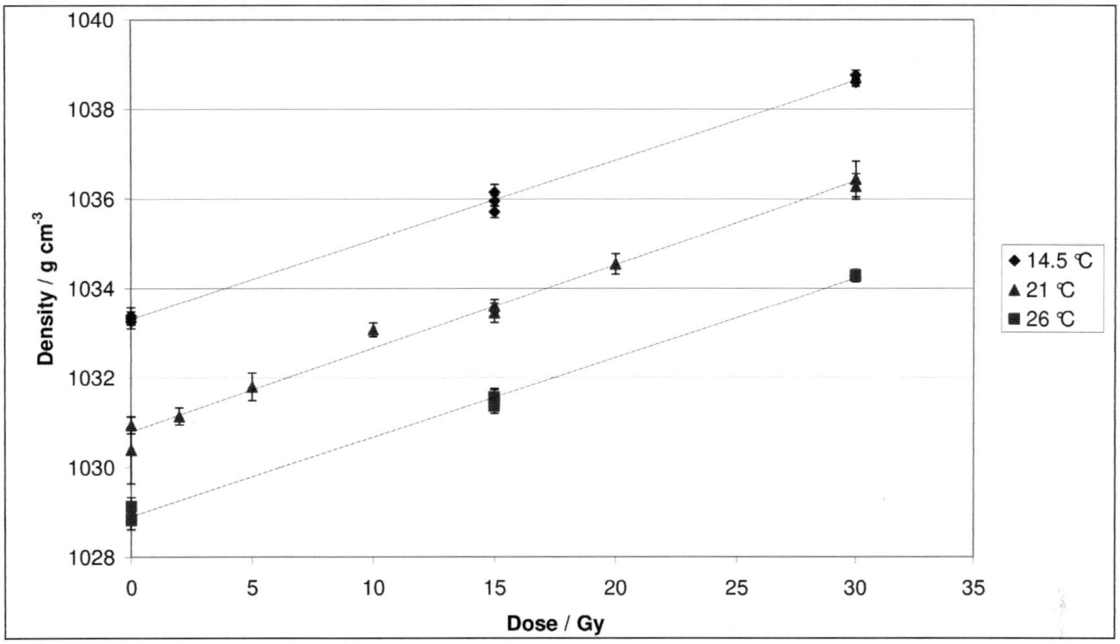

Figure 3. Variation of density of MAGIC gel with dose for three different temperatures (see key).

4. Acoustic reflector

The measurements of density and speed of sound can be used to calculate the characteristic acoustic impedance of MAGIC gel ($Z = \rho \cdot c_0$) and how this changes with radiation dose. The characteristic acoustic impedance is calculated to be 1.596×10^6 kg m^2 s^{-1} at 0 Gy and 25 °C, while it is 1.586×10^6 kg m^2 s^{-1} at 15 °C. The dose sensitivities measured indicate that changes to the characteristic acoustic impedance caused by irradiation are small. The dose sensitivity over the range of temperatures measured increases from 193 kg m^2 s^{-1} Gy^{-1} at 15 °C to 424 kg m^2 s^{-1} Gy $^{-1}$ at 25 °C. To produce an acoustic reflection that is sensitive to these small changes in acoustic impedance requires a reflector with a similar acoustic impedance to accentuate the relative change in impedance.

A polymeric material has been developed with an acoustic impedance matched to the MAGIC gel. Figure 4 shows the measured variation in reflection amplitude of an interface between the acoustic reflector and MAGIC gel averaged over regions irradiated to 0, 10, 20 and 30 Gy as a function of temperature. Each dose follows a similar pattern: starting at the lowest temperature, the reflected amplitude falls to a minimum and then increases as the temperature of the sample rises. The temperature at which the minimum occurs increases with irradiated dose. At this temperature the acoustic matching between the gel and backing is optimum. This suggests that in order to produce a practical pulse-echo dosimetry system based on this acoustic reflector a suitable temperature would need to be identified at which the reflection amplitudes showed the greatest proportional change over the dose range of interest.

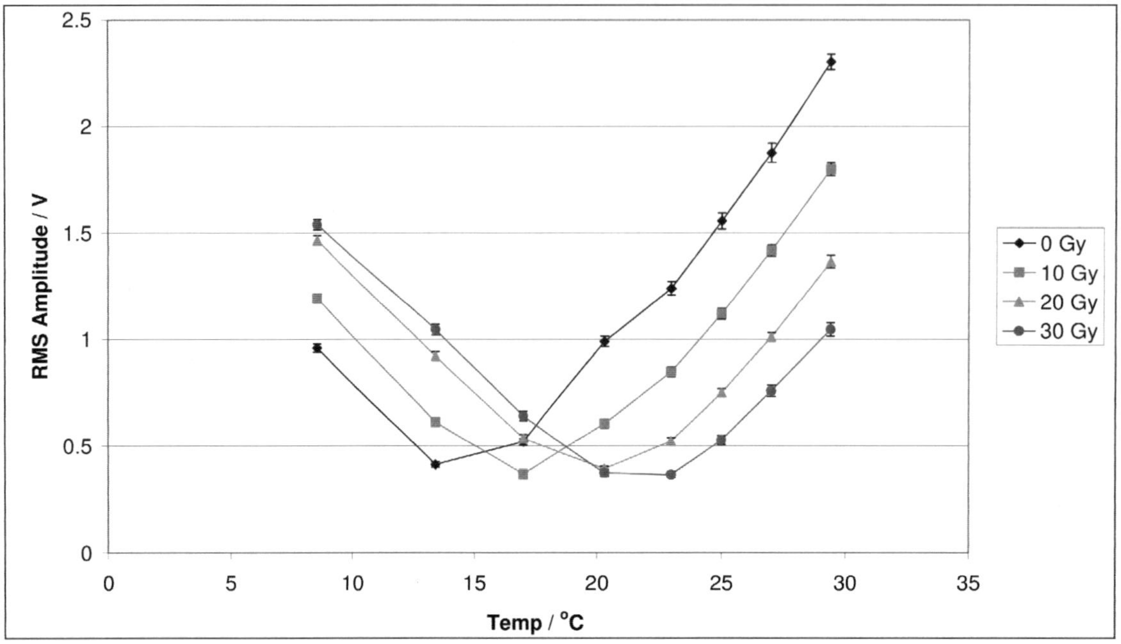

Figure 4. Temperature and dose dependence of measured acoustic reflection amplitude for doses between 0 and 30 Gy (see key).

5. Conclusions

Selected acoustic properties of MAGIC gel have been determined by measurements with the intention of producing an ultrasound backscatter based gel dosimetry system. The variation in the acoustic properties with dose and temperature have been investigated and reported. These show for the first time the variation in sound speed with temperature, and density with dose and temperature. Also the sign of the gradient of the dose dependency of sound speed changes with temperature. The density increases linearly with dose, and the gradient of this dose dependency is invariant with temperature. A material for use as a backscatterer has been developed and the acoustic reflection amplitude from a plane surface of this material has been investigated for temperature and dose response. The reflection coefficient goes through a minimum at a dose-specific temperature; the temperature at which this minimum occurs increases with dose.

The variation in the acoustic properties with temperature indicate that temperature management will be central to the development of a successful pulse-echo dosimetry system. Any such system will require a temperature dependent calibration. To give maximum sensitivity a preferred temperature may be identified.

Acknowledgements

This work was funded under the UK Department of Health NEAT programme project J008.

References

[1] ML Mather, Y De Deene, AK Whittaker, GP Simon, R Rutgers and C Baldock, 2002 *Phys. Med. Biol.*, **47**, 4397–4409.
[2] ML Mather and C Baldock, 2003 *Med. Phys.*, **30**, 2140–2148.
[3] RA Crescenti, JC Bamber, NL Bush and S Webb, 2009 *Phys. Med. Biol.*, **54**, 843–857.
[4] PM Fong, DC Keil, MD Does, and JC Gore, 2001 *Phys. Med. Biol*, **46**, 3105–3113.

Journal of Physics: Conference Series **250** (2010) 012076 doi:10.1088/1742-6596/250/1/012076

Improvement in the accuracy of polymer gel dosimeters using scintillating fibers

Nicolas M. Tremblay [1, 2], Vincent Hubert-Tremblay [2], Rachel Bujold [2], Luc Beaulieu [3], Martin Lepage [1]

1- Centre d'imagerie moléculaire de Sherbrooke, Université de Sherbrooke, Sherbrooke, QC, Canada

2- Service de radio-oncologie, Centre hospitalier universitaire de Sherbrooke, Sherbrooke, QC, Canada

3- Département de physique, de génie physique, et d'optique, Université Laval, Québec, QC, Canada and Département de radio-oncologie, Hôtel-Dieu de Québec, Centre hospitalier universitaire de Québec, QC, Canada

Martin.Lepage@USherbrooke.ca

Abstract. We propose a novel method for the absolute calibration of polyacrylamide gel (PAG) dosimeters with one or more reference scintillating fiber dosimeters inserted inside the gel. Four calibrated scintillating fibers were inserted into a cylindrical glass container filled with a PAG dosimeter irradiated with a wedge filtered 6 MV photon beam. Calibration curves using small glass vials containing the same gel as the cylindrical containers were used to obtain a first calibration curve. This calibration curve was then adjusted with the dose measured with one of the scintillating fibers in a low gradient part of the field using different approaches. Among these, it was found that a translation of the gel calibration curve yielded the highest accuracy with PAG dosimeters.

1. Introduction

The possibility of inserting scintillating dosimeters inside gel dosimeters has been suggested before [1]. We extended this work by evaluating different procedures for calibrating the gel dosimeters using the information provided by scintillating fibers as reference dose points.

2. Materials and Methods

2.1. Scintillating fiber dosimeters setup and calibration

The scintillating fiber dosimeters consisted of a 2 mm long by 1 mm in diameter BCF-12 scintillating plastic fiber (peak 435 nm). The scintillators were coupled to plastic optical fibers (length 13 m, diameter 1 mm, ESKA™ Premier polyethylene jacketed, Mitsubishi). The surface of the scintillating end of the optical fibers was manually polished and then glued using cyanoacrylate. Finally, the unjacketed end of the fibers were painted black using aerosol paint, dried at least 24 hours and inserted into a gel dosimeter. The scintillating dosimeter is thus entirely made of plastic and is considered almost water equivalent [2].

The other end of the optical fibers was held in front of a color CCD camera (Apogee Alta U2000c) inside an aluminum box following the design of Lacroix *et al.* [3]. This box enabled the precise alignment of multiple fibers in front of the CCD camera with the possibility of removing and accurately repositioning the fibers in front of the camera, thus retaining the fiber calibration. The box was placed behind the maze in the irradiation bunker away from the radiation source. No measurable noise appeared during irradiation such that filtration of the optical signal was not needed.

Dosimeters were placed in front a 6-MV photon beam (300 MU/min, Siemens Oncor Impression linac). The daily machine output was verified before every experiment using an ionization chamber in a 10 ×10 cm² square field in plastic water. Five fibers were irradiated in a water tank in a 10 × 10 cm² square field and in a 40 × 40 cm² square field with the fiber rolled in the field to maximize the production of Cerenkov light within the fibers following a published methodology [4]. The green and blue channels of the CCD camera were exploited for chromatic discrimination [5,6] of the signal such that calibration factors could be found for the subtraction of Cerenkov light. Each irradiation was repeated 5 times and the CCD pixels corresponding to each fiber were averaged to improve the signal-to-noise ratio. Black noise was removed by collecting data in the absence of radiation; the data were averaged and subtracted from the data acquired during irradiation. After completion of these calibration steps, the extremity of each fiber was placed into a custom made closed plastic box to avoid dust deposition that could alter the light transmission properties.

2.2. Gel mixing

Polyacrylamide gel (PAG) dosimeters were mixed in an oxygen-free environment. Gelatin (5%) was slowly mixed in deionized water (89%) at room temperature and then heated to 45°C in approximately 10 min with a hot plate and stabilized at this temperature. When the gelatin was completely dissolved, N,N'–methylenebisacrylamide (bis, 3%) and 3 acrylamide (aa, 3%) were added to the mixture and stirred until complete dissolution.

A borosilicate cylinder (height of 19.7 cm and inner diameter of 5.7 cm) was constructed with 4 entry ports placed at 1.2, 2.9, 5.5, and 9.2 cm from the bottom of the cylinder. Rubber tubes were connected to the ports to prevent the gel from flowing out of the cylinder. A scintillating fiber dosimeter was inserted through each port and centered inside the cylinder. The warm gel solution was transferred into the cylinder, and inside 13 glass calibration vials (height of 3.5 cm and diameter of 1.16 cm). The filled cylinder and vials were immediately immersed in water and placed in a refrigerator for approximately four hours. The samples were next transferred into a room temperature water bath for temperature stabilisation for four more hours.

2.3. Gel irradiation and imaging.

The cylinder was laid on its side on a Plexiglas support in a water tank with the gel surface at 2 cm below the water surface (SSD 100 cm). The cylinder was placed perpendicular to the linac arm and oriented such that the fibers were perpendicular to the beam. The cylinder was positioned such that the entry port at 1.2 cm was at the edge of the field.

Three measurements were made with different prescribed doses and using physical filters of 45° and 60° on a 6MV photon beam to obtain a notable variation of the dose inside the gel. A square field of 10 × 10 cm² at the surface of the water was used. The vials were irradiated with a square 10 × 10 cm² field in a water tank at various depths along the field axis. The dose given to the vials was verified with an ionization chamber and one vial was kept unirradiated.

After irradiation, the samples were stored in the MRI room for 24 hours and then scanned with a Siemens Sonata 1.5 T MRI system. The cylinder was placed in a custom made styrofoam holder designed to fit tightly and reproducibly inside the head coil. A multi-slice multi-echo sequence was used (TR = 7750 ms, TE = 40 ms, 16 echoes, 12 averages, FOV 192 mm × 96 mm, slice thickness = 4

Journal of Physics: Conference Series **250** (2010) 012076 doi:10.1088/1742-6596/250/1/012076

mm and a matrix of 256 × 128. The last 14 echoes were fitted with a monoexponential function on a pixel-by-pixel basis to produce maps of T_2 (and $R_2 = 1/ T_2$).

Finally, another calibration of the scintillating fiber dosimeters was performed for verification. Discrepancies of the dose from the gel irradiation are below 2% when comparing the dose calculated using the calibrations before and after irradiation.

The water tank and the cylinder were scanned in a Picker PQ5000 CT scan using a chest protocol in the position they were irradiated. The data was then transferred to the Pinnacle³ 8.0m planning system to simulate the dose distribution in the cylinder. Measurements inside the cylinder were made with an ionization chamber (Exradin A12) to verify the validity of the planning system. The measurements were made using water instead of gel. The measured dose was 0.8% higher than the simulated dose. Pinnacle³'s simulation is thus used as the reference dose distribution except in the high dose gradient at the field edges.

2.4. Gel calibration

A first calibration curve was generated using the R_2 values of the vials and the dose values obtained by an ionisation chamber. Then, the dose given by the scintillating fiber dosimeters are correlated to the R_2 values around the scintillating dosimeter. A margin of 1.5 mm is ignored around the fiber because of possible susceptibility artefacts or altered gel response. The calibration points (and curves) of the measurement with the 45° wedge are shown in Figure 1.

Four calibration methods were tested:

Method #1 A "vials only" calibration curve.
Method #2 A linear fit between the R_2 value of the unirradiated vial (0 Gy) and the R_2 value around a scintillating fiber.
Method #3 Adjusting the slope of the "vials only" curve to match the R_2 values around the fibers (while maintaining the 0 Gy value).
Method #4 To vertically shift the "vials only" curve to match the R_2 values around the fibers.

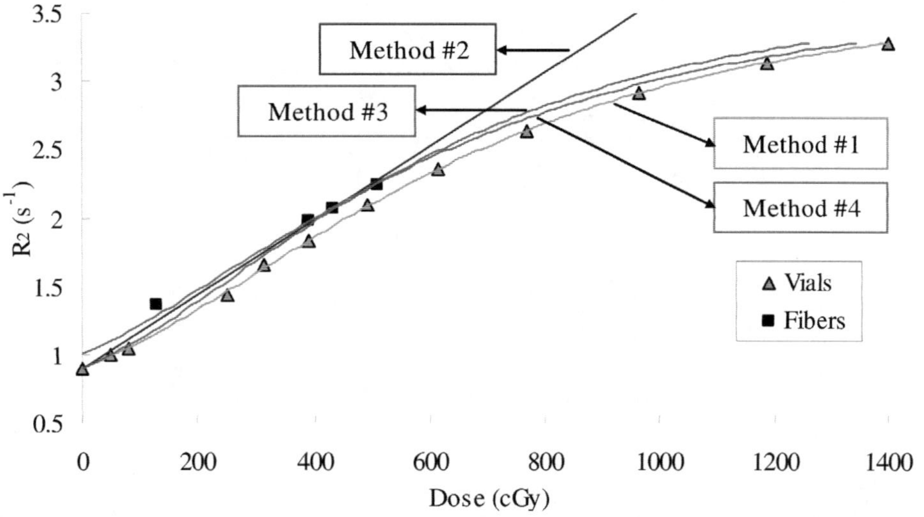

Figure 1. Calibration points and curves obtained from the calibration vials and from the fibers inserted into the gel. Only one fiber measurement (third black square from the left on the graph) is used to adjust the calibration curves from the vials for methods #2, #3 and #4.

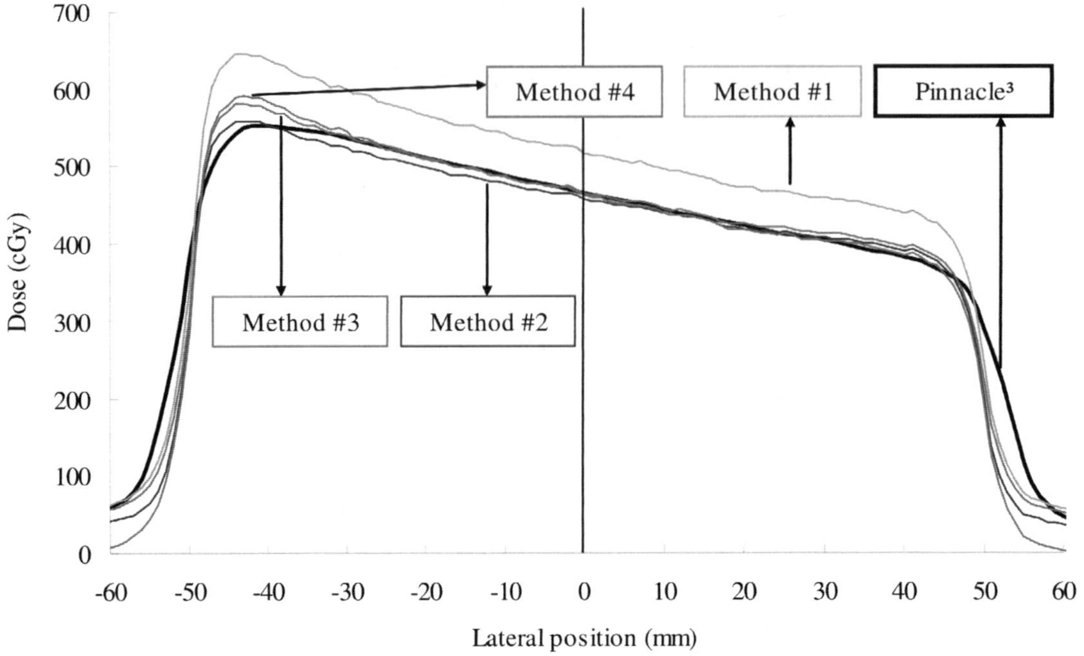

Figure 2. Lateral dose profile at 4 cm of depth from surface for the 45° physical wedge.

3. Results and Discussion

Our first observation was that the data from the fiber at the edge of the irradiation field were not useful. The large dose gradient in this location prevented a precise estimation of the R_2 value around this fiber; those data were not considered further. The dose difference between the "vials only" calibration and the fiber measurements was around 13%. The latter agreed to within 3% with Pinnacle³'s simulation. For this test, only the fiber (at 5.5 cm of depth) closer to the center of the field, was used to modify the calibration curves.

In the center part (80% of the full width at half maximum) of the profile shown in figure 2, the mean absolute difference with Pinnacle³ is 12.4%, 1.4%, 1.2% and 0.9% for methods 1 to 4, respectively. The results show that it is possible to obtain a more accurate dose when inserting a scintillating fiber inside a PAG dosimeter combined with a set of irradiated vials. The error falls below 2% compared to Pinnacle³ in the central zone of the field for the last approaches.

4. Conclusion

Inaccuracies up to 12.4% were obtained in polymer gel dosimeter phantom calibrated using small glass vials. We showed that inserting a scintillating fiber inside the radiation field away from an important dose gradient could correct these inaccuracies. A useful and practical method consists in a translation of the calibration curve from the vials to match the calibration point from the scintillating fiber. This reduced the inaccuracies to below 3%.

References

[1] L. Archambault, G. Leclerc, L. Beaulieu, M. Lepage, Absolute calibration of polymer gel dosimeter using scintillating fibers, J. Phys. Conf. Ser. **56,** 242-244 (2006).

[2] A. S. Beddar, T. R. Mackie, and F. H. Attix, Water-equivalent plastic scintillation detectors for high-energy beam dosimetry. I. Physical characteristics and theoretical considerations, Phys. Med. Biol. **37**, 1883-1900 (1992).

[3] F. Lacroix, L. Archambault, L. Gingras, M. Guillot, A. S. Beddar, L. Beaulieu, Clinical prototype of a plastic water-equivalent scintillating fiber dosimeter array for QA applications, Med. Phys. **35**, 3682-3690 (2008).

Journal of Physics: Conference Series **250** (2010) 012076 doi:10.1088/1742-6596/250/1/012076

[4] A-M. Frelin, J-M Fontbonne, G. Ban, J. Colin, M. Labalme, A. Batalla, A. Isambert, A. Vela, T. Leroux, Spectral discrimination of Čerenkov radiation in scintillating dosimeters, Med. Phys. **32**, 3000-3006 (2005).

[5] J. M. Fontbonne, G. Iltis, G. Ban, A. Batalla, J. C. Vernhes, J. Tillier, N. Bellaize, C. Lebrun, B. Tamain, K. Mercier, and J. C. Motin, Scintillating fiber dosimeter for radiation therapy accelerator, IEEE Trans. Nucl. Sci. **49**, 2223-2227 (2002).

[6] L. Archambault, A. S. Beddar, L. Gingras, R. Roy, L. Beaulieu, Measurement accuracy and Cerenkov removal for high performance, high spatial resolution scintillation dosimetry, Med. Phys. **33** , 128-135 (2006).

Journal of Physics: Conference Series **250** (2010) 012077 doi:10.1088/1742-6596/250/1/012077

Femtosecond laser pulse filamentation characterized by polymer gel dosimetry and Fricke dosimetry

Ridthee Meesat, Jean-François Allard, Daniel Houde, Luc Tremblay, Abdelouahed Khalil, Jean-Paul Jay-Gerin and Martin Lepage

Département de médecine nucléaire et de radiobiologie, Faculté de médecine et des sciences de la santé, Université de Sherbrooke, Sherbrooke (Québec) J1H 5N4, Canada

E-mail: Martin.Lepage@USherbrooke.ca

Abstract. A femtosecond laser pulse that can generate water radiolysis species was studied in view of its potential medical and biological applications. Intense ultra-short laser pulses can propagate in liquid water, leading to self-focusing and filamentation. Briefly, electrons produced by either multiphoton or tunnel ionization are further accelerated by the electric field of the pulse in an inverse Bremsstrahlung effect. If the electrons acquire enough kinetic energy, they will give rise to a second generation of electrons by impact ionization of other molecules in an avalanche-like process. The geometry and trajectory of femtosecond filaments were captured within a polymer gel dosimeter and imaged by magnetic resonance imaging (MRI) at high resolution. The results revealed that changing pulse duration modifies the penetration of the filament track in the medium. In addition, we used Fricke dosimetry to measure the absorbed dose and dose rate of the femtosecond laser pulse filamentation. A very high dose rate of 5.3×10^{12} Gy/s was calculated in filaments having a diameter of ~600 μm.

1. Introduction

We used filamentation from a femtosecond laser pulse to generate avalanche low energy electrons that is called an inverse Bremsstrahlung process (1-3). The process can lead to the creation of molecular and radical species in liquid media. For conventional radiation sources (*e.g.*, X-ray, γ-ray, accelerated electrons or heavier charge particle), the interaction with liquid water generates low-energy secondary electrons in large amounts (~5×10^4 electrons/MeV) from the absorption of the high-energy ionizing radiation. These electrons define the geometry of the radiation track (4) and their most probable kinetic energy is centered around ~10 eV (5). In this energy range, the penetration depth of electrons in water is ~10 nm (5). These low-energy electrons were shown to break plasmid DNA (6).

The physical origin of the formation of filaments from a femtosecond laser pulse is well understood (1-3). Briefly, self-focusing is an induced lens effect that results from the wavefront distortion inflicted on the beam by itself while traversing an optically nonlinear medium. Consequently, as the beam travels in the medium, the original plane wavefront of the beam gets progressively more distorted. The distortion is similar to that imposed on the beam by a positive lens. Since the direction of the optical ray propagation is perpendicular to the wavefront, the beam appears to focus by itself. This degenerative process is stabilized in the femtosecond regime by the generation of electrons forming a filament (1-3). Electrons produced by multiphoton or tunnel ionization are further accelerated by the electric field of the pulse in an inverse Bremsstrahlung effect. If they acquire enough kinetic energy (~6.5 eV in the case of water), they give rise to a second generation of electrons by impact ionization of other molecules in an avalanche-like process. Those linearly distributed electrons (in the range of ~10^{18} cm^{-3}) (1-2) transfer their excess energy to the surrounding water molecules, generating in the self-focusing region chemically reactive species such as e^-_{aq}, $H^\bullet$, $^\bullet O^\bullet$, and $^\bullet OH$, and due to the recombination of the radicals, H_2, O_2, H_2O_2 and $O_2^{\bullet -}$ (or $HO_2^\bullet$, $pK_a = 4.8$). Filaments are analogue to conventional radiation tracks, having very small diameters in condensed matter (~100-250 μm) (1-3). Demonstration of the presence of H_2 and H_2O_2 has already been published (7). Although the stabilization of the filament is most probably due to the presence of electrons, no time-resolved measurement of this presence has been reported to date.

The dose deposition from laser filamentation was measured with the Fricke (ferrous sulfate) dosimeter, a well-known radiation chemical dosimeter. Comparison with ^{137}Cs irradiation provided a dose rate for the filamentation process. The morphology of the filaments was evaluated using a polyacrylamide gel which is a 3D (8-9) and a tissue equivalent radiation dosimeter (10). The filaments were visualized and characterized in the dosimeter by magnetic resonance imaging (MRI).

2. Materials and methods

2.1. Laser instrument

A femtosecond Ti-sapphire laser beam (pulse duration 100 fs, 0.3 mJ/pulse and repetition rate 1 kHz, central wavelength 800 nm) was used in order to irradiate the sample contained in a quartz cuvette ($1 \times 1 \times 5$ cm^3) via a 3 cm focal length achromatic lens.

2.2. Fricke dosimeter

The chemicals used in the Fricke dosimeter solution are ferrous ammonium sulfate (hexahydrated salt, 99.99%, Sigma Aldrich), sulfuric acid (98%, Sigma Aldrich), and double-distilled water. The solutions were prepared from 0.342 g of $Fe(NH_4)_2(SO_4)_2.6H_2O$ dissolved in 0.4 *M* H_2SO_4. The concentration of Fe^{2+} ions in the solution was 1 m*M* and the solution density 1.024 g/cm^3. The solution was purged with oxygen for 2 h and protected from light by an aluminum foil. Special care was given to cleaning glassware according to the standard procedures used in radiation chemistry (11).

The Fricke solution was poured into 2 ml glass vials and the samples were irradiated up to 180 Gy at 25°C with ^{137}Cs γ-rays in a Gammacell (Atomic Energy of Canada Ltd.) with a dose rate of 0.2 Gy/s. For the filamentation laser irradiation, 2 ml of the Fricke solution was poured into the quartz cuvette and magnetically stirred during irradiation. The distance between the focus lens and the surface of the cell was 23 cm (Figure 1).

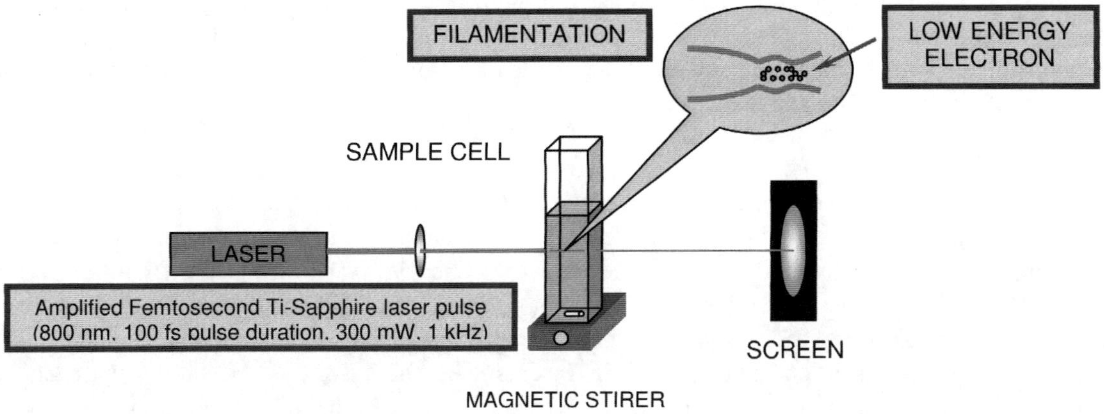

Figure 1. Schematic of the experimental setup to calculate the absorbed dose from the laser pulse using a Fricke dosimeter solution in a sample cell.

The concentration of ferric ions was calculated from the increase in the optical density at 304 nm using a molar extinction coefficient of Fe^{3+} of 2207 M^{-1} cm^{-1}. The radiation yield of Fe^{3+} ions, $G(Fe^{3+})$, was calculated from the gradient of the linear curve between the Fe^{3+} concentration and the absorbed dose. Normally, the yield of Fe^{3+} depends on the linear energy transfer (LET) of the incident radiation. For low LET values such as for gamma photons from ^{137}Cs (~0.9 keV/μm) (12), $G(Fe^{3+})$ is equal to 15.3 ± 0.3 molecules/100 eV (13).

2.3. Polymer gel dosimeter

The gel dosimeters were prepared from acrylamide (AA) and N,N'-methylene-bisacrylamide (BIS) (99+%, electrophoresis grade, Aldrich), each at 3% w/w dissolved in gelatin (300 bloom, Aldrich) at 5% w/w, and water (de-ionized). In the manufacture process, gelatin was added to water at 25°C and left to soak for 10 min. The solution was then heated and maintained at a temperature of 45°C. AA and BIS were successively added and magnetically stirred for ~15 min until complete dissolution. The gels were prepared under a controlled N_2 atmosphere inside a glove box (8,14). The gel solution was poured into glass cells (diameter 2 cm × length 5 cm) (Fig. 2a) or in rectangular plastic container (Fig. 3a). In the latter case, the gel was prepared in the glove box with the same recipe but 5 m*M* of alkaline tetrakis (hydroxymethyl) phophonium chloride (THPC) (80%, Aldrich) was added to scavenge O_2. The samples were kept inside the glove box for 24 h after manufacture (15).

2.4. Magnetic resonance imaging

The samples were imaged using a 7 T small animal magnetic resonance imaging (MRI) scanner. A fast spin-echo protocol to acquire images in the axial (sagittal) planes with the following parameters: TR: 4 400 ms (2 000 ms), echo spacing: 13 ms, echo train length: 8, FOV: 25 × 25 mm^2 (60 × 30 mm^2), matrix: 256 × 256 (512 × 256), slice thickness: 1 mm (3.5 mm), number of slices: 40 (1), number of averages: 4. The in-plane spatial resolution of the axial images was thus ~ 100 μm.

3. Results and Discussion

3.1. Morphology of the filament tracks

Laser-induced water free radicals initiate polymerization in a polymer gel dosimeter such that the tracks caused by a femtosecond pulse are captured in the gel. The size of a single track was assessed by MRI (Fig. 2). The diameter and length of a track are estimated to be ~600 μm and 4 cm, respectively.

It is possible to create arbitrary irradiation patterns with laser filamentation. Figure 3 shows a photograph of a polymer gel dosimeter irradiated with several ultra-short laser pulses. Although difficult to appreciate with a photograph, visual inspection clearly revealed that changing the laser pulse duration leads to a change in the focus position (Fig. 3b). This indicates that the depth of the deposited dose profile can be adjusted.

3.2. Dose of the filament tracks

The Fricke dosimeter has found extensive use because of its ability to provide absolute dose results with a high accuracy. The estimated kinetic energy of the electrons in filamentation in water is ~6.5 eV such that the LET is expected to be ~1 keV/μm (5). The yield of Fe^{3+} should then be 15.3 ± 0.3 molecules/100 eV (11). In other words, the same dose of gamma photons or of laser filamentation induced electrons is expected to lead to the same production of Fe^{3+} in the Fricke dosimeter.

Initial calculations based on the power of the laser pulses, the energy of the electrons produced (1-3), the expected dimension of the filament tracks (diameter 100 μm and 1 cm long (1)), and the electron density 10^{18} per cm^3 (1), predicted a dose rate of 2.4 × 10^{12} Gy/s for the filamentation process. The radiation absorbed dose of the laser filamentation and gamma radiation measured in a Fricke solution are plotted in Fig. 4. From this graph, the dose rate of the laser pulse was calculated to be 2.8

Journal of Physics: Conference Series **250** (2010) 012077 doi:10.1088/1742-6596/250/1/012077

Gy/s. This is to be compared with the dose rate of the γ-rays (0.2 Gy/s). The sample cell is irradiated uniformly with gamma radiation, but only a small fraction of the cell is absorbing the dose from the femtosecond laser. When accounting for the pulse duration (~100 fs) and the volume irradiated, the calculated dose rate is 5.3×10^{12} Gy/s, in close agreement with the predictions.

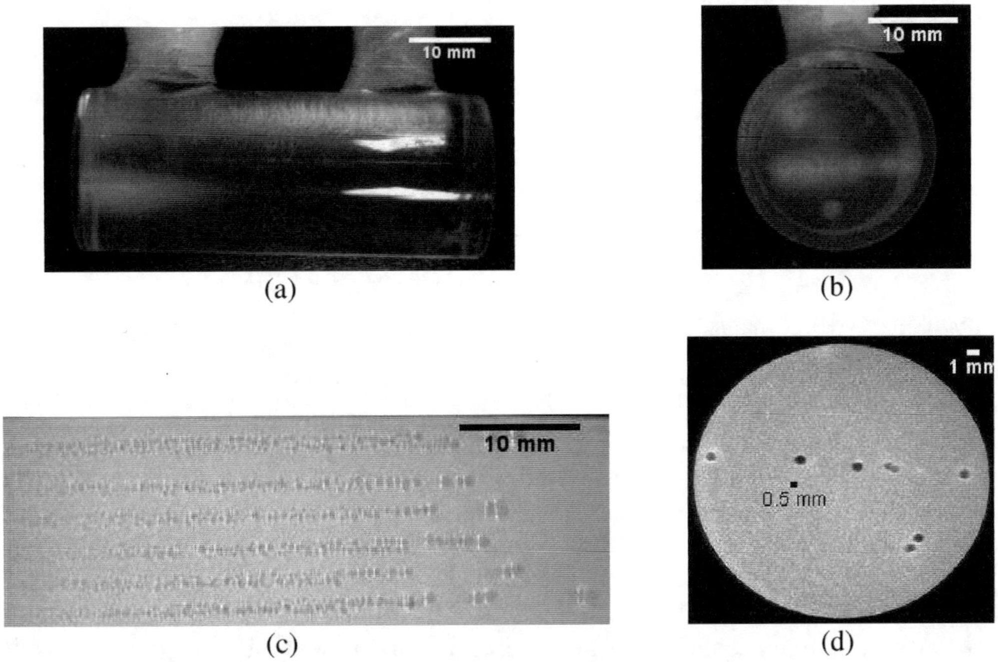

Figure 2. Femtosecond filamentation laser pulse in a polymer gel dosimeter. Photographs showing (a) the side view and (b) the top view of a glass cylinder irradiated from the top. The corresponding sagittal (c) and axial (d) magnetic resonance images were used to determine the length and the diameter of the laser filamentation tracks.

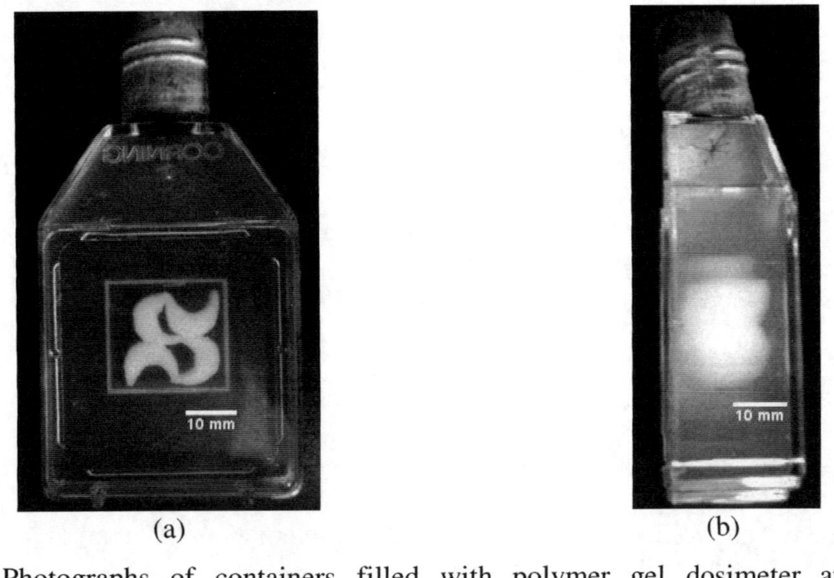

Figure 3. Photographs of containers filled with polymer gel dosimeter and irradiated with femtosecond laser pulses to draw the logo of the Université de Sherbrooke. (a) top view and (b) side view. Although not clearly visible on the photograph, the depth of the logo was modulated by changing the laser pulse duration.

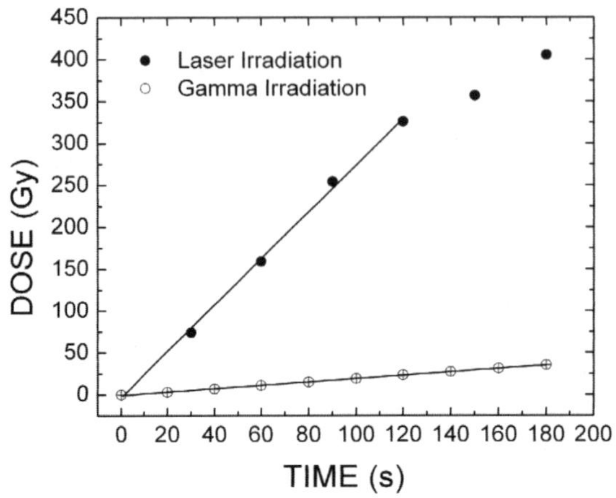

Figure 4. Concentration of Fe^{3+} calibrated in terms of an absorbed dose as a function of exposure time of a Fricke solution to (○) gamma radiation from ^{137}Cs and (●) femtosecond laser pulses.

4. Conclusion

Laser light is not normally classified as an ionizing radiation. The self-focusing process of ultra-short laser pulses, which is called filamentation, leads to the production of an avalanche of low-energy electrons. These electrons can interact with surrounding molecules, resulting in the production of molecular and radical species. We characterized the physical properties of the laser filamentation track. The diameter was evaluated by polymer gel dosimetry and MRI at ~600 μm. Its penetration in a polymer gel dosimeter was around 4 cm. The dose rate in this volume was $\sim 10^{12}$ Gy/s as determined by Fricke and polymer gel dosimetry. Thus, nonlinear effects in the femtosecond laser pulse provide a convenient and effective radiation source with a very high dose rate. In view of such properties, this powerful tool could potentially be used in medical applications such as sterilization and radiotherapy.

References

[1] S L Chin, S A Hosseini, W Liu, Q Luo, F Théberge, N Aközbek, A Becker, V P Kandidov, O G Kosareva and H Schroeder 2005 *Can. J. Phys.* **83** 863-905

[2] A Vogel J Noack, G Hüttman and G Paltauf 2005 *Appl. Phys. B* **81** 1015-1047

[3] A Couairon and A Mysyrowicz 2007 *Phys. Rep.* **441** 47-189

[4] D W Anderson 1984 *Absorption of Ionization Radiation* (Baltimore, MD:University Park Press)

[5] J Meesungnoen, J-P Jay-Gerin, A Filali-Mouhim and S Mankhetkorn 2002 *Radiat. Res.* **158** 657-660

[6] B Boudaïffa, P Cloutier, D Hunting M A Huels and L Sanche 2000 *Science* **287** 1658-1660

[7] S L Chin and S Lagace 1996 *Appl. Optics* **35** 907-911

[8] M J Maryanski, J C Gore, R P Kennan and R J Schulz 1993 *Magn. Reson. Imaging* **11** 253-258

[9] C Baldock, R P Burford, N Billingham, G S Wagner, S Patval and S F Keevil 1998 *Phys. Med. Biol.* **43** 695-702

[10] A J Venning, K N Nitschke, P J Keall and C Baldock 2005 *Med. Phys.* **32** 1047-53

[11] J W T Spinks and R J Woods 1990 *An Introduction to Radiation Chemistry* 3rd ed

[12] J Meesungnoen, M. Benrahmoune, A Filali-Mouhim, S Mankhetkorn and J-P Jay-Gerin. *Radiat. Res.* **155**, 269-278.

[13] ICRU Report 34, The Dosimetry of Pulsed Radiation. International Commission on Radiation Units and Measurements, Bethesda, MD. 1982.

[14] M Lepage, A K Whittaker, L Rintoul, S Å J Bäck and C Baldock 2001 *Phys. Med. Biol.* **46** 1061-1074

[15] R Meesat, J-P Jay-Gerin, A Khalil and M Lepage 2009 *Phys. Med. Biol.* **54** 5909-5917

Journal of Physics: Conference Series **250** (2010) 012078 doi:10.1088/1742-6596/250/1/012078

Preliminary commissioning investigations with the DMOS-RPC optical-CT Scanner

J. Newton, A. Thomas, G. Ibbott[1], M. Oldham

Dept. of Radiation Oncology, Duke University Medical Center, Durham, NC, USA
[1] Radiological Physics Center, Houston, Tx, USA

*Corresponding author: joseph.newton@duke.edu

Abstract. A midsized broad beam Optical-CT scanner is being developed for collaborative research between Duke and the Radiological Physics Center (RPC). The Duke Midsized Optical-CT Scanner (DMOS-RPC) is designed to be compatible with several of the RPC phantoms, including the head and neck, stereotactic SRS, and lung phantoms. Preliminary data investigating the basic performance of the scanner is described. Two 10 cm PRESAGE cylinders were irradiated with simple test plans. Projections of ~80 μm resolution of each dosimeter were collected at 1 degree intervals over a full 360 degrees both before and after irradiation. 3 dimensional reconstructions of attenuation coefficients throughout the dosimeter were computed with 1 mm^3 resolution. Scans were normalized to the calculated dose distribution and a 3D comparison was made with a commissioned treatment planning system. Initial results indicate DMOS-RPC can produce accurate relative dose distributions with high spatial resolution (up to 1 mm^3 in 3D) in less than 30 minutes (acquisition and reconstruction). A maximum dose of ~3.6Gy was delivered in these tests, and observed noise was ~2% for 1 mm^3 reconstructions. Good agreement is observed with the planning system in these simple distributions, indicating promising potential for this scanner.

1. Introduction

The DMOS-RPC scanner is a robust, economical scanning system which can quickly and effectively provide 3-dimensional dosimetric information on the accuracy of delivery of complex dose distributions, as determined in radiochromic dosimeters with low levels of contaminant scattered light (e.g. PRESAGE). The design of the scanner has been developed from earlier systems in our lab [1-3]. Here we describe preliminary first experiences with the scanner, carried out at Duke prior to installation at the RPC.

2. Methods

The methods section is separated in two main parts. Section 2.1 will introduce the DMOS-RPC and related components. Section 2.2 will discuss two current experiments designed to evaluate the DMOS-RPC for verification of simple radiation treatments.

2.1. The Duke Midsized Optical Scanner dedicated for the RPC

A schematic of the DMOS-RPC scanner is given in Figure 1. A telecentric lens and light source are displayed along with the central water bath which will contain the dosimeter and refractive index-matched fluid. The camera is equipped with a 1", 1608x1208 ccd array. The entire system is about 1 meter from camera (left) to light source (right). The system is designed for a dosimeter of up to 12 cm

Journal of Physics: Conference Series **250** (2010) 012078 doi:10.1088/1742-6596/250/1/012078

in diameter and 10 cm height. A 2/3", 1392x1040 ccd array camera is currently being used for preliminary testing until the 1" camera is received. The imaging lens has a magnification of 0.08X translating to isotropic reconstruction resolution down to ~80μm. All motion and image acquisition parameters are set and automated through LabView™ [www.ni.com]. For all scans presented in this work 360 projections were acquired over 360 degrees with 10 images averaged together at each projection angle for noise reduction. Each projection was flood and dark corrected. Projections were downsized from 80μm to 1 mm pixels for further noise reduction, less disk space and decreased computation time. Reconstructed 3D data cubes of isotropic resolution 1 mm^3 were created using the Matlab iradon function. Total time for data acquisition and reconstruction (including set-up time of dosimeter in the scanner) was about 30 minutes. Fuller details of the operation of Optical-CT are described in an accompanying abstract [4] and elsewhere in the literature [5-7].

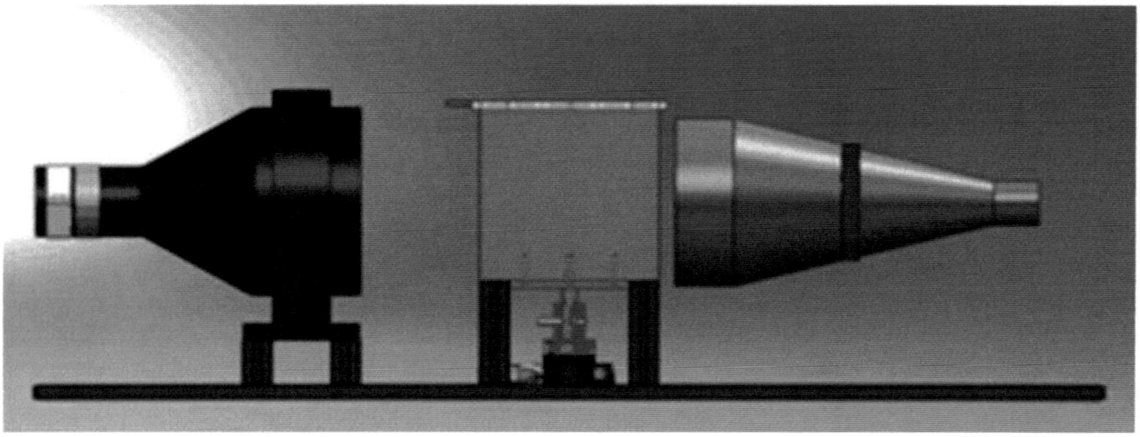

Figure 1: The DMOS-RPC system. The camera attached to a telecentric lens is shown on the left side, the water bath in the center and the light source with telecentric lens on the right. The rotation motor is displayed below the water bath. Total length is ~ 1 m.

2.2 Preliminary Tests of the DMOS-RPC
Two 10 cm diameter cylindrical PRESAGE dosimeters were irradiated with simple treatment plans where the planning system was known to provide accurate dose distributions. Both dosimeters were pre-scanned with the DMOS-RPC just before irradiation. The first dosimeter was treated with a two-field plan which consisted of orthogonal 4x4 cm fields delivering a maximum dose of ~3.6 Gy at isocenter. The second dosimeter was treated with a rotationally symmetric "spoke" plan with 9, 1cm width spokes of alternating high dose (5Gy), medium dose (3.6 Gy) and low dose (1.6 Gy) fields, with an SSD of 94 cm. The spoke plan was proposed by Sakhalkar et al [3] as a rigorous plan for evaluating optical-CT 3D dosimetry. The dosimeters were then rescanned in the DMOS-RPC scanner immediately after irradiation to find the radiation induced change in OD.

3. **Results**

3.1 The Two Field Plan
Figure 2 shows flood field corrected projections, taken with the DMOS-RPC system, of the 10 cm PRESAGE dosimeter prior to irradiation (a) and immediately after irradiation with the two field plan (b). Excellent index of refraction matching was achieved as evidenced by minimal edge artefacts. In figure 2b, increased attenuation is observed along the beam-path of one beam entering from the top of the dosimeter, orientated straight down. A bright square of increased attenuation is also observed in the center of the image, corresponding to the overlap region where the second beam is incident into the plane of the image (this projection is acquired in the beams-eye-view of the 2nd lateral beam).

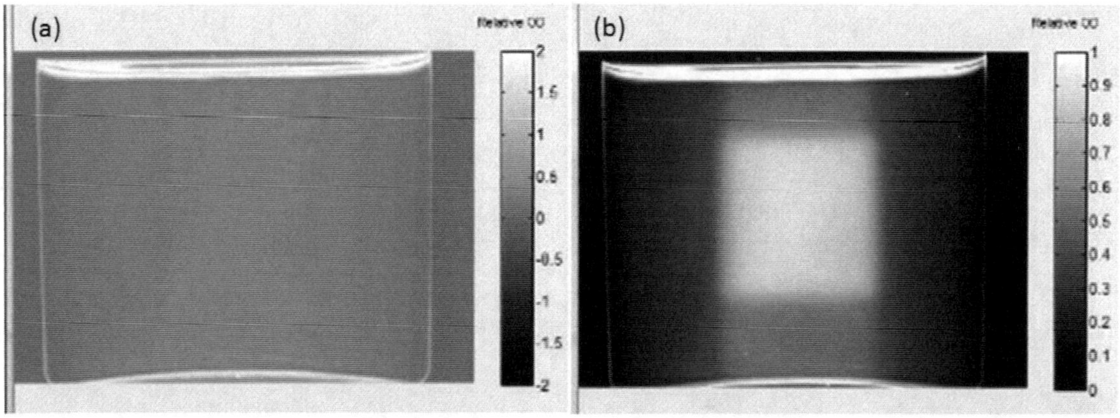

Figure 2: A single projection from the DMOS-RPC of the 10 cm PRESAGE dosimeter (a) prior to irradiation with the two field plan and (b) immediately after irradiation. In (b) one beam is incident on the top surface oriented vertically downward, and the other is incident normal to the plane of the image, resulting in the central cubed region of high attenuation (high dose).

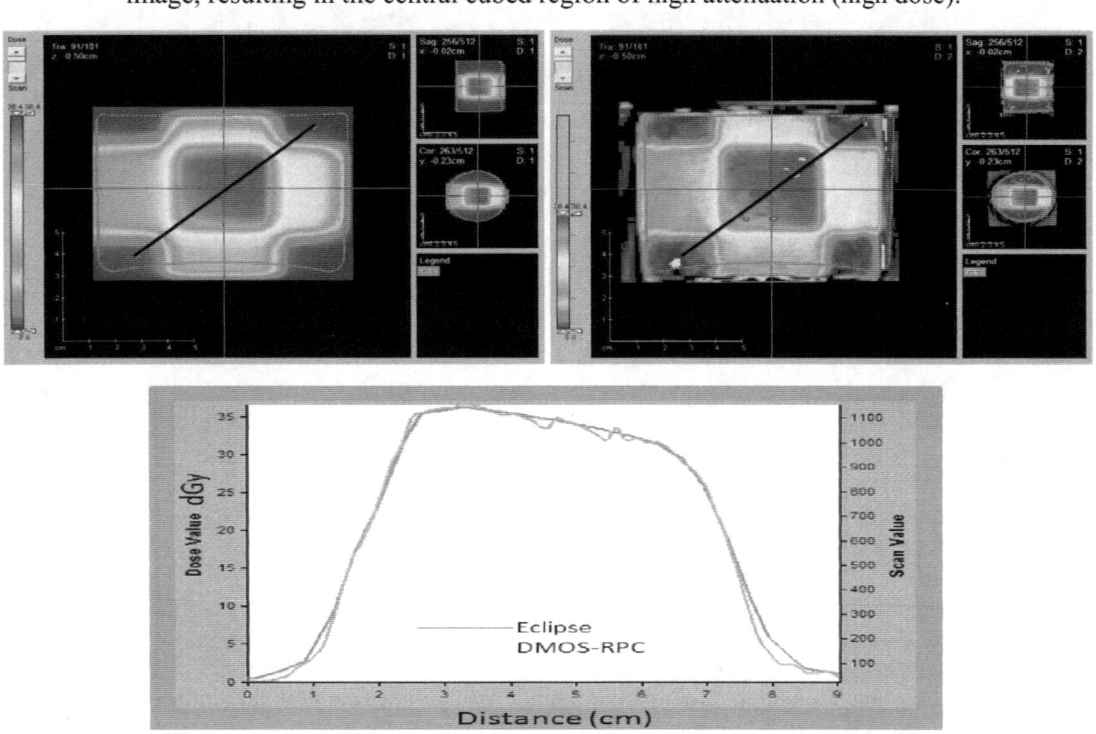

Figure 3: Comparison of dose-distributions in the central coronal slice from Eclipse (left panel) and as measured with Presage (right panel). Line profiles along the diagonal line are compared in the lower panel.

Figure 3 shows both the calculated (Eclipse) and PRESAGE measured dose distributions, in a single coronal slice towards the center of the dosimeter. A comparative line profile through the center of the two field dose is shown below. Good agreement is observed between the calculated and measured relative distributions, except for in the very low dose tails. The precise cause of the latter is under investigation.

3.2 The Spoke Plan

Figure 4 shows a reconstructed DMOS-RPC image of the OD change in the 10 cm PRESAGE dosimeter after irradiation with the spoke plan for 3 separate depths of 1cm, 3 cm and 5cm (selected from the 0.5 mm^3 isotropic 3D data of the whole dosimeter). The DMOS-RPC data have been normalized to the calculated plan at the center of the high-dose spoke. The alternating spoke intensities are clearly visible along with distinct edges to each spoke. Inter-spoke consistency between the 3 images suggests good consistency of dose response throughout the PRESAGE dosimeter. An increase in intensity is observed at the edges of the dosimeter, where the PRESAGE/fluid interface occurs. Further analysis is currently underway for a comparison of the measured DMOS-RPC data with the calculated dose distributions.

Figure 4: Individual slices of the complete 360° reconstruction of the DMOS-RPC projections for the spoke plan irradiation at depths of 1 cm (left), 3 cm (middle) and 5 cm (right). The dose map values are in units of dGy (10 dGy = 1 Gy).

4. Conclusion

Preliminary results with the DMOS-RPC system are promising. The results of the two-field plan analysis show good agreement between the DMOS-RPC system and Eclipse in the PRESAGE dosimeters, and minimal edge artifacts. The reconstructed distributions from the spoke plan showed an ability to image high resolution variations in the delivered dose, and good consistency of response throughout the volume. Further measurements and quantitative analysis are currently underway and will be presented.

Acknowledgements: Supported by NIH R01CA100835.

References

[1] H. S. Sakhalkar and M. Oldham, *Fast, high-resolution 3D dosimetry utilizing a novel optical-CT scanner incorporating tertiary telecentric collimation*. Med Phys, 2008. **35**(1): p. 101-11.

[2] H. S. Sakhalkar et al., Med Phys, 2009. **36**(7): p. 3371-7.

[3] H. S. Sakhalkar et al., *A comprehensive evaluation of the PRESAGE/optical-CT 3D dosimetry system*. Med Phys, 2009. **36**(1): p. 71-82.

[4] A. Thomas and M. Oldham, *An ivestigation into the feasibility of a fast, large field-of-view, telecentric optical-CT scanning system for 3D radiochromic dosimetry*. IC3D Dose 2010.

[5] N. Krstajic and S. J. Doran, *Focusing optics of a parallel beam CCD optical tomography apparatus for 3D radiation gel dosimetry*. Phys. Med. Biol., **51**(2006): p. 2055-2075.

[6] J. G. Wolodzko, C. Marsden and A. Appleby, *CCD imaging for optical tomography of gel radiation dosimeters*, Med. Phys., **26**(11), 1999: p. 2508-2513.

[7] J. C. Gore, M. Ranade, M. J. Maryanski and R. J. Schulz, *Radiation dose distributions in three dimensions from tomographic optical density scanning of polymer gels: I. Development of an optical scanner*. Phys. Med. Biol., **41**(1996): p. 2695-2704.

Journal of Physics: Conference Series **250** (2010) 012079 doi:10.1088/1742-6596/250/1/012079

Initial experience with optical-CT scanning of RadBall Dosimeters

Oldham M, Clift C[1], Thomas A, Farfan E[2], Foley T[2], Jannik T[2], Adamovics J[3], Holmes C[4], Stanley S[4]

Duke University Medical Center, Durham NC, USA
[1] Montefiore Medical Center, Bronx, NY, USA
[2] Savannah River National Laboratory (SRNL), Aiken, SC, USA
[3] Heuris Pharma LLC, Lawrenceville, NJ
[4] National Nuclear laboratory (NNL), Warrington, UK

Mark.Oldham@Duke.edu

Abstract. The RadBall dosimeter is a novel device for providing 3-D information on the magnitude and distribution of contaminant sources of unknown radiation in a given hot cell, glovebox, or contaminated room. The device is presently under evaluation by the National Nuclear Lab (NNL, UK) and the Savannah River National Laboratory (SRNL, US), for application as a diagnostic device for such unknown contaminants in the nuclear industry. A critical component of the technique is imaging the dose distribution recorded in the RadBall using optical-CT scanning. Here we present our initial investigations using the **D**uke **M**id-sized **O**ptical-CT **S**canner (DMOS) to image dose distributions deposited in RadBalls exposed to a variety of radiation treatments.

1. Introduction

The RadBall dosimeter, under development by the NNL, consists of a baseball-sized sphere of PRESAGE™ encased in a tungsten collimating shell (figure 1). PRESAGE™ is a radiochromic plastic material that has been investigated previously in the context of external beam radiation therapy (Sakhalkar et al [1], Adamovics et al [2]). The collimating shell is of 5-10mm thickness (depending on application) and is punctuated by an array of uniformly spaced holes that allow radiation tracks to penetrate into the dosimeter. After the RadBall is exposed in an unknown radiation field, the tracks deposited in the dosimeter can be imaged using optical-CT (first shown by Doran et al [3]), and the distribution and magnitude of unknown contaminant sources can be estimated by a back-projection of the measured radiation tracks to the surfaces containing the contaminant (walls, ceiling etc). Further details of the rationale of RadBall application are presented in an accompanying abstract (Farfan et al. [4]). The purpose of this article is to describe our initial experiences developing an optimized optical-CT scanning technique to image the radiation tracks in the RadBall dosimeters.

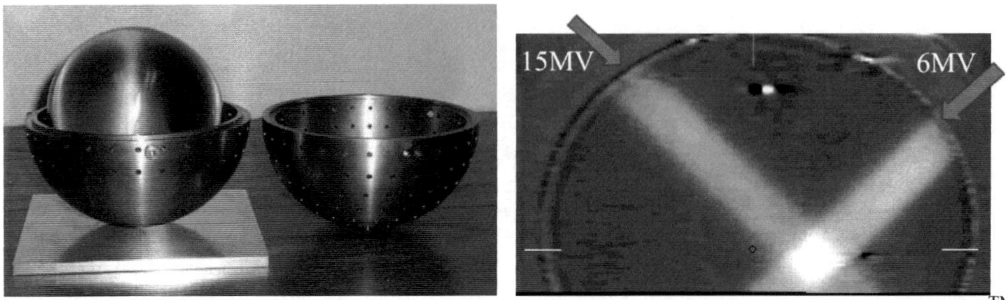

Fig 1: a) photograph of a disassembled RadBall showing the baseball sized PRESAGE™ sphere sitting inside the tungsten shell. The top part of the shell is removed for clarity. b) a coronal slice through a 3D optical-CT reconstruction of a RadBall irradiated by two 1cm^2 beams incident through the upper part of the dosimeter as indicated by red arrows.

2. Materials and Methods

RadBalls were scanned in the **D**uke **M**id-sized **O**ptical-CT **S**canner (DMOS), illustrated here in figure 2, and with the commercial MGS scanner for independent comparison. Parallel ray geometry was established using a matched telecentric light-source and lens. The telecentric light source produces a nominally parallel beam. Rays then pass through the aquarium containing refractive index (RI) matched fluid containing the RadBall (not shown in figure 2) and are collected by the telecentric lens. An aperture on the lens enables a variable acceptance angle from ~0.2-15 degrees of light rays used in image formation. Light rays deviating from the optical axis more than the acceptance angle are excluded from the image. The CCD camera mounted to a 0.08X telecentric lens and had a 1040x1392 pixel array. As the field of view is ~11cm, each pixel corresponds to ~0.08mm. Projection images were resampled to 0.5mm resolution to improve noise and maintain more manageable file sizes.

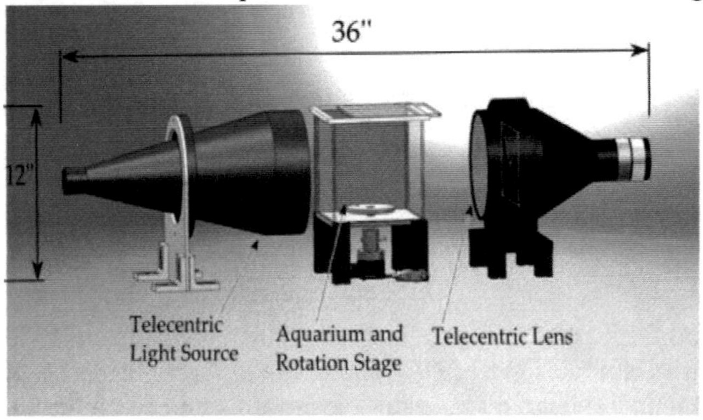

Fig 2: a) **D**uke **M**id-sized **O**ptical-CT **S**canner (DMOS)

A matched telecentric source and lens create an efficient, non-magnifying imaging beam, the projection of which through the RadBall dosimeter (not shown – but would be placed on the stage) enables measurement of the line integrals of attenuation at each pixel in the image. In principle, any scattered light is rejected from image formation in accordance with the acceptance angle of the lens.

2.1. Preliminary tests on RadBall irradiated with a Linac

Initial testing was done by irradiating two RadBalls without the collimating outer casing with beams from a medical linear accelerator. Tests were designed to determine basic feasibility, and performance capabilities regarding data quality (e.g. edge and ring artifacts), spatial resolution, noise, ability to discriminate different energies (6 and 15MV) etc. The first RadBall was irradiated with a simple two-beam arrangement as shown in figure 1b. Both beams were 1 cm^2; one was 6MV and the other 15MV. The second RadBall was irradiated with 6 – 4 mm^2 6MV beams (doses of 6, 4, 2, 1, 0.5, and 0.25 Gy) that intersected at the isocenter positioned at the base of the RadBall. The optical-CT scanning procedure for both RadBalls included acquiring 360 projection images acquired at 0.5 degree increments. An average of 10 images was acquired at each projection angle. Flood and dark correction images were created from an average of 200 images. The total acquisition time was approximately fifteen minutes.

2.2. Proof of concept RadBall irradiation

The promising results from the preliminary tests in 2.1 warranted an evaluation of a more realistic application of the RadBall dosimeter. A complete assembled RadBall (N10-2 complete with external collimating housing) was irradiated from multiple directions with a Co-60 source. The source was incident on the RadBall from 6 equi-spaced angles (60 degrees apart), and each angle delivered a dose of ~1Gy in-air at the center of the RadBall. 14 RadBalls were irradiated in various other geometries.

3. Results and Discussion

An example of a single projection of an unirradiated and irradiated RadBall, as seen through the DMOS scanner, is shown in figure 3a and b respectively. Careful matching of the refractive index of the fluid in the aquarium to that of the RadBall reduced the edge artefact to a narrow peripheral band. Although the quality of the projections is good, imperfections (particulates, scratches etc) are visible on the surface and within the RadBalls. New manufacturing processes have greatly reduced these sources of noise.

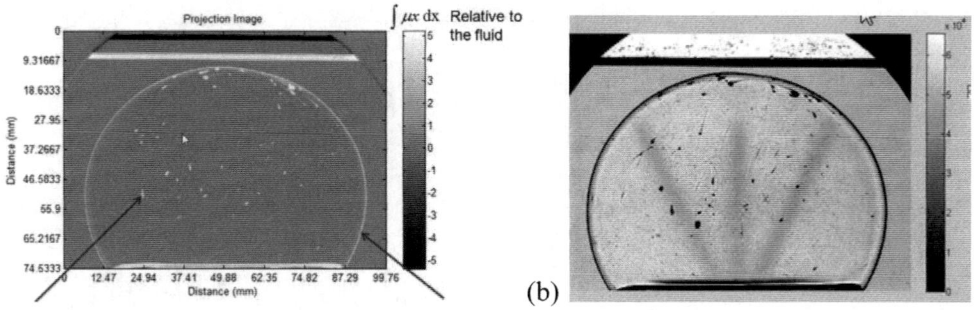

Fig 3 (a) (b)

3.1. Preliminary RadBall tests

A single coronal slice of the 2-field irradiation is shown in figure 4a. Percent depth-dose curves along the dotted red lines are shown in figure 4b. Both curves have been normalized to their respective values at the expected depth of maximum dose. Good high-resolution visibility of beams and intersection point are achieved with minimal edge effects. The lack of pronounced edge effects is a function of good RI matching between the fluid and PRESAGE^{TM}, and is an advantage of the PRESAGE^{TM} material that does not require an external container, enabling dose measurement close to the surface. Useful PDD curve data is achieved and enhanced with a simple median filter. Notable features of the PDD curves are that the dose build up is clearly visible in both beams, enabling a distinction of the different beam energies.

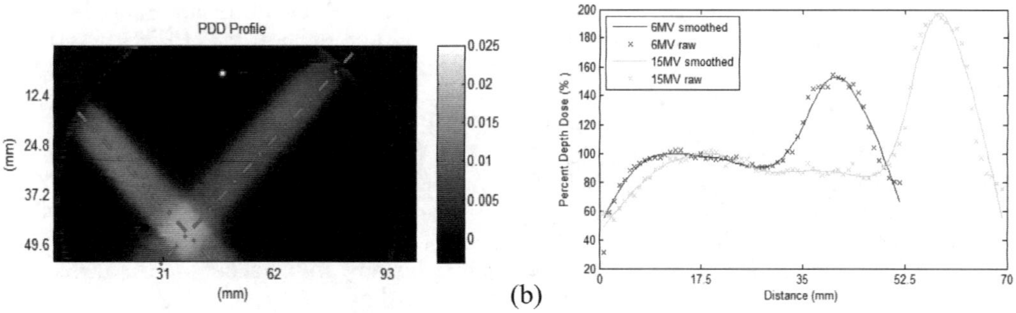

Fig 4(a) (b)

A representative and illustrative 2D reconstructed slice of the RadBall irradiated with 6 - $4mm^2$ pencil beams is shown in figure 5a, with corresponding partially transparent rendering of the entire 3D data cube in figure 5b. This data set is a subtraction of the (post)-(pre-scan) images. The high spatial resolution of the scanning system is evident from these images, as the 4x4mm beams are clearly distinguished. The 6, 4, 2, 1 and 0.5 Gy beams are easily distinguished. The 0.25 Gy beam is faint but can just be observed.

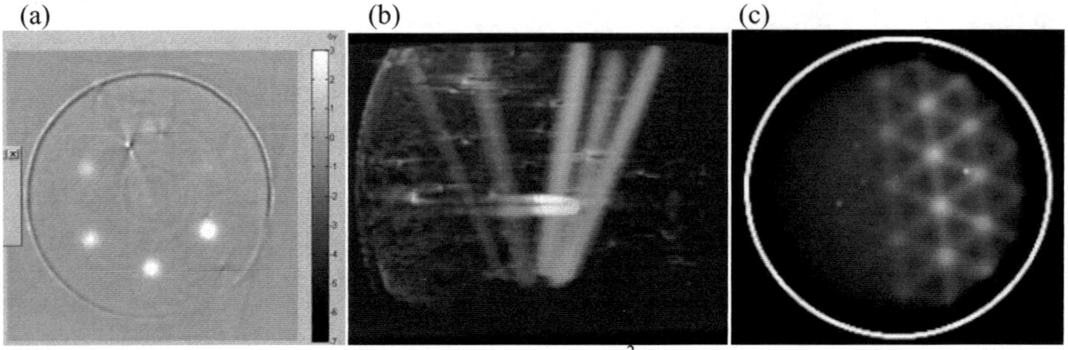

Fig 5 (a) single reconstructed axial slice of the six 4 mm^2 beam irradiation. (b) A partially transparent rendering of the entire 3D dose cube. (c) A single reconstructed slice of the proof-of-concept irradiation of a RadBall irradiated in the collimator housing.

3.2. Proof of concept RadBall irradiation

A single slice of the 3D data cube reconstructed for the RadBall irradiated in the collimator housing is shown in figure 5c. A more pronounced edge effect is observed with this RadBall presumably due to a poorer RI match. A higher background dose was also delivered to this RadBall due to its exposure in a broad high energy Co-60 beam (1.2 MeV). The increased transmission of radiation through the holes in the collimator are still clearly visible against this background, lending support for the viability of the RadBall approach.

4. Conclusions

The DMOS scanner is demonstrated to achieve good visualization of the radiation tracks deposited in the RadBall. Critical issues to achieve high quality data include good RI matching, flood and dark corrections, fine control of acceptance angle in the telecentric imaging chain, accurate modelling of spectral effects on the light source (see accompanying abstract [5]). The preliminary experiments demonstrated that energy resolution is feasible provided PDD curves can be obtained. The proof-of-concept irradiations of the complete RadBall assembly showed that the DMOS scanner has sufficient spatial resolution and sensitivity to warrant next step investigations: field studies.

References

[1] Sakhalkar HS, Adamovics J, Ibbott G, Oldham M. A comprehensive evaluation of the PRESAGE/optical-CT 3D dosimetry system. Med Phys. 2009 Jan;36(1):71-82
[2] Adamovics, J. and Maryanski, M. J., "Characterisation of PRESAGETM: A new 3-D radiochromic solid polymer dosemeter for ionising radiation", Radiat.Prot.Dosimetry, Vol. 120, pp 107-112, 2006
[3] Doran SJ, Stanley SJ, Jenneson PM, Prott E, Adamovics J. RadBall: A new departure for 3D dosimetry, J Phys: Conf Series, 164 (1), 012042
[4] Farfán EB, Foley TQ, Colemen JR , Jannik GT, Stanley SJ, Holmes CJ, Oldham M, and Adamovics J, "RadBall Technology Testing and Modeling", IC3DDose Conference Abstract, 2010
[5] A Thomas*, M Pierquet, M Oldham, "Achieving accurate optical-CT 3D using a polychromatic LED light source", IC3DDose Conference Abstract, 2010

Journal of Physics: Conference Series **250** (2010) 012080 doi:10.1088/1742-6596/250/1/012080

RadBall™ Technology Testing in the Savannah River Site's Health Physics Instrument Calibration Laboratory

Eduardo B. Farfán[1], Trevor Q. Foley[1], G. Timothy Jannik[1], Larry J. Harpring[1], John R. Gordon[1], Ronald Blessing[1], J. Rusty Coleman [1], Christopher J. Holmes[2], Mark Oldham[3], John Adamovics[4], and Steven J. Stanley[2]

[1] Savannah River National Laboratory, SRS, Aiken, SC 29808, USA.

[2] National Nuclear Laboratory, Risley, Warrington, United Kingdom.

[3] Duke University Medical Center, Durham, NC 27710, USA.

[4] Heuris Pharma LLC, Lawrenceville, NJ 08648, USA.

Email: Eduardo.Farfan@srnl.doe.gov

Abstract. The United Kingdom's National Nuclear Laboratory (NNL) has developed a radiation-mapping device that can locate and quantify radioactive hazards within contaminated areas of the nuclear industry. The device, known as RadBall™, consists of a colander-like outer collimator that houses a radiation-sensitive polymer sphere. The collimator has over two hundred small holes; thus, specific areas of the polymer sphere are exposed to radiation becoming increasingly more opaque in proportion to the absorbed dose. The polymer sphere is imaged in an optical-CT scanner that produces a high resolution 3D map of optical attenuation coefficients. Subsequent analysis of the optical attenuation data provides information on the spatial distribution of sources in a given area forming a 3D characterization of the area of interest. The RadBall™ technology has been deployed in a number of technology trials in nuclear waste reprocessing plants at Sellafield in the United Kingdom and facilities of the Savannah River National Laboratory (SRNL). This paper summarizes the tests completed at SRNL Health Physics Instrument Calibration Laboratory (HPICL).

1. Introduction

The consequences of radiological operations at various U.S. Department of Energy (DOE) sites have resulted in substantially contaminated facilities (e.g., reactors, fuel and isotope processing facilities, laboratories, hot cells, gloveboxes, etc.). These facilities are usually associated with extremely high dose rates and it is therefore imperative to use remote technologies for characterization and decommissioning to keep worker exposures as low as reasonably achievable (ALARA) in these highly contaminated environments. Although technologies might already exist that could be tested, modified, and deployed for characterization and decommissioning efforts throughout the DOE complex, development of new and innovative technologies is also beneficial. In addition, though it might be possible to complete these tasks without remote/robotic systems, the benefits of remote systems regarding ALARA and cost concerns are expected to be substantial.

The United Kingdom's National Nuclear Laboratory (NNL) has developed RadBall™, a radiation-mapping device that can locate and quantify radioactive hazards within contaminated areas of the nuclear industry. This study completed at the Savannah River National Laboratory (SRNL) addresses key aspects of the testing of the innovative RadBall™ technology. RadBall™ presents a significant opportunity to expedite initial characterization of radiologically contaminated facilities with respect to ALARA concerns, initial decontamination strategies, and costs associated with the decontamination efforts. RadBall™ will make radiation mapping safer and potentially more accurate and convenient than conventional detection devices, which are often much bigger and more cumbersome due to their electrical components and accessories. A single RadBall™ can be positioned in a highly contaminated area, glove box, or hot cell and left alone to collect data. Personnel would no longer spend valuable time carrying out manual scanning and surveying.

The device consists of a colander-like outer tungsten shell that houses a radiation-sensitive polymer sphere shown in Figure 1. The outer shell works to collimate radiation sources and those areas of the polymer sphere that are exposed react, becoming increasingly less transparent, in proportion to the absorbed dose. The polymer sphere is imaged in an optical-CT scanner developed at Duke University, which produces a high resolution 3D map of optical attenuation coefficients that are proportional to the absorbed dose.

Figure 1. Two components of a RadBall™ device: the outer collimation shell and inner polymer core.

The orientation of the opacity track provides the positional information regarding the source (achieved by using a reverse ray-tracing technique). The activity of the detected source is assessed by quantifying the magnitude of the opacity change (which follows a linear relationship with respect to the absorbed dose). NNL and SRNL have published information regarding RadBall™ development [1-3]. Information on the characterization and optical–CT scanning of the radiation sensitive polymer, PRESAGE™, has also been published [4-7].

A set of tests was performed at the Savannah River Site (SRS) Health Physics Instrument Calibration Laboratory (HPICL) using various gamma-ray sources and an x-ray machine with known radiological characteristics. The objective of these preliminary tests was to identify the optimal dose and collimator thickness of the RadBall™.

2. Experiments at SRS Health Physics Instrument Calibration Laboratory (HPICL)

The HPICL at SRS contains nine Automated Irradiator Systems (Figure 2). The primary purpose of the equipment is to calibrate radiation detection instruments and test and verify personnel dosimeters. The sources and the exposures are therefore known with a high degree of certainty. The higher activity sources were ideal for the RadBall™ experiments since these sources provided shortened periods of exposure time. The highest ^{137}Cs and ^{60}Co sources were used for the majority of the experiments at around 45.9 and 178 TBq (1,240 and 4,756 Ci), respectively. Three experimental phases were completed at the HPICL which included the exposure of 45 RadBall™ polymers. Selected experiments completed at the HPICL are presented in Table 1.

Journal of Physics: Conference Series **250** (2010) 012080 doi:10.1088/1742-6596/250/1/012080

Table 1. RadBall™ Tests at HPICL.

	Test	Total Dose (Gy)	Radiation Source(s)	RadBal™	Collimator Thickness (mm)
Phase 1	5	3.0	^{137}Cs	N-1-6	5.0
	6	5.0	^{137}Cs	N-2-4	5.0
	13	3.0	1.5 Gy ^{137}Cs + 1.5 Gy ^{60}Co (collinear)	N-3-2	5.0
	14	6.0	3 Gy ^{60}Co (no collimator) + 3 Gy ^{60}Co	N-3-3	5.0
	15	6.0	3 Gy 120 keV + 3 Gy 38keV x-ray (rotated 180°)	N-3-1	7.5
Phase 2	16	2.0	1 Gy ^{60}Co (no collimator) + 1 Gy ^{60}Co	N-3-4	5.0
	19	3.0	166 keV x-ray	N-5-5	5.0
	20	3.0	2 Gy ^{137}Cs + 1 Gy ^{60}Co (collinear)	N-4-3	5.0
	28	0.5	^{241}Am	N-4-4	5.0
	29	3.0	1.5 Gy 120 keV x-ray, rotated 90° & 1.5 Gy 38 keV	N-4-5	5.0
Phase 3	30	3.0	1.5 Gy ^{137}Cs (no collimator) + 1.5 Gy ^{137}Cs (with collimator)	N-7-2	7.5
	31	3.0	2.0 Gy ^{137}Cs (no collimator) + 1.0 Gy ^{137}Cs (with collimator)	N-8-2	7.5
	43	3.0	^{60}Co: Repeat experiment 42 using an angle of rotation of 10°.	N-8-4	7.5
	44	6.0	^{60}Co: Irradiate RadBall™ with an initial dose of 1 Gy, then rotate the RadBall™ by 60° and give another 1 Gy dose. Repeat sequence until 6 irradiations have been completed and the RadBall™ has received a total dose of 6.0 Gy.	N-10-2	10.0
	45	6.0	^{60}Co: Repeat exp. 44 with the RadBall™ tilted at a 45 degree angle.	N-10-1	10.0

Figure 2. RadBall™ positioned in front of an automated irradiator system at the HPICL.

The tests at the HPICL consisted of three phases (Table 1):

- Phase 1 experiments were primarily used to obtain information on the target dose for RadBall™. Experiments were completed with a ^{137}Cs source and a ^{60}Co source over the range of 0.5 to 5 Gy.
- Phase 2 experiments investigated the RadBall™ performance with different radiation sources and different collimator thicknesses.
- Phase 3 experiments investigated the ability of the RadBall™ technology to perform with partially exposed RadBall™ polymers. RadBall™ polymers were given a radiation dose without the shielding collimator and then a second irradiation was performed with the collimator fitted.

The ^{241}Am source was used for one exposure; however, due to its low activity and therefore long exposure times, it was not used in any other test. With the exception of experiment 28 (^{241}Am exposed at a distance of 30 cm) all experiments were completed with the RadBall™ positioned 100 cm away from the radiation source. At 30 cm, the ^{241}Am source could not be considered a point source because it is large (consisting of seven separate sources and resembling an area source at 30 cm). An x-ray source was used, which generated various photon energies with peaks at 166 keV, 120 keV, and 38 keV. The x-ray source was not ideal since it generated a distribution of photon energies; however, the peaks available were near the 60 keV gamma ray energy emitted from ^{241}Am. The x-ray machine offered a much higher rate of exposure than the ^{241}Am source and was therefore much more convenient in terms of exposure time.

After irradiation at the HPICL, the RadBall™ polymers were sent to Duke University Medical Center for optical-CT scanning. The results from the optical-CT scan were subsequently analysed at NNL, using the Image Processing and Analysis in Java (ImageJ) software (ImageJ website: http://rsbweb.nih.gov/ij/).

Phase 1 experiments produced results similar to that displayed in Figure 3, which shows a horizontal slice of a RadBall™ (N-2-4) with (a) pre- and (b) post- contrast enhancement applied (a feature within the ImageJ software, which allows the contrast between the radiation tracks and the background to be enhanced). Seven tracks are visible in the slice with the widest diameter radiation track in the middle. This is to be as expected as the collimation hole facing the source was slightly larger than the other holes on the collimation device. The diminishing intensity of the peaks to the left and right of the middle peak are due to the orientation of the hole following the curvature of the collimator geometry. The direction that each hole was drilled was always perpendicular to the surface of the sphere.

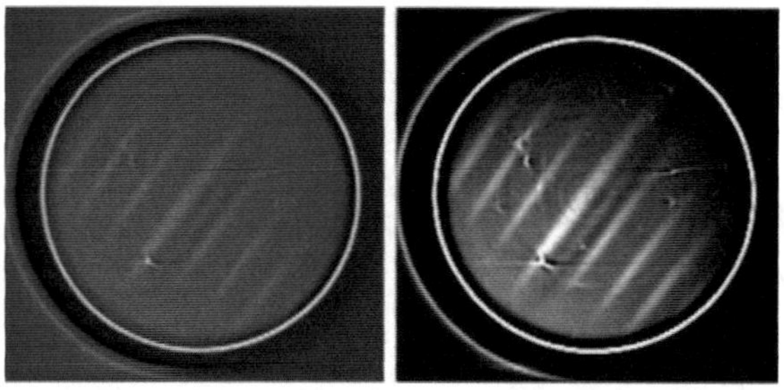

(a) **(b)**

Figure 3. Images of (a) Pre and (b) Post contrast enhanced slices of RadBall™ N-2-4

Figure 4 shows the scan images of RadBall™ N-4-5 (Table 1) irradiated with two sources (1.5 Gy 120 keV x-ray and 1.5 Gy 38 keV at 90° to the first irradiation). The scans show four radiation tracks entering from the top left and three radiation tracks entering from the top right, which crossover in the middle of the RadBall™.

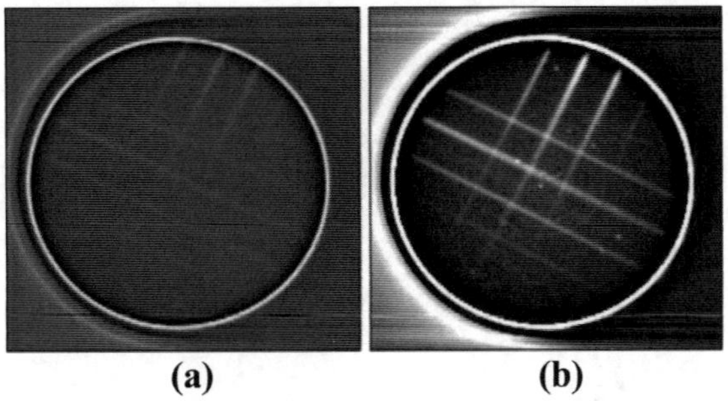

Figure 4. Images of (a) Pre- and (b) Post- contrast enhanced slices of RadBallTM N-4-5.

Figure 5 shows optical scan images for RadBall™ N-10-2 (Table 1). In this experiment the RadBall™ was given an initial dose of 1 Gy and then rotated by 60 degrees and exposed again. This exercise was repeated until six irradiations had been completed. The pattern appearing in the polymer is similar to isometric graph paper. Higher intensity bright spots appear where one or more radiation tracks intersect, increasing the radiation dose delivered to that area.

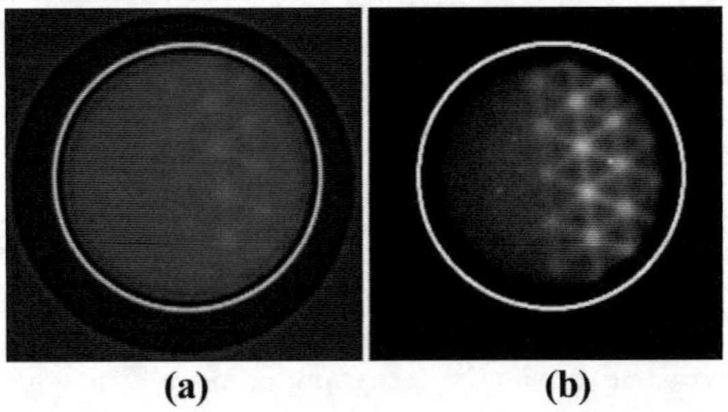

Figure 5. Images of (a) Pre- and (b) Post- contrast enhanced slices of RadBall™ N-10-2.

3. Discussions and Results

The RadBall™ technology has responded well during the HPICL calibration facility experiments. Radiation tracks were visible in all of the 45 experiments and demonstrated that the RadBall™ technology is sensitive to ^{60}Co, ^{137}Cs, and ^{241}Am sources over the radiation range of 0.5 to 8 Gy. Based on visual interpretation of the two-dimensional slice representations the target dose for the RadBall™ is 1.5 to 3.0 Gy and the optimum collimation thickness is 10 mm. Further work is ongoing to investigate whether the RadBall™ technology is able to characterize more complex radiation sources.

References

[1] Stanley SJ. RadBall: A Small Deployable Polymer Based Radiation Mapping Device for Locating, Quantifying and Characterizing Contamination in a Given Cell/Glovebox/Room, Nuclear Future, 4(6), pp 302-308; 2008.

[2] Farfán EB, Foley TQ, Jannik GT, Gordon JR, Harpring LJ, Blessing R, Stanley SJ, Holmes CJ, Oldham M, and Adamovics J. Testing of the RadBall Technology at Savannah River National Laboratory. Savannah River National Laboratory, Aiken, SC 2009 (SRNL-2009-00808).

[3] Stanley SJ, RadBall ready to roll, World Nuclear News(www.world-nuclear-news.org); June 2008.

[4] Doran SJ, Stanley SJ, Jenneson PM, Prott E, Adamovics J, RadBall™: a new field for 3D dosimetry, *DOSGEL: The 5th International Conference on Radiotherapy Gel Dosimetry,* University of Heraklion, Crete; 2008.

[5] Oldham M. *3D Dosimetry by Optical-CT scanning, Institute of Physics Publishing Journal of Physics*, Conference Series 56 58–71; 2006.

[6] Guo P, Adamovics J and Oldham M. Simple 3D validation experiments for PRESAGE™/optical-CT dosimetry *Institute of Physics Publishing Journal of Physics: Conference Series* 56187–190; 2006.

[7] Guo P, Adamovics J and Oldham M. Quality assurance in 3D dosimetry by optical-CT *Institute of Physics Publishing Journal of Physics: Conference Series* 56 191–194; 2006.

Journal of Physics: Conference Series **250** (2010) 012081 doi:10.1088/1742-6596/250/1/012081

RadBall^{TM} Technology Testing and MCNP Modeling of the Tungsten Collimator

Eduardo B. Farfán[1], Trevor Q. Foley[1], J. Rusty Coleman [1], G. Timothy Jannik[1], Christopher J. Holmes[2], Mark Oldham[3], John Adamovics[4], and Steven J. Stanley[2]

[1] Savannah River National Laboratory, SRS, Aiken, SC 29808, USA.
[2] National Nuclear Laboratory, Risley, Warrington, United Kingdom.
[3] Duke University Medical Center, Durham, NC 27710, USA.
[4] Heuris Pharma LLC, Lawrenceville, NJ 08648, USA.

Email: Eduardo.Farfan@srnl.doe.gov

Abstract. The United Kingdom's National Nuclear Laboratory (NNL) has developed a remote, non-electrical, radiation-mapping device known as RadBall^{TM}, which can locate and quantify radioactive hazards within contaminated areas of the nuclear industry. RadBall^{TM} consists of a colander-like outer shell that houses a radiation-sensitive polymer sphere. The outer shell works to collimate radiation sources and those areas of the polymer sphere that are exposed react, becoming increasingly more opaque, in proportion to the absorbed dose. The polymer sphere is imaged in an optical-CT scanner, which produces a high resolution 3D map of optical attenuation coefficients. Subsequent analysis of the optical attenuation matrix provides information on the spatial distribution of sources in a given area forming a 3D characterization of the area of interest. RadBall^{TM} has no power requirements and can be positioned in tight or hard-to reach locations. The RadBall^{TM} technology has been deployed in a number of technology trials in nuclear waste reprocessing plants at Sellafield in the United Kingdom and facilities of the Savannah River National Laboratory (SRNL). This study focuses on the RadBall^{TM} testing and modeling accomplished at SRNL.

1. Introduction

The RadBall^{TM} technology was initially developed by the United Kingdom's National Nuclear Laboratory (NNL). The U.S. Department of Energy (DOE) and Savannah River National Laboratory (SRNL) have initiated a collaborative development initiative with the NNL, to test and further develop the technology. RadBall^{TM} was developed to provide surveys of high radiation environments such as reactor spaces or hot cells. It has the advantage of being relatively inexpensive, having no electronic parts subject to radiation damage, and requiring no power. The product of a RadBall^{TM} deployment is a three dimensional projection of gamma radiation sources for the space into which it is deployed. Certain elements of the overall technology and deployment system are still under development and other components will need refinement and modification to optimize performance as additional demonstration projects are conducted. The optimized technology is expected to provide high quality survey results at modest cost and reduce worker exposure relative to existing approaches related to radiation surveys of highly contaminated facilities. This study involved the testing of the RadBall^{TM}

Journal of Physics: Conference Series **250** (2010) 012081 doi:10.1088/1742-6596/250/1/012081

technology under controlled radiation exposures at the Savannah River Site (SRS) and modeling of the tungsten collimator using a Monte Carlo computer code.

2. RadBall™ Testing

RadBall™ (a novel, non-electrical, remote radiation mapping device) offers a means to collect 3-D information regarding the intensity and distribution of radiation in a given hot cell, glovebox, or room [1, 2]. The device consists of two constituent parts shown in Figure 1: an inner core and an outer collimation shell. The inner core is made from a radiation sensitive polymer material: PRESAGE™. On manufacture, the inner core is semi-transparent; however, on exposure to radiation the material exhibits an increase in opacity. The outer collimation shell is cast from a tungsten alloy and contains over 200 individual collimation holes.

2.1. The principle of RadBall™

Once the device is deployed in a radioactive area, the collimation shell partially attenuates the incident radiation while preferentially allowing radiation to pass through the collimation holes. This radiation then creates opacity tracks within the inner core which provides information required to locate and quantify the radiation sources. The orientation of the opacity track provides the positional information regarding the source, achieved by using a reverse ray tracing technique. The activity of the detected source is assessed by quantifying the magnitude of the opacity change, which follows a linear relationship with respect to absorbed dose.

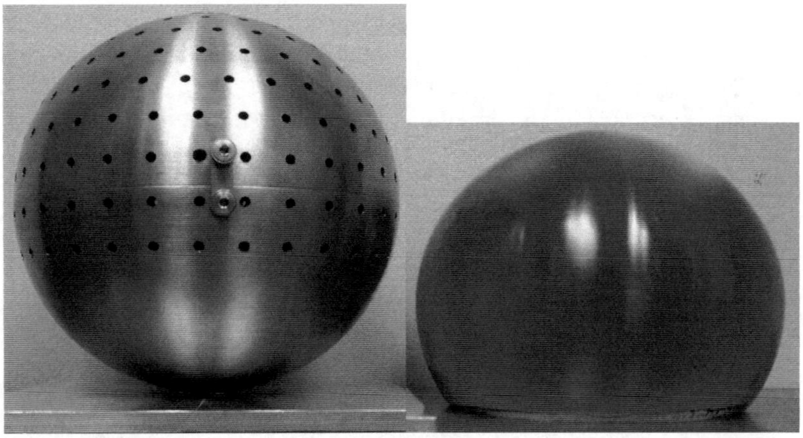

A B

Figure 1. Two components of a RadBall™ device: A) the outer collimation shell and B) inner polymer core.

2.2. RadBall™ testing at the HPICL

The RadBall™ technology was tested at the SRS Health Physics Instrument Calibration Laboratory (HPICL) in three phases involving various controlled exposure scenarios [2]. After irradiation, the RadBall™ PRESAGE™ polymers were sent to Duke University Medical Center (DUMC) for optical CT scanning. The scanning process produces a digital representation of the exposed polymer. Three experimental phases were completed at the HPICL, which included the exposure of 45 RadBall™ PRESAGE™ polymers.

2.3. Experimental phases

All experiments were completed with the RadBall™ positioned 100 cm away from the radiation source in the HPICL. The center of the RadBall™ collimator was placed at the same height as the source. An x-ray source was used to test photons with significantly less energy than photons generated

Journal of Physics: Conference Series **250** (2010) 012081 doi:10.1088/1742-6596/250/1/012081

from ^{60}Co and ^{137}Cs sources. The x-ray machine generated an energy spectrum with peaks at 38 keV, 120 keV, and 166 keV.

Three RadBall™ tungsten collimators were used at the HPICL, a 5.0, 7.5, and 10.0 mm thick collimator with 2.25, 3.0, and 4.0 mm hole diameters, respectively. The 5.0 and 7.5 mm thick collimators also had a single larger hole with a diameter of 4.0 mm. All the other collimator holes were uniform in size. All of the collimator holes were drilled with the drill bit perpendicular to the RadBall™ surface and positioned towards the center of the collimator.

Table 1 details three selected experiments completed at the HPICL (using x-ray, ^{60}Co, and ^{137}Cs sources respectively). In each of these experiments the RadBall received a total dose of 3.0 Gy.

Table 1. Selected HPICL tests.

Test	Radiation Source	RadBall	Collimator Thickness (mm)
19	166 keV x-ray	N-5-5	5.0
25	^{60}Co	N-5-3	10.0
27	^{137}Cs	N-5-4	10.0

3. RadBall™ Modeling

An isotropic point source emitting photons with a single energy was modeled in a General Monte Carlo N-Particle Transport Code (MCNP) – Version 5[1]. Multiple MCNP runs were completed varying the energy, collimator thickness, and collimator hole diameter. The center of the collimator was placed one meter from the point source. Seven of the collimator holes were included in the model. These seven holes create a section of a single row containing collimator holes spaced at 12 degree increments, as shown in Figure 2. A hole pointed directly at the source and three holes on each side were modeled. Tungsten with a density of 19.25 g cm^{-3} was entered into MCNP. Eighty cells were placed behind the collimator holes (Figure 2). Photons that entered these cells were counted to determine what effect the tungsten had on the incident photons.

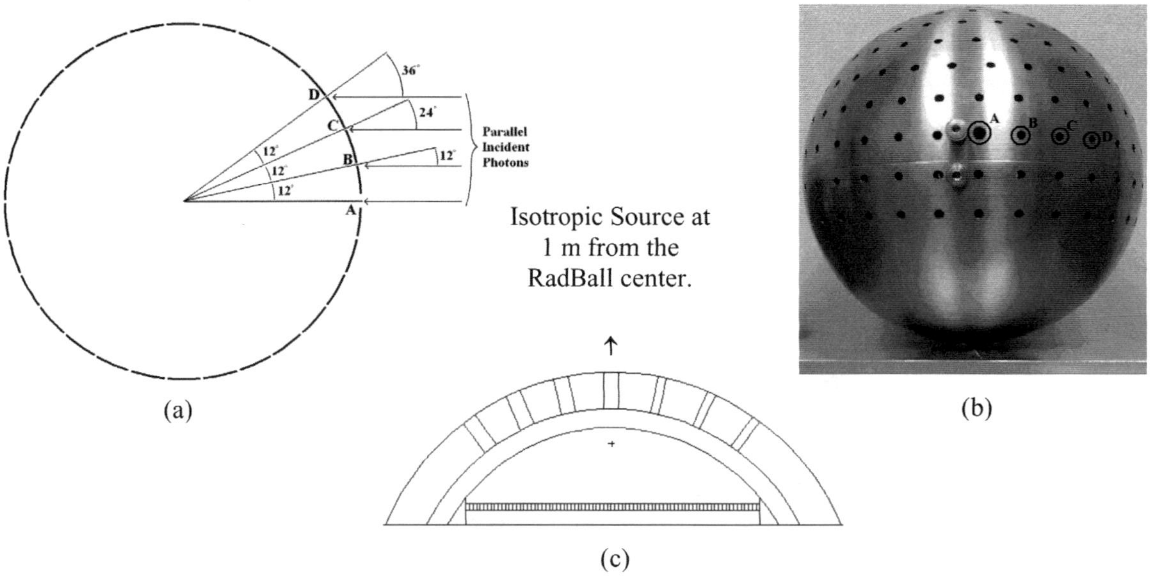

Figure 2. (a) Cross-section of the collimator with four holes labeled A,B,C, and D (b) photo of collimator showing labeled holes A,B,C, and D (c) computer model cross-section showing that only seven holes were considered and also showing the 80 MCNP tally cells.

1 MCNP is described in great detail at http://mcnp-green.lanl.gov/index.html and is available at http://www-rsicc.ornl.gov/

The 10 mm, 7.5 mm, and 5.0 mm thick collimators were modeled in MCNP. In the MCNP runs the actual hole diameter was used for the center and left side holes, different size holes were used on the right side. The three holes on the right side were all uniform in diameter size. This allowed for comparison between what was actually used in the HPICL exposures and alternative hole diameter sizes. It was hypothesized that smaller hole diameter sizes would improve the collimation effect on incoming photons. The collimation effect could not be easily shown in the experimental tests completed at HPICL as the collimators differed in thickness and hole diameter.

4. Results and conclusion

The Image Processing and Analysis in Java software (ImageJ website: http://rsbweb.nih.gov/ij/) representations for the tests in Table 1 are shown in Figure 3. These cross section images show the tracks created along a row of collimator holes. There are some high illumination points that appear in the images; these are unwanted artifacts. All the collimators provided enough shielding that tracks were visible in all 45 exposures. The dose range of 0.5 to 5.0 Gy created visible tracks with both the ^{60}Co and ^{137}Cs sources.

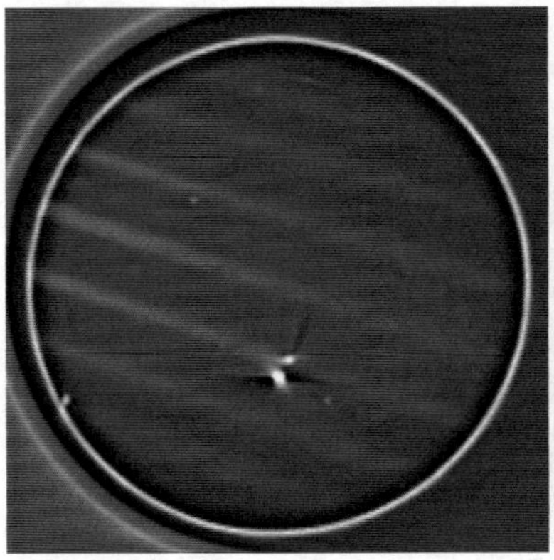

RadBall™ ID: N-5-4
Source: ^{137}Cs (662 keV)
Total Absorbed Dose: 3.0 Gy
Collimator Thickness: 10 mm
Hole Diameter: 4.0 mm
Point Source

RadBall™ ID: N-5-3
Source: ^{60}Co (1.17 & 1.33 MeV)
Total Absorbed Dose: 3.0 Gy
Collimator Thickness: 10 mm
Hole Diameter: 4.0 mm
Point Source

RadBall™ ID: N-5-5
Source: 166 keV x-ray
Total Absorbed Dose: 3.0 Gy
Collimator Thickness: 5.0 mm
Hole Diameter: 2.25 mm
Point Source

Figure 3. ImageJ representations of RadBall™ exposures at HPICL.

The collimator holes are designed to collimate the incident photons and prevent exposure to the RadBall™ dosimeter if the incident photons are not aligned with the collimation holes. The exposures with the ^{60}Co and ^{137}Cs sources demonstrate that the collimator holes allowed an angle of approximately 36 degrees into the dosimeter at higher photon energies above 600 keV. If it is deemed necessary to lower this 36 degree angle of view, it is recommended that the thickness of the collimator be increased or the diameter of the hole decreased.

As expected, the MCNP modeling represented in Figure 4 has indicated that higher energy photons are not fully shielded by the RadBall™ collimator. MCNP modeling demonstrates that higher energy photons (above 200 keV) penetrate the solid section of the collimator; thus uniformly irradiating the RadBall™. This is shown by the minimum values on the Figure 4 plots of the lines pertaining to energies above 200 keV; minimums above zero indicate photons penetrating the tungsten shield. This has to be accounted for in the analysis of the RadBall™ scan and could be used for isotopic differentiation between isotopes that emit significantly different photon energies.

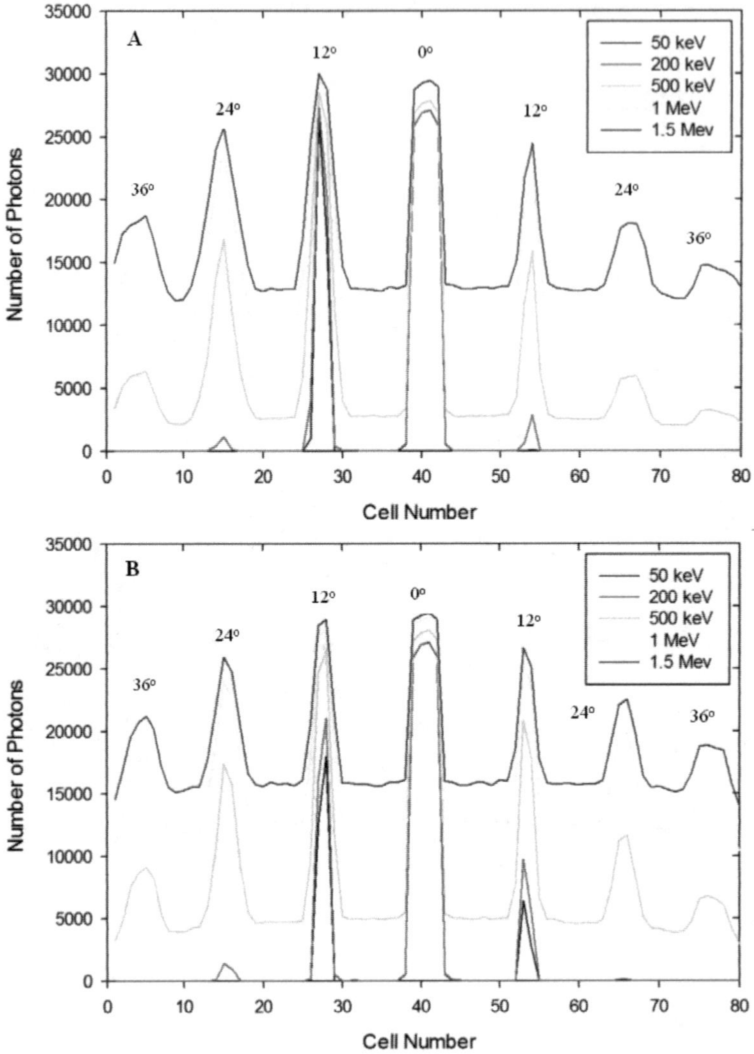

Figure 4. Number of particles crossing the cells. A) 10 mm thick tungsten collimator with 4.0 mm holes left (0 to 45 cell numbers) and 2.25 mm holes right (46 to 80 cell numbers). B) 7.5 mm thick tungsten collimator with 3.0 mm holes left (0 to 45 cell numbers) and 2.25 mm holes right (46 to 80 cell numbers).

5. References

[1] S.J. Stanley, RadBall: A Small Deployable Polymer Based Radiation Mapping Device for Locating, Quantifying and Characterizing Contamination in a Given Cell/Glovebox/Room, Nuclear Future, **4**(6), pp 302-308 (2008).

[2] E.B. Farfán, T. Q. Foley, G.T. Jannik, J. R. Gordon, L.J. Harpring, R. Blessing, S.J. Stanley, C.J. Holmes, M. Oldham, and J. Adamovics, *Testing of the RadBall Technology at Savannah River National Laboratory*. Savannah River National Laboratory, Aiken, SC 2009 (SRNL-2009-00808).

Journal of Physics: Conference Series **250** (2010) 012082 doi:10.1088/1742-6596/250/1/012082

MAGIC-*f* Gel in Nuclear Medicine Dosimetry: study in an external beam of Iodine-131

M Schwarcke, T Marques, C Garrido, P Nicolucci and O Baffa

Departamento de Física e Matemática, Universidade de São Paulo, Av. Bandeirantes, 3900, Monte Alegre, Ribeirão Preto, São Paulo, 14040-900, Brazil

E-mail: mschwarcke@usp.br

Abstract. MAGIC-*f* gel applicability in Nuclear Medicine dosimetry was investigated by exposure to a ^{131}I source. Calibration was made to provide known absorbed doses in different positions around the source. The absorbed dose in gel was compared with a Monte Carlo Simulation using PENELOPE code and a thermoluminescent dosimetry (TLD). Using MRI analysis for the gel a R2-dose sensitivity of 0.23 $s^{-1}Gy^{-1}$ was obtained. The agreement between dose-distance curves obtained with Monte Carlo simulation and TLD was better than 97% and for MAGIC-*f* and TLD was better than 98%. The results show the potential of polymer gel for application in nuclear medicine where three dimensional dose distribution is demanded.

1. Introduction

The use of radionuclides for cancer treatment led to new area of radiotherapy that is known as: targeted radionuclide therapy (TRT). This area requires specific dosimetric studies; from the theoretical point of view one possibility is to use Monte Carlo Simulation and experimentally this task can be accomplished by dosimeters, among these, polymer gel dosimetry is a relatively new method for 3D dosimetry with potential for TRT that has only a few applications in nuclear medicine [1].

On the other hand Monte Carlo (MC) Simulation is one of the most used tools assisting TRT dosimetry. The use of MC simulation codes in nuclear medicine could be more accurate through the combination with data from the Medical Internal Radiation Dosimetry (MIRD) [2]. The combination of the radiation transport in matter made by the PENELOPE code [3] and the biological information given by MIRD would make this kind of tool widely used in TRT.

Specific calibration procedures are required in nuclear medicine dosimetry considering that life times are short leading to source activity decay in a time compared with the exposition time, a situation completely different when compared with other sources used in radiotherapy consequently the dose rate also changes drastically over time. In this way, polymer gel dosimeters could be a feasible option because they are able to integrate dose in time without fluency dependency [5].

An experimental dosimetric study using ^{131}I was reported where the radionuclide was added during the gel manufacturing, getting a transversal relaxation rate (R2) value as function of the dose [4]. Our study sought to determine a relationship between R2 value and dose for an external beam, investigating the dose-response of MAGIC-*f* dosimetry for a radiotherapy treatment using ^{133}I source.

Journal of Physics: Conference Series **250** (2010) 012082 doi:10.1088/1742-6596/250/1/012082

2. Materials and Methods

Irradiation procedures were performed in the Nuclear Medicine Service (NMS) of the Hospital and Clinics of the School of Medicine of Ribeirão Preto (FMRP).

2.1. Irradiation process

In this work an iodine-131 solution with 1110 MBq of initial activity was used, the initial volume as delivered by the radionuclide supplier was diluted in final volume of 5 mL of 0.9% saline solution. Thermoluminescent dosimeters (LiF-100) were positioned radialy in front and behind the gel tubes to provide dose comparisons.

Irradiation geometry was made aiming the study of gel response in an external iodine-131 irradiation. The gel tubes and TLD were positioned in the experimental arrangement shown in Figure 1. **A** and **B** regions are symmetrical to each other and this geometry was used to study the gel dose-response with the distance and to have two measurements and estimate the uncertainty, their standard deviation. **C** region was used to calculate the attenuation coefficient for the MAGIC-*f* to ^{131}I. With this set up an efficient use the exposure time could be attained.

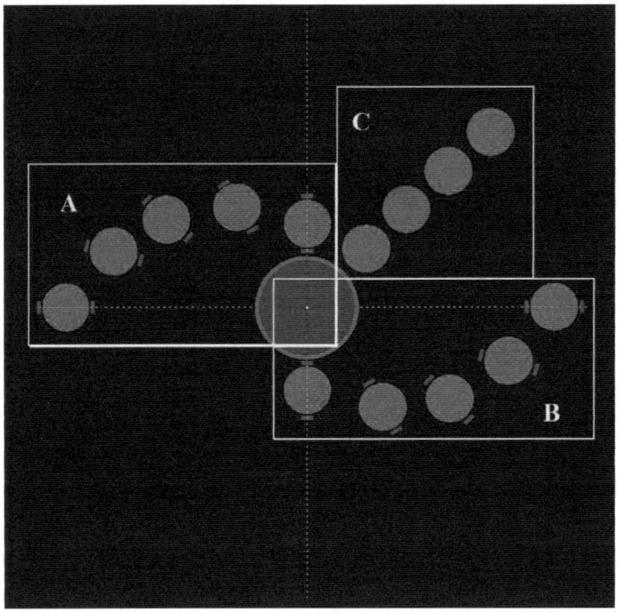

Figure 1: Experimental arrangement for gel and TLD irradiation in ^{131}I source. (A) and (B) study regions were designed to obtain the dose-response with the distance and uncertainties and (C) study region to obtain the attenuation coefficient.

2.2. Gel Manufacturing

The following composition was used for preparation of 50 ml of gel: 4.1 g of Bovine Skin Gelatin (300 Bloom), 17.6 mg of Ascorbic Acid; 2.95 g of Methacrylic Acid; 1 mg of Cooper Sulphate; 1.5 ml of Formaldehyde solution with 30% concentration stabilized in 10% of Methanol and 40.50 ml of Water Deionized [6]. The gel was stored in rubber capped glass tubes previously conditioned with a soft vacuum, commonly used for blood collection, and kept in a refrigerator at 10° during 24 hours before irradiation.

2.3. Magnetic Resonance Measurements

The gel imaging was performed in a Philips Achieva 3.0T Magnetic Resonance Imaging (MRI) Scanner, using the head coil and a multi spin echo sequence with 16 equidistant echoes, TE (echo-time) of 15-240 ms and TR (repetition-time) of 6000 ms, pixel size of 0.5 mm and a slice thickness of 2 mm. The slice position was select in order to avoid oxygen contamination effect in the image. The mean signal intensity in a region-of-interest defined in each tube (Figure 1) was extracted along the echo times. The transversal relaxation rat was calculated by fitting these mean intensities to a single-exponential decay. We report the R2 value for each tube and the standard deviation from the fitting. The relaxometry procedure is used to correlate R2 with absorbed dose [7].

2.4. Monte Carlo Simulation
The virtual source and geometry studied were faithfully represented in PENELOPE code. Photon energies considered for ^{131}I were 284.3 keV(6.2%), 364.5 keV(81.6%) and 606.0 keV(7.2%). The primary activity was 1.11×10^9 particles/s, simulating exactly the experimental source activity. Gel material was described in PENELOPE with the following weight fractions: $w_H = 10.62$, $w_C = 7.51$, $w_N = 1.39$, $w_O = 80.21$, $w_S = 2.58 \times 10^{-4}$ and $w_{Cu} = 5.08 \times 10^{-4}$. A previous work have already shown the equivalence of this simulated gel to the experimental one [8].

3. Results and Discussion

Simulated data were first compared to TLDs measurements to validate the geometry, material and source conditions. Figure 2-a presents absorbed dose values in different positions obtained with TLDs and simulated with PENELOPE. Calibration was made using an iodine-131 source with 962.6 MBq of activity with an irradiation time of 2 hours, the TLD where in air during irradiation and backscattering radiation was produced by water present inside the tubes. The concordance between these data ranged from 95.7 to 98.5%, with a maximum difference of 3%, thus a simulation was used to calibration the MAGIC-*f* gel.

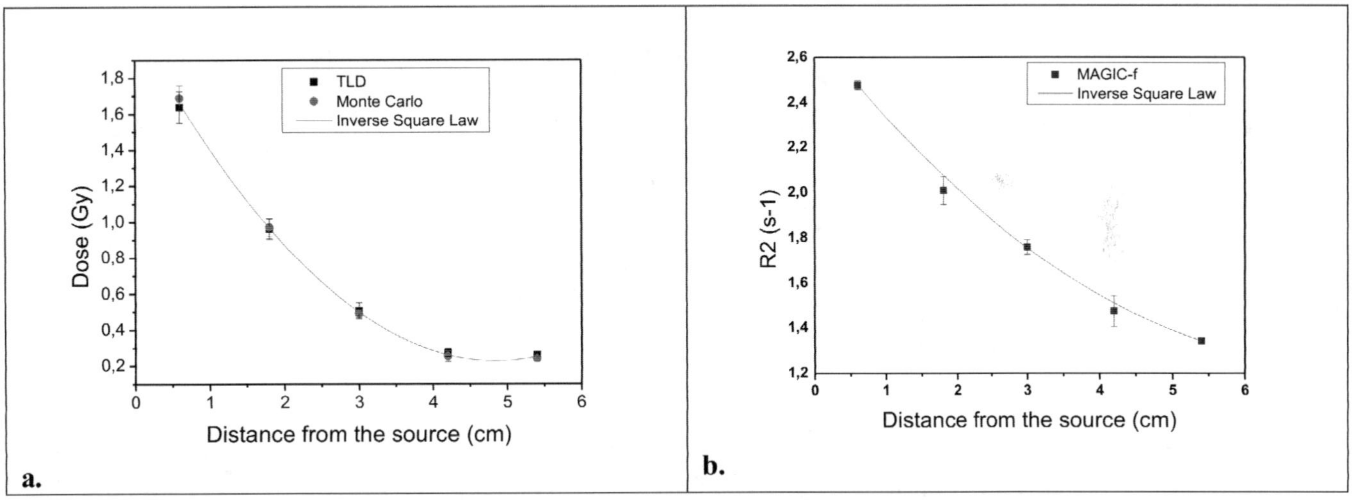

Figure 2: Absorbed dose using the geometry described in the Figure 1-a,b. (**a.**) relationship among absorbed dose in TLD and MC. (**b.**) relationship with R2 signal and distance from the source.

Figure 2-b presents the R2 value as a function of the distance from the source; it is possible to observe the agreement with the inverse square law, to a source activity of 1110 MBq. The same relation is represented in Figure 1-a, it is possible to observe a similar behavior between the both figures.

The range of dose predicted by Monte Carlo simulation for 14 hours of irradiation is 1.12 to 5.62 Gy, for 2 hours of irradiation the range of dose was 0.16 to 0.80 Gy. Figure 3 shows the linearity response obtained with gel MAGIC-*f*, sensitivity was 0.12 $s^{-1}Gy^{-1}$ for low doses, obtained with 2 hour irradiation and 0.22 $s^{-1}Gy^{-1}$ in the 14 hours irradiation. The intercept of the fitted line at zero dose shows the R2 of non-irradiated gel samples of 1.11 s^{-1} with a standard deviation of 0.1 s^{-1}.

The R2 signal for 2 hours of irradiation demonstrated that the experimental geometry and the short time of irradiation do not contribute to a stable polymerization process, probably because of oxygen contamination. Another study [4] demonstrated that for a low dose rate, the polymerization process is stable, but with an internal irradiation of the gel and a long time of irradiation condition what was different from the conditions used in the present work.

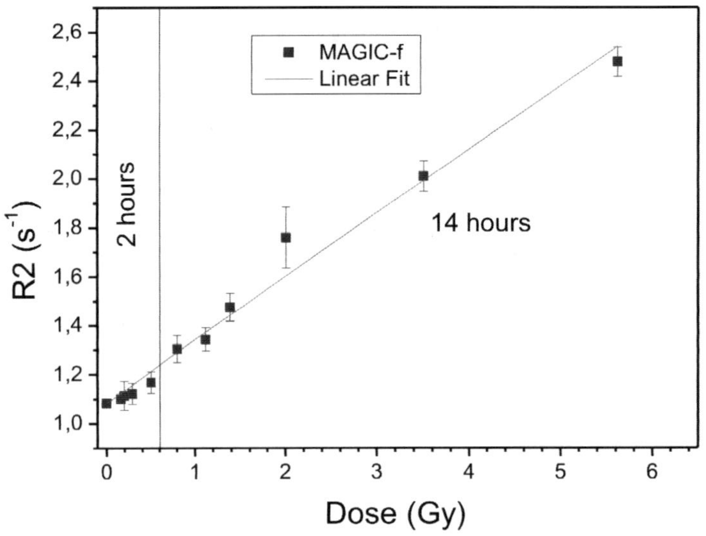

Figure 3: MAGIC-*f* sensitivity to Iodine-131 for 2 hours (Dose ≤ 0.60 Gy) and 14 hours of irradiation.

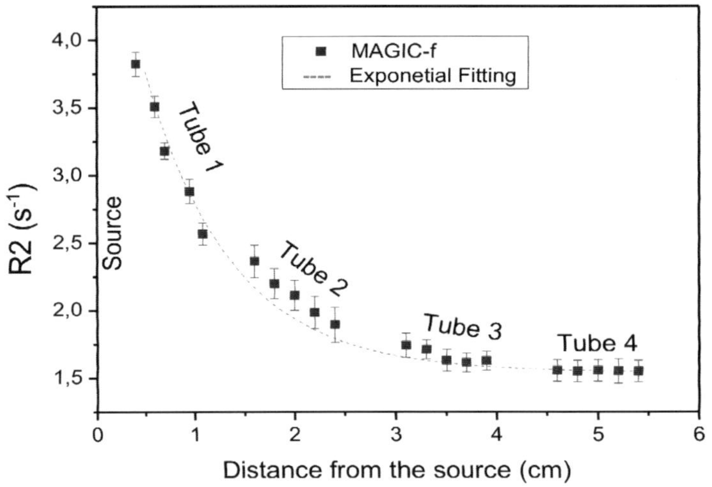

FIGURE 4: Relaxation rate R2 of the gel as function from the distance from the source.

Relaxation rates $R2_{Dose}$ distribution obtained along different distances in the polymer gel tubes at 14 hours exposure can be seen in Figure 4. This data was obtained with the arrangement show in Figure 1-c and into each tube 5 array of 3x5 pixel for each point in the Figure 4 were made. The attenuation coefficient for MAGIC-*f* was 1.28 cm^{-1} for an iodine-131 source, other studies are being studied at the present to more precisely determine the attenuation coefficient for MAGIC-*f*.

4. Conclusion

The results obtained agree well with reference values adopted in this work. MAGIC-*f* gel presents an adequate sensitivity for ^{131}I dosimetry, demonstrating the great ability of polymer gel in the study of TRT. This ability of gel includes the study of dose distribution in specific region, contributing as an accurate experimental tool to confirm the dose distribution simulated by Monte Carlo.

Acknowledgments

Authors thank the collaboration of Dr. Lauro Wichert-Ana, Dra. Léa Maciel, Dr. Thomaz Ghillardi Netto, Cecilio Jacob, Joseane Souza and to Luiz Aziani for the technical support and the CNPq, FAPESP and CAPES for partial financial support.

References

[1] Gear J I, Flux G D, Charles-Edwards E, Partridge M, Cook G and Ott R J 2006 The application of polymer gel dosimeters to dosimetry for targeted radionuclide therapy *Phys. Med. Biol.* **51** 3503-3516

[2] Snyder W.S., Ford M. R., Warner G.G. and Watson S. B. 1975 "MIRD Pamphlet 11: Absorbed Dose per Unit Cumulated Activity for Selected Radionuclides and Organs" *Society of Nuclear Medicine* Reston VA

[3] Salvat F, Fernández-Varea J M, Sempau J 2008 PENELOPE-2008: A code system for Monte Carlo Simulation of Electron and Photon Transport *Nuclear Energy Agency nº 6416.*

[4] Courbon F, Love P, Chittenden S, Flux G, Ravel P and Cook G 2006 Preparation and use of 131I MAGIC gel as a dosimeter for targeted radionuclide therapy *Cancer Biotherapy & Radiophramaceuticals* **21** (5) 427-436

[5] Baldock C, De Deene Y, Doran S, Ibbott G, Jirasek A, Lepage M, McAuley K B, Oldham M and Schreiner L J 2010 Polymer gel dosimetry – Topical Review *Physics in Medicine and Biology* **55** R1-R63

[6] Fernandes J P, Pastorell B F, de Araujo D B and Baffa O 2008 Formaldehyde increases MAGIC gel dosimeter melting point and sensitivity *Physcis in Medicine and Biology* 53 N53-N58

[7] De Deene Y, Van de Walle R, Achten E and De Wagter C 1998 Mathematical analysis and experimental investigation of noise in quantitative magnetic resonance imaging applied in polymer gel dosimetry *Signal Processing* **70** 85-101

[8] Pantelis E, Karlis A K, Kozicki M, Papagiannis P, Sakelliou L and Rosiak J M 2004 Polymer gel water equivalence and relative energy response with emphasis on low energy dosimetry in brachytherapy *Phys. Med. Biol.* **49** 3495-3514

Journal of Physics: Conference Series **250** (2010) 012083 doi:10.1088/1742-6596/250/1/012083

Verification of synchrotron microbeam radiation therapy using a purpose-built optical CT microscope

A. T. Abdul Rahman[1], Elke Bräuer-Krisch[2], Thierry Brochard[2], John Adamovics[3], David Bradley[1], Simon Doran[1,4]

[1]Department of Physics, University of Surrey, Guildford, UK

[2]European Synchrotron Radiation Facility, Grenoble, France

[3]Rider University, Lawrenceville, NJ, USA

[4]CRUK and EPSRC Cancer Imaging Centre, Institute of Cancer Research, Sutton, UK

Simon.Doran@icr.ac.uk

Abstract. This study presents an investigation of the use of 3-D dosimetry using optical computed tomography to provide verification of synchrotron microbeam radiation therapy (MRT). MRT is based on the remarkable tolerance of normal tissues to high doses of radiation when this dose is constrained to very narrow beams. At beamline ID17 of the European Synchrotron Radiation Facility, pre-clinical radiation therapy is delivered using an array of parallel microbeams of x-rays generated by a synchrotron-wiggler source. Measurement of the dose distribution around these microbeams requires a dosimeter with high spatial resolution, and the radiochromic plastic dosimeter PRESAGE™, used in conjunction with optical CT, is highly appropriate for this task. Two solid cylinders of 9.7 mm diameter PRESAGE™ were irradiated to create quality-assurance phantoms for the optical CT microscope using the dose-painting facilities at ID17. Images were analysed to ascertain the scanner linearity over the range 8 – 35 Gy and modulation-transfer function (MTF). With the initial scanner settings, MTF was found to be greater than 30% at 12 line pairs/mm and around 8% at 20.8 line pairs/mm, thus allowing individual lines of width 24 μm to be visualised. A further 9.7 mm PRESAGE™ sample was irradiated with a typical array of microbeams of FWHM 50 μm and centre-to-centre distance 400 μm. Results demonstrate how optical CT dosimetry may be capable (after further analysis) of making quantitative measurements of the peak-to-valley ratio of the microbeams. Finally, two samples of diameters 9.7 and 22 mm were irradiated from four directions using a typical MRT "cross-firing" pattern, and then imaged at two different image resolutions. The results show how optical CT dosimetry is able both to visualise the planned dose distribution and identify an incorrect treatment delivery.

1. Introduction

It is well established that normal tissues can tolerate high doses of radiation over small volumes. Despite the challenges involved in its implementation, microbeam radiation therapy (MRT) offers considerable promise in delivering lethal doses to a tumour while minimizing damage to the surrounding healthy tissue. The preclinical MRT research programme at the European Synchrotron Radiation Facility (ESRF) [1] involves very thin slice ("planar") beams of x-rays, arranged in parallel arrays with spaces in between. Specifically, the individual microbeams are 20–50 μm wide, with 100–

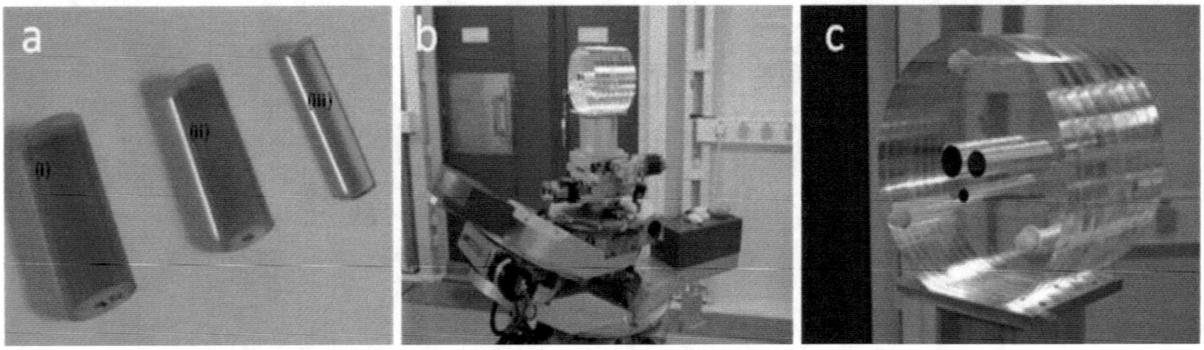

Figure 1: (a) Example of unirradiated PRESAGE™ radiochromic plastic samples; (i) 22 mm, (ii) 18 mm, and (iii) 9.7 mm; (b) the irradiation setup of microbeam radiation therapy (MRT) facility at beamline ID17, ESRF; (c) phantom used as a holder for PRESAGE™ during irradiation.

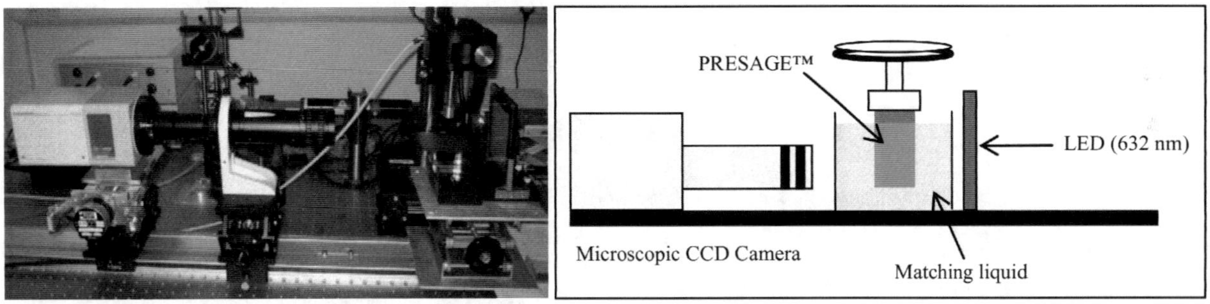

Figure 2: The optical CT microscopy scanner setup used for dose mapping of PRESAGE™ samples irradiated using MRT

400 μm centre-to-centre (c-t-c) spacing, delivering high peak dose values and relatively low valley dose values. Unlike the X-ray sources used in conventional clinical radiation therapy, a synchrotron facility can deliver extremely high dose-rates (hundreds of kGy min^{-1}). However, there is as yet no established physical dosimetric system for simultaneously providing accurate measurements of the doses in the microbeam peaks and valleys and at the same time providing 3-D spatial information on the volume as a whole. Monte Carlo simulations have been obtained [2] but these have yet to be validated by experimental measurements.

The main interest of this study is to investigate the potential of optical CT as a new candidate for dosimetry of the microbeams. Several studies have shown that the radiochromic plastic PRESAGE™ is a promising 3-D dosimeter system for radiation therapy and a dedicated optical CT microscopy scanner has been constructed [3] to evaluate the dose deposited in MRT treatments. This abstract presents the first study showing that this scanner can indeed make quantitative dose measurements at high spatial resolution.

2. Materials and Methods

PRESAGE™ was provided by Heuris Pharma (Skillman, NJ) in the form of cylinders with diameters 22, 18 and 9.7 mm. These were machined locally to give a uniform height of 60 mm, as shown in Figure 1. A Perspex phantom was constructed in-house to provide realistic scattering conditions during irradiation. The majority of investigations reported here used the 9.7 mm diameter samples.

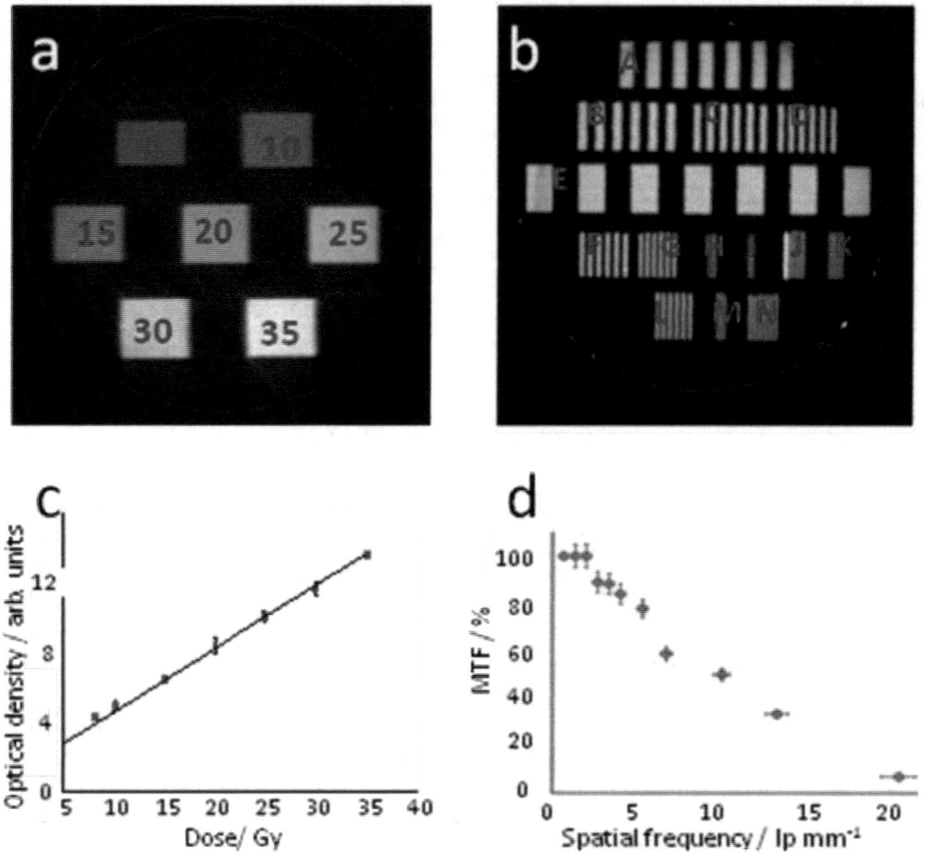

Figure 3: Reconstructed images from the 3-D datasets of (a) Sample 1, and (b) Sample 2; the numbers overlaying the square patterns in Sample 1 are the doses in Gy; the letters overlaying Sample 2 denote the line spacings in lp/mm as follows: (A) 1.37, (B) 2.04, (C) 2.73, (D) 3.43, (E) 0.68, (F) 4.10, (G) 5.43, (H) 27.8, (I) 35.7, (J) 10.2, (K) 13.5, (L) 5.43, (M) 20.8, (N) 6.85; (c) plot of deposited dose vs. measured optical density, demonstrating linearity of optical CT scanner; (d) calculated MTF for the scanner in this configuration based on the measurable results from the bar pattern.

Irradiations were carried out at the European Synchrotron Radiation Facility (ESRF) in Grenoble on the ID17 biomedical beamline. This beamline uses a wiggler source to produce an intense, highly collimated synchrotron x-ray beam from the 6 GeV electrons circulating at a beam current of approximately 200 mA. The beam has a filtered white beam spectrum, the lower energies being cut off through the use of 1.5 mm of aluminum and 1.0 mm of copper. This resulted in a maximum intensity at 83 keV, the energy spectrum covering the range 50 to 350 keV with a mean energy of 107 keV. The entrance dose on the samples was varied over several orders of magnitude, delivered at a dose rate of ~ 80 Gy per second per mA in the machine [1, 3]. The following irradiations were performed by moving the samples through the beam to provide various patterns mentioned:

- **Sample 1:** A linearity test pattern of seven 1.64 × 1.64 mm^2 squares, spanning a dose range of 8 – 35 Gy, arranged on a 9.7 mm diameter cylinder;
- **Sample 2:** An MTF test pattern, consisting of a set of near-ideal line- pairs, with a range of spacings from 0.68 – 35.7 lp/mm (732 – 14 μm), arranged on a 9.7 mm diameter cylinder;

Journal of Physics: Conference Series **250** (2010) 012083 doi:10.1088/1742-6596/250/1/012083

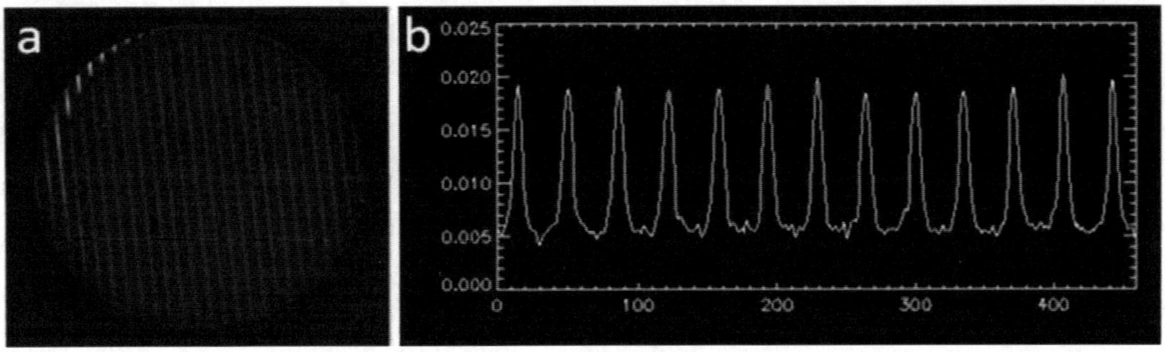

Figure 4: (a) Projection image Sample 3 (nominal FWHM 50 μm, 400 μm c-t-c); (b) profile through the dotted line in (a)

- **Sample 3:** A typical 3 × 3 mm^2 array of parallel microbeams of nominal FWHM 50 μm and c.t.c. 400 μm projected onto a centred 9.7 mm diameter PRESAGE™ cylinder, with entry dose 600 Gy;
- **Sample 4:** A typical "cross-firing" [1] MRT treatment, with beams entering the sample from four directions (0, 45, 90 and 135°), applied to a 22 mm-diameter sample, with entry dose 300 Gy;
- **Sample 5:** The same dose distribution as for Sample 4, but delivered to a 9.7 mm sample and with entry dose 140 Gy. This delivery failed for technical reasons and, although a further sample was irradiated and imaged successfully, we present these results to show the important role of 3-D dosimetry in providing quality assurance of this complex process.

Imaging was performed using the optical CT microscopy scanner described in [3]. As illustrated in Figure 2(b), the illumination arrangement is made more compact by replacing the collimated light source with a flat panel illuminator (Phlox PHLOX-LEDR-BL-100x100-S-Q-1R-24V). Spectrophotometer analysis using PRESAGE™ samples from the same batch, in standard optical cuvettes, was used as an independent measure to validate the optical CT measurements. The image field-of-view (FOV) is approximately 12 mm, and the 2-D slices were reconstructed onto matrices of 1024 × 1024 from a set of 800 projections, each of 512 × 512 pixels. Note that this enhanced pixel resolution is justified by the Nyquist Theorem at the centre of the FOV though not at the edge.

3. Results and Discussions

Figures 3(a,b) show a typical slice from the reconstructed images of Samples 1 and 2 respectively. Since we know from many previous studies that the optical response of PRESAGE™ is linear with dose, the straight line evident in Figure 3(c) provides good evidence that the new scanner has a linear response in its measurement of optical density. A similar measurement (results not shown) with squares irradiated between 20 and 80 Gy gives results that are equally linear ($r^2 > 0.994$ for both measurements). Obtaining profiles across each of the line patterns in Figure 3(b) shows that lines are resolved for all patterns up to and including (M) at 20.8 lp/mm (24 μm line width), and that the MTF at (K) 13.5 lp/mm (37 μm line width) is well over 30%.

The results of imaging Sample 3 are shown in Figure 4. One of the interests of microbeam radiation therapy (MRT) system on ID17 microbeam line in ESRF is to deliver the lethal doses to a tumour while minimizing damage to the surrounding healthy tissue. For the normal tissue sparing to be effective, the "valley" dose in the gaps between adjacent microbeams must be kept below a certain threshold. An important goal of the current research is to measure this peak/valley ratio and to compare the empirical results with Monte Carlo measurements. To do this successfully, it will be

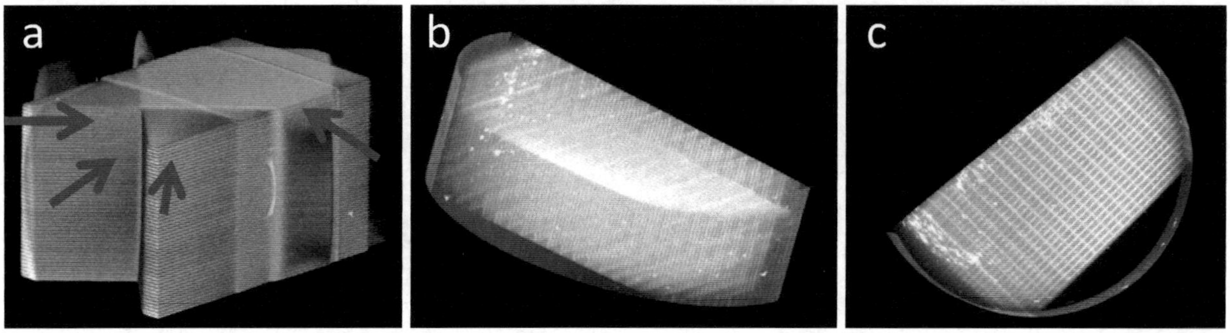

Figure 5: Surface-rendered 3-D views of (a) Sample 4, with directions of the incident radiation indicated with arrows; and (b, c) Sample 5. Note how in (a), the top of the central region shows a uniform dose has been deposited by the overlap of all four microbeam arrays, whereas in (b), it is clear that one of the four irradiations has failed, with only a single plane of radiation delivered.

necessary to deconvolve the known MTF from the profile in Figure 4(b), which should be achievable. However, even before accurate dosimetry at the microbeam lengthscale is demonstrated, the mere ability to image the microbeams at such high resolution is already proving important. Figure 5 shows the results of imaging the full MRT treatment applied to Samples 4 and 5. In particular, Figure 5(b) illustrates a failed treatment, showing graphically just why a full 3-D verification is important for such a new treatment modality.

4. Conclusions

We have demonstrated how the University of Surrey optical CT microscopy scanner can make accurate relative dose measurements at moderately high spatial resolution $(1.64\ \text{mm})^2$. Images can be formed of structures at spatial resolutions of 25 microns and possibly lower, and MTF curves generated. We anticipate that it will be possible to use this information to deconvolve the lineshape of the microbeam profile and make measurements of the doses deposited by individual microbeams. These encouraging results suggest that the methods will be applicable to the synchrotron microbeam radiation therapy application. A brief initial study has shown that the technique is well able to detect errors in delivery.

Acknowledgement

The authors gratefully acknowledge the allocation of beam time at ESRF and the associated funding for experiment MD-439.

References

[1] Bräuer-Krisch, E., et al., New irradiation geometry for microbeam radiation therapy. Physics in Medicine and Biology, 2005. **50**(13): p. 3103-3111.
[2] Siegbahn, E.A., Stepanek, J., Brauer-Krisch, E. and Bravin, A., Determination of dosimetrical quantities used in microbeam radiation therapy (MRT) with Monte Carlo simulations. Medical Physics, 2006. **33**(9): p. 3248-3259.
[3] Doran, S.J., et al., An investigation of the potential of optical computed tomography for imaging of synchrotron generated x-rays at high spatial resolution. Phys Med Biol, 2010. **55**: p. 1531-1547.

Journal of Physics: Conference Series **250** (2010) 012084 doi:10.1088/1742-6596/250/1/012084

Gel Dosimetry Analysis of Gold Nanoparticle Application in Kilovoltage Radiation Therapy

T. Marques[‡], M. Schwarcke[‡], C. Garrido[‡], V. Zucolotto[†], O. Baffa[‡], P. Nicolucci[‡]

[‡] Department of Physics and Mathematics - University of Sao Paulo.
Av. Bandeirantes, 3900 - 14040-901. Ribeirão Preto – SP, Brazil.

[†] Institute of Physics - IFSC - University of Sao Paulo.
Av. Trabalhador São Carlense,400 - 13560-970. São Carlos - SP, Brazil.

tatiana.marques@usp.br

Abstract. In this work gold nanoparticles (AuNP) were embedded in MAGIC-*f* gel and irradiated in a 250 kV x-ray clinical beam. The signal of non-irradiated gel samples containing AuNPs showed maximum difference of 0.5% related to gel without nanoparticles. Different AuNPs concentrations were studied: 0.10 mM, 0.05 mM and 0.02 mM, presenting dose enhancements of 106%, 90% and 77% respectively. Monte Carlo spectrometry was performed to quantify theoretical changes in photon energy spectrums due to AuNPs presence. Concordance between simulated dose enhancements and gel dosimetry measurements was better than 97% to all concentrations studied. This study evidences that polymer gel dosimetry as a suitable tool to perform dosimetric investigations of nanoparticle applications in Radiation Therapy.

1. Introduction

Nanoparticles have been extensively investigated in diagnostic radiology and radiation therapy [1-2]. Radiosensitization observed by several authors points out gold nanoparticles (AuNP) as the best candidates to be used as flagships of Radionanotherapy [3-4]. This technique is becoming an increasingly practical option as nanobio-interactions are mapped.

The dosimetric suitability of AuNP applications in radiation therapy is one relevant aspect that can justify its applicability in clinical environments [5]. From the dosimetric point of view, AuNP application in radiation therapy is motivated by the fact that photoelectric probability increases due to metal presence inside a biological target. Essentially, metallic nanoparticles kept inside target cells make it possible to handle effective atomic number and mass-attenuation coefficient in clinical targets improving dose delivery effectiveness [6].

Dosimetry evolved in Radionanotherapy is complex and precise investigations of changes in dose painting due to nanoparticles presence can be performed using gel dosimetry. Conditions of Radionanotherapy treatments can be experimentally simulated incorporating AuNP to tissue-equivalent dosimeters. In this work different concentrations of AuNP were embedded in MAGIC-*f* gel and dose enhancement was quantified by gel dosimetry and Monte Carlo simulation to irradiations in a clinical x-ray beam.

2. Material and Methods

2.1. Gold nanoparticle synthesis

The synthesis of dendrimer encapsulated-nanoparticles consisted in the reduction of metallic precursors in presence of dendrimer polyamidoamine generation 4 (PAMAM G4). Briefly, in a 2.0mL of 1mMol L^{-1} gold tetrachloride acid aqueous solution ($HAuCl_4$), 2.0mL of the 1mMolL^{-1} PAMAM-G4 solution was added. After 15min under agitation, 2.0mL of formic acid (10% v/v) was added, which acted as reducing agent. The system was kept in the dark. After 2h, a red coloration appeared, indicating the gold nanoparticles formation. The PAMAM AuNPs were characterized using UV-Vis and dynamic scattering light measurements (DLS), revealing an average diameter of 2–3nm.

2.2. Nanoparticles incorporation in gel

MAGIC-*f* (*Methacrylic Ascorbic acid and Gelatin Initiated by Copper with formaldehyde*) [7] was prepared from 70% dehydrated gel at 40 °C. Four samples were isolated from the same batch, in which three of them contained the AuNPs. The sample not containing AuNPs served as control. Gel homogenization was achieved by routine mechanical mix without heating. Tree different AuNPs concentration were considered: 0.10mM (A), 0.05mM (B) and 0.02(C). Figure 1a presents the gel samples without nanoparticles with different polymerization levels. In figure 1b, gel containing different AuNP concentrations can be seen as different colors.

1.a

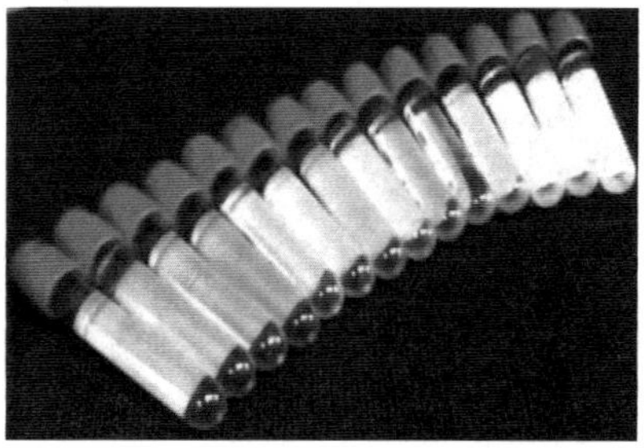
1.b

Figure 1 – Gel samples: (a) Polymerization levels of MAGIC-*f* samples irradiated with different doses; (b) Gold nanoparticles embedded in gel presenting visible color changing.

2.3. Irradiation procedures and NMR reading

Gel samples were irradiated in a Siemens Stabillipan ortovoltage clinical beam of 250kV at the Radiation Therapy facility of Hospital of Clinics of the Medical School from Ribeirão Preto - HCFMRP. MAGIC-*f* linearity of response and tissue equivalence for the beam studied were determined. A water box was positioned under the gel tubes to guarantee backscattering conditions. Control gel samples and samples containing AuNP were irradiated with 5Gy.

Dose measurement was made by relaxometry, which correlates dose and nuclear magnetic transversal relaxation rate (R2) [8]. Experiments were carried out in a Philips 3.0 Tesla Magnetic Resonance Imaging (MRI) scanner 24 hours after irradiation. A multi-spin echo sequence with 16 echoes, echo time of 15ms, repetition time of 4000ms, pixel size of 0.5x0.5mm^2 and two acquisitions averaged was used.

2.4. Monte Carlo simulations

Computational simulations of dose deposition were performed using PENELOPE Monte Carlo code [9]. Gold presence in gel was considered in the same experimental concentrations. Monte Carlo spectrometry was also performed using PENELOPE code to identify how the photon energy spectrum (PES) changes due to AuNP presence in gel. Scattered and absorbed components were identified and quantified in the PES to all nanoparticles concentrations studied.

3. Results and Discussion

MAGIC-*f* response to the 250 kV x-ray beam was characterized by comparing experimental and Monte Carlo simulated percentage depth dose curves (PDD) (see figure 2a). Maximum difference in experimental and simulated PDDs was 2.96% found at 15mm depth. Linearity of response to MAGIC-*f* was verified considering a dose range from 0 Gy to 10 Gy. Figure 2b shows the R2 signal corresponding to the evaluated doses.

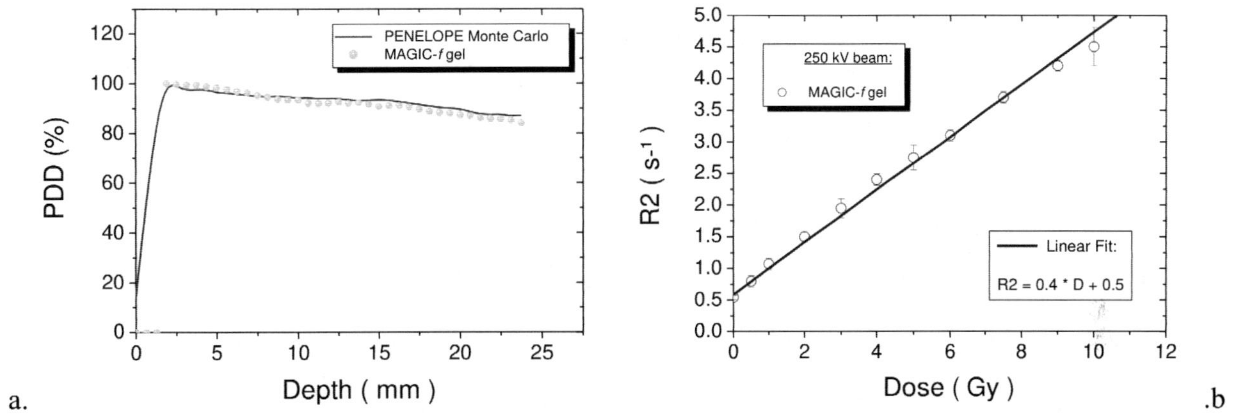

Figure 2 – Dosimetric response of MAGIC-*f* to the 250 kV x-ray beam: (a) PDD curves comparison between PENELOPE Monte Carlo simulation and gel; (b) Dose linearity response of gel.

The signal of non-irradiated tubes containing and not containing AuNPs were evaluated. Chemical interactions between PAMAM AuNPs and gel were discarded considering the low R2 signal variation found (less than 0.5%). Figure 3a presents the respective R2 averaged from 3 non-irradiated tubes considered to each nanoparticle concentration.

The dose increase due to AuNPs presence in the tissue-equivalent gel was quantified comparing the R2 of all different nanoparticles concentrations irradiated with the same dose (5Gy). Concentration A presents an increase of 106% in R2 related to gel without AuNP; concentrations B and C present increases of 90% and 77% respectively (see figure 3b).

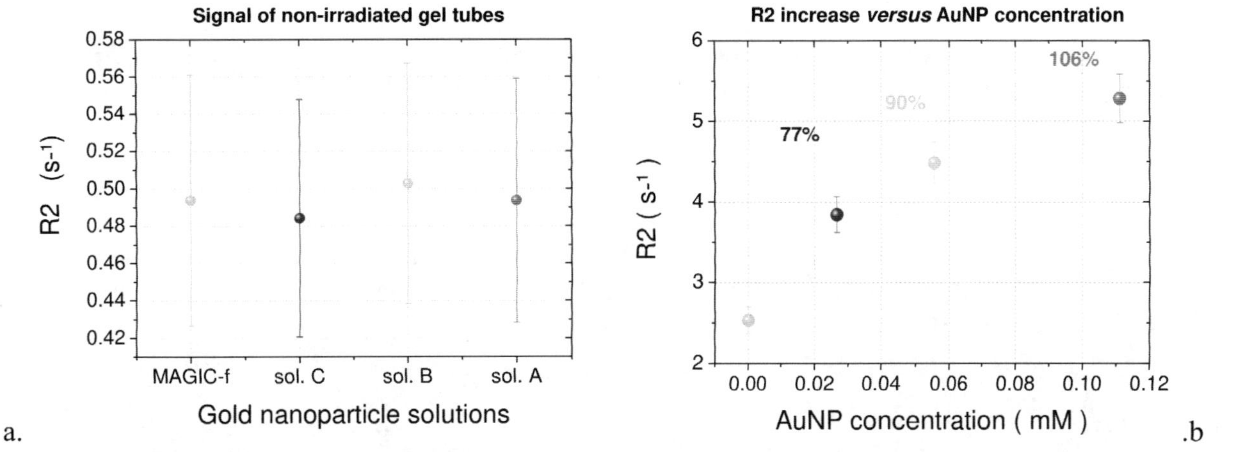

Figure 3 – (a) Signal of non-irradiated gel samples with and without nanoparticles; (b) Dose enhancement measurements performed with MAGIC-*f* gel with different concentrations of AuNP.

Monte Carlo spectrometry evidenced that the theoretical photon fluence absorption increases, resulting in 105%, 93% and 75% of dose enhancement when concentrations A, B and C of AuNP are present in gel respectively. Figure 4a shows a comparison between the PES inside gel without nanoparticles and gel containing concentration A of AuNP. It is possible to observe that the photon fluence increases between energies of 8 keV and 30 keV, indicating that incoherent scattering events occur in gold L-shells. This kind of interaction specially contributes to local dose increase because low energy photons are created with a short mean-free-path, resulting in low energy electrons created near the point where the primary photon was scattered. The absorbed part of the PES is also quantified in figure 4b. The absorption fractions in gold K-edge are referenced by the ratio between the photon fluencies in an energy range from 78 keV to 86 keV when AuNPs are not and are present in gel. That fraction is around 50% to all AuNPs concentrations studied. The absorption interactions also contribute to local dose enhancement due to photoelectron absorption.

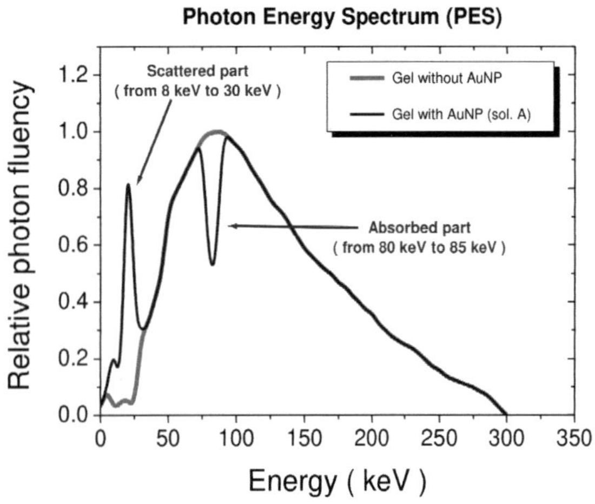

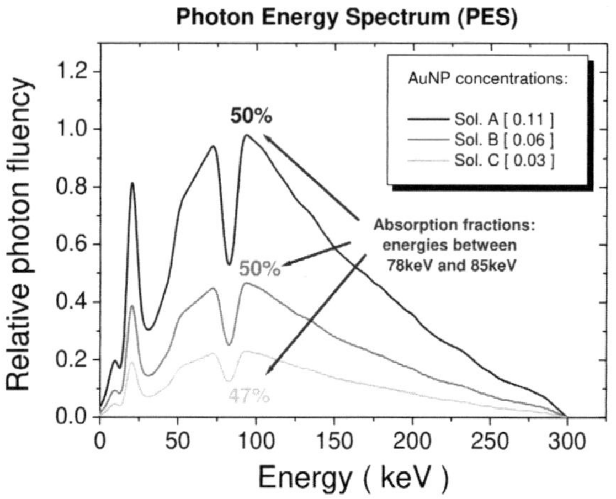

Figure 4 – Monte Carlo spectrometry inside MAGIC-*f* gel: (a) PES comparison between gel without gold nanoparticles and containing concentration A of AuNPs; (b) PES and absorption fraction inside gel containing concentrations A, B and C of AuNPs.

Dose enhancements verified by gel dosimetry and Monte Carlo simulation presented in this work instigate future investigations. Different biocompatible nanoparticles having different atomic numbers and other clinical beams can be investigated using polymer gel dosimetry aiming to optimize the application of nanoparticles in radiation therapy.

Acknowledgments

Authors are grateful to JL Aziani and to HCFMRP – USP for the essential technical contribution. This work was partially financial supported by CAPES / NANOBIOMED and CNPq. Technical and financial supports of CIDRA-USP are also appreciated.

References

[1] McMahon S.J., Mendenhall M.H., Jain S., *et al.*: "Radiotherapy in the presence of contrast agents: a general figure of merit and its application to gold nanoparticles", Phys. Med. Biol. 53, 5635-5651 (2008).

[2] Hainfeld J.F., Slatkin D.N. and Smilowitz H.M.: "The use of gold nanoparticles to enhance radiotherapy in mice", Phys. Med. Biol. 49,N309–N315 (2004).

[3] Zhang X., Gao J., Buchholz T.A., *et al.:* "Quantifying tumor-selective radiation dose enhancements using gold nanoparticles: a monte carlo simulation study", Biomed. Microdev. 11, 925–933 (2009).

[4] Zhang X., Xing J.Z., Chen J., *et al*: "Enhanced radiation sensitivity in prostate cancer by gold-nanoparticles", Clin. Invest. Med. 31, E160-E167 (2008).

[5] Cho S.H., Jones B.L., Krishnan S.: "The dosimetric feasibility of gold nanoparticle-aided radiation therapy (GNRT) via brachytherapy using low-energy gamma-/x-ray sources", Phys. Med. Biol. 54, 4889-4905 (2009).

[6] Cho S.H. "Estimation of tumor dose enhancement due to gold nanoparticles during typical radiation treatments: a preliminary Monte Carlo study", Phys. Med. Biol. 50, N163-N173 (2005).

[7] Fernandes J., Pastorello B., Araujo D., *et al.:* "Formaldehyde increases MAGIC gel dosimeter melting point and sensitivity", Phys. Med. Biol. 53, N53-N58 (2008).

[8] De Deene Y.,Van de Walle R., Achten E., *et al.:* "Mathematical analysis and experimental investigation of noise in quantitative magnetic resonance imaging applied in polymer gel dosimetry", Sig. Proces. 70, 85-101 (1998).

[9] Salvat F., Fernández-Varea J.M., Sempau J.: "PENELOPE, a code system for Monte Carlo simulation of electron and photon transport", *Facultat de Física (ECM), Universitat de Barcelona, Spain,* 2009.

Data analysis tools for 3D dosimetry: the use of CERR as a platform to integrate and compare measurements and treatment planning information

Joe Deasy and Aditya Apte

Dept. of Radiation Oncology, Washington University in St. Louis.

jdeasy@radonc.wustl.edu

Abstract. CERR, the Computational Environment for Radiotherapy Research, is a mature Matlab-based application that allows users to visualize and analyze 3D treatment planning data exported using standard protocols from clinical treatment planning systems. In this presentation we will give an in-depth discussion of the use of CERR as a tool to analyze measurements compared to expected treatment planning systems. Extensions to CERR allow for straightforward import and registration of experimental data with the planning data. These tools allow users to compare the match between measurement and treatment planning calculation in detail, as provided by profile plots and other tools. Custom Matlab scripts can also be developed, providing complete flexibility in analysis methods. In addition, several offshoot tools have been developed by our group to facilitate dosimetric data analysis, including: A film QA tool, developed under a contract for the Radiological Physics Center (RPC), and a Monte Carlo recalculation tool, also developed under the same contract for the RPC. The film QA tool is meant to facilitate the analysis of film that is irradiated in a phantom. The tool provides a simple method for registering pin-marked points on film to corresponding points in a CT-scanned phantom. Similarly, the locations of point dosimeters can be found. Once registered, data can be compared with the expected treatment plan, interpolated from the converted CERR plan. The dose-distance gamma function is available to quantify agreement. We will discuss the ways these tools can be used to support dosimetry research. All the software discussed here is being made available under open-source licensing.

Supported by NIH grant 5U10CA010953-41 (RPC).

Journal of Physics: Conference Series **250** (2010) 012086 doi:10.1088/1742-6596/250/1/012086

Imaging and 3-D dosimetry: top tips for MRI and optical CT

Simon J Doran[1,2]

[1] Department of Physics, University of Surrey, Surrey, UK

[2] CRUK and EPSRC Cancer Imaging Centre, Institute of Cancer Research, Sutton, Surrey, UK

Simon.Doran@icr.ac.uk

Abstract. The conference "refresher session" associated with this abstract reviews the main principles of the two most important imaging readout modalities for 3-D dosimetry: MRI and optical CT. Best practices for both these techniques are already described in several different places in the literature, but, for the uninitiated, there are a number of pitfalls. Here, I list some of the important considerations required to obtain good results from these methods and point to relevant prior work.

1. Introduction

Both MRI and optical CT are now well-established modalities for obtaining 3-D images of radiosensitive dosimetry phantoms. In the hands of an expert, both techniques can yield high quality results (e.g., [1, 2]). However, a number of pitfalls exist for the unwary researcher and, in order to obtain accurate results, a number of precautions (some obvious and some not-so-obvious) need to be taken. Commercial MR scanners and (to a lesser extent because the subject is some 20 years younger) optical CT scanners aim to make the acquisition of the 3-D data as simple as pushing a button. But although such developments are welcome, the data acquired can still lead to poor dose maps. The best way of avoiding this is to have a thorough understanding of the entire imaging process, and, to this end, previous "DOSGEL" conferences have included detailed articles on the background to both MRI and optical CT [3-7]. Familiarity with the dosimeter materials themselves is also important.

This article aims to distil the large body of knowledge into 10 easily digestible "top tips" for both MRI and optical CT. The suggestions are designed to encourage the reader to re-examine his or her protocol for acquisition and processing of 3-D dosimetry data, with the hope that this will lead to practical improvements in dose maps.

2. MRI

2.1 Decide whether MRI is the most appropriate technique.

As hinted at above, MRI has had much more time to mature as an imaging technology than has optical CT. For this reason, both the MRI equipment and the acquisition protocols tend to be more robust. Arguably, image artefacts have been better characterised and eliminated in MRI than optical CT. At the time of writing, MRI scanners are more widely available, although access is not always straightforward or cheap for the radiotherapy department. Quality control of the basic MR imaging process is embedded in the healthcare system in a way that is not true for optical CT. Furthermore, in the specific area of gel dosimetry, there is a longer history of research into MRI methods, and the dosimeter materials themselves have been characterised more fully. Owing to the problems that occur at interfaces when using optical CT, MRI is probably also the preferred choice if one wishes to include inhomogeneities of varying densities in the sample to create a more anatomically representative phantom.

Set against this are a number of significant reasons for choosing optical CT. Optical CT is much cheaper and it is straightforward for a planning department to buy its own scanner, which will have very low running costs. Optical CT is a much faster technique than MRI if isotropic 3-D scans are required. (If what is needed is a small number of representative slices, arbitrarily oriented in 3-D, then MRI is more competitive.) The samples used in MRI tend to be water equivalent gels; this sets an upper limit to the temperature at which scanning can take place. PRESAGE™ as used in optical CT is a very robust material that allows it to be used in harsh environments [8]. The polymer gels used in MRI can suffer from oxygen-related inhibition (see below).

2.2 Use an optimised multi-echo sequence.

In order to create an image of radiation dose, one needs to measure an MR parameter that changes with absorbed dose. The most commonly chosen of these is R_2 (the reciprocal of the NMR relaxation time T_2), and the standard method for obtaining the data required to compute an R_2 image is a multi-spin-echo pulse sequence. However, not all multi-echo sequences are created equal. In particular, the parameters chosen have a big influence on the signal-to-noise ratio and the acquisition time. De Deene and Baldock carried out an in depth study comparing different sequence designs [9].

2.3 Ensure that the 180° flip angles are properly adjusted.

The original method of multi-echo imaging was based on the Carr-Purcell-Meiboom-Gill (CPMG) pulse sequence, first proposed in the early days of NMR [10], which consists of a 90° excitation pulse followed by a string of 180° refocusing pulses. The imaging version of CPMG requires careful attention to the adjustment of RF pulse amplitudes. Modern scanners tend to use more sophisticated variants, often containing pulses with a much lower flip angle than 180°, in order to reduce RF power deposition. Furthermore, there is an increasing tendency for all the sequence adjustments to occur automatically during the "patient setup" and be hidden from the user. This "black box" approach can be dangerous. Additional complications involve the so-called "slice profile" of the RF pulses [11-13]. Since we require the final dose map to be accurate to within just a few percent, gel dosimetry is one of the most exacting quantitative tasks to use this type of multi-echo sequence. Quality assurance of the imaging results is a distinctly non-trivial exercise!

2.4 Look at the raw and fitted data, and check that they appear sensible.

Given a piece of software that claims to calculate R_2, it is very tempting simply to believe the results, particularly if the software comes from a reputable scanner manufacturer. However, even if the algorithm has been correctly coded, the non-linear fitting methods used (e.g., Levenberg-Marquardt [14]) tend to be quite sensitive to the initial guesses for the model parameters. This can lead to convergence of the algorithm at a local rather than a global minimum in parameter space, with the result that the final, fitted values are incorrect. Furthermore, as a result of the RF flip angle issues described in 2.2 above, the data may not match the assumed form of the model and this would not be obvious merely from looking at the output R_2 map.

Thus, it is very important to pick a representative sample of the voxels in the final dose map and, for each voxel, look at a graph plotting echo time vs. image intensity. Check that the fitted curve passes correctly through the points and that there are no obvious outliers. Assess visually the signal-to-noise and try to form an impression of the reliability of the data, which may be different in different regions of the phantom.

Journal of Physics: Conference Series **250** (2010) 012086 doi:10.1088/1742-6596/250/1/012086

2.5 Check that there is no spatial distortion in the MR images.

Unlike in optical CT, the image formation process in MRI is susceptible to a number of different effects that can give rise to geometric distortion (e.g., gradient non-linearity, eddy currents, chemical shift and sample susceptibility) [15, 16]. This may not be obvious, but, particularly at the edges of the field-of-view, it can be significant, certainly exceeding the "2%-2 mm gamma criteria" used in [2]. Specialised distortion assessment phantoms are available, and the types of test that need to be performed have been described by de Deene [3, 5].

2.6 Check whether the pulse sequence is heating up the sample during scanning.

NMR relaxation times are notoriously sensitive to temperature and this can have a significant effect on the results of a gel dosimetry experiment [17]. In order to be able to perform an accurate calibration, one has to ensure that the sample temperature at the time of imaging is the same for both the test sample and the calibration sample. At a basic level, this requires a constancy of the ambient temperature in the scanner room and an appropriate equilibration of the sample temperature with that of the ambient air (particularly important if the sample has been stored in a refrigerator). However, a more tricky problem is that running an MRI pulse sequence deposits energy in the sample and so heats it up. The larger the number of pulses and the higher the flip angle, the more heat is deposited. This is especially problematic for multi-echo sequences designed for the quantitative measurement of R_2 and de Deene has made measurements of this effect [3]. The simplest solution to the problem is to increase the repetition time of the MRI pulse sequence, but this may lead to extremely long imaging times.

2.7 Use a consistent and appropriate delay between irradiation and scanning.

Unfortunately, the polymerisation that is used to detect the deposition of dose in polymer gel phantoms does not happen instantly: it is an ongoing process that cannot be "frozen" at an arbitrary time. Hence, the R_2 value of the gel samples continues to evolve long after the actual irradiation. This issue was recognised very early on in the development of gel dosimetry [18], but no general solution exists to the problem. Each researcher needs to evaluate his or her own particular dosimeter and determine an appropriate time after irradiation at which to perform the imaging readout.

2.8 Ensure that the gel sample is not contaminated by oxygen.

The radiation-induced polymerisation proceeds via the creation and propagation of reactive free radicals, which become attached to the end of a growing chain. This process is very sensitive to even small quantities of oxygen [19], because, in the presence of oxygen, peroxide radicals are created and these react very readily with the growing polymer chains, thus terminating the polymerisation reaction at an early stage. Two approaches are available for avoiding these problems: (a) manufacture the gels in completely anoxic conditions (e.g., in a nitrogen-filled glove box) and pour the mixture into oxygen-impermeable containers, or (b) use a *normoxic* gel formulation. Much has been written on this subject previously [20, 21].

2.9 Use the same size containers for both calibration and measurement.

There is evidence that the response of polymer gel samples to radiation dose varies slightly depending on the size of the container used to hold the gel. The polymerisation reaction is exothermic and so the explanation proposed [22, 23] is that the different gel volumes lead to different internal temperature rises during polymerisation. These, in turn, change the reaction rates of the various steps, the lengths of polymer chains produced and hence the R_2 value and calculated dose.

2.10 Beware of potential fractionation effects.

One of the supposed benefits of polymer gels is that they are integrating dosimeters, i.e., they record the total dose to which they have been exposed, leading to the ability to verify complex multi-exposure treatments such as IMRT. However, Karlsson *et al.* discovered significant differences for some gel formulations depending on whether the dose was given to the gel in a single fraction or in multiple exposures [24].

(Aside: A fundamental disadvantage of chemical dosimetry is that the dosimeter property from which absorbed dose can be calculated (in this case R_2) is dependent not only on dose, but on a number of other factors including some relatively complicated reaction kinetics that are sensitive to the thermal history of the sample. All of problems 2.6–2.10 are linked with such chemical effects. There are also concerns relating to the ability of monomer units to diffuse within the gel, which can lead to anomalous readings in regions of very high dose deposition (ref), and finally, there can be frustrating inter- and even intra-batch variations of calibration constants (ref) between samples that are all prepared using nominally the same procedure. Medical physicists generally prefer to rely on *physical* radiation detectors like ionisation chambers, which also have the advantage of giving an instantaneous result, without having to wait for a chemical reaction to occur. These issues may perhaps be a contributing factor to the relatively slow clinical uptake of gel dosimetry compared with, say, arrays of diodes.)

3. Optical CT

3.1 Clean the apparatus well and keep sample surfaces pristine.

MRI is an imaging technique that is limited by signal-to-noise ratio (SNR). This is caused by the relatively small difference between nuclear energy levels, which leads to a low "signal" term, and the inability to cool the sample below the freezing point of gel water, which leads to a relatively high level of thermal noise from the probe. Standard clinical x-ray CT is also SNR-limited, this time because of restrictions on the amount of dose that patients can receive. By contrast, optical CT can use an effectively unlimited number of photons to improve its "counting statistics", simply by increasing the exposure time of the camera or by averaging multiple exposures. The ideal optical CT scan is thus not truly SNR-limited. However, in practice we do see noise and much of this is caused by "dirt".

On its path from the source to the camera, light has to pass through a number of media and interfaces. Each surface in the optical path has the potential to become scratched or dusty. The outside of the matching tank can also be contaminated with oily liquids that are difficult to clean and that leave smeary marks. In principle, effects like this that are time-invariant can be "calibrated out" using the "correction scan" procedure described below. However, it is always better to start with high-quality data that need little correction. Keeping equipment and samples clean and the surfaces well polished should improve image quality.

More troublesome, perhaps, are the effects of contaminants in the matching liquid and sample. Dust particles enter the matching tank from the surrounding air or are introduced when the samples are loaded, and these build up over time. The particles appear as dark specks and are particularly noticeable in high-resolution scans. They often move about, carried by convection currents or the slow turning of the sample. The reconstruction algorithm is unable correctly to take account of these unwanted time-varying features in the projections and this results in artefacts in the reconstructed images. If the impurities are embedded in the sample, then the black specks move around coherently with the sample and are reconstructed as high intensity features in the image. This may or may not be troublesome, but one should always be aware that if any feature is completely opaque, then this removes from the dataset all information from ray-paths that pass through the contaminant.

3.2 Devote time to optimising the composition of the matching liquid.

In contrast to x-rays, whose direction of travel is not affected by the orientation of boundaries between materials, light is refracted when it crosses an interface. The refractive index of the dosimeter (around 1.33 for water-based gels and 1.5 for PRESAGE^{TM}) is very different from that of air, and the light ray-

Journal of Physics: Conference Series **250** (2010) 012086 doi:10.1088/1742-6596/250/1/012086

paths deviate substantially from the parallel lines required by the standard "back-projection" CT reconstruction algorithm. Cylindrical samples act like 1-D lenses and it is difficult to generate artefact-free images from this type of data. The current generation of scanners solve this problem by placing the sample inside a tank of "index-matching" liquid. Taking the simplest example, that of PRESAGE™, which does not need a container, we require that the matching liquid has an index that is as close as possible to that of the dosimeter. The liquid is generally made from a mixture of two components (one with a higher and one a lower index than the dosimeter). These are combined in a proportion tailored to the given batch of dosimeter.

Getting the right proportions can be time-consuming, particularly if the quantity of matching liquid to be produced is large. However, the time spent pays dividends in a large reduction of the "sample edge" artefact in the final image and a corresponding increase in the volume of the dosimeter that yields accurate data. It is not always appreciated that the region of the final dose image that is affected by the edge artefact extends well beyond the bright ring seen on images.

3.3 Choose the right sample containers.

To date, PRESAGE™ is the only solid optical CT dosimeter. Others are gel-based and thus need a container. Doran *et al.* [25] showed how to model the general situation in which the gel, the container and the matching liquid all have slightly different refractive indices. Many of the clear plastics commonly used to make bottles (e.g., polyethylene terephthalate, PET) have an index that it significantly higher than that of gel, but it is possible to obtain specialist plastics (e.g., Teflon FEP®), which match well. The latter are strongly recommended, as the difference in the projection profiles near the sample edge is considerable.

3.4 Use a correction scan.

The Beer-Lambert law $I = I_0 \exp(-\alpha x)$ describes optical absorption in a dosimeter, with the term I_0 representing the incident light intensity. In an imaging context, I_0 is a spatially varying parameter (i.e., its value is different for each pixel of the projection images) that takes into account a range of intensity modulating effects that are not caused by radiation-induced changes in the dosimeter. These include: (a) directional variations in the light source intensity; (b) background absorption in the dosimeter; (c) spatial variations in the detector sensitivity — these must be time-*in*variant; and (d) static imperfections in the light transmission properties of the system components.

The easiest way to account for these properties is to scan the sample prior to irradiation and then again afterwards, keeping all the scanner settings constant. As described in [25], as simple division of the two datasets will result in a dataset with the unwanted variations removed. However, it should be noted that any intensity variations that are *additive* (e.g., a constant ambient light contribution) are not removed correctly by this procedure. These should be eliminated at source if possible, or alternatively an estimate subtracted from both the correction scan and data scan raw data before they are combined. This is the normal method for eliminating the "dark current" contribution [26].

3.5 Make sample changing easy and reproducible.

If designing one's own optical scanner, it is highly desirable to include the ability to re-load samples accurately and reproducibly with a minimum of adjustment, both in terms of the vertical and lateral position and also the rotation angle. This is particularly relevant for the correction scan described above. If the sample repositioning can be achieved to a pixel level of accuracy, then it becomes possible to correct for sample imperfections (scratches, inclusions, etc.) as well as broad variations in the light field.

Journal of Physics: Conference Series **250** (2010) 012086 doi:10.1088/1742-6596/250/1/012086

3.6 Add a dye to the matching liquid.

As described above, the correction scan method is capable of taking into account background absorption in the dosimeter. However, if the unirradiated dosimeter is absorbing to any great extent, then there will be a large intensity variation across the projections images in the correction scan, with the region outside the dosimeter being very bright. This reduces significantly the dynamic range that is available to record radiation-induced changes. The problem can be circumvented by adding a dye to the matching liquid whose absorption coefficient matches that of the unirradiated sample. Note that one should not allow the detector in the camera to saturate, as "blooming" can occur — see [27] for further details.

3.7 Beware of scattering.

It is well known that the Beer-Lambert law is only an approximation for scattering systems. At high scatterer concentrations, the exponential relationship breaks down, so that the whole basis of the optical readout (via spectrophotometry using cuvettes as well as via optical CT) is not valid. However, even at much lower concentrations, optical CT imaging can be badly affected. Although very few comparative studies have been performed to date, it seems reasonable that CCD-based techniques will suffer more than scanned laser techniques. De Jean *et al.*[28] and Bosi *et al.* [29] have presented initial data showing how scans using the Modus commercial device are affected by scatter.

3.8 Check for unwanted reflections in the apparatus.

In [30], Jordan warns of the dangers of multiple reflections at optical interfaces. In some cases, these may lead to interference fringes. In optical apparatus where there are shiny metal components, there is the potential for light to enter the detector via complex reflection paths and contaminate the desired signal.

3.9 Compare calibrations with a spectrophotometer.

There is a tendency in optical CT dosimetry to "normalise" data, on the basis that the relationship between dose and change in absorption coefficient is linear. This procedure hides changes in dose-response characteristics between batches and may obscure important effects [31]. Babic *et al.* [32] have shown that, at least for the Fricke dosimeter, careful methodology allows an absolute relation between dose and change in optical absorption coefficient to be obtained. It is recommended that the results of the CT imaging reconstruction be compared with spectrophotometer measurements on cuvettes irradiated to the same dose (but see Sec. 2.9 for a caveat if polymer gel dosimeters are being used.)

3.10 Understand the chemical and radiation physics properties of the samples.

As hinted at earlier, the number of publications characterising the radiochromic materials used in optical CT is much smaller than the equivalent body of literature for polymer gel MRI dosimetry. Two examples illustrate the need for users of optical CT to be vigilant: (i) Babic *et al.* [31] showed that when using Fricke gels, the source of xylenol orange (Aldrich, Sigma or Sigma-Aldrich) made a significant difference to the dosimeter properties. After some detailed detective work, a very interesting story emerged and led to recommendations for the procedure to be followed when using xylenol orange Fricke gels. (ii) PRESAGE^{TM} is known to be radiologically non-equivalent to water (and hence soft tissue). The first comprehensive analysis of the difference in radiation physics properties has recently been completed by Gorjiara *et al.* [33]. Looking at the results of their Monte Carlo simulations, it appears that the density-corrected depth-dose curve for PRESAGE^{TM} is different to that of water. Further work is necessary to establish whether this has implications for the calibration procedure for this dosimeter — can Oldham's depth-dose method [34] be used? — and whether simply providing a CT scan of a PRESAGE^{TM} phantom to a treatment planning system leads to an accurate calculation of the absorbed dose.

4. References

[1] Ceberg, S., RapidArc treatment verification in 3D using polymer gel dosimetry and Monte Carlo simulation. *Phys. Med. Biol.*, in press.

[2] Sakhalkar, H.S., Adamovics, J., Ibbott, G., and Oldham, M., A comprehensive evaluation of the PRESAGE/optical-CT 3D dosimetry system. *Medical Physics*, 2009. **36**(1) 71-82.

[3] de Deene, Y., Fundamentals of MRI measurements for gel dosimetry. *Journal of Physics: Conference Series*, 2004. **3**(1) 87.

[4] de Deene, Y., On the accuracy and precision of gel dosimetry. *Journal of Physics: Conference Series*, 2006. **56**(1) 72.

[5] de Deene, Y., Review of quantitative MRI principles for gel dosimetry. *Journal of Physics: Conference Series*, 2009. **164**(1) 012033.

[6] Doran, S.J., The history and principles of optical computed tomography for scanning 3-D radiation dosimeters: 2008 update. *Journal of Physics: Conference Series*, 2009. **164**(1) 012020.

[7] Oldham, M., 3D dosimetry by optical-CT scanning. *Journal of Physics: Conference Series*, 2006. **56**(1) 58.

[8] Doran, S.J. and et al., RadBall ™ : A new departure for 3-D dosimetry. *Journal of Physics: Conference Series*, 2009. **164**(1) 012042.

[9] de Deene, Y. and Baldock, C., Optimization of multiple spin–echo sequences for 3D polymer gel dosimetry. *Physics in Medicine and Biology*, 2002. **47**(17) 3117.

[10] Meiboom, S. and Gill, D., Modified Spin-Echo Method for Measuring Nuclear Relaxation Times. *Review of Scientific Instruments*, 1958. **29**(8) 688-691.

[11] Majumdar, S., Orphanoudakis, S.C., Gmitro, A., O'Donnell, M., and Gore, J.C., Errors in the measurements of T_2 using multiple-echo MRI techniques. I. Effects of radiofrequency pulse imperfections. *Magnetic Resonance in Medicine*, 1986. **3**(3) 397-417.

[12] McRobbie, D.W. and et al., Slice profile effects and their calibration and correction in quantitative NMR imaging. *Physics in Medicine and Biology*, 1987. **32**(8) 971.

[13] Wong, S.T.S. and Roos, M.S., Effects of slice selection and diffusion on T_2 measurement. *Magnetic Resonance in Medicine*, 1987. **5**(4) 358-365.

[14] Press, W.H., Teukolsky, S.A., Vetterling, W.T., and Flannery, B.P., *Numerical Recipes in FORTRAN; The Art of Scientific Computing (Ch. 15, "Modeling of Data")*. 1993: Cambridge University Press.

[15] Doran, S.J. and et al., A complete distortion correction for MR images: I. Gradient warp correction. *Physics in Medicine and Biology*, 2005. **50**(7) 1343.

[16] Reinsberg, S.A. and et al., A complete distortion correction for MR images: II. Rectification of static-field inhomogeneities by similarity-based profile mapping. *Physics in Medicine and Biology*, 2005. **50**(11) 2651.

[17] de Deene, Y., de Wagter, C., van Duyse, B., Derycke, S., de Neve, W., and Achten, E., Three-dimensional dosimetry using polymer gel and magnetic resonance imaging applied to the verification of conformal radiation therapy in head-and-neck cancer. *Radiotherapy and oncology : journal of the European Society for Therapeutic Radiology and Oncology*, 1998. **48**(3) 283-291.

[18] McJury, M. and et al., Dynamics of polymerization in polyacrylamide gel (PAG) dosimeters: (I) ageing and long-term stability. *Physics in Medicine and Biology*, 1999. **44**(8) 1863.

[19] Hepworth, S.J. and et al., Dynamics of polymerization in polyacrylamide gel (PAG) dosimeters: (II) modelling oxygen diffusion. *Physics in Medicine and Biology*, 1999. **44**(8) 1875.

[20] Baldock, C. and et al., Polymer gel dosimetry. *Physics in Medicine and Biology*, 2010. **55**(5) R1.

[21] Fong, P.M. and et al., Polymer gels for magnetic resonance imaging of radiation dose distributions at normal room atmosphere. *Physics in Medicine and Biology*, 2001. **46**(12) 3105.

[22] Dumas, E.-M. and et al., Effect of container size on the accuracy of polymer gel dosimetry. *Journal of Physics: Conference Series*, 2006. **56**(1) 239.

[23] Sedaghat, M. and et al., Volume-dependent internal temperature increase within polymer gel dosimeters during irradiation. *Journal of Physics: Conference Series*, 2009. **164**(1) 012009.

[24] Karlsson, A. and et al., Dose integration characteristics in normoxic polymer gel dosimetry investigated using sequential beam irradiation. *Physics in Medicine and Biology*, 2007. **52**(15) 4697.

[25] Doran, S.J. and et al., A CCD-based optical CT scanner for high-resolution 3D imaging of radiation dose distributions: equipment specifications, optical simulations and preliminary results. *Physics in Medicine and Biology*, 2001. **46**(12) 3191.

[26] Krstajić, N. and Doran, S.J., Characterization of a parallel-beam CCD optical-CT apparatus for 3D radiation dosimetry. *Physics in Medicine and Biology*, 2007. **52**(13) 3693.

[27] Al-Nowais, S. and Doran, S.J., CCD-based optical CT scanning of highly attenuating phantoms. *Journal of Physics: Conference Series*, 2009. **164**(1) 012023.

[28] DeJean, P. and et al., Initial experience with a commercial cone beam optical CT unit for polymer gel dosimetry I: Optical dosimetry issues. *Journal of Physics: Conference Series*, 2006. **56**(1) 179.

[29] Bosi, S. and et al., Initial investigation of a novel light-scattering gel phantom for evaluation of optical CT scanners for radiotherapy gel dosimetry. *Physics in Medicine and Biology*, 2007. **52**(10) 2893.

[30] Jordan, K., Advances in optical CT scanning for gel dosimetry. *Journal of Physics: Conference Series*, 2004. **3**(1) 115.

[31] Babic, S. and et al., An apparent threshold dose response in ferrous xylenol-orange gel dosimeters when scanned with a yellow light source. *Physics in Medicine and Biology*, 2008. **53**(6) 1637.

[32] Babic, S., Battista, J., and Jordan, K., Three-Dimensional Dose Verification for Intensity-Modulated Radiation Therapy in the Radiological Physics Centre Head-and-Neck Phantom Using Optical Computed Tomography Scans of Ferrous Xylenol–Orange Gel Dosimeters. *International journal of radiation oncology, biology, physics*, 2008. **70**(4) 1281-1291.

[33] Gorjiara, T., Hill, R., Kim, J.-H., Kuncic, Z., Adamovics, J., and Baldock, C., Study of dosimetric water equivalency of PRESAGE® for megavoltage and kilovoltage x-ray beams, in *IC3Ddose 2010*. 2010: Hilton Head Island.

[34] Oldham, M. and et al., Improving calibration accuracy in gel dosimetry. *Physics in Medicine and Biology*, 1998. **43**(10) 2709.

Journal of Physics: Conference Series **250** (2010) 012087 doi:10.1088/1742-6596/250/1/012087

Radiotherapy delivery during motion

Sofie Ceberg and Sven Å J Bäck

Department of Medical Radiation Physics, Lund University, Skåne University Hospital, Malmö, Sweden

Sofie.ceberg@med.lu.se

Abstract. This paper discusses the 3D dosimetric consequences of radiotherapy delivery during two kinds of motion, (i) the respiratory motion by the patient and (ii) the motion by the gantry while rotating around the patient.

Respiratory motion primarily compromises treatments in the thorax and abdomen regions. Several strategies to reduce respiratory motion effects have been developed or are under development. The organ motion could for instance be measured and incorporated in the treatment planning, or adapted to by using respiratory gating and tumour-tracking delivery techniques.

Gantry motion is involved in various forms of intensity-modulated arc-therapy techniques. The purpose is to increase the modulation by simultaneously varying the MLC positions, the rotation speed of the gantry, and the dose rate during the treatment.

The advantage of these techniques is the increased possibility to deliver a high absorbed dose to the target volume while minimizing the dose to normal tissues. However, the dosimetric uncertainties associated with motion, small fields and steep dose gradients, has to be evaluated in detail, and this requires adequate true 3D dose-verification tools.

1. Incorporating organ motion in treatment planning

The most widely used approach to account for organ motion is to collect a large data set of measured respiratory motion, calculate the statistical distribution, and integrate this into the treatment planning procedure as an additional margin to the PTV. The measured amplitude, applied to the internal margin (IM), can vary as a function of various factors related to the tumour (e.g. tumour site and the attachment to the chest wall), patient (e.g. respiratory capacity, oxygenation and anxiety) or treatment protocol (e.g the immobilisation and patient positioning) [1]. However, this approach, based on an addition to the geometrical margin, does not describe the uncertainties in the dose to normal tissue.

Another method is to directly introduce respiratory uncertainties into the dose calculation, by convolving the static dose distribution with a motion function [2]. This approach assumes that the target moves without deformation during treatment and that the medium is homogeneous.

2. Breathing adapted radiotherapy

2.1. Breath-hold and respiratory gating

The most obvious approach to reduce uncertainty related to movements induced by breathing is to ask the patient not to breathe during the actual beam-on time. This periods can generally be made to have a duration of 15-30 s. This breath-hold technique can either be voluntary [3] or active [4]. Another approach is to turn the beam on only when the target is in a favorable position during the patient's respiration cycle, *i.e.* the so called respiratory gating technique [5]. The advantages of these techniques are a) a reproducible immobilization, with the target volume and organs at risk (OAR) in a known position throughout the treatment, b) the irradiated lung volume will be reduced, and c) the cardiac toxicity will be reduced [6].

During deep inspiration, as with breath-hold and respiratory gating methods, the irradiated lung density will be reduced due to the 40-50% increased lung volume compared to the resting value [1]. This reduction of lung density can affect the dose to peripheral parts of the target volume due to loss of electronic equilibrium. Monte Carlo simulations have verified this impact on TPS calculations and demonstrated the risk of target under-dosage [7]. No 3D dosimetric measurements have verified this under-dosage, however, a low-density (lung tissue equivalent) polymer gel have been presented [8, 9], which may potentially enable the required measurements.

Further, depending on the size of the gating window during breath-hold or respiratory gating a certain dose smearing of the field edges due to movement during the beam-on time will remain [10]. This dose smearing effect is not accounted for in conventional treatment planning systems and dosimetric verifications is required. Experimental data have so far mostly been obtained using ionization chambers and films [10, 11]. Recently published data obtained using 3D detector systems such as PRESAGE™/optical-CT dosimetry [12] and polymer gel dosimetry [13] show very good agreement between static and gated measurements.

2.2. Tumour tracking

Another method to compensate for respiratory motion during treatment is tumour tracking. A novel promising strategy uses the dynamic multileaf collimator (DMLC) to continuously align and reshape the treatment machine aperture to follow the target motion in real time [14]. The advantage of this technique is the ability to allow for a tighter margin around the target by continuously following, and adapting the dose delivery to its motion (Figure 1, *ibid* IC3DDose proceedings, Ceberg *et al.* 2010).

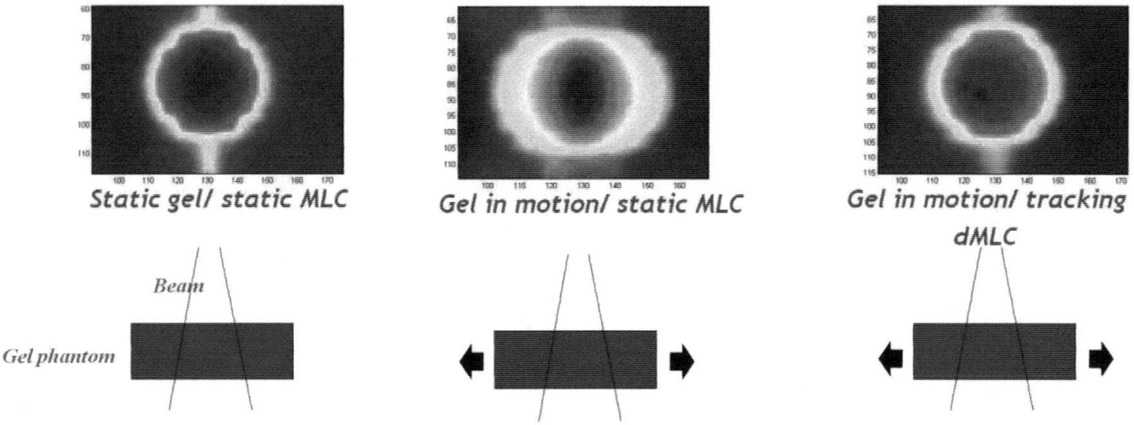

Figure 1. Dose distributions as measured using gel al dosimetry with and without tumour tracking delivery.

Compared to the breath-hold and respiratory gating methods, the tumour tracking technique potentially offers additional benefits such as higher delivery efficiency and less residual target motion. However, real-time beam adaptation is not feasible without precise real-time localisation of the tumour position in 3D, which includes parallel and perpendicular motion to the MLC leaf travel direction, in- and out-of-plane rotation as well as translation along the beam direction.

4D IMRT treatment planning methods that accounts for 3D tumour motion can be adopted [15]. The proposed method uses 4D CT and was integrated with the DMLC tumour-tracking delivery and opens up for a clinical implementation. Still, very few dosimetric measurements of the DMLC tumour-tracking system have been carried out. Dosimetric uncertainties associated with a DMLC tracking system arise from the estimation of the actual target position, possible delay between target motion detection, and beam repositioning and output variations due to e.g. dissimilar head-scatter conditions during off-axis irradiation.

A dosimetric study to investigate the ability of a DMLC-tracking system to account for target motion during radiotherapy using 3D polymer dosimetry and a bi-planar detector array (Delta4®, Scandidos) was recently undertaken (this conference). The DMLC real-time motion-tracking system was used together with the optical part of the ExacTrac system (BrainLab, Germany). The respiratory-like target motion was simulated mechanically. Good agreement in absorbed dose was found between the static and tracked measurements (ibid IC3DDose proceedings, Ceberg et al. 2010).

3. Volumetric modulated-arc therapy

Another category of motion during radiotherapy is ascribed to the gantry motion, as it rotates around the patient during a volumetric modulated-arc therapy (VMAT) delivery. The delivery could be accomplished using designated rotational therapy as Tomotherapy™ or standard linear accelerators. An early paper describing the dosimetric challenges with such techniques and the potential to use gel dosimetry was presented in 2004 [16]. RapidArc™ is a novel radiation therapy technique where the treatment is delivered during one or a few rotations of the linear accelerator gantry. The dose distribution is modulated by simultaneously varying the MLC positions, dose rate and gantry rotation speed [17]. Since the RapidArc™ optimization algorithm uses a stochastic element during the generation of the apertures it is common to see plans with apertures including single isolated leaves and disconnected small segments. Furthermore, RapidArc™ plans are delivered dynamically with leaf and gantry motion up to approximately 1 cm and 2 degrees per second, respectively. This kind of treatment delivery represents a new level of complexity, and a thorough dosimetric verification is therefore highly desirable.

In recently published work, semi-3D dosimetric verifications of RapidArc™ treatment delivery using back-projection portal dosimetry [18] and Delta4© bi-planar diode array phantom [19] have been presented. The reported results have shown very good agreement between TPS and dose measurements.

Recently a 3D verification of a RapidArc™ treatment plan and delivery using polymer gel dosimetry and Monte Carlo simulation was undertaken. Inter-comparisons between all datasets – TPS calculations, gel measurement and MC simulation - showed very good agreement [20].

4. Breathing adapted modulated-arc therapy

The combination of volumetric modulated-arc with breathing adapted methods will pose an even greater challenge to 3D dose verification.

A pre-clinical investigation, in a non-clinically released framework, evaluated the machine capability to deliver gated RapidArc™ [21]. A conventional ion chamber and a 2D ion chamber array (Seven29, PTW) were used for dosimetric verifications. Both detectors were used in combination with the Octavius phantom developed by PTW for rotational therapy verification. The results showed good agreement with TPS even for very short beam-on durations (5 s), which concluded that RapidArc™ delivery is dosimetrically accurate also when applied in combination with gating procedures.

The absorbed dose to moving targets undergoing a RapidArc™ delivery with a pre-clinical 3D DMLC tracking application has been evaluated [22]. A 2D ion chamber array (Seven29, PTW) and the bi-planar diode array Delta4© were used for dosimetric verifications. The results showed that DMLC-tracking together with RapidArc™ make a feasible combination. An additional study, using the same detector system, verified that the dosimetric accuracy was independent of the magnitude of the peak-to-peak displacement (5-25 mm) of the target and not significantly affected by the angle between the leaf trajectory and the target movements [23].

Adding an extra safety layer in the dosimetric verification of DMLC-tracked RapidArc™ delivery a true 3D dose verification using polymer gels was carried out (*ibid* IC3DDose proceedings, Ceberg et al. 2010). A high gamma (2%/1mm) pass rate was obtained comparing tracked and static measurements, demonstrating that the motion was well compensated for by the tracking performance.

5. Conclusions

True 3D detector systems with high resolution, where the response is independent of the direction of the incident radiation and can integrate the contribution from several beams, are highly desirable. Recently published result using PRESAGE©/optical-CT dosimetry and polymer gel dosimetry show a very promising future involving true 3D dosimetric verification of advanced dose deliveries during motion.

References

[1] P. Giraud, E. Yorke, S. Jiang, L. Simon, K. Rosenzweig, and G. Mageras, Cancer Radiother 10 (2006) 269.
[2] S. D. McCarter and W. A. Beckham, Phys Med Biol 45 (2000) 923.
[3] A. N. Pedersen, S. Korreman, H. Nystrom, and L. Specht, Radiother Oncol 72 (2004) 53.
[4] J. W. Wong, M. B. Sharpe, D. A. Jaffray, V. R. Kini, J. M. Robertson, J. S. Stromberg, and A. A. Martinez, Int J Radiat Oncol Biol Phys 44 (1999) 911.
[5] K. Ohara, T. Okumura, M. Akisada, T. Inada, T. Mori, H. Yokota, and M. J. Calaguas, Int J Radiat Oncol Biol Phys 17 (1989) 853.
[6] S. S. Korreman, A. N. Pedersen, L. R. Aarup, T. J. Nottrup, L. Specht, and H. Nystrom, Int J Radiat Oncol Biol Phys 65 (2006) 1375.
[7] E. D. Yorke, L. Wang, K. E. Rosenzweig, D. Mah, J. B. Paoli, and C. S. Chui, Int J Radiat Oncol Biol Phys 53 (2002) 1058.
[8] Y. De Deene, K. Vergote, C. Claeys, and C. De Wagter, Med Phys 33 (2006) 2586.
[9] P. Haraldsson, A. Karlsson, E. Wieslander, H. Gustavsson, and S. A. Back, Phys Med Biol 51 (2006) 919.
[10] X. A. Li, C. Stepaniak, and E. Gore, Med Phys 33 (2006) 145.
[11] J. Duan, S. Shen, J. B. Fiveash, R. A. Popple, and I. A. Brezovich, Med Phys 33 (2006) 1380.
[12] S. L. Brady, W. E. Brown, C. G. Clift, S. Yoo, and M. Oldham, Phys Med Biol 55 (2010) 2187.
[13] S. Ceberg, A. Karlsson, H. Gustavsson, L. Wittgren, and S. Å. J. Bäck, Physics in medicine and biology 53 (2008) N387.
[14] A. Sawant, R. Venkat, V. Srivastava, D. Carlson, S. Povzner, H. Cattell, and P. Keall, Med Phys 35 (2008) 2050.
[15] Y. Suh, A. Sawant, R. Venkat, and P. J. Keall, Phys Med Biol 54 (2009) 3821.
[16] K. Vergote, Y. De Deene, W. Duthoy, W. De Gersem, W. De Neve, E. Achten, and C. De Wagter, Phys Med Biol 49 (2004) 287.
[17] K. Otto, Medical Physics 35 (2008) 310.
[18] A. Mans, P. Remeijer, I. Olaciregui-Ruiz, M. Wendling, J. J. Sonke, B. Mijnheer, M. van Herk, and J. C. Stroom, Radiother Oncol 94 (2010) 181.
[19] S. Korreman, J. Medin, and F. Kjaer-Kristoffersen, Acta Oncologica 48 (2009) 185.
[20] S. Ceberg, I. Gagne, H. Gustafsson, J. B. Scherman, S. Korreman, F. Kjær-Kristoffersen, M. Hilts, and S. Å. J. Bäck, Phys Med Biol (2010)
[21] G. Nicolini, E. Vanetti, A. Clivio, A. Fogliata, and L. Cozzi, Phys Med Biol 55 (2010) N347.
[22] J. Zimmerman, S. Korreman, G. Persson, H. Cattell, M. Svatos, A. Sawant, R. Venkat, D. Carlson, and P. Keall, Acta Oncol 48 (2009) 245.
[23] M. Falk, P. M. af Rosenschold, P. Keall, H. Cattell, B. C. Cho, P. Poulsen, S. Povzner, A. Sawant, J. Zimmerman, and S. Korreman, Radiother Oncol 94 (2010) 218.

Journal of Physics: Conference Series 250 (2010) 012088 doi:10.1088/1742-6596/250/1/012088

3D dosimetry for brachytherapy and heterogeneities

G. S. Ibbott[1] and M. P. Heard[2]

[1]UT M. D. Anderson Cancer Center, Houston, TX
[2]Global Physics Solutions, Lexington, OH

gibbott@mdanderson.org

Abstract. Clinical applications of 3D dosimeters have been considered since the 1950s, when uses of gel dosimeters were first explored [1,2]. Since then, however, the number of investigators has increased, and the body of knowledge regarding gel dosimetry has grown considerably [3-5]. 3D dosimetry is essential for a number of clinical applications, including the complete characterisation of brachytherapy sources, and the evaluation of dose distributions around and within inhomogeneous tissues. However, each of these conditions presents several challenges which must be overcome to allow the accurate and practical determination of dose distributions.

1. Introduction

The dose distributions around brachytherapy sources and within and around heterogeneities present unusual difficulties for dose characterization. Brachytherapy sources are intended to be implanted directly in, or immediately adjacent to diseased tissue, and as a result, the most significant part of the dose distribution occurs within a few millimeters to a few centimeters from the source. In this region, the dose gradient is extremely steep and a measurement device must have both high spatial precision and low sensitivity to rapidly changing doses. In addition, a dosimeter that will be used to measure the dose distribution out to several centimeters from the source must have a dynamic range that accommodates several orders of magnitude change in dose rate.

The determination of doses within and around heterogeneous media is likewise extremely difficult [6,7]. Early measurements focused on measurements behind heterogeneities, because measurements within the heterogeneities were confounded by the effective density of the measurement device [8]. Most radiation detectors are non-unit density. Many diode detectors, for example, are largely silicon with an atomic number of 14, and have the potential to perturb the dose distribution. Inserting an air-filled ionization chamber into a low-density medium introduces electronic disequilibrium. Consequently, the perturbing effects of the dosimeter must be separated from the measurement and accounted for.

2. Brachytherapy sources

During the last several decades, low-dose rate interstitial brachytherapy using ^{125}I and ^{103}Pd encapsulated sources has become a popular treatment for prostate cancer, resulting in an increased demand for these sources. In response, manufacturers developed ^{131}Cs sources and several new models of ^{125}I and ^{103}Pd sources. The dose distributions of these low energy emitters are very sensitive to internal geometry, encapsulation, self-absorption and filtration, all of which vary with each manufacturer. The American Association of Physicist in Medicine (AAPM) Radiation Therapy Committee has recommended that the dosimetric characteristics of each new source be evaluated by at least two independent investigators [9]. The AAPM Task Group 43 protocol for interstitial brachytherapy sources details the dosimetric parameters needed to calculate the doses from these sources [10].

However, it is not only low-energy sources that present dosimetric problems. Even high-energy sources of ^{137}Cs and ^{192}Ir generate dose distributions with steep dose gradients. Measurements around high-dose rate ^{192}Ir sources are further complicated by the high dose rates and perhaps even by the heat generated by the source.

There is currently no dosimeter in routine use that provides the desired characteristics to measure these dose distributions. Most experimentalists to date have used thermoluminescent dosimetry (TLD) in water-equivalent plastics, which add uncertainty due to their energy dependence, although a technique to use TLDs in water was described recently [11]. However, the use of TLDs to measure sufficient data to fully characterize a brachytherapy source is a very tedious process. Radiochromic film allows the collection of a plane of data in a single exposure, but similarly presents issues of energy dependence, dose rate dependence, and dose linearity.

An ideal dosimeter would be tissue equivalent, have high spatial resolution in three dimensions, and be independent of energy and dose rate. In an effort to find a dosimeter that meets these requirements, the use of gels and polyurethane blocks as dosimeters has been investigated. These dosimeters often are nearly tissue equivalent at the energies emitted by the sources, provide high spatial resolution, and allow visualization of three dimensional dose distributions. Polymer gels are made up of water, gelatin, and acrylic monomers that polymerize when exposed to radiation. Ionizing radiation produces free radicals within the gel and these in turn lead to the formation of polymer microparticles that remain attached to or entangled with the gelatin [12]. A polyurethane dosimeter called PRESAGE® consists of a polyurethane matrix doped with radiochromic leuco dyes that generate a color change, and consequently a change in optical density (OD) upon exposure to ionizing radiation [13]. Both gels and PRESAGE® must be evaluated for their energy dependence, but as they allow the collection of dose information in almost a complete 3D volume, great interest is shown in these dosimeters.

2.1. Measurements around brachytherapy sources

The ability of gels to record and display dose distributions around a high-dose rate (HDR) source was first demonstrated over a decade ago [14-16]. Most radiation-sensitive gels must be protected from oxygen, as atmospheric oxygen scavenges free radicals and halts the polymerization process, reducing the sensitivity of the gel. This complicates the measurement, as a brachytherapy source must be introduced without also introducing oxygen.

Maryanski et al showed the dose distribution around a single catheter into which a high-dose-rate (HDR) remote afterloader source had been positioned [16]. The HDR unit was programmed to dwell the source at several locations in the catheter, to deliver an elliptical dose distribution. After irradiation, the gel was imaged with MR, and a map of the dose distribution was computed. Care was taken to protect the dosimeter from oxygen, through the use of a catheter material impervious to oxygen.

More recently, measurements have been made in close proximity to HDR ^{192}Ir sources [17-19]. These measurements have shown that complications occur when measurements are made in the steep dose gradients close to an HDR source. Polymerization of the gel causes an increase in the gel density and a corresponding decrease in the volume filled by the gel. The change in density causes shrinkage of the gel in the vicinity of the source, distorting the resulting measured distribution. Changes to the composition of the gel to increase the concentration of gelatin can mitigate the amount and effects of the density changes. Furthermore, there are suggestions that the high dose rates found near brachytherapy sources, particularly those of HDR afterloaders, can introduce temperature gradients which influences the polymerization of acrylamide monomer gels [20,21].

Efforts also have been made to characterize low-dose rate (LDR) sources such as prostate seeds [22-24], eye plaques [25], ^{137}Cs afterloading sources [26,27] and intravascular sources [28,29]. Studies have indicated that the diffusion of monomers (or ferrous and ferric ions in Fricke gels) across steep dose gradients can introduce errors in measurement [30]. As the use of gels to measure dose distributions from LDR sources requires long exposure times, diffusion of monomers or ions could introduce significant errors, and gels exhibiting high diffusion rates must be avoided for these measurements.

A further problem with gel dosimetry for LDR brachytherapy has been demonstrated by recent studies indicating energy dependence at lower energies. Data show that a polymer gel dosimeter under-responds to radiation in the 20 keV – 60 keV range [31]. Others have shown differences in gel response from one formulation to another, and suggest that the MAGAT gel is most water-equivalent over a wide range of energies [32]. Changes in mass attenuation coefficient of polymer gels during irradiation can also introduce errors in the dose distributions measured around low-energy sources.

The PRESAGE® dosimeter also has been used to measure dose distributions around brachytherapy sources. These investigations demonstrated some success, and Wai has shown good agreement between measurement and Monte Carlo calculations [33,34].

2.2. Intravascular brachytherapy sources

That gel dosimetry is able to visualize steep dose gradients in 3D makes it very attractive for acquiring the dose distributions of both LDR and HDR brachytherapy sources. In addition, gel dosimeters integrate the dose delivery over time so that the total dose distribution from a moving source can be measured. Different types of sources have been evaluated, as indicated above, including sources designed for intravascular therapy.

Although polymer gel dosimetry has been perceived as a suitable dosimeter for dose verification in brachytherapy, some additional difficulties may be encountered that are related to the insertion of applicators or catheters into the polymer gel. Significant dose deviations near the catheter wall have been noted [35-37]. In polymer gel dosimeters, oxygen diffusion through the wall of the cavity may contaminate the gel and result in an underestimation of absorbed dose [38]. De Deene *et al* performed a quantitative analysis of the inhibition of gel response by oxygen passing through the catheter wall of a HDR ^{192}Ir source [17]. Other sources of dose deviations related to the high dose gradients encountered with HDR point sources and measured with MRI have also been established [39].

3. Measurements Around Heterogeneities

Concern about the influence of heterogeneities has figured prominently in dosimetry applications for many decades [6]. However, the measurement of radiation dose distributions within and around heterogeneities is notoriously difficult. Most detectors themselves are non-tissue equivalent, and their introduction into non-unit density media complicates the measurement. For example, the effects of the introduction of the air cavity of an ionization chamber into a water phantom can be handled with carefully-determined replacement factors (P_{repl}). Correction for the introduction of the same chamber into a non-water equivalent medium is not as easily managed, and the changes in scattering and absorption become difficult.

As a result, efforts to measure the effects of heterogeneities have largely been limited to determinations of the transmission of radiation through the heterogeneous region, and the effects on the dose at a point or in a plane behind the heterogeneity [7].

One of the attractive features of gel dosimeters is that they are very nearly water-equivalent, particularly at megavoltage photon beam energies. However, several investigators have attempted to use gel dosimetry to measure the effects of non-unit density tissues on external beam dose distributions. Early measurements were performed to estimate the dose distribution behind high atomic number heterogeneities, to simulate the presence of bone [40-42]. More recently, measurements have been made behind or adjacent to cavities filled with air or with lung-equivalent plastic. Watanabe assembled a geometric phantom with a rectangular cavity into which air or other material could be placed [43]. A conventional gel dosimeter filled the surrounding spaces. In contrast, Vergote et al constructed an anthropomorphic thorax phantom into which they placed air cavities to simulate the lungs [40]. In both cases, the dosimetry system was a conventional polymer gel of unit density.

Attempts to construct 3D dosimeters consisting of non-unit density media were described as early as 1998, including reports by several investigators who constructed low-density versions of Fricke gels [44-46]. For these applications, Fricke gels were beaten vigorously to introduce air, or were mixed with expanded polystyrene beads to reduce density. After irradiation, the gels were examined using MR imaging. Subsequently, and to avoid the disadvantages of Fricke gels, Haraldsson et al constructed a normoxic polymer gel with a reduced density by adding expanded polystyrene spheres to a conventional mixture of polymer gel [47]. The antioxidant THP was added in various concentrations. They found the dose versus 1/T2 response to be essentially linear in the dose range of 2 to 8 Gy for all investigated THP concentrations. Good agreement between measured and Monte Carlo calculated data was obtained, in test tubes as well as in a larger 3D phantom. Similar data were shown by Borges who introduced expanded polystyrene beads into a version of MAGIC gel, in a glass vessel [48].

Construction of a low-density gel by adding expanded polystyrene spheres requires a uniform distribution of the gel and the polystyrene, and the buoyancy of the polystyrene spheres makes this difficult. However, Borges et al have shown that dense packing of truly spherical polystyrene beads results in a good approximation of lung density [48]. While these measurements showed promise, there were several potential sources of error. First, the introduction of air, or air-containing polystyrene beads introduced the possibility of oxygen contamination. Purging the polystyrene beads with nitrogen or an oxygen scavenger, or using nitrogen rather than air to foam the gel addressed this problem. The introduction of air or polystyrene eliminated the possibility of evaluating the measured dose distribution by optical scanning, and MR imaging must be used. The presence of air also may lead to partial volume imaging effects that could introduce errors into the measurement.

References

[1] Day M J and Stein G 1950 Chemical effects of ionizing radiation in some gels. *Nature* **166** 146-7.
[2] Andrews H L, Murphy R E and LeBrun E J 1957 Gel dosimeter for depth dose measurements. *Rev. Sci. Instrum.* **28** 329-332
[3] Schreiner L J 1999 Gel dosimetry: Motivation and historical foundation. In DOSGEL 1999 Proceedings of the 1st International Workshop on Radiation Therapy Gel Dosimetry (Lexington, Kentucky) ed L J Schreiner and C Audet
[4] Ibbott GS 2006 Clinical applications of gel dosimeters. *J. Phys.: Conf. Ser.* **56** 108-131.
[5] Bäck SAJ 2009 Clinical applications of gel dosimeters. J. Phys.: Conf. Ser. 164 1-4.
[6] Batho, H. F. Lung corrections in ^{60}Co beam therapy. *J. Can. Assoc. Radiol.* 1964, **15**, 79.
[7] Sontag, M. R., and Cunningham, J. R. Corrections to absorbed dose calculations for tssue inhomogeneities. *Med. Phys.* 1977; **4**: 431
[8] Young MEJ, Gaylord JD. Experimental tests of corrections for tissue inhomogeneities in radiotherapy. *Br. J. Radiol.* 1970; **43**: 349.
[9] Williamson J, Coursey BC, DeWerd LA, Hanson WF, and Nath R. Dosimetric prerequisites for routine clinical use of new low energy photon intersitial brachytherapy sources. *Med. Phys.* 1998; **25**: 2269-2270.
[10] Rivard MJ, Coursey BM, DeWerd LA, Hanson WF, Huq MS, Ibbott GS, Mitch MG, Nath R, Williamson JF. Update of AAPM Task Group No. 43 Report: A revised AAPM protocol for brachytherapy dose calculations. *Med Phys* 2004 **31**:633-74.
[11] Tailor RC, Ibbott GS, Tolani N. Thermoluminescence dosimetry measurements of brachytherapy sources in liquid water. *Med Phys* 2008 **35**:4063-9.
[12] Maryanski M J, Ibbott G S, Eastman P, Shulz R J and Gore J C Radiation therapy dosimetry

using magnetic resonance imaging of polymer gels. *Med. Phys.* 1996: **23** 699-705

[13] Adamovics J, Maryanski M. Characterisation of PRESAGE™: A New 3-D Radiochromic Solid Polymer Dosemeter for Ionising Radiation. *Rad. Protect. Dosim.* 2006;**120**:107-112.

[14] Schreiner L J, Crooks I, Evans M D C, Keller B M, and Parker W A Imaging of HDR brachytherapy dose distributions using NRM Fricke-gelatin dosimetry. *Magn. Reson. Imaging.* 1994 **12** 901-7

[15] Olsen D R and Hellesnes J Absorbed dose distribution measurements in brachytherapy using ferrous sulphate gel and magnetic resonance imaging. *Br. J. Radiol.* 1994 **67** 1121-26

[16] Maryanski M J, Ibbott G S, Xie J, Gore J C, Schulz R J Magnetic Resonance Imaging of Dose Distributions from Brachytherapy Sources Embedded in Tissue Equivalent BANG Polymer Gel Dosimeters. *Med. Phys.* 1994 **21** 919 (abstract).

[17] De Deene Y, Reynaert N and De Wagter C. On the accuracy of monomer/polymer gel dosimetry in the proximity of a high-dose-rate ^{192}Ir source. *Phys. Med. Biol.* 2001: **46** 2801–2825.

[18] Heard MP, Ibbott GS. Measurement of brachytherapy sources using MAGIC gel. *Journal of Physics: Conference Series* 2004; **3**:221-3.

[19] Heard M and Ibbott G. Gel dosimetry technique for measurements in high dose gradients. *Med. Phys.* 2004 **31** 1888 (abstract).

[20] Gelfi C and Righetti P G Polymerization kinetics of polyacrylamide gels: II. Effect of temperature. *Electrophoresis* 1981 **2** 220–8.

[21] Omidian H, Hashemi S A, Sammes P G and Meldrum I G 1998 Modified acrylic-based superabsorbent polymers. Effect of temperature and initiator concentration. *Polymer* **39**; 3459–66.

[22] Ibbott G, Maryanski M, Drogin A, Gearheart D, Painter T, Meigooni A. Characterization of a New Brachytherapy Source by BANG® Gel Dosimetry. DosGel 99: Proceedings of the 1st International Workshop on Radiation Therapy Gel Dosimetry. Canadian Organisation of Medical Physicists and the Canadian College of Physicists in Medicine: 196-8, 1999.

[23] Ibbott G S, Maryanski M J, Drogin A, Gearheart D, Painter T, Meigooni A S Characteristics of a New Brachytherapy Source by BANG® Gel Dosimetry. *Int. J. Radiat Oncol. Biol. Phys.* 1999 **45** (3S) 417.

[24] Heard M, Ibbott G, Followill D Characterizing Dose Distributions of Brachytherapy Sources Using Normoxic Gel (WIP) AAPM Annual Meeting, 2003.

[25] Chan M, Fung A Y C, Hu Y-C, Chui C-S, Amols H, Zaider M, Abramson D The measurement of three dimensional dose distribution of a ruthenium-106 ophthalmological applicator using magnetic resonance imaging of BANG polymer gels. *J. Appl. Clin. Med. Phys.* 2001 **2** 85-9

[26] Gifford KA, Horton JL, Jr, Jackson EF, Steger TR, 3rd, Heard MP, Mourtada F, Lawyer AA, Ibbott GS. Comparison of Monte Carlo calculations around a Fletcher Suit Delclos ovoid with radiochromic film and normoxic polymer gel dosimetry. Med Phys 32(7):2288-94, 7/2005.

[27] Gifford K, Horton J, Steger T, Heard M, Jackson E, Ibbott G. Verification of Monte Carlo calculations around a Fletcher Suit Delclos ovoid with normoxic polymer gel dosimetry. Journal of Physics: Conference Series 3:217-20, 2004.

[28] Maryanski MJ, Ranade M, Barry M, Rangarajan A, Nath R, Hafeli U, Amols H, Wuu C-S High-resolution 3D dosimetry for endovascular brachytherapy using optical laser CT microimaging of BANG polymer gels. *Med. Phys.* 1998 **25** A107.

[29] Wuu C-S, Schiff P, Maryanski MJ, Liu T, Borzillary S, and Weinberger J. Dosimetry study of Re-188 liquid balloon for intravascular brachytherapy using polymer gel dosimeters and laser-beam optical CT scanner. *Med. Phys.* 2003 **30** 132-7.

[30] Vergote K, De Deene Y, Vanden Bussche E and De Wagter C. On the relation between the spatial dose integrity and the temporal instability of polymer gel dosimeters *Phys. Med. Biol.* 2004 **49** 287-305.

Journal of Physics: Conference Series **250** (2010) 012088 doi:10.1088/1742-6596/250/1/012088

[31] Pantelis E, Karlis A K, Kozicki M, Papagiannis P, Sakelliou L and Rosiak J M. Polymer gel water equivalence and relative energy response with emphasis on low photon energy dosimetry in brachytherapy. *Phys. Med. Biol.* 2004 **49** 3495–3514

[32] Venning A J, Brindha S, Hill B, and Baldock C. Preliminary study of a normoxic PAG gel dosimeter with tetrakis (hydroxymethyl) phosphonium chloride as an antioxidant. *J. Phys.: Conf. Ser.* 2004 **3** 155–8.

[33] Wai P, Krstajic N, Adamovics J, Ismail A, Doran S. Dosimetry of the Amersham 6711 Oncoseed™ using PRESAGE™ and optical CT. *Journal of Physics: Conference Series* 2006; **56** 235–238.

[34] Wai P, Adamovics J, Krstajic N, Ismail A, Nisbet A, Doran S. Dosimetry of the microSelectron-HDR Ir-192 source using PRESAGE™ and optical CT. *Applied Radiation and Isotopes* 2009: **67**, 419-422 .

[35] McJury M, Oldham M, Cosgrove V P, Murphy P S, Doran S, Leach M O and Webb S Radiation dosimetry using polymer gels: methods and applications *Br. J. Radiol.* 2000: **73** 919–29.

[36] Hurley C, McLucas C, Pedrazzini G and Baldock C High-resolution gel dosimetry of a HDR brachytherapy source using normoxic polymer gel dosimeters: preliminary study *Nucl. Instrum. Methods Phys. Res.A* 2006: **565** 801–11.

[37] Massillon-JL G, Minniti R, Mitch MG, Maryanski MJ and Soares CG. The use of gel dosimetry to measure the 3D dose distribution of a $^{90}Sr/^{90}Y$ intravascular brachytherapy seed, *Phys. Med. Biol.* **54** (2009) 1661–1672

[38] Farajollahi A R, Bonnett D E, Ratcliffe A J, Aukett R J and Mills J A An investigation into the use of polymer gel dosimetry in low dose rate brachytherapy. 1999:, *Br. J. Radiol.* **72** 1085–92

[39] Amin M N, Horsfield M A, Bonnett D E, Dunn M J, Poulton M and Harding P F 2003 A comparison of polyacrylamide gels and radiochromic film for source measurements in intravascular brachytherapy *Br. J. Radiol.* **76** 824–31

[40] Vergote K, De Deene Y, Claus F, De Gersem W, Van Duyse B, Paelincka L, Achten E, De Neve W, De Wagter C 2003 Application of monomer/polymer gel dosimetry to study the effects of tissue inhomogeneities on intensity-modulated radiation therapy (IMRT) dose distributions. *Radiother. Oncol.* **67** 119–28.

[41] Vergote K, De Deene Y, Claus F, De Gersem W, Van Duyse B, and De Wagter C. Comparions between monomer/polymer gel dosimetry and dose computations for an IMRT treatment of a thorax phantom. In DOSGEL 2001 Proceedings of 2nd International Conference on Radiotherapy Gel Dosimetry (Brisbane, Australia) ed C Baldock and Y De Deene

[42] Hepworth S J, McJury M, Oldham M, Morton M J, and Doran S J 1999 Dose mapping of inhomogeneities positioned in radiosentive polymer gels. Nucl. Instrum. *Methods Phys. Res.* A **422** 756–60

[43] Watanabe Y, Mooij R B, Perera G M and and Maryanski M J 2004 Heterogeneity phantoms for visualization of 3D dose distributions by MRI-based polymer gel dosimetry. *Med. Phys.* **31** 975-84

[44] Gum F, Scherer J, Bogner L, Solleder M, Rhein B and Bock M 2002 Preliminary study on the use of an inhomogeneous anthropomorphic Fricke gel phantom and 3D magnetic resonance dosimetry for verification of IMRT treatment plans *Phys. Med. Biol.* **47** N67–N77

[45] Ølberg S, Skretting A, Bruland Ø and Olsen D R 2000 Dose distribution measurements by MRI of a phantom containing lung tissue equivalent compartments made of ferrous sulphate gel *Phys. Med. Biol.* **45** 2761–70

[46] Scherer J, Solleder M, Schiessl I, Bogner L and Herbst M 1998 3D MR-gel-dosimetry with lung equivalent gel *Z. Med. Phys.* **8** 87–95

[47] Haraldsson P, Karlsson A, Wieslander E, Gustavsson H, Bäck SAJ. Dose response evaluation of a low-density normoxic polymer gel dosimeter using MRI. *Phys. Med. Biol.* **51** (2006) 919–928.

[48] Borges JA, BenComo J, Ibbott GS. A 3 Dimensional Gel Dosimetry Lung Equivalent (WIP). Med. Phys. 30:1478, 2003.

Journal of Physics: Conference Series **250** (2010) 012089 doi:10.1088/1742-6596/250/1/012089

Dosimetry in HDR brachytherapy with Fricke-gel layers and Fricke-gel catheters

G. Gambarini[1,2], M. Carrara[3], A. Negri[1,2], M. Invernizzi[1], C. Tenconi[1], A. Scotti[1], L. Pirola[1],M. Borroni,[3], S. Tomatis[3], C. Fallai[4]

[1] Physics Department of the Università degli Studi and INFN, Milan, Italy
[2] INFN Istituto Nazionale di Fisica Nucleare, Sezione di Milano, Italy
[3] Medical Physics Unit, Fondazione IRCCS "Istituto Nazionale Tumori", Milan, Italy
[4] Radiotherapy Unit, Fondazione IRCCS "Istituto Nazionale Tumori", Milan, Italy

grazia.gambarini@mi.infn.it

Abstract. Fricke-gel layer dosimeters (FGLD) and Fricke gel dosimetric catheters (FGDC) have been designed and tested with the aim of enquiring their suitability for HDR ^{192}Ir brachytherapy source control and for in-vivo dose verification during treatment. Anisotropy function measurements have been carried out with FGLDs in which a thin plastic tube has been placed in for the ^{192}Ir source insertion. FGDCs are constituted by plastic tubes (3 mm of external diameter and 13 cm of length) filled with the dosimeter-gel. Absorbed dose images and profiles were attained by means of optical analysis. Dedicated software has been developed both for achieving anisotropy function values and for obtaining reliable results in visible light absorbance measurements across the thin cylindrical dosimeters. Preparation and analysis procedures have been optimised. The results confirm that the proposed methods are very promising for HDR brachytherapy dosimetry.

1. Introduction

HDR ^{192}Ir brachytherapy is a radiotherapy modality achieving more and more important role for cancer treatment. AAPM Task Group 43 [1,2] has outlined a protocol for dose to water calculation based on the results of air-kerma strength measurements. Conspicuous experimental activities have been carried out for dosimetry measurements in HDR ^{192}Ir brachytherapy, in particular in water phantoms [3,4], also utilising Fricke-gel [5-7].

Some Fricke-gel layer dosimeters (FGLD) were prepared containing a catheter to be jointed to the ^{192}Ir source applicator. In such a way, the brachytherapy source can enter into the gel dosimeter. This dosimeter can be placed in a water-equivalent phantom for 2-D dose measurements or piled up with FGLDs having the same size but without catheter, for 3-D measurements [8]. Each dosimeter can be previously properly calibrated, in order to achieve absolute dose results with good precision.

Besides in-phantom measurements, it is important to perform also reliable in-vivo dosimetry during HDR brachytherapy treatments, in order to verify the correctness of exposures. In particular, when the therapy is fractionated in more sessions, it is of particular importance to execute a check of the dose absorbed during each exposure. In-vivo dosimetry is particularly complex in HDR brachytherapy, because the high gradients require high spatial resolution. At present, the chosen detectors are mostly TLDs or, if immediate readout is necessary, diodes and MOSFETs. Fricke gel dosimeter sealed in thin transparent catheters (FGDC) has shown to be a good candidate for in-vivo dosimetry during a HDR radiotherapy treatment. For reliable optical analysis of such thin cylindrical dosimeters, a suitable method was set up and tested. Feasibility measurements with HDR brachytherapy unit were performed to assess the reliability of FGDCs for in-vivo dosimetry during brachytherapy treatments.

2. Material and methods

The Fricke gel dosimeters that are at the basis of the methods are laboratory-made radiochromic gels; the composition is: Agarose in the amount of 3% of the final weight, ferrous sulphate solution [1mM $Fe(NH_4)_2(SO_4)_2{\bullet}6H_2O$)]; sulphuric acid [25 mM H_2SO_4] and xylenol-orange [0.165 mM $C_{31}H_{27}N_2Na_5O_{13}S$].

For in-phantom measurements, FGLDs were prepared, consisting of gel contained between two transparent PMMA sheets, 1 mm thick. The gel thickness is of 3 mm and the frame around is usually squared, with side length of 12 cm. In some frames, a catheter for the introduction of the ^{192}Ir source is sealed. In Figure 1, a dosimeter after irradiation aimed at measuring the anisotropy function is shown. If the dosimeters are properly calibrated, absolute dose measurements are realizable, with good precision [11].

For in-vivo measurements, FGDCs were produced utilizing transparent plastic tubes with an external diameter of 3 mm and a length of 13 cm.

The dosimeter reading is based on visible light absorbance. Dosimeters are positioned on a uniform planar source of light, inserted in a special frame fitting their shape, and the transmitted light is imaged before and after irradiation. Images are acquired using a CCD camera with a band-pass filter centred at 585 nm. The measured difference in optical density (Δ(OD)) is proportional to the absorbed dose. The acquired images are analysed with the software suitably developed for each kind of dosimeter.

2.1. Dedicated software

The dedicated software (in MatLab® code) for FGLDs analysis has been suitably developed and improved [9,10], with the aim of getting better results by removing artefacts [11], effects of lack in thickness uniformity and so on.

The software for FGDCs is now in development. The dosimeter cylindrical shape implies that transmission of light occurs through a non-planar surface. Due to this feature, a special method to obtain a reliable correlation between image's gray level and absorbed dose was developed. The software finds out the position of the catheter's central axis for every row perpendicular to this axis, and then it evaluates the mean gray-level over 5 pixels centred on this position. In Figure 2, a sketch illustrating this procedure is shown. Finally, the software calculates the fitted lengthways profile of the Δ(OD) for percentage depth dose (PDD) measurements or of the absorbed dose, if the FGDCs have been calibrated, for absolute dose measurements.

Proper software has been developed also to compensate effects of non-homogeneous response within the same FGLD or FGDC. This lack in homogeneity could be due to the non-uniformity of both the CCD camera and the planar light source and/or to the dosimeter non-homogeneous sensitivity. Light transmittance images acquired before and after dosimeter irradiation in a uniform field are processed by this software, attaining a correction image for FGLD, or a correction profile for FGDC. In such a way, pixel-to-pixel correction of the dosimeter response can be performed.

2.2. Dosimeter calibration

Mostly, the dosimeters have been preliminarily irradiated in an uniform photon field, at 4 Gy, for non-homogeneity correction purpose. Moreover, for absolute dose measurements, the dosimeters were

Journal of Physics: Conference Series **250** (2010) 012089 doi:10.1088/1742-6596/250/1/012089

calibrated with the following procedure. Each dosimeter was exposed to an uniform photon beam, suitably surrounded by TE plastic, so to have a known uniform dose (4 Gy) absorbed in the gel dosimeter. Since Fricke gel dosimeter response is independent by dose rate and radiation energy, this irradiation was carried out with photons from a 6 MV clinical linear accelerator. The Δ(OD) image obtained, for each dosimeter, from the transmittance images acquired before and after such an irradiation, was utilised for attaining the dose-calibration image of the specific dosimeter. Such preliminary irradiations were always performed in the same day of the measurement. A study is in program to enquire the possibility of performing the dosimeter calibration one or two days before measurement.

2.3. Brachytherapy unit

All experiments were carried out at a Microselectron-HDR high dose rate unit, from Nucletron (Veenendaal, the Netherlands). The system is provided with a ^{192}Ir radioactive source and a remote afterloading device. The source has an active length of 3.6 mm and a diameter of 0.65 mm. The source is incapsulated at the end of a flexible cable and driven to the dwell positions evaluated by the treatment plane.

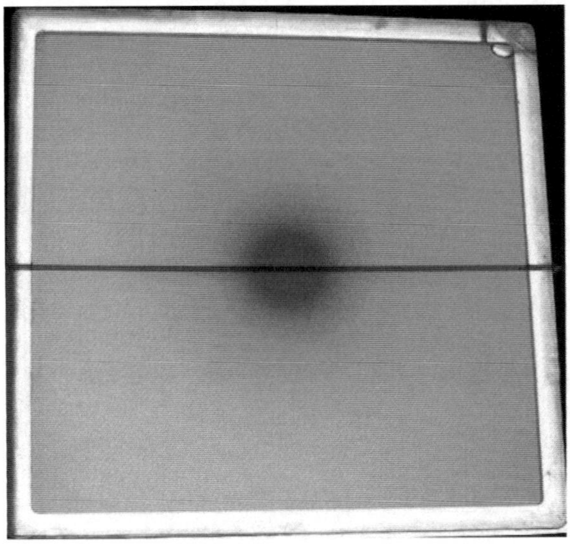

Figure 1 – FGLD with the inserted catheter for the introduction of ^{192}Ir source by remote control, after irradiation for anisotropy function measurement.

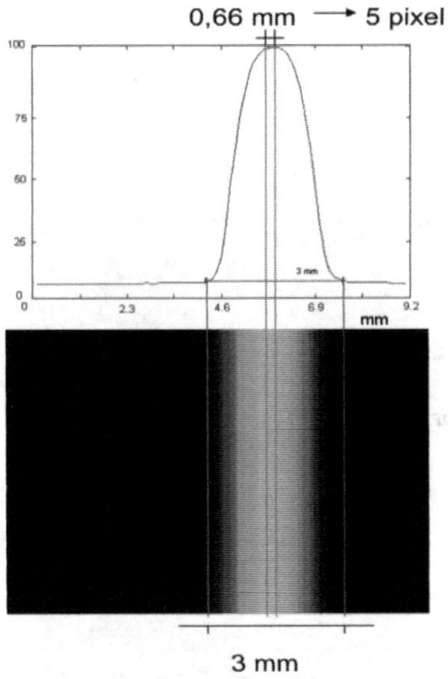

Figure 2 – Sketch of the central region of the thin cylindrical FGDC where the Δ(OD) is evaluated.

3. Results and discussion

A characterization of FGLDs irradiated at the brachytherapy source has been carried out, determining the linearity range and setting up the calibration procedure. The calibration method requires two exposures at 4 Gy. Then, the dosimeter response is linear in the range 0- 20 Gy. The dosimeter response at irradiation temperature of 37°C (temperature of human body) was investigated, for preliminary test of the feasibility of in-vivo utilization of this dosimetric compound.

Some measurements were performed irradiating FGLDs with the ^{192}Ir source stopped at the center of the dosimeter for a convenient time. After having obtained the absorbed dose images, the anisotropy function was evaluated by means of the properly developed software and the results were compared with the calculated values. The agreement is very good, within 3%, up to 30 mm of radial distance from the source. At higher distances the dispersion of data increases (remaining however within 5%), because the absorbed dose becomes very low. Some results are shown in Figure 3.

The new proposed catheters for in-vivo dosimetry (plastic tubes with inner and outer diameter of 4F and 6F, respectively) have been widely studied in order to (i) optimize the number of pixel utilized for determining the Δ(OD) all along the dosimeter, that is the dosimeter resolution, (ii) verify the linearity range, (iii) investigate the achievable precision for absolute dose measurements by means of previous calibration of each dosimeter and (iv) explore the feasibility of using thinner catheters, with external diameter of 2 mm.

FGDC measurements were also performed positioning many dosimeters in a water-equivalent phantom that was irradiated with the high dose rate remote afterloading device, with a complex treatment plane. The dosimeters were preventively calibrated following the adopted procedure, shortly described in Section 2.2 and in a previous work [13]. The results [14] show a very good consistency between measured and calculated dose profiles, showing the validity of the proposed method.

The improvement of the method aimed at using catheters with 2 mm of external diameter in now in progress. To this aim, better resolution of the optical analysis system (CCD camera plus uniform light source) is necessary. The new instrumentation is now in phase of setting up.

In conclusion, the proposed methods show to be promising for high dose rate brachytherapy dosimetry. Their characteristics suggest that such dosimeters could be proposed for clinical routine protocols.

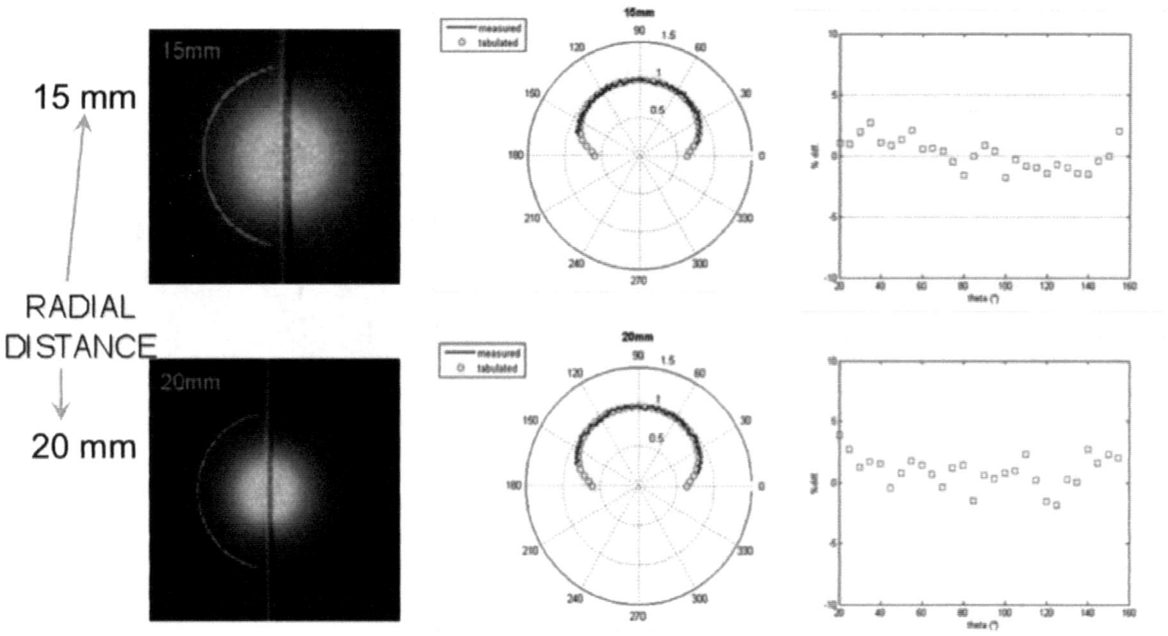

In red are the profiles where the anisotropy function was evaluated

Polar plot of measured and tabulated data

Percentage difference between measured and tabulated data

Figure 3 – Results of anisotropy function measurement performed by means of a FGLD in which the ^{192}Ir source was introduced through an inserted catheter.

Journal of Physics: Conference Series **250** (2010) 012089 doi:10.1088/1742-6596/250/1/012089

References

[1] Nath, R., Anderson, L. L., Luxton, G., Weaver, K. A., Williamson, J. F. and Meigooni, A. S., 1995, *Dosimetry of interstitial brachytherapy sources: Recommendations of the AAPM Radiation Therapy Committee Task Group No. 43. American Association of Physicists in Medicine*, Med. Phys. 22, 209–234.

[2] Rivard, M. J., Coursey, B. M., DeWerd, L. A., Hanson, W. F., Huq, M. S., Ibbott, G. S., Mitch, M. G.,Nath, R. and Williamson, J. F., 2004, *Update of AAPM Task Group No. 43 Report: A revised AAPM protocol for brachytherapy dose calculations*, Med. Phys. 31, 633–674 .

[3] Sarfehnia, A. and Seuntjens. J., 2010, *Development of a water calorimetry-based standard for absorbed dose to water in HDR 192Ir brachytherapy*., Med. Phys. 37, 1914-1923.

[4] Sarfehnia, A., Kawrakow, I., Seuntjens, J., 2010, *Direct measurement of absorbed dose to water in HDR 192Ir brachytherapy: Water calorimetry, ionization chamber, Gafchromic film, and TG-43*. Med. Phys. 37, 1924-1932.

[5] Olsen, D. R. and Hellesnes, J., 1994, *Absorbed dose distribution measurements in brachytherapy using ferrous sulphate gel and magnetic resonance imaging*, Br. J. Radiol. 67, 1121–1126.

[6] Schreiner, L. J., Crooks, I., Evans, M. D., Keller, B. M. and Parker, W. A., 1994, *Imaging of HDR brachytherapy dose distributions using NMR Fricke gelatin dosimetry*, Magn. Reson. Imaging 12, 901–907.

[7] Austerlitz, C., Mota, H. C., Sempau, J., Benhabib, S. M., Campos, D., Allison, R., DeAlmeida, C. E., Zhu, D. and Sibata, C. H., 2008 *Determination of absorbed dose in water at the reference point d(r0, theta0) for an 192Ir HDR brachytherapy source using a Fricke system*, Med. Phys. 35, 5360–5365.

[8] Carrara, M., Fallai, C., Gambarini, G., Negri, A., 2010, *Fricke gel-layer dosimetry in high dose-rate brachytherapy*. Applied Radiation and Isotopes 68, 722–725.

[9] Carrara, M., Gambarini, G., Tomatis, S., 2005. *Experimental imaging and 3D rendering of absorbed dose means of piled-up dosimetric sheets.* IEEE Nuclear Science Symposium Conference Record (IEEE Cat. No.05CH37692C) : 870-873

[10] Gambarini, G., Carrara, M., Valente, M., 2006. *3D-Reconstruction of absorbed dose obtained from gel-dosimeter-layers*. 9th ICATPP Conference on Astroparticle, Particle, Space Physics, Detectors and Medical Physics Applications, 705-709, ed. World Scientific (ISBN 981-256-798-4)

[11] Carrara, M., Gambarini, G., Tomatis, S., Valente, V., 2007. Dose distribution measurements by means of gel-layer dosimeters. Evaluation of algorithms for artifacts amendment. Nucl. Instr. Meth. A 580, 334-338

[12] Gambarini, G., Carrara, M., Gambini, I., Tomatis, S., Negri, A., Mariani, M., Fallai, C., Olmi, P., Zonca, G., *Fricke gel layer dosimetry for in-phantom absolute dose measurements in HDR brachytherapy*. Radiot. Oncol. 92, Supplement 1, (2009), S104-S105

[13] Carrara M., Gambarini G., Gambini I., Tomatis S., Fallai C., Olmi P., Zonca G., 2009, *A Novel Continuous Dosimeter for In-Vivo Absolute Dose Measurements in High Dose Rate Brachytherapy.* Int. Journ. of Radiation Oncology Biology Physics, 75, S577-S578, ISSN: 0360-3016

[14] Carrara, M., Gambarini, G., Borroni, M., Tomatis, S., Cerrotta, A., Fallai, C., Olmi P. and Zonca, G., 2010, *Feasibility study of a dosimeter for in vivo measurements of the absolute dose delivered in high dose rate brachytherapy.* Accepted for publication on Il Nuovo Cimento, DOI 10.1393/ncb/i2010-10878-2

Journal of Physics: Conference Series **250** (2010) 012090 doi:10.1088/1742-6596/250/1/012090

Electron dosimetry in the presence of small cavities

Simon Doran[1,2], Russell Thomas[3], Rachel Hollingdale[4,5], John Adamovics[6], Andrew Nisbet[1,5]

[1]Department of Physics, University of Surrey, Guildford, UK

[2]CRUK and EPSRC Cancer Imaging Centre, Institute of Cancer Research, Sutton, UK

[3]National Physical Laboratory, Teddington, UK

[4]Royal Surrey County Hospital, Guildford, UK

[5]St Thomas Hospital, London, UK

[6]Rider University, Skillman, NJ, USA

Simon.Doran@icr.ac.uk

Abstract. Tissue inhomogeneities such as bones or air cavities give rise to significant perturbations of dose during electron radiotherapy. Whilst these can be calculated using a variety of computational methods, accurate experimental verification has hitherto been difficult. In this study, we used 3-D optical computed tomography (CT) dosimetry of PRESAGE™ samples to obtain central-axis depth dose curves and to study the dose distribution around a simple air cavity. Some concerning anomalous results were obtained for the build-up region of the depth-dose curve, which are currently under investigation. However, despite this, the ability to measure rapidly varying dose distributions show great promise.

1. Introduction

Radiotherapy using electron beams has a number of attractive features, chief among which is a much more rapid fall-off of dose with depth than is available when using photon beams. By varying the electron energy and by combining electron and photon irradiations, sophisticated treatments may be given. However, this flexibility comes at a price: tissue inhomogeneities within the volume being treated and, in particular, the presence of air cavities, lead to significant perturbations of deposited dose. Methods to calculate these dose perturbations have been studied for some time [1] and commercial systems have made significant advances [2]. However, accurate *experimental* verification of complex dose distributions in 3-D has only recently become possible. Whilst it has been recognised for several years that dosimetry of unperturbed electron beams is an excellent way to calibrate and validate 3-D dosimetry scanners [3-4], optical computer tomography (CT) has not, to the best of our knowledge been widely applied to the verification of complex dose patterns around inhomogeneities.

2. Materials and Methods

Dose mapping was carried out using two groups of two PRESAGE™ dosimeters (from separate batches with different sensitivities). The samples were in the form of cylinders of diameter 60 mm and height 60 mm. The first group was irradiated with to 30 Gy at the depth of dose-maximum (D_{max}) using an electron beam of energy 6 MeV from a Varian 2100C/D linear accelerator (Varian Oncology

Journal of Physics: Conference Series **250** (2010) 012090 doi:10.1088/1742-6596/250/1/012090

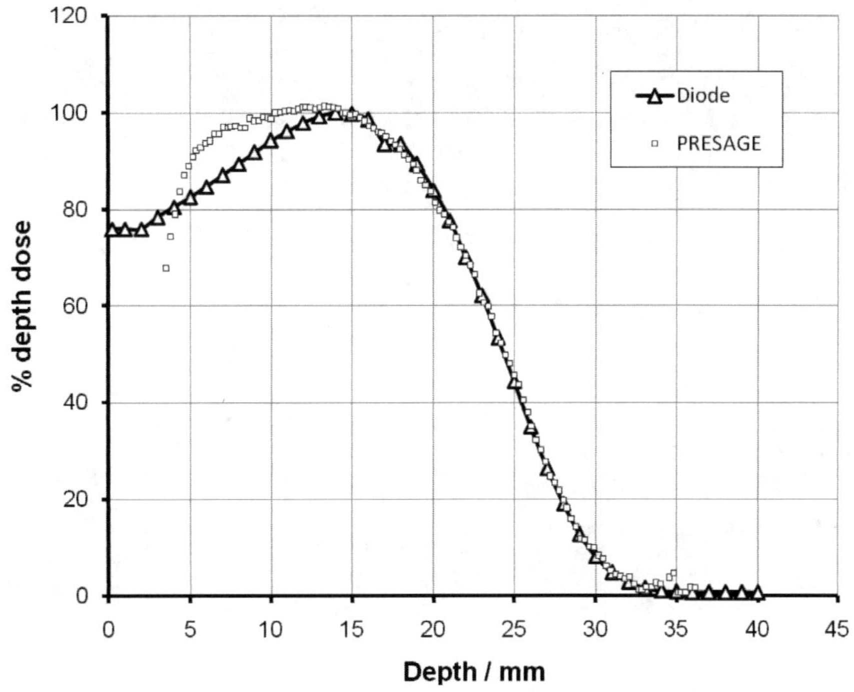

Figure 1: Comparison of diode and PRESAGE^{TM} results along the central axis for the sample with no air cavity. The poor agreement of the diode and PRESAGE^{TM} measurements between 5 and 15 mm is currently under investigation. The rapid fall-off of the PRESAGE^{TM} data below a depth of 5 mm is well understood, being indicative of image artefacts occurring at the sample-matching liquid interface at the top of the sample.

Systems, California, USA). One of the cylinders was machined with a 1cm diameter and 1 cm length cavity in one of the flat surfaces of the cylinder to determine the difference in the dose distribution in the PRESAGE^{TM} due to the cavity. A 6 cm × 6 cm electron applicator with a 4 cm diameter circle insert was used to provide an electron field at 100 cm FSD The second set of samples underwent the same procedure using a 9 MeV beam.

Given the relatively small field size of the electron applicator an electron p-type diode (Scanditronix, Uppsala) was used in a water phantom for the depth dose measurements. It was not possible to simulate the cavity in a water phantom, so an epoxy resin water-equivalent phantom (Gammex RMI457, Wisconsin, USA) was employed. Since it was impractical to use the electron diode in such a phantom, an NACP-02 plane parallel chamber (Scanditronix, Uppsala, Sweden) was used for measurements at a number of different depths in the range 0 – 30 mm. Confirmatory measurements had previously verified the consistency in depth-dose measurement between diode and plane parallel chamber beyond the build up region.

The PRESAGE^{TM} samples were placed on the treatment couch with the centre of the sample aligned with the cross hairs of the linear accelerator, and FSD was set to 100 cm at the surface of the sample. After irradiation, PRESAGE^{TM} samples were imaged with the University of Surrey optical CT scanner, using the following parameters: field-of-view 64.5 mm, projection image matrix size 256×256, acquisition time 24 mins. for 400 projections over 180°, reconstruction matrix size 256×256, nominal in-plane resolution (0.25 × 0.25) mm2.

Journal of Physics: Conference Series **250** (2010) 012090 doi:10.1088/1742-6596/250/1/012090

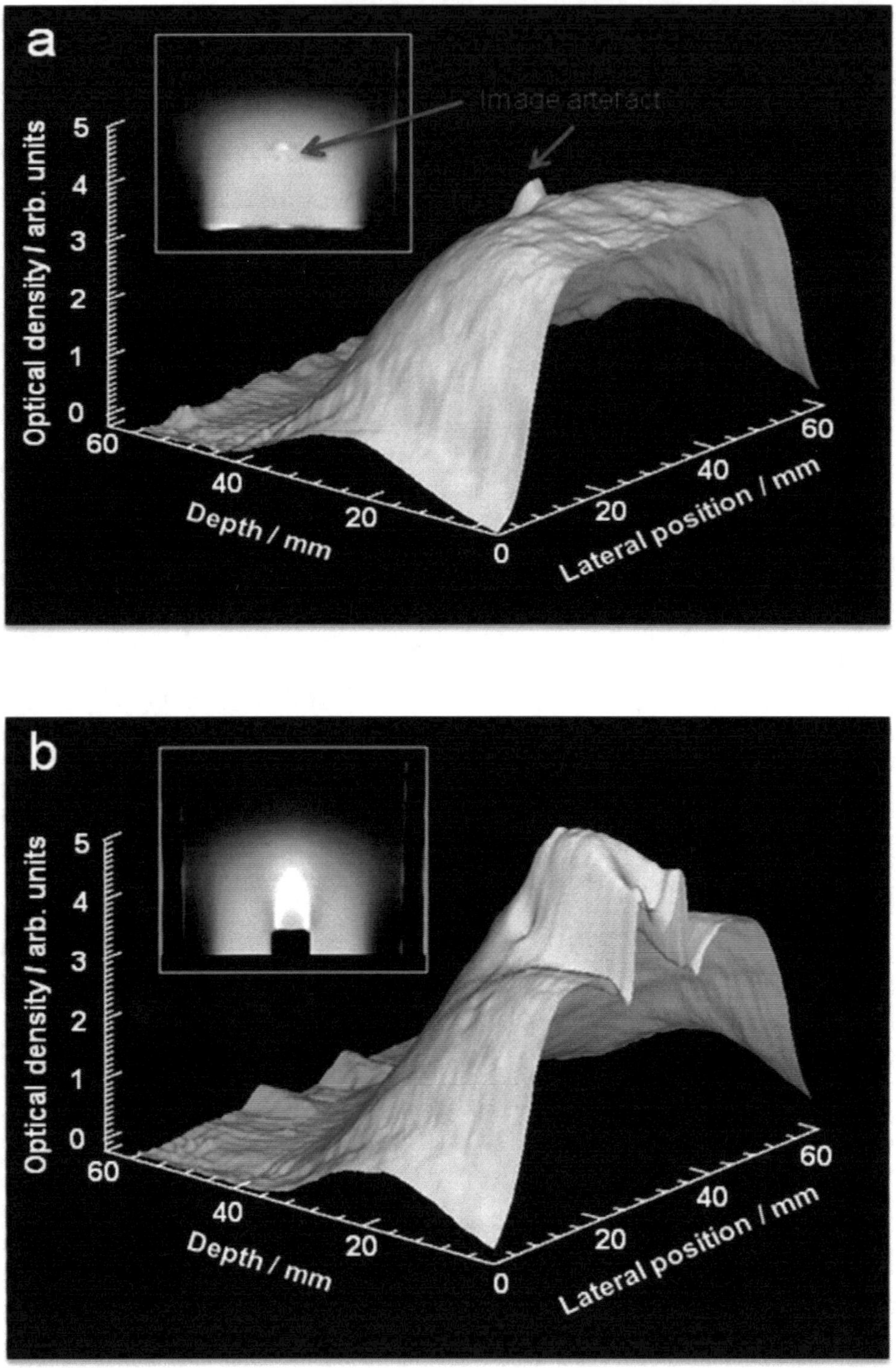

Figure 2: Optical CT data from the second set of two samples, irradiated with a 9 MeV electron beam: (a) sample with no inhomogeneity; (b) sample with an air cavity. In each case, the main graphic is the surface plot of a longitudinal cross-section through the sample approximately 2 mm off-centre. Unfortunately, an optical artefact (possibly caused by a spurious reflection from a metallic part of the apparatus) spoils the images in (a), so we attempted to avoid this where possible during the analysis.

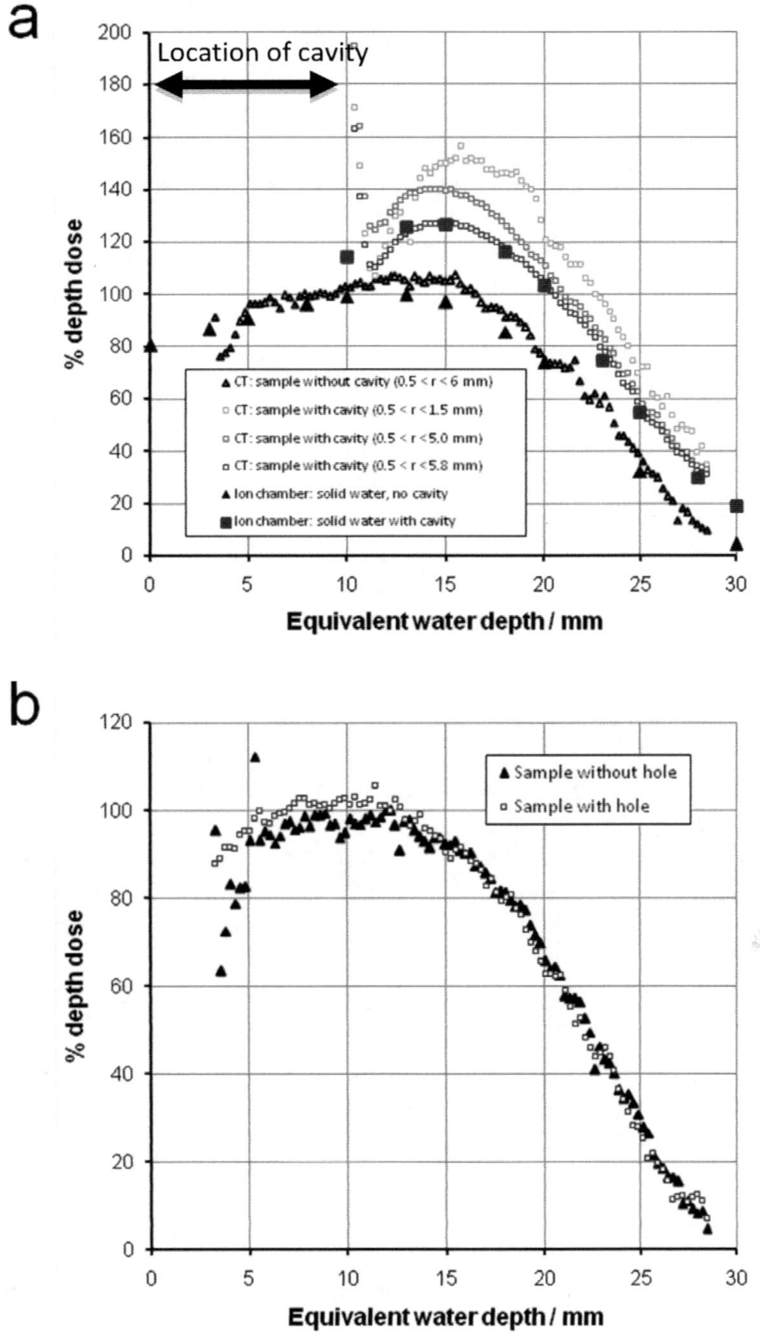

Figure 3: Comparison of data from the two 6 MeV samples: (a) depth doses averaged over annular regions of different sizes along the central axis — see main text for further explanation — together with data measured in a solid water phantom using an ionisation chamber; (b) dose at radial distance of 12.7 mm from the centre of the sample, at which point (away from the influence of the cavity) the two distributions are expected to be similar.

Two methods for increasing the signal-to-noise ratio of the measurement were tested and found to produce results of similar quality: (i) the fact that the electron dose was axially symmetric was exploited to perform a radial averaging procedure; (ii) since the dose distributions were expected to be smooth, a 3-D median filter of kernel size 7 × 7× 7 voxels was applied.

Journal of Physics: Conference Series **250** (2010) 012090 doi:10.1088/1742-6596/250/1/012090

3. Results and Discussion

Figure 1 compares diode measurements in a water tank with the depth-dose curve obtained along the central axis of the 6 MeV sample without the air cavity. The depth values for the PRESAGE™ measurements were scaled to their water equivalents by multiplying by the ratio of the densities. The agreement beyond D_{max}, is extremely good. The deviation of the optical CT results from those expected in the build-up region of Figure 1 is currently unexplained and a concern, but it is interesting to note that Figure 2(b) of [5] demonstrates a very similar effect using MRI-based polymer gel dosimetry. Repetition of all the experiments is planned.

Figure 2 shows a visualisation of the dose distribution from a longitudinal plane near the centre of the dosimeter for each of the two 9 MeV irradiations. We did not analyse the central plane because the optical artefact illustrated in Figure 2(a) was worse there. It is clear that the presence of the air cavity causes a major perturbation in the field, with the dose deposited immediately distal to the cavity being some 20 – 40% higher than that at an equivalent position the other sample. These results match well with the distributions predicted in Figure 4 of [1], but which could not, at the time of that publication, have been verified experimentally.

For a more quantitative comparison, Figure 3(a) shows PRESAGE™ data for both 6 MeV samples and the matching parallel plane ionisation chamber data. The lower curves (large and small black triangles) demonstrate that, in the sample with no cavity, the PRESAGE™/optical CT measurements agree extremely well with the ionisation chamber. It is evident from Figure 2 that the dose varies very rapidly with radial distance from the centre of the distribution. In order to compare the ionization chamber readings with the 3-D dosimetry results, we need to integrate the 3-D results over a range of radii corresponding to the size of the ionization chamber. The upper curves of Figure 3(a) depict the depth dose that results from integration over different ranges. Of interest are both the change in relative dose at D_{max} and the shift in D_{max}, both of which can be understood by looking at the distribution in Figure 2. It will be seen that an excellent match with the ionization chamber reading is obtained when we integrate over the range $0.5 < r < 5.8$ mm. This range is in reasonable agreement with [6], which found the "effective collecting radius" of a set of 6 NACP chambers to vary between 5.23 and 5.61 mm.

4. Conclusion

3-D dosimetry using optical CT of PRESAGE™ samples has the potential to be an excellent method for looking at rapidly varying electron dose distributions around inhomogeneities. However, this study is a work in progress and further experiments are needed to obtain reliable results.

References

[1] Åsell, M. and et al., Optimization of 3D conformal electron beam therapy in inhomogeneous media by concomitant fluence and energy modulation. Physics in Medicine and Biology, 1997. **42**(11): p. 2083.

[2] Schiapparelli, P., D. Zefiro, and G. Taccini, Dosimetric verification of a commercial Monte Carlo treatment planning system (VMC++) for a 9 MeV electron beam. Medical Physics, 2009. **36**(5): p. 1759-1767.

[3] Guo, P.Y., J.A. Adamovics, and M. Oldham, Characterization of a new radiochromic three-dimensional dosimeter. Medical Physics, 2006. **33**(5): p. 1338-1345.

[4] S, B. and et al., Micelle hydrogels for three-dimensional dose verification. Journal of Physics: Conference Series, 2009. **164**(1): p. 012044.

[5] Trapp, J.V., et al., The use of gel dosimetry for verification of electron and photon treatment plans in carcinoma of the scalp. Physics in Medicine and Biology, 2004. **49**(9): p. 1625-1635.

[6] Burns, D.T. and M.R. McEwen, Ion recombination corrections for the NACP parallel-plate chamber in a pulsed electron beam. Phys Med Biol, 1998. **43**: p. 2033-2045.

Journal of Physics: Conference Series **250** (2010) 012091 doi:10.1088/1742-6596/250/1/012091

On the Feasibility of Verification of 3D Dosimetry Near Brachytherapy Sources Using PRESAGE/Optical-CT

Pierquet M, Craciunescu O, Steffey B, Song H, Oldham M

Duke University Medical Center, Durham, NC, 27710

corresponding author: haijun.song@duke.edu

Purpose: The feasibility of using the PRESAGE/Optical-CT system for 3D dosimetry verification around a brachytherapy source is investigated. **Method and Materials:** Brachytherapy dose distributions were obtained by irradiation of cylindrical PRESAGE volumes 6cm in diameter by 8cm height with a GammaMed 12i Ir-192 HDR unit (Varian Medical Systems). A narrow channel on the central axis was created by setting a steel catheter in the Presage during manufacture, enabling measurements close to the source (~3mm). **Results:** Comparison of dose line profiles shows good agreement between PRESAGE and verified calculated dose calculation, in both high and low dose regions. **Conclusion:** The PRESAGE/Optical-CT shows good potential in verification of 3D dose distributions around brachytherapy sources.

1. Introduction

The goal of brachytherapy dosimetry is to determine the absorbed dose to water surrounding the brachytherapy radiation source. The current gold standard for absolute dose measurement is based on TLD dosimeters [1,2]. Absolute dosimetry with TLD dosimeters require corrections for the difference in the reference and measurement geometry as well as large corrections for the dosimeter specific energy response function. Other dosimeters such as polymer gels, radiochromic film, and diamond detectors have been used [3,4], but require further demonstration and comparison with established Monte Carlo techniques [5,6].

The PRESAGE/optical-CT 3D dosimetry system has potential to measure high resolution 3D dose distributions [7,8], and therefore potential for brachytherapy dosimetry. The lack of requirement for an external container offers potential for acquiring data closer to the source that may be achieved with other approaches [3]. This work presents a preliminary evaluation of three brachytherapy formulations with simple Ir-192 irradiations. Measured dose distributions are compared with verified calculated dose distributions from a treatment planning system.

2. Materials and Methods

2.1. The PRESAGE/Optical-CT dosimetry system for brachytherapy

The dosimetry system consists of an optical computed tomography (Optical-CT) scanner and the radio-chromic PRESAGE dosimeter. PRESAGE is a radio-chromic solid polymer for dosimetric verification of ionizing radiation [9]. The dosimeter is made up of an optically clear polyurethane matrix that is doped with a free radical initiator and a leuco dye called leucomalachite green (LMG,

Sigma-Aldrich, St Louis, MO). More detailed information can be found in previous work characterizing PRESAGE [7,8].

The optical-CT scanning was done with the **D**uke **M**id-sized **O**ptical-CT **S**canner (DMOS) an in-house CCD based telecentric system illustrated in Figure 1. This scanner has a 9 cm x 11 cm field of view (FOV) with each pixel in the projection images of size ~80microns, providing exquisite spatial resolution for small dosimeters. A telecentric system has advantages for optical scanning through acquisition of true line integrals through the dosimeter with reduced scatter. For details on telecentric scanning see Thomas et al [10].

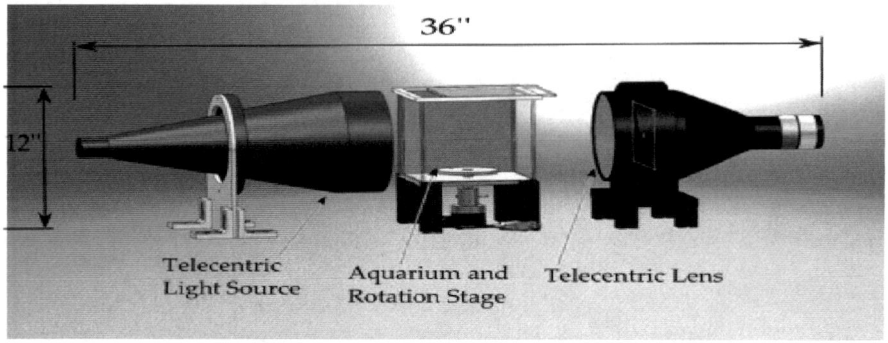

Figure 1. In-house bi-telecentric optical-CT scanner used in the PRESAGE/Optical-CT dosimetry system.

Table 1. Overview of PRESAGE Formulations and irradiation experiment studied in this work.

PRESAGE Formulation	Design Characteristics	Z_{eff}/CT #	Dose/Irradiation Type
1	Near tissue equivalent Z_{eff} and density	7.5/120±20HU	20 nominal sec
2	General formulation, temporally unstable (reusable)	8.1/130±20HU	700cGy to 1.5 cm
3	Increased sensitivity to radiation	7.8/90±20HU	500cGy to 1.5 cm distance

2.2 Brachytherapy irradiation and imaging experiments

Four cylindrical PRESAGE dosimeters, with different composition as listed in Table 1, with height 10 cm and diameter of 6 cm were supplied by Heuris Pharma LLC. Each dosimeter had 2 mm diameter cylindrical channel in the center to receive the catheters for the HDR source insertion. The channel was manufactured by setting a steel catheter in the polymer prior to curing. A release agent on the catheter allowed easy removal, leaving a channel with clean sides suitable for reducing edge artifacts.

The GammaMed 12it and Plus brachytherapy sources, both manufactured by Varian Medical Systems, were used for the simple irradiations in Table 1. Each consists of an iridium wire core surrounded by a stainless steel cylinder and both have half-lives of 73.83 days. There are only minor deviations in the geometry surrounding the iridium wire core, that result in Monte Carlo calculated dose rate data differences of 1% for $\theta<30°$ and 1% to 5% for $\theta>150°$[3].

2.3 Scanning protocol

Journal of Physics: Conference Series **250** (2010) 012091 doi:10.1088/1742-6596/250/1/012091

The central channel was filled with matching fluid for both the pre and post scans to minimize edge artifacts. The following parameters were used for optical-CT scanning: 900 projections at .2 degree spacing, 5 exposures were averaged to create each projection image used in the reconstruction. Each pixel in the projection image was of dimension 0.0774 mm. Flood field images and dark field images were averaged over 400 exposures. Each projection was flood corrected and reconstructed with an in-house FBP matlab script. Each dosimeter was scanned prior to and after irradiation. The difference between these two 3D image sets, the NOD (net optical density), represents radiation induced change in OD. The NOD is proportional to absorbed dose, and was converted using a normalization reference point based on the specific irradiation type, which is listed in Table 1.

2.4 Image Registration and Analysis

The reference dose distribution is calculated with the treatment planning system Eclipse (Varian Medical Systems) that has been commissioned for clinical use. An x-ray CT scan of each PRESAGE dosimeter was imported into Eclipse for treatment planning. The calculated 3D dose distribution was then imported in to the Computational Environment for Radiotherapy Research software (CERR) [11]. The 3D reconstructed NOD data acquired with the DMOS scanner from the Presage dosimeters were also imported into CERR for comparison and analysis. The calculated and measured distributions were registered in CERR, using the channel in the PRESAGE as a major landmark.

Eclipse and PRESAGE dose distributions for formulation 1 were normalized to 1.0 cm from the source on the transverse plane ($r = 1.0$ cm and $\theta = 90°$). Dosimeters of formulation 2 and formulation 3 were planned on Eclipse and irradiated to achieve the dose listed in Table 1 at 1.5 cm from source center on the transverse plane. NOD values were normalized to match the prescription (5Gy at 1.5 cm) consistent with using Presage as a relative dosimeter.

3. Results and Discussion

A comparison of Eclipse and PRESAGE formulation 1 coronal dose distributions is shown in Figure 2. Each image has been windowed from 0 to 10 units of dose-rate relative to dose-rate at 1.0 cm from the source center on the transverse axis.

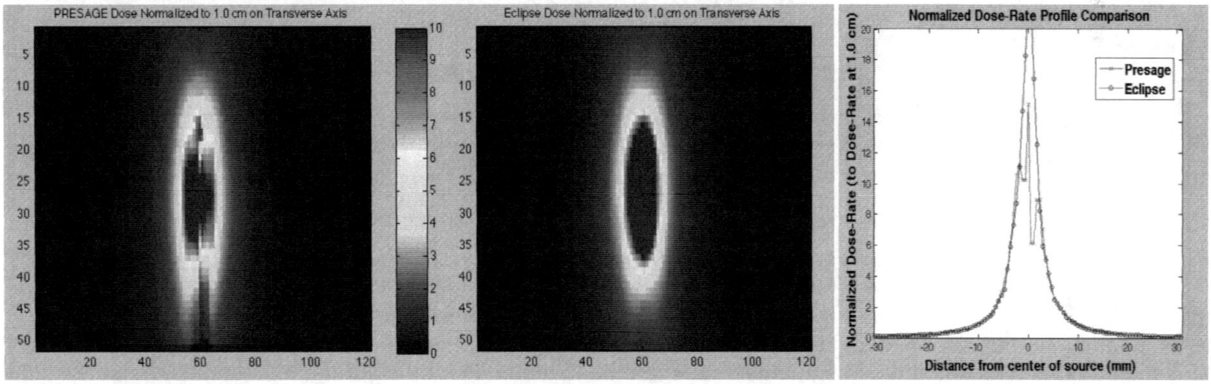

Figure 2. Coronal slice of Presage measured dose in formulation 1 (left) and the planned Eclipse dose distribution (middle). Color bar values are in Gray. Line profiles on the transverse plane are compared in right image.

Line profiles were taken along the transverse axis from the planes shown in Figure 2 are superimposed in the second image from the right in Figure 2. Good agreement is observed between the PRESAGE and Eclipse line profiles even down to very low doses. Schlerring bands, or regions of different indices of refraction within the dosimeter, were most prevalent in the dosimeter of formulation 2. These bands show up in the dosimeter projections and reconstructions, and are seen in the PRESAGE coronal view in Figure 3 on the right side of the dosimeter. In this case, the bands were about 23 mm from the center of the source as indicated by the large peaks on the right side of the profile. The schlerring bands within the polyurethane matrix appear the major contributor to the noise

Journal of Physics: Conference Series **250** (2010) 012091 doi:10.1088/1742-6596/250/1/012091

in the PRESAGE data. Despite the relatively higher noise good agreement is again observed in the areas between 5 mm and 20 mm from the center of the source. Data loss occurred in the ±3mm region from the center of source, which was expected because of artifacts from the channel. Figures 2 and 3 show that accurate dose can be measured >3mm from the source.

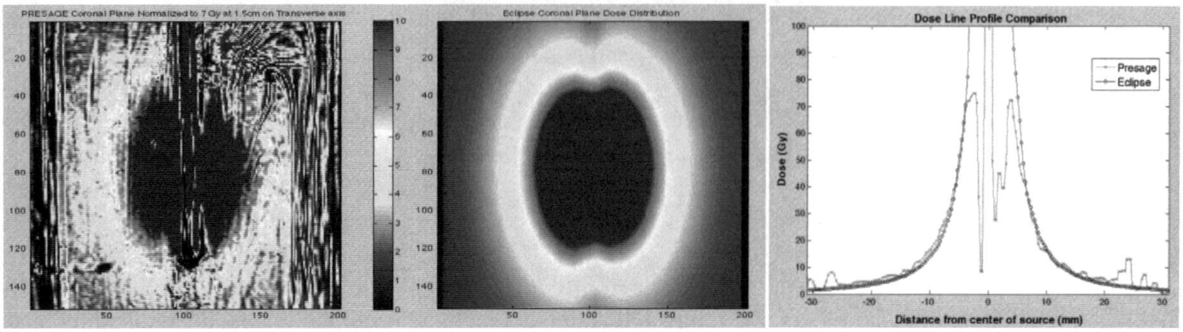

Figure 3. Coronal slice of Presage measured dose in formulation 2 (left) and the planned Eclipse dose distribution (middle). Color bar values are in Gray. Line profiles through transverse axis are shown in the right panel.

Dose distributions and profiles for profiles for formulation 3 are shown in figure 4, and are surprisingly quite different from the previous two formulations tested. Figure 4 shows the under response of this high-sensitivity formulation in the region within 1 cm of the source channel. The under response is approximately symmetric about the channel.

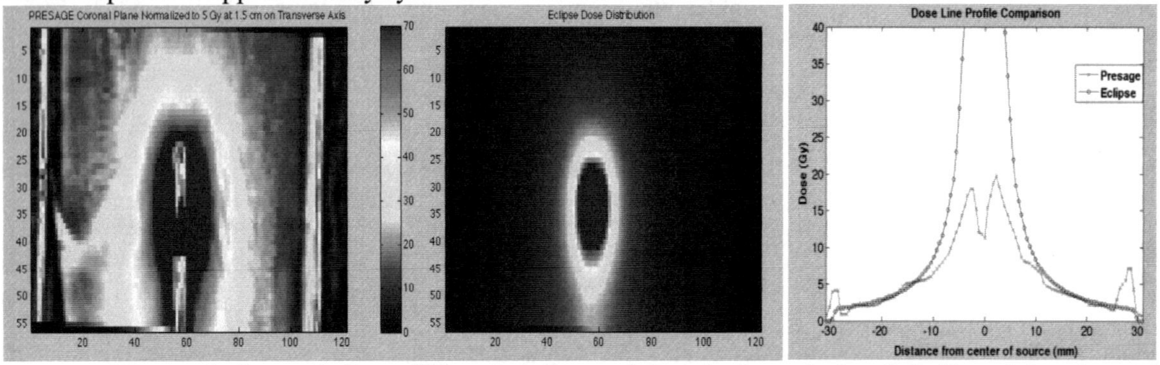

Figure 4. Coronal slice of Presage measured dose in formulation 3 (left) and the planned Eclipse dose distribution (middle). Color bar values are in Gray. Line profiles through transverse axis are shown in the right panel.

One of the main challenges, and sources of uncertainty, is that of source positioning within the narrow channel. The diameter of the channel is just large enough to allow the insertion of the brachy catheter, but the source can exhibit a slight tilt which is difficult to reproduce. In this work, the source was localized by qualitatively tracing back to the source based on identifying the distribution of dose in the PRESAGE dosimeters and knowledge of the symmetric characteristics of iridium brachy dose distributions.

4. Conclusions

High resolution (~0.5mm^3) 3D dosimetry verification around brachytherapy sources appears feasible between ~3 mm and 2 cm from the source using the Presage/optical-CT system. Presage formulations 1 and 2 both show excellent potential for brachytherapy dosimetry. No evidence of an energy response was detected with these formulations to the spectral components of the Ir source. The latter assertion needs further investigation prior to application to lower energy sources.

Acknowledgements: Supported by NIH R01 CA100835.

References

[1] Nath R, Anderson L L, Luxton G, Weaver K A, Williamson J F, Meigooni A S 1995 Dosimetry of interstitial brachytherapy sources: Recommendations of the AAPM Radiation Therapy Committee Task Group No. 43 *Medical Physics* **22**(2)

[2] Tailor R, Ibbott G, Lampe S, Warren W B, Tolani N 2008 Dosimetric characterization of a Cs-131 brachytherapy source by thermoluminescence dosimetry in liquid water Medical Physics **35** 5861-5868

[3] Baldock C, De Deene Y, Doran S, Ibbott G, Jirasek A, Lepage M, McAuley KB, Oldham M, Schreiner LJ Polymer gel dosimetry Phys Med Biol. 2010 Mar 7;55 R1-63

[4] Lin M, Huang T C, Kao M J, Wu J, Chen C L, Wu T H 2009 Three-dimensional dosimetry in brachytherapy: A MAGAT study *Applied Radiation and Isotopes* **67**(7-8): 1432-1437

[5] Ballester F, Puchades J, Lluch J L, Serrano-Andres M A, Limami Y 2001 Technical note: Monte-Carlo dosimetry of the HDR 12i and Plus Ir-192 sources *Med. Phys* **28**(12)

[6] Rivard M J, Coursey B M, DeWerd L A, Hanson W F, Huq M S, Ibbott G S, Mitch M G, Nath R, Williamson J F 2004 Update of AAPM Task Group No. 43 Report: A revised AAPM protocol for brachytherapy dose calculations *Medical Physics* **31**(3)

[7] Clift C, Thomas A, et al. 2010 Toward acquiring comprehensive radiosurgery field commissioning data using the PRESAGE/optical-CT 3D dosimetry system *Physics in Medicine and Biology* **55** 1279-1293

[8] Sakhalkar H S, Adamovics J, Ibbott G, Oldham M 2009 A comprehensive evaluation of the PRESAGE/optical-CT 3D dosimetry system *Medical Physics* **36** 71-82

[9] Adamovics J, Maryanski M J 2006 Characterisation of PRESAGE: A new 3-D radiochromic solid polymer dosemeter for ionising radiation *Radiation Protection Dosimetry* **120**(1-4) 107-112

[10] Thomas A S, Oldham M An investigation into the feasibility of a fast, large field-of-view, telecentric optical-CT scanning system for 3D radiochromic dosimetry *IC3DDose Accompanying Abstract (2010)*

[11] Deasy J O, Blanco A I, et al. 2003 CERR: a computational environment for radiotherapy research *Medical Physics* **30** 979-985

Journal of Physics: Conference Series **250** (2010) 012092 doi:10.1088/1742-6596/250/1/012092

Dose enhancement caused by gold foils on polymer gels

L.C. Afonso; M. Greiter; F. Schoefer; C. Hoeschen

Helmholtz Center Munich, Ingolstaedter Landstrasse 1, 85764 Neuherberg, Germany

luciana.afonso@helmholtz-muenchen.de

Abstract. The presence of high-Z materials adjacent to soft tissues, when submitted to irradiation, locally enhances the absorbed dose in these tissues. This effect is due to the outscattering of photoelectrons from the high-Z materials. The aim of the present work was to measure the absorbed dose enhancement caused by gold foils on polymeric gel.

1. Introduction

Previous studies showed that the presence of high-Z materials adjacent to soft tissues, when submitted to irradiation, enhances the absorbed dose in these soft tissues. Such effects happen due to the outscattering of photoelectrons from the high-Z materials. [1, 2, 3]

The concern about the effect of dose enhancement arose after discovering that oral cavity radiation therapy treatments on patients that had golden tooth replacements resulted in an increased damage of the soft tissues surrounding these gold replacements. It was found that the absorbed dose could reach an enhancement of two orders of magnitude in the microscopic vicinity of gold replacements, when exposing the patient to diagnostic X-ray examinations. [1]

In experiments with human lymphocytes covered with a thin gold film and then irradiated with diagnostic X-ray beams, it was found that the dose absorbed by the cells was increased by a factor of 45.4. The number of chromosomic aberrations was also in accordance to this dose enhancement. [2]

Thereby, the aim of the present work was to measure the absorbed dose enhancement caused by the presence of gold foils on the surface of polymeric gel samples, whose main characteristic is the soft tissue equivalence. The polymer gel dosimetry associated to the Nuclear Magnetic Resonance technique allows the absorbed dose to be measured with high three-dimensional resolution, providing the spatial distribution of the dose enhancement caused by the photoelectrons outscattered from the gold foil.

2. Dosimetric Method

In order to resolve local dose distributions in soft tissue equivalent material close to high-Z materials polymer gel dosimetry was used.

Polymeric systems were first studied in relation to their dosimetric capacity in the fifties, by evaluating the radiation effects in poly-methyl-methacrylate [4]; and studying radioinduced polymerization in liquids [5]. Later works combined these properties as the first polymer gel developed by Maryanski et al [6, 7], which used a formulation that was composed by acrylamide, bis-acrylamide and nitrous oxide diluted in an agarose aqueous matrix.

Further gel formulations were developed by changing the gel matrix (agarose to gelatin), the monomer (acrylamide to methacrylic acid, for example), etc. However, the working principle of any polymer gel dosimetry is based on the radiation-induced polymerization of the monomers present in the gel. When the gel is irradiated, water radiolysis is induced. The polymerization process is mainly initiated by free radical reactions. The polymerization degree depends on the number of free radicals created by the incident radiation, which depends on the absorbed dose; therefore, originating spatial dose resolution. [8, 9]

The polymer gel used for this work is known as MAGIC (Methacrylic Ascorbic in Gelatin Initiated by Copper), foremost developed by Fong et al [10]. Its formulation is composed by methacrylic acid, ascorbic acid, gelatin and copper sulfate. As the gel polymerization process is initiated by free radical reactions, and molecular oxygen is an efficient "scavenger" of free radicals, its presence inhibits gel polymerization. The ascorbic acid, in a process initiated by the copper sulfate, is responsible for capturing oxygen in the gel solution.

3. Methodology

3.1. Polymer Gel Preparation

The polymer gel samples used in this study were produced using deionized water, gelatin (swine skin, 300 bloom, Sigma Aldrich), methacrylic acid (99%, Sigma Aldrich), copper (II) sulfate penta-hydrate, and ascorbic acid (extra pure, Merck).

At room temperature, 140 g of gelatin were added to 700 ml of water. After the water has been absorbed by the gelatin, the mixture was heated to 55 ºC in a water bath and was let to liquefy for approximately 1 hour. The ascorbic acid (0.3522 g) was added; then, copper (II) sulfate (0.025 g); and finally the methacrylic acid. After each reagent has been added, the gel was mixtured thoroughly. The mixture was kept in the water bath for more approximately 1 hour. Afterwards the gel was poured into 12 glasses (50 ml, 5 cm of diameter), which were previously immersed in an oxygen scavenger solution (water, ascorbic acid and copper (II) sulfate) for 6 hours. The filled glasses were covered with cling film (14 ± 2 µm) and were positioned upside down in an oxygen scavenger solution, in order to prevent oxygen penetration through the cling film into the polymer gel.

3.2. Irradiation

The polymer gel samples were irradiated 24 hours after production. A Philips Industrial X-rays (now Yxlon International) system with a MCN 323 tube was used to generate x-rays at 200 kV, 15 mA filtered with 0.96 mm Cu and 4.05 mm Al. A PTW ionization chamber (type 23361, 30 cm^3 sensitive volume) coupled with a PTW Unidos electrometer was used for the dose measurement.

Gold foils (thickness about 100 nm) were placed on the cling film surface of each sample, covering half of the surface area (Figure 1). The polymer gel samples were individually positioned with the cling film surface perpendicular to the beam at the reference point at 70 cm from the x-ray focal spot (Figure 2). Irradiation doses, measured as air kerma at the reference point, were 0, 1, 2, 3, 4, and 5 Gy, attained at a constant dose rate of 4.52 mGy/s (± 1.5 % combined uncertainty; including positioning, calibration factors, instrumental stability, and corrections for air density). Two samples per dose were used.

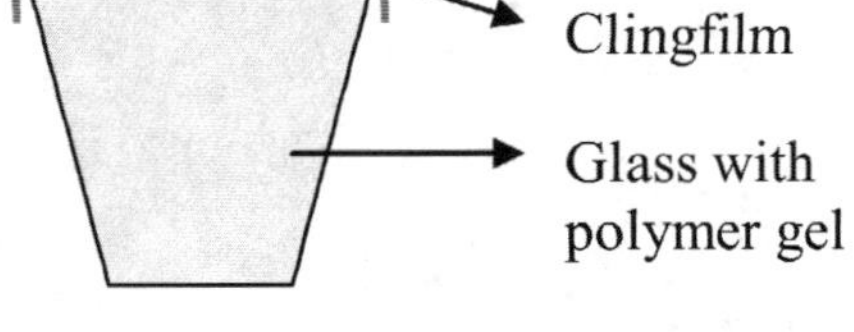

Figure 1: Schematic representation of a sample.

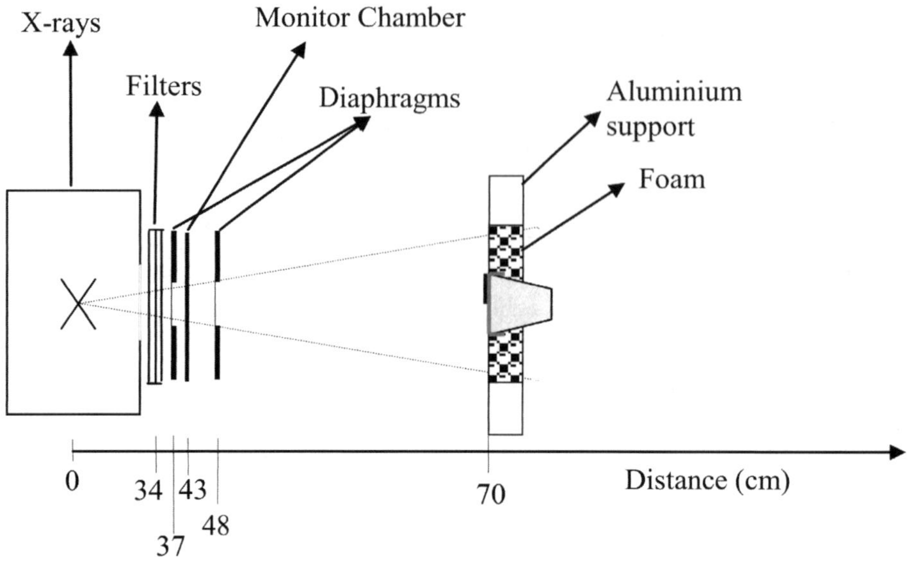

Figure 2: Irradiation set-up (not to scale).

3.3. NMR Scanning

After approximately 17 hours the samples were scanned using a Bruker BioSpin system (BioSpec 94/20 USR, 9.4 T). A Multi Slice Multi Echo sequence (MSME) was used with FOV = 6 x 6 cm^2, echo times of 20, 40, …, 320 ms and a repetition time of 2500 ms. One slice with 1mm thickness was measured in the surface of each sample. The T2 values were obtained by fitting all signal echoes to a monoexponential function. Then, the transversal relaxation rates (R2 = 1/T2) were related to the doses. Two R2 in different regions of interest (ROI) per sample were measured; one under the area covered by the gold foil and the other out of the area covered with gold. The gold foils were removed before the scanning.

4. Results and Discussion

The relation between R2 and the dose is represented in the Figure 3. It is observed that the relaxation rates measured under the gold foils are greater than those measured in the area free of gold. One can observe that the slope from the linear regression of the measurements in the area with gold is greater than the slope obtained from the measurements in the area without gold. This indicates that the presence of gold foils on the surface of the samples caused a dose enhancement in the vicinity of these gold foils by 16 %. This result can serve just as an estimate because of two important factors: the thickness of the gold foils was not known exactly; and the gold foils were very flexible and easily torn, so the difficulty of dealing with such foils caused that some small areas were not well covered.

Journal of Physics: Conference Series **250** (2010) 012092 doi:10.1088/1742-6596/250/1/012092

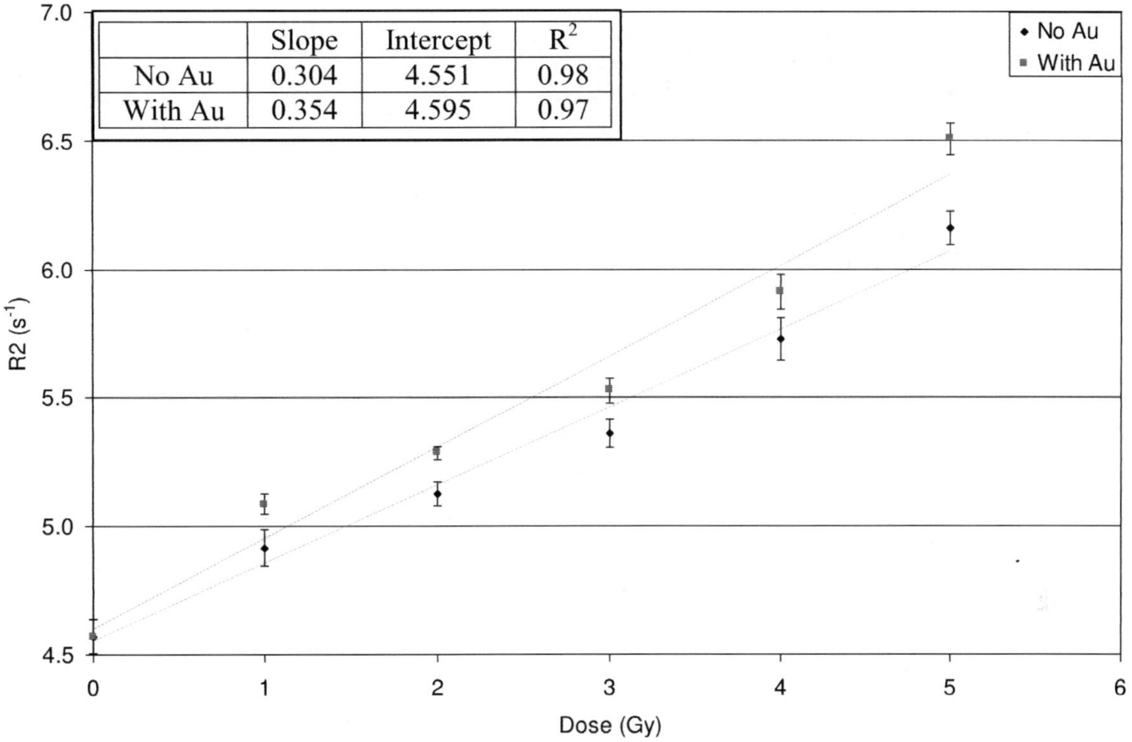

Figure 3: Relation between R2 and the dose. The red squares represent the area covered with gold foils and the black lozenges represent the area not covered with gold foils. The error bars represent the standard deviation of the R2 in the ROI.

5. Conclusion

The result of this study indicates the effect of dose enhancement in the area adjacent to the gold foils. Nevertheless, taking into account that difficulties mentioned above may play an important role, the experiment will be repeated using more resistant gold surfaces (i.e., thicker foils or plates) in order to assure this result and achieve a better quantification. The scanning will be performed with thinner slices, with a sequence of slices into the depth of the gels.

References

[1] Regulla D, Friedland W, Hieber L et al (2000) Spatially Limited Effects of Dose and LET Enhancement Near Tissue/Gold Interfaces at Diagnostic X Ray Qualities. Radiation Protection Dosimetry, 90, Nos 1–2, pp. 159–163

[2] Regulla D, Panzer W, Schmid E et al (2001) Detection of Elevated RBE in Human Lymphocytes Exposed to Secondary Electrons Released from X-Irradiated Metal Surfaces. Ratiation Research, 155, pp. 744–747

[3] Herold DM, Das IJ, Stobbe CC et al (2000) Gold Microspheres: a Selective Technique for Producing Biologically Effective Dose Enhancement. International Journal of Radiation Biology, 76, No 10, pp. 1357-1364

[4] Alexander P, Charlesby A, Ross M (1954) The Degradation of Solid Polymethyl-methacrylate by Ionizing Radiations. Proceedings of the Royal Society, pp. 392-404

[5] Hoecker FE, Watkins IW (1958) Radiation Polymerization Dosimetry. International Journal of Applied Radiation and Isotopes, 3, pp. 31-36

[6] Maryanski MJ, Gore JC, Schulz RJ (1992) 3-D Radiation Dosimetry by MRI: Solvent Proton Relaxation Enhancement by Radiation-Controlled Polymerisation and Cross-Linking in Gels. Proc. Intl. Soc. Mag. Reson. Med.

[7] Maryanski MJ, Gore JC, Kennan RP et al (1993) NMR Relaxation Enhancement in Gels Polymerized and Cross-Linked by Ionizing Radiation: a New Approach to 3D Dosimetry by MRI. Magn. Reson. Imaging, 11, pp. 253-258

[8] Berg A, Ertl A, Moser E (2001) High Resolution Polymer Gel Dosimetry by Parameter Selective MR-Microimaging on a Whole Body Scanner at 3 T. Medical Physics, 28(5), pp. 833-843

[9] Bayreder C, Georg D, Moser E et al (2006) The spatial resolution in dosimetry with normoxic polymer-gels investigated with the dose modulation transfer approach. Medical Physics, 35(5), pp. 1756-1769

[10] Fong PM, Keil DC, Does MD, Gore JC (2001) Polymer Gels for Magnetic Resonance Imaging of Radiation Dose Distributions at Normal Room Atmosphere. Physics in Medicine and Biology, 46, pp. 3105-3113

[11] Baustert IC, Oldham M, Smith TAD et al (2000) Optimized MR Imaging for Polyacrylamide Gel Dosimetry. Physics in Medicine and Biology, 45, pp. 847-858

[12] McJury M, Oldham M, Cosgrove VP et al (2000) Radiation Dosimetry Using Polymer Gels: Methods and Applications. The British Journal of Radiology, 73, pp. 919-929

Water equivalency evaluation of PRESAGE® dosimeters for dosimetry of Cs-137 and Ir-192 brachytherapy sources

Tina Gorjiara[1], Robin Hill[1, 2], Zdenka Kuncic[1] and Clive Baldock[1]

[1]Institute of Medical Physics, School of Physics, University of Sydney, NSW 2006, Australia

[2] Department of Radiation Oncology, Royal Prince Alfred Hospital, Sydney, NSW 2050, Australia

clive.baldock@sydney.edu.au

Abstract. A major challenge in brachytherapy dosimetry is the measurement of steep dose gradients. This can be achieved with a high spatial resolution three dimensional (3D) dosimeter. PRESAGE® is a polyurethane based dosimeter which is suitable for 3D dosimetry. Since an ideal dosimeter is radiologically water equivalent, we have investigated the relative dose response of three different PRESAGE® formulations, two with a lower chloride and bromide content than original one, for Cs-137 and Ir-192 brachytherapy sources. Doses were calculated using the EGSnrc Monte Carlo package. Our results indicate that PRESAGE® dosimeters are suitable for relative dose measurement of Cs-137 and Ir-192 brachytherapy sources and the lower halogen content PRESAGE® dosimeters are more water equivalent than the original formulation.

1. Introduction

Cesium-137 and Iridium-192 are two isotopes commonly used for brachytherapy treatments of gynaecologic and prostate cancer. The low dose rate (LDR) Cs-137 source has a half life of 30 years and emits photons with a mean energy of 0.662 MeV. The high dose rate (HDR) microSelectron Ir-192 source has a half life of 74 days and a spectrum of emitted photons with an average energy of 0.37 MeV. According to the American Association of Physicist in Medicine (AAPM) recommendations, the dosimetry of these sources can be performed with a thermoluminescent dosimeter (TLD) [1]. However, because of the steep dose gradients, dose measurements are difficult close to the source. In addition, the TLDs only provide point dose information and would require a large number of them to provide two or three dimensional dosimetric information. Therefore, 3D dosimetry techniques such as polymer gel dosimetry are needed to improve spatial resolution in brachytherapy dosimetry [2-5].

PRESAGE is a solid, transparent dosimeter made of polyurethane and is suitable for 3D dosimetry in modern radiation treatment techniques [6]. PRESAGE has a number of advantages over other 3D gel and polymer dosimeters including: being insensitive to oxygen contamination, being machinable and moldable to variety of shapes and sizes, and being preferentially absorbing rather than scattering which facilitates high accuracy readout by optical CT [7]. A previous study has shown that PRESAGE™ with optical CT has the potential to be used for Ir-192 brachytherapy dosimetry [8]. However, for a dosimeter to be used in brachytherapy dosimetry, it should exhibit water equivalent properties. This means that the absorption and scattering of radiation in a given thickness of material should be comparable to that of water.

PRESAGE® is another formulation of PRESAGE and it was found that the PRESAGE® original formulation has an effective atomic number (Z_{eff} = 8.65) that is 17% higher than the value of water and does not generally exhibit good water equivalent radiological properties [9]. Two new formulations of PRESAGE® dosimeters have recently been developed, with lower halogen contents (giving Z_{eff} = 7.69 and 7.74) to improve performance as a 3D dosimeter and to improve water equivalency.

In this work, we calculated the relative dose distribution in the three different PRESAGE® dosimeters for Cs-137 and Ir-192 brachytherapy sources using the EGSnrc Monte Carlo code. We have compared the relative doses in the dosimeters with the relative depth dose in water.

2. Method and Materials

The elemental composition of each PRESAGE® formulation was calculated using their molecular formulas. These are summarized in Table 1. We henceforth refer to the lower halogen content PRESAGE® formulations as (LHP1) and (LHP2), and we refer to the original PRESAGE® as (OP). All the three PRESAGE® formulations have been provided by the same manufacturer (Heuris Pharma, New Jersey, USA).

Table 1. Elemental composition and fractional weight (w_k) for materials of interest.

Material	Formula	w_H	w_C	w_N	w_O	w_S	w_{Cl}	w_{Br}
Low Halogen PRESAGE®1 (LHP1)	$C_{1758}H_{3000}N_{121}O_{442}S_4Cl_{30}Br_1$	0.08850	0.6178	0.04960	0.2069	0.003800	0.03112	0.002300
Low Halogen PRESAGE®2 (LHP2)	$C_{889}H_{1541}N_{60}O_{228}S_{10}Cl_1Br_1$	0.09050	0.6224	0.04899	0.2126	0.01870	0.002100	0.004700
Original PRESAGE® (OP)	$C_{481}H_{842}N_{30}O_{129}Cl_9Br$	0.08920	0.6074	0.04460	0.2172	-	0.03340	0.008400
Water	H_2O	0.1119	-	-	0.8881		-	-

Monte Carlo modeling was performed to determine the differences in depth doses between the different PRESAGE® dosimeters and water for Cs-137 and Ir-192 brachytherapy sources. The Monte Carlo calculations were performed using the EGSnrc package (Version 4 release 2-2-5, Research Council of Canada, Ottawa, Canada)[10, 11].

Cross section data for the three PRESAGE® dosimeters and water were generated by the PEGS4 preprocessor using the elemental compositions and mass densities of the different materials. Each PRESAGE® phantom was defined in the DOSRZnrc user code with a 20.5 cm thickness. The brachytherapy sources were modeled as cylinders with 0.05cm radius and 0.5 cm length located at 10 cm depth in the phantom. Dose was calculated as a function of radial distance from the source with 0.1 cm intervals. The electron and photon energy cutoff parameters, ECUT and PCUT, were set to 0.521 MeV and 0.001 MeV respectively. Xcom library was used to obtain the photon cross section data.

Bound Compton scattering, PE angular sampling, Rayleigh scattering and atomic relaxations were switched on during the calculation. Dose calculation was performed using 5 billion incident particles. In order to compare calculated central axis depth doses in water and in the PRESAGE® dosimeters, the water equivalent depth was calculated for each case [12].

3. Results

Figure 1 shows Monte Carlo calculated relative dose versus density corrected radial distance from a Cs-137 brachytherapy source for the three PRESAGE® dosimeters and water as well as their difference relative the value of water. In the first 0.5 cm density corrected distance from the source, the dose gradient is steeper in water than in the three PRESAGE® dosimeters. The difference in relative dose values of water and the three PRESAGE® dosimeters increases with radial distance from 0.02 for the lower halogen content PRESAGE® and 0.035 for the original PRESAGE® to 0.95 and 0.1, respectively. The maximum relative dose difference between the three PRESAGE® dosimeters and water occurs at 0.15 cm radial distance from the source.

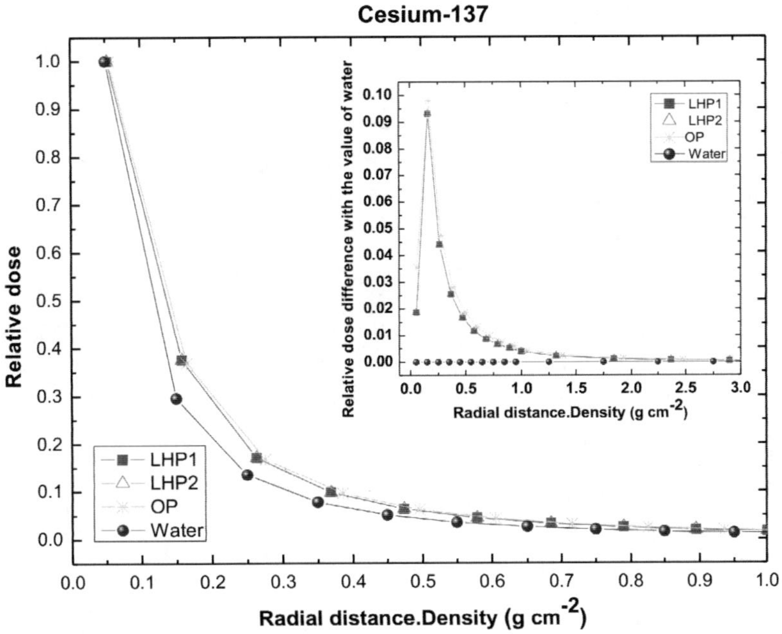

Figure 1. Monte Carlo density corrected relative dose curves for a Cs-137 source along radial distance for the three PRESAGE® dosimeters and water and their difference relative to the value in water (inset).

Figure 2 shows Monte Carlo calculated relative dose to density corrected radial distance from an Ir-192 brachytherapy source for the three PRESAGE® dosimeters and water as well as their difference relative to the value of water. At 0.5 mm radial distance from the source, the difference between the value of density corrected relative dose of the original formulation PRESAGE® with water is twice that of the low halogen content PRESAGE®. The difference with the value of water decreases gradually with radial distance. This relative dose curve suggests that the lower halogen content PRESAGE® dosimeters are only marginally more water equivalent than the original PRESAGE® in terms of dose response to Ir-192. The uncertainty of calculated dose in the all cases was below 0.1%.

Journal of Physics: Conference Series **250** (2010) 012093 doi:10.1088/1742-6596/250/1/012093

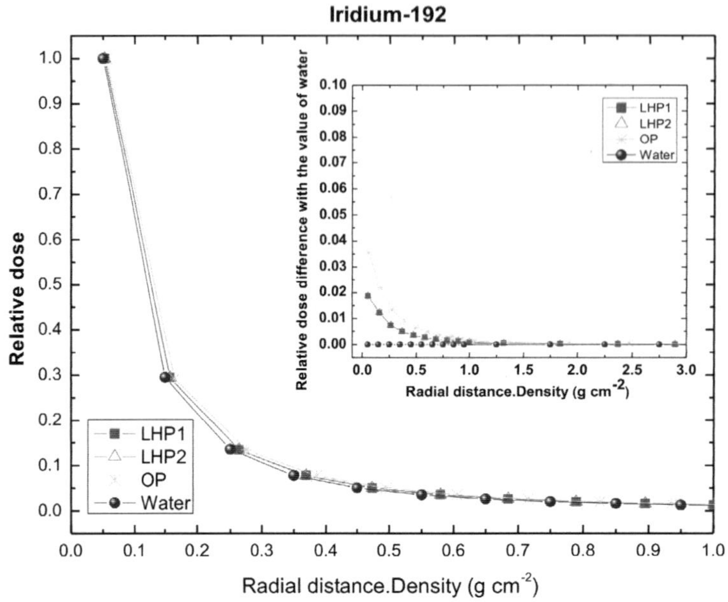

Figure 2. Monte Carlo density corrected relative dose curve for a Ir-192 source along radial distance for the three PRESAGE® dosimeters and water and their difference relative to the value of water (inset).

4. Conclusion

Results of this study indicate that like PRESAGE™, the PRESAGE® dosimeters are suitable for depth dose measurement for Cs-137 and Ir-192 brachytherapy sources. The lower halogen content PRESAGE® dosimeters exhibit a more water equivalent depth dose response than the original PRESAGE® formulation.

References

[1] Nath, R., Dosimetry of Interstitial Brachytherapy Sources - Recommendations of the AAPM Radiation-Therapy Committee Task Group No-43 - Response. Medical Physics, 1995. 22(8): p. 1351-1351.

[2] De Deene, Y., N. Reynaert, and C. De Wagter, On the accuracy of monomer/polymer gel dosimetry in the proximity of a high-dose-rate 192Ir source. Vol. 46. 2001. 2801-25.

[3] Pantelis, E., et al., Polymer gel dosimetry close to an I-125 interstitial brachytherapy seed. Physics in Medicine and Biology, 2005. 50(18): p. 4371-4384.

[4] Farajollahi, A.R., et al., An investigation into the use of polymer gel dosimetry in low dose rate brachytherapy. British Journal of Radiology, 1999. 72(863): p. 1085-1092.

[5] Hurley, C., et al., High-resolution gel dosimetry of a HDR brachytherapy source using normoxic polymer gel dosimeters: Preliminary study. Nuclear Instruments & Methods in Physics Research Section a-Accelerators Spectrometers Detectors and Associated Equipment, 2006. 565(2): p. 801-811.

[6] Baldock, C., et al., Polymer gel dosimetry. Physics in Medicine and Biology, 2010. 55(5): p. R1-R63.

[7] Guo, P.Y., J.A. Adamovics, and M. Oldham, Characterization of a new radiochromic three-dimensional dosimeter. Medical Physics, 2006. 33(5): p. 1338-1345.

[8] Wai, P., et al., Dosimetry of the microSelectron-HDR Ir-192 source using PRESAGE (TM) and optical CT. Applied Radiation and Isotopes, 2009. 67(3): p. 419-422.

[9] Brown, S., et al., Radiological properties of the PRESAGE and PAGAT polymer dosimeters.

Applied Radiation and Isotopes, 2008. 66(12): p. 1970-1974.
[10] Kawrakow, I., Accurate condensed history Monte Carlo simulation of electron transport. I. EGSnrc, the new EGS4 version. Medical Physics, 2000. 27(3): p. 485-498.
[11] Rogers, D.W.O., Fifty years of Monte Carlo simulations for medical physics. Physics in Medicine and Biology, 2006. 51(13): p. R287-R301.
[12] Khan, F.M., The physics of radiation therapy. 4th ed, ed. J. Pine. 2010, Philadelphia: Lippincott Williams & Wilkins.

Journal of Physics: Conference Series **250** (2010) 012094 doi:10.1088/1742-6596/250/1/012094

High resolution polymer gel dosimetry for small beam irradiation using a 7T micro-MRI scanner

Xuanfeng Ding [1,2] John Olsen [3] Ryan Best [1,2] Marcus Bennett [1,2] Inna McGowin [1,2] Jennifer Dorand [1,2] Kerry Link [3] J. Daniel Bourland [1,2]

[1]Department of Radiation Oncology, [2]Department of Physics, [3]Center for Biomolecular Imaging, Wake Forest University, Winston-Salem, NC, 27157, U.S.A.

Email: dingx6@wfu.edu

Abstract. The use of small field radiation beams has greatly increased with advanced radiation therapy techniques such as IMRT, rotational IMRT, and stereotactic body radiotherapy. In this work small field 3D dose distributions have been measured with high spatial resolution using polymer gels and 7T micro-MR imaging. A MAGIC (Methacrylic and Ascorbic acid in Gelatin Initiated by Copper) polymer gel [1] phantom was used to capture the 3D dose distributions for two small field (5 x 5 mm^2 and 10 x 10 mm^2) for a 6MV x-ray beam. High resolution 3D T2 maps were obtained with 7T micro-MRI (0.156mm x 0.156mm x 1mm, MSME pulse sequence). For comparison T2 maps, the gel phantom was scanned in a 3T MRI clinical scanner (0.254mm x 0.254mm x 2mm, FSE pulse sequence). Normalized 3D dose maps were calculated in Matlab. Results show that 7T micro-MRI 3D gel dosimetry measurements are much more stable, less noisy, and have higher spatial resolution than those obtained using a 3T clinical scanner for the same amount of scan time. In general, 3D gel dosimetry results also agree with simultaneously-obtained radiochromic film dosimetry. This study indicates that the MAGIC polymer gel with 7T micro-MRI for 3D dose readout could potentially be used for small radiation beams, including measurements for micro-beams (field size ~ 100um).

1. Introduction:

Dosimetric characteristics can be difficult to determine for small radiation fields used in intensity modulated radiotherapy (IMRT), rotational IMRT, and radiosurgery. In contrast to conventional dosimetry methods (ionization chambers, film, TLD etc.), polymer gel dosimetry is becoming a more important tool for evaluating the small beam 3D dose characteristics. Gel polymerization can be measured by a variety of methods including optical-computed tomography (OCT), x-ray CT, Raman Spectroscopy, Magnetic Resonance Imaging (MRI) and Ultrasound. MRI is a commonly used readout method because it can image the 3D dose distribution in both homogeneous and non-homogenous (e.g. presence of bone) media, and provides high spatial resolution compared to other 3D gel imaging modalities. Current clinical 1.5T and 3T MRI systems are capable of scanning a gel phantom in about one-half hour with finest spatial resolution of about 0.25-0.5mm[2]. However, for very small radiation fields (<1cm diameter) or micro-beams (<1mm diameter) captured in polymer gel, increased image resolution and signal-to-noise ratio (SNR) are critical and may be afforded by using MR field strengths greater than 3T. MR scan time is also important for gel dosimetry. In this initial work, we compare the performance of 3T and 7T MR imaging for the 3D dose readout of a MAGIC polymer gel for two small field (10 x 10 mm^2 and 5 x 5 mm^2). Results show that the 7T micro-MRI dose image is more stable and has less noise compared to the 3T clinical MR dose image using the same scan time.

2. Methods and Materials:

2.1. Gel preparation:

The formula of the normoxic MAGIC polymer gel is based on Fong et al in 2001 [1]. These gels are able to be prepared under normal atmosphere conditions(e.g. in-air) instead of in a special oxygen free glove box. The chemical ingredients are commercially available (gelatin swine skin[9% (w/w) 90g]; methacrylic acid [4%(w/w) 40g]; copper(II) sulfate penta-hydrate[0.025g], ascorbic acid [0.3522g (2mM)], (all available from Fisher Scientific); and deionized water). The MAGIC gel was instilled into 5cm diameter plastic tubes and sealed with plastic wrapper to prevent outside oxygen from entering. Then, the completed MAGIC gel phantom was stored in a refrigerator ($4^{o}C$) to set overnight before irradiation.

2.2. Polymer gel and radiochromic film irradiation

Small field irradiation was performed using a 6MV X-ray beam from a 2100 SCX clinical linear accelerator (Varian Medical Systems, Palo Alto, CA). Two small fields 10 x 10 mm^2 and 5 x 5 mm^2 were defined using symmetrical collimator jaws. The gel phantom was placed on the treatment table with the source-to-surface-distance (SSD) set to 100cm. In this work, we use Gafchromic EBT2 radiochromic film (ISP Technologies, Wayne, NJ) as a comparison to the gel dosimeter. Radiochromic film was handled using gloves to reduce the possibility of damage and contamination according to AAPM TG 55 [3]. The exposed radiochromic films were scanned using an Epson 10000XL scanner 20 hours after irradiation.

2.3. MR imaging and parameters:

A 7T micro-MR scanner (Bruker Corporation, Billerica, MA) was used for high resolution gel dosimeter reading (Figure 1). The high magnetic field strength provides images with very high resolution and signal to noise ratio (SNR) compared to the 1.5 T and 3 T clinical scanners typically used. However, the 7T micro-MR scanner is limited by its core space and special RF coil design, and as a result the cylindrical gel phantom size for the 7T MRI scanner could not exceed 10 cm in diameter. Moreover, due to the limited amount of B0 homogeneity in the scanner center, only a single sample can be scanned at a time. As a result, the gel container shape and size is very important consideration for use of the micro-MR high resolution scan.

The irradiated gel phantom was first scanned in the 7T MR using a linear birdcage volume coil (Bruker Corporation, Billerica, MA) with these parameters: Field of View (FOV) = 4cm Slice Thickness=2mm interslice distance = 5mm; Matrix size = 256 x 256; pixel size = 0.156mm; TE = 30, 60 … 300 ms and TR = 3330ms (Figure 2). Total scan time was about 30mins. Next, the gel phantom was scanned at the same room temperature using a 3T MRI GE signa® EXCITE™ clinical scanner (GE Healthcare, Milwaukee, WI) with 8-channel head coil. An alternative pulse sequence to MSME, Fast-Spin-Echo (FSE) was used with parameters: FOV = 6.5cm; matrix size = 256 x 256; TE = 50, 100… 500; TR = 4000 ms; NEX = 1; slice thickness = 2 mm and pixel size = 0.254mm x 0.254mm

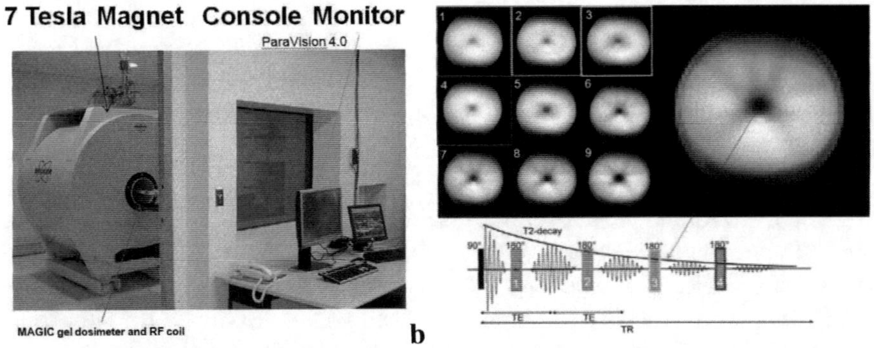

Figure 1. a) 7T Bruker micro-MR scanner setup for gel dosimeter. b) Gamma radiosurgery irradiated gel phantom scanned at 7T with the MSME pulse sequence for T2 maps calculation.

3. Results:

Polymer gel MR images are shown for 7T and 3T field strengths in Figure 2. and demonstrate higher image fidelity and less noise at 7T (Fig. 2a) vs 3T (Fig. 2b). Radiochromic film, obtained simultaneously at the time of gel irradiation, documents the field sizes (Fig. 2c). Dose profiles are shown for the 7T and 3T readouts and the radiochromic film(Figs. 2 d-e).

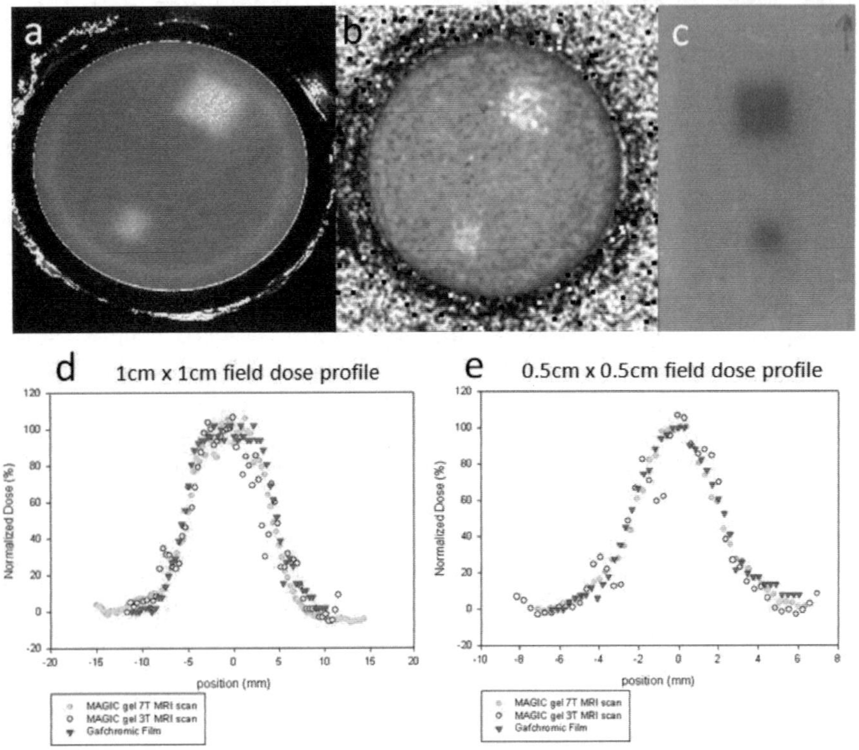

Figure 2. a) Gel phantom R2 map, axial plane, calculated from 7T MRI scan, pixel size = 0.156 mm. b) Gel phantom R2 map, axial plane, calculated from 3T MRI scan, pixel size = 0.254mm. c) Radiochromic film irradiated by 5 x 5 mm^2 and 10 x 10 mm^2 fields. d) dose profile of 10 x 10 mm^2 6MV beam at depth of 1.5cm. e) dose profile of 5 x 5mm^2 6MV beam at depth of 1.5cm.

4. Conclusions:

7T micro-MRI gel imaging greatly improves the dose measurement in both accuracy and spatial resolution compared to 3T MRI with a clinical scanner. At 7T field strength, 3D dose distribution images with voxel size of 0.156mm x 0.156mm x 1mm were achieved within 30 minutes of scan time using a 150mm inside diameter linear birdcage volume coil (max gradient 200mT/m). A higher resolution (<50um x 50um x 1mm) could be achieved within a 2 hour scan time using a 35mm inside diameter quandrature birdcage volume coil (max gradient 1000mT/m). This approach could be potentially used to measure a micro-beam (~100um) dose distribution.

References:

[1] Peter M Fong, Derek C Keil, Mark D Does and John C Gore "Polymer gels for magnetic resonance imaging of radiation dose distributions at normal room atmosphere" Phys. Med. Biol. 46 (2001) 3105–3113

[2] C Baldock, Y De Deene, S Doran, G Ibbott, A Jirasek, M Lepage, K B McAuley, M Oldham and L J Schreiner "Polymer gel dosimetry" Phys. Med. Biol. 55 (2010) R1–R63

[3] A. Niroomand-Rad Chair, C. R. Blackwell, B. M. Coursey, K. P. Gall, J. M. Galvin, W. L. McLaughlin, A. S. Meigooni, R. Nath, J. E. Rodgers, and C. G. Soares, "Radiochromic film dosimetry: Recommendations of AAPM Radiation Therapy Committee Task Group 55," Medical Physics, 25(1998) 2093-2115

Journal of Physics: Conference Series **250** (2010) 012095 doi:10.1088/1742-6596/250/1/012095

Three-Dimensional Dosimetry of a Beta-Emitting Radionuclide Using PRESAGE® Dosimeters

R L Grant[1,2,4], M L Crowder[3,4,5], G S Ibbott[1,2,4,11], J Simon[6], R K Frank[6], J Rogers[6], H M Loy[7,8], J Adamovics[9], J Newton[10], M Oldham[10], S Stearns[7,8] and R E Wendt[3,4]

[1]Radiological Physics Center and Departments of [2]Radiation Physics and [3]Imaging Physics, University of Texas MD Anderson Cancer Center, 1515 Holcombe Blvd., Houston, TX 77030 USA
[4]University of Texas at Houston Graduate School of Biomedical Sciences, Houston, TX 77030 USA
[5]Abilene Christian University, Abilene, TX 79699 USA
[6]IsoTherapeutics Group, LLC, 1004 S. Velasco St., Angleton, TX 77515 USA
[7]The Gabriel Institute, P. O. Box 55603, Houston, TX 77255 USA
[8]Valco Instruments Co., Inc., 7811 Westview Dr., Houston, TX 77055 USA
[9]Heuris Pharma, LLC, 412 Sunset Rd., Skillman, NJ 08558 USA
[10]Duke 3D Dosimetry Laboratory, Box 3085, Duke University Medical Center, Durham, NC 27710 USA

E-mail: gibbott@mdanderson.org

Abstract. Three-dimensional dose distributions from liquid brachytherapy were measured using PRESAGE® dosimeters. The dosimeters were exposed to Y-90 for 5.75 days and read by optical tomography. The distributions are consistent with estimates from beta dose kernels.

1. Background

Recent advances in radiopharmaceutical design have led to the development of a new technique called "liquid brachytherapy"[1]. Liquid brachytherapy is a proprietary formulation that is administered as an injection directly into the tissue to be treated (e.g., an osteosarcoma) and does not migrate from the point of administration. Thus, beta-emitting radionuclides can be used without the undesirable shielding effect of the encapsulation associated with sealed source brachytherapy. Intricately shaped dose distributions that exploit the limited range of beta particles are possible with this method. The work-in-progress reported here is being undertaken to assess the three-dimensional dosimetry of liquid brachytherapy using beta-emitting radiopharmaceuticals.

2. Methods

PRESAGE®[2] polyurethane radiochromic dosimeters [Heuris Pharma, LLC, Skillman, NJ] in the shape of a right circular cylinder measuring 6 cm in diameter and at least 3 cm in height were selected for three-dimensional absorbed dose measurements because of their desirable physical properties. In particular, a dosimeter for this application must be stable for at least seven halflives of the radionuclide to be studied and resistant to liquids.

Three dosimeters were prepared. The first dosimeter was designed to assess the dose distribution of a point source of a beta emitter. The second dosimeter was designed to assess our ability to measure dose distributions that overlap in two-dimensional projections but not in three dimensions. The third dosimeter was designed to assess the dose distribution from a line source and also from two point sources whose dose distributions overlap in three dimensions.

[11]To whom correspondence should be addressed.

Journal of Physics: Conference Series **250** (2010) 012095 doi:10.1088/1742-6596/250/1/012095

Holes with a diameter of 0.043" (1.09 mm) were drilled into the plastic dosimeters to allow the radionuclide to be introduced. Particular care was taken to avoid raising the temperature of the dosimeter material (which would have fogged it in the vicinity of the hole). All of the sources were solutions of Y-90 as yttrium chloride in 50 mM hydrochloric acid.

The first dosimeter contained a single hole, 1.5 cm deep, into which 1 uCi (37 kBq) of Y-90 in a volume of 1 uL was introduced in order to approximate a point source in the bottom of the drilled hole. The activity was chosen to give an expected dose of 2 Gy at a distance of 3 mm from the point source with an exposure time of 5.75 days. The second dosimeter had three holes drilled 3 cm apart in a triangular configuration. Each hole was 1.5 cm deep. The holes were filled with 0.25, 1, and 4 uCi (9.25, 37, and 148 kBq) respectively of Y-90, each in a volume of 1 uL. The radii of the 2 Gy isodose surfaces were expected to be 1.6, 3 and 4.8 mm from the respective point sources. The maximum range of the Y-90 beta emissions is about 1.1 cm and thus the dosed volumes from each point source did not overlap in three dimensions. The third dosimeter had three holes. One was 2.5 cm deep and 3 cm from the other two. The other two were 1.5 cm deep and 1 cm apart. The deep hole was filled with 2 uCi (74 kBq) of Y-90 in 9.36 uL, creating a line source with a length of 1 cm. The shallower holes were filled with 4 uCi (148 kBq) of Y-90 in 1 uL. This activity was chosen so that the 2 Gy isodose surface would have a dumbbell shape in three dimensions. After 5.75 days of exposure, the radioactivity was rinsed from the dosimeters using distilled water and hydrochloric acid.

A control dosimeter was exposed to ambient light for the same lengths of time as were the test dosimeters during their drilling, filling and rinsing.

Two calibration dosimeters were exposed to 4 cm x 4 cm external beams of 16 MeV electrons to ascertain the relationship between optical density of the PRESAGE dosimeter and absorbed dose from calibrated electron beams. The beams were designed to deliver a dose of 2 Gy to one dosimeter and 3 Gy to the other at d_{max}.

The dosimeters were then scanned with the Duke Midsized Optical Scanner dedicated for the RPC (DMOS-RPC) [Duke University, Durham, NC] using 0.5 degree per step to produce 720 projection images consisting of 798x600 pixels that were 0.140 mm on a side. Transverse images were reconstructed by filtered backprojection with a pixel size of 0.185 mm. These images were exported to the CERR software[3] [Computation Environment for Radiotherapy Research, Washington University, St. Louis] and calibrated using the electron beam data.

The calibrated radionuclide dose distributions were compared to estimates based upon the beta-ray dose kernel of Cross, et al[4].

3. Results

Irradiation of two dosimeters with electron beams yielded the calibration curve shown in figure 1. Only the pixel values at the depth of d_{max} were used for this curve. The relationship between scanner signal and dose was applied to scans of the other dosimeters to calibrate them. Figures 2-4 show isodose curves from the Y-90 dose distributions.

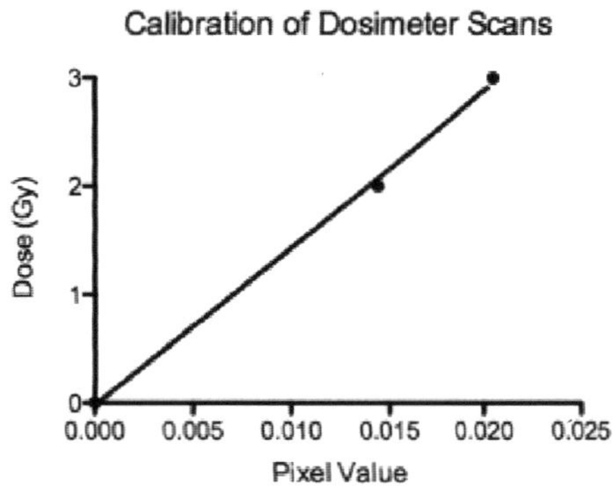

Figure 1. The calibration curve based on two points at 0 Gy, ~2 Gy and ~3 Gy. Dose (Gy) = 145.3 x Pixel Value - 0.0204.

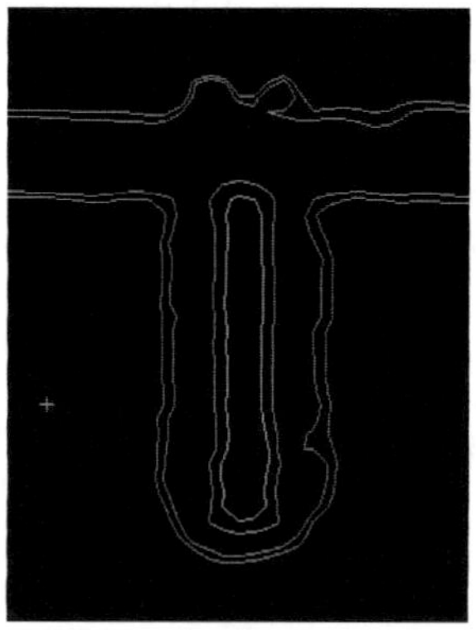

Figure 2. Coronal view of dosimeter block with single hole showing unintended uniform dose along the length of the hole. The isodose contours are plotted for 3, 4, 10 and 15 Gy.

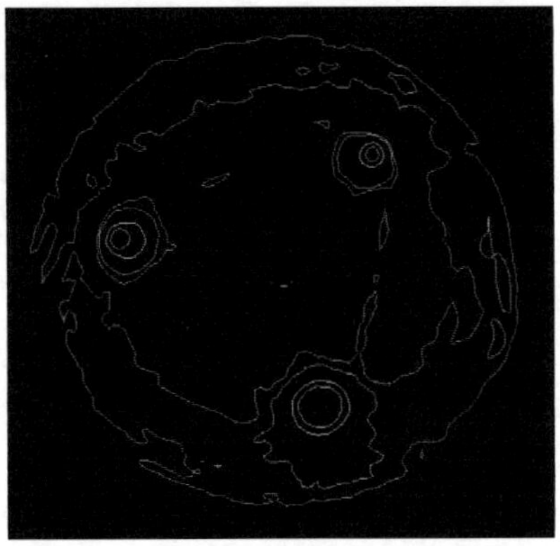

(a)

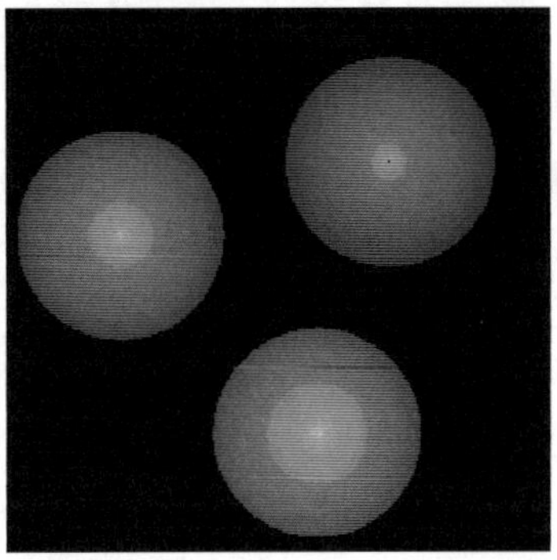

(b)

Figure 3. a) Axial view of dosimeter block with three equidistant holes showing that the diameters of the 3, 5, 10 and 15 Gy isodose lines are proportional to the activity of Y-90 (and the dose). b) Log display of dose estimate from Y-90 beta dose kernel. Gray < 3 Gy, pink >= 3 Gy.

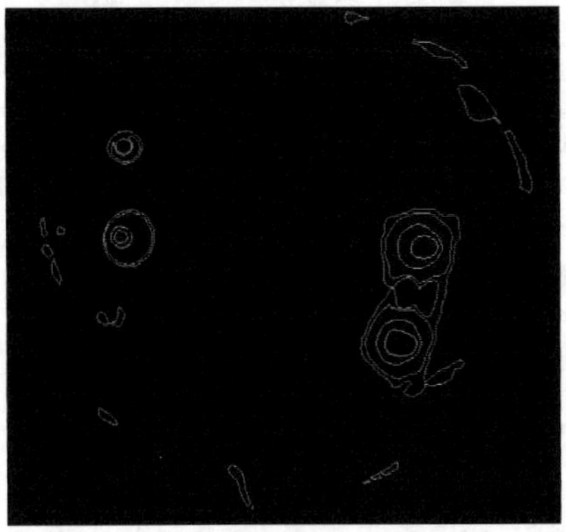

(a)

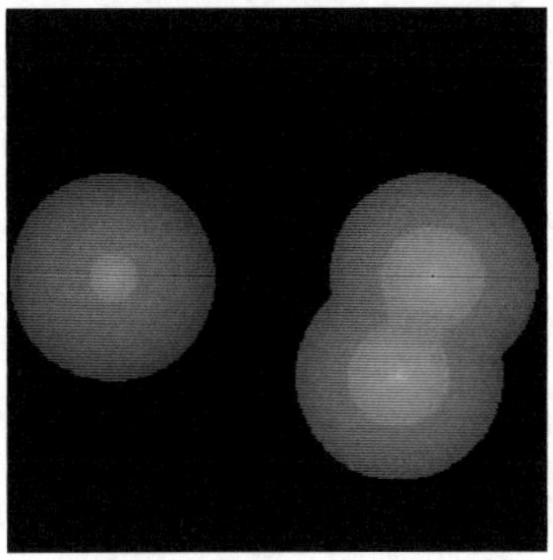

(b)

Figure 4. a) Axial view of dosimeter block with two nearby holes showing overlapping dumbbell-shaped dose distribution with isodose contours at 3, 5, 10 and 15 Gy. b) Log display of Y-90 dose kernel estimate. Gray < 3 Gy, pink >= 3 Gy.

4. Discussion and Conclusion

Good agreement was observed between a dose kernel estimate of the three-dimensional dose distribution from Y-90 and the measurements in PRESAGE dosimeters. The isodose distribution shown in figure 2 demonstrates that the Y-90 solution did not remain as a point source, but in fact was drawn up the sides of the drilled hole by capillary action or contamination during application, yielding a line source in this particular dosimeter. The dose distributions shown in figure 3 indicate clearly that the dosimeter response was proportional to the activity placed in each of the three wells. Finally, the distribution shown in figure 4 demonstrates the overlap between two nearby wells filled with Y-90 solution, and the increase in dose midway between the two holes.

References

[1] Simon J, Stearns S, McMillan K, Loy HM, Frank RK. Liquid Brachytherapy, Direct Administration of Therapeutic Isotopes into Tumors. Journal of Nuclear Medicine. 2008;49(Supplement 1):293P.
[2] Adamovics J, Maryanski M. Characterisation of Presage(TM): A New 3-D Radiochromic Solid Polymer Dosemeter for Ionising Radiation. Radiation Protection Dosimetry. 2006 Sep;120(1-4):107-112.
[3] Deasy J, Blanco A, Clark V. CERR: A Computational Environment for Radiotherapy Research. Medical Physics. 2003 May;30(5):979-985.
[4] Cross W, Freedman N, Wong P. Tables of Beta-Ray Dose Distributions in Water. Chalk River, Ontario, Canada: Atomic Energy Canada Limited; 1992.

Journal of Physics: Conference Series **250** (2010) 012096 doi:10.1088/1742-6596/250/1/012096

Fricke dosimetry of irradiated sugar by ^{1}H NMR relaxation

J Y C Lau[1,2], T Olding[1], G Wu[2] and L J Schreiner[1,3]

[1]Department of Physics, Queen's University, Kingston, Ontario, Canada, K7L 3N6
[2]Department of Chemistry, Queen's University, Kingston, Ontario, Canada, K7L 3N6
[3]Cancer Centre of South Eastern Ontario at Kingston General Hospital, 25 King Street West, Kingston, Ontario, Canada, K7L 5P9

E-Mail: John.Schreiner@krcc.on.ca

Abstract. Linear dose response in the range 0-30 Gy was observed in spin-lattice and spin-spin relaxation rates of water protons in 10 wt% sugar Fricke solutions at 300 MHz and 25 MHz. Spin-lattice dose sensitivities were higher at 300 MHz (0.058 s^{-1} Gy^{-1} with xylenol orange, 0.024 s^{-1} Gy^{-1} without) than at 25 MHz (0.047 s^{-1} Gy^{-1} with xylenol orange, 0.026 s^{-1} Gy^{-1} without), despite slightly less linearity. Spin-spin dose sensitivities at 25 MHz (0.029 s^{-1} Gy^{-1} with xylenol orange, 0.033 s^{-1} Gy^{-1} without) were slightly higher than corresponding spin-lattice dose sensitivities. Unirradiated sugar was found to accelerate background oxidation in Fricke solution, following a logistic time development curve.

1. Introduction

Electron paramagnetic resonance spectroscopy of γ-irradiated table sugar suggests the presence of at least three long-lived species of radiation-induced free radicals [1]. When dissolved in Fricke solution, these free radicals can oxidize Fe^{2+} to Fe^{3+}. Davidson and Jordan [2] demonstrated a linear relationship between optical absorbance and delivered dose in the range 1-20 Gy, using the chelating organic dye xylenol orange as an indicator for the oxidation state of iron. NMR should also be sensitive to the dose-dependent oxidation of iron since Fe^{2+} and Fe^{3+} are paramagnetic ions that have different spin relaxation enhancement effects. Dosimetry of irradiated Fricke solution [3] and gelatin matrix [4] by spin-lattice relaxation measurement of water protons has been well characterized and successfully modeled by multiple site fast exchange in the Fe^{2+} and Fe^{3+} solvation shells. The present study explores the possibilities for using ^{1}H NMR spin-lattice and spin-spin relaxation measurements in Fricke-sugar dosimetry.

2. Experiment

Food-grade granulated table sugar was purchased from a local grocery store and used without further processing. Thin plastic bags containing 3 g of sugar crystals were positioned at 100 cm source-to-axis between a 2 cm layer of solid water phantom placed on top for maximal build-up of Compton electrons and a 10 cm layer of solid water placed below for backscattering. The sugar was irradiated to doses of up to 30 Gy with 6 MV X-rays at a dose rate of 400 cGy min^{-1} in a 10×10 cm^2 field from a Varian C-Series Clinac (Varian Medical, Palo Alto, CA).

An aqueous Fricke xylenol orange (FX) solution containing 0.3 mM ferrous ammonium sulfate (Sigma-Aldrich Ltd., Oakville, Ontario, Canada), 0.05 mM xylenol orange (Sigma-Aldrich Ltd.) and 50 mM sulfuric acid (Sigma-Aldrich Ltd.) was prepared following the formulation of Davidson and

Journal of Physics: Conference Series **250** (2010) 012096 doi:10.1088/1742-6596/250/1/012096

Jordan. As NMR analysis did not require the optical properties afforded by xylenol orange, an analogous Fricke solution without xylenol orange was prepared to study the dosimetric behaviour in the absence of chelating agents. Both solutions were stored at 4ºC to minimize background development due to auto-oxidation. Sugar solutions (10 wt%, 0.32 M) were allowed to develop at room temperature for a total of 3 hours before storage at 4°C. Each sample was taken out 30 minutes before measurement to return to room temperature.

High-field NMR experiments were performed on a Bruker Avance-300 spectrometer (7.04 T). Spin-lattice relaxation times (T_1) were measured with a saturation recovery pulse sequence. A soft pulse at the water resonance frequency was applied for 4 seconds to destroy the initial magnetization, followed by 16 variable recovery delays ranging from 0 to 30 seconds. The recovered magnetization was read with a 90º pulse. Manual shimming was performed in the absence of deuterium locking. Two scans were taken following one dummy scan for each measurement. It is worth mentioning that T_1 measurement of water by inversion recovery was found to be unreliable at high field due to radiation damping, a positive feedback coupling between a strong precessing magnetization and induced currents in the coils of the probe (see Mao and Ye, 1997 [5]).

Low-field NMR experiments were performed on a MARAN 25 MHz bench-top spectrometer (Resonance Instruments Ltd., Witney, Oxon, UK). An inversion recovery pulse sequence with 16 variable recovery delays between 0 and 30 seconds was used to measure T_1. Spin-spin relaxation times (T_2) were measured using a Carr-Purcell-Meiboom-Gill echo train pulse sequence [6] with 8000 echoes at 1 ms intervals.

3. Results & Discussion

Dose sensitivity is defined as the change in relaxation rate per Gy of delivered dose, as determined by linear regression. Figures 1 and 2 compare the effect of the chelating agent xylenol orange on the dose sensitivity of the spin-lattice relaxation rate (T_1^{-1}). Sensitivities obtained at 300 MHz were higher than those at 25 MHz. Measurements at 25 MHz appear more consistent with linearity than those at 300 MHz, possibly due to residual radiation damping effects during the readout period at the higher field. The greater field-dependence of the relaxation rate in the absence of xylenol orange may partly reflect a longer correlation time (τ_c) of the bulkier chelated iron complexes.

At 25 MHz, the spin-spin dose sensitivity, figure 3, appears to be slightly higher than the spin-lattice dose sensitivity. The lower dose sensitivity observed in the presence of xylenol orange may be

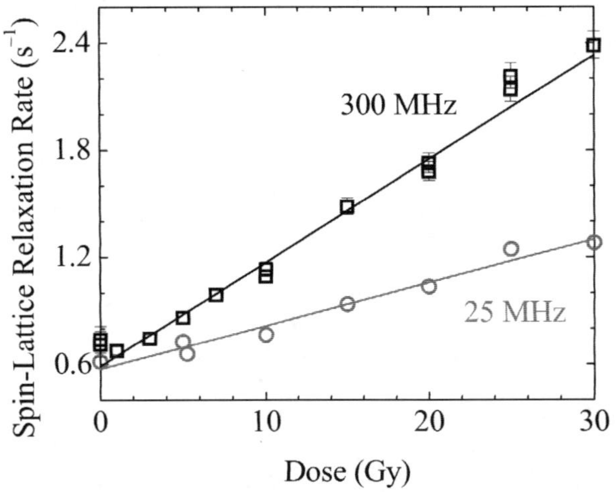

Figure 1. Dose dependence of spin-lattice relaxation rate in Fricke-xylenol-sugar system. Dose sensitivities are 0.058 s^{-1} Gy^{-1} at 300 MHz and 0.024 s^{-1} Gy^{-1} at 25 MHz.

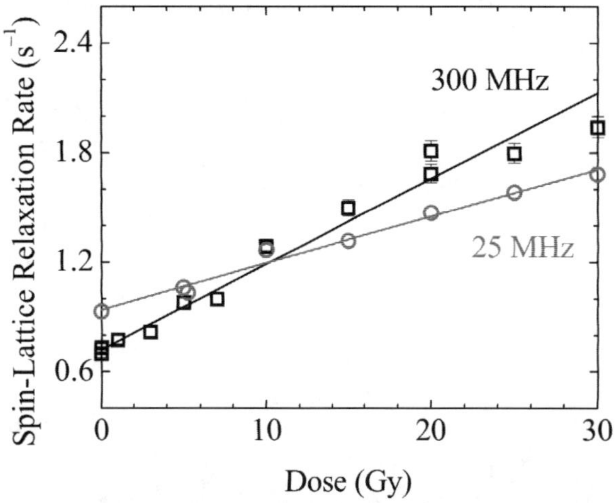

Figure 2. Dose dependence of spin-lattice relaxation rate in Fricke-sugar system. Dose sensitivities are 0.047 s^{-1} Gy^{-1} at 300 MHz and 0.026 s^{-1} Gy^{-1} at 25 MHz.

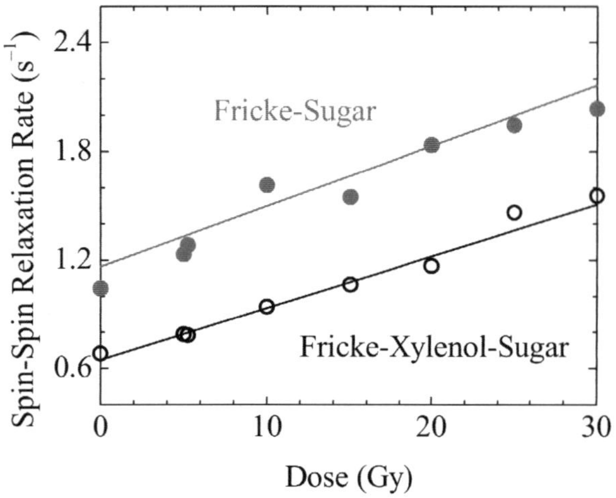

Figure 3. Dose sensitivities of the spin-spin relaxation rate are 0.029 s^{-1} Gy^{-1} for the Fricke-xylenol-sugar system and 0.033 s^{-1} Gy^{-1} for the Fricke-sugar system.

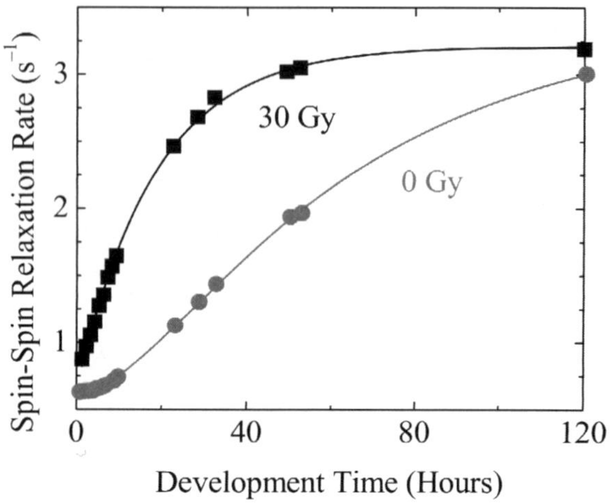

Figure 4. Time evolution of spin-spin relaxation rate in Fricke-xylenol-sugar system at 25 MHz.

attributed to competition between water and the chelating agent for coordination sites on the paramagnetic centres.

The delay between solution preparation and measurement can influence the dose sensitivity, due to the gradual signal development characteristic of Fricke dosimeters. Figure 4 shows preliminary data from our investigation of development time optimization. A logistic function modelled well the acceleration in background oxidation due to unirradiated sugar. After subtraction of this effect from a decaying exponential fit to the 30 Gy data, preliminary estimates suggest that maximum signal-to-background occurs between 16 to 48 hours of development at room temperature. Analogous experiments show a similar behaviour for the spin-lattice relaxation with and without xylenol orange.

4. Conclusions

Based on the results of this study, ^{1}H NMR relaxation can be a viable alternative to optical techniques in probing dose-dependent oxidation of Fe^{2+} in sugar-Fricke solutions. Higher dose sensitivities than those reported above may be achievable once the measurement technique is optimized. Further work is currently underway to explore the reaction kinetics in order to understand the chemical pathways connecting the free radicals from irradiated sugar with Fe^{2+} oxidation.

Acknowledgements

The authors thank Dr. Françoise Sauriol for assistance with NMR instrumentation. Experiments on the 300 MHz spectrometer are funded by the Department of Physics at Queen's University.

References

[1] Georgieva E R, Pardi L, Jeschke G, Gatteschi D, Sorace L and Yordanov N D 2006 *Free Radical Res.* **40** 553-63
[2] Davidson M T M and Jordan K J 2009 *Med. Phys.* **36** 1340-50
[3] Podgorsak M B and Schreiner L J 1992 *Med. Phys.* **19** 87-95
[4] Audet C and Schreiner L J 1997 *Med. Phys.* **24** 201-9
[5] Mao X-A and Ye C-H 1997 *Concepts Magn. Reson.* **9** 173-87
[6] Meiboom S and Gill D 1958 *Rev. Sci. Instrum.* **29** 688-91

Journal of Physics: Conference Series **250** (2010) 012097 doi:10.1088/1742-6596/250/1/012097

A 'quad-phantom' film dosimeter for use as a multi-planar verification tool for PRESAGE/optical-CT

L Stunja*, A Thomas, J Adamovics[1], J Deasy[2], M Oldham

Duke University, Durham NC
[1] Heuris Pharma LLC, Lawrenceville, NJ
[2] Washington University, Saint Louis, MO
*Corresponding author: les34@duke.edu

Abstract.
Introduction: To develop and characterize the accuracy and reproducibility of a 'quad-phantom' dosimeter which will serve as an independent verification tool during commissioning of a PRESAGE/optical-CT 3D dosimetry system. **Methods:** A 16cm x 12cm cylindrical quad-phantom was constructed from four pieces of solid polyurethane mimicking the PRESAGE material. Films were placed and anchored in orthogonal planes and the quad-phantom was fastened tightly together and placed in a water-filled Styrofoam container for irradiation. A simple, two-field plan consisting of 6x6cm anterior-posterior and right-lateral 6MV photon beams (400cGy) was delivered three times (fresh films inserted for each) with a Varian Clinac 600C. Image registration was performed in the Computational Environment for Radiological Research (CERR) and dose profiles and gamma analysis was performed in CERR and MATLAB. **Results & Discussion:** Excellent reproducibility was observed during the irradiations, with ~2.3% standard deviation between all pixels. Using a 3%, 3mm gamma criteria, excellent dosimetric accuracy was observed, with 98.8% and 96.3% passing rates in the sagittal and axial planes, respectively. **Conclusion:** The preliminary results indicate that the quad-phantom can serve as a reproducible and accurate system for high resolution dosimetry in orthogonal planes and should serve as an effective verification tool for PRESAGE/optical-CT in more challenging clinical scenarios.

1. Introduction

Recently, more and more institutions are shifting towards using Gafchromic films (e.g. EBT – ISP Corporation) for high resolution two-dimensional dosimetry. EBT films are near-tissue equivalent, structurally rugged, do not require developing nor show signs of sensitivity to room light or dose-rate [1]. These films also have the same high-resolution as their radiographic predecessors and can be digitized with flatbed scanners. EBT has been used in our group for several years as an independent tool to verify PRESAGE/optical-CT dose distributions. In prior work, PRESAGE dosimeters were cut in axial sections (after all optical-CT dosimetry had been completed) to enable EBT film verification in several axial planes [2]. It was necessary to redeliver the treatment plan to the film/PRESAGE dosimeter. While effective, this process was cumbersome. Here we report a novel quad-phantom system designed to enable efficient independent verification of PRESAGE dose-distributions without the need for slicing dosimeters.

2. Methods

2.1. Construction of the quad phantom

A standard cylindrical PRESAGE dosimeter in use in our lab is a cylinder of dimensions (12cm long x16cm diameter). The quad-phantom was designed to mimic this standard dosimeter, in both geometry and material properties (i.e. CT number). Four pieces of solid polyurethane were combined to form the cylindrical phantom as seen in Figure 1A.

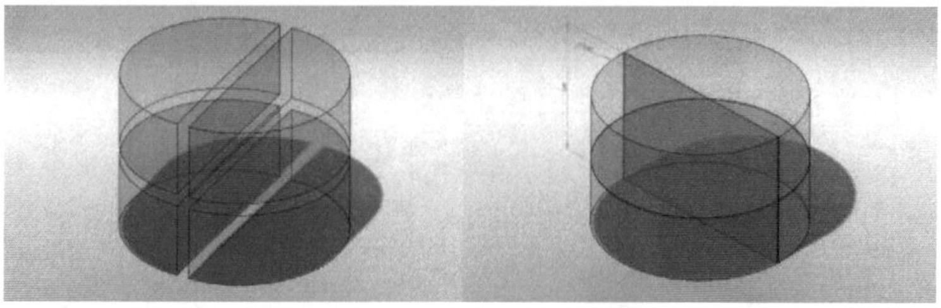

Figure 1: A) Schematic of the quadrant phantom (left). B) Four pieces of polyurethane combine with orthogonal planes of EBT2 film to form a 12x16cm cylinder (right).

This arrangement enables EBT2 film measurement in both the axial and sagittal planes. Precise alignment of the quadrants was achieved utilizing small metal pins embedded into adjoining surfaces. These markers also served as registration and alignment fiducials for treatment planning. EBT2 films were cut into rectangular and circular shapes for measurements in the sagittal and axial planes, respectively (Figure 1B). Each film was pinned to its corresponding quadrants via the metal fiducial barbs. The quadrants were then fastened tightly together with polymer tape and placed within a water-filled rectangular Styrofoam container for irradiations. This was found necessary to avoid small air-gaps present at the interface. These air gaps arose because of a slight bowing of the flat surfaces of the quadrants that occurred during curing of the polyurethane. Planning studies showed that these air gaps, although small, had a noticeable perturbation on the dosimetry in the film planes.

2.2 Irradiations, registration and image analysis

The viability of the quad-phantom approach was investigated using a simple two-field irradiation for which the planning system was known to be accurate, and could therefore be used as a gold standard. The quad-phantom was irradiated three times with the two-field plan; replacing films in both planes with each irradiation. The accuracy and reproducibility of the phantom was then determined by comparing profiles and 2D gamma maps using a 3%, 3mm gamma criteria.

The two-field plan consisted of 6x6cm anterior-posterior and right-lateral 6 MV photon beams as seen in Figure 2, using a Varian Clinac 600 C/D. Unirradiated EBT2 films were 'pre'-scanned, and then inserted into the quad-phantom prior to treatment. After 48 hrs post irradiation, the films were removed and 'post'-scanned using the same protocol as follows: Portrait mode scanning was performed in RGB mode using an EPSON Perfection 4990 flatbed scanner, which captured 48 bit images with 200 dpi resolution. It has been shown that the scanner response is most uniform at its center, so all films were positioned in this fashion. Several investigators have shown that there is a warming up effect that occurs when the scanner is first turned on which could result in dosimetric inaccuracies. Likewise, studies have shown that an average of three scans improves this further [3-4]. Following these recommendations, the first two scans were ignored and then all films were scanned three times and averaged. The radiation induced change in optical density was then calculated and

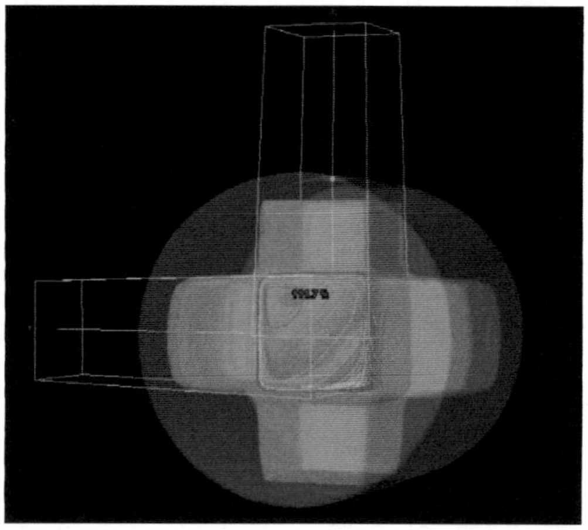

Figure 2: Screenshot of the two field plan in Eclipse. This plan consisted of two 6x6cm, 6MV photon beams in the anterior-posterior and right-lateral directions with a 400 cGy prescription dose.

converted to dose maps using a calibration curve.

The scanned film images (tiff format) were imported into MATLAB for processing purposes. Raw, red-channel pixel data was extracted from each tiff image and converted to optical-density values. Images were corrected for scanner non-uniformity on a pixel-by-pixel basis using a red-channel correction matrix generated each time by the pre-scan film. Change in optical-density was calculated and converted to absolute dose by a calibration curve. Image registration was performed in the Computational Environment for Radiological Research (CERR) using the DICOM files from Eclipse and the dose maps from MATLAB.

3. Results & Discussion

The reproducibility of the quad-phantom measurements, as determined from the three repeat independent irradiations, is shown in Figure 3. Horizontal dose profiles were taken through the central portion of the axial plane films, registered together and plotted. In addition, percent standard deviation was determined for each pixel and an average was determined over the entire surface. There was excellent precision in the results from all three irradiations, with ~ 2.3% variation between all films. This was a promising result for many reasons. Since the phantom had to be removed from the water each time and completely disassembled, there is uncertainty introduced when replacing the film and re-taping the quadrant. In addition, there is error introduced with the alignment of the phantom in the water. There are also uncertainties introduced with the scanning protocols, in terms of scattering properties from alignment and orientation on the scanner glass. It was encouraging to see that even with a combination of all of these sources of error the system seems to perform fairly consistent over time.

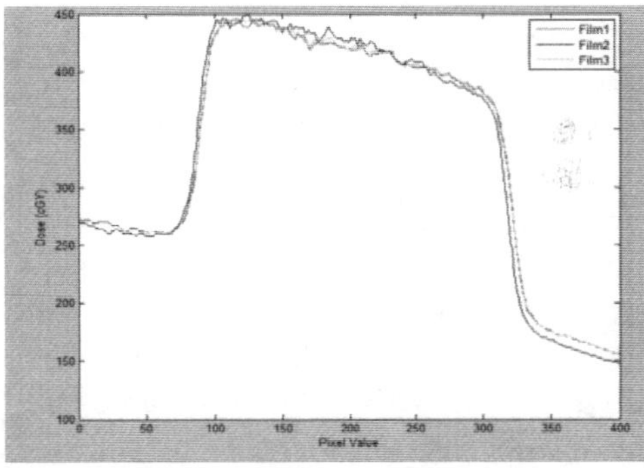

Figure 3: Horizontal Profile Comparison of 3 independent quad-phantom measurements of the Two Field Plan. This illustrates the high level of reproducibility (within 2%) in the setup and irradiation of the quadrant phantom. This required a complete re-setting of the dosimeter in the water, along with the positioning of the film and scanning of the film. These results are promising considering the sources of error involved.

The accuracy of the quad-phantom dosimetry was evaluated by comparison to the Eclipse treatment planning system. Several dose profiles were produced from fully registered film/plan combinations in both planes, as seen in Figures 4 and 5. Excellent agreement, (within 2%) was observed in both planes. The noise in the film data was slightly higher and may be attributed to the inherent scanner-induced effects listed in the previous paragraph, along with surface defects on the film. These defects were observed prior to irradiation in almost all films and may have also been created when the films were placed between the polyurethane layers.

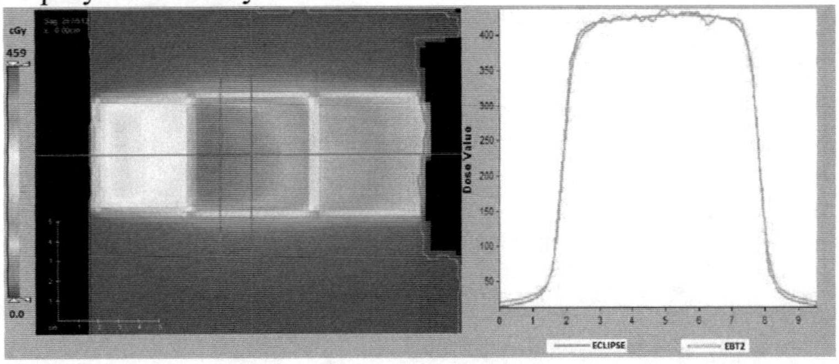

Figure 4: Vertical Profile in the Sagittal Plane. This screenshot, taken from CERR, is evidence that the dose measured with the rectangular film (green) is highly consistent with that of the Eclipse plan (orange).

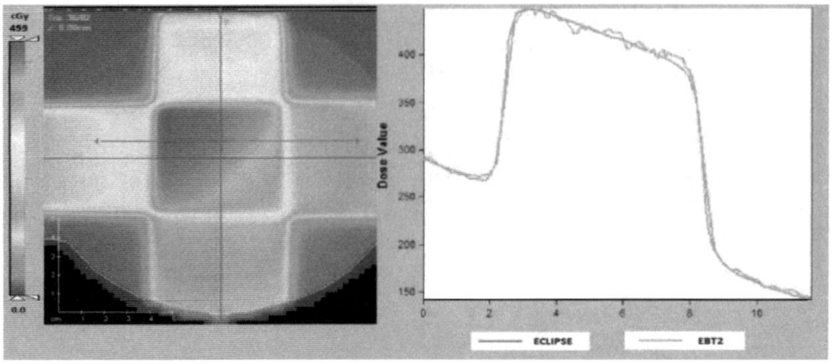

Figure 5: Horizontal Profile in the Axial Plane. Likewise, the dosimetric accuracy of the circular film (green) is high when compared to the Eclipse plan (orange).

2D gamma maps were also created using the Normalized Dose Distribution [5], and are shown in Figures 6 and 7. Using the same (3%, 3 mm) criteria that is often used in clinical scenarios, excellent agreement was seen between the films and the treatment planning system, with passing rates of 96.3% and 98.8% in the axial and sagittal planes, respectively. Visual inspection of the circular film suggests that the majority of the failing occurs at the interface between the beam and quadrant. This could be attributed to light scattering at the edges of the film during scanning. Scattered light may cause an artificially higher optical density region in this area, indicating higher dose. This would correlate with the "hotter" areas of failing that occur when the plan is used as the reference and the film as the calculated value during the NDD analysis. This may have contributed to the failing of the anterior-posterior beam interface as well, which is also witnessed in the rectangular film.

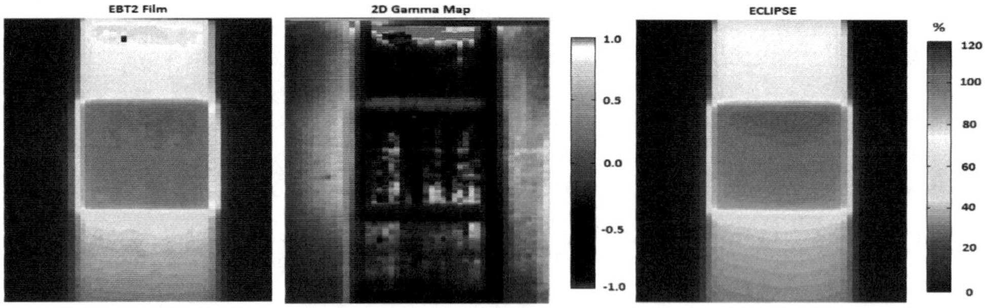

Figure 6: A 2D Gamma Map comparison of Eclipse and EBT2 in the Sagittal Plane of the quad-phantom irradiated with the 2 field plan. Using a (3%, 3mm) criteria, 2D gamma analysis was performed between the EBT2 film (left) and Eclipse plan (right), using Eclipse as the reference. A passing rate of 98.8% was observed. Of the points that failed the gamma criteria, 98.5% were hot (i.e. EBT2>Eclipse) as indicated in red, and 1.5% were cold as indicated in blue.

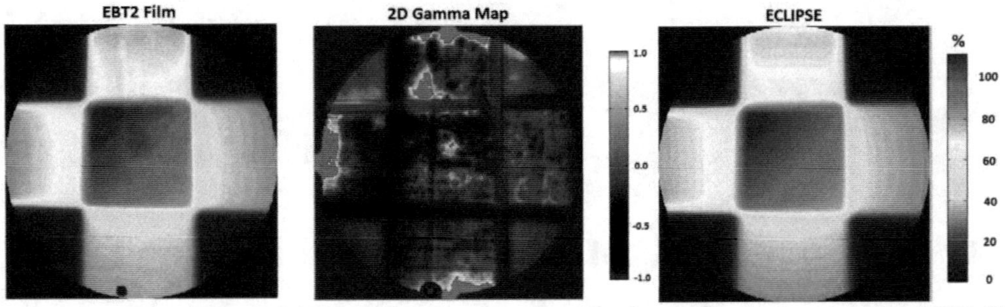

Figure 7: 2D Gamma Map of the Axial Plane. The same gamma criteria were used to compare the EBT2 film (left) and Eclipse plan (right). There was an overall passing rate of 96.3% in the axial plane, which was slightly lower than what had been observed in the sagittal. Of the points that failed, 98.9% were hot (EBT2>Eclipse) and 1.1% were cold.

4. Conclusion

The three independent quad phantom irradiations demonstrated a high level of reproducibility and accuracy when compared to the treatment planning system: with 98.8% and 96.3% passing rates in the sagittal and axial planes respectively, when (3%, 3mm) gamma criteria were used. In addition, this system provides a high level of reproducibility, with an average (~2.3%) standard deviation over all pixels. The quad phantom is anticipated to serve as an effective orthogonal plane verification tool for independent verification of our PRESAGE/optical-CT system.

Acknowledgements: Supported by NIH R01CA100835.

References

[1] Devic, S., et al. (2010). "Radiochromic film is superior to ion chamber arrays for IMRT quality assurance." Medical Physics 37(3): 959:962.

[2] Oldham, M., et al. (2008). "An investigation of the accuracy of an IMRT dose distribution using two- and three-dimensional dosimetry techniques." Medical Physics 35(5): 2072-2080.

[3] Martisikova, M., B. Ackermann, et al. (2008). "Analysis of uncertainties in Gafchromic EBT film dosimetry of photon beams." Physics in Medicine and Biology 53(24): 7013-27.

[4] Paelinck, L., W. De Neve, et al. (2007). "Precautions and strategies in using a commercial flatbed scanner for radiochromic film dosimetry." Physics in Medicine and Biology 52(1): 231-42.

[5] Jiang, S., et al. (2006). "On dose distribution comparison." Physics in Medicine and Biology 51:759.

Journal of Physics: Conference Series **250** (2010) 012098 doi:10.1088/1742-6596/250/1/012098

Absorbed dose distribution visualization for superficial treatments through the Fricke Xylenol Gel dosimeter (FXG)

M Alva[1], F G A Sampaio[1], M V Moreira[1,2], P C D Petchevist[1,2] and A de Almeida[1]

[1]Departamento de Física e Matemática, Faculdade de Filosofia Ciências e Letras de Ribeirão Preto, Universidade de São Paulo, Av. Bandeirantes 3900 CEP:14010-901 Ribeirão Preto – SP – Brazil.

[2]Instituto de Radioterapia e Megavoltagem de Ribeirão Preto - IRMEV, Rua 7 de Setembro 1150 CEP: 14015130, Ribeirão Preto, SP, Brazil.

E-mail: dalmeida@ffclrp.usp.br

Abstract. Electrons, orthovoltage X-rays and betas are used for superficial treatments. It has been shown that it is practical to measure these three types of radiation using gel dosimetry, which is an accurate dosimetric tool, from which one can infer the absorbed dose. The Fricke Xylenol Gel (FXG) dosimeter has presented adequate results due to its spatial resolution, effective atomic number and density that are near to those of soft tissue. The aim of this work is to compare three types of radiation for skin treatments like orthovoltage (X-rays), brachytherapy (beta rays) and megavoltage (electrons) using the FXG-CCD dosimetric system to determine the calibration curves (CC), beam profiles (BP) and percentage depth dose curves (PDD), evidencing why for clinical applications a specific type of radiation is selected for superficial treatment. From the results obtained we can infer that the FXG-CCD system is adequate for linear, area and volume measurements.

1. Introduction

Radiotherapy commonly has three modalities of dose delivery for superficial treatments: orthovoltage, brachytherapy and megavoltage electrons.

Gel based dosimetry has been widely investigated in some hospitals and research centers for quality control in radiotherapy, once it allows to verify the dose distributions, due to its capacity of detecting high dose gradients with high spatial resolution [1-3]. The FXG dosimeter demonstrated to be adequate for radiotherapy, because its effective atomic number (7.75) and density (1.05 g/cm^3) are equivalent to those of the soft tissue (7.42) and (1.04 g/cm^3), respectively [4,5].

In this work was shown the absorbed dose distributions from X-rays (80 keV), beta (^{90}Sr) and 5 MeV electrons. Calibration measurements, beam profiles and percentage depth dose were registered by the images through CCD detector to obtain respectively the CC, BP and PDD curves. The curves were analyzed through practical measurements done which FXG-CCD [6], using the FXG capacity to be spatially read at once. From obtained results, one can show that the FXG is appropriate for these types of measurements, which confirms that beta and electrons are better than orthovoltage for superficial irradiation.

2. Materials and methods

The measurements were done using three cuvettes for each absorbed dose value, depth and lateral distance. The FXG dosimeter was inserted in PPMA cuvettes (standard, 4.5 x 1 x 1 cm^3, and specially manufactured ones of 21 x 6 x 1 cm^3, 5 x 5 x 1 cm^3 and 7 x 5 x 1 cm^3) respectively for CC, BP and PDD measurements. From all values obtained for these measurements the average values from the non-irradiated cuvettes were subtracted.

An in-house scanner was developed to obtain FXG-CCD optical absorbance images, where all images captured with a CCD camera (NIKON/D40) were processed in a developed program with Matlab®, using the pixels intensities, to provide the absorbance values for all samples [6].

For orthovoltage measurements was used the Siemens/Stabilipan II/th. (H 250/300kV), with 80 kV and dose rate of 77.59 cGy/min. Square $^{90}Sr/^{90}Y$ planar source (Amersham England), was used for beta measurements with active area of 2 x 2 cm^2 and surface dose rate of 2.94 $cGy.s^{-1}$. The FXG dosimeter was irradiated with 5 MeV electron beams from a LINAC (Siemens/Mevatron/6MD). All field sizes were equal to 2 x 2 cm^2 for all types of radiation. To obtain the CC measurements were applied absorbed doses of 1 up to 5 Gy and for BP and PDD was applied an absorbed dose of 5 Gy, always using a 2 x 2 cm^2 field size.

For CC determinations with the $^{90}Sr/^{90}Y$ source and Stabilipan, the sample cuvettes were placed in contact with the sources varying the absorbed dose values (1-5 Gy). For 5 MeV electrons the sample cuvettes were positioned at SSD of 100 cm, for the same absorbed dose interval.

For the beam profile using the FXG, an absorbed dose of 5 Gy was given at 1mm depth for $^{90}Sr/^{90}Y$ sources [7] and orthovoltage, and 5 Gy at 100.1 cm for megavoltage, in the center of the fields.

For the PDD measurements, FXG cuvettes of 7 x 5 x 1 cm^3, were positioned with the minor side, perpendicular to the incident beams. The absorbed dose used for beta, X-rays and electron beams was 5 Gy at build-up depth (maximum dose depth).

3. Results and Discussions

The figure 1 shows calibration curves obtained from orthovoltage, beta and electrons sources. The curves show linear dependence with the absorbed dose, with correlation coefficients, R^2, of 0.999, 0.997 and 0.999 for orthovoltage, $^{90}Sr/^{90}Y$ and electrons, respectively. The standard deviations from the average readings (three measurements) for each measured point and energy not exceed 0.27%.

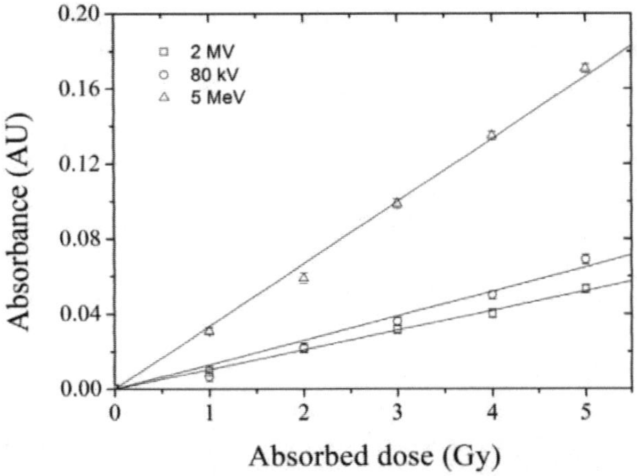

Figure 1 – Calibration curves obtained from x-rays (o), beta sources (□) and electrons (Δ)

The figure 2a) shows the beam profiles measured through FXG–CCD for X-rays, beta and electrons beams along perpendicular lines crossing the center of each radiation field. From the curves, one can see that the profiles are different due to the type of radiation and irradiation geometry, for example, the penumbra is higher for beta than for X-rays and the top of electron curve is more squared than the others, because for small field size using bigger SSD, the profiles are modified near to a Gaussian curve. Although the differences between the curves, the FWHMs are the same.

The figure 2b) shows the PDDs obtained with the FXG-CCD system for X-rays, beta and electron beams. From it one can visualize different behaviors, for example: beta shows the highest gradient, from which only very superficial lesions could be treated; orthovoltage X-rays and 5 MeV electrons could treat deeper lesions, but the electrons give homogeneous absorbed dose even for deeper distances from the surface, once the electrons have a range they would not irradiate health tissue, as do the X-rays. The FXG-CCD system permitted to evaluate how different types of radiation for superficial irradiation are selected through their absorbed dose distribution behaviors.

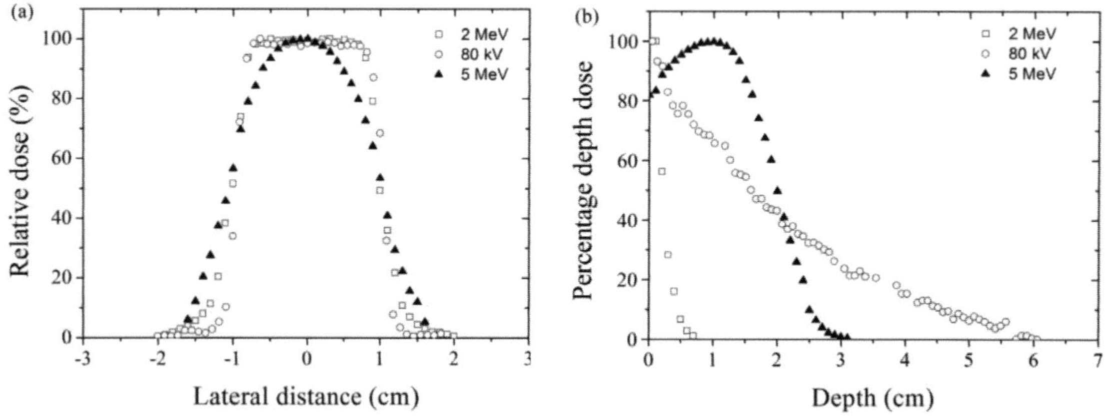

Figure 2- (a) Beam profiles obtained from X-rays (o), beta sources (□) and electrons (▲) for a 2 x 2 cm^2 field size and (b) Percentage depth dose obtained from X-rays (o), beta sources (□) and electrons (▲).

4. Conclusions

In this work, the FXG-CCD system was used for X-rays, beta and electron radiations to evaluate the behaviours of their absorbed dose distributions (calibration curve, beam profile and percentage depth dose). All calibration curves show adequate linearity with the absorbed dose, $R^2 \approx 1$). The beam profiles showed similar FWHM for all types of radiation and energy used and the PDDs showed very different absorbed dose curves, from which one can decide which type of radiation to use for a specific superficial treatment. From the results obtained we can infer that the FXG-CCD system is an adequate dosimeter for linear, area and volume measurements.

References

[1] Baldock C, Burford R P, Billinghan N C, Wagner G S, Patval S, Badawi R D and Keevil S F 1998 *Phys. Med. Biol.* **43** 695-702

[2] De Deene Y, Hurley C, Venning A, Vergote K, Mather M, Healy B J and Baldock C 2002 *Phys. Med. and Biol.* **47** 3441-63

[3] Gustavsson H, Karlsson A, Back S A J, Olsson L E, Haraldsson P, Engstrom P and Nystrom H 2003 *Med. Phys.* **30** 1264-71

[4] Calcina C S G, De Oliveira L N, De Almeida C E, De Almeida 2007 *Phys. Med. Biol.* **52** 1431 39

[5] De Oliveira L N, Zimmerman R L, Moreira M V, Ila D and De Almeida 2009 *Surf. Coating. Tech.* **203** 2367-69

[6] Sato R, De Almeida A, Moreira M V 2009 *Nu. Ins. Meth. Phys. R. B* **267** 842-5

[7] ICRU 2004 Dosimetry of Beta Rays and Low-Energy Photons for Brachytherapy with Sealed Sources *ICRU Report* 72 (Bethesda, MD: International Comission on Radiation Units and Measurements)

Journal of Physics: Conference Series **250** (2010) 012099 doi:10.1088/1742-6596/250/1/012099

Dose overshoot reduction by tetrakis hydroxymethyl phosphonium chloride in polyacrylamide gel dosimeter

K Vávrů[1], J Šemnická[2] and V Spěváček[1]

[1] Czech Technical University in Prague, Faculty of Nuclear Sciences and Physical Engineering, Břehová 7, 115 19 Prague 1, Czech Republic
[2] Na Homolce Hospital, Department of Medical Physics, Roentgenova 2, 150 30 Prague 5, Czech Republic

E-mail: katerina.vavru@fjfi.cvut.cz

Abstract. This work deals with an influence of the antioxidant tetrakis (hydroxymethyl) phosphonium chloride (THPC) on the edge enhancing effect of a polyacrylamid gel and THPC (PAGAT) dosimeter. Four batches of PAGAT gel were produced with different THPC concentrations: 0, 1, 2 and 4 mM. It was proved that antioxidant THPC reduce the edge enhancing effects on the cost of reduced sensitivity of the gel dosimeter.

1. Introduction

In the complex radiotherapy treatments, mainly radiosurgery, IMRT and brachytherapy, quite steep dose gradients often appear. For these applications a gel dosimeter is usually considered to be the only system capable of recording the full three dimensional dose distributions. In reality, however, the edge enhancing effects (dose overshoots) were observed near steep dose gradients [1], especially when the polyacrylamid gel (PAG) dosimeter is irradiated to relatively high doses. The PAG gels are spatially stable for doses up to 12 Gy [2].

Vergote et al. [3] studied the dose overshoot dependence on the gelatine concentration and proposed a mathematical model to predict the time-dependent post-irradiation evolution of the measured dose distribution in PAG dosimeters. In this study, the antioxidant tetrakis (hydroxymethyl) phosphonium chloride (THPC) was tested to reduce the edge enhancing effects.

2. Materials and Methods

2.1. *Gel manufacture*

Four batches of PAGAT gel were produced with various THPC (80% aqueous solution, Sigma-Aldrich, Switzerland) concentrations: 0, 1, 2 and 4 mM. The PAGAT components used: 3% (by weight) acrylamid (Sigma-Aldrich, Germany), 3% N,N'-methylenbisacrylamid (AppliChem, Darmstadt, Germany), and 5 % gelatine from porcine skin (300 Boom, Sigma-Aldrich, USA). All gels were made in a nitrogen environment to avoid oxygen contamination.

To prepare the PAGAT dosimeter, gelatine was added to the distilled water and left to soak over the night. In the morning, N,N'-methylenbisacrylamid and acrylamid were added to the solution and allowed to dissolve during bubbling with nitrogen gas. The gel was heated to 50°C using water bath with temperature control. When all components completely dissolved, the temperature was lowered to 35°C and the gel divided into four glass bulbs. The THPC was added according to the table 1 and each batch was again bubbled with nitrogen gas for about 30 min. Then the gel was placed in a refrigerator (4°C) to solidify over the night.

Table 1. THPC amount in PAGAT (3% AA, 3% bis, 5% gelatine)

PAGAT batch	THPC (mM)
A	0
B	1
C	2
D	4

2.2. Irradiation
Before irradiation, all the gel dosimeters were taken out of the refrigerator and allowed to equilibrate to a room temperature. Each glass bulb was placed in a purpose-built round holder and irradiated by the Leksell Gamma Knife Perfexion (Elekta, Sweden) in the Na Homolce Hospital, Prague, Czech Republic. The prescribed dose to maximum (30 Gy, dose rate: 3,504 Gy/min) was delivered by a single 16 mm shot. The irradiation took place about 52 h after the polymer gel fabrication.

2.3. Scanning
The dosimeters were scanned 6 times (1-123 h post-irradiation) at 1,5 T MR scanner Magnetom Avanto Syngo B15 (Siemens, Germany) in the Na Homolce Hospital, Prague, Czech Republic. The glass bulbs were placed in a water bath together with three calibration control vials filled with copper sulphate solutions of different concentrations (Figure 1), and scanned using a head coil (Head Matrix). Body coil acted as a transmitter. A multiple spin-echo sequence was used with 32 equidistant echoes and 1 acquisition (TE = 30,3-969,6 ms, TR = 4s, voxel size: (0,48x0,48x3)mm). The slice thickness used was 3 mm, the FOV was (260x183)mm and the matrix (384x270)px. Two slices were obtained within one TR. The measured signal-to-noise ratio (SNR) in the parametric images ranged between 80 (second echo) and 40 (32nd echo). SNR measurements were performed with the same container which was used for scanning of the gel bulbs but filled with water.

2.4. Data Processing
The R2 relaxation rate of each pixel was derived by fitting the time course of the pixel value in the consecutive base images to a mono-exponential decay using a linear fit of log values. The first base image was left out. The temperature variations were corrected by using the average values of calibration control vials filled with copper sulphate solutions of different concentrations. For data processing Matlab (The MathWorks, Inc., USA) and Microsoft Office Excel (Microsoft Corporation, USA) were used.

3. Results and Discussion
An R2 image of irradiated PAGAT gel with 4 mM THPC is shown in figure 1. The R2 profiles in figure 2 illustrate an overshoot near the edge on both sides of the 16mm shot. Each profile differs in THPC amount added to the PAGAT gel. Increasing the THPC concentration clearly restricts the edge enhancing effects. Figure 3 displays the time-dependent post-irradiation evolution of the measured dose distribution in PAG dosimeters. The edge enhancing effect in PAGAT gel with THPC concentration 4mM (b) is significantly lower than in the PAG gel without antioxidant THPC (a).

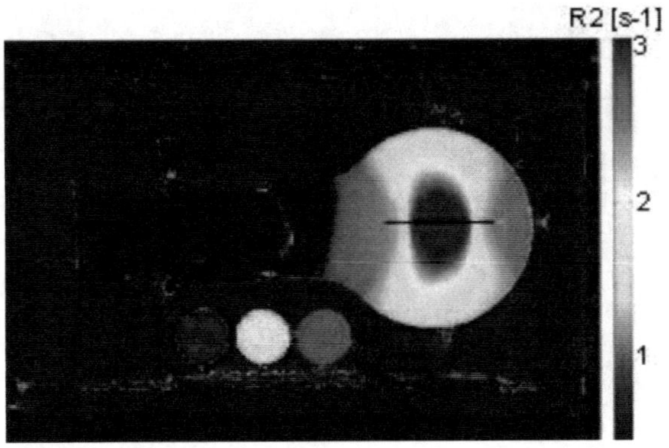

Figure 1. R2 map of PAGAT gel with 4 mM THPC. The dose to maximum 30 Gy was delivered by a single 16 mm shot on the Leksell Gamma Knife Perfexion.

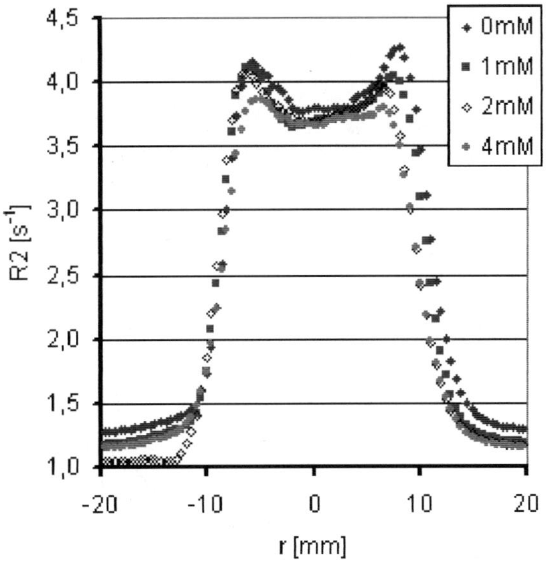

Figure 2. R2 profiles 22h post-irradiation of PAGAT gel with various THPC concentration. The profiles were obtained by average of 5 pixels along the black line in Figure 1.

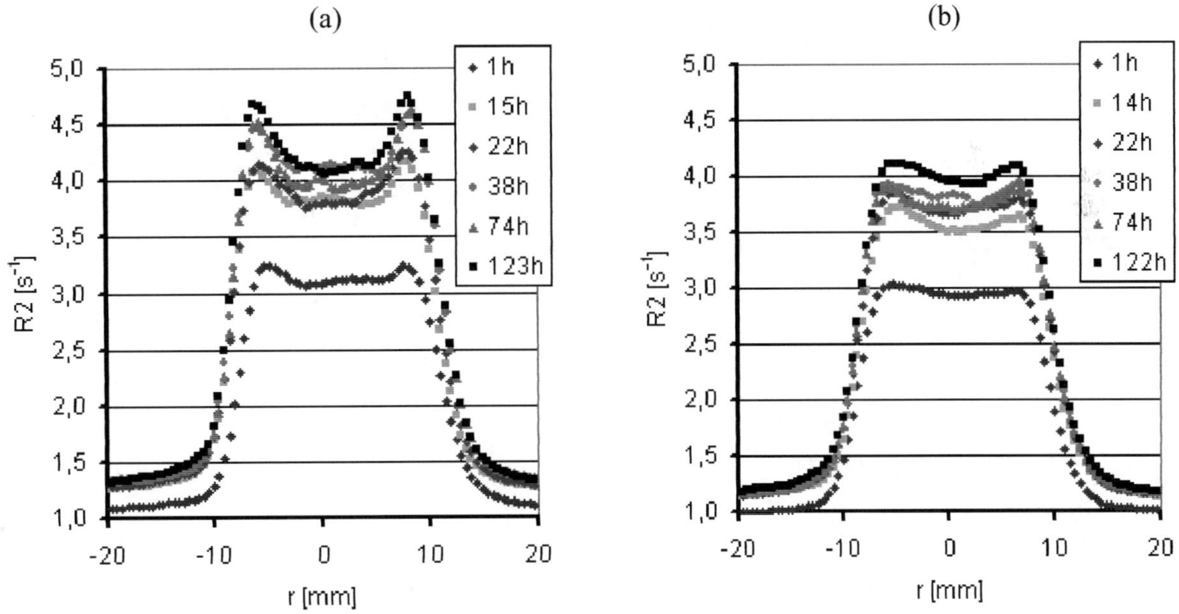

Figure 3. R2 post-irradiation profiles of PAGAT gel with THPC concentration (a) 0mM (b) 4mM.

The dose overshoot shape in this case is determined mostly by two events: First, the THPC as a scavenger of free radicals reacts both with oxygen (biradical) and with radicals at the end of polymer chain and radicals on the gelatine; thus, the polymerization process is affected. And second, irradiated volume (ca (28x19)mm ellipsoid) is too small considering the monomers diffusion velocity; therefore, the resulting edge enhancing effect is substantially influenced by monomers diffusion in 3D area. Both physical-chemical effects will be incorporated in the oncoming model.

Due to preparing the PAGAT gel in anoxic conditions, sensitivity reduction of the gel dosimeter is not so noticeable in this experiment.

Journal of Physics: Conference Series **250** (2010) 012099 doi:10.1088/1742-6596/250/1/012099

4. Conclusion

THPC dose overshoot reduction effect in PAGAT dosimeter was investigated. The comparison of PAGAT gel with various THPC concentrations (0, 1, 2 and 4 mM) proved that antioxidant THPC is useful for reducing the edge enhancing effects. Further research in dose overshoot problem is ongoing, especially for situation of 3D dose gradients such as those occurring in radiosurgery.

5. Acknowledgments

This work has been supported by Institutional research plan MSM6840770012 of The Ministry of Education, Youth and Sports.

Authors would like to express sincere gratitude to the staff of MRI department of the Na Homolce Hospital for their cooperation and for accommodating gel measurements into their busy program.

References

[1] Hrbacek J, Spevacek V, Novotny Jr J and Cechak T 2004 A comparative study of four polymer gel dosimeters *Proc. DOSGEL 2004 (Ghent)* ed Y De Deene and C Baldock pp 140–144

[2] De Deene Y, Venning A, Hurley C, Healy B J and Baldock C 2002 Dose-response stability and integrity of the dose distribution of various polymer gel dosimeters *Phys. Med. Biol.* **47** 2459–70

[3] Vergote K, De Deene Y, Vanden Bussche E and De Wagter C 2004 On the relation between the spatial dose integrity and the temporal instability of polymer gel dosimeters *Phys. Med. Biol.* **49** 4507–22

Journal of Physics: Conference Series **250** (2010) 012100 doi:10.1088/1742-6596/250/1/012100

Effect of irradiation and storage temperature on PRESAGE™ dose response

P S Skyt[1,2,3,4], P Balling[4], J B Petersen[1], E S Yates[1] and L P Muren[1,2,3]

[1] Dept of Medical Physics, Aarhus University Hospital, Aarhus, Denmark
[2] Dept of Oncology, Aarhus University Hospital, Aarhus, Denmark
[3] Institute of Clinical Medicine, Aarhus University, Aarhus, Denmark
[4] Dept of Physics and Astronomy, Aarhus University, Aarhus, Denmark

E-mail: skyt@phys.au.dk

Abstract. The temperature dependence of the PRESAGE™ dosimeter dose response has been investigated. Two series of measurements were performed. The first for measuring the temperature dependence during irradiation and the second for temperature dependence during post-irradiation storage. These measurements shows significant temperature dependence on dose response both during irradiation and storage with activation energies of respectively 1.4±0.2 eV and 1.9±0.2 eV.

1. Introduction

The development of complex radiotherapy techniques has motivated research into dosimetry materials that can measure radiation dose in three dimensions (3D) [1]. The 3D dosimeters have included both gel and solid-state materials. One such dosimeter is the commercially available PRESAGE™ material, which exhibits a radiochromic response (i.e. a change in optical density, OD) when exposed to ionizing radiation due to a leucomalachite green dye contained in the polyurethane based solid state dosimeter [2]. However, the dose response might vary with external parameters such as temperature, since the activity of the radiochromic component is likely to be temperature-dependent. In this study we have investigated: (i) The dose response of PRESAGE™ at different irradiation temperatures and (ii) The post-irradiation change in OD at a range of storage temperatures.

2. Materials and methods

During irradiation PRESAGE™ samples ($Z_{eff} = 9$, density $\rho = 1.7$ g cm^{-3}) in cuvettes were placed at 8 cm depth in a temperature controlled water phantom with a 5 cm solid water plate to ensure back scatter. The cuvettes were irradiated to 5 Gy with a 6 MV photon beam in a field size of 10x10 cm^2. Two series of measurements were performed: In the first series the cuvettes were stored at 5°C and irradiated at different temperatures extending from 5°C to 30°C. In the second series the cuvettes were irradiated at 5°C but stored in temperature controlled water baths at a range of temperatures from 4°C to 30°C. The dose response were measured with 24 hours between pre- and post-irradiation dose read-out

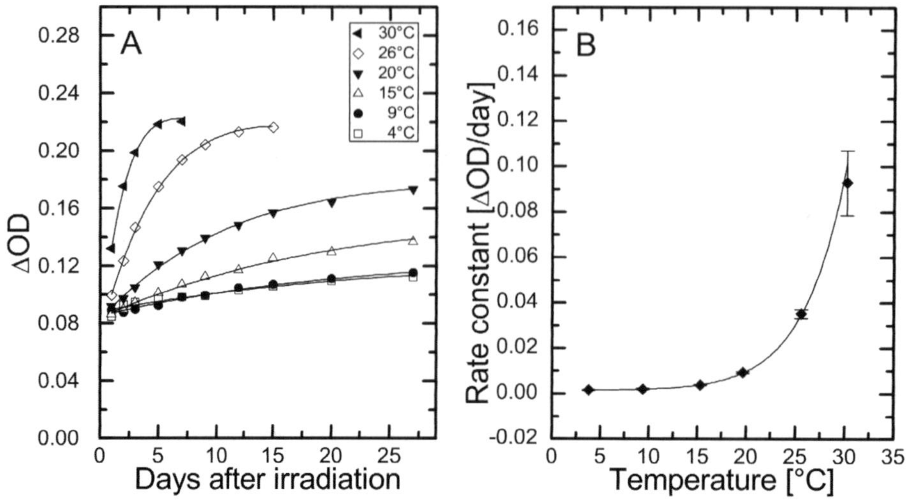

Figure 1. (A) Change in optical density of PRESAGE™ cuvettes stored at different temperatures. (B) Rate constants from (A) fitted to an Arrhenius expression.

using a standard spectrophotometer and for the second series repeated during the following month. The dose response was found by calculating the optical density difference from before to after irradiation at 632 nm and an Arrhenius expression was fitted to the data in anticipation of a thermally activated process.

3. Results

The dose read-out showed a significant change in optical density of the PRESAGE™ cuvettes when changing the irradiation temperature with a factor two increase in ΔOD from 5°C to 30°C. A fit to the Arrhenius expression reproduced the data quite well and revealed a value for the activation energy of 1.4±0.2 eV. The second series with storage temperatures from 4°C to 30°C also showed significant temperature dependence on dose response with an exponential increase in optical density (figure 1A). An Arrhenius fit to the rate constants from exponential fits of the growth in ΔOD showed again good agreement between data and fit (figure 1B), here with an activation energy of 1.9±0.2 eV.

4. Conclusion

In these experiments a significant non-linear temperature dependence on irradiation and storage temperatures of the PRESAGE™ dosimeter has been observed. These results show that the PRESAGE™ dosimeter has to be temperature stabilized both during storage and irradiation.

References

[1] Guo P, Adamovics J and Oldham M.2006 A practical three-dimensional dosimetry system for radiation therapy *Med. Phys.* **33** (10)

[2] Adamovics J and Maryanski M J 2006 Characterisation of PRESAGE™: A new 3-D radiochromic solid polymer dosemeter for ionising radiation *Radiat. Prot. Dosimetry* **120** (1-4), 107-112

Journal of Physics: Conference Series **250** (2010) 012101 doi:10.1088/1742-6596/250/1/012101

Experience Using DosimetryCheck software for IMRT and RapidArc Patient Pre-treatment QA and a New Feature for QA during Treatment.

Arthur Pinkerton, Michael Hannon, and Jae Kwag

NMMC Cancer Center Radiation Oncology 990 S. Madison Tupelo MS 38801

Wendel Dean Renner

Math Resolutions LLC. 5975 Gates Lane Columbia MD 21045

Art.pinkerton@nmhs.net

Abstract. We have used the DosimetryCheck program with the EPID's on our Varian 2100EX's to perform pre-treatment QA on more than 350 patients, between the last quarter of 2006 and the present. The software uses the EPID measured fluences of the treatment fields to reconstuct the dose distribution in the CT planning model of the patient. Since the dose calculation algorithm, is different from that used by our Eclipse planning system, this provides an independent check of planning accuracy as well as treatment delivery. 2D and 3D dose distributions, point doses, Gamma distributions, DVH statistics and MU calculations can be compared. Absolute differences of Reference Point doses between Dosimetry Check and Eclipse average 1.20%, which is similar to the ionization chamber dose differences of 1.29% for the same patient verification plans. Examples of cases for various treatment sites and delivery modes will be presented. A Special Report in Medical Physics Vol. 37 Number 6 Pg. 2638-2644 from Mans et al at The Netherlands Cancer Institute demonstrated the ability of in vivo EPID dosimetry to detect treatment errors, that escaped other QA checks. A new version of DosimetryCheck awaiting FDA approval, is capable of successfully reconstructing the dose distribution in the patient from the EPID measured exit fluences. This can also be applied to CBCT images providing actual patient dose verification for a treatment session. This should be particularly useful for monitoring hypo-fractionated treatment regimens. Examples of this method will also be presented.

1. DosimetryCheck

The DosimetryCheck software [1,2,3,4, 5], was created by Wendel Dean Renner, who also created Render-Plan 3-D, the first 3D radio-therapy planning system to receive FDA510K approval. DosimetryCheck enables one to compare the Dose Distribution created from measured field fluences in the Patient Model exported from the Treatment Planning System, with the Planned Distribution. Interfaces are provided for Varian, Elekta, and Siemens EPID's, MapCheck diode, and PTW 729 ion chamber arrays as well as Kodak CR, or film. Dicom RT and RTOG protocols are supported for importing patient plan, dose, structure, and CT image files. The default calculation grid is 0.5 cm., but may be adjusted down to 0.1 cm.. I personally prefer using 0.25 cm. which lengthens calculation time, but matches the default grid of our Eclipse TPS.

Generic deconvolution kernels are available for the various electronic devices, or may be customized from a relatively simple set of measurements. Utilities are provided for importing the treatment machine beam data from the TPS. While the beam parameters are the same as used by the TPS, the calculation algorithm is different. Input of CT# to electron density data is also required.

2. Capabilities

In addition to creating 2D isodose distribution, DVH, Gamma, point dose, and dose profile comparisons, the program offers a variety of 3D dose evaluation tools. The 3D tools are impressive, but are more applicable to specific research projects than to routine QA. There is also a MU Calculation Check Module, which in the case of dual carriage position IMRT fields provides calculations for each position. In our use at NMMC, we use only the point dose comparison at the Plan Reference Point, the 2D isodose distributions in the principal planes, through that point, and the DVH comparison.

3. Example

7 Field IMRT Prostate Case. The First Image shows the DosimetryCheck Display after the patient plan, dose, structures, CT's and EPID fluence images have been loaded, and the program run to show the comparison Isodoses and point doses.

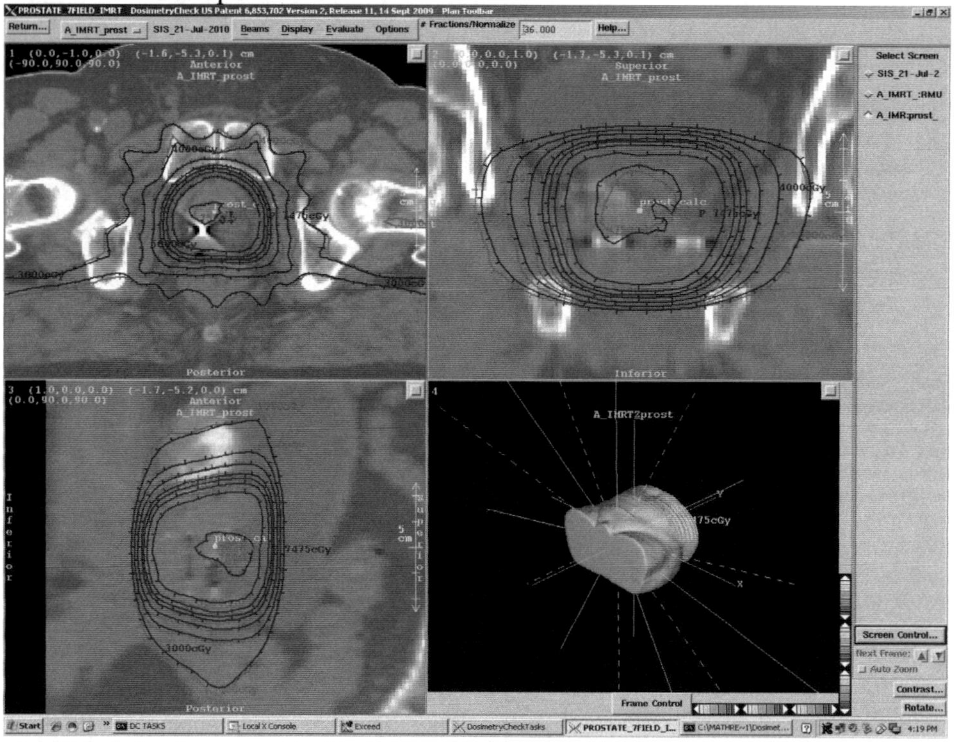

The DosimetryCheck Isodoses are in Red, and the Eclipse Isodoses are in Blue and show excellent agreement. The Reference Point dose is 7446 cGy. in DC and 7475 cGy. in Eclipse a -0.39% difference. The ion chamber dose in the verification phantom agreed at +0.76% with the Eclipse plan. The second image shows the DVH comparison for various structures. PTV, Penile Bulb, Rectum, Seminal Vesicles, Bladder an Posterior Avoidance. The DC curves are solid, the Eclipse dotted.

Journal of Physics: Conference Series **250** (2010) 012101 doi:10.1088/1742-6596/250/1/012101

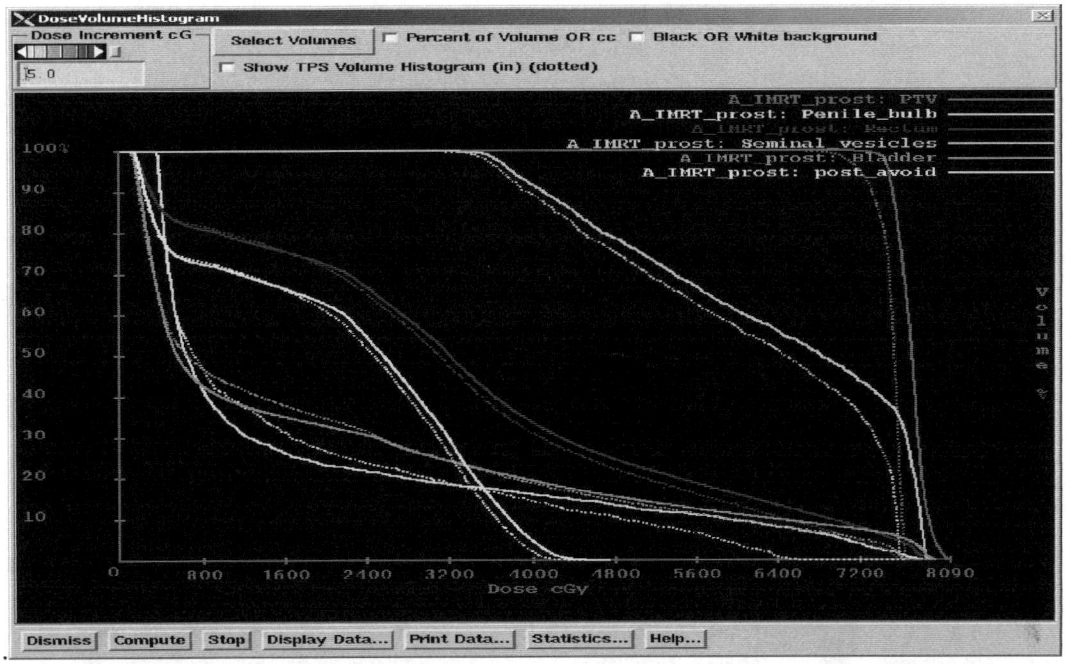

The third Image shows the DVH Statistics, Minimum, Maximum, Average Doses and Structure volumes.

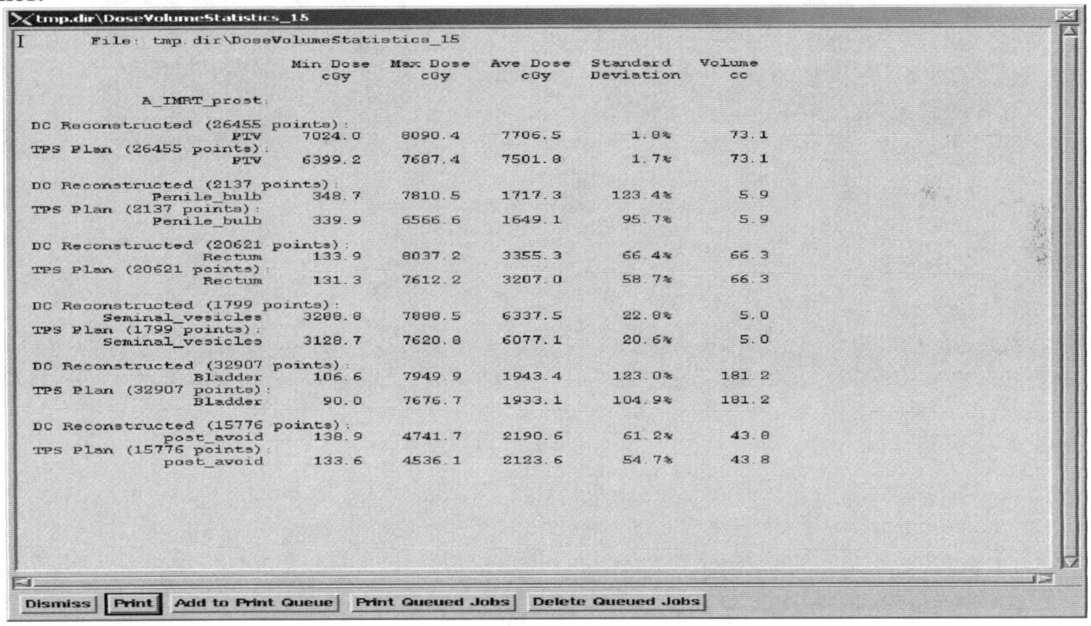

tmp.dir\DoseVolumeStatistics_15

File: tmp.dir\DoseVolumeStatistics_15

A_IMRT_prost:

	Min Dose cGy	Max Dose cGy	Ave Dose cGy	Standard Deviation	Volume cc
DC Reconstructed (26455 points): PTV	7024.0	8090.4	7706.5	1.8%	73.1
TPS Plan (26455 points): PTV	6399.2	7687.4	7501.8	1.7%	73.1
DC Reconstructed (2137 points): Penile_bulb	348.7	7810.5	1717.3	123.4%	5.9
TPS Plan (2137 points): Penile_bulb	339.9	6566.6	1649.1	95.7%	5.9
DC Reconstructed (20621 points): Rectum	133.9	8037.2	3355.3	66.4%	66.3
TPS Plan (20621 points): Rectum	131.3	7612.2	3207.0	58.7%	66.3
DC Reconstructed (1799 points): Seminal_vesicles	3288.8	7888.5	6337.5	22.8%	5.0
TPS Plan (1799 points): Seminal_vesicles	3128.7	7620.8	6077.1	20.6%	5.0
DC Reconstructed (32907 points): Bladder	106.6	7949.9	1943.4	123.0%	181.2
TPS Plan (32907 points): Bladder	90.0	7676.7	1933.1	104.9%	181.2
DC Reconstructed (15776 points): post_avoid	138.9	4741.7	2190.6	61.2%	43.8
TPS Plan (15776 points): post_avoid	133.6	4536.1	2123.6	54.7%	43.8

Dismiss | Print | Add to Print Queue | Print Queued Jobs | Delete Queued Jobs

4. Exit Dosimetry

A recent Special Report in Medical Physics [4] demonstrated the ability of in vivo EPID dosimetry to detect treatment errors, that escaped other QA checks. A new version of DosimetryCheck awaiting FDA approval, is capable of successfully reconstructing the dose distribution in the patient from the EPID measured exit fluencies[5,6]. The following three images, show the Axial, Coronal, and Sagittal Dose distributions for a Rapid Arc Prostate patient, reconstructed from the exit fluencies during an actual treatment. The DC isodoses and area are in Magenta, the Eclipse isodoses are Green. The isocenter dose is +0.47% greater than planned, exceptionally good agreement for Exit Dosimetry. The calculation Matrix in this and the following Head & Neck case is 0.5Cm.

Journal of Physics: Conference Series **250** (2010) 012101 doi:10.1088/1742-6596/250/1/012101

Rapid Arc Case

(Courtesy Advanced Radiation Centers, NY)

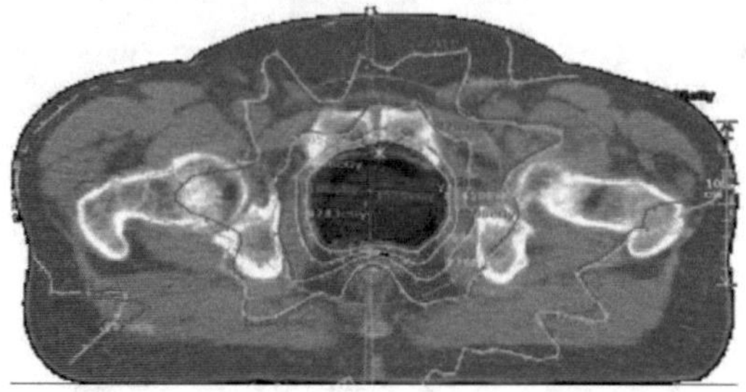

Dose Comparison in Coronal Plane

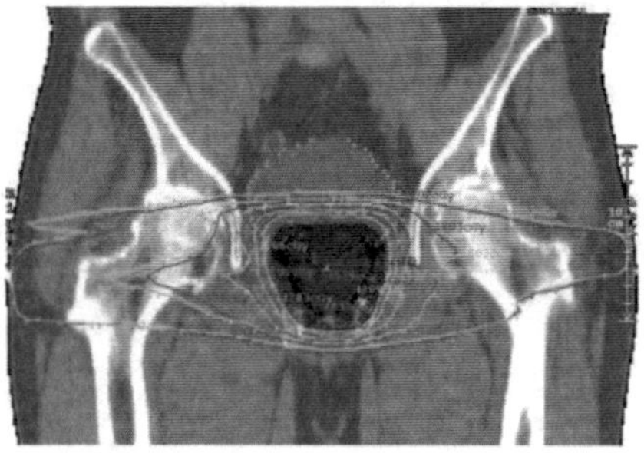

Dose Comparison in sagittal plane.

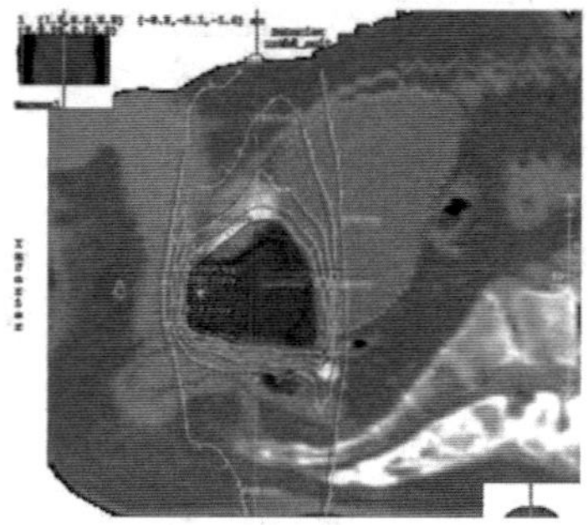

5. Exit Dosimetry Head&Neck Case

The following three images, show the Axial, Coronal, and Sagittal Dose distributions for a Head & Neck patient, reconstructed from the exit fluencies during an actual treatment, at NMMC. The DC isodoses and area are in Magenta, the Eclipse isodoses are Green. The Reference dose is +1.33% greater than the plan which is good agreement for Exit Dosimetry.

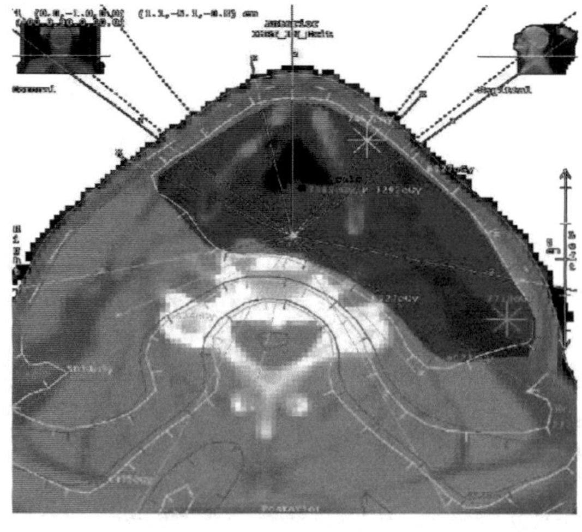

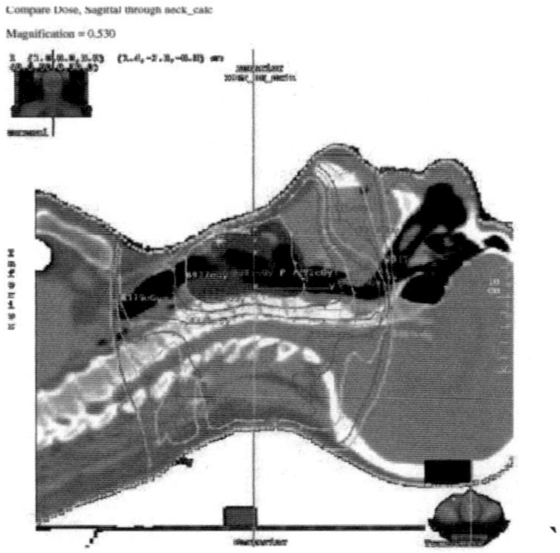

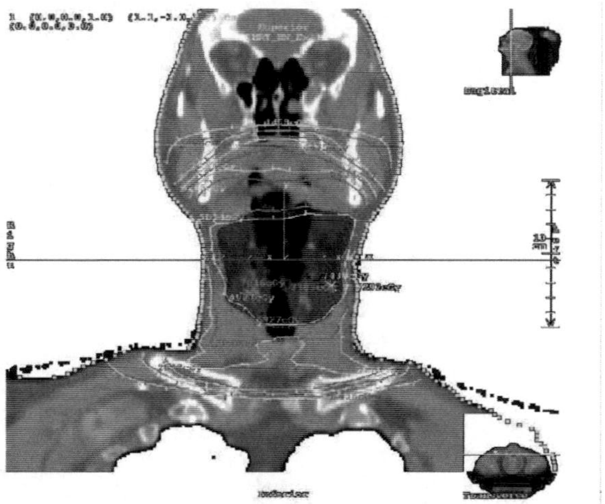

6. Conclusion

In 1992 Andre Dutreix published a paper[7], in which he wrote: "As it is not feasible to eliminate all possible errors in these complex procedures, it is highly recommended to add to the conventional programs of Quality Assurance a verification of the treated volume by portal images combined with in vivo dosimetry for each individual patient, to reach the ultimate goal: *the verification of the actually delivered treatment with the prescribed treatment.*"

References

[1] Renner W. D., Sarfaraz M., Earl M. A.., Yu C. X., A dose delivery verification method for conventional and intensity modulated radiation therapy using measured field fluence distributions. Medical Physics Vol. 30 No. 11 Nov. 2003 2996-3005

[2] Renner W. D., Norton K., Holmes T., A method for deconvolution of integrated electronic portal images to obtain incident fluences for dose reconstruction. JACMP Vol. 6 No. 4 Fall 2005 22-39

[3] Renner W. D. 3D dose reconstruction to insure correct external beam treatment of patients Medical Dosimetry Vol. 32 No. 3 Fall 2007 157-165

[4] Mans A., et al Catching errors with in vivo EPID dosimetry Medical Physics Vol. 37 No. 6 June 2010 2638-2644

[5] Kasper L. P., et al Transit dosimetry with an electronic portal imaging device (EPID) for 115 prostate cancer patients. Int. J. Radiation Oncol. Biol. Phys. Vol. 45 No. 5 1999 1297-1303

[6] Mans A., et al 3D dosimetric verification of volumetric modulated arc therapy by portal dosimetry. Radiotherapy and Oncology 94 2010 181-187

[7] Dutreix A., Leunens G., Van Dam J., and Van Der Schueren E. Quality Assurance in radiotherapy at the patient level. AMPI Medical Physics Bulletin Vol. 17 No. 4 1992 22-25

AUTHOR INDEX

Adamovics, J. 153, 158, 213, 218, 242, 303, 391, 395, 400, 410, 444, 467, 473
Afonso, L. 454
Al Nowais, S. 153
Allard, J-F. 382
Alva, M. 167, 321, 478
Ansbacher, W. 108
Appleby, A. 49
Apte, A. 420
Archambault, L. 24
Atkins, T. 372
Avery, S. 148
Back, S. 233, 284, 429
Baffa, O. 54, 167, 321, 405, 415
Baldock, C. 34, 44, 162, 242, 362, 459
Balling, P. 485
Battista, J. 299
Beaulieu, L. 24, 377
Beddar, S. 24, 171
Benedict, S. 269
Bennett, M. 464
Berman, A. 112
Best, R. 464
Bisht, R. 294
Blessing, R. 395
Borroni, M. 439
Bosi, S. 34, 44, 162
Bourland, J. 464
Bradley, D. 213, 410
Brauer-Krisch, E. 213, 410
Brochard, T. 213, 410
Brunt, J. 153
Bujold, R. 83, 88, 377
Burkhardt, K. 238
Bush, K. 112
Cai, J. 8, 269
Calvo, O. 225
Campbell, W. 117
Cao, Y. 20, 187
Carrara, M. 439
Cattell, H. 233
Ceberg, C. 284
Ceberg, S. 233, 284, 429
Chain, J. 313, 318
Chang, K. 20, 187
Chang, S. 238
Chao, K. 303
Cheeseborough, J. 303
Cho, K. Hwan 20, 187
Cho, S. Ju 20, 187
Choi, C. Heon 174
Christman, E. 49
Clift, C. 391
Clowes, S. 213
Coleman, J. 395, 400
Constantinou, C 289
Craciunescu, O. 449
Crowder, M. 467
Cullip, T. 238
Czelusniak, C. 17
Damilakis, J. 122
Darko, J. 12, 127
Das, I. 158
Das, S. 229
Davies, J. 34
De Almeida, A. 17, 478
De Deene, Y. 39, 69
Deasy, J. 280, 420, 473
Del Lama, L. 17
Ding, X. 464
Doran, S. 153, 213, 410, 421, 444
Dorand, J. 464
Duck, F. 372
Falk, M. 233
Fallai, C. 439
Farfan, E. 391, 395, 400
Foley, T. 391, 395, 400
Frank, R. 467
Freeman, C. 148
Fulton, R. 362
Gambarini, G. 439
Garrido, C. 54, 405, 415
Georgiou, E 289
Ghysel, S. 39
Gingras, L. 24
Gopishankar, N. 294
Gordon, J. 395
Gorjiara, T. 162, 242, 459
Grant, R. 467
Greer, P. 108
Greiter, M. 454
Guillot, M. 24
Gustafsson, H. 233
Gutierrez, A. 225
Gutman, G. 330
Han, H. Soo 174
Haneda, K. 79
Hannon, M. 487
Harpring, L. 395
Hayashi, S. 79
Heard, M. 433
Hill, R. 162, 242, 459
Hilts, M. 357
Hindmarsh, J. 362
Hoeschen, C. 454
Hollingdale, R. 444
Holmes, C. 391, 395, 400
Houde, D. 382
Hsi, W. 143
Hubert-Tremblay, V. 377

AUTHOR INDEX

Huh, H. Do....20, 187
Humphrey, V....372
Ibbott, G....1, 280, 387, 433, 467
Invernizzi, M....439
Jaffray, D....126
Jannik, G....395, 400
Jannik, T....391
Jay-Gerin, J-P....382
Jiang, Q....330
Jirasek, A....117, 335, 357
Jordan, K....136, 140, 191, 222, 251, 299
Kacperek, A....153
Kale, S....294
Karaiskos, P....289
Keall, P....233
Khalil, A....382
Khoei, S....93
Kim, C. Yong....20, 187
Kim, J-H....242
Kondo, T....79
Korreman, S....233
Kumar, D....177, 182, 367
Kumaran, S....294
Kuncic, Z....162, 242, 459
Kupelian, P....143
Kwag, J....487
Langton, C....93
Lau, J....470
Laviraj, M....294
Lawrence, M....238
Lee, S....20, 187
Lee, S. Hoon....20, 187
Lepage, M....83, 88, 251, 377, 382
Letourneau, D....126
Li, S....330
Li, Z....143
Lian, J....238
Libby, B....269
Link, K....464
Liu, Y....330
Lopatiuk-Tirpak, O....143
Lovelock, M....269
Low, D....346
Loy, H....467
Mackie, T....125
Mahapatra, A....294
Mans, A....97
Maris, T....122
Marques, T....54, 321, 405, 415
Marsden, C....49
McAuley, K....64, 313, 318, 357
McGowin, I....464
Medin, J....233
Meeks, S....143
Meesat, R....382
Meng, J....247
Mijnheer, B....97
Min, C. Kee....20, 187
Moorrees, J....93
Moran, J....103
Moreira, M....17, 478
Moutsatsos, A....289
Muren, L....485
Nasr, A....318
Negri, A....439
Newton, J....49, 280, 387, 467
Nicolucci, P....54, 167, 321, 405, 415
Nilsson, P....284
Nisbet, A....153, 444
Nordstrom, F....233, 284
Numasaki, H....79
O'Daniel, J....229
Olaciregui-Ruiz, I....97
Oldham, M....8, 29, 49, 208, 218, 280, 308, 387, 391, 395, 400, 449, 467, 473
Olding, T....12, 64, 127, 470
Oldman, M....158
Oliver, L....362
Olsen, J....464
Palta, J....143
Pantelis, E....289
Papadakis, A....122
Park, J....20, 187
Park, U. Jae....174
Park, Y. Je....20, 187
Peraticou, A....289
Perrotte, L....325
Persson, G....233
Petchevist, P....478
Petersen, J....485
Petrokokkinos, L....289
Pianoschi, T....167, 321
Pierquet, M....208, 218, 449
Pinkerton, A....487
Pirola, L....439
Potter, L....238
Rahman, A. Abdul....213, 410
Rath, G....294
Renner, W....487
Rink, A....126
Ripoll, J....122
Rogers, J....467
Rosenschold, P....233
Safronov, V....330
Sakelliou, L....289
Salomons, G....12
Sampaio, F....478
Samuel, E....177, 182, 367
Santanna, M....167, 321
Saupin, G....325

AUTHOR INDEX

Sawant, A. 233
Schlesinger, D. 269
Schoefer, F. 454
Schreiner, L. 12, 64, 127, 313, 318, 470
Schwarcke, M. 54, 405, 415
Scotti, A. 439
Sedaghat, M. 83, 88
Seimenis, I. 289
Sekimoto, M. 136, 140, 222
Sellakumar, P. 367
Semnicka, J. 481
Shahnazi, K. 148
Sharpe, M. 126
Sheng, K. 269
Shim, J. Bo 20, 187
Simon, J. 467
Skyt, P. 485
Smee, R. 44
Snir, J. 299
Sochor, V. 59
Solc, J. 59
Son, K-J. 174
Song, H. 449
Sonke, J-J. 97
Sozontov, E. 330
Spevacek, V. 59, 481
Sriprisan, S. 143
Stanley, S. 391, 395, 400
Stathakis, Q. 132
Stearns, S. 467
Steffey, B. 449
Stroom, J. 97
Strumban, E. 330
Stunja, L. 473
Subramani, V. 294
Svatos, M. 233
Swift, C-L. 108
Tenconi, C. 439
Teshima, T. 79
Thomas, A. 29, 49, 158, 208, 218, 280, 387, 391, 473
Thomas, R. 444
Thulkar, S. 294
Tielenburg, R. 97
Tomatis, S. 439
Tominaga, T. 79
Tooley, M. 372
Townson, R. 112
Trapp, J. 93
Tremblay, L. 382
Tremblay, N. 377
Usui, S. 79
Van Herk, M. 97
Vandecasteele, J. 39
Varveris, C. 122
Vavru, K. 481
Vijlbrief, R. 97
Vivekanandhan, S. 294
Wang, S. 238
Wang, Z. 280
Wells, D. 117
Wendt, R. 467
Wetterstedt, S. 284
Wolodzko, J. 49
Wu, G. 470
Wu, J. 229
Wu, Q. 8
Wuu, C. 303
Wuu, C-S. 198
Xu, Y. 198, 303
Xu, Z. 238
Yang, D. Sik 20, 187
Yang, W. 269
Yates, E. 485
Ye, S-J. 174
Yewondwossen, M. 247
Yi, J. 20, 187
Yin, F-F. 8, 229
Yoon, W. Seob 20, 187
Yoshioka, M. 79
Zacharakis, G. 122
Zavgorodni, S. 112
Zeidan, O. 143
Zhao, L. 158
Zhao, Q. 158
Zhu, T. 259
Zimmerman, J. 233
Zourari, K 289
Zucolot, V. 415